ENCYCLOPEDIA OF

Construction Methods & Materials

William P. Spence

Sterling Publishing Co., Inc.
New York

DEDICATION

To my wife, Bettye Margaret Spence, for her steady encouragement,
editorial assistance, and the many hours she spent at her computer
preparing the manuscript.

ACKNOWLEDGMENTS

Thanks are due to the manufacturers, professional and trade
organizations, governmental agencies, and architectural and
engineering associations that have so generously contributed
to this book.

Library of Congress Cataloging-in-Publication Data
Spence, William Perkins, 1925–
    Encyclopedia of construction methods & materials / William P. Spence.
      p.   cm.
    Includes index.
    ISBN 0-8069-6851-6
    1.Building—Encyclopedias. I. Title.
TH9.S64 2000
690'.03—dc21   00-037289

Book design by Judy Morgan
Editing and layout by Rodman Pilgrim Neumann

10 9 8 7 6 5 4 3 2 1

Published by Sterling Publishing Company, Inc.
387 Park Avenue South, New York, N.Y. 10016
© 2000 by William P. Spence
Distributed in Canada by Sterling Publishing
c/o Canadian Manda Group, One Atlantic Avenue, Suite 105
Toronto, Ontario, Canada M6K 3E7
Distributed in Great Britain and Europe by Cassell PLC
Wellington House, 125 Strand, London WC2R 0BB, England
Distributed in Australia by Capricorn Link (Australia) Pty Ltd.
P.O. Box 6651, Baulkham Hills, Business Centre, NSW 2153, Australia
Manufactured in the United States of America
All rights reserved

Sterling ISBN 0-8069-6851-6

DISCLAIMER

The materials included in this book were obtained from a large number
of sources such as manufacturers, professional and trade associations,
governmental organizations, and architectural and engineering firms.
While every effort has been made to report material accurately, the publisher and author do not warrant and assume no liability for its accuracy
or completeness or the suitability of the material for any particular application. The publisher and author shall not be liable for damages of any
type resulting from the reader's use of this material. The reader is expected to seek the advice of experts in each field to verify and utilize the
information in this book.

CONTENTS

Division 4

Masonry   95

Division 5

Metals   139

Division 8

Division 9

Division 10

Division 11

Division 12

Division 13

Division 14

Division 15

Division 16

PREFACE

This book provides a detailed look at the major methods of construction, building systems, and the vast array of materials and products provided by the manufacturers supplying the construction industry. It is organized following the MasterFormat™ developed by The Construction Specifications Institute, 601 Madison St., Alexandria, Va., 22314, and Construction Specifications Canada, 100 Lombard St., Toronto, Ontario, Canada, M5C 1M3. This system of organization consists of 16 divisions that are divided into hundreds of topics. The complete MasterFormat is detailed in a 300-page manual available from the above organizations.

The first division, **General Requirements**, reviews briefly the areas included in this section of the MasterFormat™. While they pertain to administrative work, it is important for architects, engineers, and contractors to be aware of these materials. In some cases these requirements will relate directly to their work with contracts, payment procedures, administrative requirements, and other management activities.

In addition, the first division contains extensive information about trade and professional organizations in the United States and Canada, various publications, and a review of the metric system of measurement.

The second division, **Site Construction,** pertains to a study of the building site and analysis of various types of soil, and commonly used foundation systems. The study of soils is critical to the ultimate design of the foundation.

The third division, **Concrete,** presents a detailed study of the manufacture, types, characteristics, and properties of concrete. Consideration of the use of admixtures, proportions, water, mixing, and placing is included.

Division 4, Masonry, includes detailed information on mortar, the key to satisfactory masonry construction. The materials and techniques involved with clay brick and tile, concrete masonry, and stone construction are explained.

Ferrous and nonferrous metals are covered in detail in **Division 5, Metals.** Their characteristics, mechanical properties, and practical applications are discussed. Steel-frame construction systems are illustrated.

One of the largest divisions in the book is **Division 6, Wood & Plastics,** covering the vast array of wood and plastic materials. Their properties, characteristics, and recommended applications are explained. Illustrated are wood structural framing systems and methods and materials of light wood-frame construction. The remainder of this division is used to present information on plastic materials useful in building construction.

Insulation, waterproofing, and sealing buildings against the weather are covered in **Division 7, Thermal & Moisture Protection.** Various bonding agents, sealers, and sealants available are covered. The roofing systems for residential and commercial building conclude this division.

The types, styles, methods of operation, and materials used for doors and windows are extensive. Many of the products available are illustrated in **Division 8, Doors & Windows,** as well as factory stock and custom-made store fronts. Many of these use some form of glass, so the types, properties, and uses for the various glass products available are discussed. The types and installation of cladding systems commonly used on large commercial buildings conclude this division.

Finishing the interior of the building involves possibly the largest and most diverse range of products in any area of construction. **Division 9, Finishes,** includes interior finishes, decorative and protective coatings, gypsum, lime and plaster materials and products, acoustical finishes and materials, and all types of finish flooring.

Division 10, Specialties, covers some of the specialty products, such as visual displays, screens, grilles, service walls, and identifying signs such as building directories and interior and exterior signs. A host of other products included as specialty items are shelving, partitions, fire-protection devices, telephone enclosures, and toilet and bath accessories.

Division 11, Equipment, includes items specified as equipment. Examples of some of the many products so specified are discussed. Some include ecclesiastical equipment, library equipment, and vending equipment.

Furnishings form the final touch to a completed building, and **Division 12, Furnishings,** classifies these into seven major divisions. Artwork, window treatments, and rugs and mats offer a variety of types and styles. Casework and furniture for use in commercial buildings form the largest group.

A most interesting assembly of features is found in **Division 13, Special Construction,** which discusses and illustrates a diverse offering including air-supported structures, prefabricated and pre-engineered assemblies, sound and seismic control devices, nuclear reactors, and radiation-shielding components. Other diverse constructions are in this section.

Division 14, Conveying Systems, discusses and illustrates the variety of conveying systems available such as conveyors, elevators, escalators, moving walks, and material-handling systems.

Division 15, Mechanical Systems, is one of the more complex divisions of this book. It covers the mechanical systems used in residential, commercial, and industrial buildings. Extensive information is presented on fire protection systems, plumbing systems, and heating, air-conditioning, ventilation (HVAC), and refrigeration. The types of systems and range of products are great and improvements are occurring regularly.

Division 16, Electrical Systems carries the discussion from the generation and transmission of electrical power to the service entrances in residential and commercial buildings. Various internal power distribution systems are illustrated and extensive information on lighting is available. Equipment for controlling and operating the electrical system as well as equipment used for communication such as alarm, television, public address, and other communication systems are presented.

The book concludes with the extensive Appendices, Glossary, Selected Bibliography, and Index.

—William P. Spence

DIVISION

GENERAL REQUIREMENTS

CSI MasterFormat™

Courtesy H.H. Robertson, A United Dominion Company

A description of level-two titles, as listed on the previous page, for Division 1, General Requirements, of the CSI MasterFormat™ is provided below. A detailed listing of the level-three titles is available in the 1995 edition of the MasterFormat™, published jointly by the Construction Specifications Institute (CSI), Alexandria, Virginia, and the Construction Specifications Canada (CSC), Toronto, Canada.

Following the description of the nine second-level titles, this division also contains extensive information about codes and the various model-building-code-defining organizations as well as other trade and professional organizations in the United States and Canada. Information about various publications and a review of the metric system of measurement are also provided.

SUMMARY

The first level-two title within Division 1, General Requirements, of the CSI MasterFormat™, is **Summary,** which includes Summary of Work, Multiple Contract Summary, Work Restrictions, and Project Utility Sources.

▼ **Summary of Work** identifies the work to be covered in a single contract and describes provisions for future construction. It may include work by the owner, work covered in the contract documents, and specific things such as products ordered in advance, owner-furnished and installed products, and the use of salvaged material and products.

▼ **Multiple Contract Summary** describes the construction delivered under more than one construction contract such as a construction management contract and a multiple-prime contract. It includes the sequence of construction, contract interface, any construction by the owner, and a summary of the contracts.

▼ **Work Restrictions** pertains to restrictions that affect construction operations such as the location of construction, occupancy requirements, and the use of the building premises and site during construction. If the construction involves a building already occupied, procedures for coordination with the occupants are detailed.

▼ **Project Utility Sources** includes the identity of utility companies that are to provide permanent services to the project.

PRICE & PAYMENT PROCEDURES

The second level-two title is **Price & Payment Procedures.** It details the allowance adjusting procedures for cash and quantity allowances for products, installation, testing, and contingencies.

▼ **Allowances** includes the adjusting procedures for cash and quantity allowances for products, installation, inspection, and testing contingencies.

▼ **Alternates** provides for the submission and acceptance procedures for alternate bids including, when desired, a list and description of each alternate.

▼ **Value Analysis** includes the procedures and submittal requirements for value analysis, value engineering, application for consideration, and consideration of proposals.

▼ **Contraction Modification Procedures** details the procedures for making clarifications and proposals for changes to the contract, as well as making those changes to the contract.

▼ **Unit Prices** establishes the procedures associated with unit prices and measurement and payment. It may include a list and descriptions of the actual unit price items.

▼ **Payment Procedures** describes the procedures for submitting schedules of values and applications for payment.

ADMINISTRATIVE REQUIREMENTS

The third level-two title, **Administrative Requirements,** encompasses the areas of Project Management & Coordination, Construction Progress Documentation, Submittal Procedures, and Special Procedures.

▼ **Project Management & Coordination** includes the administration of subcontractors and coordination with other contractors and the owner. This includes project meeting and on-site administration.

▼ **Construction Progress Documentation** covers the requirements for scheduling, recording, and reporting progress. This includes such things as construction photographs, progress reports, site observation, purchase order tracking, scheduling of construction, and survey and layout data.

▼ **Submittal Procedures** include the general procedures and requirements that are necessary for making submittals during the course of the construction. Submittal procedures can include items such as certificates, design data, field test reports, shop drawings, product data, samples when required, and source quality control reports.

▼ **Special Procedures** includes procedures for any of the project situations that require special procedures such as a historic restoration, renovation or alteration, preservation, or hazardous material abatement.

QUALITY REQUIREMENTS

The fourth level-two title of Division 1 is **Quality Requirements.** This covers four areas: Regulatory Requirements, References, Quality Assurance, and Quality Control.

▼ **Regulatory Requirements** provides information that is required for conformance to regulatory requirements such as building codes, mechanical codes, and electrical codes. This also includes information for other regulations and fees applicable to the project.

▼ **References** includes lists of reference standards cited in the contract documents and the organizations whose standards are cited. This includes items such as symbols, abbreviations, and acronyms used.

▼ **Quality Assurance** references are used to provide for procedures that are necessary to assure the quality of construction. This includes field observations and tests performed by manufacturers' representatives during installation. Quality assurance is provided by fabricators, installers, manufacturers, suppliers, and testing agencies.

▼ **Quality Control** establishes procedures for measuring and reporting the quality and performance of construction. It may require field samples and mock-ups assembled at the site. The contractor provides quality control plans. Quality control can be provided by inspections, inspection services, and testing laboratories.

TEMPORARY FACILITIES & CONTROLS

The fifth level-two title, **Temporary Facilities & Controls,** provides requirements for installation, maintenance, and removal of temporary utilities, controls, facilities, and construction aids during construction. This includes Temporary Utilities, Construction Facilities, Temporary Construction, Construction Aids, Vehicular Access & Parking, Temporary Barriers & Enclosures, Temporary Controls, and Project Identification.

▼ **Temporary Utilities** includes all such services used during construction such as electrical, water, lighting, gas, telephone, fire protection, fuel oil, gasoline, and diesel fuel.

▼ **Construction Facilities** includes any temporary facilities built on the site for use during construction. Typical examples include field offices, storage buildings, sanitary facilities, and first aid stations.

▼ **Temporary Construction** includes those facilities built to provide access to various parts of the site and project so as to facilitate the construction process or to accommodate the needs of the owner and occupants. Typical examples include temporary bridges, ramps, overpasses, decking, and turnarounds.

▼ **Construction Aids** includes all requirements and procedures related to tools and equipment used during construction such as scaffolding, cranes, hoists and construction elevators.

▼ **Vehicular Access & Parking** provides requirements for and procedures related to access to the site and parking facilities to meet the needs of the construction and owner's operations. Typically this can include access roads, haul routes, parking areas, temporary roads, and control of traffic.

▼ **Temporary Barriers & Enclosures** includes all facilities and procedures for the protection of the occupants or existing spaces during construction. This can include air barriers, barricades, dust barriers, fences, noise barriers, pollution control, security, and the protection of trees and other vegetation.

▼ **Temporary Controls** includes site or environment controls required to allow construction to proceed, including such things as erosion and sediment control and pest control.

▼ **Project Identification** includes any signs used to identify the construction site and any particular areas on the site.

PRODUCT REQUIREMENTS

The sixth level-two title, **Product Requirements,** includes a comprehensive series of requirements.

- ▼ **Basic Product Requirements** includes the basic requirements for new, salvaged, and reused products used in construction.
- ▼ **Product Options** includes the basic requirements for options the contractor may have in selecting products and how the determination is made for equal products.
- ▼ **Product Substitution Procedures** includes the basic requirements and procedures when proposals for the substitution of a product are made.
- ▼ **Owner-Furnished Products** includes the basic requirements for products that are to be furnished by the owner. This also involves scheduling, coordinating, handling, and storing owner-furnished products.
- ▼ **Product Delivery Requirements** specifies the basic requirements for packing, shipping, delivery, and acceptance of products at the site.
- ▼ **Product Storage & Handling Requirements** sets forth the basic requirements for storing and handling products on the site.

EXECUTION REQUIREMENTS

The seventh level-two title, **Execution Requirements,** has eight sections.

- ▼ **Examination** involves the acceptance of conditions, existing conditions, and the basic requirements for determining acceptable conditions for installation.
- ▼ **Preparation** details the requirements for preparing to install, erect, or apply products, including activities as field engineering, protection of adjacent construction, surveying, and construction layout.
- ▼ **Execution** includes the basic requirements for installing, applying, or erecting products that are new, prepurchased, salvaged, or owner-furnished.
- ▼ **Cleaning** sets the requirements for maintaining the site in a neat condition during construction and the final cleaning in preparation for turning the project over to the owner.

- ▼ **Starting & Adjusting** involves establishing the initial checkout and startup procedures and any adjustments needed to ensure safe operation during the acceptance testing and commissioning.
- ▼ **Protecting Installed Construction** provides the requirements and procedures for protecting installed construction.
- ▼ **Closeout Procedures** includes the administrative procedures for substantial completion and final completion of the work.
- ▼ **Closeout Submittals** sets the procedures for closeout submittals, revised project documents, and delivery and distribution of spare parts and maintenance materials.

FACILITY OPERATION

The eighth level-two title, **Facility Operation,** has to do with the final requirements for preparing the facility for decommissioning.

- ▼ **Commissioning** requirements include the commissioning of summaries, system performance evaluations, as well as testing, adjusting, and balancing procedures.
- ▼ **Demonstration & Training** includes the requirements and procedures for the demonstration of the products and systems within the facility, including training the owner's operating and maintenance personnel.
- ▼ **Operation & Maintenance** sets forth the requirements and procedures for operating the facility after commissioning.
- ▼ **Reconstruction** involves any renovation or reconstruction of the existing facilities that may be required.

FACILITY DECOMMISSIONING

Finally, the last and ninth level-two title, **Facility Decommissioning,** includes the basic requirements for deactivating a facility or a portion of it from operation. This includes activities such as facility demolition and removal, hazardous materials abatement, removal and disposal, and protection of deactivated facilities.

TABLE 1-1 MODEL BUILDING CODES

BOCA National Building Code	Building Officials and Code Administrators International (BOCA) 4051 West Fossmore Rd. Country Club Hills, IL 60478
Standard Building Code (SBCC)	Southern Building Code Congress International (SBCC) 900 Montclair Rd. Birmingham, AL 35213-1206
Uniform Building Code (ICBO)	International Conference of Building Officials (ICBO) 5360 Workmen Mill Rd. Whittier, CA 90601-2298
National Building Code of Canada	Associate Committee on The National Building Code National Research Council Ottawa, Ontario, Canada K1A 0R6

1-1 *The Unified Development Ordinance is approved by the local governing body and becomes part of the codes administered by the building officials. It includes requirements on land use and zoning districts.*

CODES

The architect and engineer are constantly aware of the codes that regulate the minimum standards pertaining to the design and construction of buildings and other facilities in the area in which the project will be built. The codes control all aspects of the project from space utilization and structural design to plumbing, mechanical, electrical, and fire protection. Codes apply to the alteration, moving, repair, and demolition of buildings and other structures. Construction is subject to local and state codes as well as adapted national codes.

Local & State Codes

The city may have codes to regulate work within its incorporated limits while counties may have codes to regulate the work in their jurisdictions not already regulated by city codes. These are legislated by local governing bodies and become municipal codes. The local building official is responsible for checking projects to see they meet codes before a building permit is issued. This person inspects the job as it proceeds to see that the work is performed according to the approved design.

National Codes

There are several model codes which are used in various parts of the country. The United States does not have a single national code. These regional codes have no power of enforcement individually. They must be adopted by the governing body of a city, county, or state as part of its overall building code. The local authorities then enforce the code. Some of the model building codes available are listed in Table 1-1.

Other Codes

Within the local government administrative structure are committees, bureaus, or elected bodies that have the authority to review, regulate, approve, and disapprove construction projects. The range of activities is great but may include such things as an **architectural review board** to pass an opinion on the design of the project, and a **planning and zoning board** that considers the use of the land and controls through ordinances the type of structures that can be built. A unified development ordinance is shown in 1-1. A zoning board of adjustment, which considers appeals to the zoning ordinances or rulings and can grant a variance, and the office of the **building official** (inspector), where salaried staff review proposed projects to see if they meet the codes and zoning regulations, issue building permits, inspect the job while under construction, and finally issue a certificate of occupancy when it is complete and has satisfied all regulations.

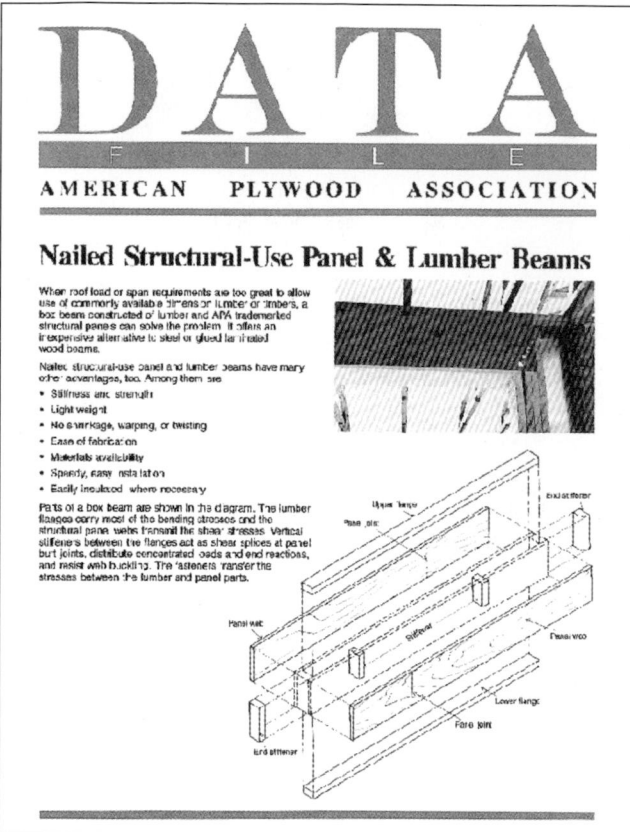

1-2 *A technical publication providing design information and technical data.*

Courtesy APA-The Engineered Wood Association

1 Panel grade
2 Span rating
3 [not represented]
4 Exposure durability classification
5 Product standard
6 Thickness
7 Mil number
8 APA's performance-rated panel standard
10 [not represented]
11 HUD/FHA recognition
12 Panel grade, Canadian standard
13 Panel mark—rating and end-use designation, Canadian standard
14 Canadian performance-rated panel standard
15 Panel face orientation indicator

1-3 *This stamp on plywood products indicates that APA-The Engineered Wood Association has certified that it meets their specifications.*

Courtesy APA-The Engineered Wood Association

Various professional organizations prepare model codes for specialized areas such as electrical, plumbing, and fire. Some examples of a few of these include The Uniform Plumbing Code and The Uniform Mechanical Code available from the International Association of Plumbing and Mechanical Officials and The National Electrical Code and the National Fire Code (both on CD-Rom) available from The National Fire Protection Association. The Southern Building Code Congress International publishes a series of 10 codes including the Standard Gas Code, Standard Mechanical Code, Standard Plumbing Code, and Standard Housing Code. In addition they have a series of code-related publications. Other United States professional organizations also have specialty codes.

In Canada, in addition to their National Building Code, they have National Housing, National Fire, National Plumbing, and Canadian Farm Building Codes.

Some projects require other approvals such as an environmental impact study and approval, air pollution controls, and local or state health department approval.

TRADE ASSOCIATIONS

Trade associations are involved with the advancement of knowledge about materials and methods used in the construction industry. Some have helped develop codes related to products within their area of interest.

Trade associates are organizations having a membership composed of manufacturers or businesses involved in the production or supply of materials or services. Some support research into the use and improvement of construction materials and better construction methods. As research produces results this often leads to the development of material **specifications** and **performance procedures** for the trade area. Frequently these become nationally accepted and become national standards. Some associations operate programs certifying the quality and performance of products.

There are a large number of active trade associations. One example is **APA-The Engineered Wood Association.** Among their activities is the establishment of standards for plywood and various other wood products (see 1-2). Their plywood certification seal is shown in 1-3. Another example is the **American Institute of Steel Construction,** which serves the needs of those involved in structural steel work including publication of manuals, textbooks, and specification.

The list of trade associations is long and they form a very important part of the construction industry (see Appendix A for trade and professional associations detailed for each division). Materials and techniques would not be at their current state of development without the work of these associations.

Other Organizations

Those involved in the construction industry find their work constantly influenced by other organizations working to improve materials, techniques, safety, and other aspects of the industry.

One such organization is the **American Society for Testing & Materials.** It is a non-profit corporation formed to develop standards on characteristics and performance of materials, products, systems, and services and the promotion of related knowledge. It has over the years developed an extensive list of published works and influences not only the construction industry but all of the industrial United States.

Underwriters Laboratories Inc. (UL) is a non-profit organization that utilizes techniques of scientific investigation, study, experiments, and tests on materials and a wide range of equipment to establish standards and specifications for these so as to reduce or prevent bodily injury and loss of life and property from associated hazards. Many products used in construction, such as electrical, heating, air conditioning, refrigeration, and fire protection are tested and if certified carry the UL symbol (see 1-4).

One other major independent organization that has influenced standards for years and continues as a major factor in standard development is the **American National Standards Institute (ANSI).** One of the hundreds of ANSI standards is shown in 1-5. ANSI is a coordinating organization for the national standards system in the United States. It coordinates the voluntary development of national standards. It helps identify standards that are needed and arranges for competent organizations to develop them. ANSI works to establish a national consensus on standards leading to its acceptance across the country. Another important responsibility is that ANSI represents the United States interests in international standardization. ANSI is the U.S. representative to the **International Organization for Standardization (ISO),** which is the world body coordinating and producing engineering and product standards for worldwide use.

1-4 *Manufacturers whose products meet the safety standards of Underwriters Laboratories, Inc. (UL) may be authorized by UL, subject to testing and evaluation, to use this UL mark on those products.*

Reproduced with the permission of the Underwriters Laboratories, Inc.

1-5 *This American National Standard was developed by the Hardwood Plywood and Veneer Association.*

Courtesy Hardwood Plywood and Veneer Association

CANADIAN ORGANIZATIONS

The following are a few of the Canadian organizations active in various aspects of the construction industry.

The Canada Mortgage and Housing Corporation (CMHC) is an agency of the Federal Government and is responsible for administering the National Housing Act. This legislation is designed to improve the housing and living conditions in Canada. Among its responsibilities it conducts research into the social, economic, and technical aspects of housing and related fields. It publishes the results of these activities.

The Canadian Housing Information Centre (CHIC) is a part of the Canadian Mortgage and Housing Corporation. It provides thousands of publications related to all aspects of housing, building, and community development.

The Canadian Wood Council (CWC) is a national federation of forest products associations. It is involved in the development and promotion of technical product information to assist members of the specifying and regulatory community. It has an extensive listing of technical manuals and books. The **Canadian Standards Association (CSA)** prepares standards related to design and materials.

CONSTRUCTION SPECIFICATIONS INSTITUTE (CSI)

The Construction Specifications Institute's members include architects, engineers, specification writers, contractors, and others involved with the preparation and use of construction specifications.

▼ The Institute has five major goals.
1. To improve specifications writing
2. To simplify specifications
3. To standardize building codes
4. To standardize specifications for government work
5. To study new materials and processes to substitute for those becoming more costly or scarce.

CSI has over 17,000 members and 140 local chapters nationwide, plus chapters in Mexico. Following are some of the major publications. MasterFormat™ is a master list of numbers and titles for organizing information about construction requirements, products, and activities into a standard sequence. Master-Format™ facilitates the organization of standard filing and retrieval systems related to the construction industry. In addition it can be used to organize information in project manuals, file product information, and organize cost data; the McGraw-Hill Sweet's General Building and Renovation Catalog File and the R.S. Means Company construction cost data publications are organized using this format. You will also notice that the content of this book is organized following the 16 divisions of the MasterFormat™.

MasterFormat™ is published jointly by the Construction Specifications Institute (CSI) (a U.S. organization) and Construction Specifications Canada (CSC).

SectionFormat® is a three-part format for the organization of a specification section.

PageFormat® presents a recommended format for uniform presentation of information on pages of a specification.

Spectext® is a master specification system providing a listing of paragraphs and statements arranged for the purpose of developing a project specification and covers many of the sections in the MasterFormat™.

The Spec-Data® presents product data from manufacturers in a format enabling objective comparisons to be made. They are available in three-ring binders and CD-ROM. Spec-Data II® is a microfilm library detailing products from thousands of manufacturers and containing several hundred thousand complete catalog pages.

The Construction Specifications Institute also has other publications pertaining to the production of specifications.

SWEET'S CATALOG FILES

Sweet's Catalog Files contain catalog information from hundreds of manufacturers. There are five sets of catalogs available. These are General Building and Renovation, SweetSource CD-ROM, Engineering and Retrofit, Homebuilding and Remodeling, and Contract Interiors.

These catalog files are produced and distributed by Sweet's Group, a Division of the McGraw-Hill Companies, 1221 Avenue of the Americas, New York, NY 10020. They are sold only to those in construction doing a certain level of business. Those desiring to secure a set should file an application giving information about their business, and from that it is determined which catalog file the applicant is eligible to purchase.

The General Building and Renovation Catalog File consists of 15 volumes organized in a manner that is compatible with the MasterFormat™. It provides product descriptions and specifications for hundreds of building products. The SweetSource CD-ROM provides access to a wide range of building products and specifications and is updated quarterly. The Engineering and Retrofit Catalog File, a three-volume set, is a source of product data and specifications of interest to engineers, building material suppliers, contractors, building owners, and developers. The Homebuilding and Remodeling Catalog File is a two-volume set and contains catalogs of products commonly used for residential work such as doors, windows, kitchens, baths and finishes. The Contract Interiors two-volume set provides information and specifications from manufacturers of interior building materials such as floors, tiles, and wallcoverings.

MANUFACTURERS' PUBLICATIONS

Companies manufacturing products used in building construction have available an extensive range of materials, most available at no cost. Their sales literature usually gives detailed product descriptions and often considerable technical information about the product. It is up to the person reading this material to follow up on the claims to see if the product is accurately described. For example, if certain electrical products included do contain the Underwriters Laboratory certification, you can feel assured they meet UL requirements. If the material is silent on any certification, you should make inquiries to get additional information.

Manufacturers also publish considerable information about how to assemble, test, install, maintain, and otherwise operate the equipment. This is essential information needed by the installation crew and later by the maintenance staff of the finished building.

It is important to keep your files of catalogs up to date so you are not designing or constructing using obsolete information. A service as described by the Sweet's Catalog Files is one way to do this.

Table 1–2 SI METRIC BASE UNITS

Quantity	Name of Unit	Symbol
Length	meter	m
Mass	kilogram	kg
Time	second	s
Electric current	ampere	A
Thermodynamic temperature	kelvin	K
Amount of substance	mole	mol
Luminous intensity	candela	cd

U.S. FEDERAL OCCUPATIONAL SAFETY & HEALTH ACT

The United States Federal Occupational Safety & Health Act (OSHA) sets standards that must be observed to protect the health and safety of construction employees as well as those in other industries. Construction sites are subject to periodic inspection by an OSHA inspector. The publication related to the construction industry, **Construction Industry, OSHA Safety and Health Standards** can be purchased from the Superintendent of Documents, U.S. Government Printing Office, Washington, D.C. 20402. Other publications available include **Personal Protective Equipment** and **Hand and Power Tool Safety.**

METRIC SYSTEM (SI METRIC)

The metric system of measure is used over the entire world. The United States construction industry is slowly moving into the use of metric measures. It is a simple system and when fully adopted will make the measuring tasks of the construction trades much simpler and less prone to error.

The International System of Units is also known as the SI metric system. The "SI" stands for **Système International,** the French name of the system. For those in the construction industry working outside the United States, a knowledge of the metric system is essential.

SI Metric Base Units

The seven base units of measurement in the SI metric system are length, mass, electric current, time, thermodynamic temperature, quantity of substance, and luminous intensity. These are indicated by standard symbols as shown in Table 1-2. Notice some are capital letters and some are lower case. It is important to show the correct case of the symbol.

▼ On the following page is a brief discussion of each of the units in Table 1-2.

The seven base units of measurement in the SI metric system are described along with their standard symbols. The case of the symbol is always important.

1. **Length**—The basic unit of length is the meter (m). Subunits are decimeter, centimeter, and millimeter. A decimeter is one-tenth of a meter, a centimeter is one-hundredth of a meter, and a millimeter is one-thousandth of a meter.
2. **Mass**—Basically this is considered as weight. Very small weights are given in grams (g) and heavier weights in kilograms (one thousand grams) (kg).
3. **Time**—Time is measured in seconds (s) in the same manner as the Customary System.
4. **Electric current**—Amperes (A) are used to measure electric current. An ampere is the unit of the rate of flow of electric current.
5. **Thermodynamic temperature**—The metric unit for temperature is the kelvin (K).
6. **Amount of substance**—The quantity of substance is the mole. It is the amount of atoms in carbon 12 and is used mostly in chemistry.
7. **Luminous intensity**—The metric unit of luminous intensity is the candela (cd).

Prefixes

A prefixes is a word added to the beginning of a word to alter its meaning. Metric prefixes are added to SI metric units to change the multiples of the unit. For example, the prefix "kilo" means "1000," so a kilogram is 1000 grams. A prefix like "milli" means 1/1000 of the unit. For example, a millimeter is one thousandth of a meter. It takes 1000 millimeters to equal a meter (see Appendix B, Metric Information, for acomplete listing).

Derived SI Units

Derived SI units are combinations of base or other metric units generated by multiplying or dividing. For example, area is derived by multiplying the length and width of a rectangle (in meters), producing a result in square meters (m^2). This is a unit derived from the base unit, meter (see Appendix B, Metric Information, for complete listing of SI-derived units).

▼ Following are some of the SI units derived from SI base units that are commonly used in construction work:

Volume—Small volumes are specified in cubic millimeters (mm^3) or cubic centimeters (cm^3) and larger volumes (as concrete) in cubic meters (m^3).

A cubic meter is the volume of a cube the edges of which are one meter. The liter (L) is the measurement for liquid volume.

Area—The basic unit for area is the square meter (m^2) while smaller areas are specified in square centimeters (cm^2) or square millimeters (mm^2). A square meter is the area of a square with edges of one meter. The hectare (ha) is the metric measurement used in surveying.

Pressure—The metric unit for pressure (replacing pounds per square inch) is kilopascals (kPa) and megapascals (MPa) for stress.

Heavy loads—The metric ton (t) is used to denote large loads such as those used in excavation.

Plane and solid angles—The radian (rad) and steradian (sr) denote plane and solid angles used in lighting work and some engineering calculations. The units degree (°), minute ('), and second (") are used in surveying.

Conversion Factors

Metric projects are designed in metric units; ordinarily this means that conversion to Customary units should not be necessary. There may be times, however, when a project is designed in Customary units and certain measurements or specifications might have to be converted to metric units. See Appendix B, Metric Information, for a listing of some of the most frequently required conversion factors.

U.S. Metric Policy

The Metric Conversion Act of 1975 as amended by the Omnibus Trade and Competitiveness Act of 1988 established the modern metric system (Système International, or SI) as the preferred system of measurement in the United States. It required that, to the extent feasible, the metric system be used in all federal procurement, grants, and business-related activities by September 30, 1992. The U.S. Department of Commerce established the Interagency Council on Metric Policy (ICMP). Its subcommittee, the Metrication Operating Committee, established a Construction Subcommittee to facilitate the metrication of all federal construction. The Construction Subcommittee established the goal of instituting the use of metric design in all new federal facilities by January 1994.

DIVISION 2

SITE CONSTRUCTION

CSI MasterFormat™

02050 Basic Site Materials & Methods

02100 Site Remediation

02200 Site Preparation

02300 Earthwork

02400 Tunneling, Boring & Jacking

02450 Foundation & Load-Bearing Elements

02500 Utility Services

02600 Drainage & Containment

02700 Bases, Ballasis, Pavements & Appurtenances

02800 Site Improvements & Amenies

02900 Planting

02950 Site Resoration & Rehabilitation

Courtesy GOMACO Corporation, Ida Grove, Iowa

THE BUILDING SITE

Site work planning requires the services of architects, landscape architects, engineers, and specialists. The on-site work includes a wide range of activities, including subsurface investigation, demolition, preparing the site for construction, dewatering excavations, shoring and underpinning, various types of earthwork, tunneling, constructing piles and caissons, paving, installation of various types of piping, water distribution, sewerage and drainage systems, construction of ponds and reservoirs, site improvements (fences, walks, etc.), and landscaping (see 2-1).

Details are shown on site plans developed by the architect with the assistance of engineers specializing in the activities required. Additional information is contained in reports of subsurface investigations, including a geotechnical report and soil test boring data.

Building codes have sections devoted to site work and demolition.

Site Plans

The site plan will show the results of the land survey, the existing contours of the land, and the desired finished contours. The buildings, roads, wells, and other features are located. The location of utilities (gas, electric service, public water and sewer lines) are indicated. Set backs and easements must be located. The location and size of parking areas, vehicle access routes, streets, curbs, and other details such as fences, walks, and retaining walls are shown. A typical site plan is shown in 2-2.

Environmental Considerations

As a part of the preliminary work the project in the United States must address environment considerations by making an impact study as required by the passage of the *National Environmental Policy Act*. This determines how the project and site work will affect the natural features in the surrounding area. When the plan is found to be sound site work can proceed. However, the general contractor must be constantly vigilant to maintain conditions friendly to the environment. For example, soil runoff into a nearby stream or lake must be curtailed. Discharge of toxic waste materials must be prohibited. The existence of flood zones must be acknowledged and decisions be made concerning how they impact on the proposed building site.

2-1 *Site preparation involves the activities of a number of construction trades.*

Preparing the Site

Preparing the site includes removal of trees and brush in areas to be occupied by the structure (see 2-1), parking areas, roads and driveways, removal of existing improvements above and below grade, stripping and saving existing soil, removal and replanting of existing trees, and demolition or removal of existing structures. The site plan and landscape plan give the information needed to accomplish these.

Subsurface Investigation

A key factor in the design of a structure is the determination of the characteristics of the soil. This requires field tests made by a soils engineer and tests made on soil samples in a soil testing laboratory or by field test procedures. The soil report is used by the architect and design engineers to determine what must be done as the building is designed and constructed. This can influence the footing design, the need for retaining walls, paving, dewatering requirements, and other factors.

The soil tests are made by the results of subsurface test borings which are detailed on a soil test report: The report includes a detailed geologic description of the site, subsurface conditions, and recommendations based on the findings. Recommendations may pertain to site grading, foundation support, paving design, retaining walls, and other aspects of the project that are to be built.

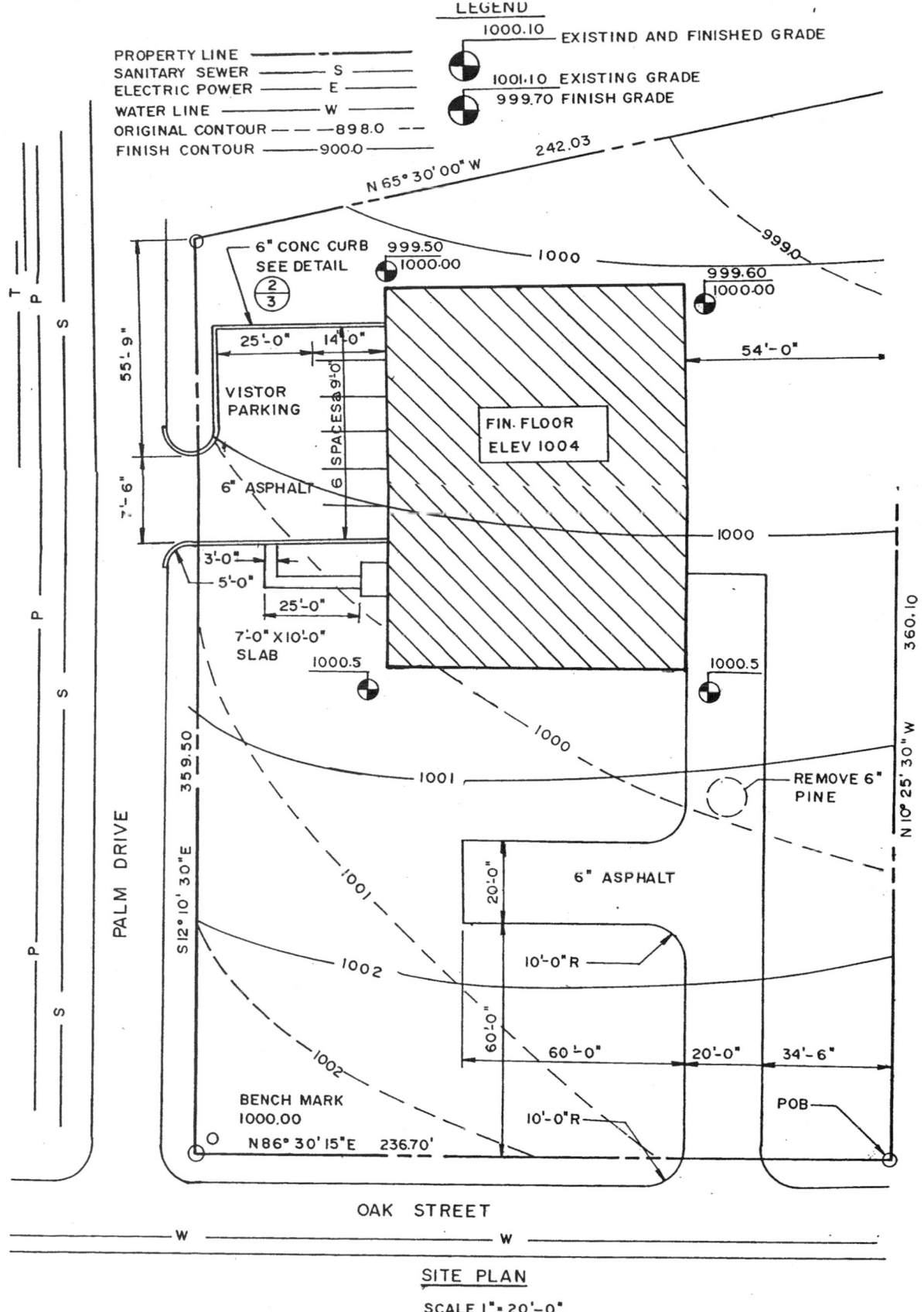

2-2 *A site plan details information about the many site features.*

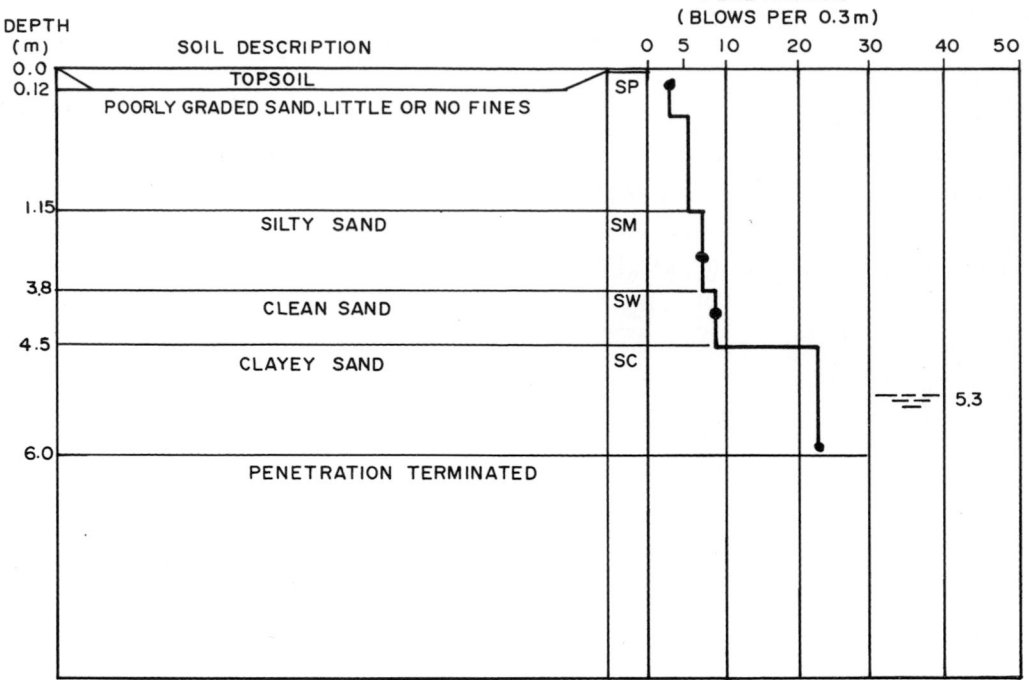

PENETRATION TEST RECORD

2-3 *The results of a typical soil investigation using the penetration method.*

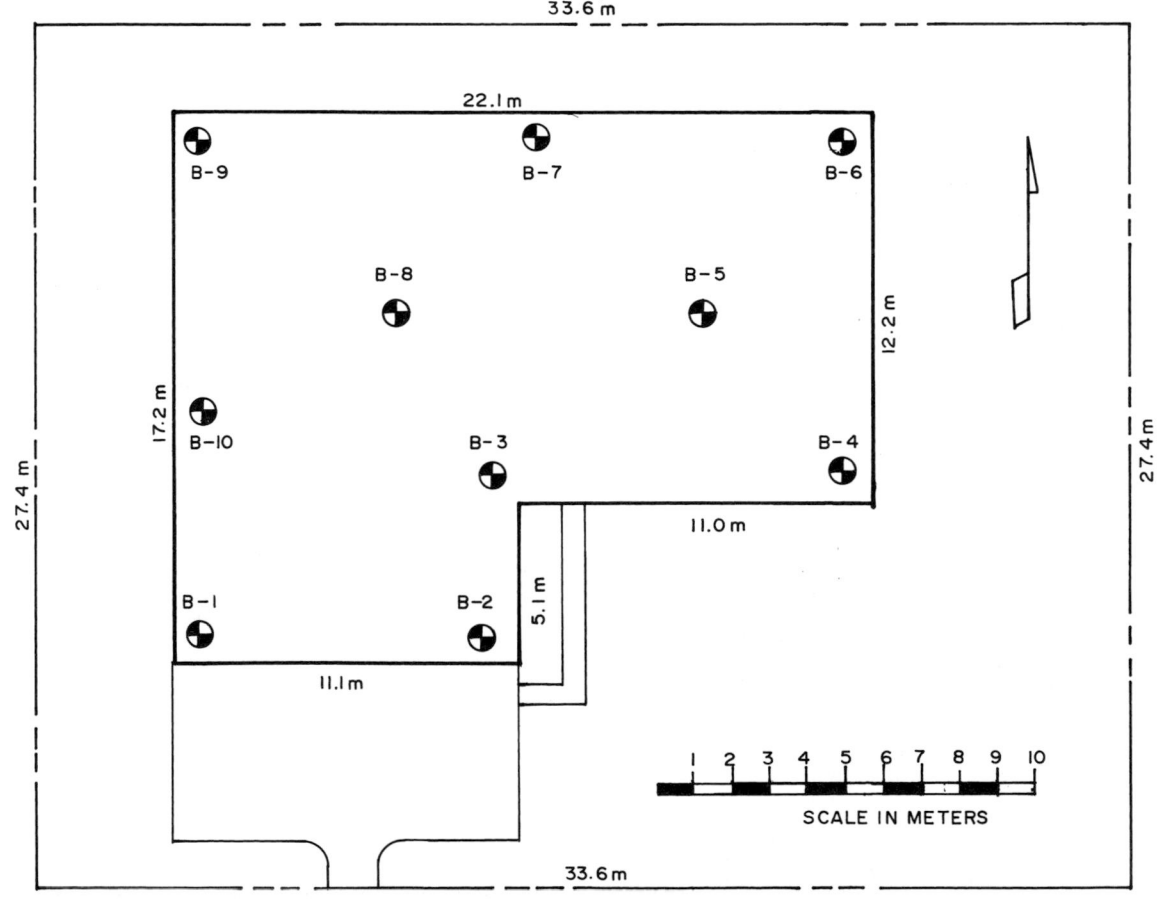

2-4 *A site plan locating the boreholes.*

Soil investigations for shallow depths (around 8 ft. or 2440mm) can be made by digging an excavation and taking soil samples. Investigations can be made using a **penetration method,** which involves driving a test rod into the soil and recording the number of blows it takes with a standard driving hammer to penetrate the soil a predetermined distance. A typical report is shown in 2-3.

Soil investigations are also made by **drilling test holes** using a hollow stem auger and wash boring techniques. The auger is mounted on the end of a drill rod that is rotated causing the auger to penetrate the soil. Water is run in the pipes under pressure. It washes away waste material and lubricates the drilling process. The location, depth, and number of borings varies with site conditions and is based on the judgment of the soil engineer. The location of boreholes is shown on a drawing of the site (see 2-4). The findings are reported on some type of drawing as shown in 2-5. The height of the water table can sometimes be noted in the borehole.

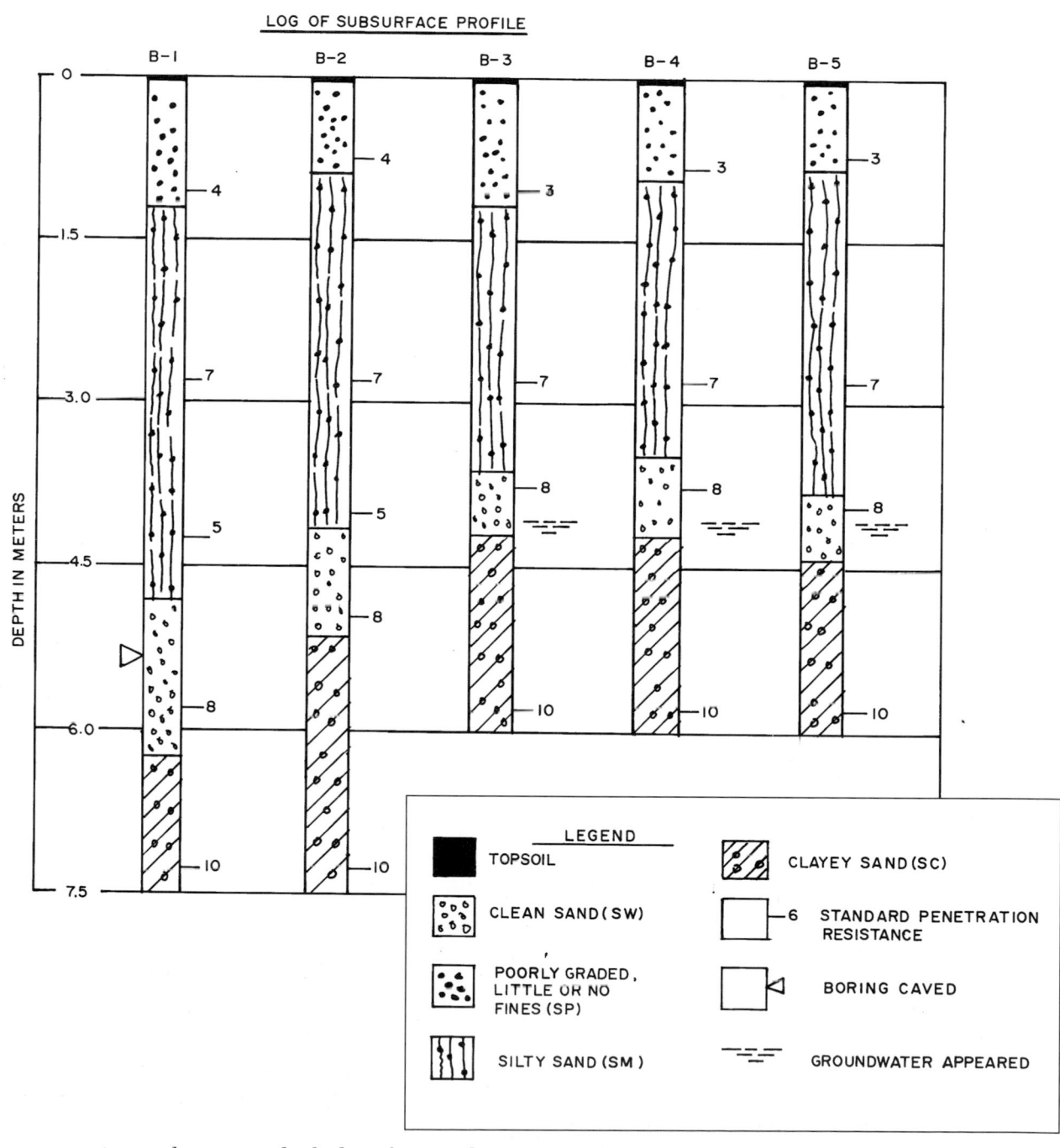

2-5 *A typical report on the findings from test borings at predetermined locations on the site.*

EARTHWORK

Earthwork includes grading, excavating, backfilling, compacting, laying base courses, stabilizing soil, slope protection, erosion control, carrying out soil treatment, and building earth dams.

Grading & Excavation

Rough grading involves adjusting the level of the ground so that the excavation and construction of the building is facilitated. Usually the top soil is removed and stockpiled for use during finish grading. Finish grading occurs after the building is complete. It brings the site elevations up to the level shown on the site or grading plan.

Excavations are required for footings, foundations, and other aspects of a project such as underground utilities. Excavations are made with power shovels, backhoes, draglines, clamshells, trenching machines, wheel-mounted belt loaders, and tractors with a bulldozer.

Usually the excavation begins with the removal of the topsoil, which is piled out of the way of the construction area. If the footings are shallow they are usually dug with a backhoe. In cold climates they will have to be several feet deep to get below the frost line. Another consideration is the need to excavate to a depth where the soil tests have shown the soil to have the required load-bearing capacity. A building with a basement will have a large excavation and a considerable amount of soil will have to be removed. Sometimes excess soil has to be removed from the site. Multistory buildings require extensive excavations, which in some cases go several stories into the ground to get to bedrock or other adequate supporting soil. If rock must be removed the cost of excavation increases rapidly. Weak and thin layered rock can be loosened with power shovels, tractor-mounted rippers, backhoes, and pneumatic hammers. Other rock formations can be fractured by explosives. The fractured rock can then be removed from the excavation by a front-end loader.

Deep excavations present an ever present danger of collapse of the sides. The depth and type of soil influence what must be done. If the excavation is not too deep and the site is large enough the side of the excavation can be sloped as shown in 2-6.

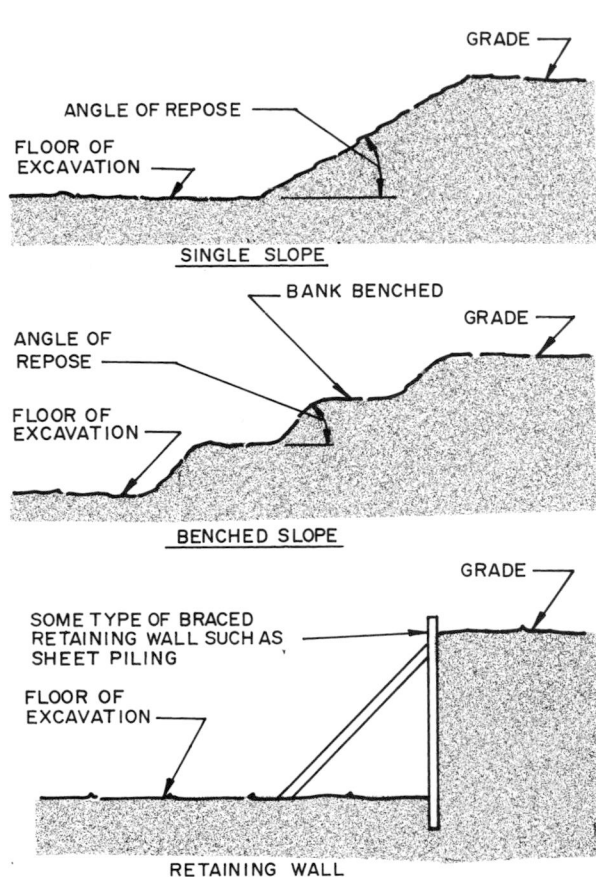

2-6 *Ways to contain the sides of excavations to prevent collapse. The angle of repose varies depending on the characteristics of the soil. Maximum slopes are specified by the* Occupational Safety and Health Act *(OSHA) regulations.*

2-7 *On shallow excavations steel pilings are driven before the excavation begins.*

Sheeting

If the sides of the excavation cannot be sloped to provide protection from slides some form of sheeting must be used, such as sheet piling, lagging, or slurry walls.

Sheet piling may be wood, aluminum, steel, or precast concrete, placed vertically (see 2-7). Sometimes steel and concrete sheet pilings are left in place after construction is finished. Wood pilings are usually removed.

Some subsoils permit the use of tiebacks. Tiebacks are steel cables or tenons that are inserted into holes drilled through the sheeting and into the rock or subsoil. The drilled hole is filled with concrete grout. When this has set the cables are tightened with hydraulic jacks and fastened to the walls. This provides an excavation free of barriers.

Another sheeting technique uses a **slurry wall**. A **slurry wall** serves as sheeting protecting the excavated area and becomes a part of the permanent foundation. It may be cast in place or built from precast concrete units. The excavation is dug with a narrow clamshell bucket, which establishes the width of the wall. The type of soil present is important to the success of this method. In order to help prevent collapse of the excavation walls the excavation is filled with a slurry composed of bentonite clay and water which stabilizes the wall. The clamshell bucket moves through the slurry to continue digging

After the required depth is reached a welded cage of steel reinforcing is lowered into the cavity. Concrete is poured from the bottom of the excavation filling it to the top. The slurry rises above the concrete and is pumped off (see 2-8). Once the entire wall is poured and has reached sufficient strength the soil can be excavated from one side. Steel tiebacks are set in holes drilled through the wall and into the subsoil as the excavation deepens.

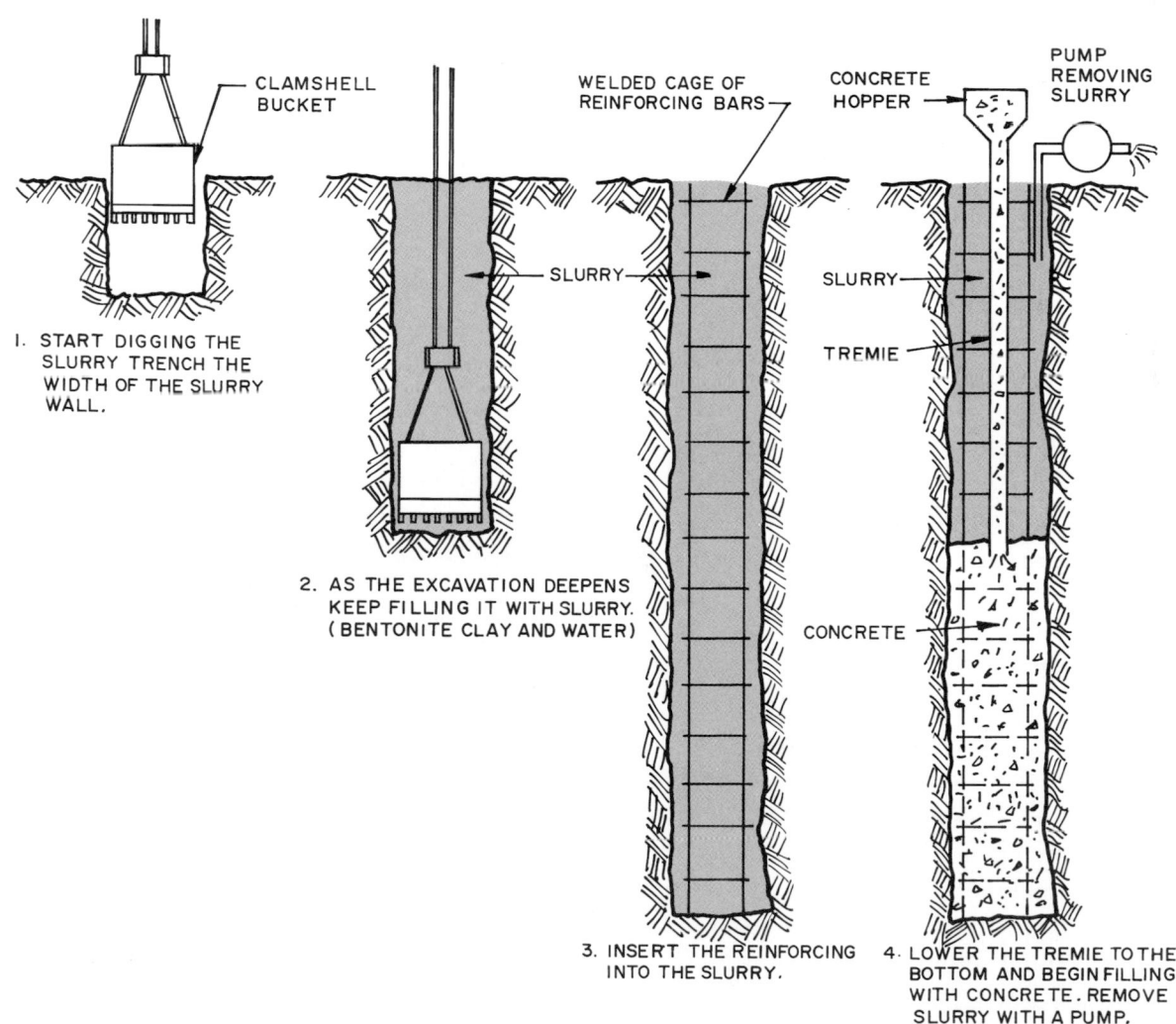

2-8 *A typical procedure for constructing a slurry wall.*

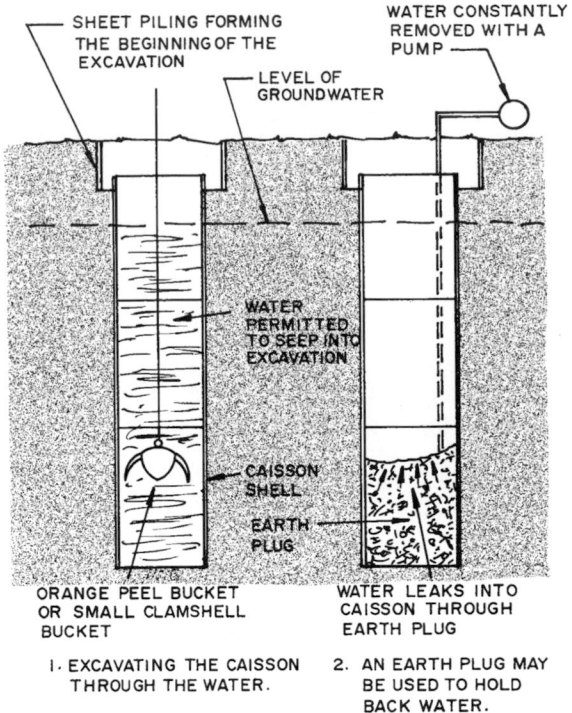

1. EXCAVATING THE CAISSON THROUGH THE WATER.

2. AN EARTH PLUG MAY BE USED TO HOLD BACK WATER.

2-9 *A typical open-air caisson.*

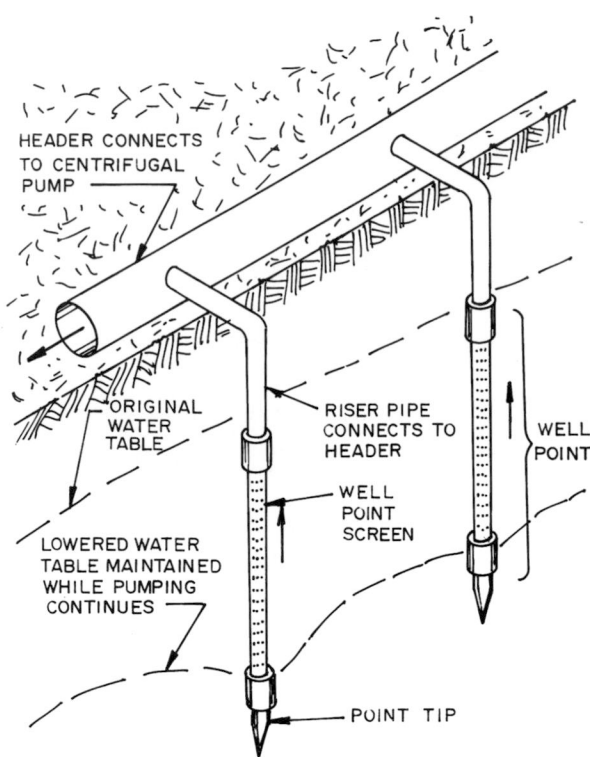

2-10 *Well points are driven into the water-bearing soil.*

Cofferdams & Caissons

Cofferdams are temporary watertight enclosures built in an area of water-bearing soil or directly in the water. They hold out the water from the area inside so that construction can proceed in a dry, stable environment. Water is pumped from within the cofferdam and pumps are maintained during construction to remove any leakage that may occur. Typically they are built using sheet piling, soldier beams with lagging, or as a double wall structure. The cofferdam extends through the areas of permeable water-bearing soil formations, through any impervious rock formations that have low-bearing capacity, and on to solid bedrock, which is used to support the foundation.

A **caisson** is a watertight shell in which construction work is carried on below water level. They may be open or pneumatic. The top of the open caisson is exposed to the weather and work is performed under normal atmospheric pressure (2-9). Pneumatic caissons are air and water tight. They are open on the bottom so soil excavation can be accomplished. The caisson is kept filled with air under pressure to keep water from entering as soil is excavated at the bottom. Workers must go through decompression when they leave to prevent suffering from the bends.

Dewatering Techniques

Dewatering is the process used to lower the subsurface water on a site so that the excavation remains dry and stable. It is usually started before excavation is started and continues as the excavation proceeds. Subsurface water is removed using a system of pumps, pipes, and well points. Basically a well point is a perforated unit placed at the bottom of a pipe driven into the soil (see 2-10). A series of well points are driven in and around the area to be excavated. They are connected to horizontal pipes above the ground called headers. The headers are connected to centrifugal pumps. The water is drawn up into the header pipes and ejected onto the side of the site where it can be drained away from the excavation. More than one series of well point, header, pump assemblies are usually required, especially as the excavation is dug deeper.

Underpinning Techniques

Excavation work involves the protection of buildings next to the excavated area. Occasionally the excavation for a new building is so close to an existing building it becomes necessary to support the building during construction of the neighboring building. This requires considerable engineering work and special considera-

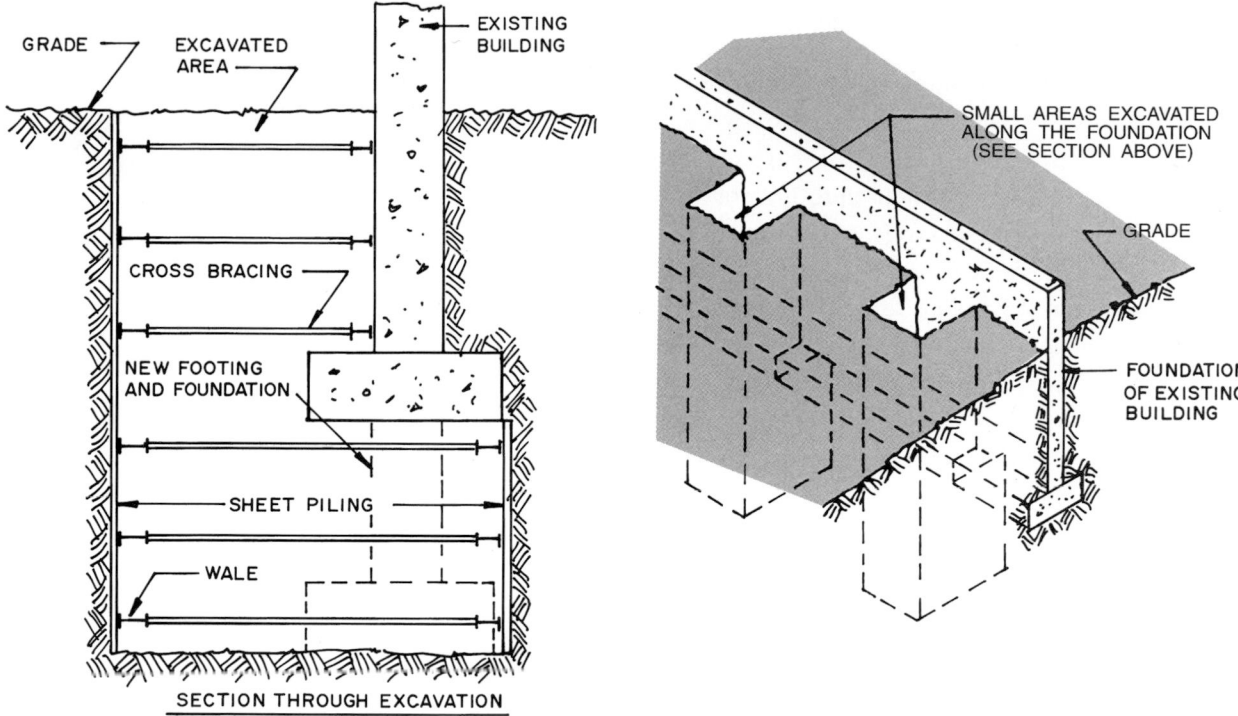

2-11 *An existing building can be supported by using the trench underpinning technique if the remaining soil can carry the weight of the building.*

tions are needed for each situation. It is important to take soil samples and have the test results evaluated by trained engineers. The soil should be evaluated to the depths required to ascertain if the new footings will be bearing upon a suitable stratum. Two frequently used techniques are to use trenches dug below the foundation of the existing building or to use needles.

Trench technique—uses trenches dug at intervals beneath the foundation. The building is supported by the remaining undisturbed soil. The required underpinning is installed in the trenched area and additional trenches are dug through the unexcavated area. This is repeated until the existing building has adequate underpinning (see 2-11).

Needles—are heavy wood timbers or steel beams that are run horizontally through the wall of the building and are supported on each end. The spacing between needles in a wall is determined by an engineer. In 2-12 the building is supported with needles run through the foundation, the area under it is excavated, and a new footing and foundation are built. Notice the sheeting used under the building to prevent cave in of the excavation wall.

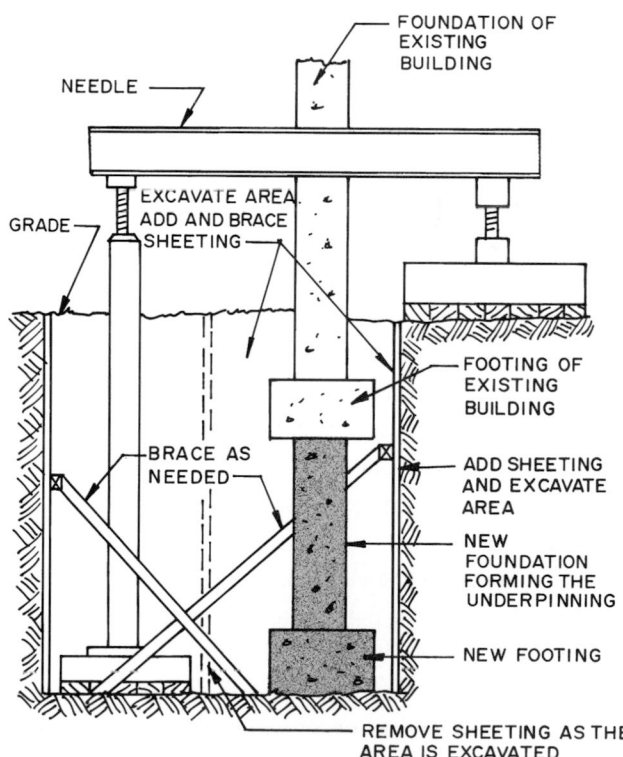

2-12 *A typical underpinning installation using needles to support the building during construction.*

TYPES OF SOIL

Soils are the end result of mechanical and chemical weathering of rock. Naturally occurring abrasive and mechanical forces wear down large rock masses into smaller particles due to thermal and gravitational forces. This produces gravels, sands, and fine silts.

The less stable minerals in rocks produce very small, flake-like particles in crystalline form. These are produced by chemical action in clayey soils.

A knowledge of soils, their characteristics, and properties is essential for those who design foundations. Building codes specify maximum design loads for various types of soils. Soil investigations on the site are required to produce the information needed.

Systems of Soil Classification

Building codes have provisions pertaining to soils. They are concerned with activities such as test borings, seismic studies, the taking of soil samples, and procedures for soil testing. Usually they specify the need for a soils report. You can find requirements for the types of loadbearing materials and loadbearing values of soils to be used to support foundations in the building code.

Important to understanding soil classification are the liquid limit, plastic limit, plasticity index, and the shrinkage limit of the soil. These states of soil consistency are indicated in terms of water content (see 2-13).

Liquid limit (LL)—of a soil is the water content expressed as a percentage of the dry weight at which the soil will start to flow when tested by the shaking test.

Plastic limit (PL)—of a soil is the percent moisture content at which the soil begins to crumble when it is rolled into a thread ³ in. (3mm) in diameter.

Plasticity index (PL)—is the difference between the liquid limit and the plastic limit. This indicates the range in moisture content over which the soil would remain in a plastic condition.

Shrinkage limit (SL)—is the water content at which the soil volume is at its minimum.

Test methods, practices, guides for the classification of soils, test methods, and preparation of test specimens are detailed in the American Society of Testing and Materials publication, *ASTM Standards on Soil Stabilization with Admixtures.*

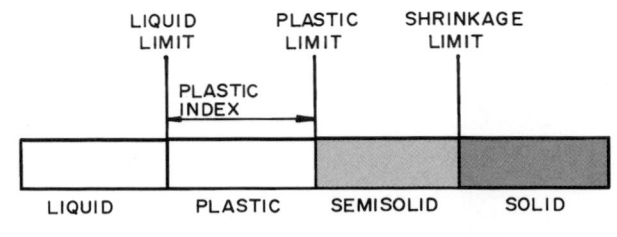

2-13 *When the water content in a soil increases it becomes more fluid.*

SOIL CLASSIFICATION BY PARTICLE SIZE

Soils are classified by the sizes of particles and their physical properties. Soil tested is typically found to be a mixture of the following five types of particle.

▼ **Gravel** is a hard rock material with particles larger than ¼ in. (6.4mm) in diameter but smaller than 3 in. (76mm).

▼ **Sand** is fine rock particles below ¼ in. (6.4mm) in diameter to 0.002 in. (0.05mm).

▼ **Silt** is fine sand with particles smaller than 0.002 in. (0.05mm) and larger than 0.00008 in. (0.002 mm).

▼ **Clay** is a material with microscopic particles (less than 0.00008 in. or 0.002mm) that is very cohesive.

▼ **Organic matter** is partly decomposed vegetable matter.

Rock particles larger than 3 in. (76mm) are called cobbles or boulders.

The commonly used soil classification systems are the Unified Soil Classification System (USCS) and the system of the American Association of State Highway and Transportation Officials (AASHTO).

UNIFIED SOIL CLASSIFICATION SYSTEM

The **USCS** was developed by the U.S. Corps of Engineers and is used to classify soils for use in roads, embankments, and foundations. Soils are classified according to grain size distribution of the fraction less than 3 inches in diameter, plasticity index, liquid limit, and organic matter content. Soils are grouped in 15 classes. Eight of these are coarse-grained soils and are identified by the letter symbols GW, GP, GM, GC, SW, SP, SM, and SC.

Six classes are fine grained and are identified as ML, CL, OL, MH, CH, and OH. The one class of highly organic soils is identified as PT. Soils on the borderline between two classes are given a dual classification such as GP-GW. Descriptive details are in Table 2-1.

Table 2-1   CLASSIFICATION OF SOILS USING THE UNIFIED CLASSIFICATION SYSTEM

Type	Letter Symbol	Description	Rating as Subgrade Material	Rating as Surfacing Material
Gravel and gravelly soils	GW	Well-graded gravel; gravel-sand mixture; little or no fines	Excellent	Good
	GP	Poorly graded gravel; gravel-sand mixture; little or no fines	Good	Poor
	GM	Gravel with silt; gravel-sand-silt mixtures	Good	Fair
	GC	Clayey gravels; gravelly sands; little or no fines	Good	Excellent
Sand and sandy soils	SW	Well-graded sands; gravelly sands; little or no fines	Good	Good
	SP	Poorly graded sands; gravelly sands; little or no fines	Fair	Poor
	SM	Silty sands; sand-silt mixtures	Fair	Fair
	SC	Clayey sands; sand-clay mixtures	Fair	Excellent
	ML	Inorganic salts; fine sands; rock flour; silty and clayey fine sands with slight plasticity	Fair	Poor
Silts and clays with liquid limit greater than 50[a]	CL	Inorganic clays of low to medium plasticity; gravelly clays; silty clays; lean clays	Fair	Fair
	OL	Organic silts of low plasticity	Poor	Poor
Silts and clays with liquid limit less than 50[a]	MH	Inorganic silts; micaceous or diatomaceous fine sandy or silty soils; elastic silts	Poor	Poor
	CH	Inorganic clays of high plasticity	Very poor	Poor
	OH	Organic clays of medium to high plasticity; organic silts	Very poor	Poor
Highly organic soils	Pt	Peat and other highly organic soils	Unsuited for subgrade material	Unsuited for surfacing

[a] The liquid limit is the water content, expressed as a percentage of the weight of the oven-dried soil, at the boundary between the liquid and plastic states of the soil.

Courtesy U.S. Department of Interior, Bureau of Reclamation.

Table 2-2 THE AASHTO SYSTEM OF SOIL CLASSIFICATION

| Typical Material | A-1 Sand and Gravel | | A-2 Gravel or Silty or Clayey Sand | | | | A-3 Fine Sand | A-4 Silt | A-5 Silt | A-6 Clay | A-7 Clay |
	A-1-a	A-1-b	A-2-4	A-2-5	A-2-6	A-2-7					
No. 10 sieve max.	50%										
No. 40 sieve	30% max.	50% max.					51% min.				
No. 200 sieve	15% max.	25% max.	35% max.	35% max.	35% max.	35% max.	10% max.	36% min.	36% min.	36% min.	36% min.
Fraction passing No. 40 sieve											
Liquid limit	6% max.	6% max.	40% max.	41% min.	40% max.	41% min.	—	40% max.	41% min.	40% max.	41% min.
Plasticity index	—	—	10% max.	10% max.	11% min.	11% min.	—	10% max.	10% max.	11% min.	11% min.

From *Standard Specifications for Transportation Materials and Methods of Sampling and Testing,* Fifteenth Edition,

Copyright 1990 by the American Association of State Highway and Transportation Officials, Washington, D.C.

Used by permission.

AMERICAN ASSOCIATION OF STATE HIGHWAY & TRANSPORTATION OFFICIALS SYSTEM

The **AASHTO soil classification system** classifies soils according to those properties that affect their use in highway construction and maintenance. In this system the mineral soil is classified in one of seven basic groups ranging from A-1 through A-7 on the basis of grain size distribution, liquid limit, and plasticity index. Soils in the A-1 through A-3 groups are sands and gravels and have a low content of fines. Soils in the A-4 through A-7 groups are silts and clays and are therefore fine grained soils. A special grade, A-8, is reserved for highly organic soils.

When laboratory tests are made on soil samples the A-1, A-2, and A-7 groups can be further classified into groupings as A-1-a, A-1-b, A-2-4, A-2-5, A-2-7, A-7-5, and A-7-6. Details are shown in Table 2-2.

FIELD CLASSIFICATION OF SOIL UNDER THE UNIFIED SOIL CLASSIFICATION SYSTEM

Sometimes it is not possible to get soil samples tested in a laboratory and field tests are made. The soil particles are separated by size by screening them through sieves with various size openings.

Table 2-3 shows the division of soil into various fractions based on the size of the soil particles. While this is important when considering the properties of a soil other properties such as the shape of the particles (angular or sharp distinct edges versus subangular or rounded edges) influence the ability of a soil to interlock particles.

Table 2-3 SOIL SIZE FRACTIONS

Constituent	U.S. Standard Sieve No.
Cobbles	Above 3 in.
Gravel	
Coarse	3–¾ in.
Fine	¾ in.–No. 4
Sand	
Coarse	No. 4–No. 10
Medium	No. 10–No. 40
Fine	No. 40–No. 200
Silts & clays	Below No. 200

Table 2-4 TYPICAL WEIGHTS, SWELL & SHRINKAGE FOR SELECTED TYPES OF SOIL

	Loose		Weight[a] Bank Measure		Compacted			
	lb./yd.³	kg/m³	lb./yd.³	kg/m³	lb./yd.³	kg/m³	Swell	Shrinkage
Clay, natural	2300	1300	2900	1720	3750	2220	38%	20%
Common earth, dry	2100	1245	2600	1540	3250	1925	24%	10%
Sand with gravel, dry	2900	1720	3200	1900	3650	2160	12%	12%
Sand, dry	2400	1400	2700	1600	2665	1580	15%	12%

[a]Weights, swell & shrinkage of actual samples may be larger or smaller.

Soil Volume Characteristics

When soil is excavated, hauled, placed, and compacted its volume changes. When soil is excavated and moved it increases in volume. This is called **swell.** The grains loosen and air fills the voids between them. The amount of swell varies in different types of soil. It is measured as a percentage of the original volume.

When the soil is placed and compacted the air is forced out between the grains and its volume decreases. This is called **shrinkage.** The amount of shrinkage varies with the type of soil and is measured as a percentage of the original volume.

The volume of one cubic yard of soil **in situ** (undisturbed soil) is referred to as **bank measure.** It will, when disturbed, produce more than one cubic yard and when compacted produce less than one cubic yard of soil. The weight and volume changes for typical types of soil are shown in Table 2-4.

Soil volume also changes due to a change in the moisture content. Clays are especially susceptible to changes in moisture. As moisture content increases the soil expands sometimes to the point of lifting a footing or slab. If the moisture decreases the soil will shrink sometimes causing the footing or slab to settle. Notice these actions as shown in 2-14.

Soil Water

▼ The two major types of water present in soils are capillary water and gravitational water.

Capillary water—is retained in the microscopic pores of the individual sand particles and enables them to bind together.

Gravitational water—is water flowing within the soil in the voids between particles. It can be removed by pumping.

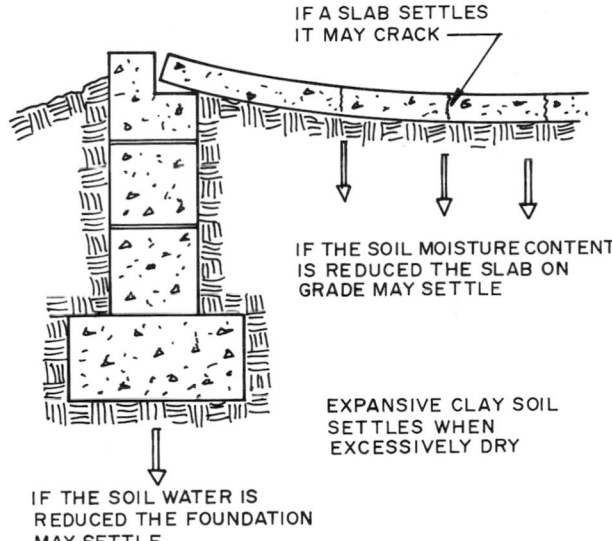

IF A SLAB SETTLES IT MAY CRACK

IF THE SOIL MOISTURE CONTENT IS REDUCED THE SLAB ON GRADE MAY SETTLE

EXPANSIVE CLAY SOIL SETTLES WHEN EXCESSIVELY DRY

IF THE SOIL WATER IS REDUCED THE FOUNDATION MAY SETTLE

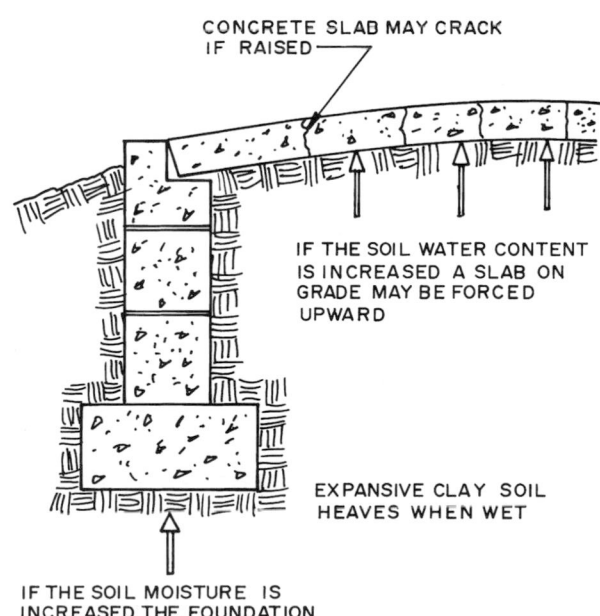

CONCRETE SLAB MAY CRACK IF RAISED

IF THE SOIL WATER CONTENT IS INCREASED A SLAB ON GRADE MAY BE FORCED UPWARD

EXPANSIVE CLAY SOIL HEAVES WHEN WET

IF THE SOIL MOISTURE IS INCREASED THE FOUNDATION MAY BE FORCED UPWARD

2-14 *Expansive clay soils heave when wet and settle when excessively dry.*

The **water table** or **groundwater level** is the water below the surface. Since it is free flowing the water level below the soil surface can be varied by pumping water from wells, by a dry or rainy season, and the installation of drain pipes to remove it from the construction area.

The moisture content of soil is an important factor when compacting it. The density varies with the moisture content. The moisture in proper amounts enables the soil to be compacted to its maximum density. This is called the **optimum moisture content** and is determined by laboratory tests of soil samples.

Soil-Bearing Capacities

The acceptable load-bearing capacities of the various types of soils are specified by the local building code. In Table 2-5 are the maximum allowable load-bearing values for supporting soils under spread footings placed at or near the surface of the site. In many cases it is necessary to test soil samples to determine the maximum allowable pressure.

If the soil in the area of the structure is uniform in composition and the footing design is adequate the building will remain stable. Even if it settles a little over time the settlement will be uniform. However, if the soil composition varies and is not compensated for in the footing design the building could begin to settle unevenly. Actions of others could also affect existing structures. For example, extensive pumping of subsurface water will cause soil to settle because of a lowering of the water table.

Soil Modification

After the soil tests have been made and the properties determined the soil can be modified to improve its characteristics. Commonly used processes to alter the properties include compaction, stabilization, and dewatering.

COMPACTION

Compaction of soil refers to increasing its density by mechanically forcing the soil particles closer together. This expels the air in the voids between the particles. Soil density can also be increased by **consolidation** which involves removing the water in the voids between soil particles permitting the particles to come closer together. Compaction produces immediate results. Consolidation requires a much longer period of time to increase the soil density.

Table 2-5 PRESUMPTIVE LOAD-BEARING VALUES OF FOUNDATION MATERIALS

Class of Material	Load-Bearing Pressure lb./ft.^{2}	kg/m^2
Crystalline bedrock	12,000	58 560
Sedimentary rock	6,000	29 280
Sandy gravel or gravel	5,000	24 400
Sand, silty sand, clayey sand, silty gravel, and clayey gravel	3,000	14 640
Clay, sandy clay, silty clay, and clayey silt	2,000	9760

From *The BOCA National Building Code/1993*
Courtesy Building Officials and Code Administrators International, Inc.
Copyright 1993. All rights reserved.

The amount of compaction that can be obtained depends upon the physical and chemical properties of the soil, its moisture content, the compaction method used, and the thickness of the soil layer being compacted. As compaction occurs the air and water between the soil particles is forced out of the soil. Excess water must be drained away. The soil engineer will use the results of the soil tests to decide the best way to achieve the required compaction density.

There are many ways compaction can occur. These include a kneading action, vibration, static weight, explosives, and impact.

Pneumatic-tired rollers are multitired units that effect compaction of the soil by providing a kneading action. The rows of tires are staggered to give complete coverage and some units have wheels mounted to give a wobbly effect increasing the kneading action. The weight of the unit can be varied by adding ballast. They are effective on most soils but least effective on sands and gravels.

Tamping foot rollers, such as the sheep's-foot roller in 2-15, use static weight to compact the soil. The tamping foot rollers are available in a range of foot sizes and shapes. As they pass over the soil the feet sink into it, compacting it below the surface. With repeated passes the feet do not penetrate as deep and eventually walk on the top surface. Tamping rollers are most effective on cohesive soils.

2-15 *This compactor has sheep's-foot rollers that provide a tamping action. The front-mounted bulldozer is used to rough grade the surface.*

Courtesy Caterpillar, Inc.

2-16 *Smooth-wheeled rollers are used to compact granular materials and asphalt pavements.*

Another type of static-weight compactor has smooth **steel wheels** or **drums.** It is best when used to compact granular bases, asphalt bases, and asphalt pavements. It is not effective on cohesive soils because it tends to compact the surface forming a crust over a loose, noncompacted subsurface (see 2-16).

There are a variety of compaction devices utilizing **vibration.** Small hand-operated vibratory compactors are used on compacting areas where it may be difficult to get larger equipment, such as preparing the base for a concrete floor or sidewalk. Larger vibratory compactors include tamping foot rollers and smooth drum rollers. In addition to static weight the vibratory action provides additional compaction.

Impact rammers are used to provide compaction in tight areas. The vertical movement of the rammer provides considerable compaction. Units are also available for mounting on the end of a backhoe boom.

Loose, saturated, granular soil can be compacted by subjecting it to a sudden shock and vibration. This causes the soil particles to fill into a denser pattern displacing the water existing between the particles. The weight of the particles forces the water to flow from the soil. Explosives are typically used for this purpose. The spacing, depths, and sizes of the explosive charges are determined by experienced soil engineers.

Another way to increase the density of cohesionless soils is to use **vibrocompaction.** This uses a vibratory probe, which is a large diameter tube with a vibrating device mounted on one end. It is lifted by a crane and driven into the soil by the weight of the tube and the

vibrations. This process using a patented compacting probe is shown in 2-17. The probe is inserted in the soil by a powerful vibrator. The frequency and duration of vibration can be monitored to achieve maximum compaction for the prevailing soil conditions. The degree of compaction depends upon the soil type, the spacing of the probes, and the duration of vibration.

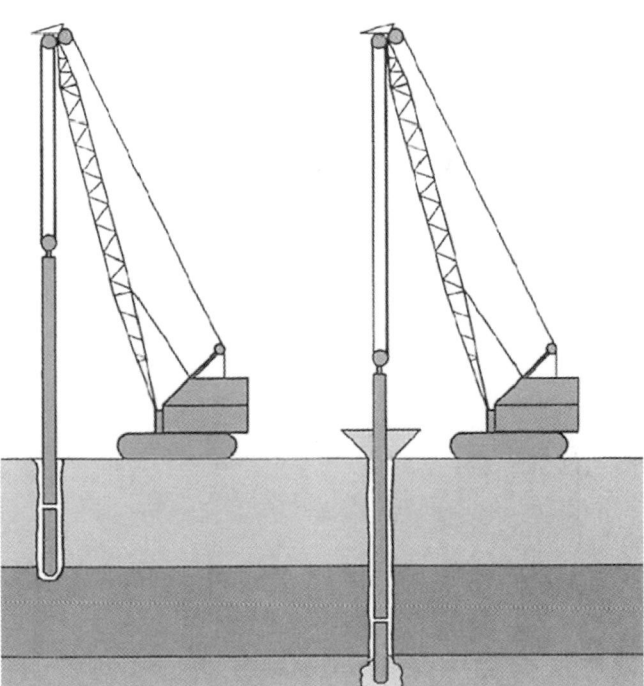

2-17 *Vibrocompaction consolidates the soil by inserting a patented compaction probe into the soil and using a powerful vibrator to drive the probe into the soil.*

Courtesy Franki International

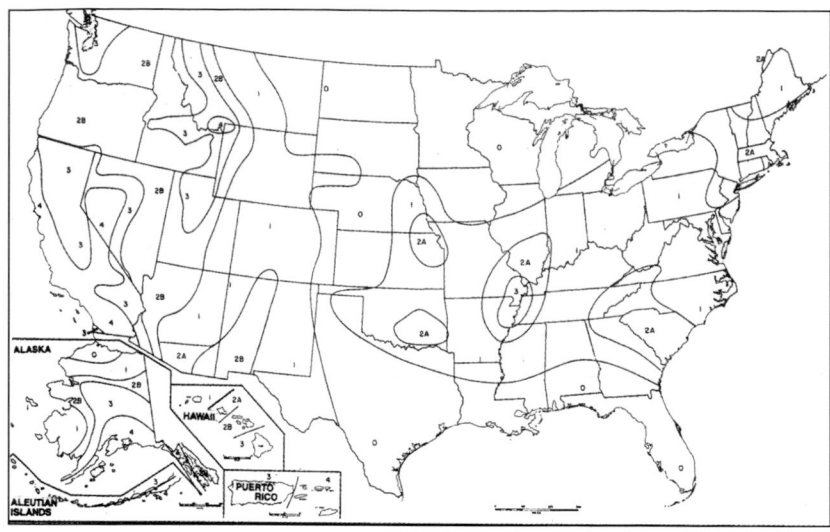

ZONE 0 — No damage.

ZONE 1 — Minor damage. Distant earthquakes may cause damage to structures with fundamental periods greater than 1.0 seconds. Corresponds to intensities V and VI of the Modified Mercalli Intensity Scale.

ZONE 2 — Moderate damage. Corresponds to intensity VII of the Modified Mercalli Intensity Scale.

ZONE 3 — Major damage. Corresponds to intensity VIII and higher on the Modified Mercalli Intensity Scale.

2-18 *Seismic zone map of the United States.*
Courtesy International Conference of Building Officials

Soil Stabilization with Admixtures

Test methods and specifications on soil stabilization with admixtures are detailed in the publication, *ASTM Standards on Soil Stabilization with Admixtures.* Following are explanations of some of these.

BLENDING SOILS

Soils may be blended by bringing in soil and mixing it with the original with a motor grader or disc. Power shovels or deep cutting belt loaders can be used to cut deeply through the soil mixing that below the surface with that on top.

LIME-SOIL STABILIZATION

Clays and silty clay soils can be stabilized by mixing lime, which produces a chemical reaction. Clays expand when wet causing damage to concrete surfaces over them.

The concrete slabs heave up and crack due to the expansion of the clay. The lime reduces the amount of expansion and forms a moisture barrier protecting the expansive clay from subsurface water. In some cases it can remove the need to excavate unsatisfactory soil and replace it.

Usually slaked or hydrated lime or quicklime is spread over the base material and blended into the base with a pulverizer type machine. Water may be added if conditions warrant. The blending is generally 12 to 18 in. (300 to 450mm) deep and compacted to the final thickness. The compacted layer will continue to gain strength for many months.

ASPHALT-SOIL STABILIZATION

Asphalts blended with granular soils produce a durable, stable soil and may be used as a finished surface for low traffic roads or a stabilized base course for a higher quality pavement. Some soils require the addition of fines with the asphalt to fill the spaces between the soil particles.

CEMENT-SOIL STABILIZATION

Predominately granular soils having minute amounts of clay particles can be stabilized by blending with portland cement. In some cases fly ash is used to replace some of the portland cement. The portland cement is uniformly spread over the surface, and blended with the soil using a pulverizer type machine. After this has been accomplished to the specified depth the surface is fine graded and compacted. Water may be sprinkled on the surface during blending if the soil moisture content is low. Since portland cement sets up rapidly compaction immediately after blending is required.

SALT-SOIL STABILIZATION

Coarse crushed rock salt can be used to stabilize well graded clay or loamy soil having some limestone fines. The salt can be blended in the soil in dry form or mixed with water to form a brine. The salt and water form a bond with soil particles and thus form a stable soil.

The base soil is pulverized to the specified depth, the salt added, and the mixture blended. This layer is then compacted, fine graded, and watered and finished with a steel smooth roller. It can take up to several weeks before the soil is cured.

SEISMIC CONSIDERATIONS

Destructive forces caused by earthquakes are called **seismic forces.** An earthquake is the oscillatory movement of the earth's surface that follows a release of energy in the earth's crust. When subjected to deep-seated forces the crust may first bend and when the stress exceeds the strength of the rocks they break and "snap" to a new position. This produces vibrations called seismic waves. The waves travel along the surface and through the earth. Some of these vibrations have a high enough frequency to be audible. Others are of low frequency. While accurate determination and prediction of earthquakes is difficult certain sections of the country are known to have subsurface conditions that make them possible earthquake zones as shown in 2-18. These areas have special building code requirements regulating the design and construction of buildings and other structures.

An earthquake is caused by the slippage of the earth along a fault plane. A **fault** is a fracture in the earth's crust along which two blocks of crust have slipped with respect to each other. Common movements include one crustal block moving horizontally in one direction while the block facing it moves horizontally in the other direction. This is referred to as **strike-slip.** In other cases one block may move up vertically while the abutting block moves downward. This is referred to as **vertical slip.** In some cases the movement along the fault may have both vertical and horizontal movement. A **thrust fault** results when sections of rock press together forcing one side up over the other. A **blind thrust** raises the surface into folded hills without breaking the surface (see 2-19). Frequently these slippages of strata occur deep in the earth and cause no surface rupture but do cause surface vibration and shaking. These fault slippages generate ground motions that radiate vertically and horizontally in all directions. Earthquakes frequently produce ruptures of the surface as well as a rolling, waving motion of the ground surface. The ground shakes as this occurs and the shaking is what causes the most damage to buildings and other structures rather than the more visible and dramatic ground rupture.

The magnitude of an earthquake has been measured by the **Richter scale,** which is based on the maximum single movement recorded on a seismograph. The most widely used measurement used by seismologists is the **moment magnitude.** It is based on the size of the fault on which the earthquake occurs and the amount of earth slippage. The larger the fault and the larger the actual slip the higher the moment magnitude of the earthquake.

The design of footings for buildings in seismic areas are specified by local codes. The design varies with the soil classification. Following are some examples. Spread footings and pier foundations are connected with ties capable of resisting specified forces in tension and compression. Piles are designed to resist the maximum imposed curvatures from seismic forces. Piles and pile caps are interconnected by ties.

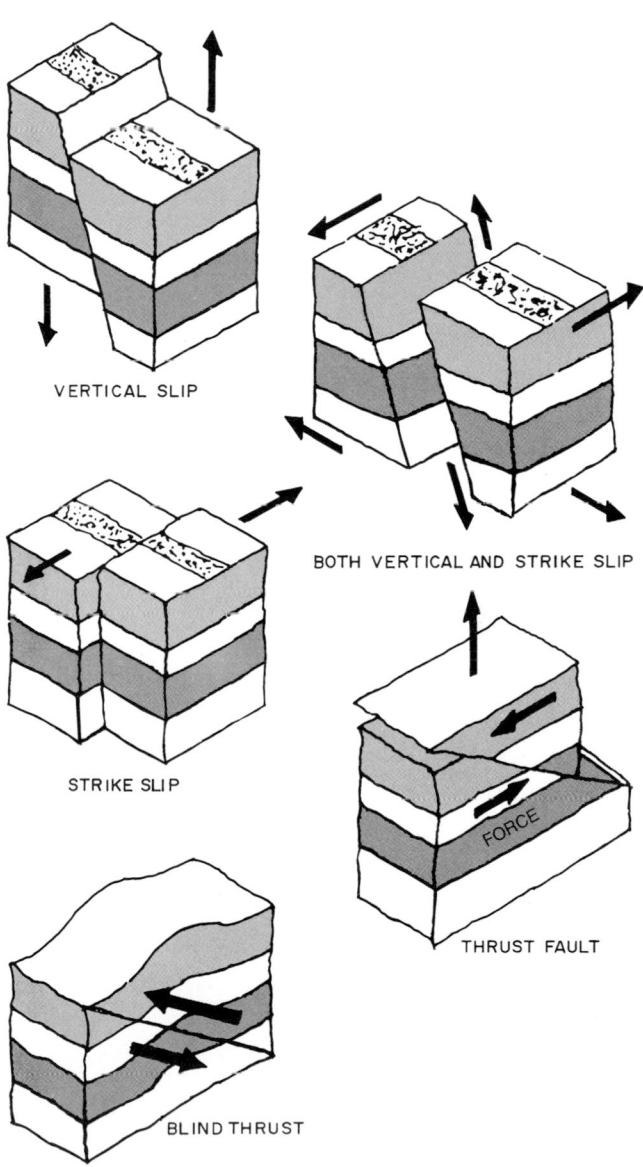

VERTICAL SLIP

BOTH VERTICAL AND STRIKE SLIP

STRIKE SLIP

FORCE

THRUST FAULT

BLIND THRUST

2-19 *Typical types of fault.*
Courtesy U.S. Geological Survey

Connections between structural steel piles and the pile cap must be designed for a specified tensile force. Concrete-filled steel and tube piles must have the specified minimum reinforcement and longitudinal reinforcement is also required. Cast-in-place piles also must have the minimum reinforcement specified by the code and the reinforcing must be sized and placed according to the code.

Masonry construction is usually required to meet the requirements for seismic design set forth by the American Concrete Institute, American Society of Civil Engineers, and the Masonry Council. Structural steel seismic requirements are usually those set forth by the American Institute of Steel Construction, Inc. in their publication *AISC Seismic Provisions for Structural Steel Buildings*, plus other requirements specified by the local code. Seismic requirements for wood and timber construction follow the recommendations of the National Forest and Paper Association and additional local code specifications.

SURFACING MATERIALS

Surfacing materials can be divided into two major types: flexible and rigid. Flexible pavement includes those using some form of bituminous or rubber modi-fied material. Rigid pavement includes various materials utilizing portland cement. You can see typical construction for a flexible pavement in 2-20 and a rigid pavement in 2-21. In both cases the preparation of the base for the pavement is vital. The lowest strata is the natural subgrade, which is consolidated by compaction.

The **subbase** course is laid on top of the natural subgrade. It is a natural soil of a higher quality than the natural subgrade. It provides additional support as required to distribute the loads imposed on the finished surface. The **base course** is laid over the subbase and is the surface upon which the **finished wearing surface** is laid. It is a granular base of high quality such as crushed stone, gravel, slag, or some combination of these. Sand is also sometimes mixed with these. It is often treated with a bitumen to bind the materials. The subbase and base courses are extended beyond the edge of the next layer to provide a cone dispersal of the loads.

Asphalt Materials

Asphalt is a dark brown to black cementitious material in semisolid or solid form made up of bitumens found in deposits of natural asphalt. Asphalt is also refined from petroleum. Following are the types of asphalts that find some use in paving.

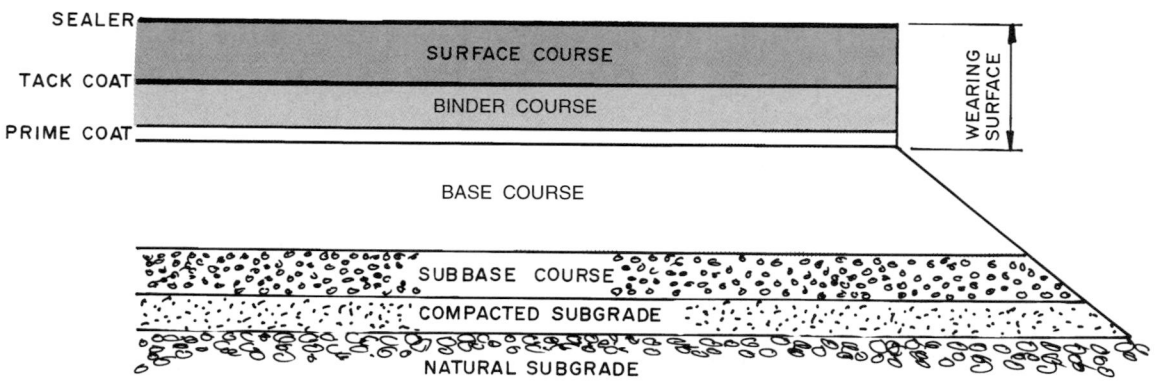

2-20 *The base courses and wearing surface course for a typical flexible pavement.*

2-21 *The base courses and concrete paving for a typical rigid pavement.*

Cutback asphalt is a broad classification of residual asphalt materials left after the petroleum has been processed, which produces gasoline, kerosene, diesel oil, and lubricating oil.

The **residual asphalt** is blended with various solvents producing cutback asphalt. There are three classifications: rapid curing (RC), medium curing (MC), and slow curing (SC). They are mixed with granular soil to stabilize a road bed before paving or to serve as a binder for a finished road surface for light traffic.

The **emulsion** group of asphalts consists of emulsified asphalt. An emulsion is a liquid dispersed within an immiscible liquid in the form of droplets. The emulsion is made by adding heated, fluid asphaltic cement into water to which an emulsifying agent, such as soap or bentonite clay, has been added. A stabilization agent, protein, is added to prevent the particles from blending together within the mix. The suspended particles of asphalt blend with the aggregate or soil particles as the water drains away and evaporates.

2-22 *An asphalt plant.*
Courtesy The Asphalt Institute

Emulsified asphalts are either anionic or cationic depending upon the emulsifying agents used. Anionic emulsions are those in which the asphalt globules in the mix are negatively charged. Cationic emulsions have asphalt globules that are positively charged. Each comes in three grades. They are used for the same purposes as cutback asphalts.

ASPHALT PAVING

Four types of asphalt paving are in common use. They are made with either cutback or emulsified asphalt.

Asphalt concrete—consists of asphalt cement and graded aggregates carefully proportioned and mixed in an asphalt plant at controlled temperatures (see 2-22). The mix is transported to the site, spread by an asphalt paving machine, and rolled while it is still hot (see 2-23).

Cold-laid asphalt—is made in the same way as asphalt concrete but the asphalt liquid is cold.

Asphalt macadam—is laid using a penetration method. A coarse aggregate is laid over the base, compacted to a smooth surface, and sprayed with an asphalt emulsion or hot asphalt cement. This is covered with fine aggregates and rolled, forcing the fine aggregate into the voids between the larger aggregate.

Concrete Paving

Roads, driveways, and parking lots are often paved with concrete. After the subbase and base courses are laid, side forms are placed along the edges of the paving to contain the plastic concrete and are used as rails upon which the concrete placer spreader rides (see 2-24). Some placing spreaders are designed to ride inside the side forms while others ride outside the forms.

2-23 *Laying asphalt concrete paving.*
Courtesy The Asphalt Institute

2-24 *Laying concrete paving.*
Courtesy Bid-Well Division of CMI Corporation

2-25 *The concrete paving is leveled, consolidated, and finished.*

Courtesy Bid-Well Division of CMI Corporation

2-26 *This concrete curb and gutter are slip formed in a continuous casting process.*

The pavers may be form or slip-form types. The **form-type paver** rides on the metal side forms, places the concrete, strikes it off, and consolidates (see 2-25). Some types of pavers can be used for form or slip-form placement. **Slip-form pavers** place a slab without the side forms. The concrete is of a consistency that develops sufficient strength to be self-supporting as it leaves the paver. The slip-form paver spreads, consolidates, and finishes the concrete slab (see 2-26). (Information on concrete is in Division 3.)

FOUNDATIONS, PILES & CAISSONS

The foundation is the part of a structure that serves to transmit the weight of the structure to the earth or rock at or below ground level. Residential and small commercial buildings consist of two major parts, a superstructure, which is the portion above grade, and the foundation, which is mostly below grade. Large commercial buildings will often have a third part, the substructure, which is habitable space below grade and a foundation, typically caisson piles that penetrate the soil well below the substructure.

Foundation Design

The design of the foundation begins as soon as the basic architectural concepts are developed. Since there are so many factors to consider the actual foundation

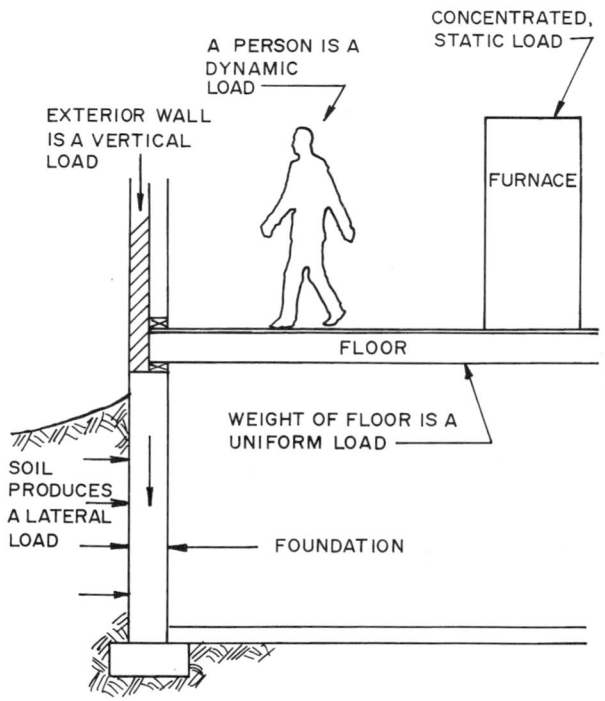

2-27 *Loads on the foundation may be vertical, lateral, static, dynamic, concentrated or uniform.*

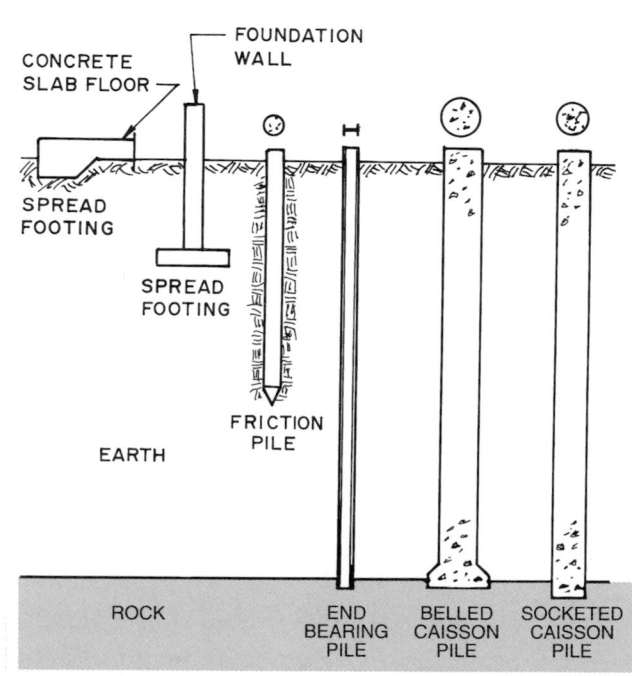

2-28 *Spread, pile and caisson foundations.*

may be in a period of flux during the preliminary stages. Codes require an investigation of subsoil conditions. These data may dictate the choice of foundation used. The size of the building, loads to be carried, conditions on the site, climate, type of structural system to be used, and other such factors are considered by the foundation engineer.

The final building is the result of the cooperative effort of these three engineering staffs:

Architect—is involved with factors such as the shape and size of the building, orientation, space utilization and material selection.

Structural engineer—designs the structural aspects of the building.

Foundation engineer—must work with both as foundation designs are developed.

Foundation Loads

Foundations are designed to carry the loads imposed by the structure and any additional loads produced by occupancy. These are divided into dead loads and live loads.

Dead loads—are those, such as the weight of materials used to construct the building, that act continuously on the foundation.

Live loads—are those not constant, such as the weight of furniture, people, wind or snow.

Loads on the foundation (see 2-27) may be: **vertical** or **lateral** (horizontal), **static** (fixed) or **dynamic** (moving), **concentrated** (on one spot) or **uniform** (spread evenly over a surface).

Types of Foundation

Foundations are classified into three major types: spread, pile, and caisson pile foundations.

Spread foundations—transfer the loads to the soil through the footings at the bottom of a foundation wall or column.

Pile foundations—use long wood, concrete, or steel piles driven into the earth. They get their load carrying ability by friction on the sides of the pile or from the end resting on load-bearing strain.

Caisson pile foundations— are formed by drilling holes in the earth and filling them with concrete. They get their load-carrying ability in the same manner as piles (see 2-28).

SPREAD FOUNDATIONS

Spread foundations utilize spread footings to distribute the building loads over a large enough area of soil to provide adequate bearing capacity. They are usually near the surface of the site and are considered shallow footings. Following are the most commonly used spread footings.

Continuous wall footing—(strip footing) is the most common spread footing. Columns and piers are placed on **independent footings,** which are usually square or rectangular. Both types generally are reinforced with steel reinforcing bars (see 2-29). Spread footings must be placed on undisturbed soil. If the excavation for footing is dug too deep, it must be filled with concrete. In some cases compaction is allowed but only if carefully engineered to provide a known load-bearing capacity. Wall footings on sloped sites are usually stepped using horizontal steps.

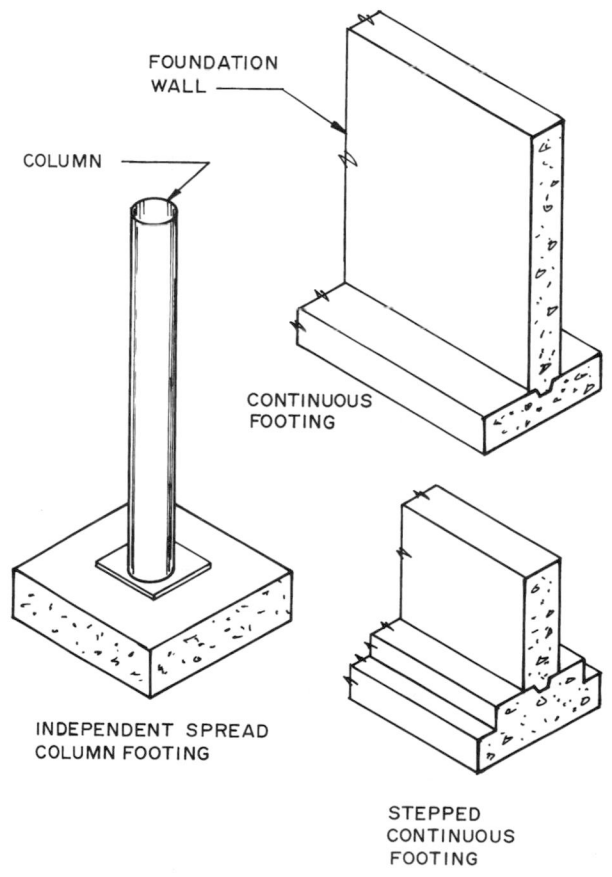

2-29 *Spread footings are used under foundation walls, piers, and columns.*

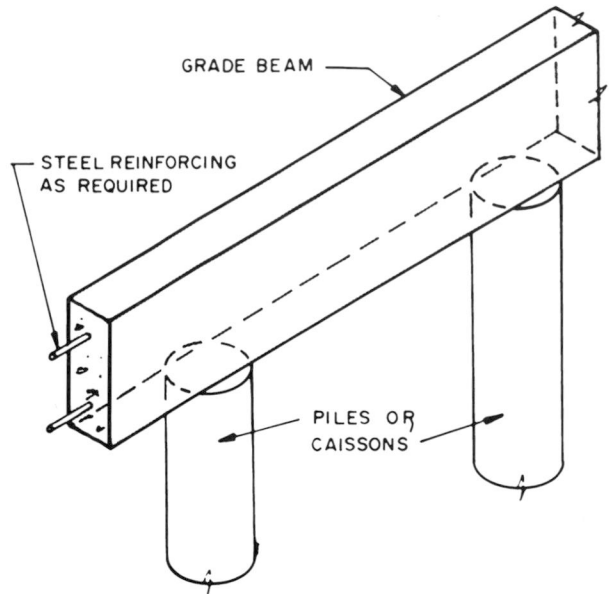

2-30 *Grade beams are cast on piles or piers and serve as the foundation.*

Grade beam foundations—support the building on a grade beam that rests on a series of piles or piers. A grade beam is a reinforced concrete beam serving as the foundation wall (see 2-30).

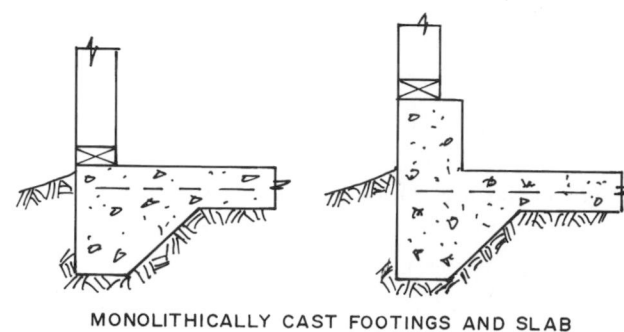

MONOLITHICALLY CAST FOOTINGS AND SLAB
USED IN WARM CLIMATES

2-31 *These shallow monolithically cast footings and floor slabs are used in warm climates.*

Slab on grade—is a widely used shallow foundation, as shown in 2-31. There are many designs for this type of footing. It is used in warm climates where the surface soil has the necessary bearing capacity. The slab can be thickened within the building to support interior load-bearing partitions.

Raft or **mat foundations**—are reinforced concrete slabs several feet thick covering the entire area of the foundation (see 2-32). This slab spreads the weight of the building over the entire area below the building and reduces the load per square foot on the soil.

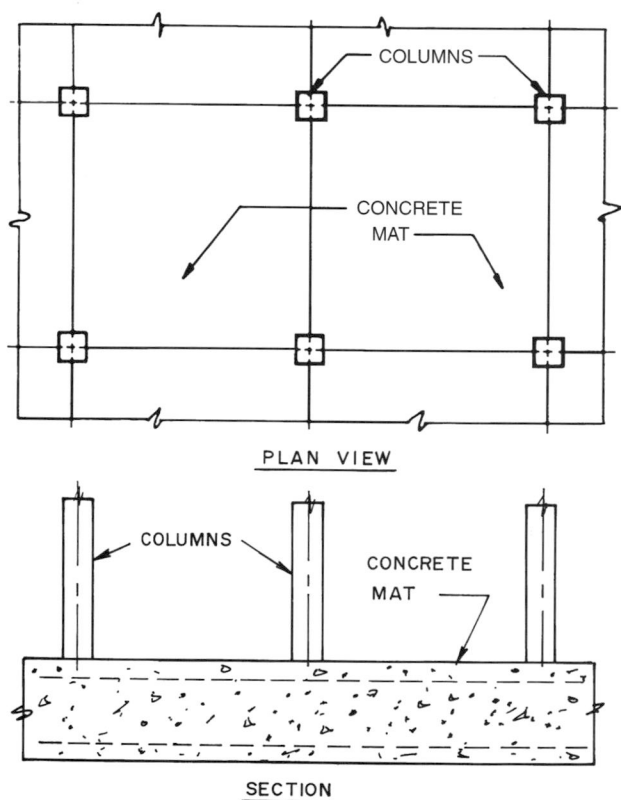

2-32 *This is a solid poured concrete matt foundation supporting several structural columns.*

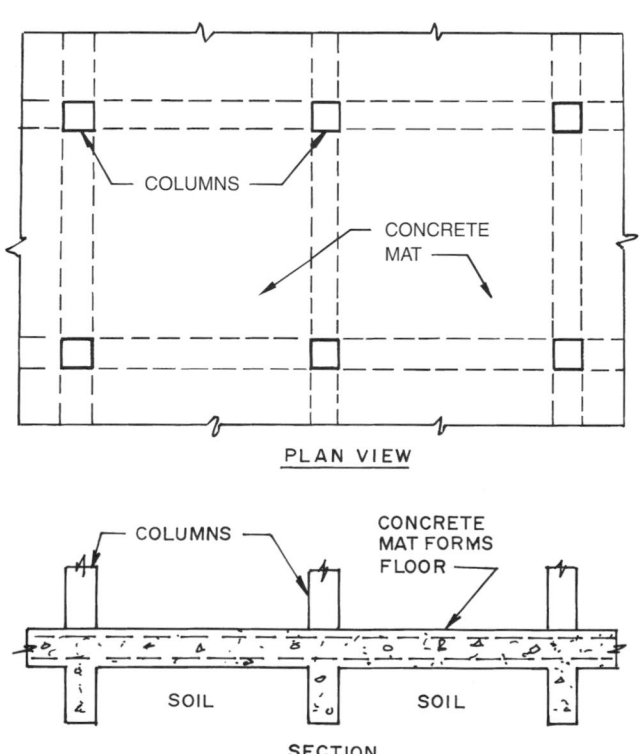

2-33 *One type of T-beam matt foundation.*

This is especially useful when the soil characteristics vary across the area. In order to avoid construction joints the entire slab must be poured continuously. Another raft design uses a T-beam construction that has a solid slab poured and two directional beams poured above or below the slab (see 2-33 and 2-34). When the beam is inverted the slab becomes the basement floor. The beams cavities are dug in the soil, which must be able to stand without caving in as the beams and slab are poured. When the beams are on top of the slab the spaces between them must be filled before pouring a concrete floor.

Steel grillage footings—use steel beams to reinforce the footing. The base layer is sized to the area needed to support the load. The upper layer of beams are joined at right angles to the lower level and the entire thing encased in concrete (see 2-35).

Combined footings—are used to support walls and columns near the property line where it is not possible to allow the footing to cross the line onto the neighboring property. The combined footing ties the outside row of columns to the next row within the building. If the outer row of columns were to be placed on the edge of individual footings the loads would not be symmetrical and the footings would tend to settle unevenly or rotate (see 2-36).

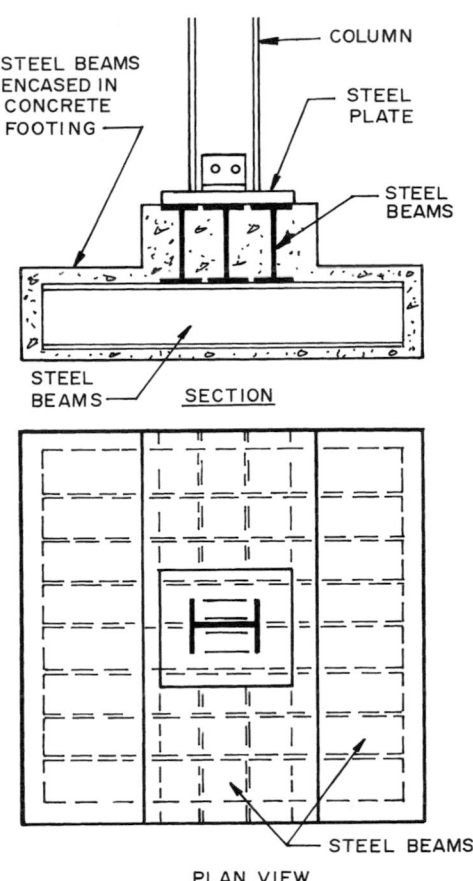

2-35 *A steel grillage column footing.*

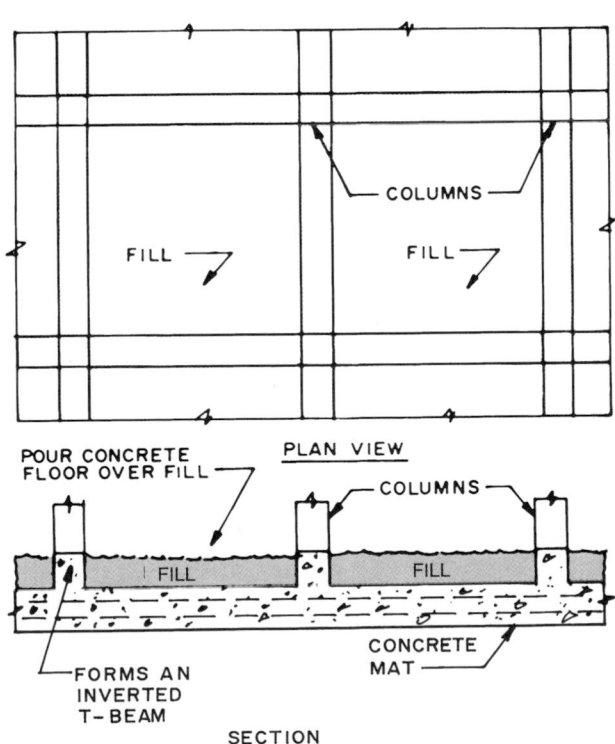

2-34 *Another type of T-beam matt foundation.*

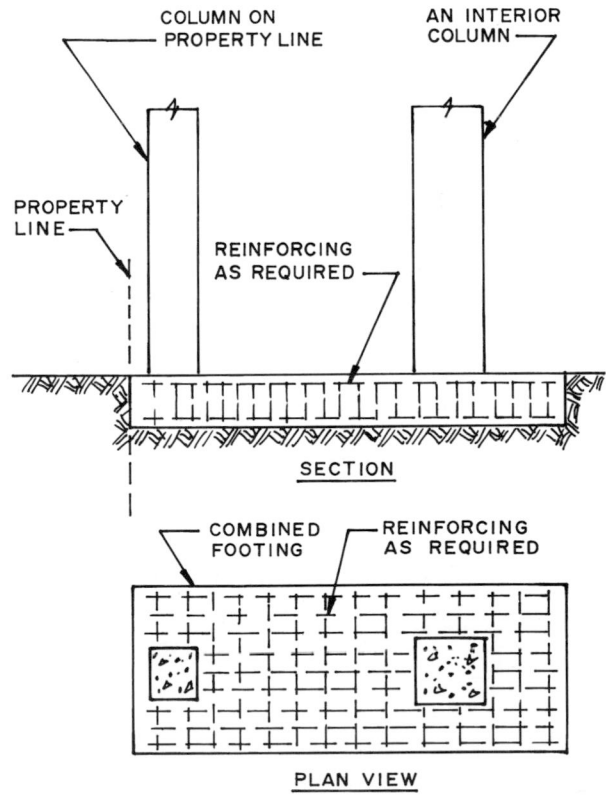

2-36 *Columns that must be located on the property line can be supported with a combined footing.*

Cantilever footings—are a form of combined footing that permits erection of wall columns near the edge of the footing. A beam rests on independent footings and cantilevers over the outer footing to support the column (see 2-37). The footings may be reinforced concrete or steel grillage.

Continuous footings—are reinforced concrete members extending continuously under several columns. They are especially useful if the soil varies in the area and reduces uneven settlement between the columns. They can be a flat or T-footing as shown in 2-38.

PILE FOUNDATIONS

Pile foundations use long wood, steel, and concrete piles driven into the earth with a pile driver. This delivers a series of blows to the top of the pile, driving it into the earth. The commonly used driven piles are shown in 2-39. These include timber, pipe, precast concrete and steel shells, and steel H piles.

Timber piles—are tree trunks trimmed and stripped of bark and treated with preservatives. They are driven with the small end down and may have a pointed metal shoe on this tip. The minimum tip is 6 in. (152mm) in diameter while the top or butt end minimum diameter is 12 in. (305mm) and has a metal pile ring to protect it during driving. These work best in soils relatively free of rock. Parts below the water level will last for years while the part above the ground water level may decay.

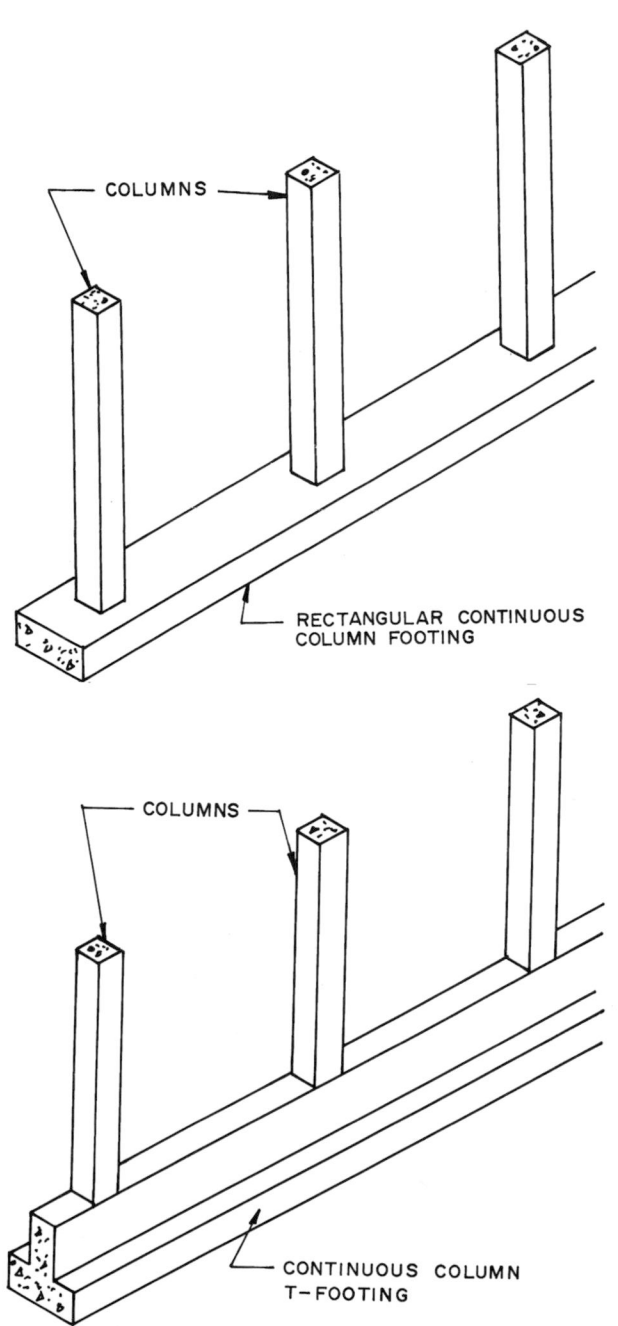

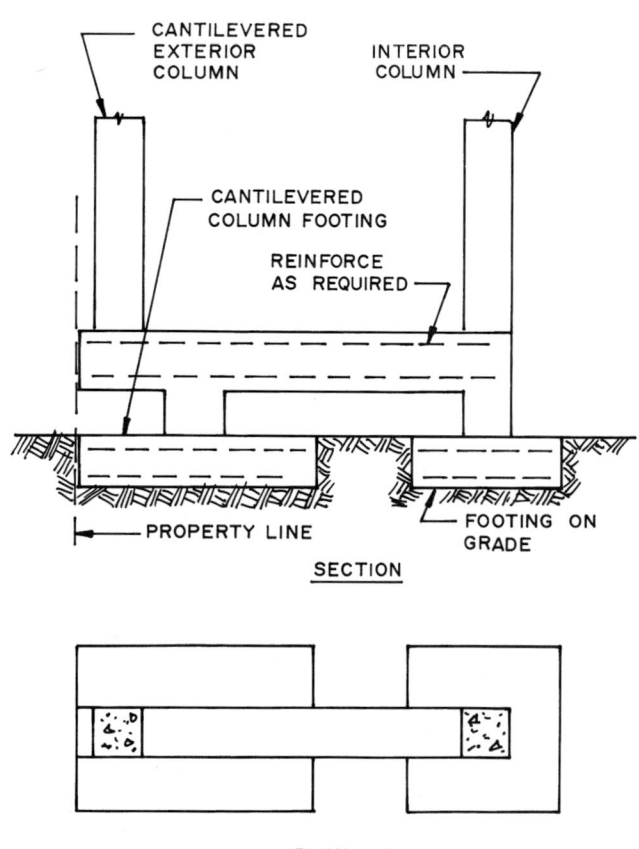

2-37 *Columns on the property line can be supported with a cantilevered footing.*

2-38 *Columns can be supported with continuous footings.*

Wood piles that reach rock or other load-bearing strata transmit the load through the end and are called **end-bearing piles.** Those that get their load-bearing capacity from friction between the sides of the pile and the earth are called **friction piles.**

H-piles—are HP structural steel shapes and are especially good for driving in soil with rocks or thin rock strata. Since they have a small surface area they cannot rely solely on friction but must be driven to end bearing on rock. They are available in widths from 8 to 14 in. (200 to 355mm) and can be driven to depths of 150 to 200 ft. (45 to 61m). Long lengths are developed by driving shorter sections and welding on another section.

Pipe piles—are heavy-gauge steel pipe that may be driven with an open end. However, small-diameter piles generally have the end closed. As open-end piles are driven the earth inside is removed to decrease resistance. The soil can be removed by inserting a pipe into it and blowing it out. Water may be mixed with the soil to help loosen it for removal. Concrete is placed inside the pipe starting the fill from the bottom. If the pipe fills with water a concrete seal is inserted at the bottom and the water pumped out. Pipe diameters may be any practical size with 8 to 18 in. (203 to 457mm) frequently used, and wall thicknesses from ¼ to ½in. (6 to 12mm) are common (see 2-40).

TIMBER   PIPE, OPEN END   PIPE, CLOSED END   H-PILE

METAL DRIVING POINT

CAST-IN-PLACE CONCRETE   STRAIGHT STEEL SHELL DRIVEN WITH A MANDREL, FILLED WITH CONCRETE   TAPERED SECTION STEEL SHELL   PRECAST AND PRESTRESSED CONCRETE   COMPOSITE

STEEL SHELL   TIMBER

2-39 *Some of the commonly used driven piles.*

2-40 *These pipe piles are ready to be driven by the pile drivers in the background.*
Courtesy Franki Canada Limited

Precast and pretensioned concrete piles—are
made in a variety of ways. They may be cylindrical, square, or octagonal and are available tapered
and uniform in size their entire length. While
most are solid in cross section some cylindrical
types are hollow. The piles are reinforced with
steel to withstand the installation stresses as well
as the building loads. Piles driven in clay soils
have a blunt point while a tapered point is used in
sand and gravel. Precast concrete piles can be
lengthened by welding together the reinforcing
bars where the two sections meet. Some are installed by jetting with water. A pipe is cast in the
center of the pile and high pressure water is
forced through it blasting the soil as the pile penetrates it (see 2-41).

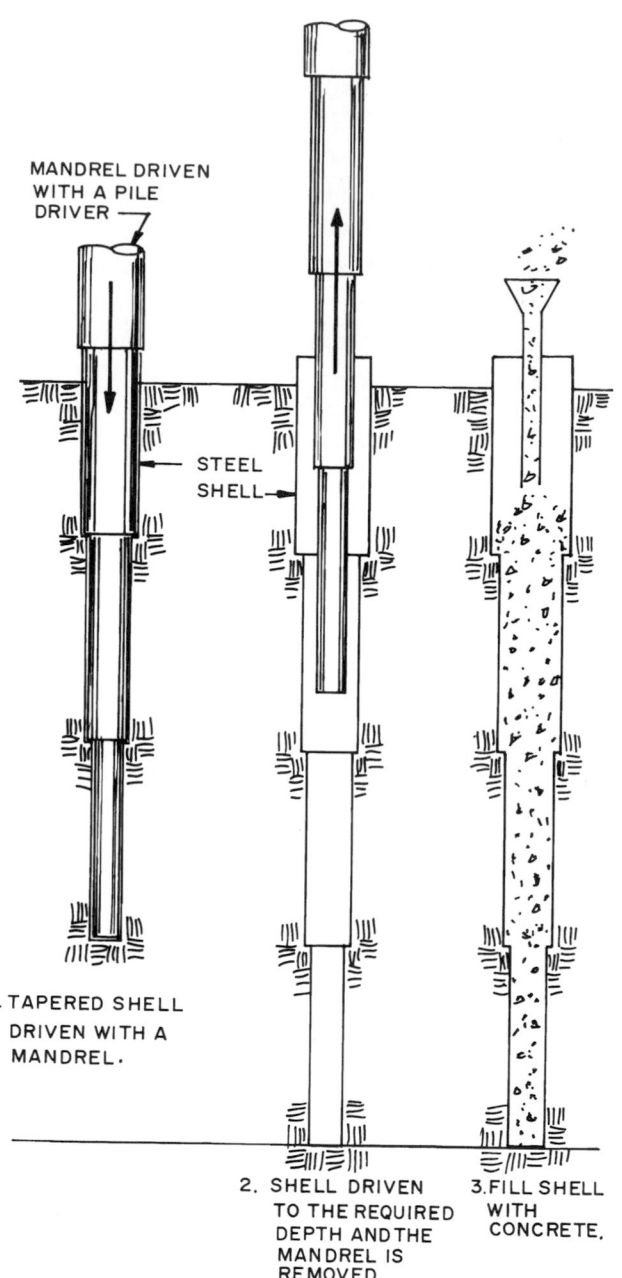

1. TAPERED SHELL
DRIVEN WITH A
MANDREL.

2. SHELL DRIVEN
TO THE REQUIRED
DEPTH AND THE
MANDREL IS
REMOVED.

3. FILL SHELL
WITH
CONCRETE.

2-41 *These precast concrete hollow cylinder piles have
been set to support a bridge.*

Courtesy Precast/Prestressed Concrete Institute

2-42 *Steel shells used for cast-in-place concrete pile can
be driven with a mandrel and then filled with concrete after
the required depth has been reached and the mandrel
removed.*

Cast-in-place piles—are cased with a metal shell or can be uncased and shell-less. One type of pile uses a tapered steel shell driven with an internal steel mandrel. A mandrel is a steel core inserted into a hollow pile to reinforce the pile shell while it is being driven into the earth. When the required depth has been reached the mandrel is removed and the shell is filled with concrete (see 2-42). The shells are in 8 ft. lengths and sections are added to increase the length as it is driven.

Shell-less uncased pile—is formed by driving a steel shell with an interior driving core, removing the core, and filling the shell with concrete as the steel shell is removed from the earth (see 2-43).

The Franki pressure-injected footing construction procedure is shown in 2-44. It uses a steel casing into which zero-slump concrete is rammed to produce a pedestal footing.

Composite piles—are used when it reduces cost or has other advantages. For example, a wood pile (low in cost) could be driven into the area below groundwater and a concrete pile could be placed on top.

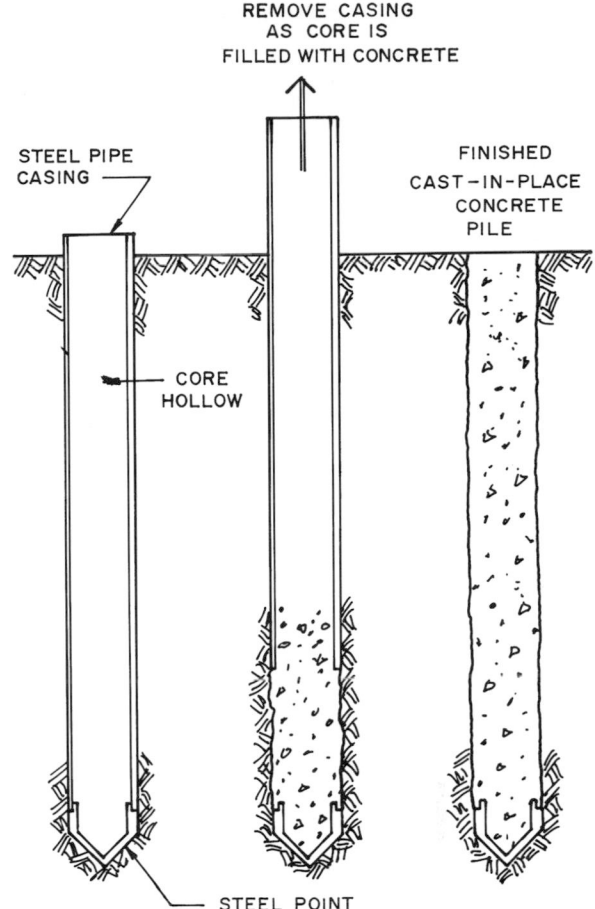

2-43 *A typical cast-in-place concrete pile when the steel shell is removed.*

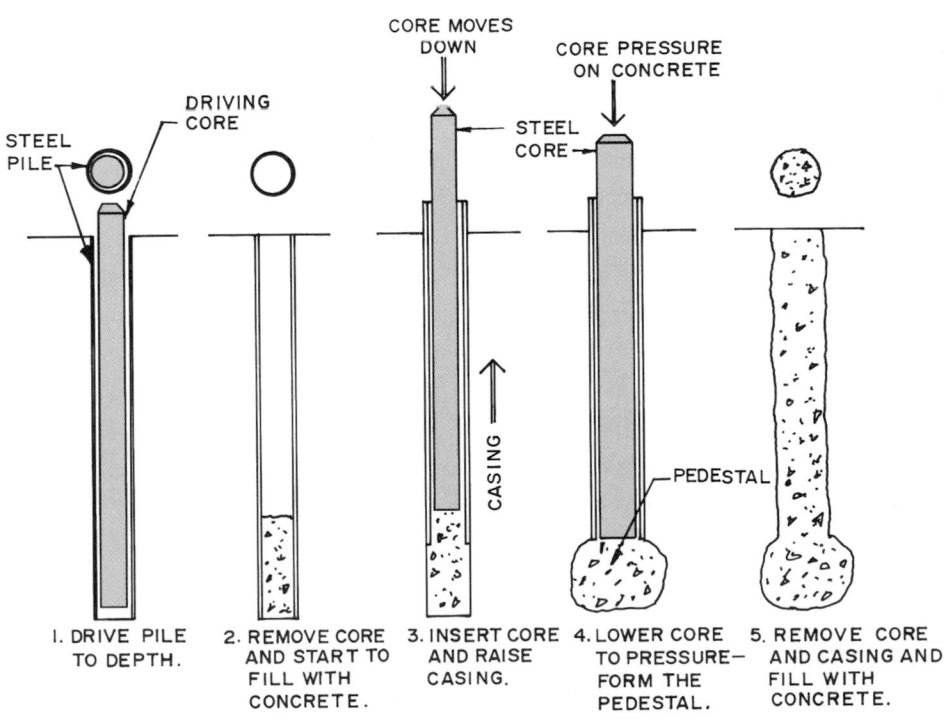

2-44 *The Franki pressure-injected process produces a pile with a pedestal footing.*

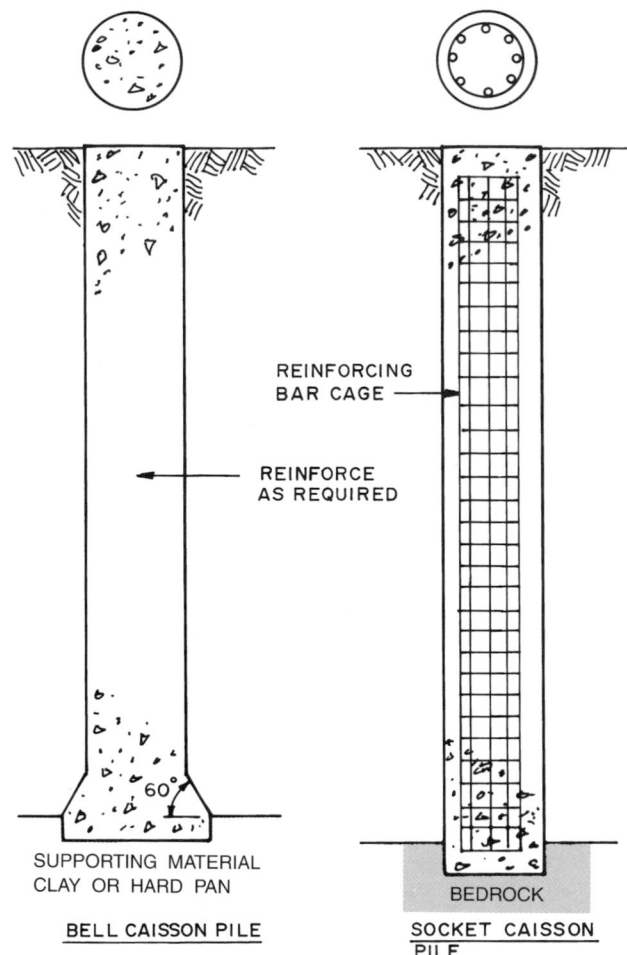

REINFORCING
BAR CAGE

REINFORCE
AS REQUIRED

60°

SUPPORTING MATERIAL
CLAY OR HARD PAN

BELL CAISSON PILE

BEDROCK

SOCKET CAISSON
PILE

2-45 *Caisson piles may rest on bedrock or other approved bearing material as specified by local codes.*

2-46 *The bottom of caisson piles is belled with a belling bucket.*

Courtesy Franki Canada International

CAISSON PILES

A caisson is a water-tight structure used to provide a dry working area in building foundations or structures below water. A **caisson pile** is a pile cast-in-place in the earth by driving a tube into the soil, emptying the earth from inside the tube, and filling it with concrete and required reinforcing steel (see 2-45).

Caisson piles are end-bearing units that are drilled using large-diameter augers. A steel casing is lowered into the hole as drilling continues. The soil must be able to stand without collapsing when the casing is removed from the hole. Caisson piles are bored through the weaker soil until they strike the more resistant rock or other acceptable bearing soil. When hard rock is not reached, the bottom of the hole is belled using a belling bucket on the end of the drill. This belling of the hole provides a larger end-bearing surface (see 2-46). The bottoms of large-diameter bor-

ings are inspected by a person lowered into them. If all is as expected the required reinforcing steel is lowered into the hole and it is filled with concrete. As it is filled the steel casing is removed. However, the casings are sometimes left in place.

UTILITY SERVICES

Site work involves the installation of utility services, such as water, gas, steam, oil, sanitary sewerage, electrical, and telephone service. Each of these requires the services of engineers as the system is planned and careful planning by the construction manager so work flows along with the other construction activities.

Temporary utility services most needed during the construction process are also required.

SITE IMPROVEMENTS

Finally, as the project nears completion and the site is brought to finished grade, attention is given to the vast array of site improvements. As with other parts of the project these involve the architect and specialists such as a landscape architect. The range of installations varies but includes such things as plantings, walks, fences, retaining walls, and signs.

DIVISION 3

CONCRETE
CSI MASTERFORMAT™

Courtesy Precast/Prestressed Concrete Institute

Concrete is a major metal of construction used all over the world. The raw materials needed are available in most parts of the world. It does not require complex or expensive equipment to make concrete. In many parts of the world wood resources are meager and what is available is too expensive for use in construction as is done in the United States. Concrete is an inexpensive construction material but has no form of its own or tensile strength. It requires forms be built to shape it into useful form and reinforcing steel be added to provide tensile strength

Concrete is a solid, hard material produced by combining in proper proportions **portland cement,** coarse and fine **aggregates** (sand and stone), and water. The chemical reaction between the cement and water produces heat and a hardening of the mass.

COMPOSITION OF CONCRETE

Concrete is a major construction material used all over the world. It is a hard, solid, durable material produced by combining in proper proportions **portland cement,** coarse and fine **aggregates** (sand and stone), **water** and sometimes a**dmixtures.** An admixture is a material added to the concrete to produce specific reactions such as retarding or accelerating setting time.

PORTLAND CEMENT

Portland cements are fine, pulverized materials consisting of compounds of lime, iron, silica, and alumina. They are **hydraulic** cements, which means they set and harden due to a chemical reaction when combined with water. This process is called hydration and combines the cement and water into a hard, durable mass.

Manufacture of Portland Cement

The manufacture of portland cement requires the combination of the various elements in proper proportion under carefully controlled conditions. The basic process is shown in 3-1. The lime is commonly derived from limestone, marble, marl, or seashells. Iron, silica and alumina are obtained from mining clays containing these elements. In some areas iron furnace slag and flue dust is used, as is sand, chalk, and bauxite. These ingredients are crushed in the primary crusher and sent to a vibrating screen. Pieces falling through the screen go to storage. Those that are too big go to a secondary crusher and on to storage. After this either the

wet or dry process is used. The **dry process** grinds the raw material to a powder and blends it in mixing silos. From here it goes to the kiln. In the **wet process** the ground materials are mixed with water to form a slurry, blended, and moved to a kiln. The **kiln** is a rotating cylinder operating at 2600 to 3000°F (1600 to 1780°C). It burns the materials into a clinker. The clinker is cooled, has a small amount of gypsum added, and ground into a powder so fine it would pass through a sieve with 40,000 openings per square inch. The gypsum is added to retard the curing process. Most portland cement is shipped in bulk in railroad cars or large trailers designed especially for this material. Some is shipped in bags containing one cubic foot of volume weighing 94 pounds.

Types of Portland Cement

Various types of portland cement are manufactured. These have different chemical and physical properties and are designed to serve different purposes. The eight types of portland cement are specified in the American Society for Testing and Materials (ASTM) designation C150.

▼ The eight types of portland cement include:

Type 1	Normal
Type IA	Normal with air entering
Type II	Moderate
Type IIA	Moderate with air entering
Type III	High-early strength
Type IIIA	High-early strength with air entering
Type IV	Low heat
Type V	Sulfate resisting

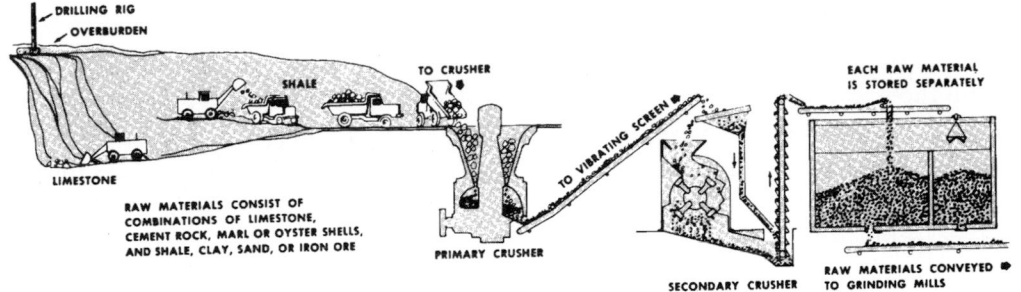

1. Stone is first reduced to 5-in. size, then to ¾ in., and stored.

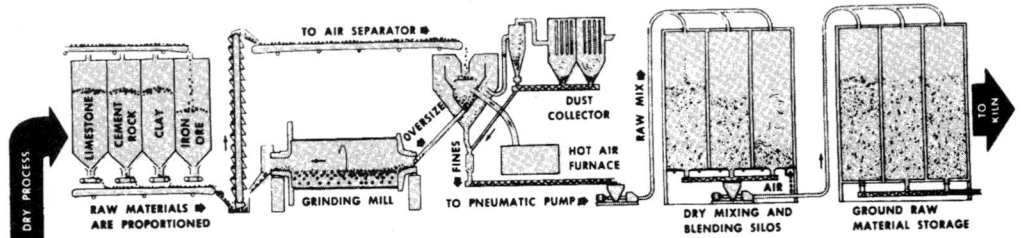

OR 2. Raw materials are ground to powder and blended.

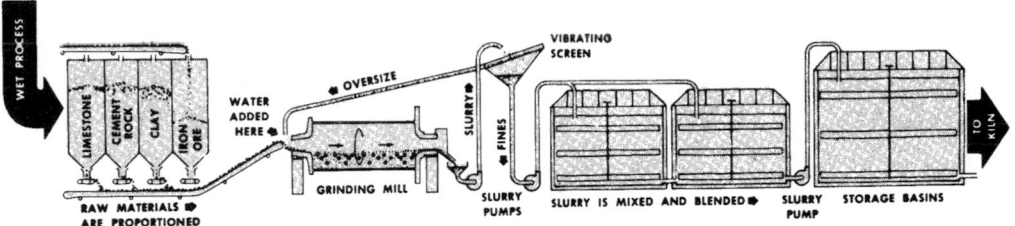

2. Raw materials are ground, mixed with water to form slurry, and blended.

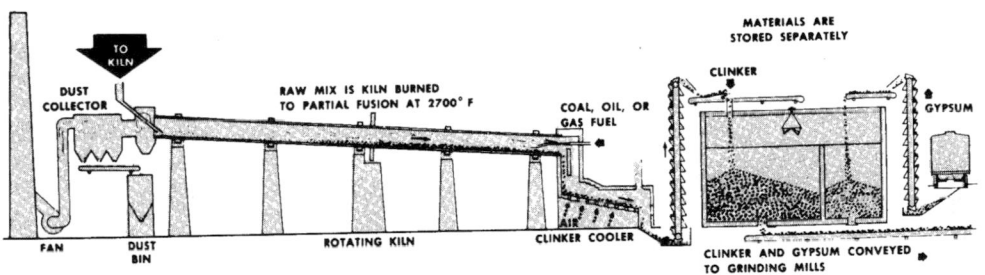

3. Burning changes raw mix chemically into cement clinker.

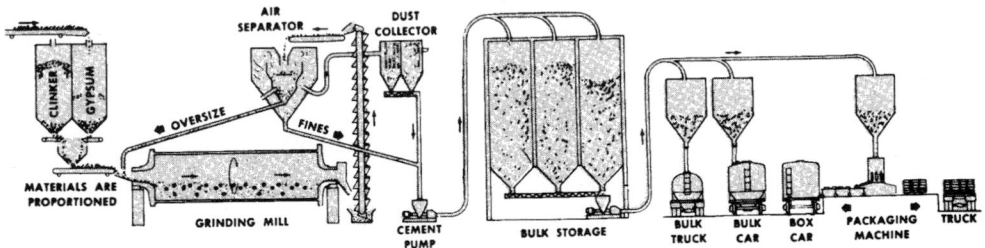

4. Clinker with gypsum is ground into portland cement and shipped.

3-1 *The steps in the manufacture of portland cement.*
Courtesy The Portland Cement Association

TYPE I NORMAL

Type I is a general purpose portland cement used for all purposes where the special properties of the other types are not necessary such as sulfate attack from soil or water or the temperature rise in the concrete due to hydration is not objectionable. It is used for purposes such as pavements, sidewalks, reinforced concrete structural members, bridges, tanks, water pipes, and masonry building units.

TYPE II MODERATE

Type II is used where protection against moderate sulfate attack such as found in some soils and groundwaters is necessary. It also generates less heat by hydrating at a slower rate than Type I. This makes it useful in structures having larger masses of concrete such as abutments, piers, or large retaining walls.

TYPE III HIGH EARLY STRENGTH

Type III portland cement provides high strength faster than Type I or II. It is used when forms are to be removed as soon as possible or to get strength faster in cold weather, thus reducing the curing period when special protection from freezing is necessary.

TYPE IV LOW HEAT OF HYDRATION

Type IV is used where it is necessary to reduce the rate and amount of heat generated by hydration. It develops strength slower than Type I. Its major use is for structures having large concrete masses such as dams and nuclear plants.

TYPE V SULFATE RESISTING

Type V portland cement is used only in concrete that is exposed to severe sulfate conditions. This occurs most frequently in soils and groundwater in areas where sulfate concentrations are high.

AIR-ENTRAINED PORTLAND CEMENT

There are three types of **air-entrained** portland cement, IA, IIA, and IIIA. These are the same as Types I, II, and III except small quantities of air-entraining materials are interground with the clinker and gypsum. The primary active ingredient used for air-entraining admixtures is alkylbenzene sulfonate. However, polyethylene oxide, detergents, and salts of fatty acids are sometimes used.

Air-entraining entrains millions of microscopic air bubbles into the concrete. This greatly improves the durability of the concrete exposed to moisture and the freezing and thawing cycles in the winter. It also increases the concrete's resistance to surface scaling, which is often caused by salts used to remove ice on the surface. Air-entraining improves the workability of concrete, reduces segregation of the aggregate, bleeding, and the amount of water needed.

WHITE PORTLAND CEMENT

White portland cement is a true portland cement that differs from regular portland cement in that during the manufacturing process the components are controlled so the finished products will be white. It is used mainly for architectural purposes such as precast curtain wall and facing panels, terrazzo floors, stucco, finish-coat plaster, and tile grout.

BLENDED HYDRAULIC CEMENTS

Hydraulic cements are cements that are capable of hardening under water. There are two blended hydraulic cements, portland blast-furnace-slag cements and portland pozzolan cements.

ASTM C595 sets requirements for two types of portland blast-furnace-slag cements, IS and IS-A (air-entrained). Blast furnace slag is interground with the portland cement clinker or separately ground and blended with portland cement.

There are four types of **portland pozzolan cements,** IP, IP-A, P, and P-A. The "A" indicates air-entraining. They are manufactured by intergrinding portland cement clinker with a suitable pozzolan. **Pozzolan** is a siliceous or siliceous and aluminous material that chemically reacts with calcium hydroxide to form compounds possessing cementitious properties. Pozzolans may be manufactured or natural. Natural pozzolans include pumicite, volcanic ash, and volcanic tuff. Tuff is a porous rock formed by the consolidation of volcanic ashes and dust. Processed natural pozzolans include clay and shale that is burned or calcined in a kiln, crushed and ground. All four types can be used for general concrete construction. Types P and P-A are used in massive concrete structures where high early strengths are not required.

MASONRY CEMENTS

Masonry cements are used for the manufacture of masonry mortars. They are covered under ASTM C91. They may contain some of the following: portland cement, slag cement, and hydraulic lime. In addition they may contain one or more of the following: hydrated lime, limestone, chalk, calcareous shell, talc, slag, or clay. It is important that mortar cements have excellent workability, plasticity, and water retention properties.

OTHER CEMENTS

There are other types of cements that are not covered by ASTM specifications. These include some special types of portland cement.

Waterproof portland cement—has a small amount of calcium stearate or aluminum stearate added to the portland cement clinker during grinding.

Plastic cements—have plasticizing agents added to Type I or Type II portland cement during the milling operation. They are used in stucco and plaster.

Expansive cement—is a hydraulic cement that expands during the early hardening period. The three types available are K, M, and S. They can be used to compensate for the effects of drying shrinkage.

Oil-well cements—are used to seal oil wells and are a blend of portland cement and hydraulic cement. They must be slow setting and withstand high pressures and temperatures.

Regulated set cement—is a hydraulic cement that can be compounded to have a set time from 1 to 2 minutes to as long as 50 or 60 minutes.

Properties of Portland Cement

The specifications for portland cement place limits on its chemical composition and physical properties. A knowledge of these properties is necessary to evaluate the results of tests on the cement itself and on concrete made with the cement.

FINENESS

The rate of hydration increases as the cement increases in fineness, therefore accelerating the strength development of the concrete. The effects of greater fineness on strength are most apparent during the first 5 to 7 days.

SOUNDNESS

Soundness is the ability of a hardened paste to retain its volume after setting. The major problem is a delayed destructive expansion after the paste has hardened. Excessive amounts of hard-burned free lime or magnesia are the main cause. Cement can be tested following ASTM standards to determine soundness.

CONSISTENCY

Consistency is the relative mobility of a fresh mixture, in other words its ability to flow. Cement paste tests for consistency are made using the Vicat-needle technique.

SETTING TIME

Setting time tests are made to determine if the cement paste undergoes setting and hardening during the first few hours. The initial set must not occur before the concrete finishing operations can be completed. The final set should occur without unnecessary delay. Setting times of cement pastes and concrete do not correlate directly because of water loss, the surface upon which concrete has been poured, and temperature differences.

FALSE SET

False set results in the loss of plasticity without the development of much heat shortly after the concrete has been mixed. False set will cause no difficulty if the concrete is given additional water. Chemical admixtures can be added to delay the occurrence of false set.

COMPRESSIVE STRENGTH

The compressive strength of cement is a major physical property. It is found by testing 2 in. (50 mm) cubes of mortar as specified by ASTM C150. Compressive strength is influenced by the type of cement (its compound composition and fineness). Cement compressive strengths cannot be used to determine concrete compressive strengths because of variances in concrete mixtures. Actual concrete samples must be tested to find the compressive strength of concrete.

HEAT OF HYDRATION

Heat of hydration is the heat generated when the cement and water chemically react. The amount of heat generated depends upon the chemical composition of the cement. The rate of heat generated is affected by the fineness of the cement, the chemical composition, and temperature during hydration. Structures with a large concrete mass may experience a significant rise in temperature unless the heat can be rapidly dissipated. If not controlled it may create undesirable stresses. In cold weather the excessive heat may be beneficial because it will help maintain favorable curing temperatures.

LOSS OF IGNITION

Loss of ignition is a test to find out how much weight a sample of portland cement will lose when heated to 1652 to 1832°F (900 to 1000°C). A loss of ignition may be caused by improper and prolonged storage or adulteration during transfer and transport.

SPECIFIC GRAVITY

The specific gravity of portland cement is about 3.15. Portland-blast-furnace and portland pozzolan cements may be as low as 2.90. The specific gravity is not an indication of the quality of the cement. It is used when calculating mix designs.

CONSISTENCY

Consistency is an indication of the ability of a fresh mix to flow. Cement pastes are tested using the Vicat plunger technique. Mortars are mixed and tested for flow and the findings used to regulate the water content. The consistency of concrete is measured by a slump test.

WEIGHT OF CEMENT

Bulk cement is sold by the ton (2000 lb. Or 907.2 kg.). Smaller quantities are bagged with a volume of one cubic foot (0.028m) weighing 94 pounds (42 kg).

Storing Portland Cement

Since portland cement is moisture-sensitive it must be protected from dampness. Sacked material must be stored on pallets regardless of whether it is in a warehouse with a concrete floor or temporarily on the site in the open. Warehouse storage must be watertight and bags must not touch the exterior walls. Pack the bags closely together to reduce air flow. Cover with plastic or tarpaulins if they are to be stored for long periods.

Cement in bags if stored a long time tends to pack. To correct this roll the bags on the floor before opening. If lumps remain test the cement before using it on jobs where properties are important.

Bulk cement is stored in water-tight bins or silos. Dry low-pressure aeration or vibration should be used to make the cement flow better. When cement is loaded into bins or silos it swells so the unit will store only about 80% of its rated capacity.

WATER

Water used in making concrete should be clear and free of sulfates, acids, alkalis, and humus. Most munic-

Table 3-1 ROCK & MINERAL CONSTITUENTS IN AGGREGATES

Minerals	Igneous Rocks	Metamorphic Rocks
Silica	Granite	
Quartz	Syenite	Marble
Opal	Diorite	Metaquartzite
Chalcedony	Gabbro	Slate
Tridymite	Peridotite	Phyllite
Cristobalite	Pegmatite	Schist
Silicates	Volcanic Glass	Amphibolite
Feldspars	Obsidian	Hornfels
Ferromagnesian	Pumice	Gneiss
Hornblende	Tuff	Serpentinite
Augite	Scoria	
Clay	Perlite	
Illites	Pitchstone	
Kaolins	Felsite	
Chlorites	Basalt	
Montmorillonites		
Mica		
Zeolite	**Sedimentary Rocks**	
Carbonate	Conglomerate	
Calcite	Sandstone	
Dolomite	Quartzite	
Sulfate	Graywacke	
Gypsum	Subgraywacke	
Anhydrite	Arkose	
Iron Sulfide	Claystone, Siltstone,	
Pyrite	Argillite and Shale	
Marcasite	Carbonates	
Pyrrhotite	Limestone	
Iron Oxide	Dolomite	
Magnetite	Marl	
Hematite	Chalk	
Goethite	Chert	
Ilmenite		
Limonite		

For brief descriptions, see *Standard Descriptive Nomenclature of Constituents of Natural Mineral Aggregates* (ASTM C294).

Courtesy Portland Cement Association

ipal water systems provide water suitable for use. Potable water from wells is also usually acceptable. Water from lakes, ponds, or rivers should be carefully checked for its suitability before using it.

Water of questionable quality can be used for concrete if test mortar cubes have 7 and 28 day strengths equal to 90% of samples made with drinkable water. These tests should be made following ASTM C109 specifications. ASTM C191 specifies how to test the samples to see if impurities in the water adversely shorten or lengthen the setting time.

Seawater containing up to 35,000 ppm (parts per million) of dissolved salts is generally suitable as mixing water for unreinforced concrete. Concrete made with seawater will have a higher early strength than normal concrete and will usually have lower strength after 28 days. Seawater may be used for reinforced concrete but the reinforcement must be protected from corrosion and the concrete must contain entrained air.

AGGREGATES

Approximately 60% to 80% of concrete is made up of the aggregates. The cost of concrete and its properties are directly related to the aggregates. Natural aggregates are composed of rocks and minerals. **Minerals** are naturally occurring inorganic substances that have distinctive physical properties and a composition that can be expressed by a chemical formula. **Rocks** are usually composed of several minerals. For example, limestone is basically calcite (a mineral) with small amounts of quartz, feldspar, and clay (all minerals). The crushing and weathering of rock produces stone, gravel, sand, silt, and clay. Rocks and minerals commonly found in aggregates are shown in Table 3-1. Aggregates must conform to ASTM specifications. They must be clean, hard, strong, free of absorbed chemicals, and free of coatings of clay, humus, and other fine materials. Aggregates containing some shale, shaly rocks, soft or porous rocks, and some types of chert are not to be used because they do not weather well and cause popouts on the exposed concrete surface.

Normal-weight concrete weighs about 135 to 160 lb. per cubic foot (2150 to 2550 kg/m^3). It uses sand, gravel, crushed stone, and air-cooled blast furnace slag aggregates. They are specified in ASTM C33. Structural lightweight concrete weighs about 85 to 115 lb. per cubic foot (1350 to 1850 kg/ml) and uses expanded shale, clay, slate, and slag aggregates. They are specified in ASTM C330. Insulating concrete weighs about 15 to 90 lbs. per cubic foot (250 to 1450 kg/m^3) and uses pumic, seoria, perlite, vermiculite, and diatomite aggregates. They are specified in ASTM C332. Heavyweight concrete will weigh 175 to 400 lb. per cubic foot (2800 to 6400 kg/m^3) and uses barite, limonite, magnetite, ilmenite, hematite, iron, and steel slugs as aggregate. They are specified in ASTM C637.

Characteristics of Aggregates

The characteristics of aggregates influence their use in concrete mixes to be used for various applications.

Table 3-2 FINE & COARSE SIEVE SIZES

Fine Sieve Sizes
⅜ in. (9.5 mm)
No. 4 (4.75 mm)
No. 8 (2.36 mm)
No. 16 (1.18 mm)
No. 30 (600 µm)
No. 50 (300 µm)
No. 100 (150 µm)

Coarse Sieve Sizes	
Size Number	Nominal Size (sieves with square openings)
1	3½ to 1½ in. (90 to 37.5 mm)
2	2½ to 1½ in. (63 to 37.5 mm)
357	2 in. to No. 4 (50 to 4.75 mm)
467	1½ in. to No. 4 (37.5 to 4.75 mm)
57	1 in. to No. 4 (25.0 to 4.75 mm)
67	¾ in. to No. 4 (19.0 to 4.75 mm)
7	½ in. to No. 4 (12.5 to 4.75 mm)
8	⅜ in. to No. 8 (9.5 to 2.36 mm)
3	2 to 1 in. (50.0 to 25.0 mm)
4	1½ to ¾ in. (37.5 to 19.0 mm)

Courtesy Portland Cement Association

▼ The characteristics that are considered follow.

Abrasion resistance—is important when the aggregate is to be used in an area such as a factory floor, subject to heavy abrasive use.

Freeze–thaw properties—of an aggregate are important in concrete to be exposed to a wide range of temperatures.

Strength—of aggregates under compression is an important factor to consider especially when compressive strength of the concrete is important.

Shape and texture—of aggregate particles influence the properties of fresh concrete more than cured concrete. Rough textured, angular, elongated particles require more water than do smooth, rounded aggregates. The cement tends to bond better to angular rather than smooth particles.

Aggregates are graded into standardized sizes by passing them through a sieve as specified by ASTM C136 and CSA A23.2.2. The sieves used for fine aggregates are in seven sizes and there are ten sizes for coarse aggregates (Table 3-2).

Specific gravity is a measure of the relative density of an aggregate. It is a ratio of its weight to the weight of an equal volume of water. It is used to determine the absolute volume occupied by the aggregate. Most natural aggregates have specific gravities between 2.4 and 2.9.

The absorption and surface moisture of an aggregate is determined so the net water content of the concrete can be controlled and the correct batch weights determined.

▼ Moisture conditions are designed in four categories.

Oven dry—completely dry and fully absorbent.

Air dry—dry on the surface but have some interior moisture.

Saturated surface dry—neither absorbing water or contributing water to the mix. The surface is dry but the voids and interior of the aggregate are fully saturated.

Damp—contains an excess of surface moisture.

Chemically stable aggregates do not react chemically with cement causing harmful reactions. Some aggregates contain minerals that do react with alkalies in cement causing abnormal expansion and cracking in the concrete.

HARMFUL MATERIALS

Intermixed with aggregates may be a number of **harmful materials** such as silt, organic materials, coal, and soft rock particles. Aggregate specifications limit the amount of these materials that may be present. There are a series of ASTM tests that are used to identify these harmful materials in aggregate samples.

STRUCTURAL LIGHTWEIGHT AGGREGATES

Lightweight aggregates are used to produce lightweight structural concrete. Lightweight concrete has a density ranging from 90 to 115 pounds per cubic foot (1440 to 1850 kg/m^3) depending upon the aggregate. Shale, slate, clay, and slag are generally used.

INSULATING LIGHTWEIGHT AGGREGATES

Perlite, diatomite, vermiculite, pumice, and scoria are commonly used as insulating lightweight aggregate.

Perlite—is siliceous volcanic rock that contains moisture. It is crushed and heated causing it to expand forming a honeycomb structure. It is used as loose fill insulation and plaster and concrete aggregate.

Diatomite—is a diatomaceous earth (which is composed of silicified skeletons of microscopic one-celled animals).

Vermiculite—is a hydrated magnesium-aluminum-iron silicate that occurs in thin layers having moisture between them. When it is crushed and heated the layers expand forming dead air cells.

Pumice—is a type of volcanic rock that is porous and light in weight.

Scoria—is a type of volcanic slag and also used to describe a form of blast furnace slag. It is a cellular material and is crushed to form aggregate of various sizes.

HEAVYWEIGHT AGGREGATES

Heavyweight aggregates are used to produce heavyweight concretes that have densities up to 400 pounds per cubic foot (6400 kg/m^3).

Heavyweight aggregates frequently used include ferrophosphorus, barite, goethite, hematite, ilmenite, limonite, magnetite, and steel shot and punchings.

Handling & Storing Aggregates

Store aggregates in layers of uniform thickness. When forming stockpiles discharge each truck load tight against the previous load. Remove it from stockpiles with a front end loader. If removing from a tall, conical pile, load by moving back and forth across the face of the pile so as to reblend sizes. Give washed aggregates sufficient time in storage to drain to a uniform moisture content. Do not drop fine aggregates from a bucket or conveyor because the wind tends to blow away the very fine particles. Build dividers to keep various grades from intermingling or store in completely separate locations. Above all, store and handle aggregates so they are not contaminated by unwanted substances, such as earth and leaves.

ADMIXTURES

Admixtures are ingredients added to concrete, other than portland cement, aggregates, and water. They may be added before or during mixing. Admixtures change the properties of concrete so should be used sparingly and only on the advice of a concrete specialist.

Table 3-3 ADMIXTURES BY CLASSIFICATION

Desired Effect	Type of Admixture	Material
Improve durability	Air entraining (ASTM C260)	Salts of wood resins Some synthetic detergents Salts of sulfonated lignin Salts of petroleum acids Salts of proteinaceous material Fatty and resinous acids and their salts Alkylbenzene sulfonates
Reduce water required for given consistency	Water reducer (ASTM C494, Type A)	Lignosulfonates Hydroxylated carboxylic acids (Also tend to retard set so accelerator is added)
Retard setting time	Retarder (ASTM C494, Type B)	Lignin Borax Sugars Tartaric acid and salts
Accelerate setting and early strength development	Accelerator (ASTM C494, Type C)	Calcium chloride (ASTM D98) Triethanolamine
Reduce water and retard set	Water reducer and retarder (ASTM C494, Type D)	(See water reducer, Type A, above)
Reduce water and accelerate set	Water reducer and accelerator (ASTM C494, Type E)	(See water reducer, Type A, above. More accelerator is added)
Improve workability and plasticity	Pozzolan (ASTM C618)	Natural pozzolans (Class N) Fly ash (Class F and G) Other materials (Class S)
Cause expansion on setting	Gas former	Aluminum powder Resin soap and vegetable or animal glue Saponin Hydrolyzed protein
Decrease permeability	Dampproofing and waterproofing agents	Stearate of calcium, aluminum, ammonium, or butyl Petroleum greases or oils Soluble chlorides
Improve pumpability	Pumping aids	Pozzolans Organic polymers
Decrease air content	Air detrainer	Tributyl phosphate
High flow	Superplasticizers	Sulfonated melamine formaldehyde condensates Sulfonated naphthalene formaldehyde condensates

Courtesy Portland Cement Association

Concrete should be workable, finishable, strong, durable, watertight, and wear resistant. Whenever possible, these properties should be obtained by careful selection of suitable types of aggregate, portland cement, and the water-cement ratio. If this is not possible, or special circumstances exist, as freezing weather, admixtures will be of benefit. A summary of admixtures and their uses is in Table 3-3.

▼ Admixtures are of the following types:

> Air-entrainment
> Retarders
> Water reduction
> Accelerators
> Pozzolans
> Workability agents
> Superplasticizers
> Dampproofing & permeability-reducing
> agents
> Bonding admixtures
> Concrete coloring agents
> Hardeners
> Grouting agents
> Gas-forming agents

AIR-ENTRAINMENT

Air-entraining admixtures are used to entrain microscopic air bubbles in concrete. The entrained air bubbles are distributed uniformly throughout the cement paste. Entrainment can be produced by using air-entrained portland cement or adding an air-entraining admixture to the concrete as it is being mixed.

The active ingredients used in air-entraining admixtures include polyethylene oxide polymers, some fats and oils, sulfonated compounds, and detergents. Air-entraining admixtures are specified by ASTM C226.

Entrained air bubbles improve the durability of the concrete increasing the resistance to damage due to the freeze-thaw cycle and to damage from deicers, which cause scaling. It gives improved workability during placement and has superior watertightness. It also improves resistance to sulfate attack from soil water and seawater. An important feature is that properly proportioned air-entrained concrete requires less water per cubic yard than non-air-entrained concrete of the same slump. This improves the water-cement ratio and is responsible for some of the advantages mentioned earlier.

RETARDERS

Retarding admixtures are used to retard the setting time of the cement paste in the concrete. They are often used to retard set time in hot weather because hydration is accelerated by the heat, which uses up some of the water needed to give the plasticity required for placement and finishing. Without a retarder in hot weather more water is required to get the desired slump, which produces lower strength concrete. Retarding admixtures tend to reduce the water required producing a better water-cement ratio and increasing the ultimate strength. Retarders also help when it is necessary to pour large amounts of concrete or where placement is difficult.

WATER-REDUCING ADMIXTURES

Water-reducing admixtures reduce the amount of water needed to produce a concrete of a given consistency. It may also be used to increase the amount of slump without requiring additional water. This means a lower water-cement ratio can be used. The lower the water-cement ratio the greater the strength of the concrete. Lignin solfonic acids and metallic salts are commonly used.

ACCELERATORS

An accelerating admixture accelerates the strength development of concrete. Strength development can also be accelerated by using Type III high-early-strength portland cement, lowering the water-cement ratio by increasing the amount of cement, or curing at higher temperatures. Accelerators are used in cold weather to develop strength faster in order to offset freeze damage.

POZZOLANS

A pozzolan is a siliceous or siliceous and aluminous material that will, when finely ground and with the presence of moisture, chemically react with calcium hydroxide at ordinary temperatures to form compounds possessing cementitious properties.

Pozzolans affect concrete in many different ways. The concrete should be tested using the specific cement and aggregates to see if it is suitable.

WORKABILITY AGENTS

If fresh concrete is harsh because of improper aggregate grading or incorrect mix proportions workability agents can be added to improve workability. **Workability** is a term used to describe the ease with which concrete can be placed and consolidated. If concrete is to have a troweled finish workability is important. Improved workability may be needed if the concrete has to be placed in forms containing considerable reinforcing or if it must be pumped.

SUPERPLASTICIZERS

When cement and water mix the wet cement particles form into small clumps which inhabit the proper mixing of the cement and water reducing workability and inhibiting hydration.

Superplasticizers are admixtures that, when added to the portland cement and water, coat the cement particles causing them to break away from the lumps and disperse in the water. Superplasticizers give each cement particle a negative charge, which causes them to repel each other providing a more thorough dispersement.

DAMPPROOFING & PERMEABILITY-REDUCING AGENTS

Sound, dense concrete having a water-cement ratio of 0.50 by weight and properly placed and cured will be watertight.

Concretes that have low cement contents, a deficiency in fines, or a high water-cement ratio can have the permeability reduced by using **permeability-reducing agents** such as certain soaps, stearates, and petroleum products.

Permeability—is a measure of the amount of water that passes through channels running between the outer faces of the concrete. The permeability-reducing agent reduces the flow of water through these channels.

Dampproofing—admixtures are used to reduce the transmission of moisture that is transferred by capillary action. This occurs when one side of the concrete is exposed to moisture, as a slab on the ground, and the other to the air, which tends to dry that surface. Capillary action occurs as moisture flows to a dry surface. The effectiveness of dampproofing admixtures varies and manufacturers test data needed to be studied.

BONDING ADMIXTURES

Often it is necessary to place fresh concrete over a concrete surface that has already set. It is important to get a good bond between them. When fresh concrete is poured over hardened concrete the fresh concrete shrinks, breaking its bond to the hardened concrete. The hardened surface must be prepared so the fresh concrete will firmly bond to the aggregate in the hardened surface.

Of great importance is the condition of the surface. It must be dry, clean (free of dirt, dust, grease, paint, etc.), and at the proper temperature. Bonding admixtures can be added to portland cement mixtures or applied to the surface of old concrete to increase the bond strength.

CONCRETE COLORING AGENTS

Concrete can be colored by thoroughly mixing pure, finely ground **mineral oxides** with the dry portland cement. When added to the mix, thorough mixing is necessary to produce a uniform color. Mixing time may sometimes be longer than normal. Oxides added to normal portland cement are usually limited to earthy colors and pastels because of the cost and graying effect of the cement. White portland cement will produce clearer, brighter colors and is preferred. Some color agents are compounded to give water-reducing and set-controlling properties.

HARDENERS

When a concrete surface is to be subjected to heavy wear, as a factory or warehouse floor, its life can be extended by using a liquid **chemical** or **dry-powder hardener.**

One form of **chemical hardener** is a colorless, nontoxic, nonflammable liquid containing magnesium and zinc fluosilicates and a wetting agent. The wetting agent reduces the surface tension of the liquid hardener, which makes it easier for it to enter the pores of the concrete. The hardening agent produces a chemical reaction with the free lime and calcium carbonates in the portland cement. Concrete surfaces are also hardened using dry-powder hardeners. One product uses quartz silice aggregates and alkali-fast inorganic oxides that color the concrete. These two are mixed with portland cement and plasticizing agents giving a dry shake that is applied to the freshly poured concrete.

GROUTING AGENTS

Portland cement grouts are widely used for many purposes, such as stabilizing foundations, filling cracks in concrete walls, filling joints, grouting tenons and anchor bolts, and other such applications. Grout properties can be altered by using the various admixtures discussed in this chapter.

GAS-FORMING AGENTS

Gas-forming agents are added to concrete or grout to cause a slight expansion in it before it hardens. Aluminum powder is one of several gas-forming agents in use. It reacts with the hydroxides in hydrating cement and produces small hydrogen gas bubbles. This also helps eliminate voids caused by settlement of the concrete or grout.

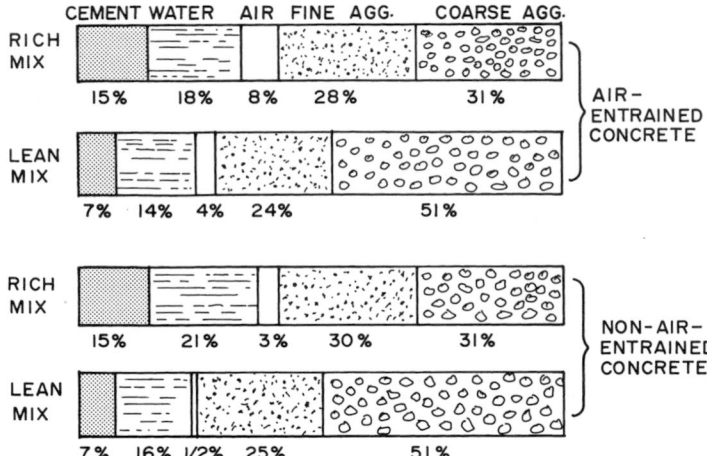

CEMENT WATER AIR FINE AGG. COARSE AGG.

RICH MIX
15% 18% 8% 28% 31%

LEAN MIX
7% 14% 4% 24% 51%

AIR-ENTRAINED CONCRETE

RICH MIX
15% 21% 3% 30% 31%

LEAN MIX
7% 16% 1/2% 25% 51%

NON-AIR-ENTRAINED CONCRETE

3-2 *The range in proportions of materials used in concrete by volume.*
Courtesy Portland Cement Association

BASICS OF CONCRETE

Concrete is made up of two parts, aggregates and a paste. The paste is a portland cement and water mixture with some entrapped air. The paste coats the particles of aggregate, hardens and bonds them together. The paste hardens due to a chemical reaction between the portland cement and water. This process is called **hydration.** The aggregates used range from fine to coarse with various degrees of size within each.

In 3-2 are shown the range in proportions of materials used in typical designs for rich and lean mixes. Notice that the volume of cement varies from 7% to 15% and water from 14% to 21%. A small percentage is air and the rest, the major volume, is aggregate. This means that the selection of aggregates is vital for the production of quality concrete. It is essential that a continuous graduation of particle size be maintained.

In 3-3 is a section cut through a sample of hardened concrete. The paste must completely fill the spaces between the particles and coat each particle.

WATER-CEMENT RATIO

If all conditions are held constant the quality of the hardened concrete is determined by the water-cement ratio. The water-cement ratio is the ratio by weight between the water and cement used to make the paste. For example, a mix that uses 0.62 lbs. of water per one pound of cement has a water cement ratio of 0.62/1.00 or 0.62. In metric measure this would be 620 g of water per kilogram of cement for a metric water-cement ratio of 620. Stated another way, it is 62% of the weight of the cement. The water-cement ratio should be the lowest value required to meet the design considerations. If too much

water is used giving a high water-cement ratio, the paste is thin and will be porous and weak when hardened. The effect of the water-cement ratio on the strength of concrete is illustrated by the graph in 3-4. Only a very small amount of water is needed for hydration to occur. A water-cement ratio of 0.31 will product hydration but usually more water is added so the concrete is workable and can be properly placed. The lower the water-cement ratio, the stronger the concrete.

▼ Concrete exposed to the elements must have the following for durability:
  Air entrainment
  Low water-cement ratio
  Quality cement and aggregate
  Proper curing
  Proper construction practices

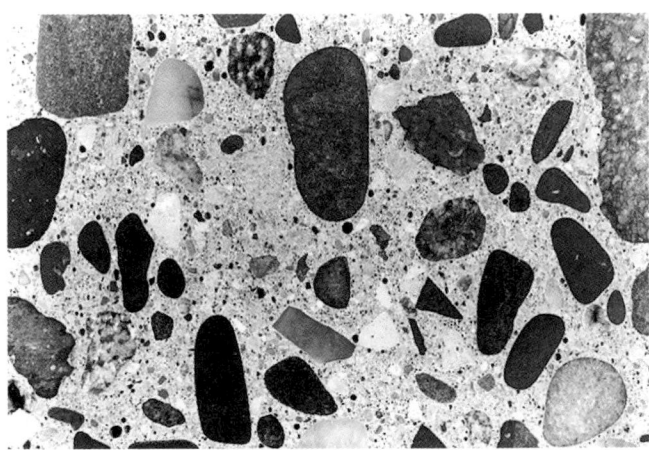

3-3 *A cross section of hardened concrete showing how the cement and water paste coats each particle of aggregate and fills the voids between the particles.*
Courtesy The Portland Cement Association

Concrete Construction	Slump, in. Maximum[a]	Minimum
Reinforced foundation walls and footings	3	1
Plain footings, caissons, and substructure walls	3	1
Beams and reinforced walls	4	1
Building columns	4	1
Pavement and slabs	3	1
Mass concrete	2	1

[a]Slumps shown are for consolidation by mechanical vibrations. may be increased 1 in. for consolidation by hand methods, such as rodding and spading

Courtesy Portland Cement Association

MINIMUM CEMENT CONTENT

In addition to the specification of the water-cement ratio, the minimum cement content is also given. This insures that the concrete will have good finishability, wear resistance, and appearance.

AGGREGATES

Aggregates must be properly graded and of proper quality. Grading pertains to the particle size and the distribution of particles. The properly graded aggregate produces the most economical concrete because it allows the use of the maximum size coarse aggregate. This reduces the amount of water and cement required. The reduction in cement reduces the cost.

The maximum size of aggregate used depends upon the size and shape of the members being formed and the amount of reinforcing steel. The maximum size aggregate acceptable can be no more than **one-fifth** the narrowest dimension between the sides of the forms or **three-fourths** the clear space between steel, as reinforcing bars, ducts, conduit, or bundles of bars. Aggregate in unreinforced slabs on the ground should not exceed one-third the slab thickness. For leaner concrete mixes a finer grade of sand is used to improve workability. Richer mixes use a coarse grade of sand for greater economy.

ENTRAINED AIR

Entrained air should be used in all concrete exposed to freezing and thawing cycles. It should be used in all concrete paving regardless of the temperatures.

Slump

Slump is a measure of the **consistency** of concrete. Consistency is the ability of fresh concrete to flow. This is measured by the slump test that is explained later in this chapter.

Slump is the decrease in height of a molded mass of fresh concrete that occurs immediately after it is removed from a standard metal slump cone. The higher the slump (the more the sample lowers), the wetter the concrete mixture. A measure of slump can only be used to compare mixes of identical design. Slump is usually specified in the concrete specifications. Recommended slumps are shown in Table 3-4. These slumps are for concrete consolidated by mechanical vibration.

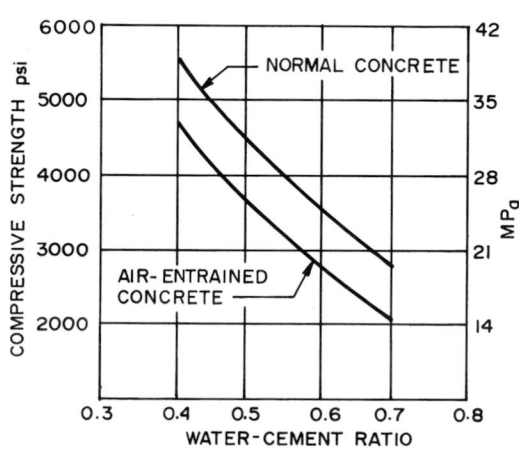

3-4 *This shows the effect of the water-cement ratio on the strength of concrete.*

Courtesy Portland Cement Association

Table 3-5   MAXIMUM WATER–CEMENT RATIOS FOR VARIOUS EXPOSURE CONDITIONS

Exposure Condition	Normal-Weight Concrete, Absolute Water-Cement Ratio by Weight
Concrete protected from exposure to freezing and thawing of application of deicer chemicals	Selct water–cement ration basis of strength, workability, and finishing needs
Watertight concrete	
In fresh water	0.50
In seawater	0.45
Frost-resistant concrete	
Thin sections; any section with less than 2-in. cover over reinforcement and any concrete exposed to deicing salts	0.50
All other structures	0.45
Exposure to sulfates	
Moderate	0.50
Severe	0.45
Placing concrete under water	Not less than 650 lb. of cement per cubic yard (386 kg/m^3)
Floors on grade	Select water-cement ratio for strength, plus minimum cement requirements

Courtesy Portland Cement Association

Normal-Weight Concrete Design

The design of normal weight concrete depends first upon the required strength and durability. Consideration of workability, plasticity, and cost are factored. The goal is to produce a concrete that meets these requirements.

As discussed earlier, the **water-cement ratio** is a basic premise used when designing normal weight concrete. Based on research and experience water-cement ratios for various applications can be recommended as shown in Table 3-5.

Another way mix design is accomplished is by using the absolute volume of material amounts. The absolute volume of a loose material, as aggregate, is the total volume including the particles and air spaces between them. This includes the absolute volume of the cement, aggregate, water, and trapped air.

$$\frac{\text{Absolute}}{\text{volume}} = \frac{\text{weight of dry material}}{\text{Specific gravity} \times \text{unit weight of water}}$$

For example, the absolute volume of 100 pounds of aggregate having a specific gravity of 2.5 would be:

$$\frac{\text{Absolute}}{\text{volume}} = \frac{100 \text{ lb.}}{2.5 \times 62.5 \text{ lb. (one cubic foot of water)}}$$
$$= 0.64 \text{ ft.}^3$$

Publications of the Portland Cement Association give detailed instructions on concrete design by water-cement ratio and absolute volume.

Lightweight Insulating Concrete Design

The design of lightweight insulating concrete depends upon the aggregate used and the desired compressive strength. The amount of water required varies greatly depending upon the circumstances. An air-entraining agent is recommended for some mixes. Detailed information is available from the Portland Cement Association.

Lightweight Structural Concrete Design

Lightweight structural concrete can be designed to produce structural members 25% to 35% lighter than the same member made using normal weight concrete and have no loss in strength. Since the aggregate is cellular and weights are different from normal aggregates, its design is usually derived from testing trial batches, experience, and other reliable test data.

CONCRETE TESTS

Hardened, cured concrete and freshly mixed concrete are tested to make certain they meet the specifications written for the concrete. This is especially important when working in various parts of the country because aggregates and water differ and the mix must be adjusted accordingly.

Tests with Fresh Concrete

When testing fresh concrete it is essential to take samples that are representative of the batch. Samples must be taken and handled following the specifications set forth in ASTM C172. Except for slump and air-content test, the sample must be at least 1 cu. ft. (0.030 m³). The sample must be used within 15 minutes after it has been taken from the batch and it must also be protected from sources of rapid evaporation during the test. Samples taken at the very first and last of a batch are not representative.

Each load of transit-mixed concrete has a certificate listing its ingredients and their proportions. An on-site slump test is made to see if it has the required consistency. A mix with a high slump may be too wet or if it has a low slump it may be too stiff.

The slump test is made following ASTM C143 specifications. A standard slump cone is 8 in. (200 mm) in diameter at the bottom and 12 in. (305 mm) high. The cone is placed on a flat surface and is held in place by standing on the foot pieces. It is filled about ⅓ full and rodded 25 times with a ⅝ in. (16 mm) diameter, 24 in. (600 mm) long rod with a rounded tip. The second ⅓ is poured and rodded as above being certain the rod penetrates the surface of the layer below. After the top layer has been rodded, the excess concrete is struck off leveling the top surface and the mold is carefully lifted. The amount the concrete will slump is measured from the top of the cone—e.g., if the top of the concrete is 4 in. below the top of the cone, the slump is 4 in. (see 3-5).

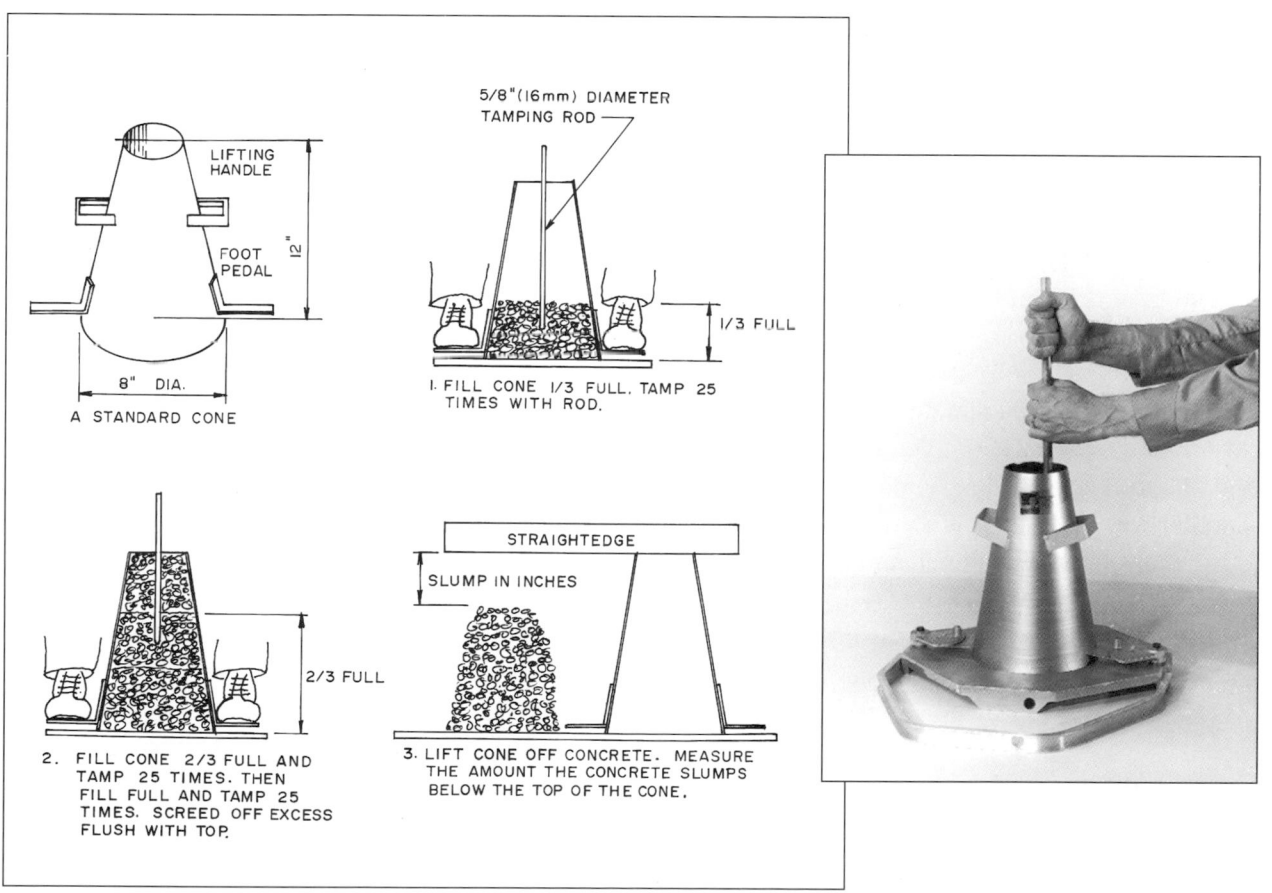

3-5 *The steps to make a slump test.*

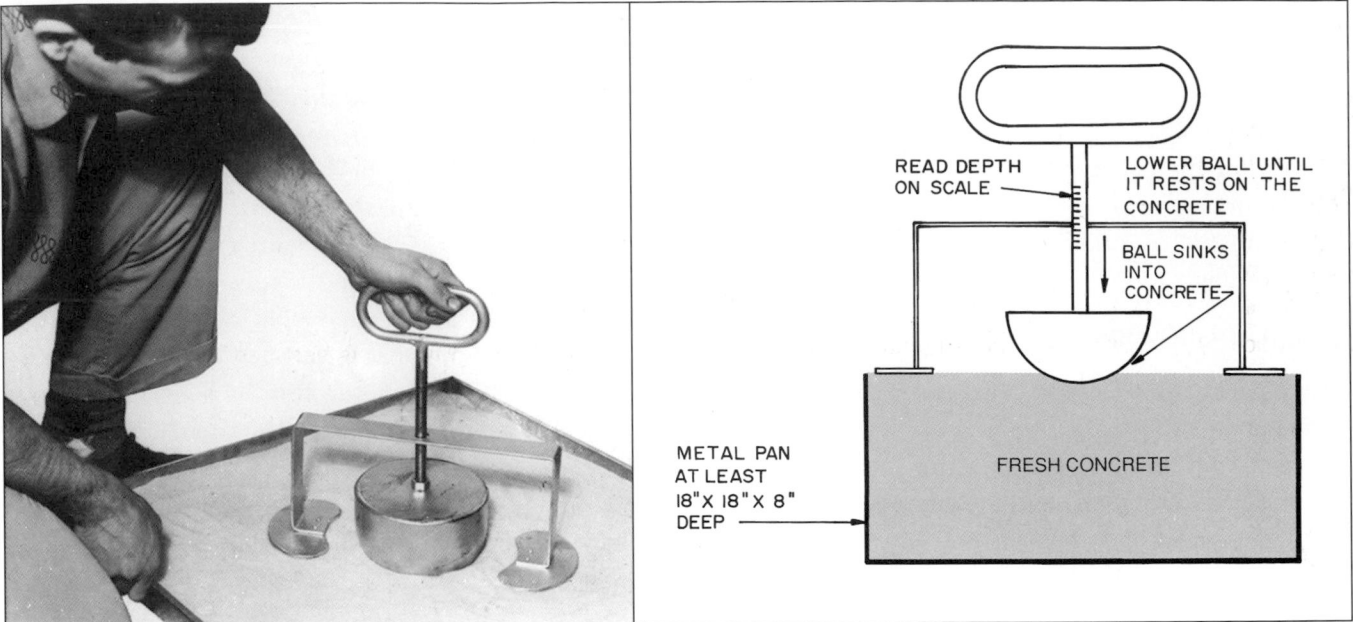

3-6 *The ball penetration test is used to test slump.*
Courtesy Portland Cement Association

Another method for testing slump is the ball penetration test as specified by ASTM C360. The depth to which a 30 lb. (13.6 kg), 6 in. (150 mm) diameter hemisphere will sink into the fresh concrete is measured. When calibrated for a particular set of materials, the results can be related to the slump. The concrete is placed in a container at least 18 in. (450 mm) square and at least 8 in. (200 mm) deep (see 3-6).

UNIT WEIGHT TEST

The unit weight test involves weighing a properly consolidated specimum in a calibrated container following ASTM C138 standards. It can determine the quantity of concrete produced per batch and also indications of air content.

AIR CONTENT TEST

Methods for measuring **air content** include the pressure method, ASTM C231, the volumetric method, ASTM C173, and the gravimetric method, ASTM C138.

Pressure method—requires the sample be placed in a pressure air meter and subjected to an applied pressure. The air content can be read directly (see 3-7).
Volumetric method—measures air content by agitating a known volume of concrete in an excess of water. This method is suitable for concrete containing all types of aggregate.

Gravimetric method—uses the same test as used for the test for unit weight of concrete. The actual unit weight of the sample is subtracted from the theoretical unit weight as determined from the absolute volumes of the ingredients, assuming no air is present. The mix proportions and specific gravities of the ingredients must be known. The difference in weight is given as a percentage and is the air content. This method requires laboratory control and is therefore not suitable for on-site use.

TEMPERATURE TEST

Sometimes concrete specifications place a limit on the **temperature** of the fresh concrete because it influences the properties. There is no standard method for measuring temperature. Armored thermometers are placed into the fresh concrete sample and left until the reading becomes stable (see 3-8).

CEMENT CONTENT TEST

The cement content test is used to determine what the water and cement content is of fresh concrete. The water–cement ratio has a major influence upon the strength of the concrete. Therefore this test will give an estimate of the strength potential without waiting for samples to harden and cure, a process that usually takes 7 to 28 days.

3-7 *This is a pressure type air meter used to ascertain air content of concrete.*
Courtesy Portland Cement Association

3-8 *This armored thermometer is placed in the fresh concrete to take its temperature.*
Courtesy Portland Cement Association

3-9 *Test cylinders being cast on the job site. Notice the cones of concrete on the left from a slump test.*
Courtesy Portland Cement Association

3-10 *The cured test cylinders are capped in preparation for the compression test.*
Courtesy Portland Cement Association

Tests with Hardened Concrete

Concrete specimens for strength tests of hardened concrete are made and cured according to ASTM C31 (in the field) and ASTM C192 (in the laboratory).

COMPRESSIVE STRENGTH TEST

The compressive strength test is made following ASTM C31 specifications. It is one of the most frequently required tests. Specifications indicate the required curing times before tests are made but 7 to 28 days are commonly used.

Concrete specimens are cast in cylinders 6 in. (152 mm) in diameter and 12 in. (305 mm) high if the coarse aggregate is 2 in. (50 mm) or smaller (3-9). Larger cylinders are specified for aggregate above 2 in. (50 mm). The cylinder is filled in three equal layers. Each layer is rodded 25 times with a ⅝ in. (16 mm) diameter rod with a round end. After the last layer is rodded and the mold is struck off full the ends are covered with a glass or metal plate to prevent evaporation. After 24 hours the hardened specimen is removed from the mold and placed in a curing location. Some are stored under controlled con-

3-11 *Standard, cured concrete cylinders are tested in a compression testing machine.*
Courtesy Portland Cement Association

ditions in a test laboratory while others may be stored on the site and cured under actual field conditions.

After the specified curing period has passed the ends of the samples are capped with a commercially available capping material (ASTM C617) (see 3-10). The capped specimen is placed in a compression testing machine and loaded until it fractures. Cylindrical specimens are tested according to ASTM C39 (3-11).

FLEXURAL STRENGTH TEST

The flexural strength test is used to determine the flexural or bending strength of the concrete. The concrete sample is formed in a mold in the shape of a beam. Samples with aggregates up to 2 in. (50 mm) should have a minimum cross section of 6 × 6 in. (150 × 150 mm). Large aggregate samples should have a minimum cross section dimension of three times the maximum size of the aggregate. The span of the test beam should be three times the depth of the beam plus 2 additional inches. A 6 × 6 in. (150 × 150 mm) beam would be 20 in. (508 mm) long.

The mold is filled in two layers with one rodded stroke for every 2 in.^{2} (13 cm^2) of area. The top is struck flush with the mold and the sample is cured with controlled temperature and moisture (see 3-12).

3-12 *A test beam is cast in a mold and cured.*
Courtesy Portland Cement Association

3-13 *The cured test beam is loaded to the breaking point to find the ultimate flexural strength.*
Courtesy Portland Cement Association

The cured specimen is tested as shown in 3-13. It is supported on each end and pressure is applied to the midpoint until the specimen breaks. The ultimate flexural strength is read on a dial in pounds per square inch (kilopascals).

ABRASION TEST

The abrasion test is used to ascertain the resistance to wear of hardened concrete samples. A hardening admixture or surface coating is used with the sample concrete mix. The test is made on a machine that rolls steel balls under pressure in a circular motion on the surface of the specimen. The specimen is weighed before and after the test. The loss in weight determines the ability to resist abrasion (see 3-14).

FREEZE-THAW TEST

The cured specimens are placed in a freeze-thaw tester, which is a cabinet much like a freezer. It is run through a series of freeze-thaw cycles. The loss between the original weight and final weight of the specimen is used to determine which samples withstand the freeze-thaw cycle best.

3-14 *This abrasion test machine determines the ability of a cured concrete sample to resist wear.*
Courtesy Portland Cement Association

Accelerated-Curing Tests

Accelerated-curing tests are used when it is desired to get acceptance of structural concrete without the usual 28-day curing period. ASTM C684 has three methods for making accelerated strength tests.

Nondestructive Tests

These tests are used to evaluate the strength and durability of hardened concrete. Commonly used tests are rebound, penetrating pullout, and dynamic or vibration tests.

Rebound tests are made with a Schmidt rebound hammer (3-15). It measures the distance of rebound of a spring-loaded plunger after it has struck the concrete surface. The reading is related to the compressive strength of concrete.

The **penetration** method uses a Windsor probe, which is a power activated gun that drives a hardened alloy probe into the concrete. The exposed length of the probe is measured and related by a calibration table to the compressive strength of the concrete. This leaves a small indention in the concrete surface.

The **pull-out test** requires that a steel rod with an enlarged end be cast in the concrete. A device used to pull the rod from the concrete measures the force required. This gives the shear strength of the concrete. It has the disadvantage of damaging the surface of the concrete.

3-15 *A rebound hammer can be used to ascertain the compressive strength of cured concrete.*
Courtesy Portland Cement Association

CAST-IN-PLACE CONCRETE CONSTRUCTION

Cast-in-place concrete utilizes wood, metal, molded plastic, and wood fiber forms that are set in place, reinforcing is placed in the forms, and the concrete is placed over the reinforcing filling the form. Cast-in-place concrete enables an engineer to design concrete members in a wide range of sizes and shapes and they can have a variety of surface textures and colors. Cast-in-place concrete members include spreadfootings, foundations, caissons, pilings, piers, flat slabs on grade and a wide range of columns, beams, girders, and above-grade floor and roof slabs.

Cast-in-place concrete structural members are usually heavier than steel, wood, or precast members. This increases the load on the foundation and footings and is a major design factor.

A wide range of forms are available as are various equipment such as concrete pumps and power finishing machines.

BUILDING CODES

Reinforced concrete structural members and prestressed concrete must be designed and constructed according to the provisions in the building code. This includes provisions to resist seismic forces if they are a factor in the area. Codes specify provisions for bending the reinforcement, the surface conditions, placing it in the forms, and required coverage of reinforcing with concrete.

ADVANCE PREPARATION

Concrete cast on the site involves much planning. Among the things to be considered is access to the site, preparation of temporary roads, and sources of suitable water and aggregates. Areas must be cleared and prepared for stockpiling materials and protecting them from the weather and contamination, location of the batching plant or other sources of concrete, and an overall safety plan for the site, including scaffold and form construction. Following is a discussion of the major activities in preparing and placing concrete for cast-in-place construction.

Concrete Production

There are various ways to produce, deliver, finish, and cure concrete. Those used depend on the circumstances under which the job is operating.

BATCHING

Concrete is usually prepared in batches. A **batch** is the amount of concrete mixed at one time. The quantities of each dry ingredient are usually weighed. Water and admixtures are specified by either weight or volume. The use of volume measurements is discouraged because it is inaccurate. Moisture in the aggregate will change the weight but this is not accounted for in volume measures. Aggregates, especially sand, tend to fluff when handled so the actual volume of sand can vary from batch to batch.

MIXING

Concrete is mixed until it is uniform in appearance. All ingredients must be evenly distributed. If an increased amount of concrete is needed get an additional mixer rather than overloading or speeding up those in operation. Follow the manufacturers recommendations. Keep the mixing blades clean and replace them if they become bent or worn.

3-16 *Concrete from a stationary mixer is being loaded into a transit truck for movement to the place on the site where it is needed.*

Courtesy Tennessee Valley Association

STATIONARY MIXING

On a large job the concrete is often mixed on the job using a stationary mixer (3-16). They may be a tilting or nontilting type and may be manual, semiautomatic, or automatically controlled. Some have data for various mix designs stored on computer tape. Generally the batch is mixed one minute for the first cubic yard and an additional 15 seconds for each additional ½ cubic yard (0.35 m^3) or fraction thereof. The mixing time is measured from the time when all ingredients are placed in the mixer. All water must be added before one-fourth of the mixing time has elapsed.

Generally about 10% of the mixing water is placed in the drum before the dry ingredients are added. Then water is added uniformly with the dry ingredients saving 10% to be added after all dry ingredients are in the drum.

READY-MIX CONCRETE

Ready-mix concrete may be fully mixed in a central mixing plant and delivered to the site in a truck mixer operating at agitating speed or a special nonagitating truck. The concrete may be partially mixed in a central mixer and completed in a truck mixer as it is moved to the site or the dry ingredients may be placed in a transit truck mixer, water added, and the entire mixing is done by the truck.

REMIXING

Fresh concrete in the drum tends to stiffen even before the concrete has hydrated to initial set. It may be used if remixing will restore sufficient plasticity so it can be compacted in the forms. Under special conditions a small amount of water can be added but it must not exceed the allowable water-cement ratio, designated slump, allowable drum revolutions, and must be remixed at least half the minimum required mixing time or number of revolutions.

Transporting, Handling & Placing Concrete

Before the fresh concrete arrives on the job preparations for moving it to its point of placement must be complete. Delays in placing the concrete can cause a loss of plasticity. Also the method of moving the concrete must not result in the segregation of concrete materials. **Segregation** is the tendency of the coarse aggregate to separate from the sand-cement mortar. In some cases the heavy aggregate may settle to the bottom and the sand-cement rises to the top. This produces unsatisfactory results.

3-17 *Many jobs permit the concrete to be poured directly from the transit truck onto the area of placement.*
Courtesy Portland Cement Association

3-19 *A concrete pump moves concrete for placement.*
Courtesy Portland Cement Association

3-18 *A pour being made using a concrete bucket.*

3-20 *Concrete can be placed with a telescopic conveyor. This permits a long range of placement. Notice the down flow discharge tube.*
Courtesy ROTEC Industries

There are many ways to move and place concrete. Concrete is moved about the site to the point of placement with cranes using concrete buckets, barrows, buggies, chutes, belt conveyors, pneumatic guns, and concrete pumps. Often the pour can be made directly from the transit-mix or ready-mix truck as shown in 3-17.

Multistory slabs, beams, and columns are poured from buckets lifted by cranes or by concrete pumps. The buckets are filled on the ground, lifted over the point of pour, and opened to deposit the concrete as shown in 3-18.

Concrete pumps are heavy duty piston pumps that force concrete through a pipe ranging from 6 to 8 in. (152 to 203 mm) in diameter. They can place concrete over long distances ranging up to about 100 ft. (30.5 m) vertically and 800 ft. (244 m) horizontally. A pump is shown in 3-19.

Concrete is also placed with a telescopic conveyor as shown in 3-20. Notice the concrete truck discharging onto a belt that raises the mix to the telescopic conveyor.

3-21 *Concrete is being consolidated with a mechanical vibrator.*

Courtesy Portland Cement Association

3-22 *After the concrete has been placed it is screeded.*

Courtesy Portland Cement Association

3-23 *A bull-float is used to lower high spots and fill in low spots.*

Courtesy Portland Cement Association

Concrete should be placed continuously as near as possible to its final location. In slab construction, work is started along one end and each batch is discharged against the one previously placed. If the concrete is to be thick, as in a foundation wall, it should be placed in layers 6 to 20 in. (150 to 500 mm) thick for reinforced members and 15 to 20 in. (400 to 500 mm) thick for mass work. Each layer should be consolidated before a second layer is placed on it.

As the concrete is placed it is consolidated. **Consolidation** is the process of compacting the freshly placed concrete to the forms and around reinforcing steel to remove air pockets and pockets of stone. It may be done by hand by pushing a rod into the concrete or with a mechanical vibrator (see 3-21).

PLACING CONCRETE IN COLD WEATHER

Concrete placed in cold weather gains strength slowly. Fresh concrete must be protected from freezing. The critical period after which concrete is not seriously damaged by several freezing cycles depends upon the ingredients, conditions of mixing, placing, curing, and long-term drying.

PLACING IN HOT WEATHER

In addition to maintaining low concrete temperature by cooling the aggregates and water, precautions must be made to maintain a low temperature while placing. This could involve shading and painting white the mixers, chutes, hoppers, pump lines, and other concrete handling equipment. Forms can be cooled with water and the subgrade moistened before the concrete is placed. The concrete must be transported from the mixing station to the point of placement as quickly as possible. In hot weather the mix should be in place 45 to 60 minutes after mixing.

Finishing Concrete

After the concrete has been placed and consolidated it is screeded (see 3-22). **Screeding** (also called strikeoff) involves striking off the excess concrete to bring it flush with the tops of the forms. Then the surface is **bull-floated** and **darbed.** This is done immediately after strikeoff to lower high spots, fill low spots, and embed large aggregate that may be on the surface. A bullfloat has a long handle connected to a float (see 3-23). A darby has a shorter handle and is used for shorter distances. This work must be done before any bleed water appears on the surface. **Bleed** refers to the water that

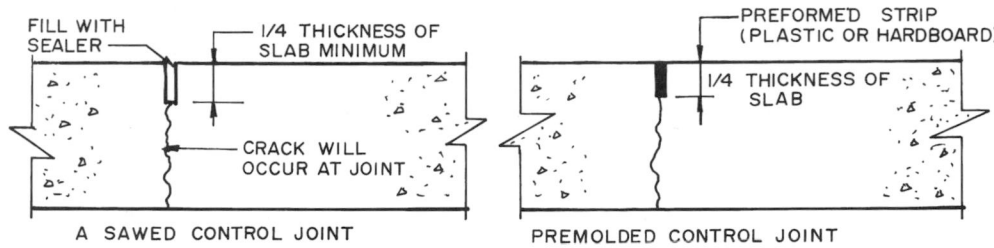

3-24 *Control joints provide a place for the slab to crack without being visible or running at angles across the slab.*

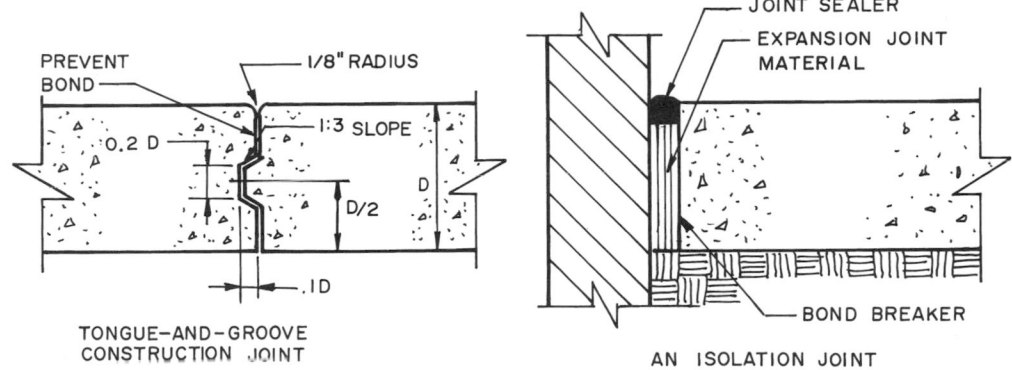

3-25 *Construction joints occur where two pours will meet. Isolation joints separate a pour from a wall, column, or other abutting form.*

rises to the surface very soon after the concrete is placed on the forms. Any finishing operation performed on the surface of the concrete while bleed water is present will cause it to scale and dust.

When the bleed water sheen has evaporated the surface is ready for final finishing. Final finishing includes one or more of the following: edging, jointing, floating, troweling, and brooming.

Edging rounds off the edge of the slab to prevent chipping. **Jointing** forms control joints in the slab. A groove ¼ the thickness of the slab is formed across the slab at intervals specified by the architect. Its purpose is to provide a weak spot where the slab can crack when stresses exceed the strength of the concrete. Control joints can be formed in the wet concrete, sawed after the concrete has hardened, or be formed by inserting plastic or hardwood slices in the concrete (see 3-24). Other joints used are isolation and construction. **Isolation** joints provide a space between a slab and a wall allowing each to move without disturbing the other. **Construction** joints are formed where one pour is to end and a joining one will meet it (see 3-25).

After the concrete is edged and jointed it can be floated. This is done with a wood or metal hand held float or a finishing machine with float blades (see 3-26). **Floating** embeds aggregate slightly below the surface, removes imperfections, compacts the mortar at the surface for final finishing, and keeps the surface open,

allowing excess moisture to escape. Marks left by edging and jointing are removed by floating. If they are wanted for decorative purposes they need to be rerun after floating.

The final finish might be accomplished by troweling, brooming, or forming a pattern or texture in the surface. **Troweling** produces a hard, dense, smooth surface. The trowel is a steel blade, handheld tool. **Brooming** involves roughing the surface with a steel-wire or coarse-fiber broom. It provides a slip-resistant surface. Various patterns can be formed in the surface by placing divider strips in the concrete.

3-26 *The slab can be floated with a power finishing machine.*
Courtesy Portland Cement Association

Curing

As explained in an earlier chapter the addition of water to portland cement forms a water–cement paste that produces a chemical reaction called hydration. This reaction produces a hard cement paste that bonds the aggregate into a solid mass. Hydration continues for an indefinite period at a decreasing rate as long as water is in the mix and the temperature is favorable (73°F (23°C) is recommended). Therefore, concrete should be protected so moisture remains in the mix during the early hardening period and the temperature is maintained. Concrete that is protected and kept moist for 7 days will have about twice the compressive strength of unprotected concrete (exposed to air with no attempt to keep moisture in the mix). The curing process is essential to producing concrete members with the expected compressive strength.

▼ Following are some of the curing methods used.
   Forms—may be left in place and the top exposed concrete surfaces kept moist.
   Curing compounds—are sprayed on the surface. They retard the evaporation of moisture.
   Sheet materials—are used to cover the exposed concrete to hold moisture in. Among those in use are waterproof curing paper, and plastic film.
   Wet covering materials—such as burlap or moisture retaining fabrics can be placed over the concrete. They should be kept wet over the entire curing period.
   Sprinkling—the surface continuously is an excellent way of curing.

PNEUMATICALLY PLACED CONCRETE

Concrete can be placed by pneumatically forcing a dry mixture of sand, aggregate, and cement through a hose and mixing it with water at a nozzle. It is referred to as pneumatically placed concrete or shotcrete. It is used where concrete has to be placed in difficult locations to form thin sections, and to cover large areas. It is used to place concrete in free-form shapes, such as coating a dome, for applying protective coatings, and for repairing concrete surfaces.

Formwork

There are a variety of standard metal forms (steel and aluminum) manufactured. These are available in a range of sizes and have assembly and bracing systems (see 3-27). Plywood is used for carpenter-built forms. While these are used for general wall construction they are the main material for the construction of more intricate and complex forms. An excellent material for carpenter-built forms is exterior high-density overlay plywood. Waferboard and solid wood are also used for form construction. There are a number of molded plastic and waxed cardboard forms available.

The finish produced by forms ranges from the untreated surface left by the forms to one produced by **form liners** (see 3-28). Form liners are molded plastic sheets that have been molded from actual concrete, masonry, or wood patterns. They are bonded to plywood sheets and are secured inside the form, producing a textured surface (see 3-29).

3-27 *Foundation walls can be rapidly formed using standard metal forms.*
Courtesy Precise Forms, Inc.

3-28 *Form liners are moldable plastic sheets with a textured surface that are secured to the inside of the form.*
Courtesy Symons Corporation

There are forms used to construct cast-in-place structural members. Columns are usually round, square, or rectangular. Round columns are formed with circular forms such as shown in 3-30, using a molded fiberglass column form.

Circular formwork is shown in 3-31. The horizontal wales are curved metal units and the form is supported with round, steel braces that are bolted to the concrete floor.

3-30 *This fiberglass column form has the reinforcing steel in place. The workers are installing a metal ring to which the bracing will be connected.*
Courtesy Molded Fiberglass Concrete Forms Co.

3-29 *Form liners produce a wide range of surface textures on poured concrete.*
Courtesy Symons Corporation

3-31 *A plywood skin is used on this circular formwork. Notice the curved wales and braces bolted into the concrete floor.*
Courtesy PERI Formwork Systems, Inc.

3-32 *The climbing formwork is raised to the next level and used to pour the next exterior wall panel.*

Courtesy PERI Formwork Systems, Inc.

3-33 *These pans with steel reinforcing in place are typical of those used to form cast-in-place concrete floors and roofs on multistory buildings.*

Courtesy U.S. Corps of Engineers

3-34 *This unit of flying formwork has molded fiberglass pans supported by a steel framework. It is being set in place by a crane to be used to pour the floor directly below it.*

Courtesy Molded Fiberglass Concrete Forms Co.

3-35 *The polystyrene forms are spaced with plastic inserts and braced with wood posts and wales.*

Courtesy American ConForm Industries

An example of climbing formwork is shown in 3-32. After the section is poured and has reached sufficient strength the form is raised to the next level, positioned, reinforcing set in place, and the pour repeated.

There are a number of ways to construct cast-in-place concrete floors and roofs. In 3-33 are pans set for pouring a waffle floor. Between the pans are the reinforcing bars needed in the ribs. The reinforcing for the floor is placed over the pans. The large areas without pans are column heads cast over each column below the floor.

Flying formwork is built in large sections moved up by a crane to the next floor and reused (see 3-34). The formwork is supported by metal trusses providing a strong, rigid unit.

Another type of form used for residential and light commercial foundation construction uses an insulated stay-in-place form. The form is made from expanded polystyrene and the units are notched to facilitate stacking them, forming the cavity for the cast in place concrete wall. As you can see in 3-35 the forms remain in place after the wall is poured.

Large projects use a variety of forms—3-36 shows the forms being installed for part of the construction of a dam. Straight wall forms, horizontal slabs, and forms for slanting concrete surfaces are seen.

Concrete Reinforcing Materials

Since concrete has no useful tensile strength steel reinforcing is added. Steel has high tensile strength. Since concrete and steel have about the same coefficient of thermal expansion, concrete bonds to steel, and steel is not corroded by concrete, they work together to provide a satisfactory structural system.

The two commonly used steel-reinforcing materials are reinforcing bars with associated hooks and stands and welded-wire reinforcement. A wide range of fibers also find use as concrete reinforcing.

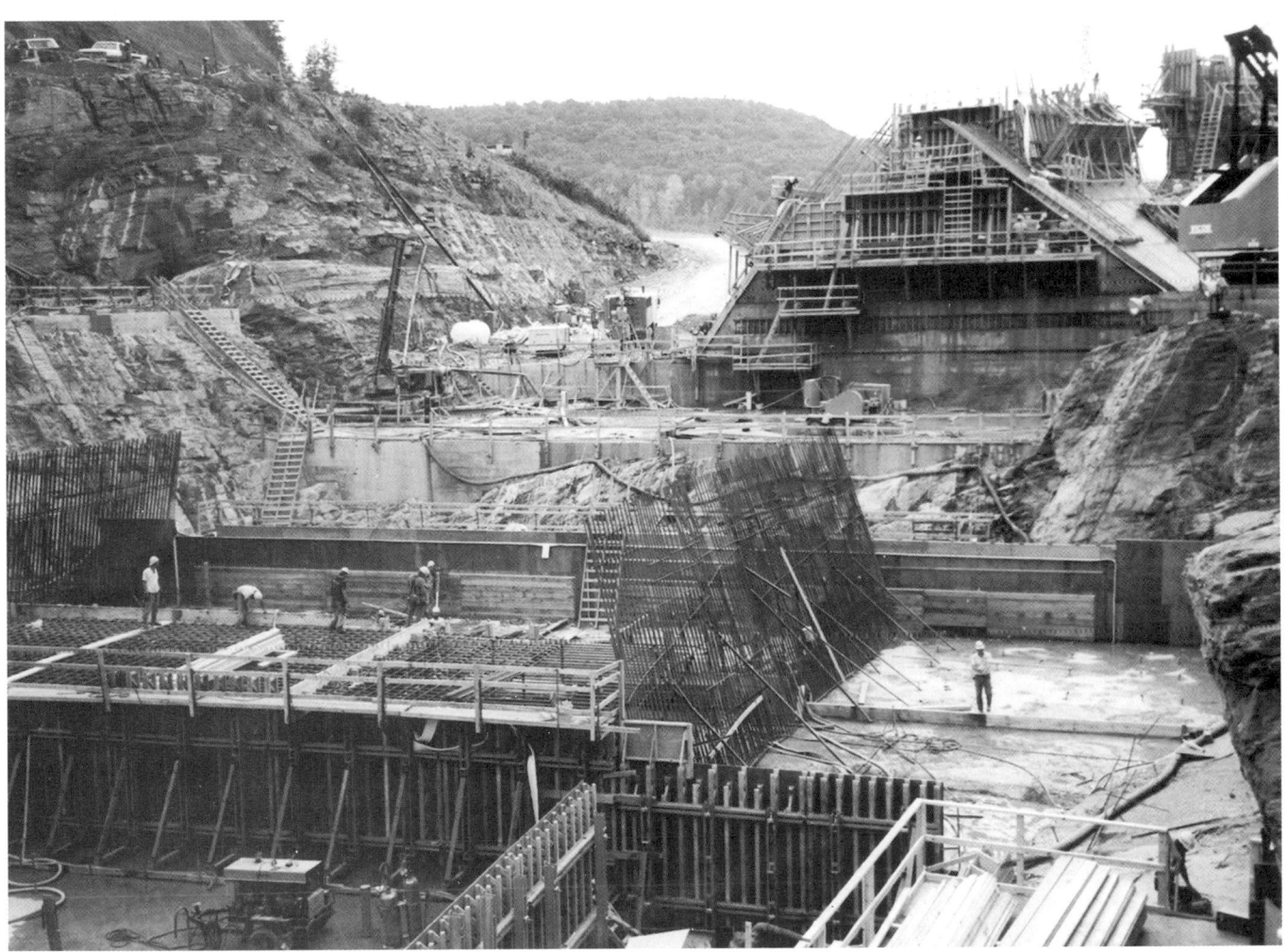

3-36 *Notice the variety of concrete forms required on this dam construction project.*

Courtesy PERI Formwork Systems, Inc.

Table 3-6 STEEL-REINFORCING BAR SIZES

ASTM Inch-Size Steel-Reinforcing Bars

Bar Size Designation	Weight (lbs./ft.)	Nominal Dimensions Diameter (in.)	Cross-Sectional Area (in.^{2})
#3	0.376	0.375	0.11
#4	0.668	0.500	0.20
#5	1.043	0.625	0.31
#6	1.502	0.750	0.44
#7	2.044	0.875	0.60
#8	2.670	1.000	0.79
#9	3.400	1.128	1.00
#10	4.303	1.270	1.27
#11	5.313	1.410	1.56
#14	7.650	1.693	2.25
#18	13.60	2.257	4.00

ASTM Metric-Size Steel-Reinforcing Bars

Bar Size Designation	Mass (kg/m)	Nominal Dimensions Diameter(mm)	Area (mm^2)
#10M	0.785	11.3	100
#15M	1.570	16.0	200
#20M	2.355	19.5	300
#25M	3.925	25.2	500
#30M	5.495	29.9	700
#35M	7.850	35.7	1000
#45M	11.775	43.7	1500
#55M	19.625	56.4	2500

Copyright American Society for Testing and Materials.

Reprinted with permission.

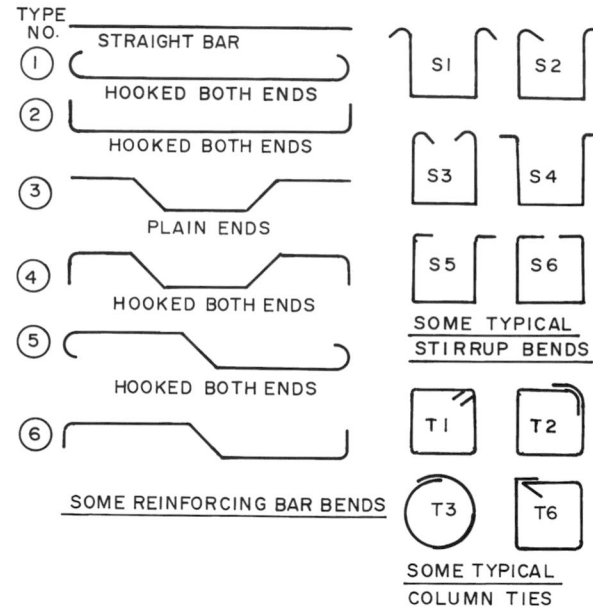

3-38 *Some of the standard bends used with steel-reinforcing bars.*

Courtesy American Concrete Institute

STEEL-REINFORCING BARS

Reinforcing bars (also called **rebars**) are hot-rolled steel rods that may be plain (smooth) or deformed. The deformed type has surface ridges giving better bonding to the concrete. The smooth type is used for special applications.

The inch-size bars are available in eleven standard diameters. The metric bars are made in eight diameters (Table 3-6).

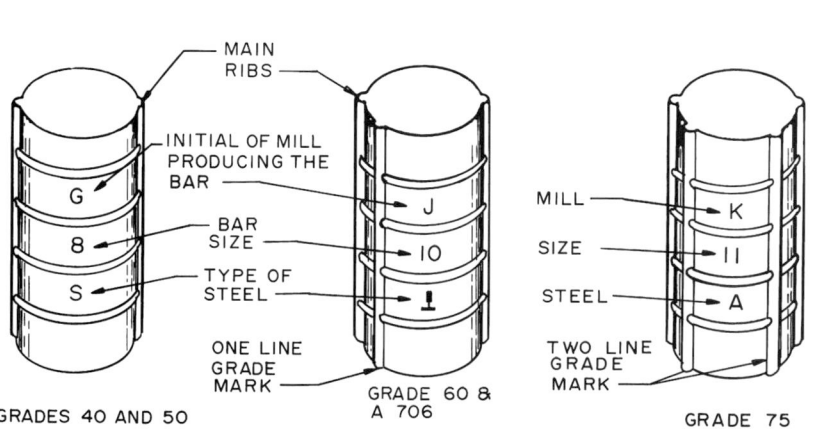

3-37 *Markings stamped into reinforcing bars tell the bar size, type of steel, grade mark, and identify the mill that made them.*

Courtesy Concrete Reinforcing Steel Institute

Table 3-7 SPECIFICATIONS COVERING WELDED-WIRE FABRIC

U.S. Specification	Canadian Standard	Title[a]
ASTM A82	CSA G30.3	Cold-Drawn Plain Steel Wire for Concrete Reinforcement
ASTM A185	CSA G30.5	Welded Plain Steel Wire Fabric for Concrete Reinforcement
ASTM A496	CSA G30.14	Deformed Steel Wire for Concrete Reinforcement
ASTM A497	CSA G30.15	Welded Deformed Steel Wire Fabric for Concrete Reinforcement

[a]The titles of the American Society for Testing and Materials specifications and the Canadian Standards Association are identical.

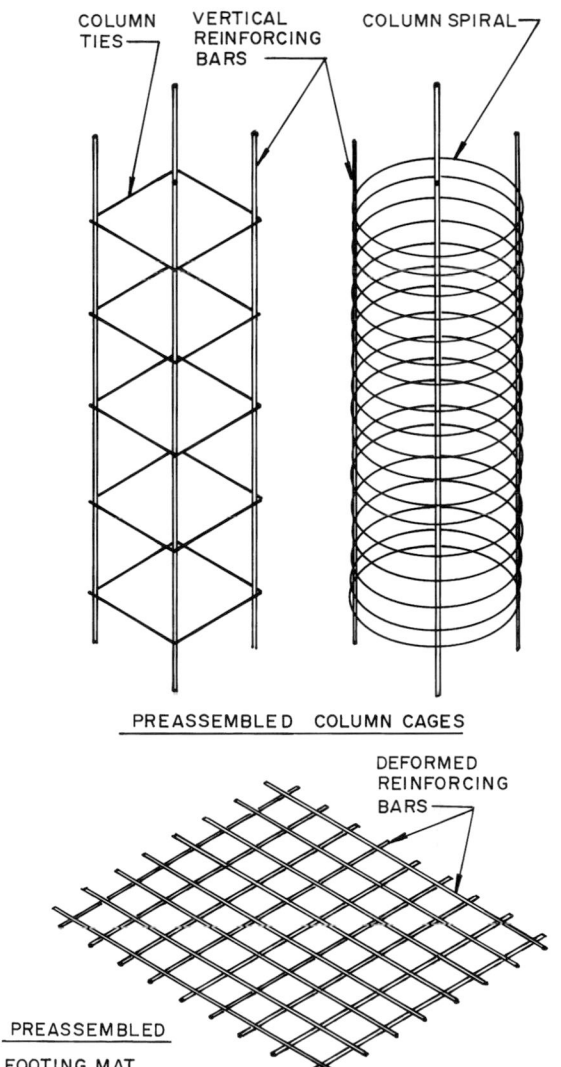

3-39 *Some reinforcing is preassembled and placed in the form.*

The diameters are identified by the bar size designation. Inch-size bar designations represent ⅛th of an inch of bar diameter. For example, a No. 4 bar is 4/8 or ½inch in diameter. Metric designations give the diameter in millimeters. A No. 20 M bar has a diameter of 20 mm (actually 19.5 mm as shown in Table 3-6).

Reinforcing bars are made in grades 40, 50, 60 and 75. These refer to the minimum yield strength of the steel—10,000, 50,000, 60,000, and 75,000 pounds per square inch (276, 345, 414, 517 MPa). In Canada metric reinforcing bars are available in grades 300 (43,600 psi or 300 MPa), 350 (50,800 psi or 350 MPa), and 400 (58,000 psi or 400 MPa). Grade 400W is a weldable bar.

Examples of the markings stamped into the rebar are in 3-37. Shown are the type of steel, bar size, grade mark, and identification of the mill making the bar.

The types of bends are standardized and identified by number. A few of these are shown in 3-38.

Some preassembly of reinforcing bars is necessary before they are placed in the forms. Examples are column spirals, column ties, and footing bars (3-39). The steel in beams usually involves a set of bottom bars and stirrups. The bottom bars resist tension forces that exist in the bottom of the beam. The stress is dissipated from the bars into the concrete through the bond between the concrete and the bars.

The bottom bars in the beam are raised above the bottom of the form with one of the several types of bar supports. These could be bolsters or chairs. Bars in the top of the slab are supported with high chairs. Bar supports are available in wire, precast concrete, cementitous fiber-reinforced, and all-plastic types.

WELDED-WIRE REINFORCEMENT (WWR)

Welded-wire reinforcement (sometimes called welded-wire fabric WWF) is an assembly of steel-reinforcing wires made from rod that is worked either cold-drawn or cold-rolled or both. The material is available in rolls and sheets. The wires may be plain (W) or deformed (D). The plain welded wire bonds to concrete by the positive mechanical anchorage at each welded intersection.

In addition to uncoated wire there are two coatings available. One is a hot-dipped galvanized coating and the other is an epoxy coating. Note the information in Table 3-7.

The transverse wires are electrically resistance welded to each of the longitudinal wires and form squares or rectangular grids (see 3-40).

Welded-wire reinforcement is specified by listing the longitudinal wire spacing, the transverse wire spacing, longitudinal wire size, and the transverse wire size.

▼ An example for WWR style is 12 × 12 – W12 × W5. This means:

Longitudinal wire spacing 12 in. apart

Transverse wire spacing 12 in. apart

Longtudinal wire type and area-W means plain wire, 12 means area of 12 in.^{2}/ft.

Transverse wire type and area-W means plan wire, 5 means area of .05 in.^{2}/ft.

▼ Following are examples of metric specifications:

Structural WWR-305 × 305-MD71 × MD71

This is equivalent to inch size 12 × 12-D11 × D11. The spacings between wires are in millimeters (mm) and wire sizes are in mm^2. Metric plain wire is MW and deformed wire is MD.

Detailed information relating to the manufacture, specifications, properties, design, information, and building code requirements are available from the Wire Reinforcement Institute, Inc. The American Concrete Institute publications AC1 318, *Building Code Requirements for Reinforced Concrete*, contains design data information on welded wire.

Welded-wire sheets are shipped in bundles in quantities varying with the size and weight of the sheets. Typical bundles weigh 500 to 5000 lbs. (227 to 2268 kg). They are bound together with steel strapping. The bundles should never be lifted off the truck by the steel strapping. Lifting eyes can be specified when bundles are to be lifted with cranes.

FIBER-REINFORCED CONCRETE

In addition to the use of steel-reinforcing bars and welded-wire fabric a number of fibers are being used to reinforce concrete. They are mixed with the concrete as it is prepared in the mixer. A number of manufacturers produce these various products and research will probably increase their effectiveness and use in the future. They may under some conditions reduce the amount of reinforcing bars required and replace welded-wire fabric in some installations. They also are used along with these to produce more desirable properties in concrete.

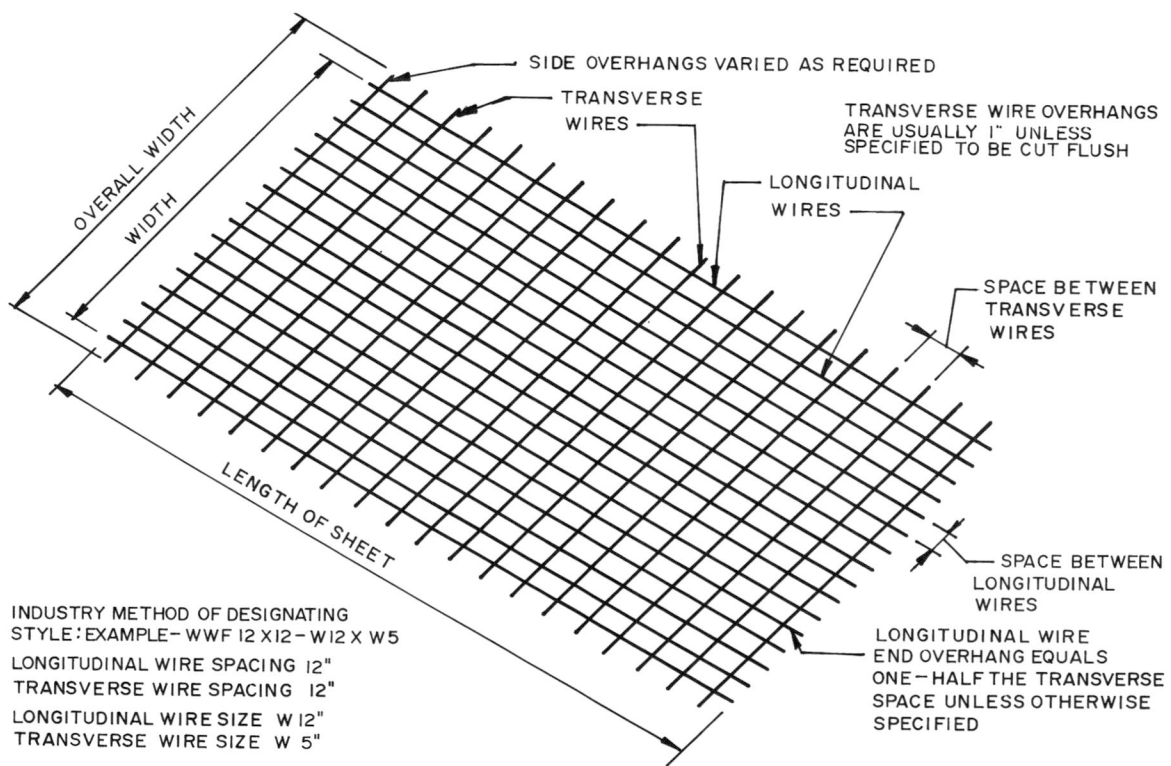

3-40 *Identification of the parts of a sheet of welded-wire fabric.*
Courtesy American Society for Testing and Materials

The types of fibers available include glass fibers, polymeric (polypropylene, polyethylene, polyester, acrylic, and aramid), steel, asbestos, carbon, and natural fibers (wood, sisal, coconut, bamboo, jute, okwara, and elephant grass).

It should be noted that each of these fibers has different characteristics and consultation with a concrete specialist should occur before using them. The Portland Cement Association has information on fiber-reinforced concrete.

Cast-in-Place On-Grade Concrete Slabs

On-grade concrete slabs require preparation of the slab bed. While this varies with the type of soil often a base of compacted gravel is required over the soil. Some soils require only compacting before pouring the slab. To control moisture penetration a plastic sheet can be laid over the base. Some place 2 or 3 in. (50 to 75 mm) of compacted sand over the plastic sheet (3-41). Finally the reinforcing is placed over this base and held above the surface the required amount with concrete brick.

The common type of slabs include **unreinforced, lightly reinforced,** and **structurally reinforced. Unreinforced** slabs rely entirely upon the earth for support against tension forces. If subjected to excess loads they could crack. As the slab cures the surface may also have cracks open due to shrinkage. **Lightly reinforced** slabs also depend upon the earth for total support. However, they are reinforced with welded-wire fabric or one of the fiber reinforcements. In some cases both are used. This helps hold together surface cracks that occur during curing. **Structurally reinforced** slabs contain steel reinforcing bars and often welded-wire fabric and/or fiber reinforcing. The design of the reinforcing and thickness of the slab varies with the loads to be carried. Some designs depend upon the soil for support while other designs have sufficient reinforcing so the slab can extend from one support, as a foundation, to another and not depend upon the earth for support. These slabs are designed by a professional engineer. Selected examples are shown in 3-42.

Joints are necessary when building concrete slabs on grade to help control cracking, reduce the size of the pour, and separate the slab from surfaces where it should not bond. **Control joints** are used to provide a weakened place in the slab where it can crack (if necessary) and be relatively hidden.

Construction joints are used to separate a large area to be poured into smaller more manageable areas.

Isolation joints are used to keep a slab from bonding to some abutting part of the building.

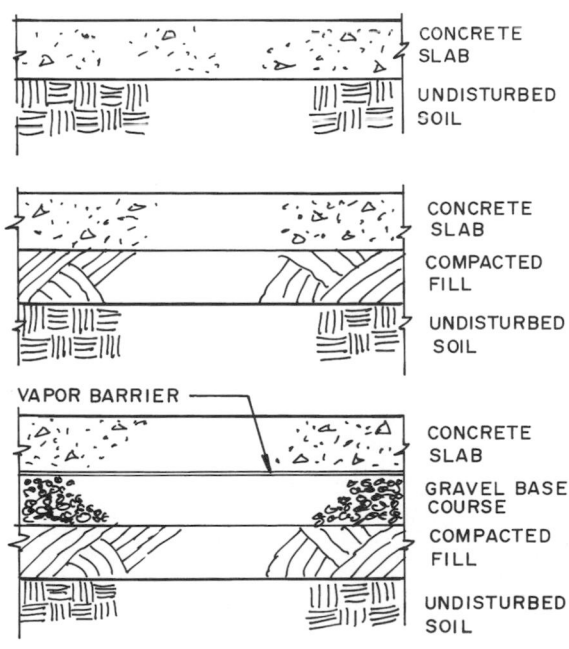

3-41 *Various methods for constructing on-grade concrete slabs.*

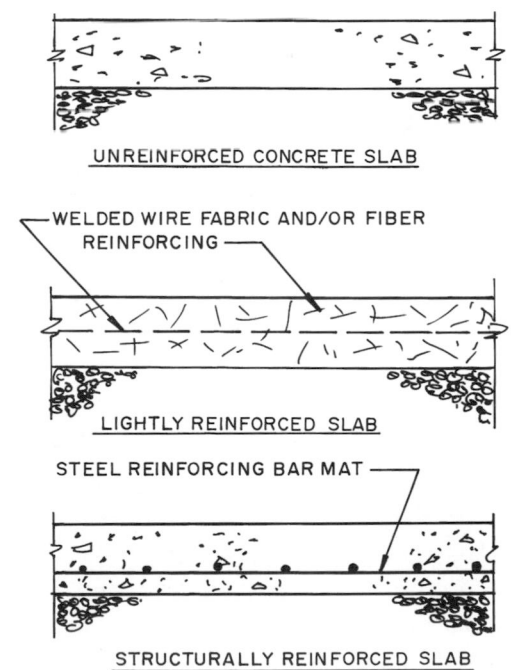

3-42 *Various degrees of reinforcing used on concrete slabs.*

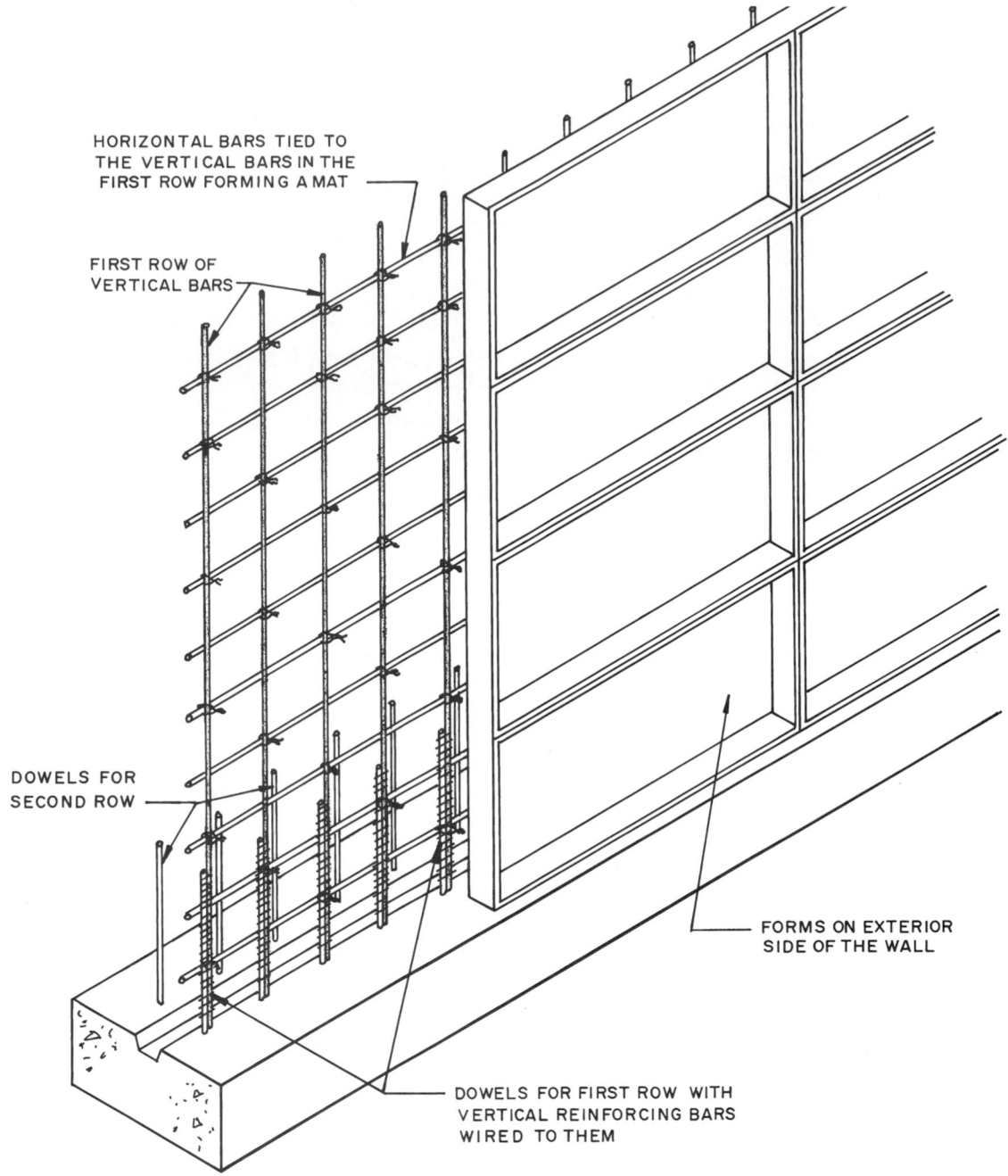

HORIZONTAL BARS TIED TO
THE VERTICAL BARS IN THE
FIRST ROW FORMING A MAT

FIRST ROW OF
VERTICAL BARS

DOWELS FOR
SECOND ROW

FORMS ON EXTERIOR
SIDE OF THE WALL

DOWELS FOR FIRST ROW WITH
VERTICAL REINFORCING BARS
WIRED TO THEM

3-43 *An example of typical wall forming and reinforcing. This example shows the first vertical matt in place.*

Reinforcing Cast-in-Place Concrete Walls

Reinforced concrete cast-in-place walls may rest on a continuous concrete footing and be below grade, as a basement wall, or extend above grade forming the exterior or interior walls of a building. When the footing is poured metal dowels are inserted that project above the top of the footing. Some footings have a key cast into them which forms a tie at the bottom of the wall. The reinforcing bars are formed into a mat and tied together inside the form as you can see in 3-43.

If the wall is to be topped with a concrete slab floor or roof the working drawings will show the required reinforcing for the connection. Typically vertical reinforcing will extend above the top of the wall and be bent to be cast into the slab (3-44). Corners between walls also have to be tied together. Usually some form of a hook or elbow bar is used. After all steel is in place it should be rechecked before the other side of the form is set in place and braced. The wall is now ready to be poured (3-45).

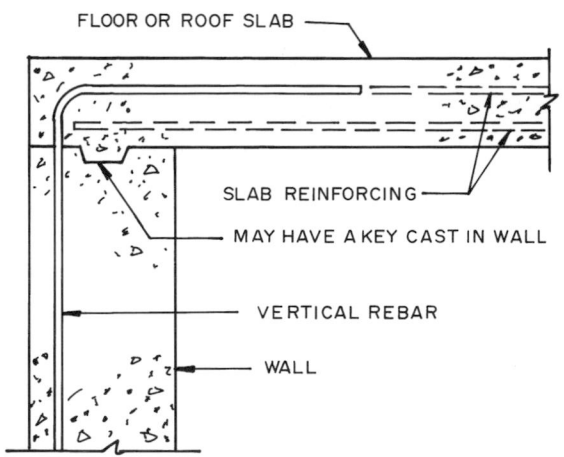

3-44 *A typical wall to slab connection.*

3-45 *A straight wall form with reinforcing in place. Notice the form supports a working deck and the wall is being poured using a concrete pump.*
Courtesy PERI Formwork Systems, Inc.

Reinforcing Cast-in-Place Concrete Beams

A simple single span cast-in-place beam rests on end supports. One possible arrangement for placing the reinforcing is shown in 3-46. The stirrups are held in place with a small diameter top bar or the top member of a truss bar.

Continuous beam casting is common in structures with a cast-in-place concrete structural system. The beam ties together the structure from column to column. The bottom of the beam at midspan is in maximum tension and this force dissipates toward the ends of the beam. The stirrups at the end transfer these tension forces to the concrete. Over the column the top of the beam is subject to tension forces due to bending. Therefore appropriate top bars are placed over each column.

Reinforcing Cast-in-Place Concrete Columns

Cast-in-place concrete columns are typically round, square, or rectangular. They are reinforced with vertical bars, which help carry compressive loads and resist tension forces due to lateral loads as caused by a high wind. The vertical bars may be arranged in a square or rectangular pattern or a circular pattern. Either pattern can be used for square and circular columns.

Cast-in-Place Concrete Framing Systems

A typical cast-in-place concrete framing system utilizes cast-in-place columns, one-way or two-way concrete slabs, and cast-in-place joists, beams, and girders.

There are several types of reinforced concrete framing systems used with cast-in-place concrete construction. In some cases the beams and girders are cast with

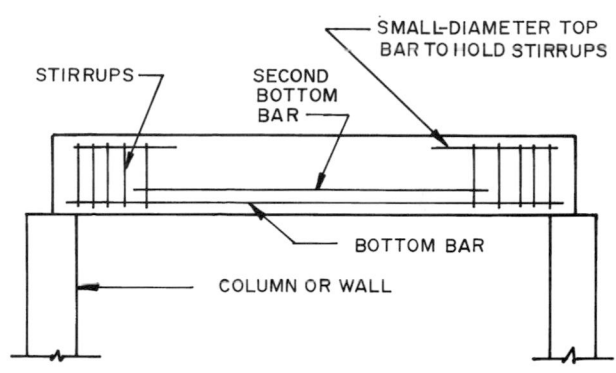

3-46 *Typical reinforcing for a single span cast-in-place beam.*

the floor or roof slab. The two commonly used reinforced concrete floor and roof systems are the flat slab and the flat plate. The flat slab is a concrete slab reinforced in two or 'more' directions and supported by columns with dropped panels and capitals that enlarge the columns at their top or by beams or joists. When two-way reinforcement is used in the slab it is called a two-way slab. A **flat plate floor** or roof is much like the flat slab with reinforcing running in two directions and supported by columns that are not enlarged where they meet the slab.

One-Way Flat Slab Floor & Roof Construction

The one-way flat slab systems in common use include a one-way solid slab, one-way flat slab with beams and girders, and one-way flat slab with joists and beams.

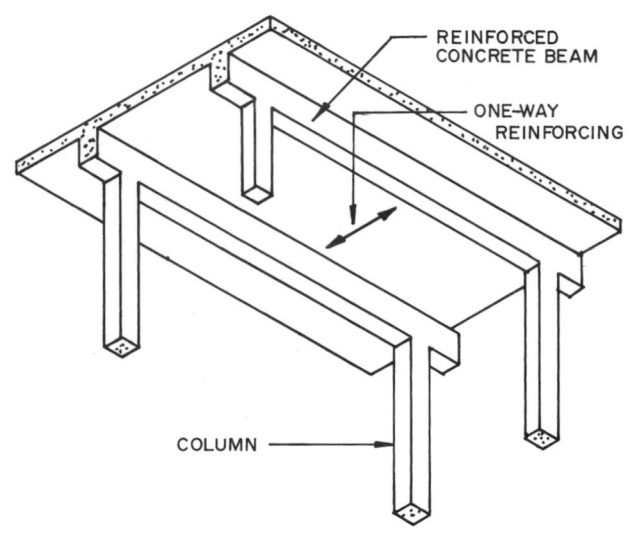

3-47 *One-way solid slab with beam construction has the main beams on one axis and the slab spans the distance between them.*

REINFORCED CONCRETE BEAM

ONE-WAY REINFORCING

COLUMN

ONE-WAY SOLID SLAB WITH BEAMS

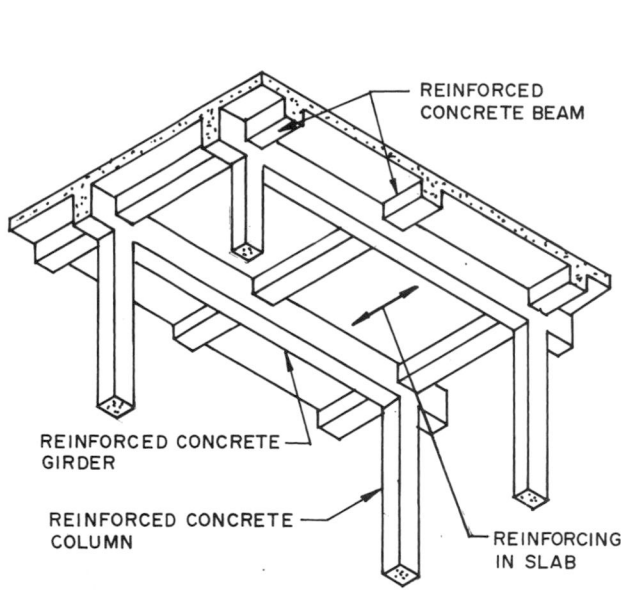

REINFORCED CONCRETE BEAM

REINFORCED CONCRETE GIRDER

REINFORCED CONCRETE COLUMN

REINFORCING IN SLAB

ONE-WAY FLAT SLAB WITH BEAMS AND GIRDERS

3-48 *A one-way flat slab with beams and girders has a slab that spans between the beams that are supported by the girders.*

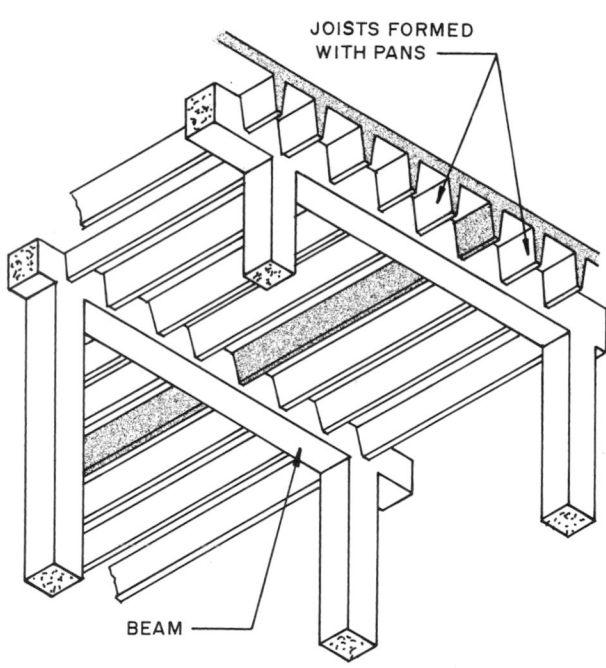

JOISTS FORMED WITH PANS

BEAM

ONE-WAY FLAT SLAB WITH JOISTS AND BEAMS

3-49 *A one-way flat slab with joists and beams has the joists span the distance between the beams.*

ONE-WAY SOLID SLAB CONSTRUCTION

One-way solid slab construction has the slab supported on two sides by beams or load-bearing walls. An example of one-way solid slab construction is shown in 3-47. The main beams are on one axis and the slab spans the distance between them. The reinforcing bars or pre-stressed tendons are placed perpendicular to the supporting walls or beams.

ONE-WAY FLAT SLAB WITH BEAMS & GIRDERS

Another one-way slab construction is shown in 3-48. It utilizes a one-way flat slab with beams. The construction follows the same procedures described earlier for one-way solid slab construction except this slab is supported by cast-in-place concrete beams that are supported by cast-in-place girders. Usually the girders, beams, and slab are cast monolithically.

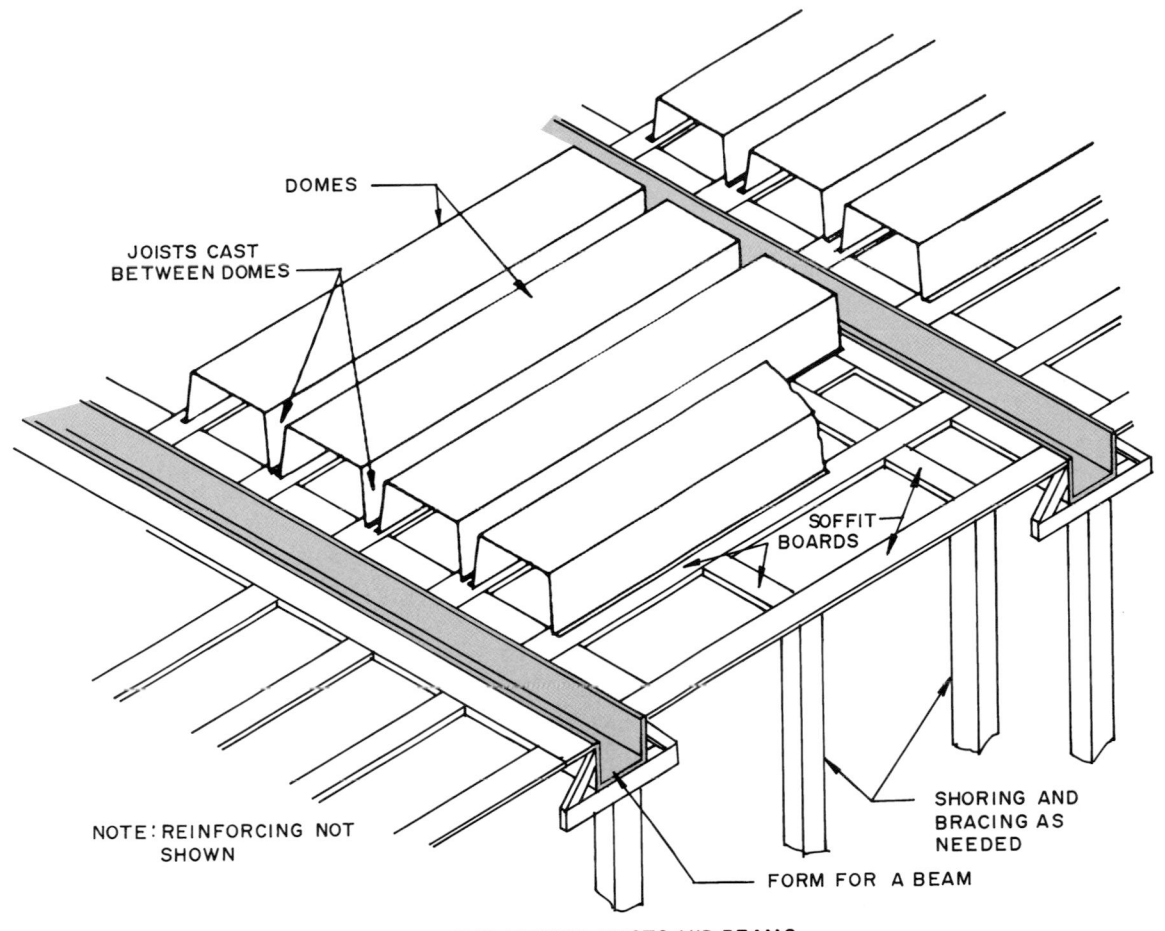

DOMES

JOISTS CAST
BETWEEN DOMES

SOFFIT
BOARDS

NOTE: REINFORCING NOT
SHOWN

SHORING AND
BRACING AS
NEEDED

FORM FOR A BEAM

ONE—WAY FLAT SLAB WITH JOISTS AND BEAMS

3-50 *The beams, joists, and flat slab are cast monolithically using domes to form the joists and the base for the floor.*

ONE-WAY FLAT SLAB WITH JOISTS & BEAMS

Another widely used system uses a one-way flat slab with joists and beams as shown in 3-49. The beams, joists, and slab are cast monolithically. The joists are formed with molded fiberglass or metal domes resting on the temporary framing and shoring (3-50). When the forms are removed the exposed ceiling appears as shown in 3-51.

Two-Way Flat Slab Floor & Roof Construction

As mentioned earlier, the two-way flat slab has reinforcing running in two (perpendicular) or more directions. The systems in general use are the two-way flat slab, two-way solid slab, two-way flat plate, and the two-way joist (or waffle) slab. This construction enables a building structure to be designed in nearly square bays.

3-51 *The ceiling formed with a cast-in-place flat slab with joists and beam construction.*
Courtesy Molded Fiberglass Concrete Forms Co.

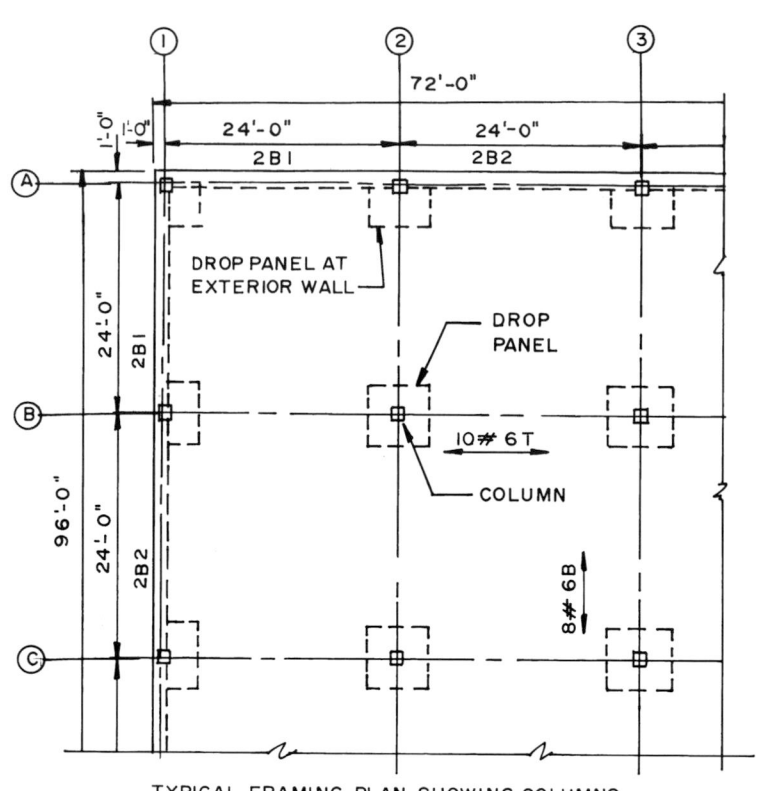

DROP PANEL AT EXTERIOR WALL

DROP PANEL

10 # 6 T

COLUMN

8 # 6 B

NOTE: DIMENSIONS ON THESE DRAWINGS ARE FOR ILLUSTRATION PURPOSES ONLY.

TYPICAL FRAMING PLAN SHOWING COLUMNS AND DROP PANELS

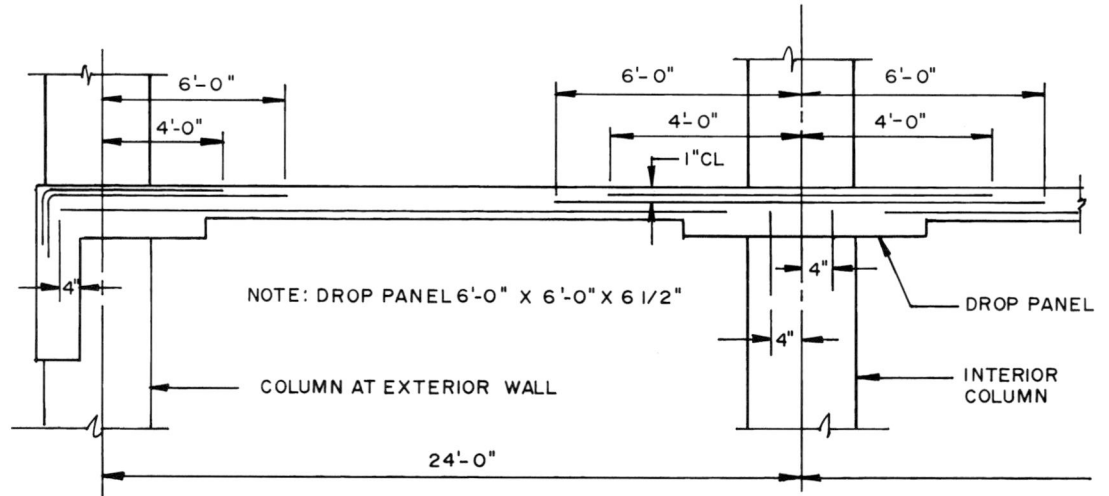

NOTE: DROP PANEL 6'-0" X 6'-0" X 6 1/2"

COLUMN AT EXTERIOR WALL

DROP PANEL

INTERIOR COLUMN

TYPICAL DETAIL FOR DROP PANEL AND FLAT SLAB CONSTRUCTION

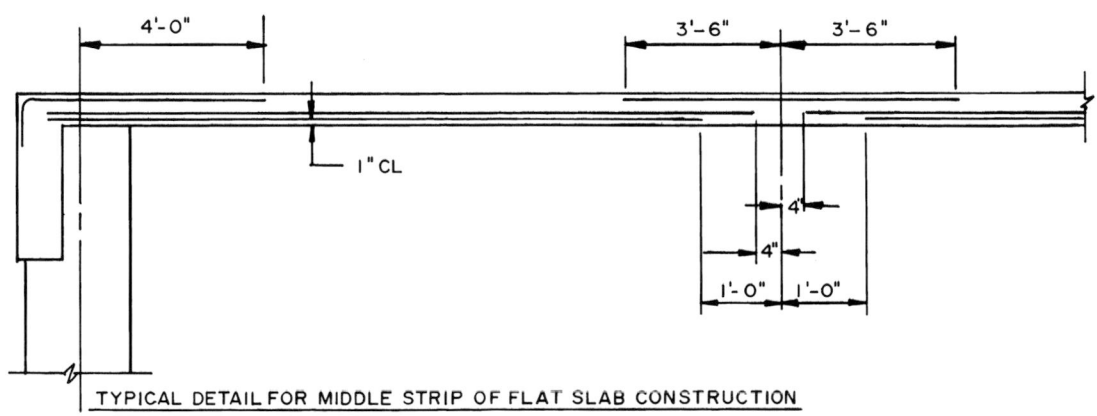

1" CL

TYPICAL DETAIL FOR MIDDLE STRIP OF FLAT SLAB CONSTRUCTION

3-53 *Two-way flat slab construction using columns with drop panels and drop panels with a column capital.*

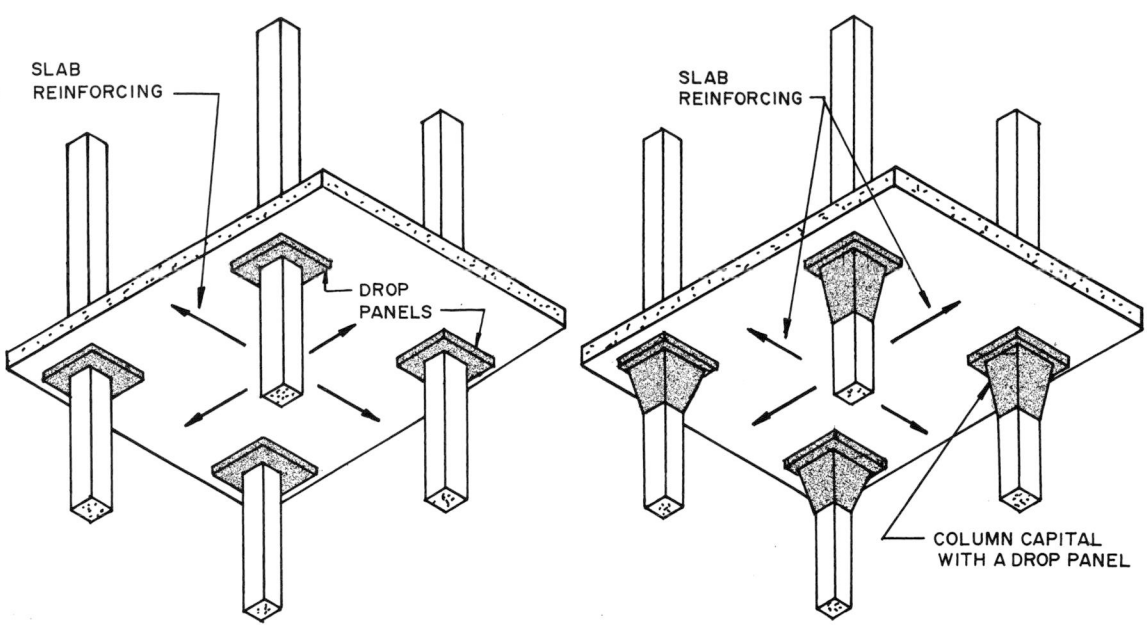

TWO-WAY FLAT SLAB WITH DROP PANELS

TWO-WAY FLAT SLAB WITH DROP PANELS AND COLUMN CAPITALS

TWO-WAY FLAT SLAB CONSTRUCTION

The two-way flat slab construction system is useful for buildings that will have heavy loads. A simplified partial plan view and sections are in 3-52. The system utilizes a flat slab supported by thickening the slab at each column with a drop panel or a drop panel and a capital (3-53). The slab reinforcing runs in two directions.

TWO-WAY SOLID SLAB

The two-way solid slab has the flat slab supported by beams running between columns. The slab and beams are cast monolithically (3-54). This construction can be used for buildings designed to carry heavy loads. The slab is reinforced in two directions.

TWO-WAY FLAT PLATE

The two-way flat plate uses the same general construction as the flat slab but has no drop panel or capital on the top of each column (3-55).

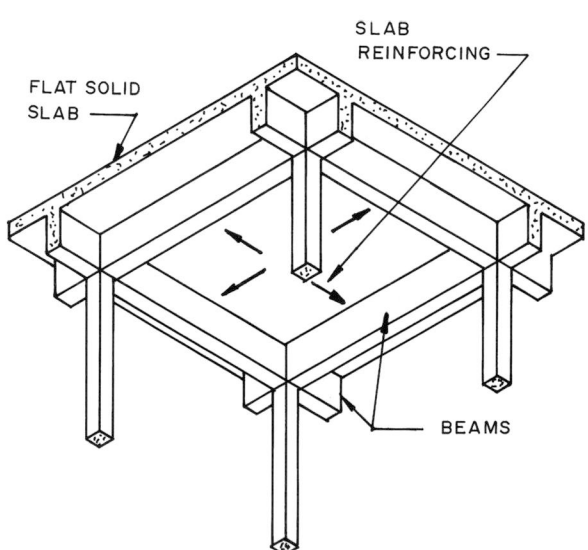

TWO-WAY SOLID SLAB AND BEAM

3-54 *A two-way solid slab and beam is supported by beams running between the columns.*

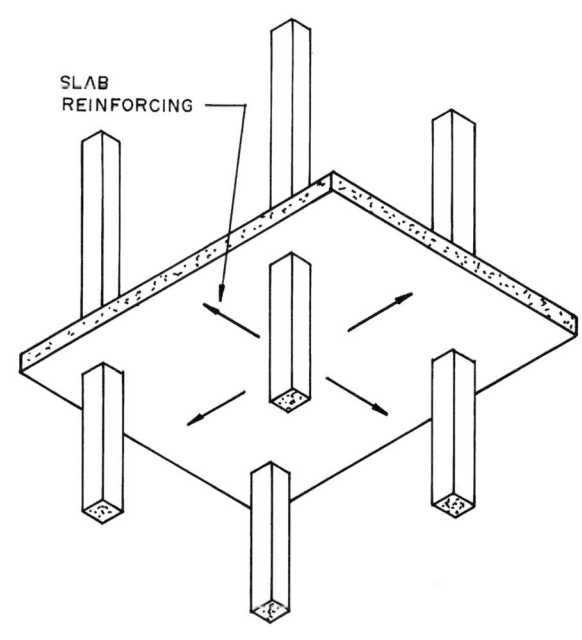

TWO-WAY FLAT PLATE

3-55 *This two-way flat plate does not have drop panels or column capitals.*

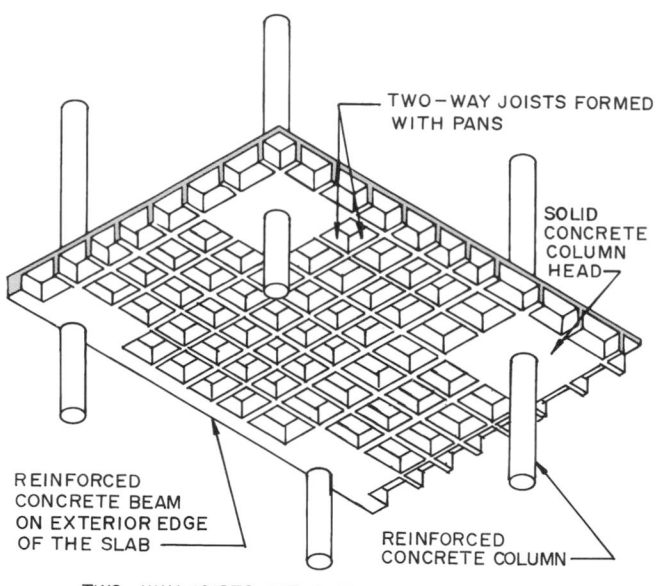

TWO—WAY JOISTS FORMED WITH PANS

SOLID CONCRETE COLUMN HEAD

3-56 *The two-way joist framing system has cast-in-place joists at right angles to each other forming a waffle-like ceiling.*

REINFORCED CONCRETE BEAM ON EXTERIOR EDGE OF THE SLAB

REINFORCED CONCRETE COLUMN

TWO—WAY JOISTS WITHOUT BEAMS (WAFFLE FLAT PLATE CONSTRUCTION)

TWO-WAY JOIST (WAFFLE) FLAT SLAB

The two-way joist framing system consists of a series of concrete joists cast at right angles to each other forming a grid. The slab spans the distances between the joists and they are poured monolithically (3-56).

The system is designed around a series of columns usually equally spaced. The joists are formed by domes placed on a wood deck supported by shoring. Areas over the columns are left open so a solid concrete slab area called a column head can be poured (3-57).

3-57 *These fiberglass domes are set on a wood platform supported by shoring.*
Courtesy Molded Fiberglass Concrete Forms Co.

TILT-UP WALL CONSTRUCTION

Tilt-up concrete construction is a form of site cast concrete. Wall panels are cast in a horizontal position near the building or most frequently on the concrete on-grade floor slab. A bond breaking agent is placed on the slab so the wall panel will not adhere to it. The surfaces of the panel can be finished in many ways, such as texture from a form liner, aggregate surface, trowelled smooth, or painted. After the wall panel has developed sufficient strength it is lifted into position with a crane and secured to the footing (3-58). The panels are temporarily braced until all wall and roof structural members are in place. The panels are secured to columns, which may be precast concrete, steel, or cast-in-place after the walls are up.

LIFT-SLAB CONSTRUCTION

Lift-slab construction involves casting a reinforced concrete slab for each floor of the building and the roof one on top of the other. The ground floor on-grade slab is cast first. It is coated with a bond release compound that prevents the next slab that is cast on top of it from

bonding to it. Slabs are cast for each floor and the roof if it is to be a concrete slab. Each is coated with a bond release compound. The slabs can be reinforced concrete flat slabs or prestressed. They are generally two-way flat slabs.

The slabs are lifted by hydraulic jacks mounted on the top of the columns. They slide up to the desired elevation and are secured to each column by welding, bolting, or a series of pins (3-59). This is repeated for each floor.

SLIP FORMING

Slip forming is used to cast-in-place tall structures such as grain elevators and stairwells. It involves constructing the formwork, which is then slowly lifted upward by jacks supported by vertical steel rods. The forms are filled with concrete and reinforcing bars as the form is raised. The concrete gains sufficient strength to carry the weight of the newly poured concrete as the form slowly moves upward. A key to success is a proper concrete mix and the equipment to lift it to high levels as the structure increases in height.

3-58 *This site cast concrete wall is being tilted up into a vertical position.*

Courtesy Portland Cement Association

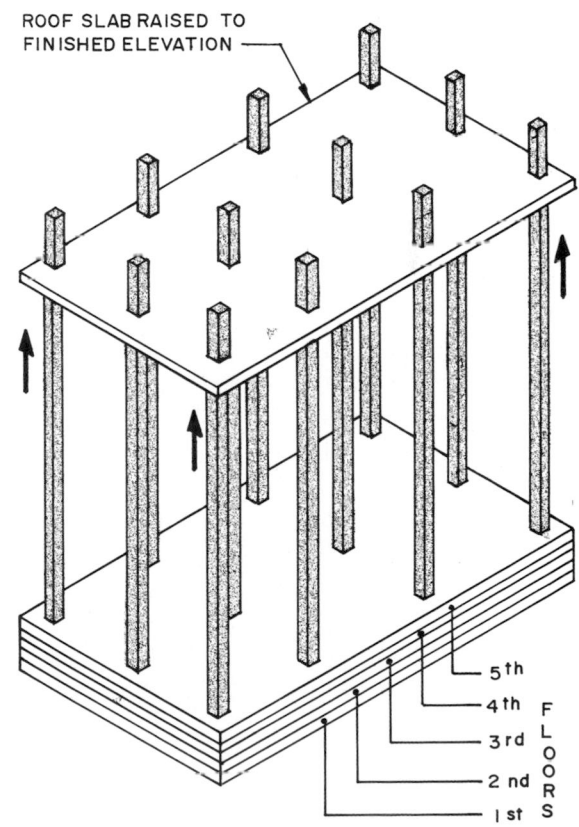

3-59 *Lift-slab construction casts the floor and roof slabs on grade and lifts them to their required elevation.*

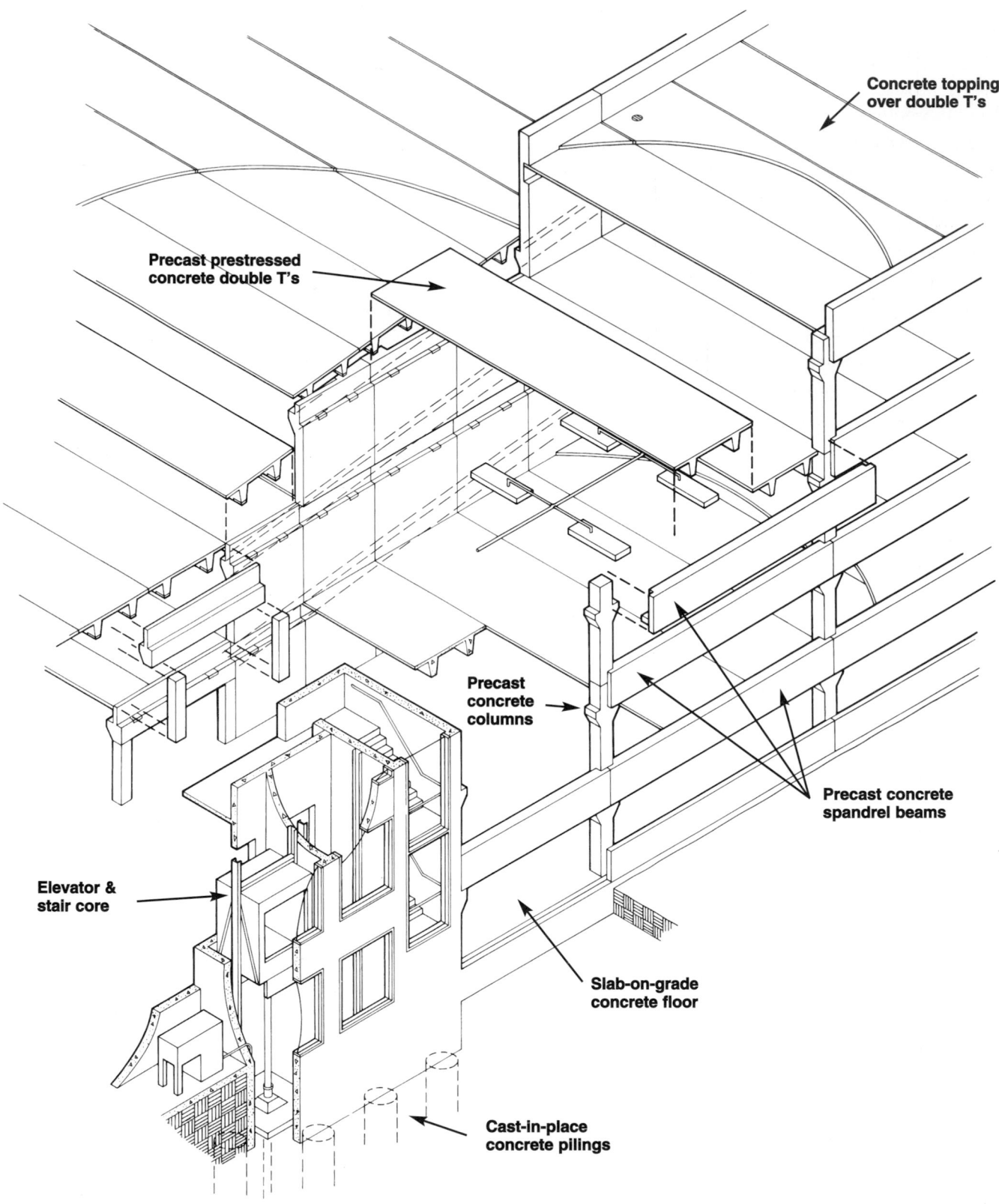

Concrete topping over double T's

Precast prestressed concrete double T's

Precast concrete columns

Precast concrete spandrel beams

Elevator & stair core

Slab-on-grade concrete floor

Cast-in-place concrete pilings

3-60 *A structure system of precast concrete members.*

Reproduced with permission from *The Building Systems Integration Handbook,* Richard Rush, editor,
Butterworth-Heinmann Publishers, Newton, MA. 02158, l986.

PRECAST CONCRETE CONSTRUCTION

Structural precast concrete units are used to form the structural system of a building. These include columns, beams, girders, various floor and roof slabs, and exterior and interior wall panels (3-60). They are cast under factory controlled conditions and moved to the job site.

Precast concrete units can be classified into two major groups, precast structural concrete units and precast architectural units. Typical precast structural units include floor and roof slabs, beams, girders, columns, and wall panels. Precast architectural concrete includes wall panels and other features and may or may not be structural.

Precast units are made in a plant under controlled conditions. They are cast in beds, which are permanent forms made of metal, wood, or glass-fiber. The quality of the surface of the form produces the finished surface of the precast unit. Both pretensioned and posttensioned units can be produced (3-61).

Precast pretensioned and posttensioned units can span longer distances, and they are smaller and lighter than typical cast-in-place concrete members. Generally the quality of the concrete is better on precast units because it is produced in computer-controlled facilities and used at the point of production. The quality of materials used can also be more carefully controlled. Samples of the concrete are taken regularly and tested as per ASTM standards. Typically 5000 psi (35 MN/m^3) concrete is used. It is mechanically vibrated in the form and cured under carefully controlled conditions. To speed up production the units are covered with insulation blankets and are steam cured (autoclaved). When Type III, high early strength portland cement is used, the units can be removed from the form after 24 hours. After it has been determined the units have achieved adequate strength they are delivered to the job site, usually by truck. They are erected with a crane much the same as structural steel.

Building Codes

Building codes have extensive specifications pertaining to the design and installation of precast concrete members. In addition to engineering design data, the codes refer to areas such as reinforcements, connections, lifting devices, fabrication, shop drawings, tendon anchorage, and grout. The codes also specify the requirements for meeting fire codes.

Nonprestressed Precast Units

Nonprestressed units are cast in molds in a plant, cured, and shipped to the job site. They are reinforced in the same manner as cast-in-place concrete and are not under tension forces.

Prestressed Precast Concrete

Prestressed concrete units have stresses introduced before they are placed under a load. There are two types, pretensioned and posttensioned.

PRETENSIONED PRECAST UNITS

Pretensioned precast concrete units are cast in forms, some as long as 800 ft. (250 m) or more. High tensile strength steel reinforcing strands are stretched from one end of the form to the other. One end is anchored to a large fixed abutment. The other end passes through the abutment on the other end and has hydraulic stressing equipment fastened to it. The desired design stress is applied to the cable by pulling it in tension.

After stressing the strand the concrete is placed in the form. As it hardens it bonds to the strands. Generally after 24 hours samples that were taken of the concrete are tested. If they have attained the required strength the strands are cut from the abutments and the resulting forces produced by the strands prestress the concrete. The units are lifted from the bed and moved to storage from which they are shipped to job sites.

3-61 *A cast concrete unit being removed from the casting bed.*

Courtesy Portland Cement Association

A comparison of nonprestressed (3-62) and prestressed (3-63) concrete shows that the nonprestressed unit has no camber (arch) and deflects slightly under load while the pretensioned member has some camber due to the tension in the strands. When under load it will tend to move toward a level position.

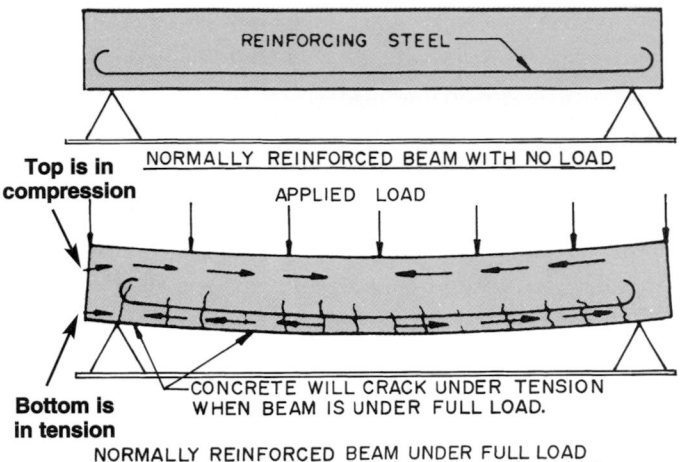

Top is in compression

Bottom is in tension

REINFORCING STEEL

NORMALLY REINFORCED BEAM WITH NO LOAD

APPLIED LOAD

CONCRETE WILL CRACK UNDER TENSION WHEN BEAM IS UNDER FULL LOAD.

NORMALLY REINFORCED BEAM UNDER FULL LOAD

3-62 *A cast-in-place concrete beam will deflect under load and possibly crack on the bottom because concrete cannot resist the tension forces.*

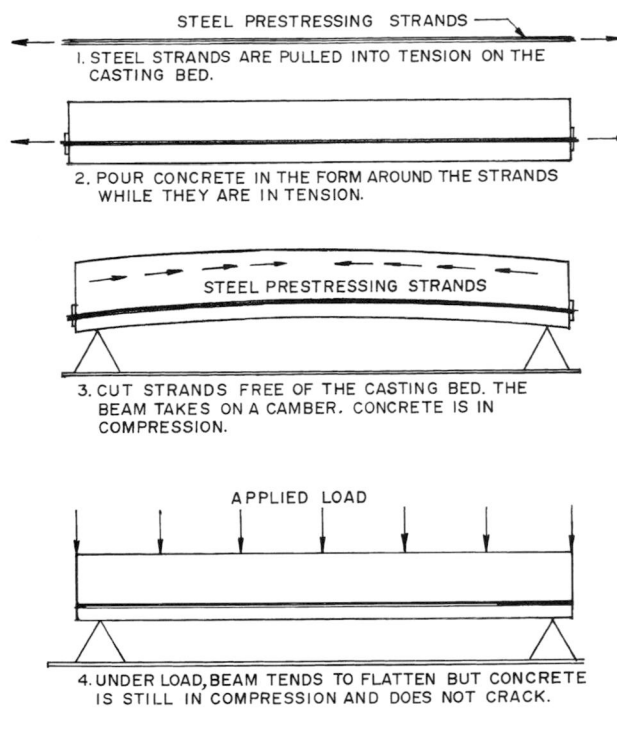

STEEL PRESTRESSING STRANDS

1. STEEL STRANDS ARE PULLED INTO TENSION ON THE CASTING BED.

2. POUR CONCRETE IN THE FORM AROUND THE STRANDS WHILE THEY ARE IN TENSION.

STEEL PRESTRESSING STRANDS

3. CUT STRANDS FREE OF THE CASTING BED. THE BEAM TAKES ON A CAMBER. CONCRETE IS IN COMPRESSION.

APPLIED LOAD

4. UNDER LOAD, BEAM TENDS TO FLATTEN BUT CONCRETE IS STILL IN COMPRESSION AND DOES NOT CRACK.

PRESTRESSED PRECAST CONCRETE BEAM

3-63 *A prestressed concrete beam has a slight camber that places the concrete in compression. When under load the concrete is still in compression.*

POSTTENSIONED UNITS

Posttensioning involves applying stresses to the concrete unit after it has been cast and hardened. It is used primarily with cast-in-place concrete but finds some use with precast concrete units.

Posttensioning involves casting the unit that contains normal reinforcement and in addition running several posttensioning strands through the unit in such a way that the concrete does not bond to them as it sets. The strands may be greased or placed in thin wall metal tubes to keep the concrete from bonding with them. After the concrete has hardened the posttensioning strands are anchored firmly on one end and a hydraulic jack is used on the other to apply tension to the strand. Once the desired tension is reached this end is also anchored (3-64). The steel strands may be left unbonded. If they are in steel tubes they can be bonded by filling the space between the strands and the tube with high pressure grout. The structural framing and tenons in tubes on a posttensioned concrete structure is shown in 3-65.

Precast Concrete Slabs

The major types of slabs used for floor and roof construction are solid flat slabs, channel slabs, hollow core, double tees, and single tees. Some are prestressed and others precast with untensioned reinforcing.

Solid and channel slabs are used for short spans and minimum slab depth. Spans of 25 ft. (7.6 m), and thicknesses of 2 to 8 in. (50 to 200 mm) and widths of 8 to 12 ft. (2.4 to 3.7 m) are common.

3-64 *This concrete beam is being posttensioned with a hydraulic jack tensioning the cables in each tendon. The steel protruding on the top of the beam will be imbedded in the concrete deck to be poured on top of the beam.*
Courtesy Portland Cement Association

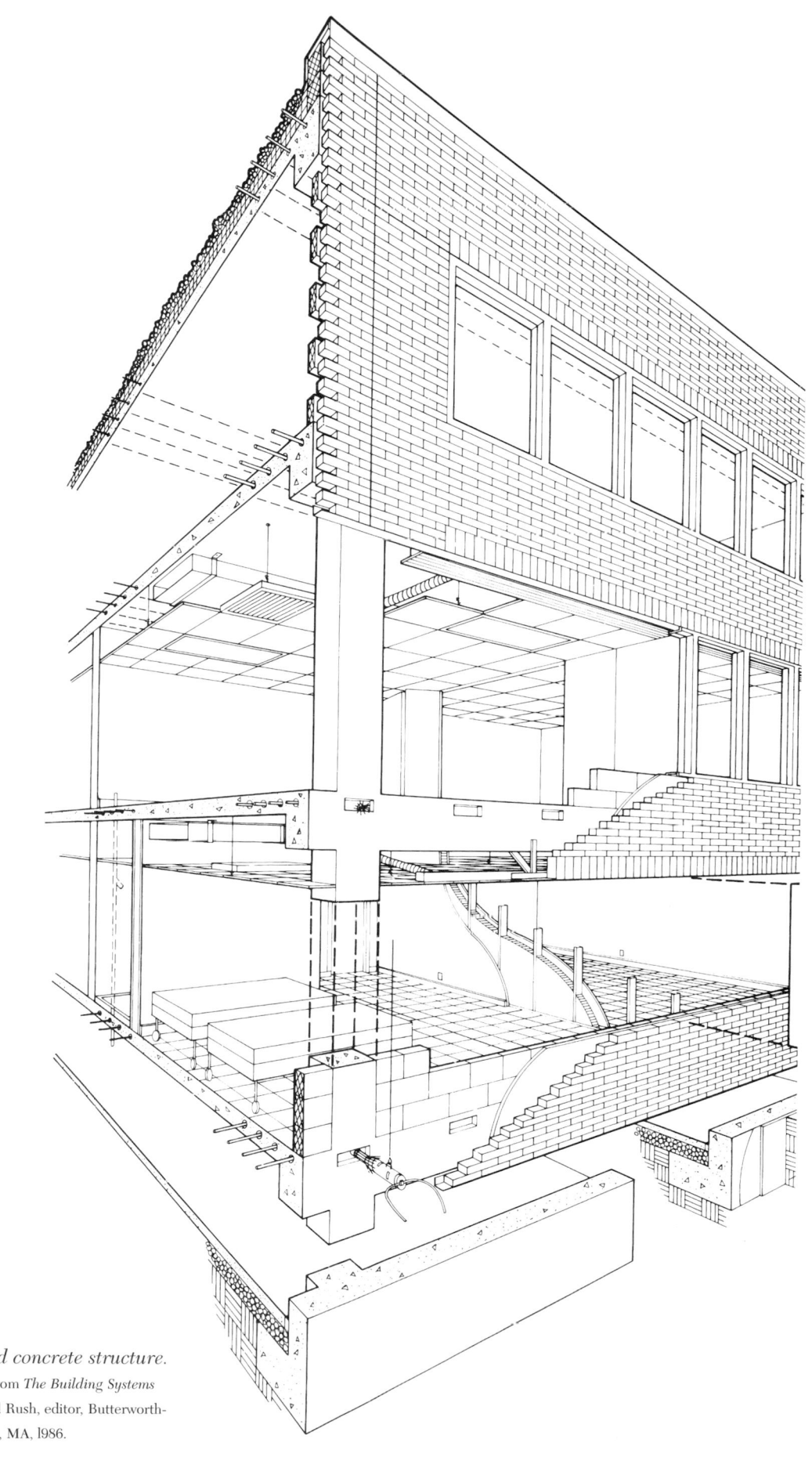

3-65 *A posttensioned concrete structure.*
Reproduced with permission from *The Building Systems Integration Handbook,* Richard Rush, editor, Butterworth-Heinmann Publishers, Newton, MA, l986.

Hollow-core slabs are used for spans ranging up to 40 ft. (12 m) in length and are 6 to 12 in. (150 to 300 mm) thick. The hollow core removes concrete from an area of the slab that has little influence on strength and lightens the member. A topping of 2 in. (50 mm) thick concrete is often applied after installation to increase structural performance and fire rating.

Double tees and **single tees** are used to span the longest distances. Standard lengths range to over 100 ft. (30.5 m). Lengths over 60 ft. (18.3 m) present special problems in transporting them.

Tees are cast with a rough or smooth surface on top. The rough surface is used when a concrete topping is to be applied over them. This topping is usually 2 in. (50 mm).

Normal weight and lightweight concrete is used to cast these units. Lightweight concrete is more expensive than normal weight but reduces the weight on the structure.

Precast Concrete Columns

Precast columns are usually combined with precast concrete beams forming a post and beam structural framework. Most precast concrete columns are reinforced conventionally and if prestressed this is mainly to reduce stresses on the column during transportation and handling. Columns may be rectangular or square.

Beams & Girders

Precast prestressed concrete beams and girders can be used in any building where precast construction is desired. They are made in several standard shapes and special designs can be produced (3-66). The L-shaped and inverted tee beams provide a bearing surface for precast floor units as single and double tee units. This helps reduce the overall height when compared with resting floor and roof units on top of rectangular beams.

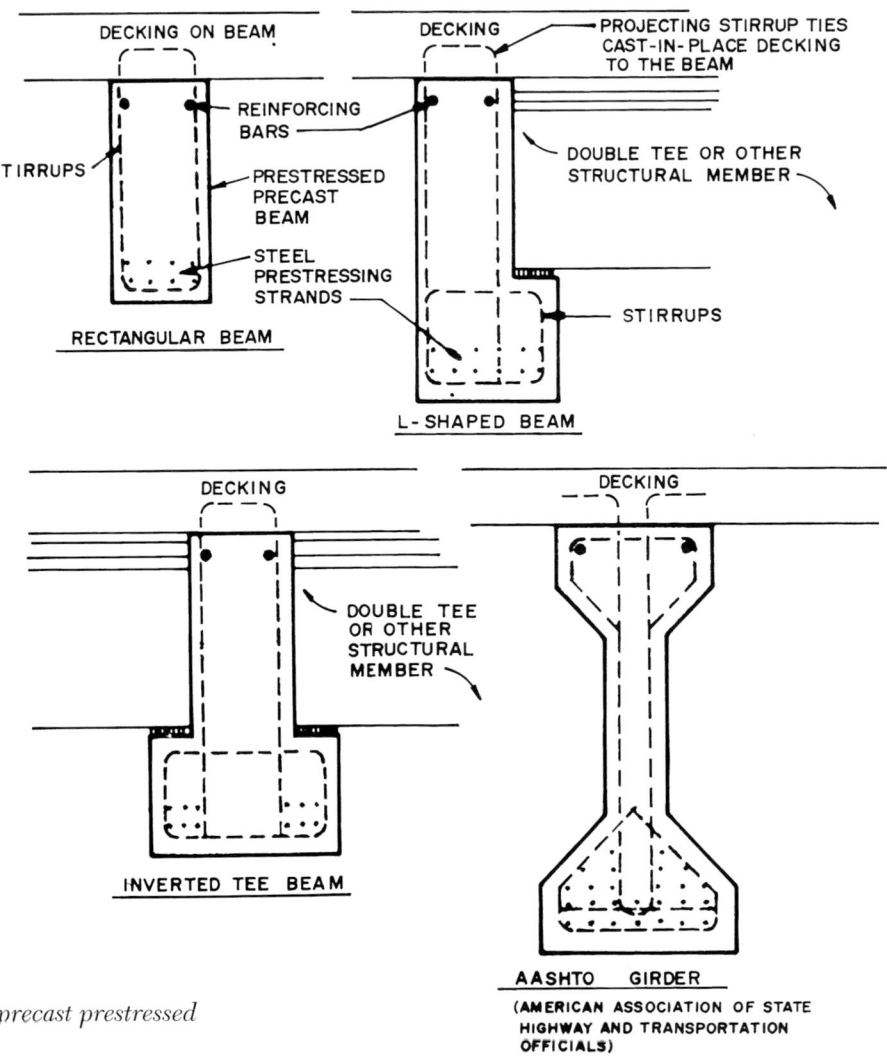

3-66 *Typical precast prestressed concrete beams.*

Precast Concrete Wall Panels

Precast concrete wall panels may be prestressed or conventionally reinforced. They may be load or non-load bearing; have a flat, ribbed, or other surface configuration; and be hollow, solid, or of a sandwich construction. They may be used in connection with a precast concrete framing system, cast-in-place concrete, or steel framing systems. Flat panels can be two stories high while ribbed panels can be cast up to four stories high. Panels with openings are not prestressed.

Erection of Precast Concrete

Consideration of the erection procedures begins before the units are cast. The design engineer must discuss with the erector the use of lifting hardware and connections. The manufacturer may have specific details for the design and manufacture of the units that the engineer should observe.

Precast units are erected and connected as shown on the erection drawings. The erector must comply with the specified dimensional tolerances, hardware sizes, weld lengths, and torque on bolts. Connections should be planned so workers can make them from a stable work platform. Planning should include provision for activities such as welding, posttensioning, and grouting. The erector should install temporary bracing during construction. Final adjustment and alignment of precast structural units can then be made without the need for a crane to support them.

TYPES OF CONNECTION

The types of connections include bolt connections, welded connections, posttensioned connections, and dowelled connections.

Bolt connections—speed up erection and allow final alignments to be made later after the crane has placed the member. Generally ½, ¾, and 1 in. (12, 18, 25 mm) bolts with national coarse threads are used. Steel washers may be required under certain conditions. Bolts should be tightened to the recommended torque. A bolted column-to-column connection can be seen in 3-67.

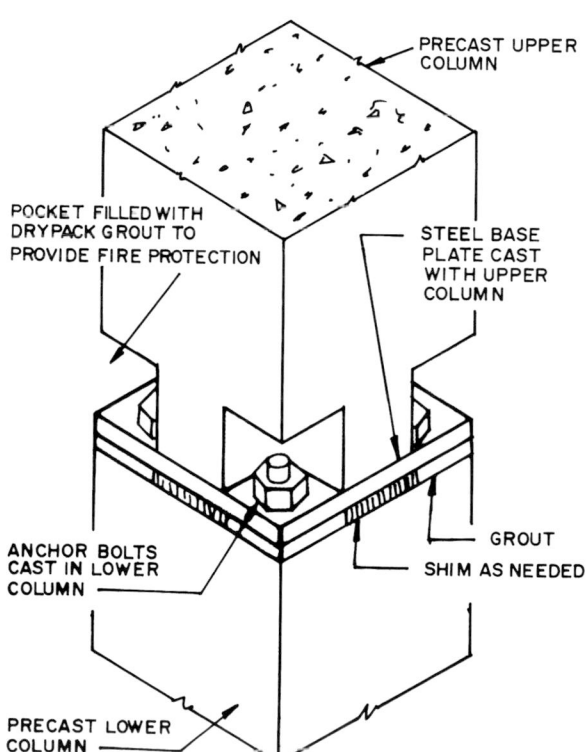

3-67 *A typical column-to-column connection.*

Welded connections—are strong and easily made on the site. They should be made following the details on the erection drawings, which include the size, type, length of the weld, type of electrode, and required temperatures. Welded connections are made before the unit is released by the hoisting device. A welded beam to column connection is shown in 3-68.

Posttensioned connections—are made using either bonded or unbonded tendons. Bonded tendons are installed in holes cast in the member and are bonded to the member after tensioning by filling the hole with grout. Unbonded tendons are not grouted in place but are coated with an organic coating to prevent corrosion.

Dowelled connections—use reinforcing bars grouted into dowel holes in the precast member. The strength of the connection depends upon the dowel diameter, the depth it is in the member, and the bond developed by the grout. A doweled connection can be seen in 3-69.

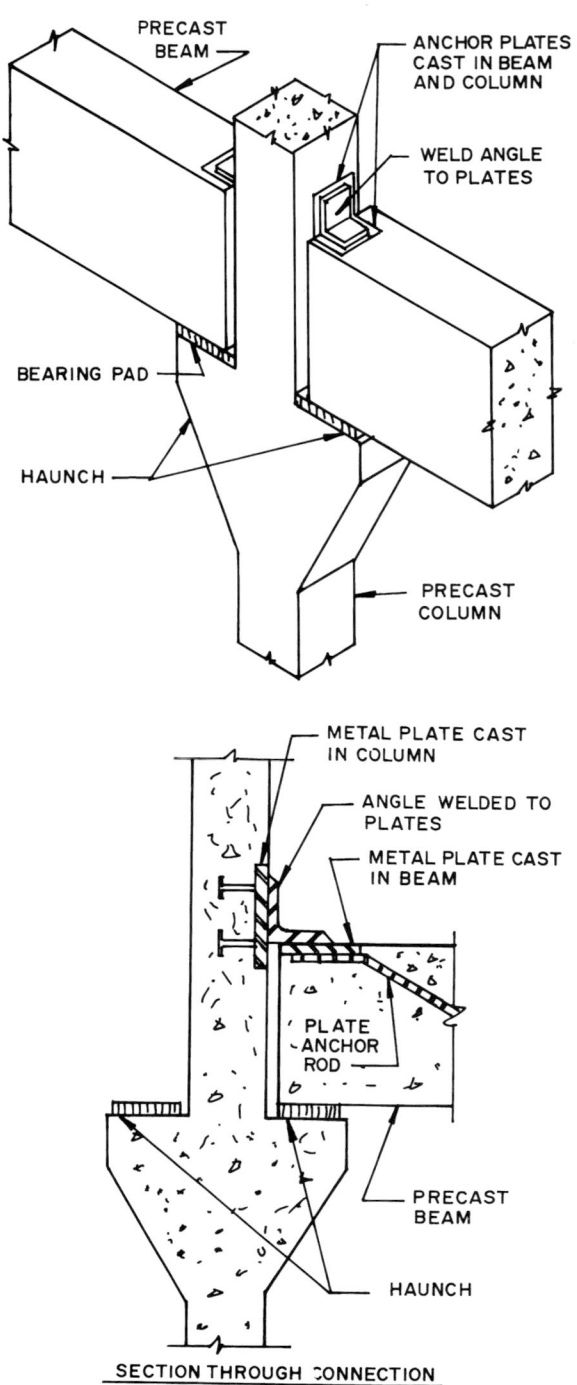

3-68 *Precast rectangular beams are supported on column haunches.*

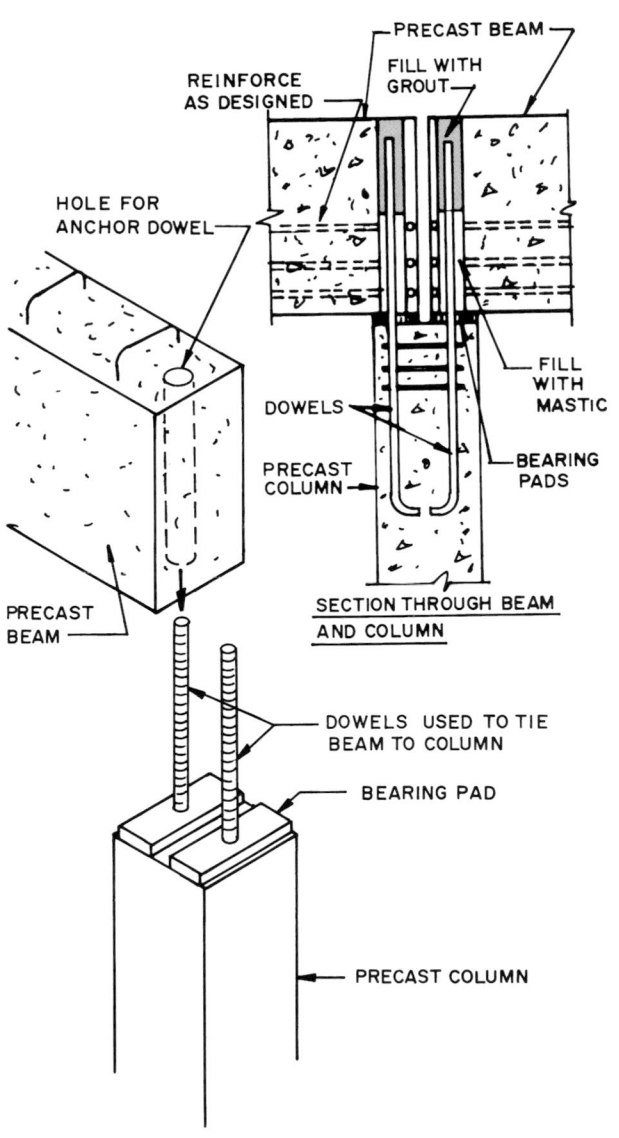

3-69 *A typical connection for a precast rectangular beam to a precast column.*

DIVISION 4

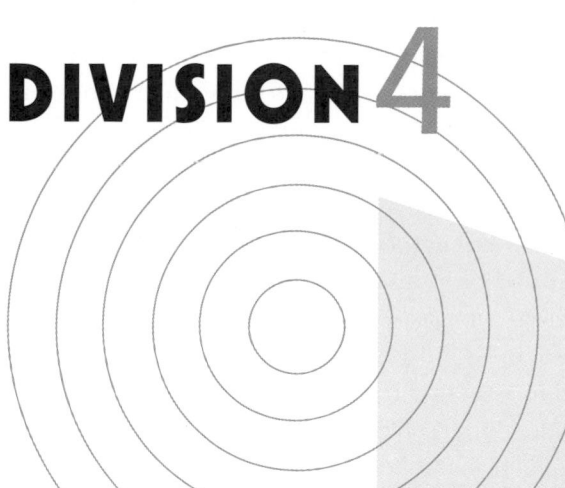

MASONRY

CSI MASTERFORMAT™

Courtesy Italian Government Travel Office

MORTAR FOR MASONRY WALLS

Mortar is the bonding agent used to join masonry units into an integral structure. While it bonds the units it also must (1) seal the spaces between the units so they are not penetrated by air or moisture, (2) tie steel reinforcement, ties, and anchor bolts into the walls, (3) provide a design function by providing lines of color and shadows, and (4) allow for the adjustment of the slight variations that occur in masonry units.

Types of Mortar

There are four types of mortars specified for use in the United States. These are M, S, N, and O and are specified by ASTM C270. The Canadian Standards Association Standard A179 specifies two types, S and N.

Mortar types are identified by either proportion or property specifications, but only by one of these, not both. In Table 4-1 you can find the specifications for mortar **proportion** indicated by the combination of portland cement and masonry cement. Mortar types, classified under **property specifications,** are based upon the compressive strength, water retention, and air content. This data is developed by testing 2 in. (50.8 mm) cubes of hardened mortar to find the compressive strength.

TYPE M MORTAR

Type M Mortar is a high-strength mortar with a compressive strength of 2500 psi (17 MPa) and has better durability than the other types. It is recommended for masonry below grade and in contact with earth and for conditions of severe frost action. It is also used for reinforced masonry.

TYPE S MORTAR

Type S Mortar is a medium high strength mortar with a compressive strength of 1800 psi (12.5 MPa). It is often permitted in wall construction instead of Type M because it has almost the same allowable strength and can be used above or below grade. Type S has better workability and more water retention than Type M. It is used for reinforced and unreinforced masonry and has high tensile bond strength.

TYPE N MORTAR

Type N Mortar is a medium-strength mortar with a compressive strength of 750 psi (5 MPa) used for above-grade general construction involving exposed masonry load-bearing walls where requirements for compressive strength and lateral strength are not high.

TYPE O MORTAR

Type O Mortar is a medium-strength mortar with a compressive strength of 350 psi (2.5 MPa) used for general interior purposes such as non-load bearing walls where compressive strength does not exceed 100 lb. per sq. in. It must not be in contact with the soil or exposed to saturated freezing conditions.

A guide for the selection of masonry mortars for various uses above and below grade is in Table 4-2.

Mortar Composition

Mortar is composed of cementitious materials, clean, carefully graded mortar sand, clean water, and sometimes a coloring agent of an admixture. The actual composition of mortars varies according to the intended use. For general masonry applications they may contain portland cement or masonry cement, sand, hydrated lime or lime putty, and water. Mortars for special applications such as prefabricated masonry units which must be moved into place require additives that increase compressive and tensile strength and have greater bonding capabilities.

CEMENTITIOUS MATERIALS

Cementitious materials provide the bonding ingredient in mortars. They are made according to ASTM (American Society for Testing and Materials) and CSA (Canadian Standards Association) as shown in the following listing:

▼ Bonding ingredient standards:
Masonry cement
    ASTM C91 (Types M, S, or N)
    CSA A8 (Types S or N)
Portland cement
    ASTM C150 (Types I, IA, II, IIA, III or IIIA)
    CSA A5 (10, 20, 30 or 50)
Blended hydraulic cement
    ASTM C595 (Types IS, IS-A, IP, IP-A, I (PM), or I(PM)-A). (Pozzolan modified portland cement)
    CSA A362 (Type 10S)
Hydrated lime for masonry purposes
    ASTM C207 (Types S, SA, N or NA)

You can find detailed information on various cements in Division 3.

Table 4-1 PROPORTION SPECIFICATIONS FOR MORTAR

United States–ASTM C270

| Mortar type | Portland cement or blended cement | Masonry cement type | | | Hydrated lime or lime putty | Aggregate[a] |
		M	S	N		
M	1	—	—	1	—	4½ to 6
	—	1	—	—	—	2¼ to 3
	1	—	—	—	¼	2¹³/₁₆ to 3¾
S	½	—	—	1	—	3⅜ to 4½
	—	—	1	—	—	2¼ to 3
	1	—	—	—	Over ¼ to ½	.
N	—	—	—	1	—	2¼ to 3
	1	—	—	—	Over ½ to 1¼	.
	—	—	—	1	—	2¼ to 3
O	1	—	—	—	Over 1¼ to ½	.

Canada—CSA A179M

| Mortar type | Portland cement | Masonry cement type | | Hydrated lime or lime putty | Aggregate[a] |
		S	N		
S	—	—	1	—	2¼ to 3
	½	—	1	—	3½ to 4½
	1	—	—	½	3½ to 4½
N	—	—	1	—	2¼ to 3
	1	—	—	1	4½ to 6

[a]The total aggregate shall be equal to not less than 2¼ and not more than 3 times the sum of the volumes of the cement and lime used.
Notes: 1. Under both ASTM C270, Standard Specification for Mortar for Unit Masonry, and CSA A179, Mortar and Grout for Unit Masonry, aggregate is measured in a damp, loose condition and 1 cu ft. of masonry sand by damp, loose volume is considered equal to 80 lb. of dry sand (in SI units 1 cu m of damp, loose sand is considered equal to 1280 kg of dry sand).
2. Mortar should not contain more than one air-entraining material.
Courtesy Portland Cement Association and the Canadian Standards Association. Canadian material presented with the permission of the Canadian Standards Association; material is reproduced from CSA Standard A179-94 (Mortar and Grout for Unit Masonry), which is copyrighted by CSA, 178 Rexdale Blvd., Etobicoke, Ontario M9W 1R3. Although use of this material has been authorized, CSA shall not be responsible for the manner in which the information is presented, nor for any interpretations thereof. This material may not be updated to reflect amendments made to the original content. For up-to-date information, contact CSA.

Table 4-2 GUIDE FOR THE SELECTION OF MASONRY MORTARS (UNITED STATES)[a]

| Location | Building segment | Mortar type | |
		Recommended	Alternative
Exterior, above grade	Load-bearing walls	N	S or M
	Non-load-bearing walls	O[b]	N or S
	Parapet walls	N	S
Exterior, at or below grade	Foundation walls, retaining walls, manholes, sewers, pavements, walks, and patios	S[c]	M or N[c]
Interior	Load-bearing walls	N	S or M
	Non-load-bearing partitions	O	N

[a]Adapted from ASTM C270. This table does not provide for specialized mortar uses, such as chimney, reinforced masonry, and acid-resistant mortars.
[b]Type O mortar is recommended for use where the masonry is unlikely to be frozen when saturated or unlikely to be subjected to high winds or other significant lateral loads. Type N or S mortar should be used in other cases.
[c]Masonry exposed to weather in a nominally horizontal surface is extremely vulnerable to weathering. Mortar for such masonry should be selected with due caution.
Courtesy Portland Cement Association

Table 4-3 ADMIXTURES—BENEFITS AND CONCERNS

Admixture	Primary Benefits	Possible Concerns
Air-entraining	Freeze-thaw durability, workability	Effect on compressive and bond strengths
Bonding	Wall tensile (and flexural) bond strength	Reduced workability, bond strength regression upon wetting, corrosive properties
Plasticizer	Workability, economy	Effect on hardened physical properties under field conditions
Set accelerator	Early strength development	Effectiveness at cold temperatures, corrosive properties, effect on efflorescence potential of masonry
Set retarder	Workability retention	Effect on strength development, effect on efflorescence potential of masonry
Water reducer	Strength, workability	Effect on strength development under field conditions with absorptive units
Water repellent	Weather resistance	Effectiveness over time
Pozzolanic	Increase density and strength	Effect on plastic and hardened physical properties under field conditions
Color	Esthetic versatility	Effect on physical properties, color stability over time

Courtesy Portland Cement Association

Most mortars are now made with mortar cement. Masonry cement mortars are made by combining masonry cement, clean, carefully graded sand, and enough clean water to produce a plastic, workable mix.

MASONRY CEMENTS

Masonry cements are hydraulic cements used in mortars for masonry construction. They produce a mortar that has greater plasticity and water retention than if portland cement was used. Typically they may include portland cement, slag cement, blastfurnace slag cement, portland-pozzolan cement, natural cement, and hydraulic lime. Manufacturers may include chalk, clay, talc, careous shell, or limestone to add the properties desired.

Masonry cements meet American Society for Testing and Materials specifications found in ASTM C91. This report specifies three types of **masonry cement,** Type N, Type S, and Type M. They are used to produce mortars Types N, O, S, and M as specified in ASTM C270.

Type N masonry cement is used in Type N and O mortars. It can be blended with portland cements for use in Type S and M mortars.

OTHER MORTAR INGREDIENTS

Lime is used to help stabilize the volume of the mortar by controlling shrinking and expansion.

Masonry sand may be natural or a manufactured product that meets the requirements of ASTM C144, *The Standard Specification for Aggregate for Masonry Mortar.* In Canada the standard is CSA A82.56, *Aggregate for Masonry Mortar.* Since sand makes up the major portion of a mortar, the quality of the finished mortar depends heavily on clean, quality sand.

Water used to produce mortar must be clean, free of acids, alkalies, and other organic materials. Water containing soluble salts will cause the mortar to effloresce. **Efflorescence** appears on the finished masonry wall as a white, powdery substance that discolors the mortar and masonry units. The water used should be potable.

Mortar can be **colored** by adding various organic pigments. White mortar is made with white masonry cement or white portland cement, lime, and white sand. Colored mortars are made with white masonry cement and pigments, typically some type of mineral oxide compound such as iron, chromium, cobalt, or manganese oxides. Carbon black is used to produce a dark gray or black mortar.

Admixtures may sometimes be used to alter the properties of the mortar. However, the possibility of their creating unforeseen problems is great. Admixtures are not typically used unless laboratory tests have been made to verify the changes they will have on the mortar. For example, certain admixtures may affect the hydration process and hardening properties. Consultants such as those at the Portland Cement Association can provide helpful information.

In Table 4-3 you can see some of the commonly available admixtures (modifiers) and their benefits and possible problems they may cause. You can find more information on admixtures in Division 3.

Sources of Mortar

Mortars are available ready-mixed, dry-batch, and on-site mixed.

READY-MIXED MORTAR

Ready-mixed mortar is mixed at a central batch plant and transported to the job site where a slump test is made. Additional water is added if required to get the desired slump. This mortar contains a retarding, set-controlling admixture that keeps the mortar workable and in a plastic condition for more than 24 hours. The mix is delivered by a ready-mix truck and is stored on the site in large tubs (see 4-1). Ready-mixed mortar must meet the requirements of ASTM C1142, Specifications for Ready-Mixed Mortar for Unit Masonry.

DRY-BATCH MORTAR

Dry-batch mortar has the required ingredients blended and packaged in bags or delivered to the site in a sealed truck where it is stored in a sealed hopper. When mortar is needed the dry ingredients are moved to an on-site mixer and the required water is added. Dry-batching provides control over the proportions of the various ingredients in the mortar.

ON-SITE MIXED MORTAR

On-site mixed mortar can be made by storing the dry ingredients in a silo-type mixer where an auger-type mixing device receives the correct proportions of dry ingredients and blends them together. The dry ingredients are stored in separate chambers from which they are fed to the mixer. Then the water is injected from a pressurized water source, as a city water main, and the mixer produces the finished mortar. The process is usually computer controlled. The mortar is discharged into tubs or wheelbarrows for distribution on the site. This produces a very accurately proportioned mortar (4-2).

Small jobs will typically deliver the ingredients to the site as separate items. The mortar cement is in bags, and the sand is in bulk. The sand must be protected from moisture and dirt. The mortar is often proportioned by the worker putting so many shovelfulls of sand into a small power-operated mixer (see 4-3, on page 100) and adding the required number of bags of mortar cement. Water may be added by a certain number of bucketfulls.

The gasoline-powered mixer produces the mortar which is moved to the masons, usually by wheelbarrow. Obviously this is often done on small jobs and the mix is based on experience. The proportions of the ingredients can vary considerably and this can often be seen after the mortar has cured and the

4-1 *Ready-mixed mortar may be batched at a central location and delivered to the job site ready to use. It is stored in mortar containers (tubs) until it is needed.*
Courtesy Portland Cement Association

4-2 *This is a storage bin/batcher used at a central plant in which the dry mortar ingredients (masonry cement, portland cement, lime, and sand) are stored in separate compartments. A computer batches the dry ingredients by weight and dry blends them prior to their being sent to the site or to a mixing silo where water is added.*
Courtesy Portland Cement Association

joints over the wall do not have a uniform shade. To be of greatest effectiveness the proportions should be carefully measured and the power mixer run from 3 to 5 minutes. Shorter times produce a nonuniform mortar and overmixing may reduce the strength of the mortar.

Properties of Plastic Mortar

The properties that affect mortar in a plastic condition are workability and water retention.

4-3 *Small quantities of on-site mixed mortar should be batched in a portable mixer.*
Courtesy Portland Cement Association

4-4 *Mortar must be workable, cling to vertical surfaces, extrude from joints, and not drop off.*
Courtesy Portland Cement Association

Workability is a term used to describe the condition of a mortar that will spread easily, cling to vertical surfaces, extrude easily from joints but not drop off, (4-4), and permit easy positioning of masonry units. It is a condition of the mortar that is influenced by other properties, such as consistency, water retention, setting time, weight, adhesion, and penetrability. Masons judge workability by observing how the mortar slides or adheres to their trowel.

WATER RETENTIVITY

Mortar must have good **water retention,** which means it resists rapid loss of the mixing water to the air or to an absorptive masonry unit. Rapid loss of mix water causes the mortar to stiffen, making it difficult to get a good bond or a watertight joint. Mortar that has good water retentivity remains workable, enabling the mason to properly place the masonry units. Low absorption masonry units may float when placed on a mortar with too much water retentivity. This will cause the mortar joint to bleed.

Water retentivity is increased by entrained air, very fine aggregate or cementitious materials. Water retention is measured by a flow test.

MORTAR FLOW TEST

The water retention limit of mortar is measured by initial flow and flow after suction tests made in a laboratory as described in ASTM C91.

The **initial flow test** is made by placing a truncated cone of mortar with a 4 in. (100 mm) diameter on a metal flow table. The table is mechanically dropped ½in. (12 mm) 25 times in 15 seconds. The mortar will flow into an enlarged circular shape. The diameter of this shape is measured and compared with the original diameter. Allowable initial flow should be in the range of 100% to 115% (4-5).

The **flow after suction** test is used to determine flow after loss of water to an absorbent masonry unit. The mortar sample is placed for one minute in a vacuum device that removes some water. The mortar is then tested as described above and the flow measured. The **flow after suction** should be in a range of 70% to 75%.

Properties of Hardened Mortar

The properties essential to quality hardened mortar are bond strength, durability, compressive strength, volume change, appearance, and rate of hardening.

BOND STRENGTH

Bond strength refers to the **degree of contact** between the mortar and the masonry units and also the **tensile bond strength** available to resist forces, tending to pull the masonry units apart.

The degree of contact between masonry units is essential to watertight joints and tensile bond strength. Good bond strength requires good, workable, water retentive mortar, good workmanship, full joints, and masonry units with a medium rate of absorption (suction).

Tensile bond strength is necessary to withstand forces such as wind, structural movement, expansion of clay masonry units, shrinkage of mortar or concrete masonry units, and temperature changes. Tensile bond strength is tested by bonding together samples of the concrete masonry units with the mortar, curing the mortar, and pulling them apart on a tensile machine (4-6).

The major factors affecting the bond strength are (1) the characteristics (strength) of the masonry units, (2) quality of the mortar, (3) workmanship of the mason, and (4) curing conditions.

Bond is high on **textured surfaces** and low on smooth surfaces. **Suction rates** of masonry units influence bond. Concrete masonry units tend to retain moisture after they are cured and have relatively low suction rates. Some clay bricks have very high suction rates and unless wetted before using will pull water from the mortar resulting in a poor bond. After the bricks are wetted the surface should be permitted to dry before they are used.

Mortar flow influences tensile bond strength. As the water content increases the **bond strength increases.** This indicates it is wise to use the highest water content possible and yet retain a workable mortar. As water content increases, mortar **compressive strength decreases.** Bond strength takes precedence over compressive strength.

Good **workmanship** requires a minimum of time elapse between the spreading of the mortar and the placing of the masonry unit. Some water in the mortar will evaporate and some will be sucked away by the masonry unit on which it is placed, leaving insufficient water to form a good bond on the next masonry unit. After placing a unit on the mortar and getting its initial alignment, it should not be moved, tapped, or slip in any way. This would break the initial bond and it cannot be reestablished. The mortar must be replaced if this happens.

Good curing conditions require the maximum amount of water possible to be in the mortar because it is needed for hydration. The laid units should be covered with plastic to retain moisture while curing. Under severe dry conditions it may be necessary to keep the wall wet with a fine mist spray for several days. It must also be protected from freezing by using insulating blankets.

4-5 *The mortar flow test is used to measure the water retention limit.*

Courtesy Portland Cement Association

4-6 *The tensile bond strength test measures the extent of the bond between the mortar and the masonry unit.*

Courtesy Portland Cement Association

Table 4-4   RECOMMENDATIONS FOR ALL-WEATHER MASONRY CONSTRUCTION[a]

Air temperature, °F	Construction Requiements	
	Heating of materials	**Protection**
Above 100 or above 90 with wind velocity greater than 8 mph	Limit open mortar beds to no longer than 4 ft. and set units within one minute of spreading mortar. Store materials in cool or shaded area.	Protect wall from rapid evaporation by covering, fogging, damp curing, or other means.
Above 40	Normal masonry procedures. No heating required.	Cover walls with plastic or canvas at end of work day to prevent water entering masonry
Below 40	Heat mixing water. Maintain mortar temperatures between 40°F and 120°F until placed.	Cover walls and materials to prevent wetting and freezing. Cover should be plastic or canvas.
Below 32	In addition to the above, heat the sand. Frozen sand and frozen wet masonry units must be thawed.	With wind velocities over 15 mph, provide windbreaks during the work day, and cover walls and materials at the end of the work day to prevent wetting and freezing. Maintain masonry above 32°F using auxiliary heat or insulated blankets for 24 hours after laying units.
Below 20	In addition to the above, dry masonry units must be heated to 20°F.	Provide enclosure and supply sufficient heat to maintain masonry enclosure above 32°F for 24 hours after laying units.

[a]Adapted from recommendations of the International Masonry Industry All Weather Council and requirements of ACT530 1/ ASCE 6/TMS 602. (References 4 & 13).
Courtesy American Concrete Institute

DURABILITY

Mortar must have the durability to withstand the forces of weathering. Frost or freezing will not damage mortar joints unless they leak and are water soaked. High compressive strength mortar usually has good durability. Air-entrained mortar also provides protection against the freeze-thaw cycle.

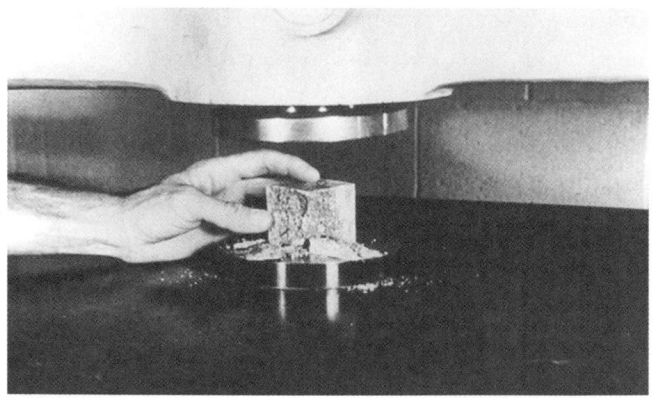

4-7   *The cube compressive strength test measures the compressive strength of the mortar.*
Courtesy Portland Cement Association

COMPRESSIVE STRENGTH

The compressive strength of mortar depends mainly upon the type and quantity of cementitious material used in the mortar. It increases as cement content increases and decreases as air-entrainment, lime, or water content increases. The compressive strength of mortar is found by testing standard cured 2 in. (50.8 mm) square cubes in a laboratory compression testing machine (4-7) following ASTM C270 standards. Mortar can be tested in the field using ASTM C780 standards.

The compressive strength of a wall depends not only upon the mortar but the masonry unit, design of the structure, workmanship, and curing.

LOW VOLUME CHANGE

Mortars have low volume change. The actual shrinkage during curing of a mortar joint is miniscule.

APPEARANCE

The mortar joints in a wall should have a uniform color or shade. Each batch of mortar should have exactly the same proportions of each ingredient. The time of tooling also causes variances in the shade of the joint. If

Table 4-5 CANADIAN PROTECTION REQUIREMENTS[a]

Mean daily air temperature, °C	Protection
0 to 4	Masonry shall be protected from rain or snow for 24 hours.
−4 to 0	Masonry shall be completely covered for 24 hours.
−7 to −4	Masonry shall be completely covered with insulating blankets for 24 hours.
−7 to below	The masonry temperature shall be maintained above 0°C for 24 hours by enclosure and supplementary heat.

[a]The amount of insulation required to properly cure masonry in cold weather shall be determined on the basis of the expected air temperature and wind velocity (wind-chill factor) and the size and shape of the structure.
This material is presented with the permission of the Canadian Standards Association; material is reproduced from CSA Standard A179-94 (Mortar and Grout for Unit Masonry), which is copyrighted by CSA, 178 Rexdale Blvd., Etobicoke, Ontario M9W 1R3. Although use of this material has been authorized, CSA shall not be responsible for the manner in which the information is presented, nor for any interpretations thereof. This material may not be updated to reflect amendments made to the original content. For up-to-date information, contact CSA.

tooled when fairly hard, a darker shade will occur than if it is tooled when fairly soft. White mortar cement should be tooled with a glass or plastic joint tool because a metal tool will darken the joint.

Pigments, if added to color the mortar, must be carefully measured. It is best to premix the color with enough mortar cement to do the entire job rather than one batch at a time.

Other things affecting joint color and shade are atmospheric conditions, admixtures, and the moisture content of the masonry units.

Colored Mortar

Architects frequently specify the color of mortar. White mortar is made using white masonry cement or white portland cement, lime, and sand. Colored mortars are made using white masonry cement or white portland cement and get colored by using color pigments, colored sand, or colored masonry cements. The final color achieved is the result of the blending of these ingredients. Trial batches are made and cured until the desired color is developed.

Mortar in Cold Weather

When mortar is placed and cured the temperature of the mortar should be kept in the range of 60°F (15.7C) to 80°F (26.9C). The water in the mortar is needed for hydration, a chemical reaction between the masonry cement and the water. This leads to the hardening of the mortar, which will be slowed or stopped if the cement paste in the mortar drops below 40°F (4.5°C). The cement paste is the mixture formed by water and the mortar cement.

When laying mortar when the air temperature is above 40°F (4.5°C), use normal procedures and cover the finished work at the end of the day to keep water from penetrating the uncured joints. When laying with the air temperature below 40°F (4.5°C) the **mortar water** will usually have to be heated. However, the temperature of the mortar should never go above 120°F (50°C) because higher temperatures will cause the mortar to set up too fast and the mortar will have a loss of compressive and bond strength. If the temperature of the air is falling as the mortar is laid, the **minimum mortar temperature** should be 70°F (21°C). You can find the recommended mortar temperatures for various air temperatures in Table 4-4 and Table 4-5.

The finished laid masonry should be covered when work is stopped. When air temperatures are above 40°F (4.5°C) the walls should be covered with plastic sheets to protect from rain or light snow. When below 40°F (4.5°C) you cover the walls with plastic or canvas to prevent their freezing or becoming wet. Some use insulated plastic or canvas covered blankets. When air temperatures fall below 32°F (0°C) the wall must be covered with insulated blankets and a source of heat provided to keep the mortar temperature from freezing for at least 24 hours. You can find protection recommendations in Tables 4-4 and 4-5.

Surface-Bonding Mortars

There are a variety of surface-bonding mortars available from various manufacturers. These are applied by trowel, brush or spray to any masonry surface and present a new surface to which plaster, stucco, concrete, and cement-based paints will adhere.

Another form of surface bonding is the application of a cement mortar that is reinforced with glass fibers to both the surfaces of concrete block walls that were laid up without mortar. This bonds them into a solid wall and provides a waterproof coating. Tests indicate it is as strong in bending flexure as walls laid with conventional mortar joints. If the surface between the blocks is not flat and smooth, vertical compressive strength is reduced.

CLAY BRICK

Clay brick is made from surface or deep mined clays that have the necessary plasticity when mixed with water to permit its being molded to the desired shape. It also must have the tensile strength to hold this shape while in a plastic condition and contain clay particles that will fuse together when subjected to high temperatures. Properly manufactured brick is fire resistant. It is made in small units, providing the designer the opportunity to use a variety of designs and patterns. The commonly manufactured bricks are shown in 4-8.

Clays

Clay is found in three forms: (1) surface clay, (2) shales, and (3) fireclays.

Surface clays are found near the surface of the earth and are strip mined. Most bricks are made from surface mined clays. **Shales** are clays that have been subjected to high pressures causing them to be relatively hard. **Fireclays** are found at deeper levels than the first two and have more uniform physical and chemical properties. Fireclays can withstand higher temperatures and are used where these will occur such as the lining of a fireplace firebox.

Manufacturing Clay Bricks

The manufacture of clay bricks requires seven steps: (1) winning and storage, (2) processing the raw materials, (3) forming the bricks, (4) drying, (5) glazing (if required), (6) burning and cooling, and (7) drawing and storage of the finished bricks.

WINNING AND STORAGE

Winning is a term used to describe the mining of the clay. Most bricks are made from surface mined clays dug from open pits. Fireclays are obtained from underground mines. The clays are moved by truck or rail to the plant where they are crushed and moved to storage piles.

PROCESSING THE RAW MATERIALS

Clays are removed from the various storage piles and blended to produce the desired chemical composition and physical properties. The blended clays are then moved to crushers where stones are removed and the clay lumps reduced to a maximum of about 2 in. (50 mm) in diameter.

This material is then moved by conveyor to grinders. Here it is ground to a fine powder and passed over vibrating screens. Material too large to go through the screen is sent back to the grinder. The fine material is placed in storage.

FORMING THE BRICKS

The three major methods used to form bricks are (1) the soft-mud process, (2) the stiff-mud process, and (3) the dry-press process.

The **stiff-mud process** is most widely used. It is a high-production process where the clay, which contains 12 to 15 percent moisture, is passed through a deairing machine (a vacuum), which removes air pockets. The clay is then forced by an auger through a die, producing a continuous column of clay of the desired size and shape.

The column then passes through a cutter that cuts the bricks to size. The cut bricks are then moved by conveyor to the inspection area. Imperfect bricks are removed and sent back to be reprocessed. The good bricks are placed on drier cars for transfer to a drier kiln.

The **soft-mud process** is used for making clay bricks from clays having too much natural water to permit the use of the stiff-mud process. The bricks are formed in molds lubricated with water or sand. Brick formed in water-lubricated molds are called **water struck** while those made in sand-lubricated molds are called **sand struck.** Water-struck bricks have a relatively smooth surface. Sand-struck bricks have a matte textured surface.

The **dry-press process** is used with clays having 10 percent or less moisture. The mix is formed into bricks in steel molds under high pressure.

DRYING

The moisture content of the green bricks (unfired, newly formed bricks) will vary depending upon the clay and the process used. Once formed they are placed in a low-temperature drier kiln for one to two days. The temperature and humidity are carefully controlled to

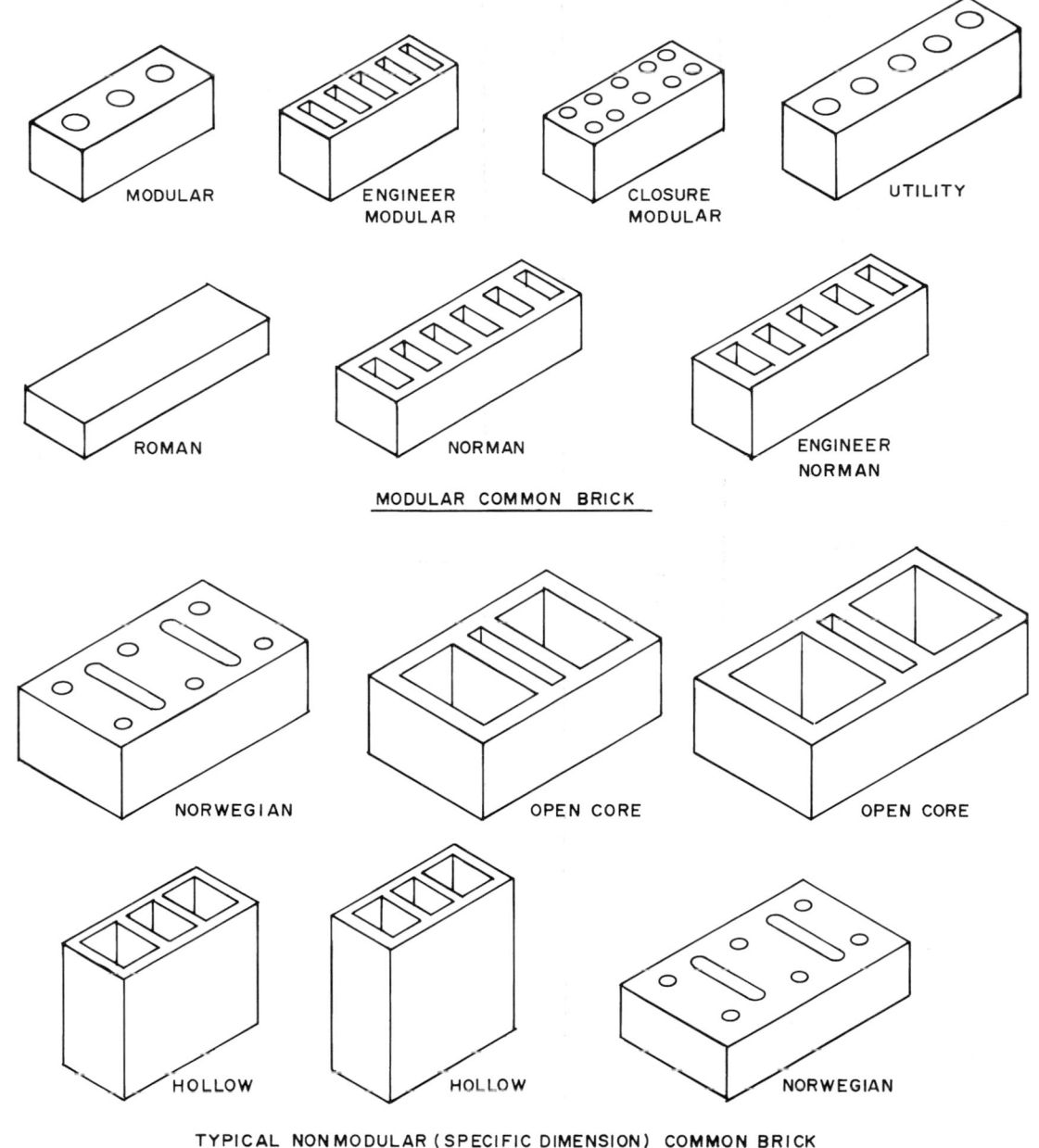

MODULAR   ENGINEER MODULAR   CLOSURE MODULAR   UTILITY

ROMAN   NORMAN   ENGINEER NORMAN

MODULAR COMMON BRICK

NORWEGIAN   OPEN CORE   OPEN CORE

HOLLOW   HOLLOW   NORWEGIAN

TYPICAL NONMODULAR (SPECIFIC DIMENSION) COMMON BRICK

4-8 *Some of the more commonly used types of modular brick.*

prevent rapid shrinkage and possible cracking. From here they may be glazed if required or go directly to high-temperature kilns.

GLAZING

Some bricks have a ceramic glaze applied to one or more surfaces. This is applied after the brick has been dried. **Glaze** is a sprayed coating of mineral ingredients that, when subjected to the required temperature, melts and fuses to the brick. The glaze forms a smooth, glasslike coating and is available in a wide range of colors.

BURNING & COOLING

Burning involves raising the temperature of the dried bricks to a predetermined level. The two types of kilns in use are a periodic kiln and a tunnel kiln. The **periodic kiln** is filled with bricks stacked so air can circulate between them. The temperature in the kiln is raised, held, and lowered. The **tunnel kiln** is a long, narrow building through which the bricks are moved on cars. When the bricks reach the other end of the kiln they have been fired and cooled. The burning process for both methods takes from 40 to 150 hours, depending on desired end results.

Table 4-6 MODULAR INCH-SIZE COMMON BRICK SIZES

Unit Name	Actual Dimensions[a]		Nominal Dimensions[b]	Joint Thickness	Modular Courses
Modular	w	3½	4		
	h	2¼	2⅔	½	3C = 8"
	l	7½	8		
Engineer modular	w	3½	4		
	h	2¾	3⅕	½	5C = 16"
	l	7½	8		
Closure modular	w	3½	4		
	h	3½	4	½	1C = 4"
	l	7½	8		
Roman	w	3½	4		
	h	1⅝	2	½	2C = 4"
	l	11½	12		
Norman	w	3½	4		
	h	2¼	2⅔	½	3C = 8"
	l	11½	12		
Engineer norman	w	3½	4		
	h	2¾	3⅕	½	5C = 16"
	l	11½	12		
Utility	w	3½	4		
	h	3½	4	½	1C = 4"
	l	11½	12		

[a]Actual size unit as manufactured.
[b]Specified unit size plus intended joint size.

Courtesy Brick Institute of America

The burning process involves water-smoking, dehydration, oxidation, vitrification, flashing, and cooling.

Water-smoking removes free water by evaporation and requires temperatures up to about 400°F (204°C). **Dehydration** removes additional moisture and temperatures ranging from 300 to 1800°F (150 to 980°C). **Oxidation** temperatures range from 1000 to 1800°F (540 to 980°C) and **vitrification** from 1600 to 2400°F (870 to 1315°C). These last processes transform the clay into a solid, ceramic material. **Flashing,** if required, follows at this point and is accomplished by adjusting the fire to reduce the atmosphere in the kiln (insufficient oxygen to support combustion). This produces a variation in the colors and color shading of the bricks.

After the bricks have been burned and flashed as required, the cooling period begins. This will take from 48 to 72 hours. The rate of cooling affects the color of the brick and controls cracking and checking.

The color of brick is related to the chemical composition of the clay or shale used and the temperature during the burn. The iron in clay turns red in an oxidizing fire

and purple in a reducing fire. As mentioned earlier, the higher the temperature of the burn, the darker the color.

DRAWING AND STORAGE

After the cooling stage is complete the bricks are removed from the kiln, sorted, graded and moved to storage. Often they are stacked on wood pallets for loading by a forklift. Each pallet load is wrapped in plastic to keep the bricks dry.

Structural Clay Masonry Units

Structural clay masonry units are classified as solid masonry and hollow masonry.

SOLID MASONRY

Bricks are classified as solid masonry if they have cores whose area does not exceed 25 percent of the gross cross sectional area of the brick. The cores (holes) help the drying and burning of the unit and reduce its weight.

Bricks are made in a wide range of types and sizes and a check with a brick supplier is sometimes neces-

Table 4-7 MODULAR METRIC COMMON BRICK SIZES[a]

Unit Name	Actual Dimensions (comparable inch size)		mm[b]	Metric Modular Size (mm)[c]	Metric Nominal Size (mm)[d]	Vertical Coursing
Modular	w	3½	89	90	100	
	h	2¼	57	57	67	3:200 mm
	l	7½	190	190	200	
Engineer modular	w	3½	89	90	100	
	h	2¾	70	70	80	5:400 mm
	l	7½	190	190	200	
Closure modular	w	3½	89	90	100	
	h	3½	89	90	100	2:200 mm
	l	7½	190	190	200	
Roman	w	3½	89	90	100	
	h	1⅝	41	40	50	4:200 mm
	l	11½	292	290	300	
Norman	w	3½	89	90	100	
	h	2¼	57	57	67	3:200 mm
	l	11½	292	290	300	
Engineer norman	w	3½	89	90	100	
	h	2¾	70	70	80	5:400 mm
	l	11½	292	290	300	
Utility	w	3½	89	90	100	
	h	3½	89	90	100	2:200 mm
	l	11½	292	290	300	

[a]Metric sizes are hard conversions.
[b]Actual size unit as manufactured.
[c]Size used when specifying metric brick.
[d]Modular size plus 10 mm mortar joint.

Courtesy Brick Institute of America

sary to verify the size of the unit to be supplied. Both modular and nonmodular units are shown in 4-8.

Modular bricks are those whose **actual size** plus a mortar joint can be assembled on a standard unit or module. The modular unit for inch-size bricks is 4 in. The actual size plus the thickness of a mortar joint (½ in.) is the **nominal size.** The actual and nominal sizes for modular inch bricks are shown in Table 4-6. Modular metric brick sizes are in Table 4-7. Brick sizes are specified by three dimensions, **width, thickness,** and **length,** given in that order. The application of the modular size of a standard inch modular brick is shown in 4-9. Using the 4 in. module, three courses of brick produce an 8 in. module.

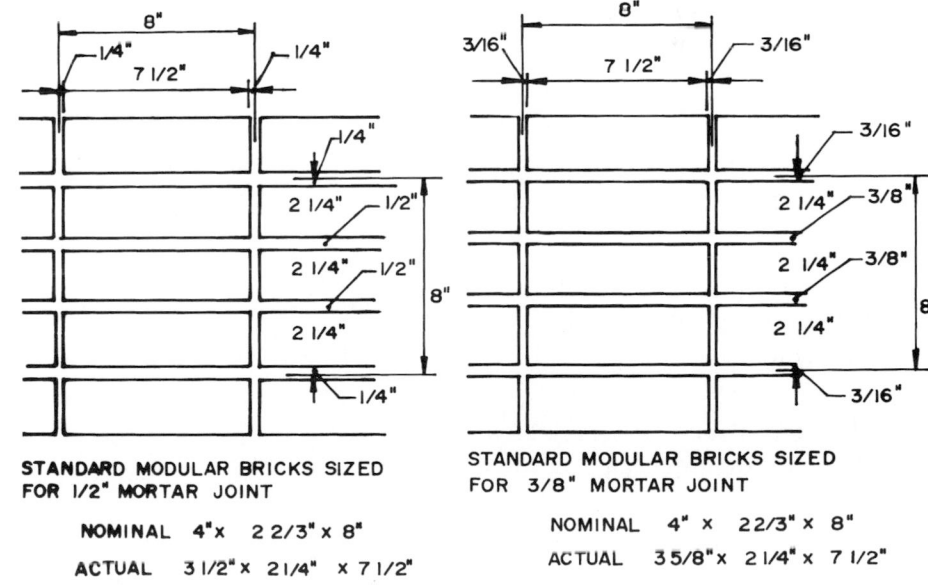

STANDARD MODULAR BRICKS SIZED FOR 1/2" MORTAR JOINT

NOMINAL 4" x 2 2/3" x 8"

ACTUAL 3 1/2" x 2 1/4" x 7 1/2"

STANDARD MODULAR BRICKS SIZED FOR 3/8" MORTAR JOINT

NOMINAL 4" x 2 2/3" x 8"

ACTUAL 3 5/8" x 2 1/4" x 7 1/2"

4-9 *Modular bricks are sized for an 8 inch module with ½ or ⅜ inch mortar joints.*

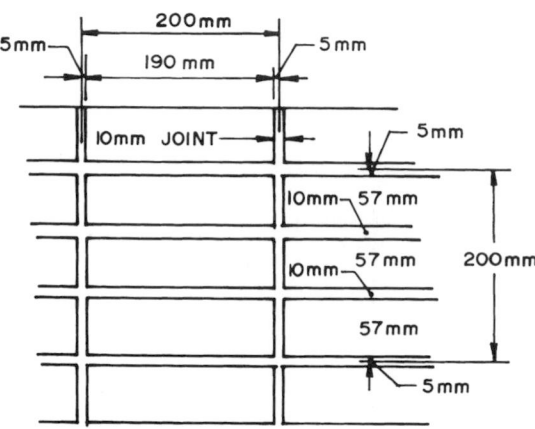

ACTUAL SIZE METRIC MODULAR BRICK
89 X 57 X 190 mm
MORTAR JOINT 10 mm

4-10 *Standard metric modular bricks are laid on a 200 mm multiple.*

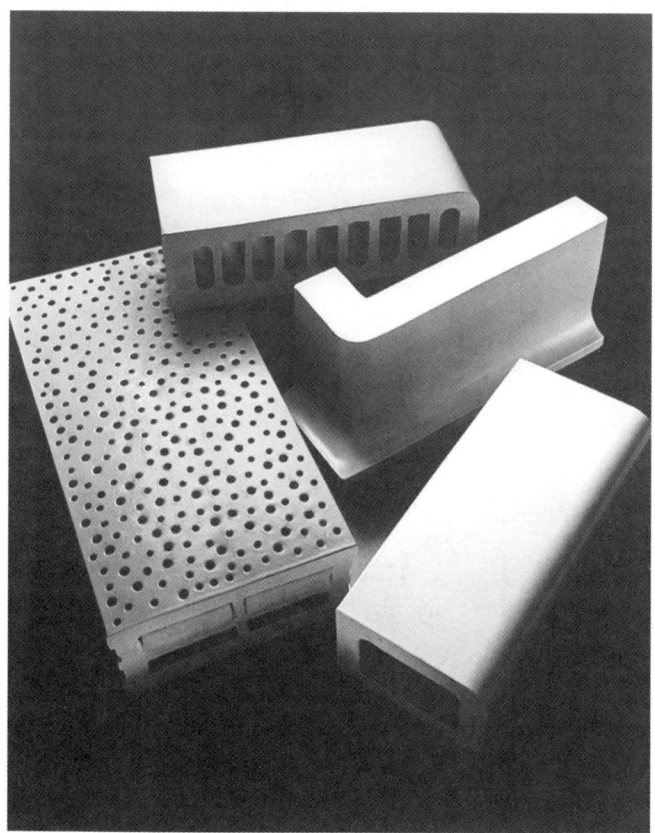

4-11 *There are a wide variety of structural clay facing tile available.*

Courtesy Stark Ceramics, Inc.

The **actual size** of modular metric brick is the size of the manufactured product. The **modular size** is the size stated when specifying the brick. For example, a 89 × 57 × 190 mm actual size modular metric brick is specified when ordering as 90 × 57 × 190 mm. The **nominal size** is the modular size plus the 10 mm mortar joint. The modular metric mortar joint is 10 mm. A standard metric modular brick can be laid on a 200 mm multiple as shown in 4-10.

HOLLOW MASONRY

Hollow brick is a clay masonry unit whose net cross sectional area in the plane of the bearing surface is not less than 60 percent of the gross cross sectional area of that face. Therefore a hollow brick may have a cored area from 25 to 40 percent of the gross cross sectional area of the bearing surface.

Structural clay facing tile are hollow clay units having cores exceeding 25 percent of the gross cross sectional area (4-11). They are used in load-bearing and nonload-bearing walls. They are available with glazed or non-glazed finished surface. A smooth, colored glaze is the most frequently used surface. However, they are available in matt, speckled, and mottled finishes. Selected sizes of structural clay facing tile are in 4-12.

Material Specifications for Clay Masonry

The specifications for clay masonry units have been developed by the American Society for Testing and Materials. When specifying clay masonry units the appropriate ASTM specifications should be included. The ASTM specifications for brick, hollow brick, and structural clay tile are based on the weathering index.

WEATHERING INDEX

The weathering index reflects the ability of clay masonry units to resist the effects of weathering. It is the product of the number of **freezing cycle days** and the annual **winter rainfall.**

A freezing cycle day is a day when the temperature of the air rises above or falls below 32°F (0°C). Winter rainfall is measured in inches between the first and last killing frosts in the fall and spring.

The weathering index ranges from 0 to over 500. Regions rated over 500 are called **severe-weathering regions,** those between 50 and 500 are **moderate regions,** and areas below 50 are termed **negligible regions** (4-13).

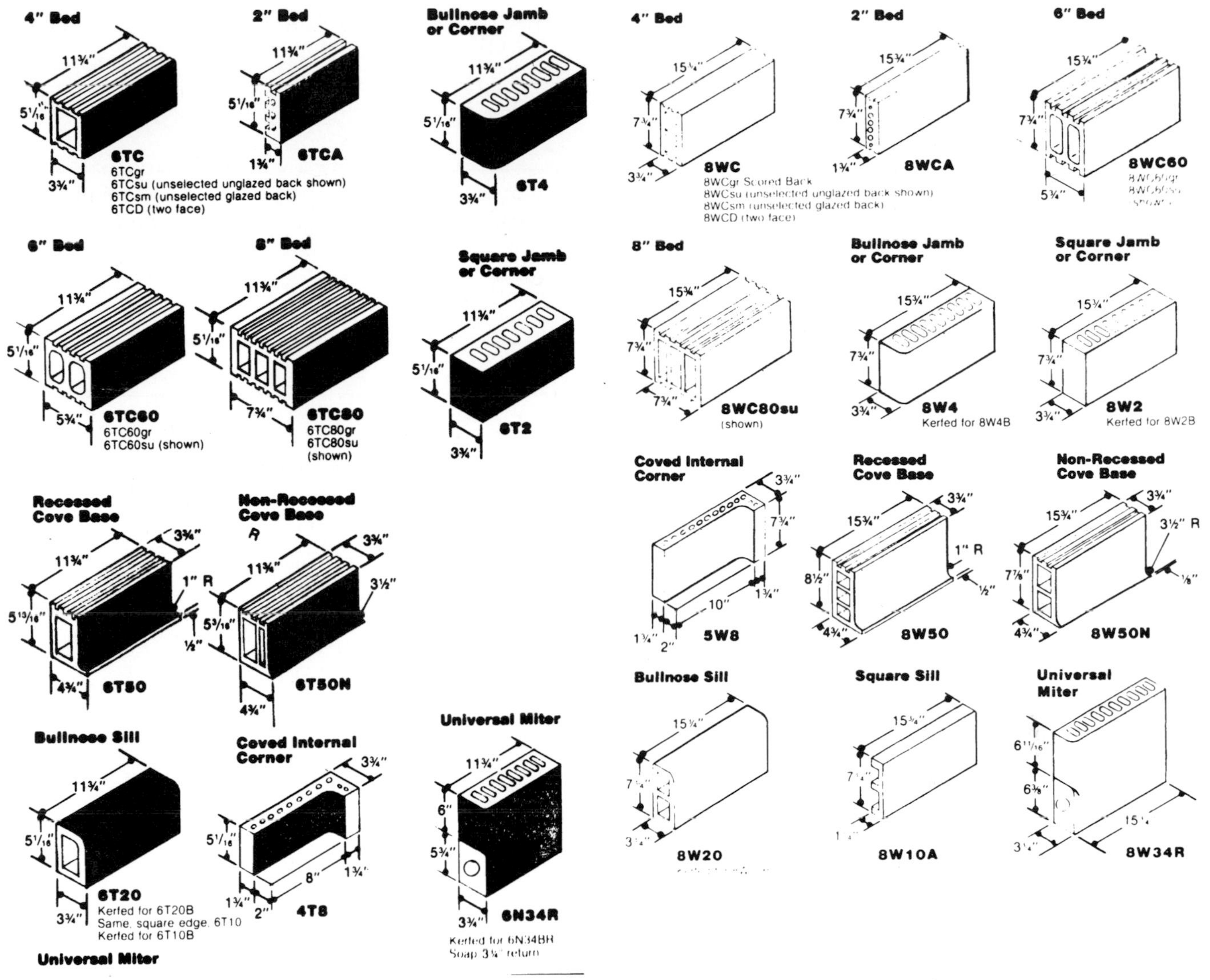

4" Bed — 11¾", 5¹/₁₆", 3¾" — **6TC**
6TCgr
6TCsu (unselected unglazed back shown)
6TCsm (unselected glazed back)
6TCD (two face)

2" Bed — 11¾", 5¹/₁₆", 1¾" — **6TCA**

Bullnose Jamb or Corner — 11¾", 5¹/₁₆", 3¾" — **6T4**

4" Bed — 15¾", 7¾", 3¾" — **8WC**
8WCgr Scored Back
8WCsu (unselected unglazed back shown)
8WCsm (unselected glazed back)
8WCD (two face)

2" Bed — 15¾", 7¾", 1¾" — **8WCA**

6" Bed — 15¾", 7¾", 5¾" — **8WC60**
8WC60gr
8WC60su
(shown)

6" Bed — 11¾", 5¹/₁₆", 5¾" — **6TC60**
6TC60gr
6TC60su (shown)

8" Bed — 11¾", 5¹/₁₆", 7¾" — **6TC80**
6TC80gr
6TC80su
(shown)

Square Jamb or Corner — 11¾", 5¹/₁₆", 3¾" — **6T2**

8" Bed — 15¾", 7¾", 7¾" — **8WC80su** (shown)

Bullnose Jamb or Corner — 15¾", 7¾", 3¾" — **8W4**
Kerfed for 8W4B

Square Jamb or Corner — 15¾", 7¾", 3¾" — **8W2**
Kerfed for 8W2B

Recessed Cove Base — 11¾", 5¹³/₁₆", 3¾", 1" R, ½", 4¾" — **6T50**

Non-Recessed Cove Base — 11¾", 5³/₁₆", 3½", ½", 4¾" — **6T50N**

Coved Internal Corner — 3¾", 7¾", 10", 1¾", 1¼", 2" — **5W8**

Recessed Cove Base — 15¾", 3¾", 8½", 1" R, ½", 4¾" — **8W50**

Non-Recessed Cove Base — 15¾", 3¾", 7⅛", 3½" R, ⅛", ½", 4¾" — **8W50N**

Bullnose Sill — 11¾", 5¹/₁₆", 3¾" — **6T20**
Kerfed for 6T20B
Same, square edge: 6T10
Kerfed for 6T10B

Coved Internal Corner — 3¾", 5¹/₁₆", 8", 1¾", 1¾", 2" — **4T8**

Universal Miter — 11¾", 6", 5¾", 3¾" — **6N34R**
Kerfed for 6N34BR
Soap 3¾" return

Bullnose Sill — 15¾", 7⅛", 3¾" — **8W20**

Square Sill — 15¾", 7⅛", 3¾", 1" — **8W10A**

Universal Miter — 6¹¹/₁₆", 6⅜", 15¾", 3¾" — **8W34R**

4-12 *Selected sizes of structural clay facing tile.*
Courtesy Stark Ceramics, Inc.

4-13 *The weathering index is an indication of the ability of clay masonry units to resist the effects of the weather.*
Courtesy Brick Institute of America

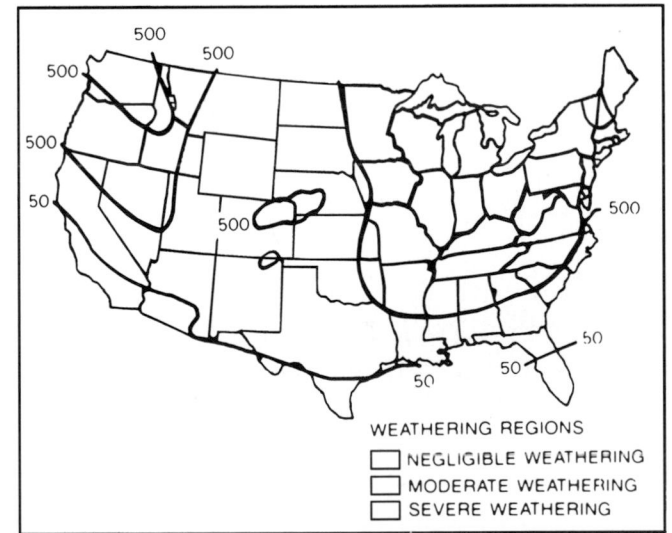

Weathering Indexes in the United States

WEATHERING REGIONS
☐ NEGLIGIBLE WEATHERING
☐ MODERATE WEATHERING
☐ SEVERE WEATHERING

Table 4-8 GRADES AND USES OF SOLID BUILDING BRICK (ASTM C62)

Grade (based on weathering index)	Use
SW Severe weathering	For wet locations below grade where bricks may be frozen, such as in foundations
MW Moderate weathering	For vertical masonry surfaces exposed to the weather in relatively dry conditions where freezing can occur
NW Negligible weathering	For use as backup or interior masonry where no freezing occurs

Courtesy American Society for Testing and Materials

Table 4-9 GRADES, TYPES, AND USES OF SOLID FACING BRICK (ASTM C216)

Grade (based on weathering index)	Use
SW Severe weathering	Masonry in wet locations, in contact with the ground, and subject to freezing
MW Moderate weathering	Exterior walls and other exposed masonry above grade where freezing can occur

Type (based on appearance of the finished wall)	Use
FBX	High degree of physical perfection, minimum variation in color, minimum variation in size
FBS	Wider color range and size variations than permitted in Type FBX
FBA	Nonuniform in size, color, and texture

Courtesy American Society for Testing and Materials

GRADES, TYPES, & CLASSES

Building (common) brick is available in three grades, SW, MW, and NW as specified by ASTM C62. See Table 4-8 for details.

Facing brick is available in two grades, SW and MW and three types, FBX, FBS, and FBA as specified by ASTM C216 as shown in 4-9.

Hollow brick is available in two grades, SW and MW, and four types, HBS, HBX, HBA, and HBB as specified by ASTM C652 (Table 4-10).

Table 4-10 GRADES, TYPES, & USES OF HOLLOW BRICK (ASTM C652)

Grade (based on weathering index)	Use
SW Severe weathering	High degree of resistance to disintegration by weathering when brick may be permeated with water and frozen
MW Moderate weathering	Moderate degree of resistance to frost action where brick is not likely to be permeated with water when it is exposed to freezing temperatures

Type (based on appearance of the finished wall)	Use
HBS	Visible interior and exterior walls where graded variations in color and size than specified for HBX are acceptable
HBX	Visible interior and exterior walls where a small variation in color and size are acceptable
HBA	Nonuniform in size, color, and texture
HBB	Color & texture are not a consideration Size variation greater than specified for HBX is acceptable

Courtesy American Society for Testing and Materials

Table 4-11 GRADES AND USES OF STRUCTURAL CLAY LOAD-BEARING TILE (ASTM C34)

Grade (based on weathering index)	Use
LBX	Tile exposed to the weather and as a base for the applications of stucco
LB	Tile not exposed to frost or earth May be used in exposed masonry if covered with 3 in. or more of other masonry

Courtesy American Society for Testing and Materials

Load-bearing structural clay tile is available in two grades, LBX and LB, as specified by ASTM C34 (Table 4-11). Load-bearing tile will carry building loads in addition to its own weight.

Nonload-bearing structural clay tile is available in one grade, NB, as specified by ASTM C56 (Table 4-12). Nonload-bearing tile carries only its own weight.

Table 4-12 GRADE AND USE OF STRUC-
TURAL CLAY NONLOAD-BEARING TILE
(ASTM C56)

Grade	Use
NB	Nonload-bearing walls, partitions, fireproofing, and furring

Courtesy American Society for Testing and Materials

Table 4-13 GRADES AND USES OF UN-
GLAZED STRUCTURAL CLAY FACING TILE
(ASTM C212)

Grade	Use
FTX	Exposed masonry with minimum variation in color and dimensions, smooth face, mechanically perfect
FTS	Smooth or rough textured, with moderate absorption and variation in dimensions, medium color range, and minor surface finish defects

Courtesy American Society for Testing andc Materials

Table 4-14 GRADES, TYPES & USES OF
CERAMIC GLAZED STRUCTURAL CLAY
FACING TILE (ASTM C126)

Grade	Use
S Select	For masonry with narrow mortar joints (¼")
SS Select sized or ground edge	For masonry where the face dimension variation is small
Type	**Use**
I Single-face units	Where only one finished face is to be exposed
II Two-faced units	Where two opposite finished faces are to be exposed

Courtesy American Society for Testing and Materials

Table 4-15 CLASSES, TYPES & USES OF
PEDESTRIAN & LIGHT TRAFFIC PAVING
BLOCK (ASTM C902)

Class (based on weathering index)	Use
SX	Where brick may be frozen when saturated with water
MX	Where resistance to freezing is not necessary
NX	Interior use when an effective sealer or water-resistant surface coating will be applied
Type (based on traffic)	**Uses of the brick**
I	Where brick will be exposed to extensive abrasion, as in driveways
II	Where brick will be exposed to inter-mediate traffic, as on floors in stores
III	Where bricks will be exposed to low traffic, as in residences

Courtesy American Society for Testing and Materials

Unglazed structural facing tile is available in two classes, FTX and FTS, as specified by ASTM C212. These are based on face shell thickness and factors affecting the appearance of the finished wall. (Table 4-13).

Ceramic glazed facing brick and structural clay facing tile are available in two grades, S and SS, and two types, 1 and 11 as specified by ASTM C126. Details are in Table 4-14.

Pedestrian and light traffic paving bricks are available in three classes, SX, MX and NX, and three types. I, II, and III as specified by ASTM C902 (Table 4-15). These are used on patios, walkways, floors and driveways.

Properties of Clay Brick & Tile

Finished clay units vary considerably in their physical properties. The properties of the raw materials used and the effects of how they are processed as the units are made greatly influence these properties. The major properties include compressive strength, durability, absorption, color and texture.

COMPRESSIVE STRENGTH

The compressive strength developed depends upon the clay, how the units are made, and the temperature and length of the burn. In general, plastic clays used in the stiff-mud process produce units with the higher compressive strengths. Higher burn temperatures will produce higher compressive strengths in almost any clay or process used. Bricks vary in compressive strength from 1600 psi (10.35 MN/m²) to over 5,000 psi (34.50 MN/m²).

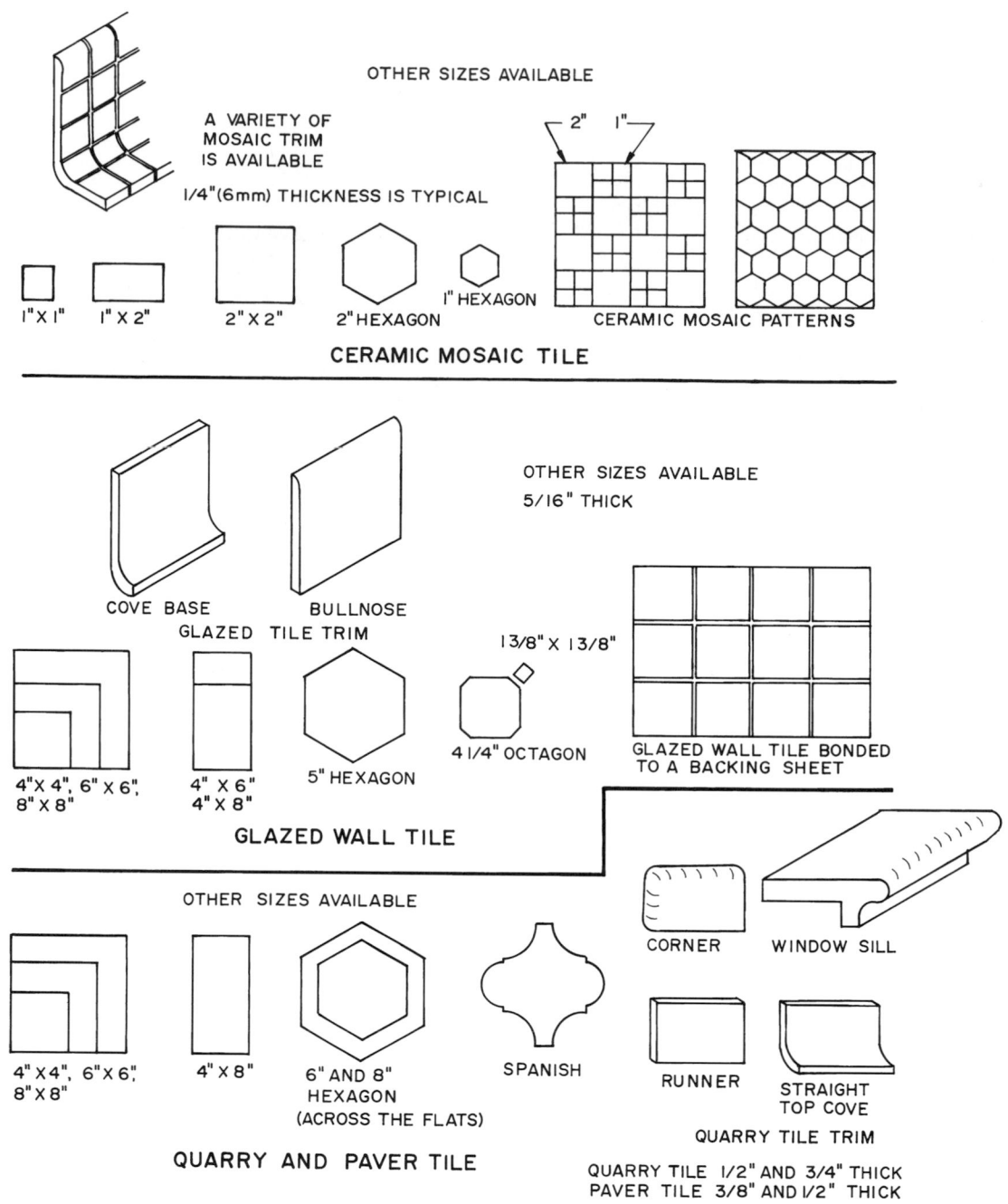

CERAMIC MOSAIC TILE

OTHER SIZES AVAILABLE

A VARIETY OF MOSAIC TRIM IS AVAILABLE

1/4"(6mm) THICKNESS IS TYPICAL

1" HEXAGON

2" 1"

CERAMIC MOSAIC PATTERNS

1"X 1"   1" X 2"   2" X 2"   2" HEXAGON

GLAZED WALL TILE

COVE BASE   BULLNOSE
GLAZED TILE TRIM

OTHER SIZES AVAILABLE
5/16" THICK

13/8" X 13/8"

4 1/4" OCTAGON

GLAZED WALL TILE BONDED TO A BACKING SHEET

4"X 4", 6"X 6", 8"X 8"   4" X 6", 4" X 8"   5" HEXAGON

QUARRY AND PAVER TILE

OTHER SIZES AVAILABLE

4"X 4", 6"X 6", 8"X 8"   4" X 8"   6" AND 8" HEXAGON (ACROSS THE FLATS)   SPANISH

CORNER   WINDOW SILL

RUNNER   STRAIGHT TOP COVE

QUARRY TILE TRIM

QUARRY TILE 1/2" AND 3/4" THICK
PAVER TILE 3/8" AND 1/2" THICK

4-14 *Commonly manufactured glazed wall, ceramic mosaic, quarry, and paver tile.*

DURABILITY

Durability refers to the ability of a clay masonry unit to resist damage due to freezing and thawing cycles while the units are subjected to moisture. Durability is a result of fusion of the clays during burning. High burn temperatures tend to produce a harder more durable unit.

ABSORPTION

Clay units will **absorb** a certain amount of water. This property affects the bond strength of the mortar to the brick. The properties of the clays, the process used to make the brick, and the burn temperature affect absorption. The rate at which a clay unit absorbs moisture is called **suction.** Suction refers to the tendency of a brick to take up moisture in pores and small openings in the surface by capillary action. It does not refer to moisture penetrating the brick itself but just surface water. Suction effects the bond strength of the mortar to the brick. The strongest bond is achieved when the unit has a suction that does not exceed 0.7 ounces (20 grams).

Table 4-16 FIRE RESISTANCE OF
SELECTED BRICK AND TILE WALLS

Solid Brick	Fire Rating in Hours
4" (100 mm) brick	1
6" (150 mm) brick	3
8" (200 mm) brick	4
Hollow Core Brick	
8" (200 mm) brick wall	2
8" (200 mm) brick wall plastered both sides	4
Structural Facing Tile	
4" (100 mm) tile wall plastered one side	1
6" (150 mm) tile wall	2
8" (200 mm) tile wall plastered both sides	4

COLOR

The color of clay masonry units depends upon the clays used, the burning temperature, and the method of controlling color during the burn. Burned clays vary widely in color, ranging from creams and buffs through reds and purples. Since the composition of clays varies, the color of the units produced has some variation within each unit. The application of surface coatings, as glazes, enables units to be produced with almost any color desired. Chemicals can be introduced into the clay producing, when vaporizing, a range of colors.

TEXTURE

The **surface texture** of a finished clay masonry unit is produced by the surface of the die or mold used to form the unit or by attachments that cut, scratch, roll, or in some other way alter the surface as the clay unit leaves the die. Some of the standard textures are smooth, matt (with horizontal or vertical markings), barks, rugs, sand-mold, stippled, waterstruck, and sandstruck.

HEAT TRANSMISSION

The ability of clay brick walls to transmit or resist the transmission of heat directly influences the surface temperature of interior walls. In most applications resistance to heat transmission is very important. In others, as the storage and transmission of solar heat, the ability to transmit heat is important.

SOUND TRANSMISSION

A major factor when designing walls is the ability of the material used to block the transmission of sound through the wall. Walls can reduce sound transmission by absorbing or reflecting sound or reducing the diaphragm action of the materials.

FIRE RESISTANCE

The fire resistance of a material is an indication of its ability to prevent materials behind it from igniting. Fire resistance is stated in hours and is usually called the **fire rating.** These ratings are developed by testing the units under actual fire conditions set up in a test laboratory. Most floors, walls, and ceilings are made up of several materials so fire ratings of these assemblies are also known. In Table 4-16 are the fire ratings for selected clay masonry units.

Ceramic Tile

Ceramic tile includes wall tile, mosaic tile, quarry tile, and paver tile. Wall tiles are 5/16 in. thick; mosaic tiles ¼ in. thick; paver tiles ¼, ⅜, and ½ in. thick; and quarry tiles, ⅜, ½ and ¾ in. thick. They are made in a variety of trim types and flat tiles. Refer to manufacturers catalogs for information about the extensive designs and flat tiles.

CERAMIC TILE
GRADE MARKING & CERTIFICATION

▼ The three grades of ceramic tile are:

Standard Grade—is the best and meets all of the specifications.

Second Grade—meets all specifications except it has facial defects noticeable from a distance of 10 feet.

Decorative Thin Wall Tile Grade—meets all specifications except breaking strength.

Ceramic wall tile, mosaic tile, quarry tile, paver tile, and special purpose tile are shipped in sealed cartons that have the grade indicated by a label glued to the carton.

▼ The color of the label also indicates the grade.
> **Standard Grade**—labels are blue
> **Second Grade**—yellow
> **Decorative Thin Wall Tile**—orange

TYPES OF CERAMIC TILE

Ceramic wall tiles are fired clay tiles that are usually glazed and widely used on interior surfaces such as walls, floors, showers, and countertops. They are available in a variety of shapes and sizes (4-14). They are not expected to withstand excessive impact or be subject to freezing and thawing.

High gloss and matt finishes are commonly used and the surface may be smooth or textured. A wide range of colors are available including solid, multishaded, and speckled variations. Much tile is imported from foreign manufacturers. Distributors catalogs should be consulted for sizes available. These tiles are often supplied mounted on a backing sheet so a number of tile, properly spaced, can be applied to the wall at one time (4-15).

Ceramic mosaic tile may be either natural clay or porcelain in composition. They are smaller than wall tiles and usually square or hexagon in shape. Standard sizes are in 4-14, on page 112. These individual tiles are assembled on a backing material forming sheets from 9 × 9 in. (225 × 225 mm) to 12 × 24 in. (300 × 600 mm). The entire sheet is embedded in the adhesive on the wall or floor as if it were one big tile. Ceramic tile can be installed over wood or metal studs or furring or any masonry or concrete wall.

To get metric sizes the industry has converted inch sizes to the nearest millimeter (e.g., 4¼ × 4¼ in. tile is 108 × 108 mm; 5/16 in. thick tile is 8 mm).

Quarry tiles are hard, burnt unglazed or glazed clay tiles which are red, brown or buff in color, depending upon the clay. They may be extruded or dry-pressed. They are made in a variety of shapes. They generally are from ½ to ¾ in. (12 to 20 mm) thick. A number of trim tiles are made for use with quarry tiles. These include straight base, cover base, double bullnose, and internal and exterior base units.

Paver tiles are the same as quarry tile but thinner, ranging from ⅜ to ⅝ in. (9.5 to 16 mm). They are used in areas where the loads and traffic are less.

Other Clay Tile Products

Clay tiles are vitrified clay products, as clay pipe, used for drainage and sewer systems.

Roof tiles are burnt clay units used for finished roofing. They are made in flat, plain shingles, single-lap tiles, and interlocking tiles. They are made in various styles reflecting the countries where they have been used for centuries. They include Spanish, French, Roman, and English. Dimensions will vary, so manufacturers should be consulted for design data.

ARCHITECTURAL TERRA-COTTA & CERAMIC VENEER

Terra-cotta is a hard fired clay that has been used for hundreds of years for architectural decorative purposes. Modern terra-cotta products are extruded and molded or pressed to shape. Most of these products are custom made to the designs and specifications of the architect. A machine-made unit having either a flat or ribbed back and a flat face is called ceramic veneer. Molded or pressed units are called architectural terra-cotta.

4-15 *Glazed wall tile are assembled with a backing sheet permitting the installation of a large area at one time.*
Courtesy American Olean Tile Company

Ceramic veneer may be an earth red if unglazed but may have a transparent ceramic glaze, a nonlustrous glaze with a satin or matt finish, or a ceramic color glaze in either a solid color or a mottled blend of colors. A polychrome finish is used, which means that two or more colors are applied to separate areas and each color is burned separately. The surface may be smooth, scored, combed, or roughened.

Architectural terra-cotta is also a red earth color when unglazed. It is available with a smooth, plane surface that may be unglazed, have a transparent glaze, a nonlustrous glaze, giving a satin or matte finish, a ceramic color glaze, or a polychrome finish. Sculptural and decorative reproductions of classic architectural ornamentation can be made from terra-cotta and are generally handmade in special molds.

CONCRETE MASONRY

Concrete masonry units are manufactured in a wide range of standard sizes and custom-designed architectural units. They are one of the most widely used construction materials and find use in structural and nonstructural applications. For example, they are used for foundations, piers, columns, pilasters, and other structural purposes. They also are used for nonload-bearing walls, for fire protection of steel, for sidewalks, drives, and patios, and as a backing for other materials, such as clay brick.

Manufacturing of Concrete Masonry Units

Concrete masonry units are made of a relatively dry mix of portland cement, aggregates, water, and in some cases admixtures. The dry materials are carefully weighed and moved to a mixer. The mixer adds the required water and mixes the batch for a predetermined time. The mixed batch is discharged into the hopper of the block machine. Here it is fed into molds and consolidated by pressure and vibration. The freshly molded blocks are called green blocks. They are moved from the block machine on steel pallets to a curing rack.

The curing rack full of green units is moved to a low-pressure steam kiln of an autoclave for hardening. In a **low-pressure steam kiln** the green units are allowed to attain an initial set before steam is introduced. This takes 1 to 3 hours and the temperature is kept at 70°F to 100°F (22° to 38°C). Then the steam is introduced which provides heat and moisture. The temperature is gradually raised to 150° to 180°F (66° to 82°C) depending upon the composition of the concrete. This condition is maintained for 10 to 20 hours until the units reach the required strength. After the blocks are removed they are stored in a protected condition and attain almost full strength in 2 to 4 more days.

The **autoclave** uses high-pressure steam. The molded units are placed in the autoclave and allowed to set for 2 to 5 hours. Then they are gradually heated with saturated steam under a pressure of 150 psi (1035 kPa). This takes 2 to 3 hours. Once the maximum temperature, 350°F (178°C), and pressure are reached, the units soak for 5 to 10 hours. Finally the pressure is gradually released over a 30-minute period and the units go to storage. The can be used 24 hours after removing from the autoclave. These units have greater stabilization against volume changes caused by moisture than the low-pressure units (4-16).

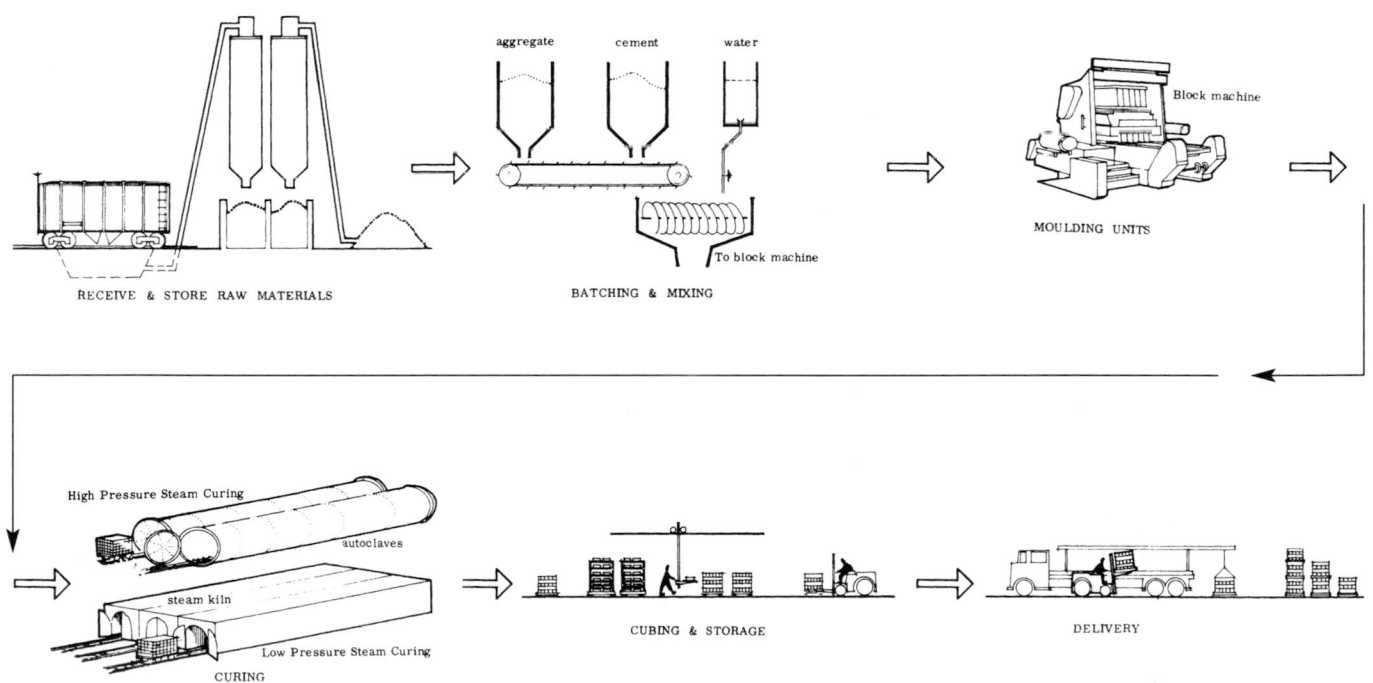

4-16 *Manufacturing process used to produce concrete masonry units.*
Courtesy National Concrete Masonry Association

Physical Properties of Concrete Masonry Units

The physical properties of concrete in Division 3 apply to concrete masonry units. The properties of the units are determined by cement paste and aggregates. Concrete masonry units are generally made with less cement per cubic yard and a lower water-cement ratio. The aggregate is finer with ⅜ in. (10 mm) being the largest size. Concrete masonry units have a large volume of void spaces between aggregate particles while normal concrete should have no voids.

The properties of importance when considering the use of concrete masonry units are weight, compressive strength, water absorption, and the coefficient of thermal expansion. The required properties are established by national building codes and ASTM standards.

WEIGHT

The weight of a concrete masonry unit varies with the design of the block and the mix used to make it. It is necessary to know weights so the dead loads of the structure can be calculated.

COMPRESSIVE STRENGTH

Compressive strength data gives a means of determining a unit's ability to carry loads and withstand structural stresses. The strength will vary depending upon the wetness of the mix. The wetter mixes give the highest strength but cause difficulties in making the units.

TENSILE STRENGTH

Tensile strength is about 5 to 10 percent of compressive strength. **Flexural strength** is about 15 to 20 percent of compressive strength and the **modulus of elasticity** ranges from 300 to 1200 times the compressive strength.

WATER ABSORPTION

Water absorption varies with the density of the concrete masonry unit. Units made with a dense aggregate have much lower absorption rates than those made with lightweight aggregate.

Units used for exterior walls that are not to be painted should be low absorption units. Painting or other waterproof coatings greatly reduces water absorption.

COEFFICIENT OF THERMAL EXPANSION

The coefficient of thermal expansion is used to calculate the amount of expansion that can be expected as temperatures change. Concrete masonry units expand when heated and contract when cooled. These changes are reversible and the unit returns to its original size after the temperature returns to the point at which the change started.

Other properties that are a result of those just discussed include insulation value, coefficient of heat transmission, sound absorbing properties, and fire resistance.

INSULATION VALUE & COEFFICIENT OF HEAT TRANSMISSION

The **insulation value** of units made with porous, lightweight aggregate is better than those made using denser material. The units with lightweight aggregate will also have a lower **coefficient of heat transmission.** Heat transmission, U, is stated in British thermal units, BTUs, per hour per square foot per degree Fahrenheit for each degree difference in temperature between the air on the cool and warm sides of a wall. Concrete masonry walls are poor insulators and good heat transmitters. Therefore they require the addition of insulation materials to be energy efficient.

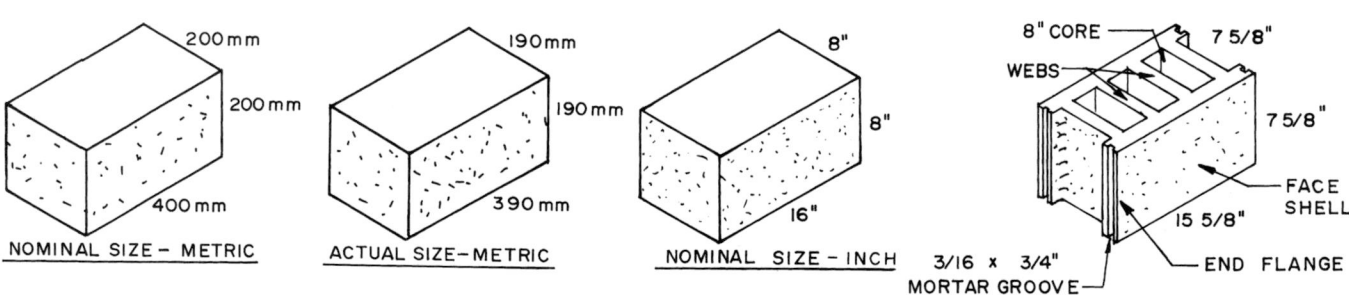

4-17 *Standard inch and metric concrete blocks.*

SOUND-ABSORBING PROPERTIES

Concrete masonry units resist the **transmission of sound.** Hollow block made with lightweight aggregate are recommended. The addition of a plaster interior or exterior finish increases this property. Concrete masonry units that have an open, porous surface will absorb sound better than denser units with smooth surfaces. Painting the surface fills these pores and reduces sound-absorbing properties. Acoustical concrete block units are manufactured combining a sound-deadening liner panel with a concrete unit with open slots on the face.

FIRE RESISTANCE

Building codes are very strict on required **fire resistance ratings** for various parts of a building. Products, such as concrete masonry units, must be carefully tested before they are given a fire resistance rating. The rating is the number of hours the material can be exposed to a flame before it fails. The rating will vary depending upon the aggregate. Plaster on the concrete unit is an effective way of increasing its fire resistance.

Types of Concrete Masonry Unit

Concrete masonry units, often called concrete blocks, are made in a wide range of sizes and types. Many are many based on a special design prepared by an architect. They can be divided into several classifications: (1) concrete brick (solid units), (2) concrete block (hollow and solid units), and (3) special units.

MODULAR SIZES

Concrete masonry units are made in modular sizes and are specified by their nominal size. They are made on a 4 in. module. The metric modular unit is 100 mm.

The **nominal** or **modular** size is the theoretical size without allowance for a mortar joint. The actual size is ⅜ in. less than the nominal size, so the actual unit plus a ⅜ in. mortar joint gives the modular size. See 4-17. Concrete units are also related to standard bricks. For example, in 4-18, the unit is two bricks wide, three brick courses high, and two bricks long. This allows for the mortar joints between the standard modular bricks. A standard modular brick is 4 × 2⅔ × 8 in. nominal.

True metric modular concrete blocks modular size is 200 × 200 × 400 mm. Actual modular size is 190 × 190 × 390 mm (7½ × 7½ × 15⅜ in.). American metric modular blocks soft converted from inch sizes are (actual size) 194 × 194 × 397 mm (7⅝ × 7⅝ × 15⅝ in.), which is quite similar to the true metric block. The mortar bed is 10 mm. Refer back to 4-17 and 4-18.

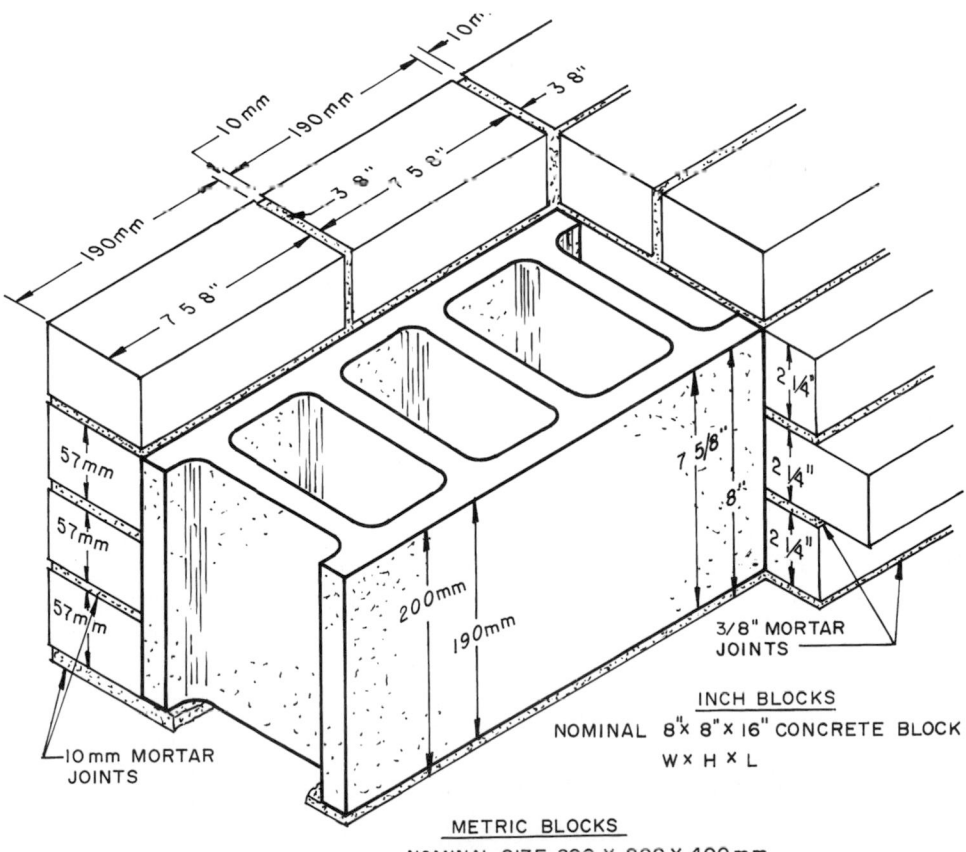

4-18 *The size of standard concrete blocks and bricks are related.*

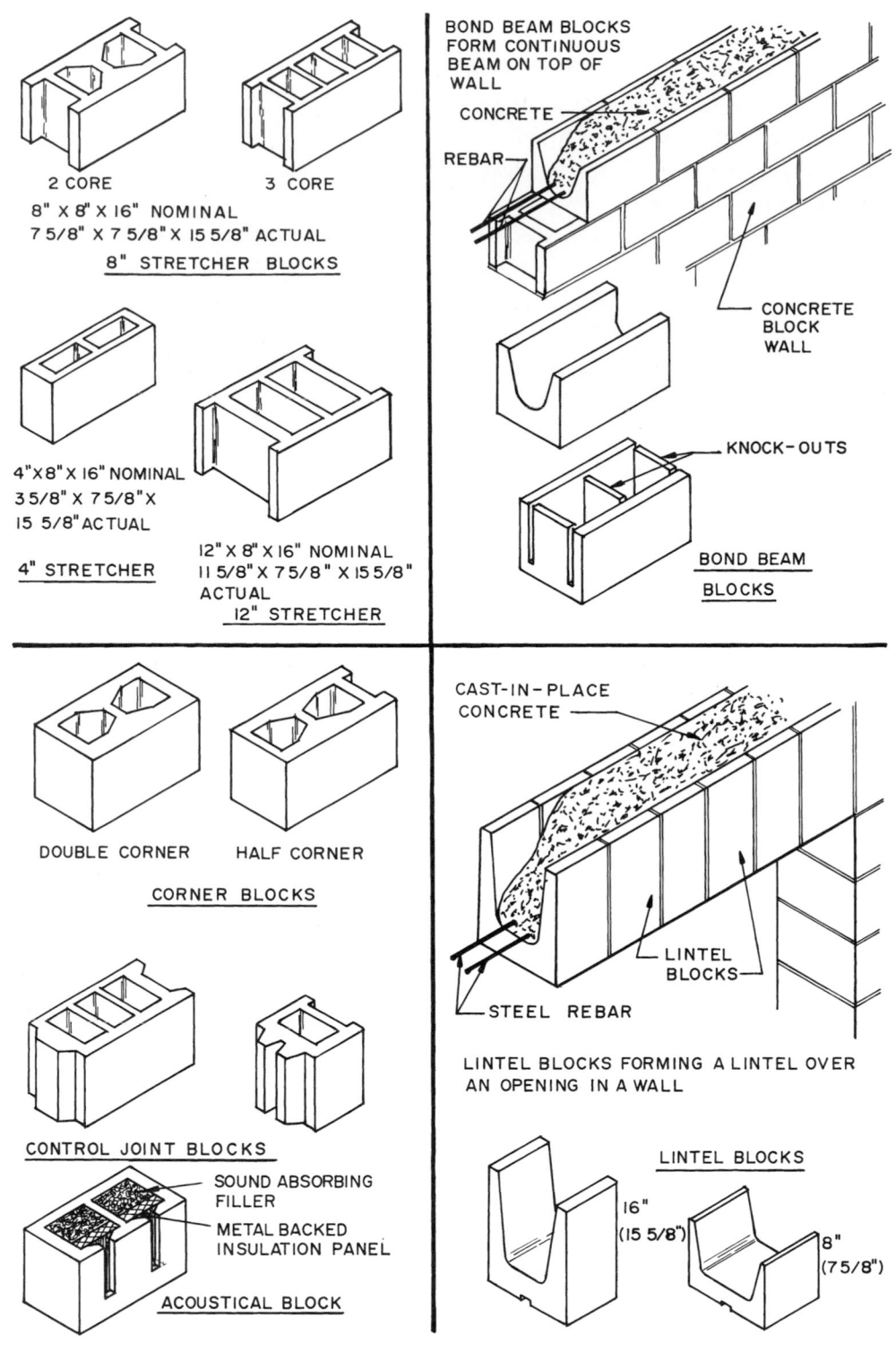

4-19 *Shapes of concrete masonry units that are commonly available.*

CONCRETE BRICK

Concrete brick are made either solid or with a depressed area called a frog. The frog reduces the weight. They are laid with a ⅜ in. mortar joint. A standard modular brick is 4 × 2⅔ × 8 in. nominal. Other sizes are made by some block manufacturers.

SLUMP BRICK AND BLOCK

Slump brick and block are made with a concrete mix that permits the unit to slump a bit when it is removed from the mold, producing units having irregular faces and some differences in height and surface texture. They give an unusual appearance when laid into a wall.

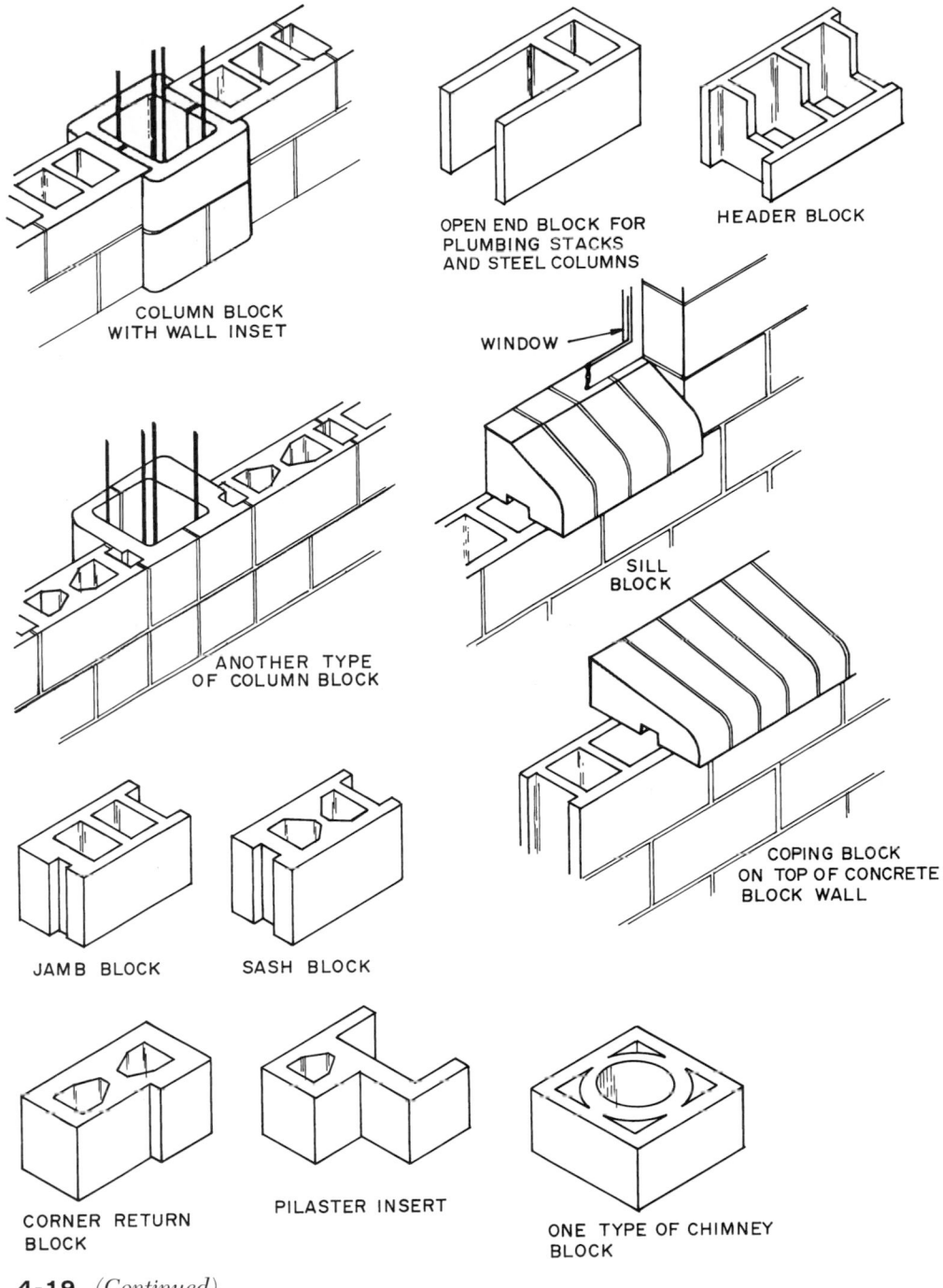

COLUMN BLOCK
WITH WALL INSET

OPEN END BLOCK FOR
PLUMBING STACKS
AND STEEL COLUMNS

HEADER BLOCK

WINDOW

SILL
BLOCK

ANOTHER TYPE
OF COLUMN BLOCK

COPING BLOCK
ON TOP OF CONCRETE
BLOCK WALL

JAMB BLOCK

SASH BLOCK

CORNER RETURN
BLOCK

PILASTER INSERT

ONE TYPE OF CHIMNEY
BLOCK

4-19 *(Continued)*

CONCRETE BLOCK

There are a wide variety of types of concrete blocks available, many designed for a special use. Some of these are shown in 4-19. They are produced in three major groups: (1) solid load-bearing, (2) hollow load-bearing, and (3) non-load-bearing units. A **solid load-bearing unit** is one whose cross-sectional area in every lane parallel to the bearing surface is not less than 75% of the gross cross-sectional area measured in the same plane. A **hollow concrete block** is one whose cross-sectional area in every plane parallel to the bearing surface is less than 75% of the gross cross-sectional area measured in the same plane.

Many units are made with two rather than three cores. The two-core unit has the advantages of having an increase of the face shell thickness and the center web. This increases strength, reduces cracking due to shrinkage when being cured, is lighter, reduces heat concentration in the wall, and permits the cores to line up vertically so plumbing and electrical runs can be made inside the wall.

The stretcher unit is most commonly used in foundation and wall construction. Half blocks are also available.

Corner blocks are used to make an exposed corner and for piers and pilasters. They are made with one or both ends flush.

Corner return block permits a full block face to appear on one wall with a recess to help turn the corner.

Header blocks have a recess that holds the header unit in a masonry bonded wall.

Pilaster blocks are used where the design calls for pilasters to reinforce a wall. They also provide a bearing surface for beams. Some are made in halves and others are a single unit. They may have rebar and concrete in the core if greater strength is necessary.

Control joint blocks are used where vertical shear-type control joints occur in a wall. This relieves stresses in long masonry walls.

Bond beam blocks are filled with concrete and reinforcing bar to form a continuous reinforced concrete beam along the top of a wall. They also are used to span above lintels and below sills.

Sash blocks and jamb blocks are used where windows are to be installed. One style is used for wood windows and the other for metal windows.

Joist block permits a floor joist to rest on the wall and the end is covered with the concrete wing. From the exterior it appears as a normal block wall.

Open-end units are used to provide a vertical cavity for running plumbing, electrical conduit, or to enclose a steel beam.

Lintel blocks are used to form a steel reinforced, concrete lintel over an opening in a wall, as a window opening. They can also be used to form bond beams. More commonly, precast concrete lintels are used over and below openings in concrete block walls.

Sill and coping blocks are used to provide a finished cap. Sill blocks are placed below the window and direct water away from the wall. Coping blocks are placed on top of the exterior wall to seal out moisture.

Partition blocks are available in 4 and 6 in. thickness and are used for nonload-bearing walls.

Chimney blocks are used to quickly form a concrete surround around a fire clay flue lining.

Solid top blocks are used where a solid, coreless surface is needed, such as on the top of a foundation.

Table 4-17 ASTM[a] STANDARDS FOR CONCRETE MASONRY UNITS

Standard	Type of Unit
ASTM C55	Concrete building brick, solid veneer, and split block
ASTM C90	Hollow load-bearing concrete masonry units
ASTM C129	Hollow nonload-bearing concrete masonry units
ASTM C145	Solid[b] load-bearing concrete masonry units

[a]Metric sizes are hard conversions.
[b]Actual size unit as manufactured.

▼ There are a number of types of blocks used to form floors. One type called a **soffit block** is shown in 4-20. Notice it as reinforcing bars and a poured concrete deck on top of it.

Split-face blocks are concrete units that are split in half lengthwise after they have hardened. This produces a rough, irregular surface that is to be the exposed face. They may be hollow or solid units. Color variations are produced by using aggregates of various colors and by putting mineral colors in the mix.

Faced blocks have the exposed surface covered with a ceramic glaze, a plastic overlay, or polished by grinding the surface smooth.

Decorative blocks are made with many different face designs (4-21). Typically they are pierced or recessed to produce an unusual wall surface. A common use is for a decorative, privacy wall. Common sizes are 8 × 16 in. and 12 × 12 in. Special designs can be made by the block manufacturer to meet the requirements of the architect.

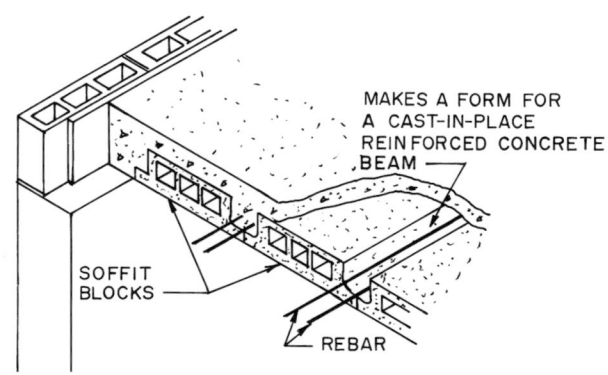

MAKES A FORM FOR A CAST-IN-PLACE REINFORCED CONCRETE BEAM

SOFFIT BLOCKS

REBAR

4-20 *Soffit blocks are used to form floors.*

SPECIAL UNITS

Special units are custom-made blocks designed by an architect to face a project where a special surface treatment is desired. The design possibilities are unlimited. Block manufacturers can produce almost any design provided a mold can be made to produce the unit.

Decorative walls can be produced by using standard concrete masonry units and varying the bond pattern.

Material Specifications

The American Society for Testing and Materials has specifications for four types of concrete masonry units. These are used by block manufacturers to produce quality blocks as shown in Table 4-17.

GRADES

Load-bearing concrete masonry units are available in two grades, S and N (Table 4-18). Grade S units are restricted to above-grade applications and if used in exterior walls must have a protective coating. Grade N units can be used above and below grade and may be exposed to the weather or moisture penetration. When used below grade protection coatings are recommended and are often required by building codes.

Table 4-18 GRADES, TYPES, WEIGHTS & USES OF CONCRETE MASONRY UNITS (ASTM C90)

Grade	Use
N	General use above and below grade
S	Above grade only in walls not exposed to weather or with weather-protective coating if exposed to the weather

Type	Use
I	Moisture controlled during manufacture
II	Nonmoisture-controlled units

Weight	Use
Normal weight	Manufactured using concrete weighing more than 125 pcf (2000 kg/m³)
Medium weight	Manufactured using concrete weighing 105 to 125 pcf (1680 to 2000 kg/m³)
Lightweight	Manufactured using concrete weighing 105 pcf (1680 kg/m³) or less

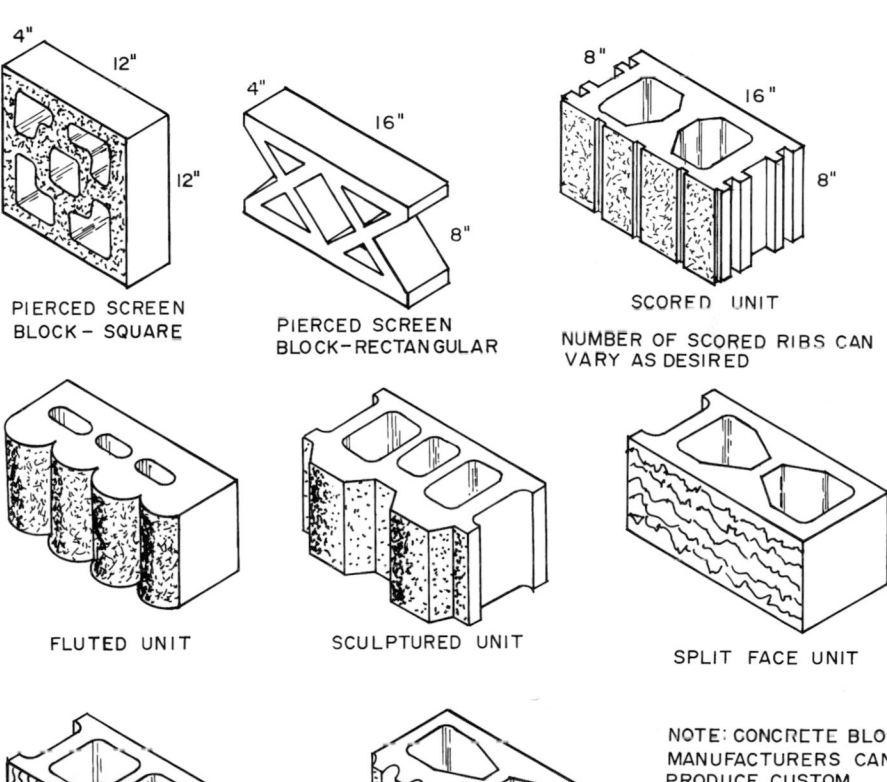

4" 12"
12"
PIERCED SCREEN BLOCK – SQUARE

4" 16"
8"
PIERCED SCREEN BLOCK – RECTANGULAR

8" 16"
8"
SCORED UNIT
NUMBER OF SCORED RIBS CAN VARY AS DESIRED

FLUTED UNIT

SCULPTURED UNIT

SPLIT FACE UNIT

NOTE: CONCRETE BLOCK MANUFACTURERS CAN PRODUCE CUSTOM DESIGNED BLOCKS TO MEET THE DESIGN REQUIREMENTS OF THE ARCHITECT.

STRIATED FACE UNIT

RIBBED UNIT

4-21 *Selected examples of decorative concrete masonry units.*

TYPES

There are two types of concrete masonry units specified by ASTM C90, Type I and Type II. Type I units are manufactured with specific limits on moisture content. Type II has no moisture control limits. Moisture limits on block manufacturing have the advantage of minimizing shrinkage of the units after they have been laid. Shrinkage could cause cracking of the wall. Applications of the various grades and types are in Table 4-19.

WEIGHTS

ASTM C90 establishes three weights of concrete masonry units—normal, medium, and lightweight. Normal weight units are made from concrete weighing more than 125 pcf (2000 kg/m³), medium weight uses concrete weighing 105 to 125 pfc (1680-2000 kg/m³), and lightweight uses concrete 105 pfc (1680 kg/m³) or less.

STONE

While stone was used for centuries as a material to build structural load-bearing walls, it is now mainly used as a veneer or facing material. This greatly reduces the weight of the building, yet permits the designer to use the beauty of the stone as a finish material (4-22).

Table 4-19 GRADES & USES OF SOLID LOAD-BEARING CONCRETE BLOCK, ASTM C145 & HOLLOW LOAD-BEARING CONCRETE BLOCK, ASTM C90

Grade	Use
N-1, N-2	Exterior walls above or below grade that may be exposed to moisture and weather, and for interior walls and backup
S-1, S-2	Above-grade walls not exposed to weather or exterior walls if covered with weather-protective coatings

Rock is a solid mineral material, occurring in individual pieces or large masses, as a hill. **Stone** is a rock selected or processed by shaping to size for building purposes.

Basic Classifications of Rock

Rock is divided into three categories, depending upon its origin—igneous, sedimentary, and metamorphic.

IGNEOUS ROCK

Igneous rock is formed by change of a molten material, usually deep in the earth, to a solid. Commonly used forms include granite, serpentine, and basalt.

Stone on the interior of the African Pavilion in the Smithsonian Institution, Washington, D.C.

Stone used on the exterior of a building

4-22 *Stone is used as a finish material on the interior and exterior of buildings.*

Courtesy Bybee Stone Company, Inc. and Cold Spring Granite Co.

Granite is an igneous rock having crystals or grains of visible size. It consists mainly of quartz, feldspar, mica, and other colored minerals. Colors include black, gray, red, pink, brown, buff, and green. It is hard, strong, nonporous, and durable. It is one of the most permanent building stones and can be used under severe weather conditions and in contact with the ground. Granite stones are finished with a range of textures including a highly polished surface. Granite is used for window sills, cornices, columns, floors, and wall veneers.

Serpentine is an igneous rock named after its major ingredient, serpentine. It ranges from an olive green to a greenish black. It has a fine grain and is dense. Since some types deteriorate due to weathering, its major uses are on interiors. Since it can be cut into thin sections, it can be used as paneling, windowsills, stools, stair treads and risers, and landings.

Basalt is an igneous rock that ranges in color from gray to black, is fine grained, and is used mainly for paving stones and retaining walls.

SEDIMENTARY ROCK

Sedimentary rock is formed of materials, such as sediments, deposited on the bottom of bodies of water or on the surface of the earth. Major types include sandstone, shale, and limestone.

Sandstone is a sedimentary rock composed of sand-sized grains cemented together by naturally occurring mineral materials such as silica, iron oxide, and clay. Quartz grains predominate in the sandstone used for building construction. The two most familiar forms are brownstone, used mainly in wall construction, and bluestone, used for paving and wall copings. Colors include gray, brown, light brown, buff, russet, red, copper, and purple.

Shale is a sedimentary rock derived from clays and silts. It is weak along planes and is in thin laminations. It is not suitable as a concrete aggregate. Shale that is high in limestone is ground into small particles used in making cement, bricks, and tiles. It is basically gray in color but is found ranging from black to red, yellow and blue.

Limestone is a sedimentary rock composed mainly from calcite and dolomite. There are three types of limestone. **Oolitic** is a calcite-cemented calcareous stone formed from shells. It is very uniform in

4-23 *Stone is a popular material for interior use such as this fireplace.*
Courtesy Mr. and Mrs. Roy Register

composition and structure. **Dolomitic** limestone consists mainly of magnesium carbonate. It has a greater compressive strength than oolitic. **Crystalline** limestone consists mainly of calcium carbonate crystals. It has a high tensile and compressive strength.

Limestone is used for building stones and is available as dimension (cut), ashlar, and rubble stones. It is used for paneling, veneer, window stools, and sills, flagstone, mantels, copings and facings (4-23). It is also crushed to form crushed-stone aggregate or burnt to produce lime.

METAMORPHIC ROCK

Metamorphic rock is either igneous or sedimentary rock that has been altered in appearance, density, and crystalline structure by high temperature and/or high pressure. Major types used in construction include marble, quartzite, shist, and slate.

Marble is a metamorphic rock made up largely of calcite or dolomite that has been recrystallized. The colors vary from white through gray and black. Other colors include red, violet, pink, yellow, and green.

Marble is used as wall panels and column facings as well as window stools and sills, and floors. Some types are used on the exterior of buildings while others are limited to interior applications. The surface can be ground to a fine, polished condition.

4-24 *Slate is a beautiful roofing material.*
Courtesy Evergreen Slate Co., Inc.

4-25 *Fieldstone are laid in random design forming textured, attractive walls.*

Quartzite is a metamorphic rock that is often confused with granite. It is a variety of sandstone composed mainly of granular quartz that is cemented by silica producing a coarse, crystalline appearance. It has a high tensile and crushing strength. It is used for building stone, gravel, and aggregate in concrete. In color it could be brown, buff, tan, ivory, red, or gray.

Shist is a metamorphic rock generally made up of silica with smaller amounts of iron oxide and magnesium oxide. The color depends upon the mineral make-up but blue, green, brown, gold, white, gray, and red are common. It is commonly available in rubble veneer and flagstone and finds use as interior and exterior wall facing, patios, and walks.

Slate is a hard, brittle metamorphic rock consisting mainly of clays and shales. The major ingredients include silicon dioxide, aluminum oxide, iron oxide, potassium oxide, magnesium oxide, and sometimes titanium, calcium, and sulfur. It is found in parallel layers, which enable it to be cut into thin sheets.

Slate is produced in three textures, sand rubbed, honed, and natural cleft. It is cut into three types, roof tiles, random flagging and dimension slate (cut to size) (4-24). It is commonly used for interior and exterior wall facing, flooring, flagstones, countertops, coping, and windowsills and stools.

Types & Uses of Stone

Commercially, stone is used in several types: rubble stone, rough stone, monumental stone, dimension stone, flagstone, broken and crushed stone, and stonepowder and dust.

Rubble stone are irregular pieces that have one good face that are fragments from a quarry. They are irregular in shape and size, usually in pieces 12 in. (300 mm) by 24 in. (600 mm). They are cut and fitted by the mason.

Rough building stone, sometimes called fieldstone, are naturally found rock masses. They are generally used in the shapes as found (4-25).

Monumental stone is used for monuments, gravestones, and similar purposes.

Dimension stone, also referred to as cut stone, is cut to size at a stone mill and shipped to the site. The surface may be rough as occurs when it is split or polished. It is used as veneers on interior and exterior walls, floors, copings, stair treads, and other similar uses. **Ashlar stone** is a cut rectangular stone that is smaller than dimension stone and is generally rectangular with square corners and faces. It is a form of dimension stone.

Flagstone are thin, rather flat pieces of stone from ½ to 4 in. (12 to 100 mm) in thickness. Those laid over a concrete base are usually ¾ to 1 in. (1I to

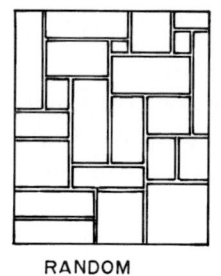
RANDOM
RECTANGULAR

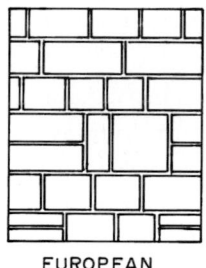
EUROPEAN

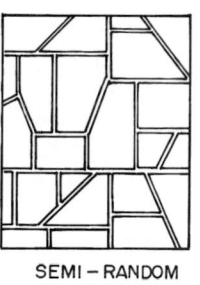
SEMI – RANDOM
RECTANGULAR

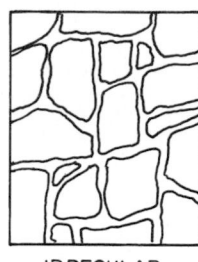
IRREGULAR

4-26 *Some of the commonly used patterns for laying flagstone.*

25 mm) thick. If laid over a sand or loam base 1¼ in. to 1½in. (31 to 37 mm) thick pieces are required. The surface may be left rough or polished. Random flagstone are the natural shape of the pieces with minor shaping. Trimmed flagstone are random shaped pieces with several edges sawed straight. Trimmed rectangular flagstone are pieces with four sides sawed, forming rectangles. They are cut into square and rectangular pieces. Typical patterns for laying flagstones are in 4-26.

Broken and **crushed stone** include irregular shapes and crushed pieces of stone of one type of rock that are graded for hardness and size and used as aggregate in concrete, surfacing roads and driveways and aggregate for surfacing fiberglass asphalt shingles and built-up roofing.

Stone powder and **dust** is used as fill in paints and asphalt paving surfaces.

Quarrying & Producing Building Stone

A quarry is an excavation from which stone used for building is taken by blasting or cutting (4-27). Broken stone is produced by blasting the rock. The larger pieces can be rebroken or cut into smaller units for use as an exterior finish material. The rest is crushed and sorted into various sizes for use as an aggregate.

Most stone used in building construction is dimensional stone produced by cutting large blocks in the quarry. These are often cut by a channeling machine which makes a cut one to three inches wide. Some machines use a rotating chisel-like cutter and others use a wire that runs over pulleys and moves a quartz sand cutting agent over the stone. This produces a saw-type cut in the stone.

Large blocks are removed from the quarry to a mill where they are cut to the sizes and thicknesses needed. The architectural drawings specify the shape and size of each stone. Holes are drilled in each block as indicated for lifting and anchoring it in place (4-28).

Choosing Stone

When selecting the stone to be used in a building the architect must consider the following: (1) cost, (2) strength, (3) durability, (4) hardness, (5) grain and color, and (60 texture and porosity.

4-27 *A typical quarry showing a cut shelf of rock. The rock-sawing equipment is on top of the shelf.*
Courtesy Bybee Stone Company, Inc.

4-28 *Finished pieces of stone cut to the size and shape indicated on the architectural drawings. Notice the markings on the ends indicating the location of each on the building.*
Courtesy Bybee Stone Company, Inc.

The cost requirement is influenced by the ease with which the stone can be shaped. Soft stones have lower production costs. Another factor is accessibility. Cost increases as the distances the cut stone must be transported increase. Some stone is easier to quarry so quarrying costs are another factor that must be considered.

In general, the commonly used stone has sufficient (strength) for the purposes for which it is to be used. However, this must be a design consideration when choosing stone. If the stone is exposed to the weather it must have sufficient **durability** to withstand the freeze-thaw cycles and erosive conditions. **Hardness** is very important for stones that are to be used on floors, steps, patios, and other areas exposed to traffic. **Grain** and **color** are considered when the appearance of the stone in its finished position is decided. Colors vary widely and must be specified. The **texture** of the stone has a great influence on the finished appearance. Fine-grained stones can have a smoother, polished surface while coarse-grained stones will present a more open face. The **porosity** pertains to the ability of the stone to resist penetration by moisture. Porous rock tends to permit some of the minerals to dissolve and stain the exposed face. It also is not durable and will be damaged by the freeze-thaw cycle.

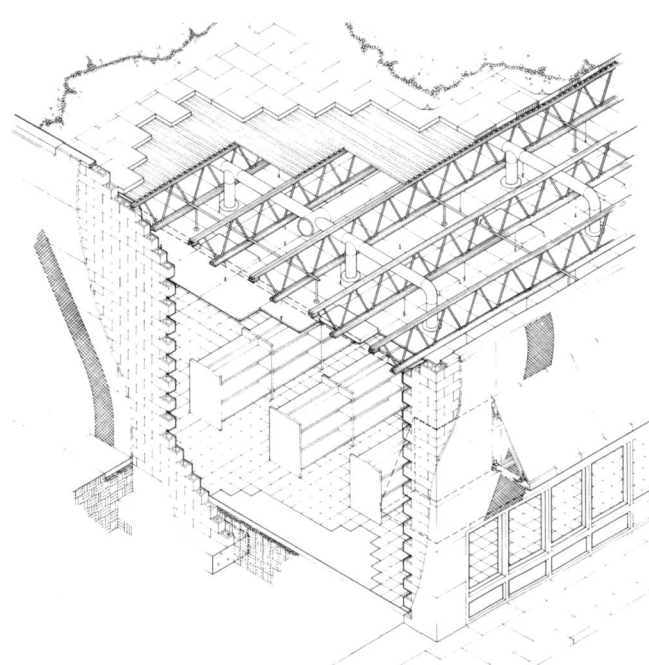

4-29 *Masonry is used to provide structural support as well as finished interior and exterior walls. This building has masonry load-bearing walls and a bar joist roof.*
Reproduced with permission from *The Building Systems Integration Handbook*, Richard Rush, editor, Butterworth-Heinmann Publishers, Newton, MA, 1986.

MASONRY CONSTRUCTION

Masonry wall construction can be of clay brick, structural tile, concrete masonry units, or stone. If they are load-bearing they carry floor and roof loads on the exterior and interior of a building (4-29). Non-load-bearing masonry walls are used to enclose a building to protect it from the elements and for interior partitions. Masonry units provide accoustical control, fireproof construction, and are highly resistant to damage. Masonry walls are easy to design and construct and often produce a more economical building than other methods. They can be used to construct one-story low rise buildings without lateral bracing. Walls above certain heights require lateral bracing. Masonry is a heavy material and this limits its use in high-rise buildings. This weight requires extensive footings and produces compressive stresses and can cause possible buckling. Various design features are used to control lateral forces, such as building internally reinforced masonry walls.

Masonry Bearing Walls

A masonry load-bearing wall may be unreinforced or reinforced, solid masonry or a cavity type wall, and may be a composite wall built using several different types of units such as clay brick over concrete blocks.

Reinforced masonry is used where the flexural, compressive, and shear stresses exceed those permitted for partially or unreinforced masonry. The amount of steel reinforcing is required by building codes.

Cavity wall construction is generally used on exterior walls because they can be built to control moisture penetration and can be insulated. Solid masonry walls do not have these advantages and are therefore generally used for interior partitions.

Composite masonry walls will have an exterior veneer of a quality masonry unit such as brick, tile, or stone and the hidden interior is built from a cheaper unit such as concrete block. The engineer designing the wall must consider how differences in thermal expansion, moisture absorption, and load-bearing capabilities affect the wall.

Control & Expansion Joints

Movements occur within masonry walls that produce stresses that may cause cracking. These stresses can be created by expansion and contraction due to changes in temperature or moisture content, structural movements due to settling of the foundation, and concentration of stresses at openings in the wall. Masonry walls often use units of different materials. These expand and contract at different rates and can create stresses. Examples include brick veneer over concrete block or the use of metal lintels over openings. The location of control and expansion joints must be carefully determined as part of the wall design.

Expansion joints provide a space between adjacent parts of masonry construction and permit a limited amount of movement. The actual spacing varies with the type of masonry unit, size of the area, and the reinforcing used. Vertical expansion joints are placed near the corners or where the walls change direction. Horizontal joints are placed above masonry walls that butt structural frames or the bottom of floor or roof structures (4-30).

Control joints provide tension relief between parts of a masonry wall that may change from their original dimensions. One major contributor is the fact that masonry walls tend to contract as they dry after laying. The tensile strength of the wall tries to resist these stresses and the wall cracks if the strength of the wall is exceeded. This cracking is controlled by using carefully placed control joints and adequate steel reinforcing in the wall.

Control joints are generally located where there are changes in wall height, wall thickness, at openings, wall intersections, returns in U-, T-, and L-shaped buildings, and junctions between walls and columns. These are illustrated in 4-31.

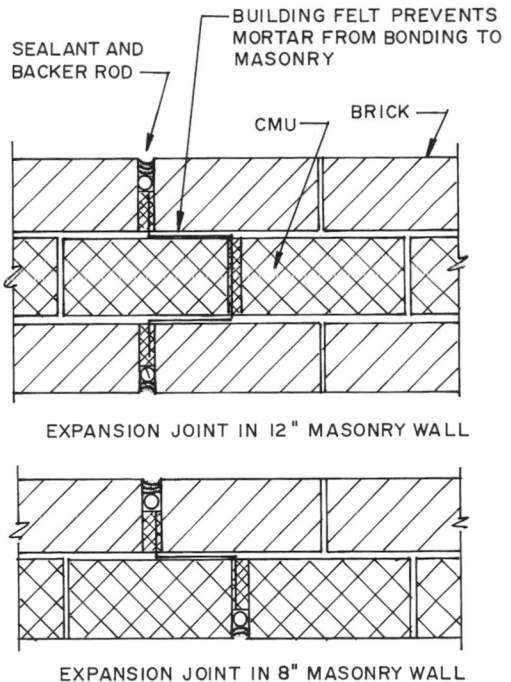

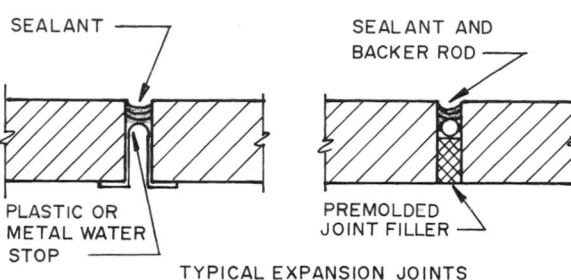

4-30 *Expansion joints used in masonry construction.*

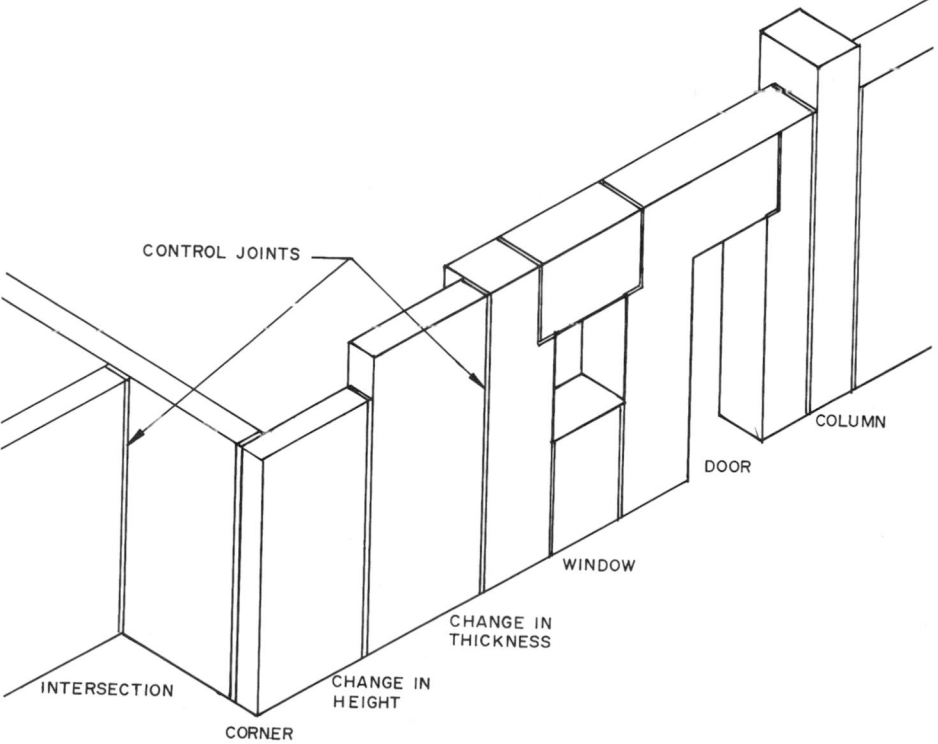

4-31 *Commonly used locations for control joints in masonry construction.*

Brick Masonry Construction

Some of the commonly used brick masonry wall constructions are shown in 4-32. The fire resistance of these walls will vary with the composition of the unit. This varies because of the clays used in various parts of the country. The interior space in cavity walls will tend to collect moisture, which can be controlled by providing for the drainage of this moisture. This construction uses flashing at the base and weep holes (openings in the mortar) located just above the bottom of the flashing (4-33). Moisture that enters the cavity moves to the bottom of the wall and is directed out of the building through weep holes. **Weep holes** are openings left in the mortar joint.

Some examples of brick bearing walls supporting floor and roof joists and also those supporting decking are in 4-34 through 4-37.

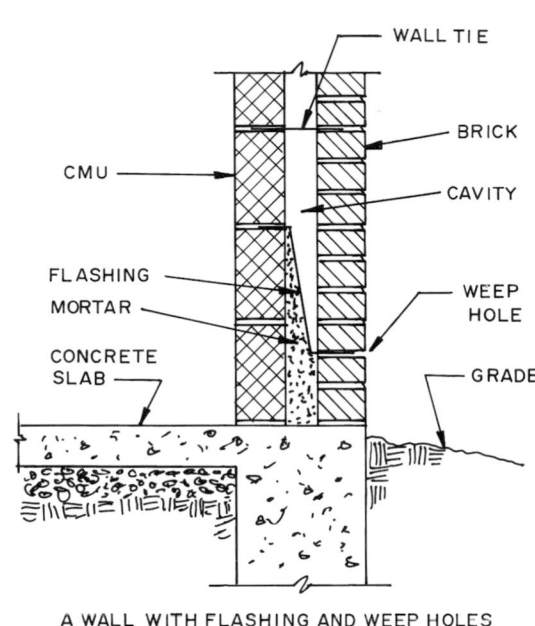

A WALL WITH FLASHING AND WEEP HOLES

A PARGED WALL

4-33 *Two techniques used to control moisture inside a brick cavity wall.*

4" SOLID BRICK WALL

6" "SCR" BRICK WALL

8" SOLID BRICK WALL

8" HOLLOW BRICK WALL

CAVITY — CMU — TIE

COMPOSITE CAVITY WALL. WIDTH CAN VARY.

CMU — REBAR IN CONCRETE CORE

8" COMPOSITE WALL

REINFORCED BRICK WALL

4-32 *Typical brick masonry wall construction.*

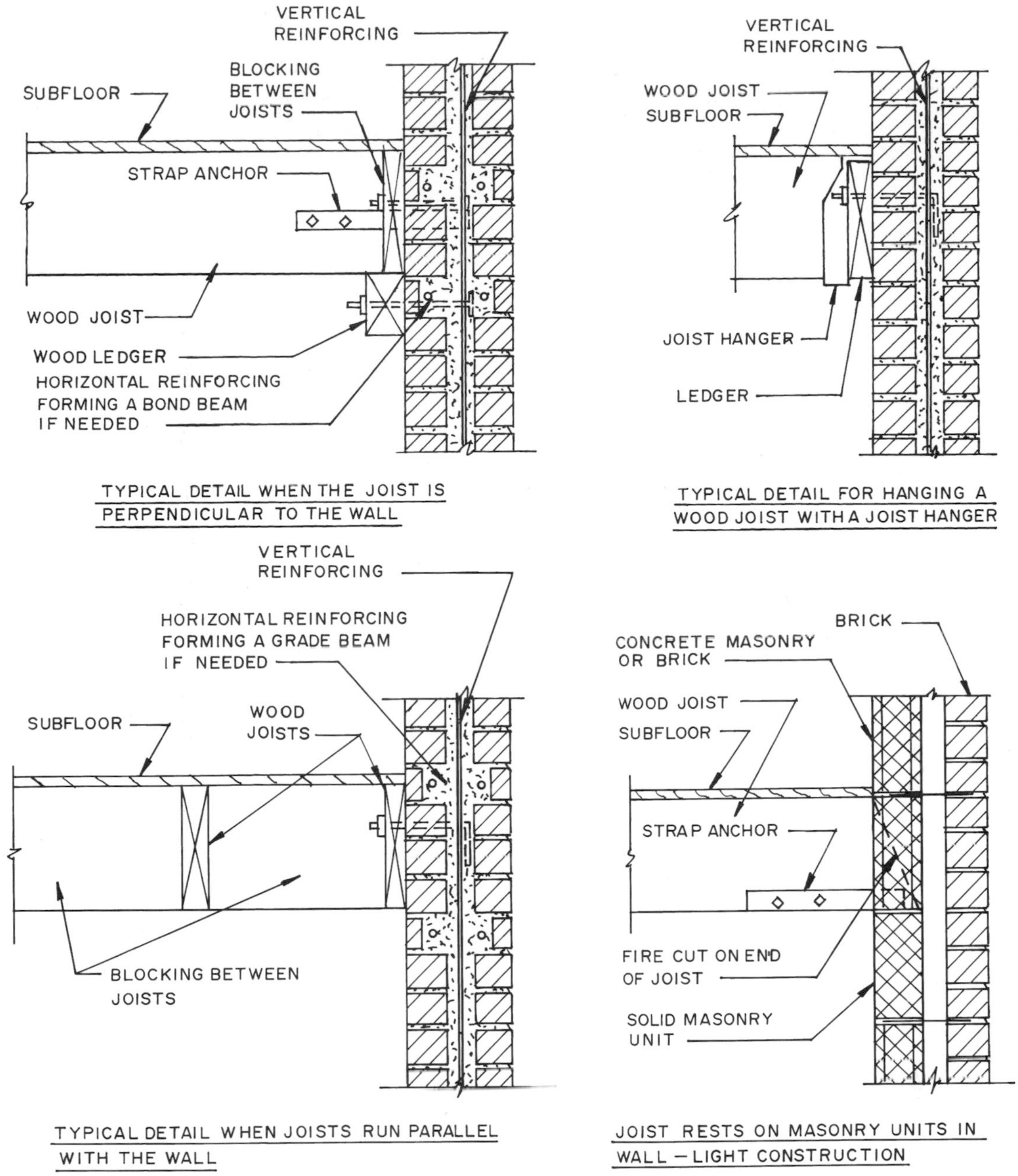

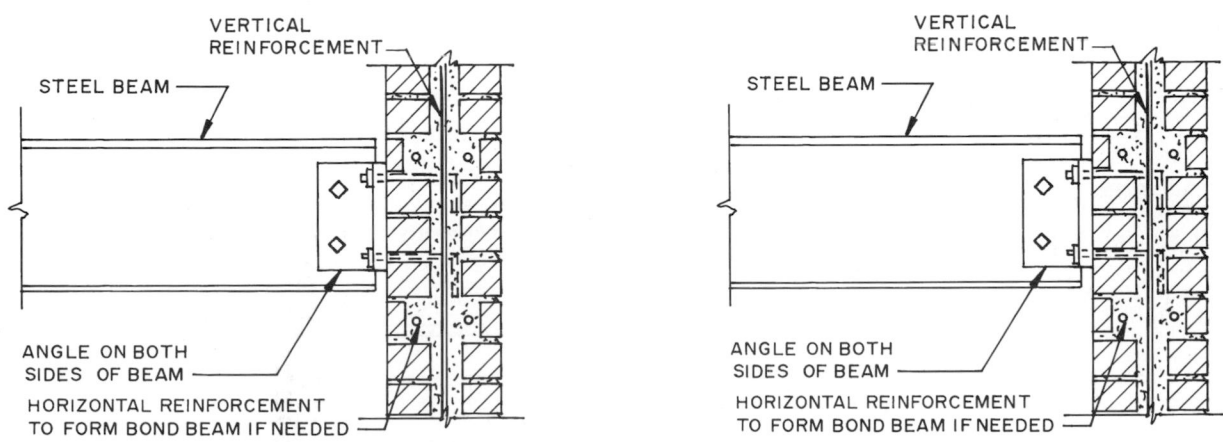

4-34 *Commonly used ways to frame wood joists and beams to brick masonry walls.*

4-35 *Several ways to frame steel joists or beams to brick masonry walls.*

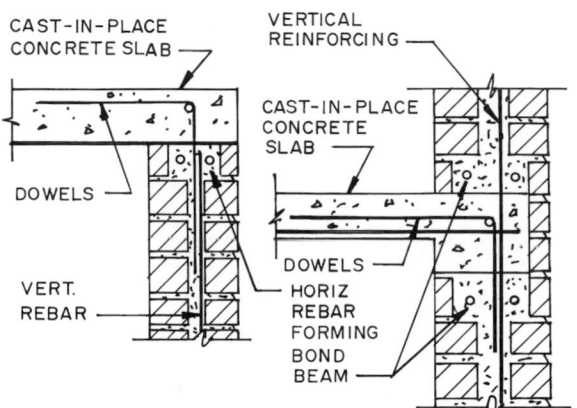

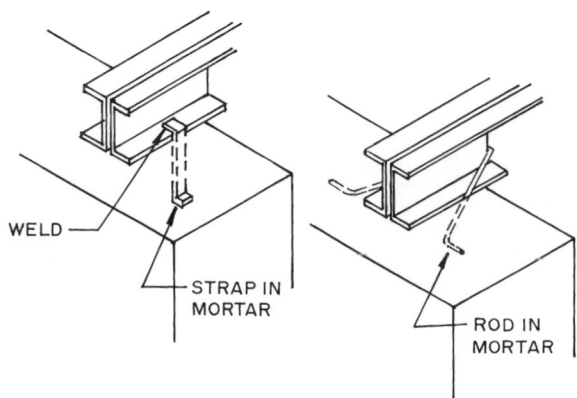

4-36 *Two ways to support cast-in-place concrete decks with brick masonry walls.*

4-37 *Open-web joists are anchored to masonry walls with metal straps set in the mortar joint.*

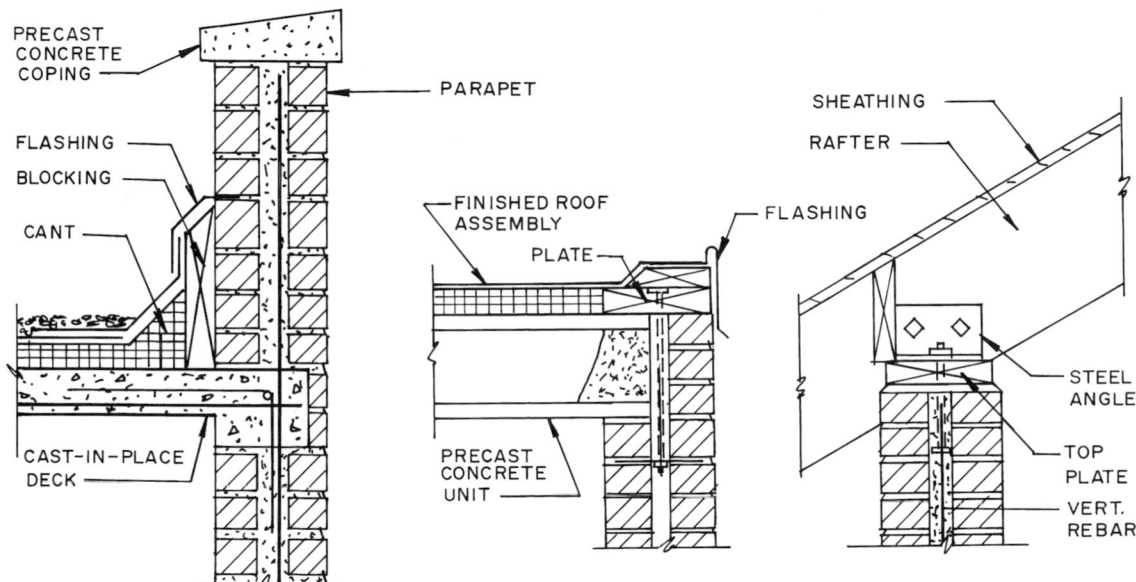

4-38 *Examples showing roof construction with brick masonry walls.*

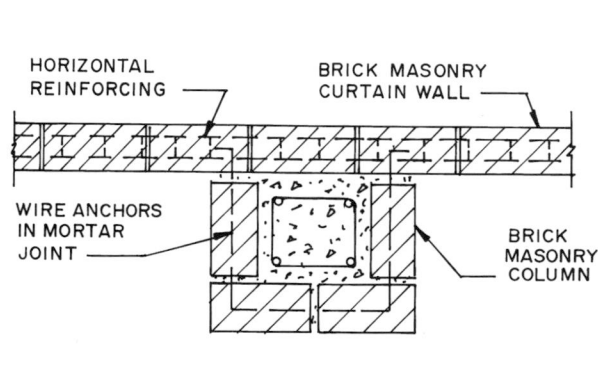

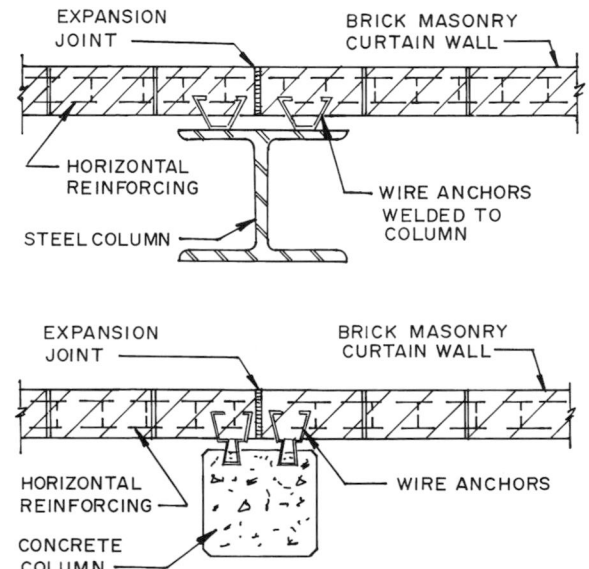

4-39 *Typical brick masonry curtain wall construction.*

Examples of roof construction with brick masonry walls are shown in 4-38.

Details for typical brick **curtain wall** construction are in 4-39. The curtain wall is tied to columns with metal reinforcing. A **curtain wall** is an exterior non-bearing wall between columns not supported by the beams or girders of the structural frame.

Openings in brick masonry walls are spanned with a **lintel.** These are typically steel angles; however, precast concrete or concrete masonry lintel blocks are used. In some cases reinforced brick lintels can be used. The choice and design depend upon the span of the opening and the load to be carried by the lintel (4-40). The precast concrete and concrete masonry lintels generally carry the loads imposed on the wall and a steel lintel carries the face brick. Some classic architectural styles use exposed concrete or stone lintels.

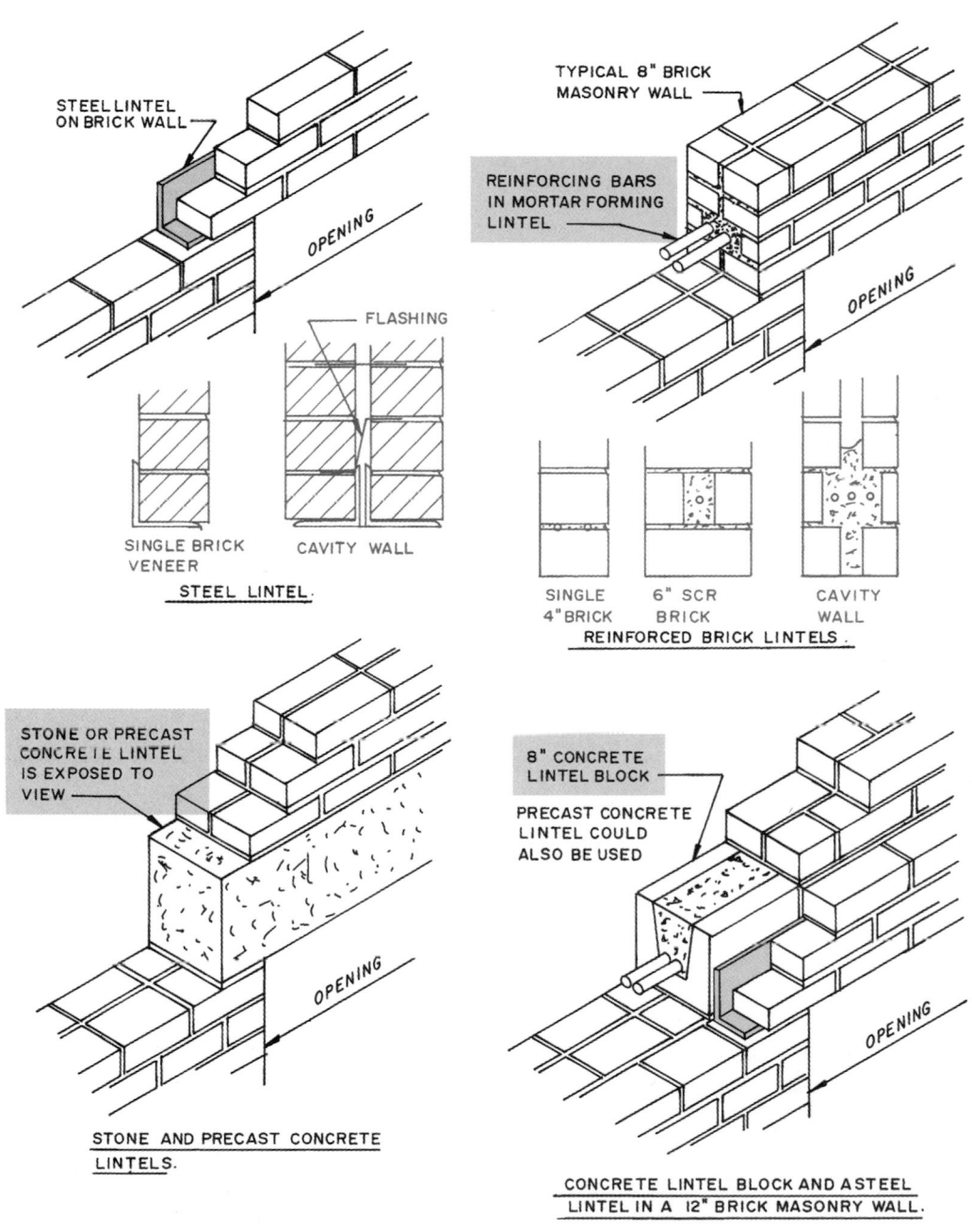

4-40 *Types of lintels used with brick masonry construction.*

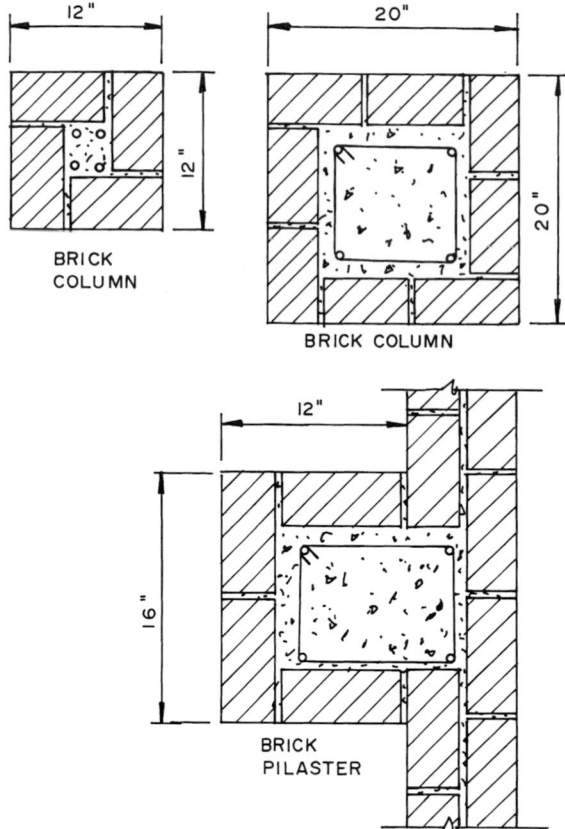

4-41 *Reinforced brick masonry pilaster and column construction.*

Brick masonry **columns** are typically designed to be built without requiring the bricks to be cut. The interior core has reinforcing bars and is filled with concrete (4-41). Pilasters are built the same as columns but are an integral part of the load-bearing wall.

INSULATION

Solid masonry walls are poor insulators. They conduct heat and cold. To improve the energy efficiency of a masonry wall, insulation can be placed on the exterior face, within the wall, or on the interior surface. Rigid insulation consists of rigid foam plastic sheets adhered to the masonry and layers of acrylic polymer stucco material reinforced with fiberglass mesh is applied over it. This forms an attractive hard exterior finish.

Cavity walls permit the insulation to be placed inside the wall. If rigid insulation is used it is bonded to the inside face of the masonry units, forming the inside wall (4-42).

BRICK BOND PATTERNS

Bond patterns commonly used for standard and oversize bricks are shown in 4-43. When the longer, narrower Norman bricks are used the bond patterns in 4-44 are used. Notice the running bond consists entirely of stretchers. The English bond alternates courses of headers and stretchers. The common bond has a header course every sixth course.

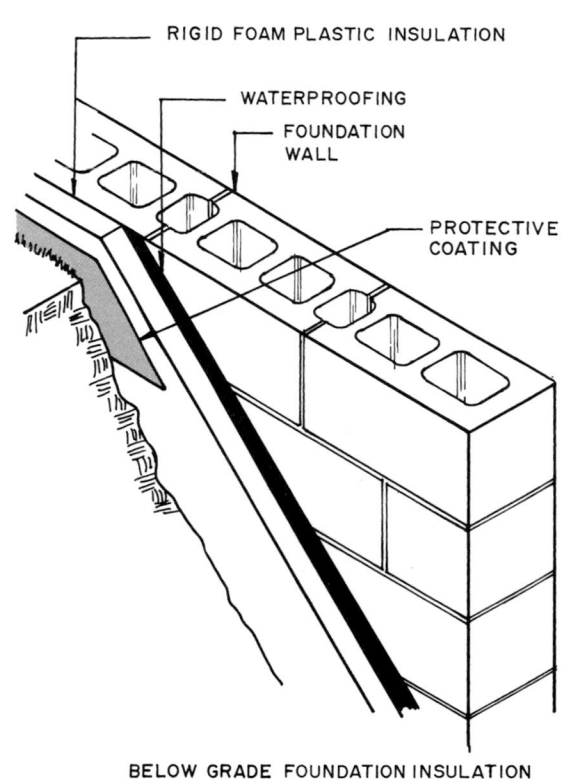

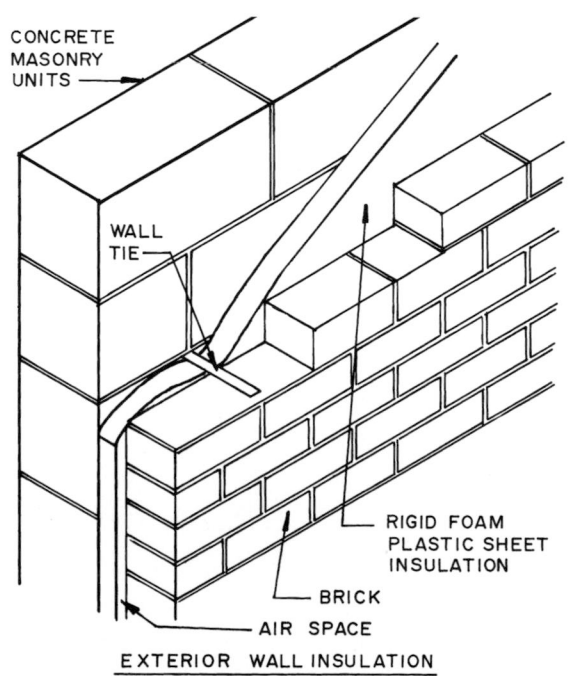

4-42 *Rigid plastic foam insulation is used on all types of masonry construction.*

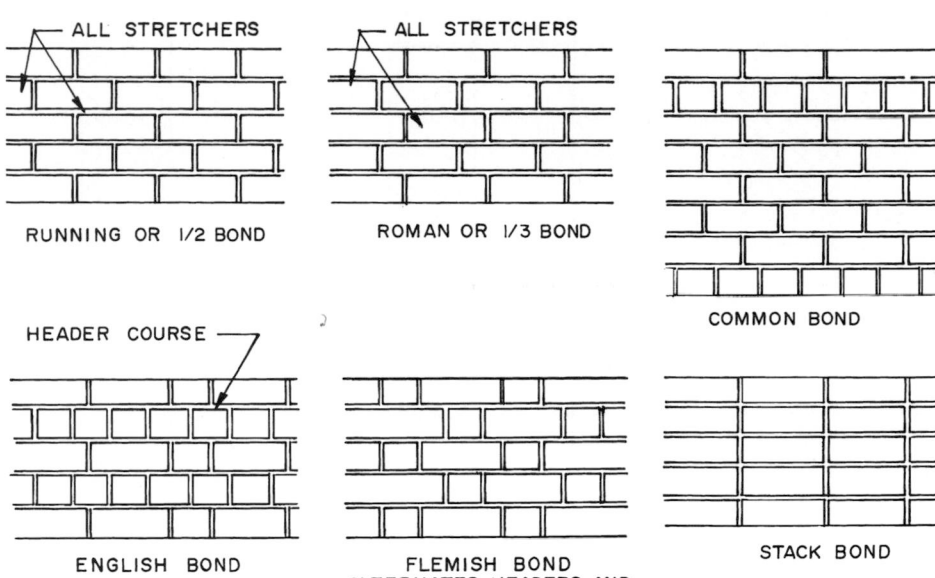

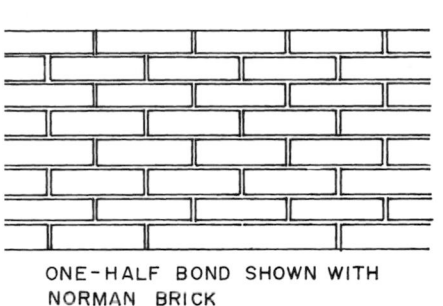

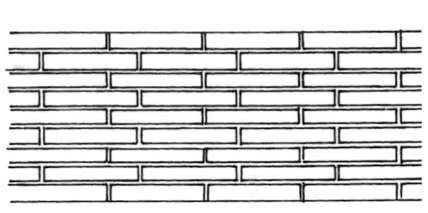

4-43 *Brick bond patterns used with standard and oversize brick masonry units.*

4-44 *Bond patterns used with Norman and Roman brick masonry units.*

TYPES OF MORTAR JOINT

Mortar joints can take several forms. The appearance of the finished wall depends upon the type of joint. Those in common use are in 4-45. The mortar joint is finished by troweling or tooling. **Troweling** involves striking off the excess mortar using the mason's trowel. This does not produce the most water-tight joint. The weathered, concave, and V-joints resist leakage the best, and are called tooled joints. Tooled joints are the result of compressing the mortar into the joint and against the face of adjacent units. Ruled, flush, and flush and rodded joints are not as watertight. Extruded, beaded, struck, and raked joints are most likely to eventually let moisture penetrate the wall.

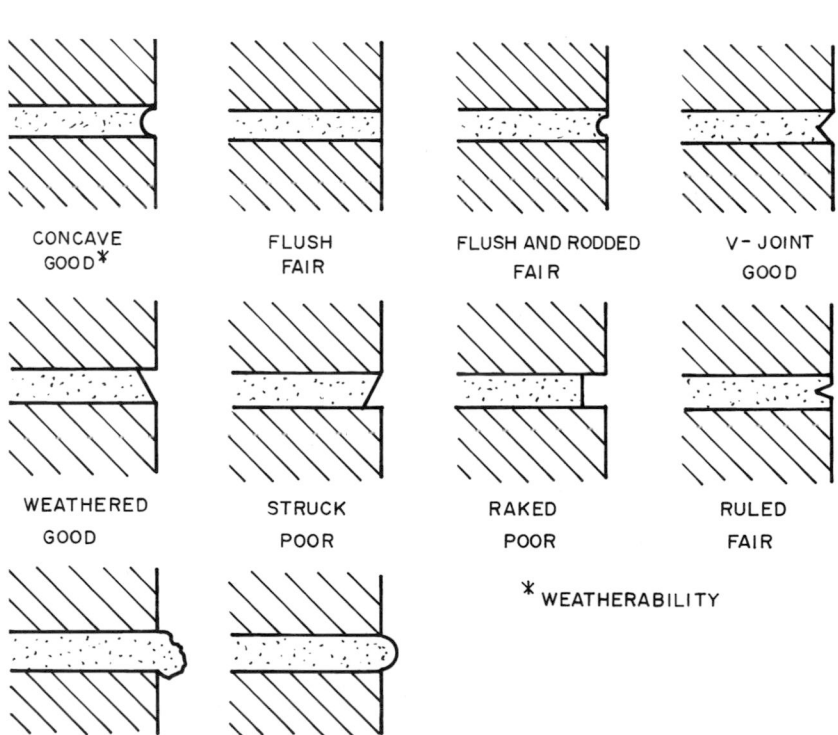

4-45 *Commonly used mortar joints.*

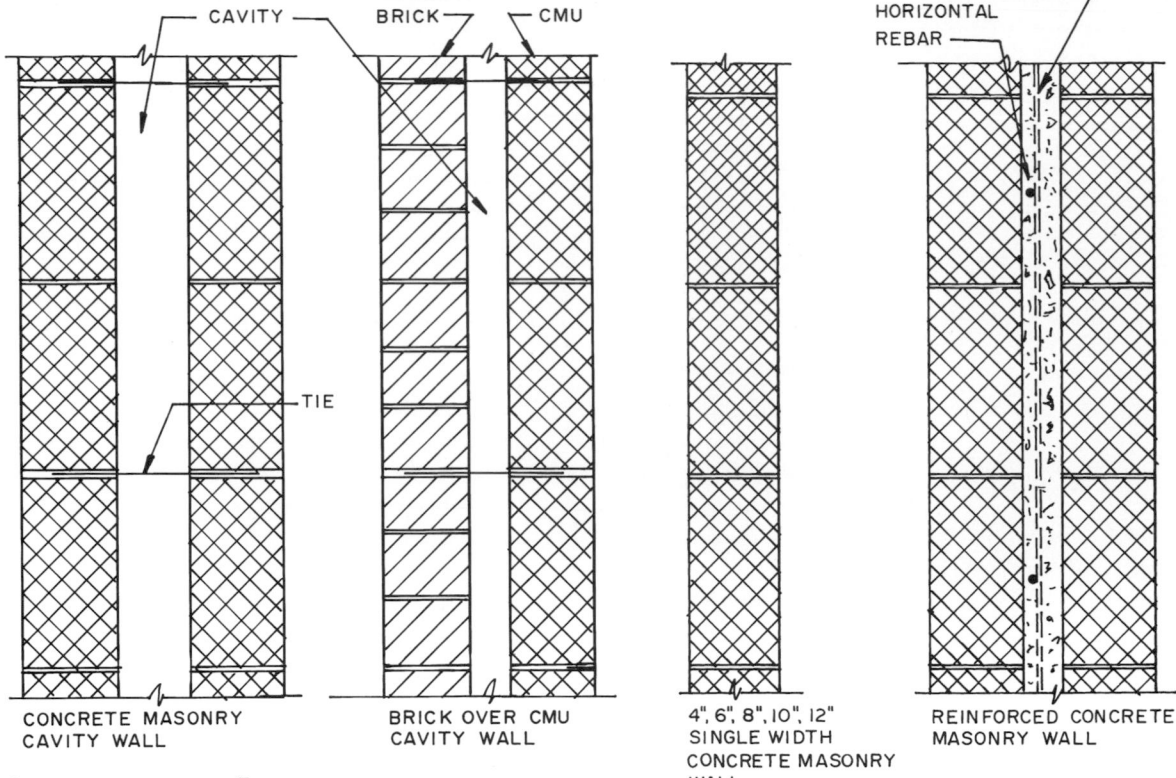

4-46 *Concrete masonry wall construction.*

Concrete Masonry Construction

Some of the frequently used concrete masonry wall constructions are shown in 4-46. Since the composition of materials used to make concrete masonry varies, their compression strength, fire resistance, coefficient of thermal expansion and other properties must be known so the designer can specify the type of construction required. Typical floor and roof construction details can be seen in 4-47 and 4-48.

Stone Masonry Construction

Stone laid in mortar may be rubble or ashlar masonry. **Rubble** is stone found naturally and is in irregular shapes and sizes. **Ashlar masonry** uses units cut into squared shapes. Stone may be laid in a random or coursed pattern. The coursed pattern maintains horizontal lines while the random pattern courses fall in many directions (4-49). One form of rubble stone masonry, ledgerock, is a rock formed naturally into thin layers of varying thicknesses.

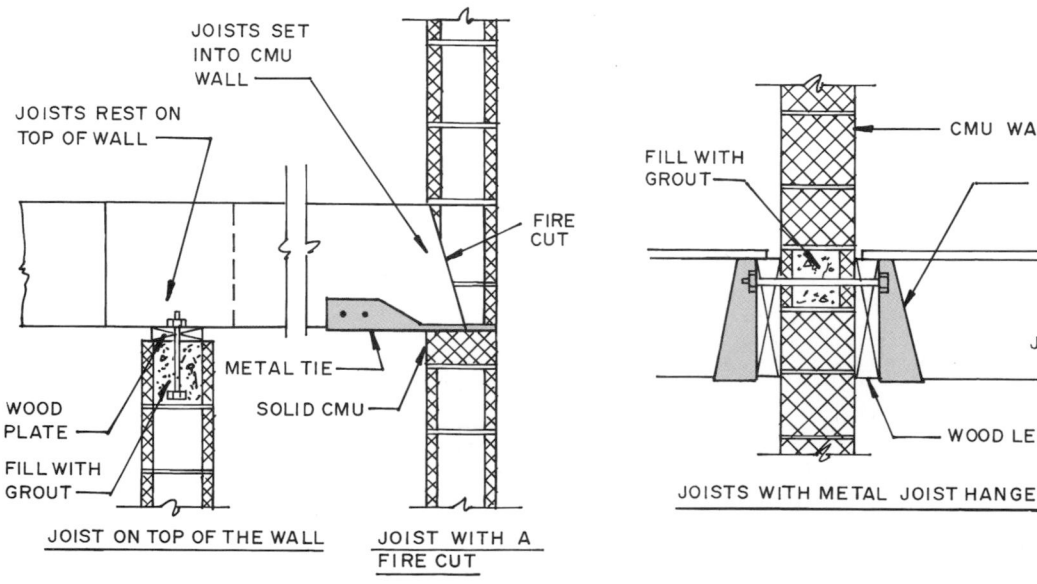

4-47 *Ways to support floor joists using concrete masonry foundations and walls.*

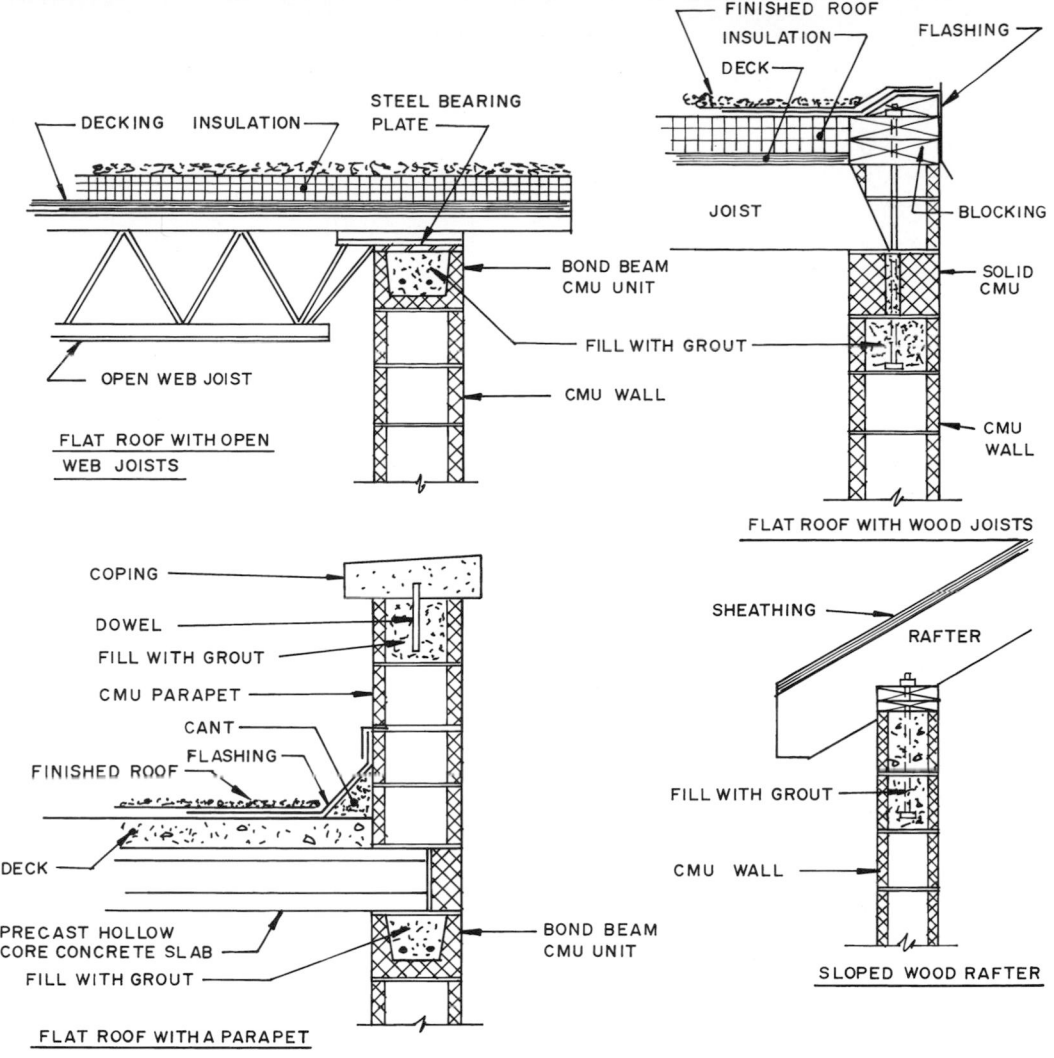

DECKING INSULATION STEEL BEARING PLATE

OPEN WEB JOIST

BOND BEAM CMU UNIT

FILL WITH GROUT

CMU WALL

FLAT ROOF WITH OPEN WEB JOISTS

FINISHED ROOF INSULATION
DECK FLASHING

JOIST

BLOCKING

SOLID CMU

CMU WALL

FLAT ROOF WITH WOOD JOISTS

COPING

DOWEL

FILL WITH GROUT

CMU PARAPET

CANT

FINISHED ROOF FLASHING

DECK

PRECAST HOLLOW CORE CONCRETE SLAB

FILL WITH GROUT

BOND BEAM CMU UNIT

FLAT ROOF WITH A PARAPET

SHEATHING RAFTER

FILL WITH GROUT

CMU WALL

SLOPED WOOD RAFTER

4-48 *Typical roof details for concrete masonry wall construction*

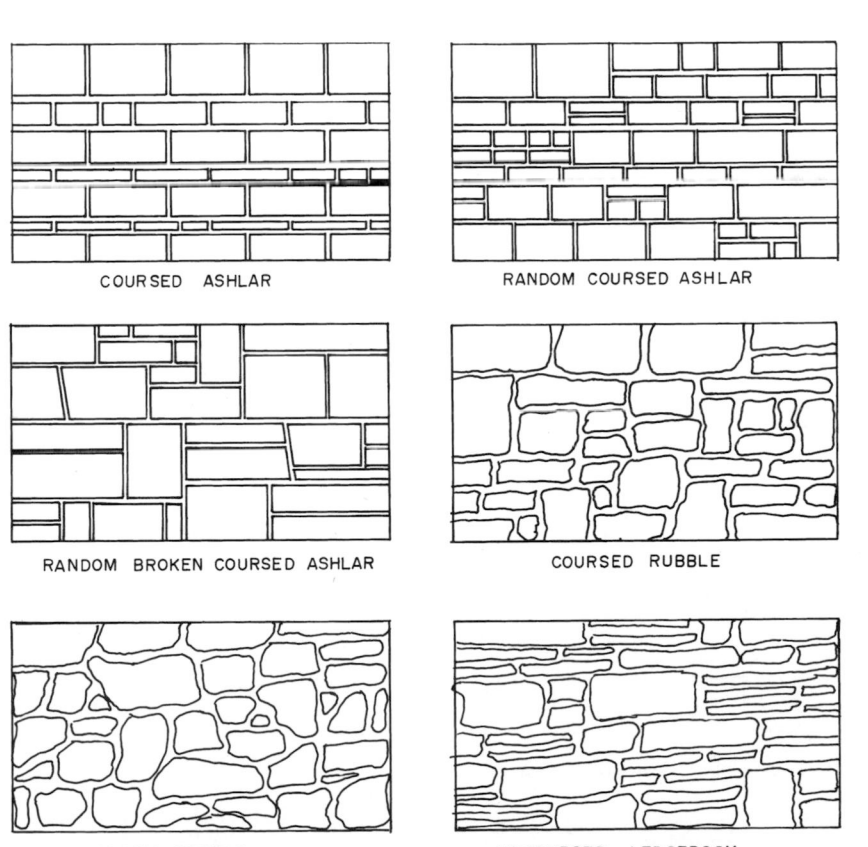

COURSED ASHLAR

RANDOM COURSED ASHLAR

RANDOM BROKEN COURSED ASHLAR

COURSED RUBBLE

RANDOM RUBBLE

UNCOURSED LEDGEROCK

4-49 *Patterns for laying ashlar and rubble stone.*

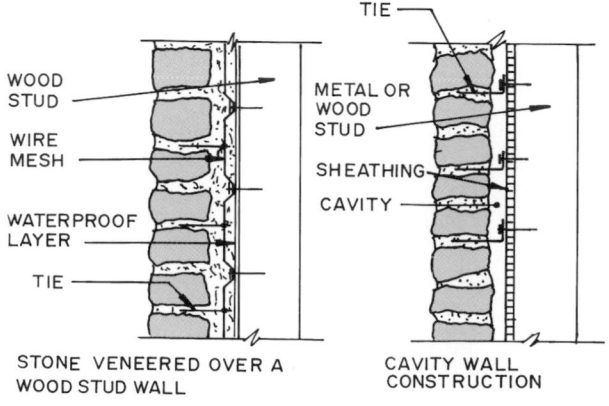

4-50 *Some typical stone veneer wall constructions*

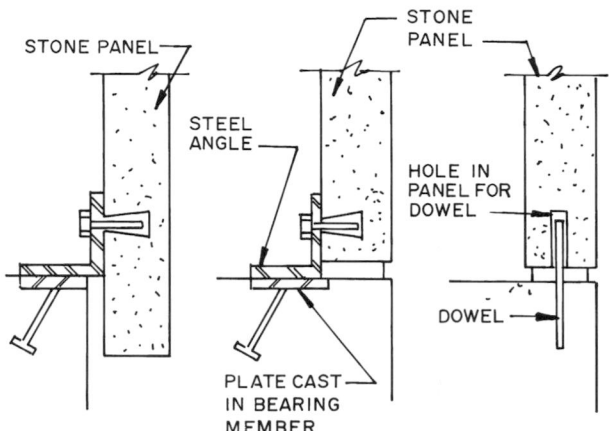

4-51 *Vertical stone panels may rest on the bearing member or overlap it.*

Rubble stone is difficult to lay because the mason must select pieces to fit the open space and work in with the shapes of adjoining stones. Sometimes it is necessary to trim the stone with a hammer. The mortar joints are irregular and often quite large. The squared ashlar stones are easier to lay than rubble but it is still necessary for the mason to select stones that will fit together as needed. Since the stones are squared the mortar joint can be of relatively uniform width over the entire wall. Joints in ashlar masonry are usually ⅜ to ¾ in. Large squared blocks are called **dimension stone** and usually require a hoist to set them in place. Mortar joints in ashlar masonry are usually raked after setting and allowed to thoroughly set. Then the raked space is filled with mortar (called pointing) and tooled to the specified shape. Great care is taken to avoid getting mortar on the face of the stone. The face is cleaned with a mild soap and soft brush and flushed clean with water.

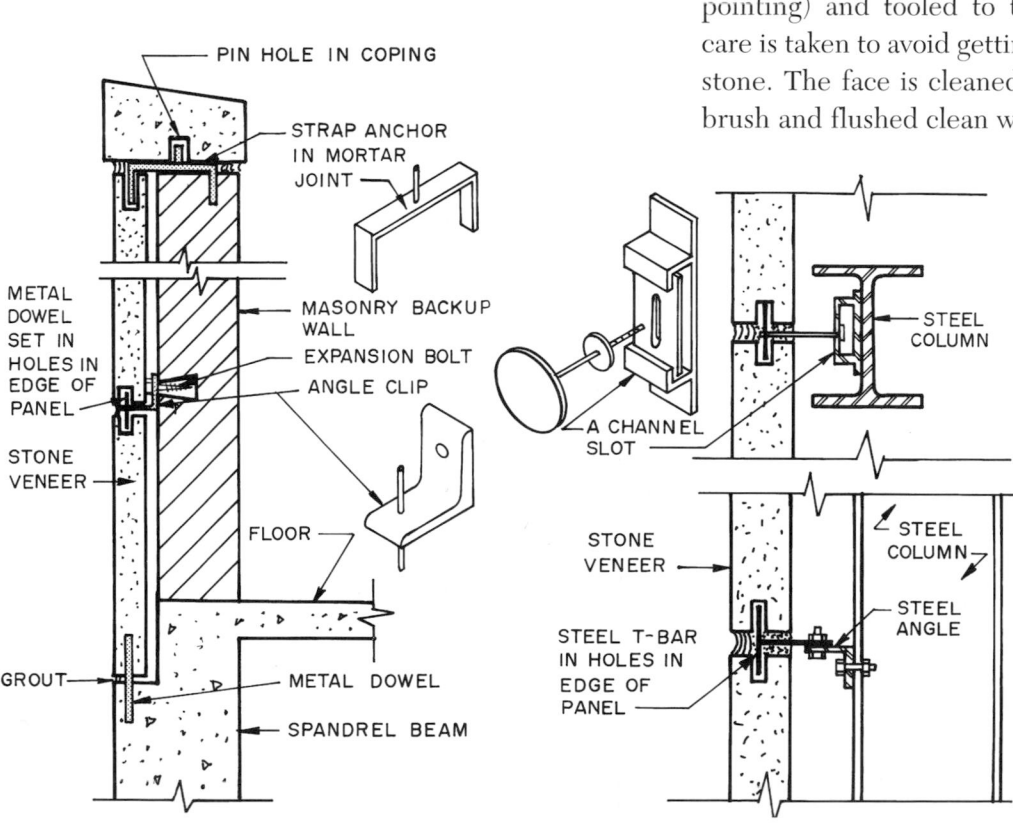

4-52 *Typical details for installing vertical stone panels over masonry, concrete, and steel framing.*

When stone is laid the grain in each piece should run in a horizontal direction. The stone is stronger in this position and tends to better resist weathering.

Stone masonry is usually a veneer over a backup wall. Concrete masonry units are frequently used. Some form of metal tie is used to bond the stone veneer to the backup wall much the same as is done with brick masonry construction.

Anchors should be chromium-nickle, stainless steel, or a zinc alloy. Building codes often ban the use of galvanized steel. Copper, brass, and bronze ties may cause some staining. Construction details for several types of stone masonry veneered wall construction are shown in 4-50. The manufacturers of various types of stone panels have engineered systems for securing the panels in place.

Some stone is cut into large thin panels and these are scored to accept some form of metal tie to secure it in place to the backup wall. A number of connection methods are used. The first veneer panel on the base can be set on the foundation or other supporting masonry or can overhang it and be secured with metal connectors as shown in 4-51. As the panels are placed up the wall, some form of metal connector is used to secure it to the backup wall. Connections to masonry walls and structural steel frames are shown in 4-52. Connections to a coping are also shown in this figure. Often columns are cased with stone veneer. Typically metal connectors tie the stone veneer to the column. Preassembled column covers are available and are bonded with high-strength epoxy adhesives.

Clay Structural Tile Construction

Structural clay tile may be load-bearing or nonload-bearing. Units that are load-bearing may form an unfaced wall or be used as a backup wall and faced with another material. Non-load-bearing tile walls are used as interior partitions, fireproofing barriers, and masonry screens.

Examples of commonly used structural tile construction are in 4-53.

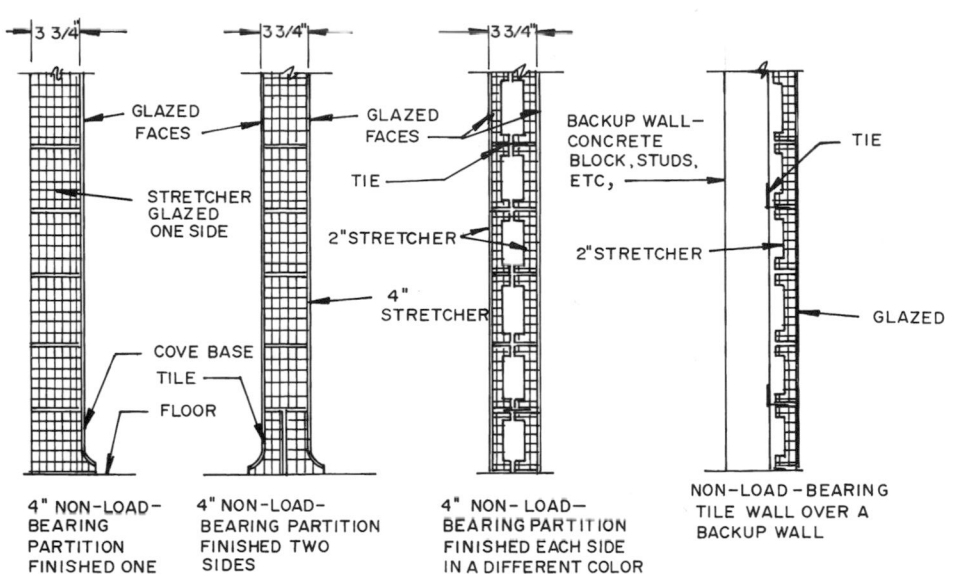

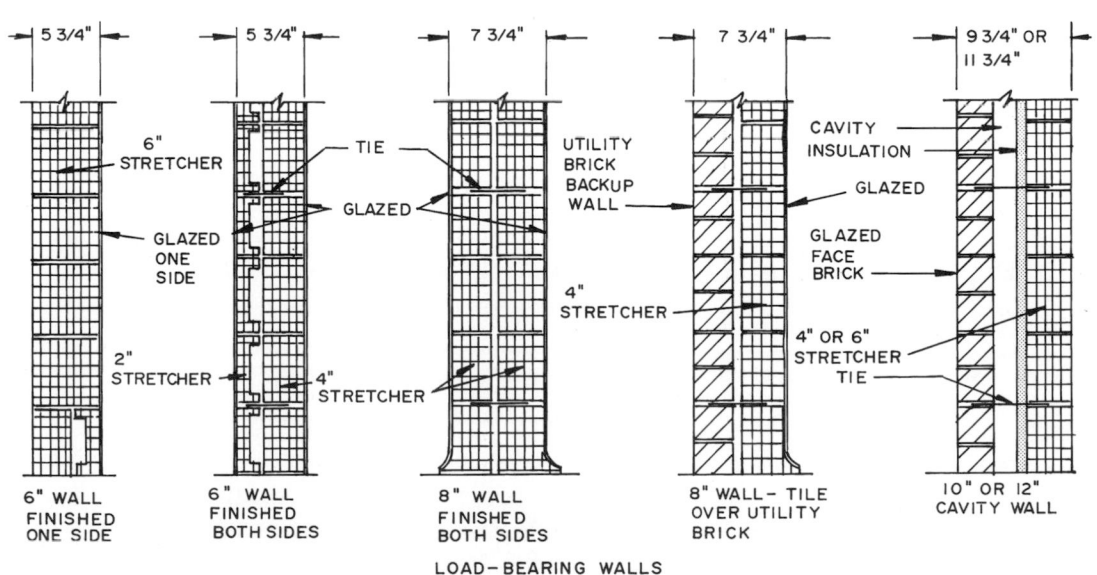

4-53 *Typical wall and partition construction using structural tile.*

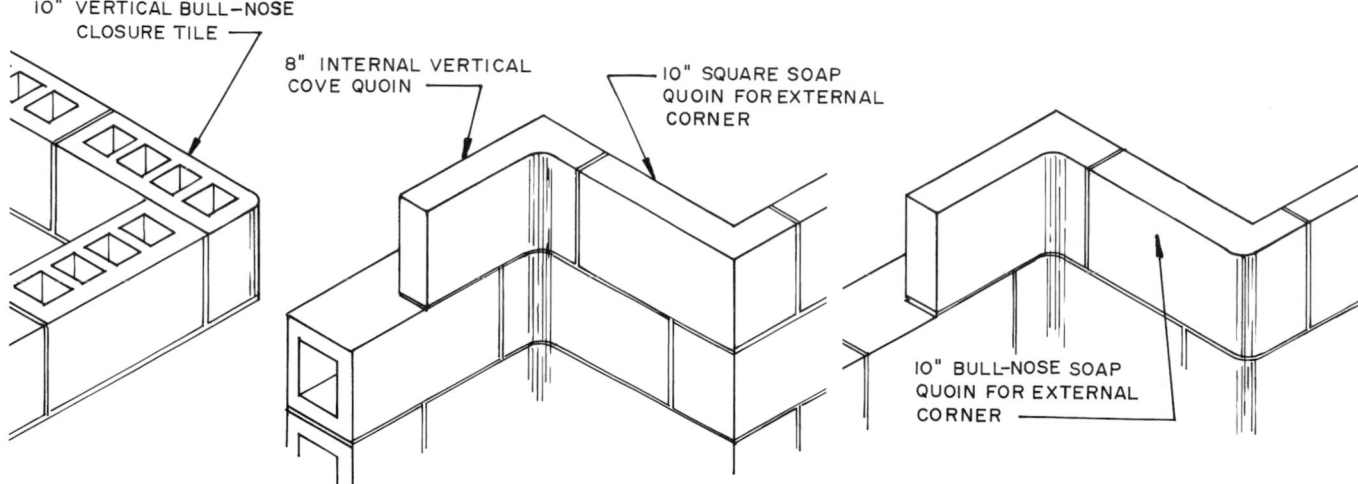

10" VERTICAL BULL-NOSE CLOSURE TILE

8" INTERNAL VERTICAL COVE QUOIN

10" SQUARE SOAP QUOIN FOR EXTERNAL CORNER

10" BULL-NOSE SOAP QUOIN FOR EXTERNAL CORNER

4-54 *Several ways to construct corners with structural tile.*

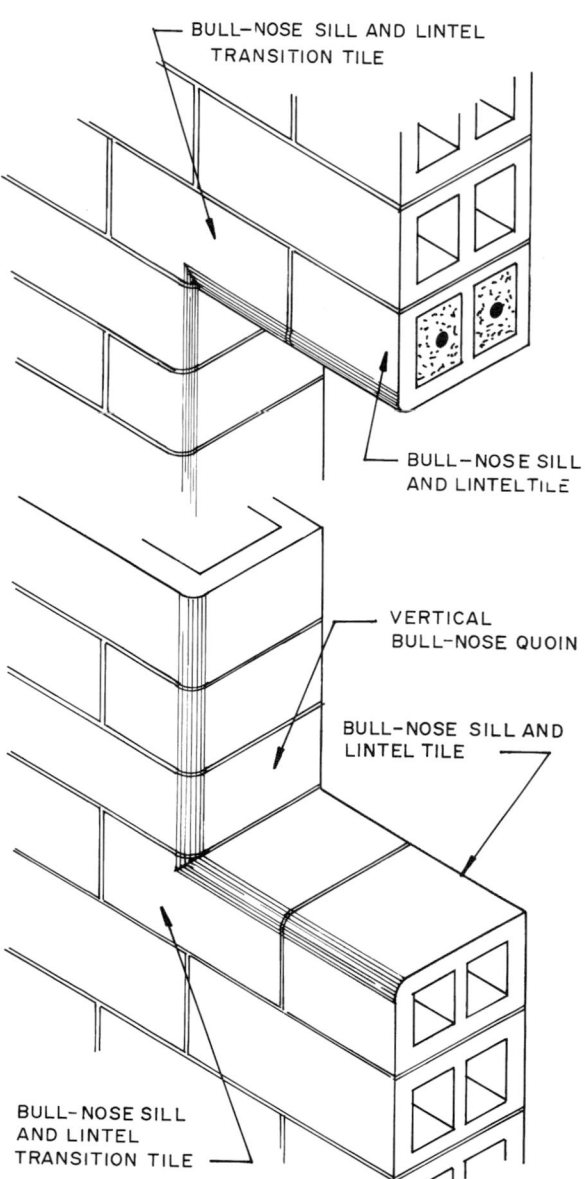

BULL-NOSE SILL AND LINTEL TRANSITION TILE

BULL-NOSE SILL AND LINTEL TILE

VERTICAL BULL-NOSE QUOIN

BULL-NOSE SILL AND LINTEL TILE

BULL-NOSE SILL AND LINTEL TRANSITION TILE

4-55 *One way to frame openings in structural tile walls.*

It should be noted there are many possible variations depending upon the expected end results, loads, and exposed face. Some tile units are utility block used to construct a backup wall, while others have glazed exposed faces available in a variety of colors and textures.

The cavity walls use flashing and weep holes as described for brick construction. They may be insulated with rigid plastic foam insulation. Floor joists and rafters also are installed much like those on brick and concrete masonry. Tile can be used as a veneer over a backup wall of a different material, such as a concrete masonry wall or a stud wall.

BUILDING STRUCTURAL CLAY MASONRY WALLS

As with brick and concrete, masonry structural clay tile walls should be designed to use standard size units to avoid having to cut them. The face size of standard stretcher blocks is $5\frac{1}{16}$ in. high and $11\frac{3}{4}$ in. long ($5\frac{1}{2} \times 12$ in. nominal) and $7\frac{3}{4}$ in. high and $15\frac{3}{4}$ in. long (8×16 in. nominal). The designer must have knowledge of these sizes and other special units available.

High-strength mortars are used with structural tile. The local codes and information from the tile manufacturer should be used when specifying the mortar to be used. The construction requires the use of metal ties as discussed for brick and concrete masonry. Expansion joints must be planned.

Corners can be built in several ways. Typical internal and external corner constructions are shown in 4-54. Sills on openings can be constructed using horizontal bullnose units as shown in 4-55.

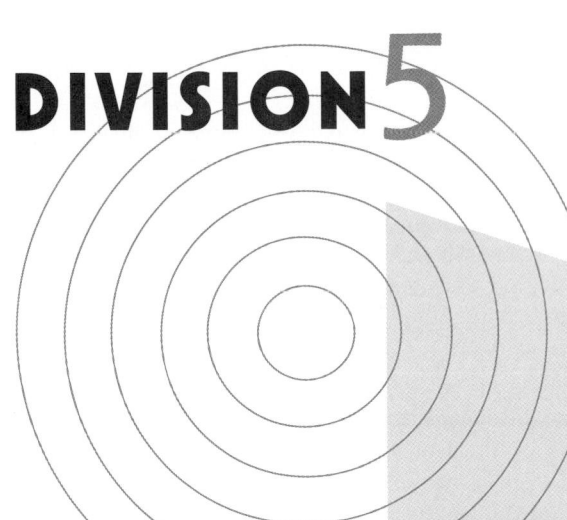

DIVISION 5

METALS
CSI MasterFormat™

05050 Basic Metal Materials & Methods

05100 Structural Metal Framing

05200 Metal Joists

05300 Metal Deck

05400 Cold-Formed Metal Framing

05500 Metal Fabrications

05600 Hydraulic Fabrications

05650 Railroad Track & Accessories

05700 Ornamental Metal

05800 Expansion Control

05900 Metal Restoration & Cleaning

Courtesy H.H. Robertson

FERROUS METALS

Ferrous metals are those in which the chief ingredient is the chemical element iron (ferrum). Iron (symbol Fe) is found in the earth's crust in large quantities mixed with other minerals. To be useful it must be extracted from the mined ore, have impurities removed and ingredients added to alter its properties, and then be formed into useful products.

Ferrous metal products are widely used in the construction industry. They are a major construction material and architects, engineers, and contractors should be familiar with the various types, their properties, and the proper applications for each. Although a ferrous metal product may fail, it generally is not because it is a poor material but because the choice of the type of ferrous metal for a particular application was incorrect. It is a material such that, when the properties are known, its performance can be accurately determined during the engineering design process.

Iron

Iron is found in large quantities in the earth's crust. Pure iron, free from impurities and other elements, is ductile, soft, but not strong enough for structural purposes. It has high magnetic properties, oxidizes (rusts) easily, and does not resist attack by acids and some chemicals. For commercial purposes iron must have alloying elements added to improve its characteristics. Iron can be hardened by heating and then cooling rapidly and can be made more workable by heating and allowing it to cool slowly.

The typical commercial iron product produced is called pig iron. It will contain 3 to 5 percent carbon and traces of other elements such as manganese, sulfur, silicon, and phosphorus. Pig iron is the base material used to produce various types of iron and steel.

Iron is used for a wide range of purposes in construction. For example, iron particles may be used as the abrasive for sandblasting and the main ingredient in cast iron, steel, stainless steel, and iron alloys. It finds some use as an aggregate for specialized concretes, and as the basis for some color pigments.

MINING & PROCESSING IRON ORE

Iron is found in rock, gravel, sand, clay, and mud. It is mined in open pit and underground mines. Common iron-bearing minerals are pyrite (FeS), siderite

($FeCo_3$), and hematote (Fe_2O_3). These contain up to 70 percent iron. Jasper and taconite rock contain 20 to 30 percent iron.

The surface ores are blasted and loaded on to trucks, trains, and ships and moved to blast furnaces. If the iron content of the ore is less than about 50 percent it is too costly to ship any distance. These lower content ores are processed at the site by a **beneficiation**

5-1 *This very large blast furnace is capable of producing 8,000 tons of molten iron per day.*

Courtesy Bethlehem Steel Corporation

process, which removes some of the unwanted elements leaving an ore with a higher iron content. **Beneficiation** involves grinding the ore to remove the unwanted elements and then increasing the size of the ground particles by a process called agglomeration. **Agglomeration** involves pelletizing the ore particles into ball-like pellets, which are easier to ship than the finely ground ore. Iron ore dust produced is recycled by sintering. **Sintering** involves fusing the iron-ore dust with coke and fluxes into a clinker, which is high in iron content.

Ores that have a very high iron content are made into pellets and briquetts containing over 90 percent iron. These pellets are so pure they are not used in producing pig iron but go directly to the steel-making processes. The processes of producing these pure pellets is called **direct reduction.**

PRODUCING THE IRON

The iron ore is converted into pig iron in a blast furnace by smelting and reduction.

Smelting is a process in which the ore is heated permitting the iron to be separated from impurities, which may be chemically or physically mixed. **Reduction** is a process that separates the iron from oxygen with which it is chemically mixed.

BLAST FURNACE FUELS

Coal, oil, and natural gas are the commonly used fuels in a blast furnace. Most operate on **coke,** which is produced from coal. The coke not only provides heat to melt the ore but helps separate the iron from its oxides.

FLUXES

A flux is a mineral added to the molten ore in the blast furnace, which combines with impurities and forms a **slag** that floats on top of the molten material. Limestone and dolomite are typical basic fluxes. Sand, gravel, and quartz are acid fluxes.

THE BLAST FURNACE

The blast furnace is used to process iron ore into iron. It separates the iron from the waste materials and sinters the ore and flue dust. A very large blast furnace is shown in 5-1. It operates 24 hours a day and is computer controlled and fed by a continuous conveyor.

The heart of the blast furnace is a tall, cylindrical shaft 150–200 feet (45.8 to 61m) tall and about 30 feet (9.2m) in diameter at the base and is usually smaller in diameter at the top. It is lined with a refractory brick made of a material, such as magnesia, that will withstand temperatures approaching 3000°F (1662°C) (See 5-2).

A charge of ore, hot coke (fuel), and limestone (flux) are loaded at the top. A blast of hot air is injected at the bottom. As the air works its way up through the charge the oxygen in the hot air combines with the hot coke causing combustion, which produces the

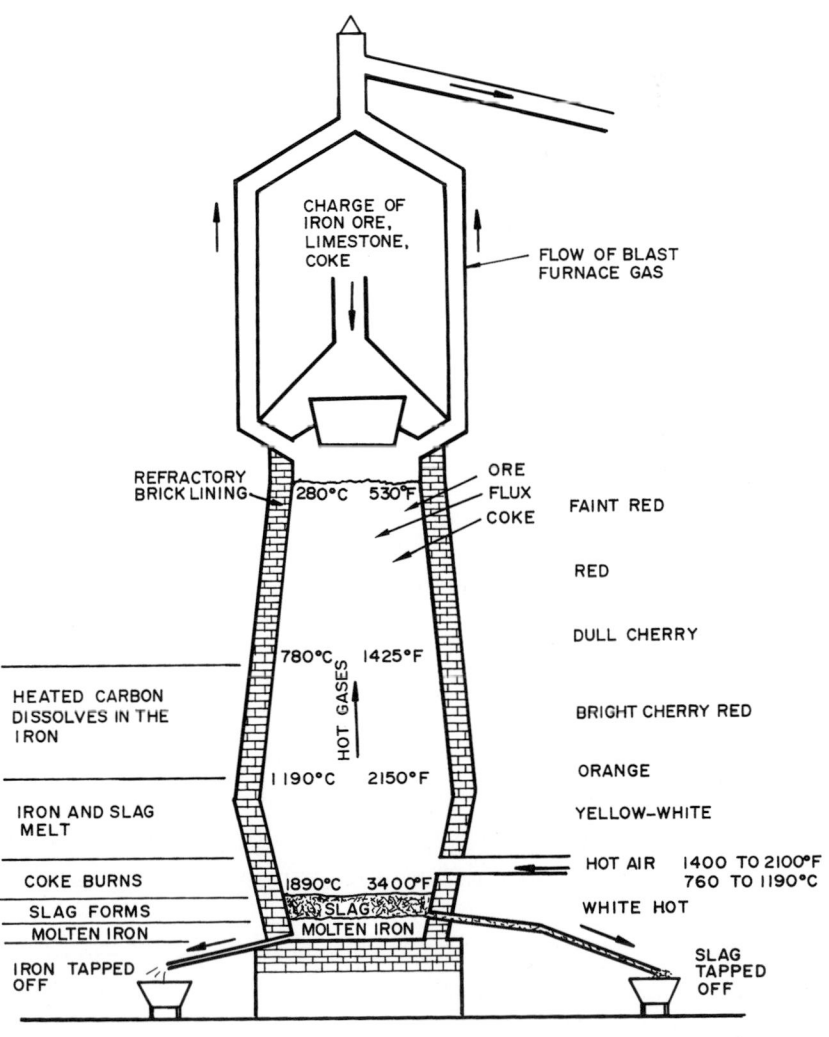

5-2 A blast furnace receives the charge at the top and hot air at the bottom causing combustion, which frees the iron from the oxide-forming pig iron.

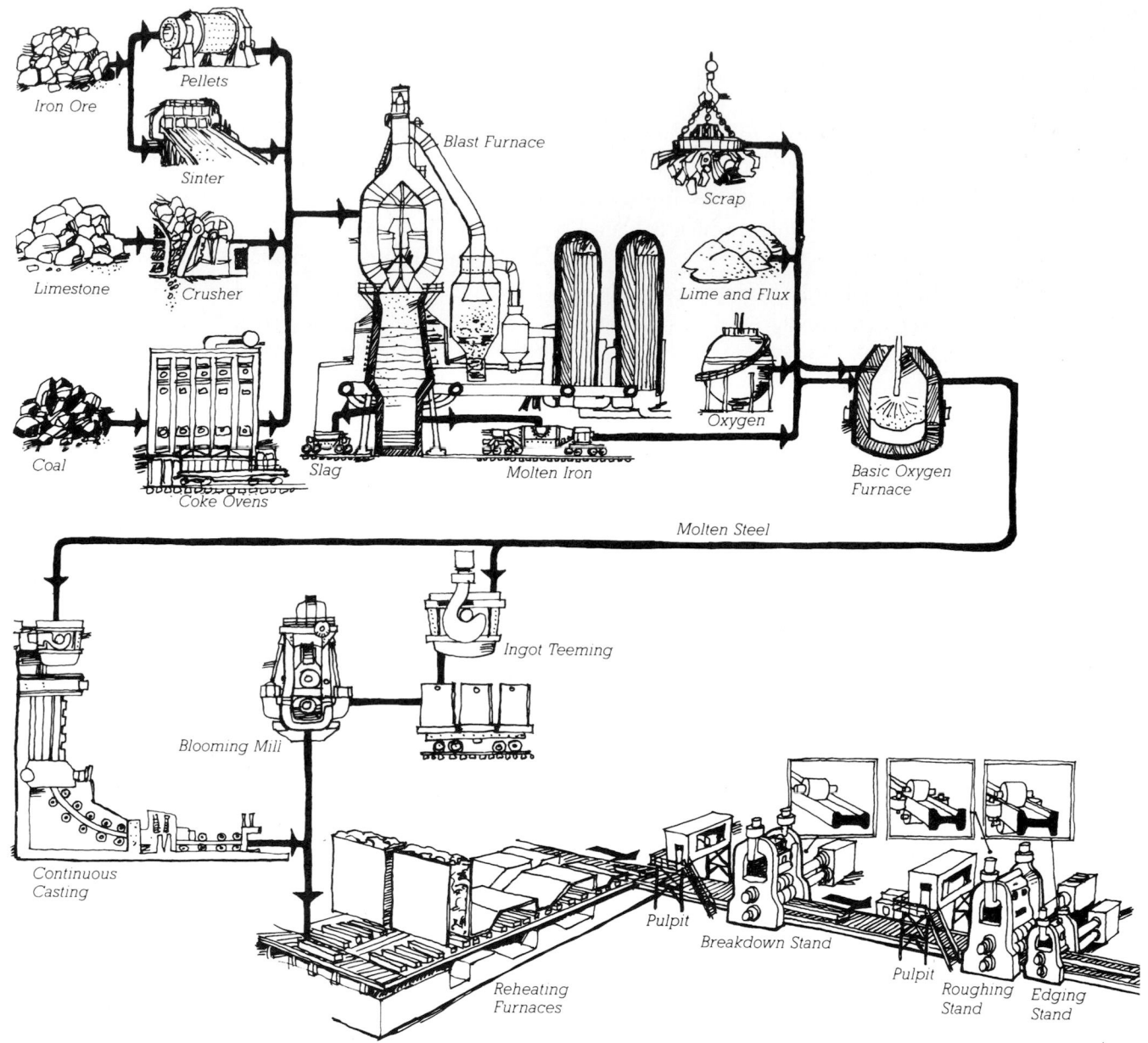

5-3 *The steelmaking process from preparation of the ore through processing into steel to the production of steel beams.*
From *Steelmaking Flowline* by permission of the American Iron and Steel Institute

heat necessary to melt the ore. As the gases from combustion pass through the heated ore a chemical reaction occurs, which frees the iron from its oxide. The gases are removed from the top of the furnace and pass through cleaners that retain the dust. The cleaned heated air is passed through brick heaters that heat the new incoming air. A blast furnace must be operated continuously.

The molted pig iron settles to the bottom of the furnace and is tapped off every few hours. It is at this point impure and brittle. It contains about 4 to 5 percent carbon. This high carbon content is what makes pig iron brittle and not as useful as steel. It has no ductility and cannot be rolled into useful products such as beams or studs. It must have additional processing in a steel-making furnace to reduce the carbon content. The impurities, flux, and oxygen combine to form a slag on top of the molten iron. The slag is tapped off every few hours. Slag is used as an aggregate for concrete, loose fill, and road surfacing.

The molten pig iron is moved to furnaces called mixers. Here it is mixed to equalize the chemical composition and kept liquid until it is to be used. The molten iron sometimes goes into a casting machine where it is cast into ingots (pigs) and is allowed to cool. In other operations it is moved directly to basic oxygen steelmaking furnaces. In this case energy is saved because the iron is already in molten form. You can see the total process in 5-3.

The pig iron ingots are usually made to specific specifications depending upon their end use. The elements that are carefully controlled include carbon, sulfur, silicon, phosphorus, and manganese.

Cast Irons

Iron containing almost no carbon is identified as a **wrought iron.** It is a mixture of low carbon iron and a large amount of slag. It is soft, tough, and ductile and easily worked. **Ingot** iron is a very low carbon iron that has no slab and is also tough, ductile, and soft.

Cast irons have carbon contents above 1.7 percent and include white, gray, and malleable types.

White cast irons have a low silicon content and are cooled rapidly. They are hard, brittle, and have few applications for construction uses. **Malleable cast iron** is produced by reheating white cast iron and holding the required temperature for a long time and then cooling it slowly.

It is not as brittle as white and gray cast iron because it has a lower carbon content and has greater ductility. It is used for hardware and other cast items requiring toughness and resistance to breakage.

Gray cast irons are produced by increasing the silicon content and cooling the molten metal slowly. It is tougher and softer than white cast iron and is gray in color. It may have other elements, such as nickel, copper and chromium, added to it. Gray cast irons are widely used for all types of castings such as sewer pipe, ornamental railings, and decorative lamp posts.

Steelmaking

The production of steel involves (1) controlling impurities in the pig iron, (2) adding alloying elements as needed to get the required properties, and (3) lowering the carbon content of the pig iron. This involves combining molten pig iron, scrap metal, and fluxes in a steelmaking furnace from which the molten steel is cast into ingots or from a strand casting machine. Steel is produced using the basic oxygen process or with an electric furnace.

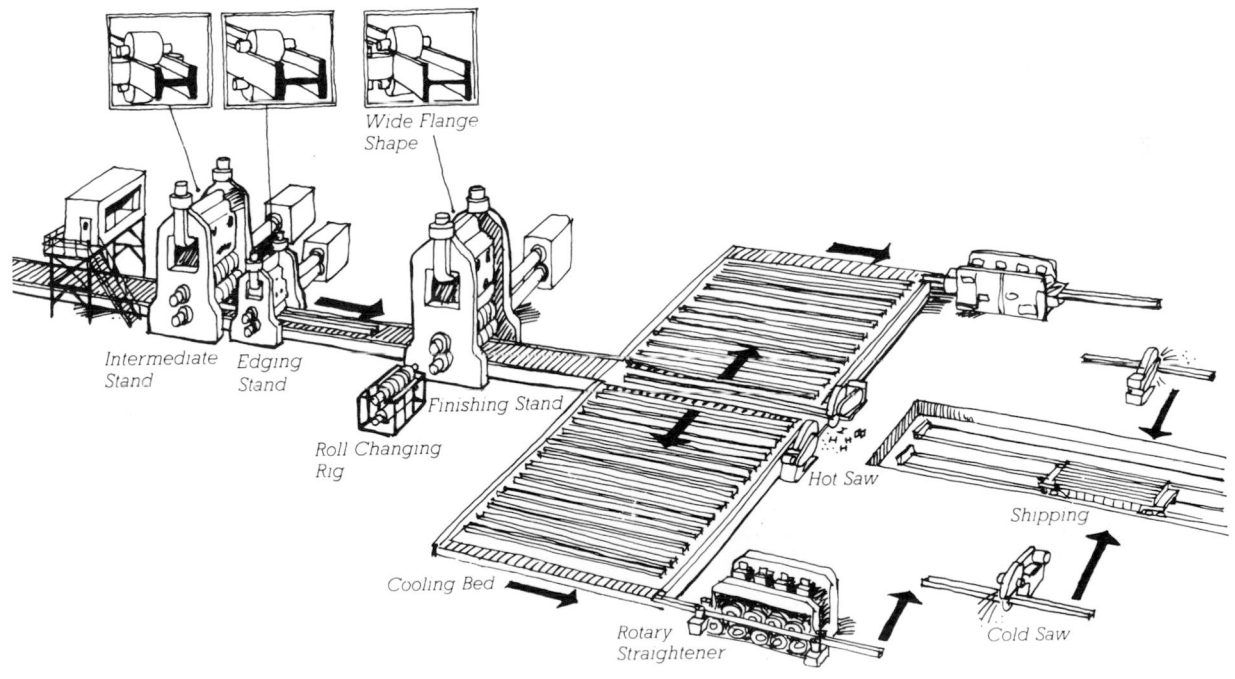

Wide Flange Shape

Intermediate Stand   Edging Stand

Roll Changing Rig

Finishing Stand

Cooling Bed

Rotary Straightener

Hot Saw

Shipping

Cold Saw

5-3 *(Continued)*

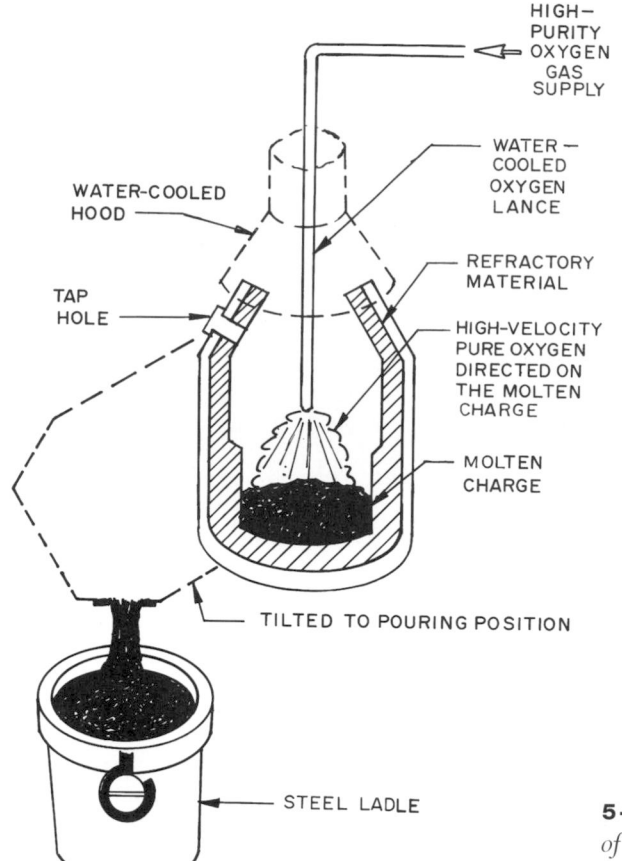

HIGH–
PURITY
OXYGEN
GAS
SUPPLY

WATER–
COOLED
OXYGEN
LANCE

REFRACTORY
MATERIAL

HIGH-VELOCITY
PURE OXYGEN
DIRECTED ON
THE MOLTEN
CHARGE

MOLTEN
CHARGE

WATER-COOLED
HOOD

TAP
HOLE

TILTED TO POURING POSITION

STEEL LADLE

5-4 *The basic oxygen process uses a jet of high purity oxygen plus the heat of the molten pig iron to start the process.*

5-5 *This is a basic oxygen steelmaking furnace. Notice its size compared to the man in the picture. It is operating behind closed pollution control doors, which allow dust and gases to be collected by the pollution control system. An interior view showing a pour is in 5-6.*

Courtesy Bethlehem Steel Corporation

5-6 *Internal view of a basic oxygen furnace during a pour.*

Courtesy Bethlehem Steel Corporation

BASIC OXYGEN PROCESS

This furnace has a large pear-shaped vessel lined with refractory material as you can see in 5-4. The charge consisting of molten pig iron, metal, scraps, and fluxes is added at the top. A jet of high-purity oxygen is shot into the vessel through a water-cooled lance. The heat of the molten pig iron is great enough to start a chemical reaction among the oxygen, carbon, and other impurities. This oxidation produces the heat necessary to melt the charge without additional heat. The slag and molten steel are tapped off. A modern basic oxygen furnace can be seen in 5-5. It operates behind huge pollution control doors which remain closed during the steelmaking process. This enables dust and gases generated during the process to be collected by the pollution control system. A pour of the molten steel can be seen in 5-6.

ELECTRIC STEELMAKING PROCESSES

High-grade steels, such as stainless, tool, heat resisting, and alloy, are generally produced in electric furnaces. These furnaces can develop the high temperatures needed to produce the reducing conditions required to produce these metals. These furnaces use arc radiation and electric resistance to current flow to produce the high temperatures needed. High-purity oxygen is injected when needed, which oxidizes impurities. Since it can be carefully controlled there is less loss of the alloying elements that have been added.

The furnaces are charged with iron, scrap metal, fluxes, and alloying elements.

The two types of electric furnaces are the electric arc furnace and the induction furnace. The electric arc furnace can produce steels in large quantities while the induction furnace is used to produce smaller quantities of special grades of steel that require the use of expensive alloying elements.

THE ELECTRIC ARC FURNACE

The electric arc furnace has a circular steel shell somewhat like a large cooking pot. Several cylindrical electrodes project into the furnace from the top. A high voltage electric current is passed through the electrodes causing an arc to pass between them. This generates the heat needed to melt the charge. The furnace is tilted and the steel is poured off under the slag. The electric arc furnace is shown in operation in the photo in 5-7.

THE INDUCTION FURNACE

This furnace has a cylindrical vessel made of magnesia and is insulated with refractory materials. Outside of this are windings of copper tubing through which high voltage alternating current is passed. This creates an induction current, which is resisted by the charge. This resistance generates the heat needed for melting the charge. The vessel is tilted and the steel is poured off below the slag.

5-7 *An electric arc steelmaking furnace.*
Courtesy Bethlehem Steel Corporation

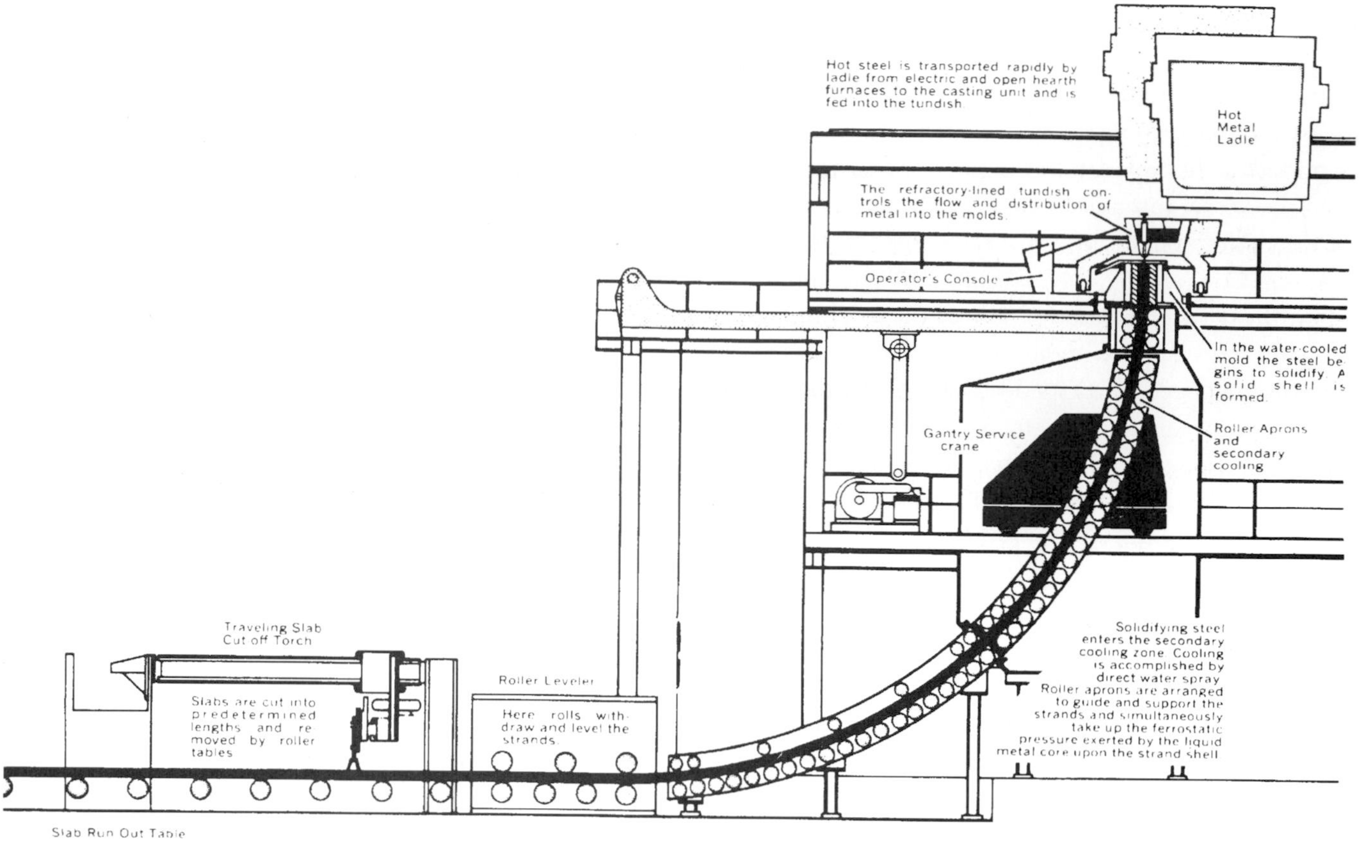

Hot steel is transported rapidly by ladle from electric and open hearth furnaces to the casting unit and is fed into the tundish.

Hot Metal Ladle

The refractory-lined tundish controls the flow and distribution of metal into the molds.

Operator's Console

In the water-cooled mold the steel begins to solidify. A solid shell is formed.

Gantry Service crane

Roller Aprons and secondary cooling

Traveling Slab Cut off Torch

Slabs are cut into predetermined lengths and removed by roller tables

Roller Leveler

Here rolls withdraw and level the strands.

Solidifying steel enters the secondary cooling zone. Cooling is accomplished by direct water spray. Roller aprons are arranged to guide and support the strands and simultaneously take up the ferrostatic pressure exerted by the liquid metal core upon the strand shell.

Slab Run Out Table

5-8 *The molten steel can be cast into a continuous slab by a strand-casting machine.*

Courtesy American Iron and Steel Institute

5-9 *This is the steel strand produced by the continuous casting machine. The slab is automatically cut into predetermined length by an automatic torch machine.*

Courtesy Bethlehem Steel Corporation

When the steel is ready to be poured from the furnace it is either cast into **ingot molds** or run through a **strand-casting machine.**

Ingot molds can be several feet in diameter and six to eight feet high. They are coated inside to prevent the steel from damaging the surface. After the ingot has begun to harden it is removed from the mold and taken to a soaking pit. Here the inside, which is still molten, solidifies.

Strand-casting machines produce a continuous ribbon of steel, which begins to harden as it passes through a series of rollers. When it reaches a horizontal conveyor it has hardened and is cut to the required lengths. This process takes about 30 minutes (5-8). The steel strand can be seen coming from the casting machine in 5-9.

Manufacture of Steel Products

Steel products are manufactured by rolling, extruding, cold drawing, forging, and casting. Those items produced by casting are referred to as cast-steel products and all of the others are called **wrought** products. The most frequently used products are produced by hot rolling and cold finishing.

When the steel is produced in the form of slabs, billets, or blooms it is called semifinished. When **semi-finished** steel is fabricated into a product that is used to produce a finished product or fabricated directly to produce a finished product it is **finished steel.**

Cast-steel products are made by pouring molten steel into sand molds. The steel used will have gone through a process to alter its properties to suit the purpose served by the finished product.

Most wrought products are produced by hot and cold rolling. The ingots produced in the steel mill are used to produce slabs, blooms and billets. **Slabs** are large rectangular semifinished steel pieces and are made in a number of sizes. Their width is typically more than twice the thickness. The thickness is usually around 10 in. (254mm) or less. **Blooms** are square shapes of semifinished steel that are usually larger than 6 in. (152mm) square. **Billets** are also square shapes of semifinished steel but smaller and longer than blooms.

All types of structural shapes, rods, rails, strip, and seamless pipe are manufactured using blooms. Finished structural beams are shown in 5-10 where one has just left the hot rolling process and will be moved to the cooling beds on the side.

Billets are used to produce solid steel bars and rounds, wire rods and wire, seamless pipe, and cold drawn bars.

Steel Identification Systems

There are a number of nationally used metal and alloy numbering systems administered by societies, trade associations, and individual users and producers of metals and alloys. Among these are numbering systems by the American Society for Testing and Materials (ASTM), SAE/Aerospace Materials Specifications

5-10 *These structural steel beams have been roll formed hot and are being transferred to a "walking" cooling bed before they are cut to the desired lengths. Notice the beam in the center is just out of the machine and is very hot.*
Courtesy Bethlehem Steel Corporation

Table 5-1 IDENTIFICATION NUMBER, TYPE & YIELD POINT OF SELECTED STRUCTURAL STEELS

Unified Number	ASTM Number	Type	Yield Point, ksi[a]
K02600	A36	Carbon steel	36
K11510	A242	High-strength low-alloy, corrosion resistance 5 to 8 times carbon steel	42–50
K11630	A514	High-yield-strength, quenched and tempered alloy steel	90–100
K02703	A529	Carbon steel	42
K02303	A572	High-strength low-alloy, niobium-vanadium, structural quality, corrosion resistance four times carbon steel	42–65
K11430	A588	High-strength low-alloy, corrosion resistance four times carbon steel	42–50
K01803	A633	Normalized high-strength low-alloy steel	42–60
K01600	A678	Quenched and tempered steel	50–75
K12043	A852	High-strength quenched and tempered alloy steel	70

[a]Kips per square inch (ksi). One ksi equals 1000 lb. per square inch.

(AMS), American Welding Society (AWS), American Iron and Steel Institute (AISI), Society of Automotive Engineers (SAE), Federal Specifications, Military Specifications (MIL) and the American Society of Mechanical Engineers (ASME). Following are brief descriptions of several of these systems.

The American Iron and Steel Institute (AISI) and the Society of Automotive Engineers (SAE) series of identifying numbers for carbon and alloy steels uses a four digit number and in some cases five digits are used. The first two digits indicate the type of steel (carbon or alloy) and the last two digits indicate the carbon content. For example in AISI 1030 the 10 indicates a carbon content of about 0.30 (actual range is 0.28 to 0.34). An AISI 4012 indicates a molybdenum steel alloy with 0.15 to 0.25 molybdenum. It also has 0.09–0.14 carbon, 0.75–1.00 manganese, 0.035 phosphorus maximum, 0.040 sulfur maximum, and 0.15–0.35 silicon.

The American Society for Testing and Materials (ASTM) sets standards for steel designations using an arbitrary number to indicate chemical compositions and to specify minimums for strength and ductility. These specifications regulate specific chemical elements that directly affect the fabrication and erection of the steel. This makes it possible for various proprietary steels of different chemical compositions to conform to ASTM performance standards. Structural steels used in construction are designated by these designations. Typical steels used for construction purposes are identified in Table 5-1. Detailed information on these and other systems is available from the sponsoring organization.

One system, the Unified Numbering System for Metals and Alloys (UNS), was developed jointly by the Society of Automotive Engineers, Inc., and the American Society for Testing and Materials. It provides a means for describing the composition of several thousand metals designations and cross-references those of the organizations mentioned earlier in this chapter.

UNIFIED NUMBERING SYSTEM
FOR METALS AND ALLOYS (UNS)

The Unified Numbering System (UNS) provides the uniformity required for indexing, record keeping, data storage and retrieval, and cross-referencing. The UNS is not a specification but is used as an identifier of a metal or alloy for which controlling limits have been established in specifications published elsewhere.

Table 5-2 THE UNIFIED NUMBERING SYSTEM (UNS) FOR METALS AND ALLOYS

Series Designation	Family of Metals
Axxxxx	Aluminum & aluminum alloys
Cxxxxx	Copper & copper alloys
Exxxxx	Rare earth & similar metals & alloys
Fxxxxx	Cast irons
Gxxxxx	AISI & SAE carbon & alloy steels
Hxxxxx	AISI & SAE H-steels
Jxxxxx	Cast steels (except tool steels)
Kxxxxx	Miscellaneous steels & ferrous alloys
Lxxxxx	Low melting metals & alloys
Mxxxxx	Misc. nonferrous metals & alloys
Nxxxxx	Nickel & nickel alloys
Pxxxxx	Precious metals & alloys
Rxxxxx	Reactive & refractory metals & alloys
Sxxxxx	Heat & corrosion-resistant steels (including stainless), valve steels, & iron-base "superalloys"
Txxxxx	Tool steels, wrought & cast
Wxxxxx	Welding filler metals
Zxxxxx	Zinc & zinc alloys

Courtesy Society of Automotive Engineers

Table 5-3 NUMBERING SYSTEMS CROSS-REFERENCED TO UNS

Cross-Reference Prefix	Specifying Organization
AA	(Aluminum Association) numbers
ACI	(Steel Founders Society of America) numbers
AISI	(American Iron and Steel Institute) including SAE (Society of Automotive Engineers) numbers (carbon and low-alloy steels)
AMS	(SAE/Aerospace Materials Specification) numbers
ASME	(American Society of Mechanical Engineers) numbers
ASTM	(American Society for Testing and Materials) numbers
AWS	(American Welding Society) numbers Federal Specification Numbers
MIL	(Military Specification) numbers
SAE	(Society of Automotive Engineers) "J" numbers

Courtesy Society of Automotive Engineers

The UNS has 18 series of designations for metals and alloys. Each UNS designation has a single-letter prefix followed by five digits. The letter is used to identify the family of metals, such as S for stainless steels and A for aluminum. The prefixes can be seen in Table 5-2.

The significance of the digits can vary with the UNS series. Therefore their meaning for each series can serve a different purpose. Effort has been made to relate UNS digits to those established by related trade organizations. In the published UNS manual, *Metals and Alloys in the Unified Numbering System,* the UNS numbers are cross indexed with the systems of other organizations whose documents describe materials that are the same as or similar to those covered by the UNS numbers. These are identified by the letters for each organization shown in Table 5-3.

Steel & Steel Alloys

Steel is a term generally applied to plain carbon steels that are alloys of iron and carbon with a carbon content below 2 percent. Other steel products include stainless and alloy steels.

PLAIN CARBON STEELS

Plain carbon steels have iron as the major element (over 95 percent) but will have impurities present such as sulfur, nitrogen, and oxygen. Other elements may be present as residual impurities or by being added to change the properties of the steel. These could include phosphorus, nickel, aluminum, copper, silicon, and manganese.

The properties of carbon steel are varied not only by the elements that are added to alter the chemical composition but by the type of mechanical and heat treatment used as it is produced. For example, during production carbon steel could be hot or cold rolled, cast, cooled slowly or cooled rapidly. All of these will influence the final properties.

Control of the **carbon content** is the major factor in establishing the properties of carbon steel. As the carbon content increases the strength and hardness increases and ductility decreases.

The differences in the amount of carbon between the various types of steel is very small, usually in terms of hundredths of a percent. The percentage of carbon for several types of carbon steel are shown in Table 5-4.

Carbon steels also will contain other elements such as manganese, phosphorus, and sulfur. The amount varies with the chemical design of the metal. For example AISI/SAE 1008 carbon steel will contain 0.10% maximum carbon, 0.30% to 0.50% manganese, 0.040% maximum phosphorus, and 0.050% maximum sulfur.

Carbon steels are used for many products used in construction, including structural shapes, bars, sheet and strip products, plate, pipe, tubing, wire, nails, rivets, and screws. It is also used to produce cast products which are typically from a medium grade carbon steel. The castings are often heat treated to relieve internal strain developed during the casting process.

ALLOY STEELS

An alloy steel contains one or more alloying elements other than carbon (such as chromium, nickel, molybdenum), which have been added in an amount exceeding a specified minimum that produce properties not obtained in carbon steels. These elements give particular physical, mechanical, and chemical properties to the steel. Stainless steels, specialty steels, and tool steels are not considered alloy steels although they do contain alloying elements.

Alloy steels are used where their special properties are required. For example, a nickel steel has great toughness and is used on bits for rock drilling machines and air hammers. Silicon gives steel good magnetic permeability and is used in making transformers, motors, and generators.

Standard alloy steels generally have elements of carbon, manganese, and silicon occurring naturally. Alloy steel that has one additional alloying element is identified as a **single alloy steel** while two alloying elements produce a **double** or **binary alloy steel** and three elements a **triple** or **ternary steel.** The various alloying elements produce changes in the properties of the steel. The major alloying elements and the property changes they produce can be found in Table 5-5.

As can be seen from reviewing Table 5-5 the alloying elements are used to improve properties such as hardness, performance of the material at high and low temperatures, strength, ductility, workability, wear resistance, electromagnetic properties, and electrical

Table 5-4 CARBON CONTENT OF CARBON STEELS

Type of Steel	Percent Carbon Content	Characteristics
Extra soft grade	0.05–0.15	Ductile, soft, tough Used for wire, rivets, pipe, sheets
Mild structural grade	0.15–0.25	Ductile, strong Used for boilers, bridges, buildings
Medium grade	0.25–0.35	Harder than mild structural Used for machinery & general structural purposes
Medium hard grade	0.35–0.65	Harder than medium grade Resists abrasion & wear
Spring grade	0.85–1.05	Strong & hard Used to make springs
Tool steel	1.05–1.20	Hardest & strongest of carbon steel

resistance or conductivity. The architect and structural engineer are especially involved with those alloys having increased strength, wear resistance, and resistance to corrosion, expansion, contraction, and ductility.

There are many alloy steels specified by the various organizations mentioned earlier. The uses range widely including such types as alloy steel electrode, alloy steel welding wire, alloy steel, high-strength low-alloy steel, and others. Some of these have a direct application to products used in construction. Examples include high silicon content, which improves magnetic permeability making it useful in transformers, motors, and generators. Nickel improves toughness and nickel alloys are used on the cutting tools used for rock drilling and air hammers. Since alloy steels are more expensive than carbon steels they typically are not used for structural members unless special requirements exist such as functioning in areas with high temperatures or very high strengths are required. They are also used in the manufacture of power tools and heavy construction equipment.

Some common uses for carbon and alloy steels are shown in Table 5-6.

Table 5-5 PROPERTIES IMPARTED TO STEEL ALLOYS BY ALLOYING ELEMENTS

Element	Properties
Aluminum	A deoxidizer used to control the grain size within the structure of the steel, promotes surface hardening
Boron	Increases the depth of hardness
Carbon	Increases hardness, strength High percentages contribute to brittleness
Chromium	Increases hardness, corrosion, and wear resistance
Cobalt	Hardens or strengthens the ferrite, resists softening at high temperatures Used in high-speed tool steels
Copper	Increases resistance to atmospheric corrosion and increases yield strength
Manganese	Increases strength and resistance to wear and abrasion
Molybdenum	Increases corrosion resistance, raises tensile strength and elastic limit, reduces creep, improves impact resistance
Nickel	Increases elastic limit and internal strength Increases strength and toughness in heat-treated steel In some steels increases hardness, fatigue, and corrosion resistance
Niobium	Retards softening during tempering operations, increases resistance to creep at high temperatures, increases ductility and impact strength.
Phosphorus	Increases corrosion resistance and strength
Silicon	Increases hardening ability, strength, and magnetic permeability in low-alloy steels
Sulfur	Improves machining properties Especially useful in mild steels
Titanium	Prevents intergranular corrosion of stainless steels, is a deoxidizer, increases strength in low-carbon steels
Tungsten	Used in tool steels to promote hardness In stainless steels helps maintain strength at high temperatures
Vanadium	Improves resistance to thermal fatigue and shock
Zirconium	Inhibits grain growth and is a deoxidizer

Table 5-6 CONSTRUCTION USES FOR SELECTED CARBON & ALLOY STEELS

UNS Designation	ASTM Designation	Construction Use
K02600	A36	Structural steel members
K20504	A53	Welded and seamless pipe
K11510	A242	Structural members
K02303	A572	Structural members
K11430	A588	Structural members
K02703	A529	Structural members
K01803	A633	Structural members
K03000	A500	Cold-formed, welded, and seamless tubing
K10600	A678	Quenched and tempered steel plate
K03000	A501	Hot-formed, welded, and seamless tubing
K02706	A325	High-strength bolts
K03900	A490	High-strength alloy steel bolts
	A307	Machine bolts and nuts
	A328	Sheet piling
KJ0300	A27	Cast-steel items
J31575	A148	High-strength cast-steel items
	615	Billet steel bars for reinforcing concrete
	616	Rail-steel bars for reinforcing concrete
	606, 607	Sheet steel

Courtesy Society of Automotive Engineers

Types of Structural Steel

Of major importance to the architect and engineer is the steel specified for structural purposes. They have low to medium carbon content. The American Institute of Steel Construction (AISC) publication *Code of Standard Practice for Steel Buildings and Bridges,* includes those things required in construction documents, including such things as columns, beams, trusses, bearing plates, and various fastening devices and connectors. The American Institute of Steel Construction publication *Specification for Structural Steel Buildings* details information on structural steels for use in building construction.

▼ **Structural steels** fall into four major classifications:

1. **Carbon steel**—ASTM A36, A529, UNS K02600
2. **Heat-treated construction alloy**—ASTM A514, UNS K11630
3. **Heat-treated high-strength carbon**—ASTM A633, A678, A852, UNS K01803, K01600, K12043.
4. **High-strength low-alloy**—ASTM A242, A572, A588 UNS K11510, K02303, K11430

Carbon steels used must meet maximum content requirements for manganese, silicone. Copper requirements have minimum and maximum specifications. There are no other minimums specified for other alloying elements.

Heat-treated construction alloy steels have alloying element specifications in excess of those for carbon steel and produce the strongest general use structural steel.

Heat-treated high-strength carbon steels are brought to the desired strength and toughness by heat treating. Heat treating refers to the process of heating and cooling metals to produce changes in the physical and mechanical properties.

High-strength low-alloy steels are a group of steels to which alloying elements have been added to produce improved mechanical properties and greater resistance to atmospheric corrosion. The carbon range is typically from 0.12 to 0.22 percent. The percent of alloying agents varies with the manufacturer of the steel. These are available from manufacturers and listed under specific trade names. ASTM has specifications to cover all the trade name steels. One type that is of special interest is known by the general name, weathering steels. These are designed for architectural exterior applications where the exposure to the atmosphere causes the steel to form a natural, rust-colored, self-healing oxide coating. It is never painted but left with the oxide coating protecting and coloring the exterior material. These are the strongest steels made for general use.

High-strength low-alloy steels can be worked by many metal processing operations, including hot and cold forming, punching, shearing, gas cutting, and welding. In construction alloy steels find use in high strength bolts, cables used in prestressed concrete, sheet and strip material, plates, various bars and structural shapes, wire, tubing, and pipe.

STAINLESS & HEAT-RESISTING STEELS

Stainless steels have outstanding corrosion and oxidation resistance at a wide range of temperatures. **Heat-resisting steels** maintain their basic mechanical and physical properties when subjected to high temperatures. Typically standard stainless steels have 10 to 25 percent chromium. The chromium gives the stainless steel its corrosion resistance qualities and produces a thin, hard, invisible film over the surface which inhibits corrosion. Nickel and manganese increase strength and toughness and make the material easier to fabricate.

Other alloying elements that may be used to produce desired characteristics include zirconium, titanium, sulfur, silicon, selenium, phosphorus, molybdenum, and Columbium.

Classifying Stainless & Heat-Resisting Steels

Stainless and heat-resisting steels are classified into three major groups based on their chemical composition and their reaction when heat treated. These groups are **ferritic steels, martensitic steels,** and **austenitic steels** (Table 5-7.)

Ferritic stainless steels have a chromium content of about 16 to 18 percent and a low carbon content of 0.12 maximum. These are nonhardenable steels which means they cannot be hardened by heat treating. Ferritic steels have excellent corrosion resistance, greater even than the martensitic steels. While they have especially good corrosion resist-

Table 5-7 AISI/SAE & UNS DESIGNATIONS & CHARACTERISTICS
OF HEAT-RESISTING & STAINLESS STEELS

Grain Structure	Chief Alloying Elements	Series	Designation		Characteristics	Steel Type
			AISI/SAE	UNS		
Martensitic (ferromagnetic and hardenable by heat treatment)	Chromium 4%–6%	500	502	S50200	Retain their mechanical properties at high temperatures	Heat-Resisting steels
	Chromium 11.50–13.50%	400	410	S41000	Moderate corrosion resistance, high strength & hardness	
Ferritic (ferromagnetic and nonhardenable	Chromium 16%–18%	400	430	S43000	Very good corrosion resistance, particularly at high temperatures	
Austenitic (nonmagnetic and hardenable by cold working)	Chromium 17%–19% Nickel 8%–10% Manganese 2.0% Max.	300	302	S30200	Excellent corrosion resistance, high strength & ductility Suitable for many fabrication techniques	Stainless steels
	Chromium 17%–19% Nickel 4%–6% Manganese 7.5%–10%	200	202	S20200	Excellent corrosion resistance at high temperatures, high strength and toughness	

Courtesy American Iron and Steel Institute

ance at high temperatures they do lose strength under these conditions. They are used in conditions where high temperatures exist and corrosion resistance is important and where a low coefficient of thermal expansion is helpful.

Martensite stainless steels have a chromium content of 11.50 to 13.50 percent and a carbon content of 0.15 percent maximum. They can be hardened by heat treating and can be worked hot, forged, or formed cold. They are typically used where hardness, strength, and abrasion resistance are critical, such as in a steam turbine.

Austenitic stainless steels have a chromium content of about 17 to 19 percent and a nickel alloying element of 8 to 10 percent. They typically will have a maximum carbon content of 0.15 percent. They are nonmagnetic and harden when worked cold. At high temperatures they have high strength, are very tough, and are corrosion resistant. They have

a high coefficient of thermal expansion, which must be taken into consideration by the engineer specifying the use of this material. Of all of the stainless steels they have the least resistance to corrosion by attack of sulfur gases. Typical uses include curtain wall panels, railings, and door- and window-finished surfaces. Some types are used in the manufacture of food preparation equipment and various appliances in hospitals.

Designations for Heat-Resisting & Stainless Steels

The American Iron and Steel Institute (AISI) designations for heat resisting and stainless steels are based on a three-digit numbering system ranging from 200 to 500 as shown in Table 5-7. The first number indicates the group, and the last two numbers indicate the type within the group. Some types have the prefix, TP, which refers to tubular grades.

Table 5-8 STAINLESS STEELS USED IN CONSTRUCTION

UNS Designation	AISI Designation	Typical Applications
S30100	301	Gutters, trim, flashing, household & industrial appliances
S30200, S30400	302, 304	Storefronts, curtain walls, doors, windows, railings, household & industrial appliances
S31600	316	All exterior marine uses, on seacoast building exteriors, in chemical, petroleum, & paper manufacturing facilities
S43000	430	Exterior applications in areas exposed to salt water atmosphere, column covers, trim, grills, gutters
S20100	201	Gutters, trim, flashing, household & industrial appliances
S20200	202	Storefronts, curtain walls, doors, windows, railings, household & industrial appliances

STAINLESS STEEL USES IN CONSTRUCTION

Typical applications for stainless steels frequently specified by architects and engineers are shown in Table 5-8. Of the ferritic stainless steels Type 430 is most commonly used in construction for such applications as column covers, trim, grilles, and gutters. Of the austenitic stainless steels Type 301, 302, and 304 are used for applications such as store fronts, doors, windows, and railings. Type 201 and 202 are used for the same applications as 301 and 302. Other uses include drinking fountains, kitchen sinks, flatware, and cooking utensils

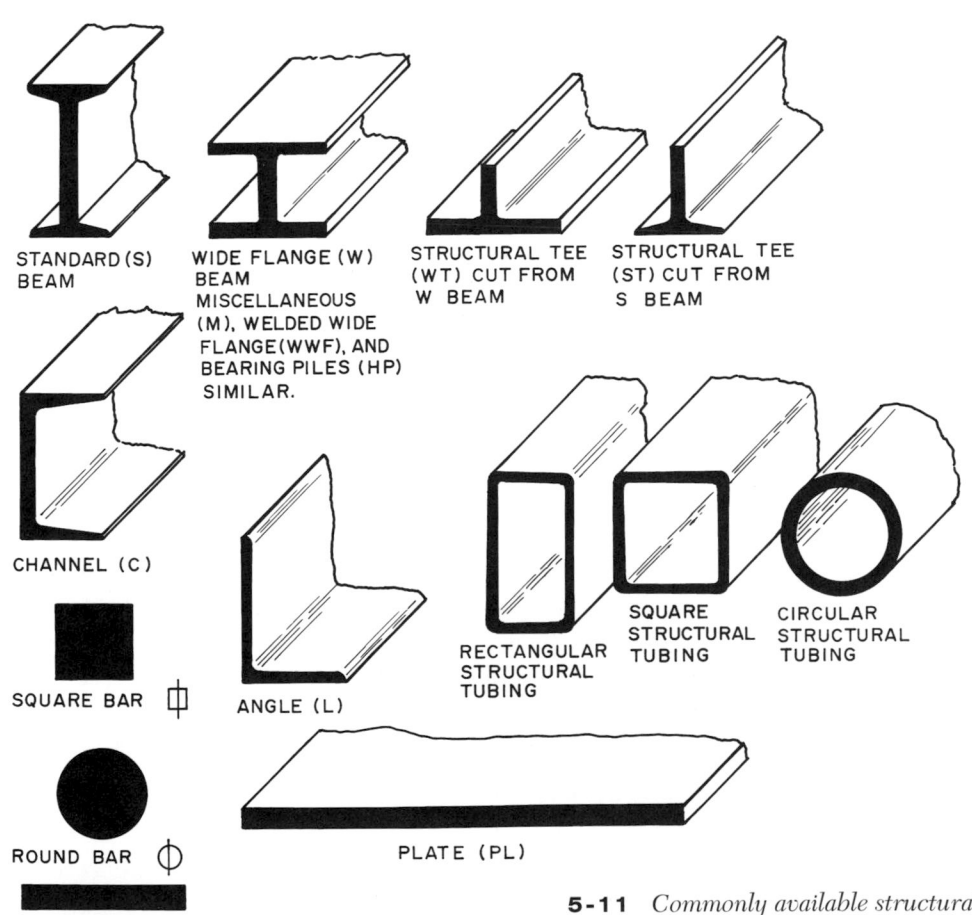

STANDARD (S) BEAM

WIDE FLANGE (W) BEAM MISCELLANEOUS (M), WELDED WIDE FLANGE (WWF), AND BEARING PILES (HP) SIMILAR.

STRUCTURAL TEE (WT) CUT FROM W BEAM

STRUCTURAL TEE (ST) CUT FROM S BEAM

CHANNEL (C)

SQUARE BAR

ANGLE (L)

RECTANGULAR STRUCTURAL TUBING

SQUARE STRUCTURAL TUBING

CIRCULAR STRUCTURAL TUBING

ROUND BAR

PLATE (PL)

FLAT BAR

5-11 *Commonly available structural steel shapes.*

5-12 *Structural steel H beams are available in a wide range of sizes.*
Courtesy Bethlehem Steel Corporation

Steel Products

Steel of various types is used extensively in construction. It ranges from something as small as a nail to hugh beams and columns. Following are a few of the products in common use.

STRUCTURAL STEEL PRODUCTS

Rolled structural steel shapes are extensively used in construction. They carry the live and dead loads imposed on the structure. The commonly used shapes are in 5-11. These include S (standard beam) shapes, W (wide-flange) shapes, M (miscellaneous) shapes, HP (bearing pile) shapes, C (channel) shapes, T (structural T) shapes cut from M, W, and WWF shapes, L (angle) shapes with equal and unequal legs, HSS (hollow structural section), square and rectangular shapes, pipe, bar (flat, round and square), and plate (see 5-12). The S beam has narrow flanges that taper at a 17 per cent slope. The W beam has wide flanges that have very little slope. The WWF shapes are made by welding the flanges to the web while the S and W beams are rolled to shape from a single piece of steel.

Open-web steel joists are a widely used steel product. They are lightweight and produced by welding structural steel shapes, such as angles and bars, into a Warren-type truss unit. They are manufactured in shortspan, longspan, and deep longspan series. The shortspan joists are manufactured to span clear openings from 8 ft. (2.4m) to 60 ft. (18.3m) and in depths from 8in. (203mm) to 30in.(762mm). The longspan joist series is a heavier joist and is made to span clear openings from 25 ft. (7.6m) to 96 ft.(29.2m) and depths from 18 in.(457mm) and 48 in. (1219mm). The deep longspan

Table 5-9 STANDARD THICKNESSES FOR SOME BASIC STEEL SHEETS

Minimum Thickness	
Equivalent Inches	Millimeters
0.3937	10.0
0.3543	9.0
0.3150	8.0
0.2756	7.0
0.2362	6.0
0.1969	5.0
0.1575	4.0
0.1378	3.5
0.1181	3.0
0.0984	2.5
0.0787	2.0
0.0630	1.6
0.0551	1.4
0.0472	1.2
0.0394	1.0
0.0354	0.90
0.0276	0.70
0.0236	0.60
0.0197	0.50
0.0157	0.40
0.0138	0.35

series spans a clear opening from 89 ft. (27m) to 144 ft.(43.9m) and in depths from 52in.(13mm) to 72in.(1829mm). They are used to support floor and roof loads. The openings in the web permit the passage of plumbing, heating ducts, and electrical runs (see 5-13).

SHEET STEEL PRODUCTS

Many products are made by rolling flat steel sheets to shape. The thickness of steel sheets is given by gauge numbers as shown in Table 5-9. Common among these are roofing, siding, decking, and light gauge steel framing systems. Steel decking can be seen in 5-13.

5-13 *Open-web joists with corrugated steel decking.*
Courtesy Vulcraft Division of Nucor Corporation

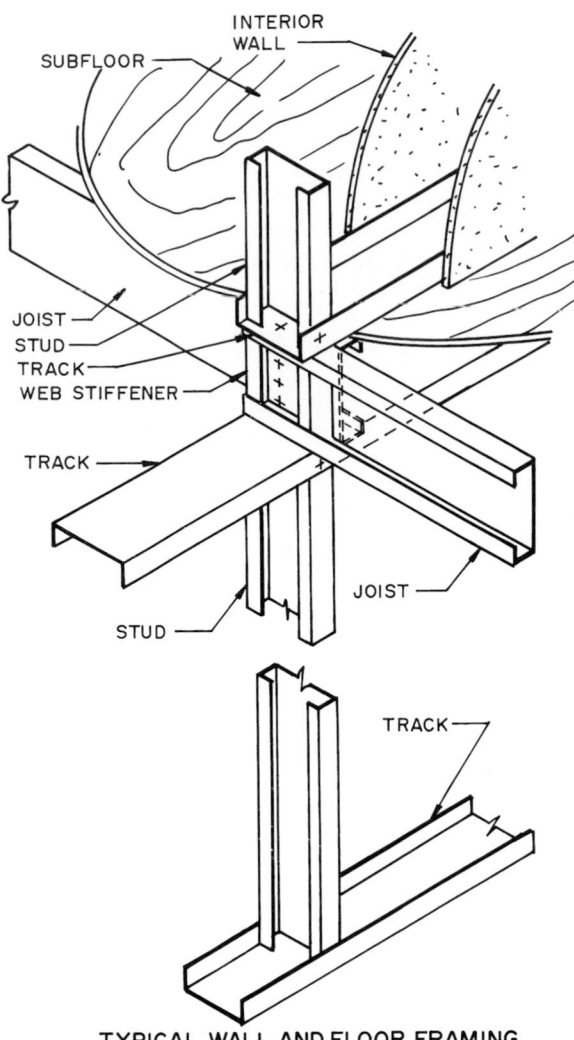

INTERIOR WALL

SUBFLOOR

JOIST
STUD
TRACK
WEB STIFFENER

TRACK

STUD

JOIST

TRACK

TYPICAL WALL AND FLOOR FRAMING

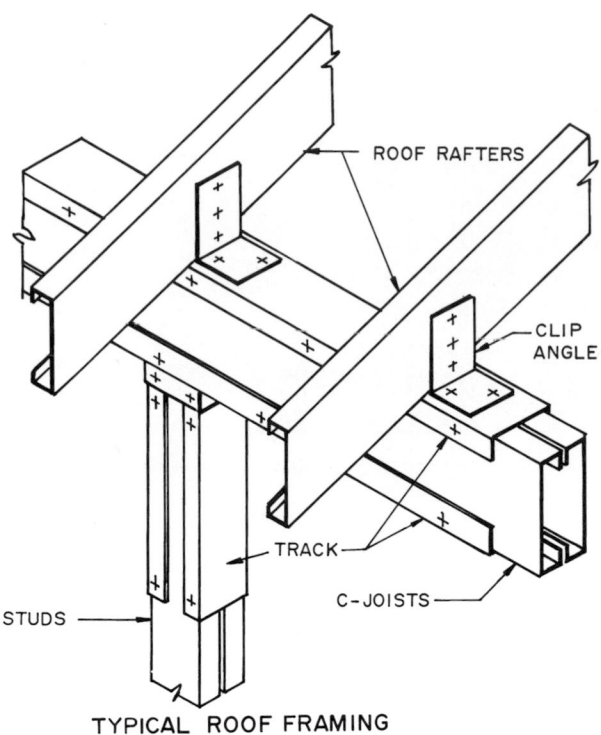

ROOF RAFTERS

CLIP ANGLE

TRACK

STUDS

C-JOISTS

TYPICAL ROOF FRAMING

5-14 *Construction details of lightweight steel framing*
Courtesy Marino Industries Corporation

There are many different styles of steel roof and siding systems available. A variety of coatings are used to protect the steel surface. Typical coatings in use include galvanizing, zinc-aluminum alloy with a paint over it, siliconized polyester in a variety of colors, a fluoroplymer paint finish, and a weathering copper coating.

Corrugated **steel floor decking** is used with structural steel framing. It will have a sitecast concrete topping. This decking will usually span 6 to 15 ft. (1.8 to 4.6m). Cellular floor decking provides openings in the floor slab for running electrical systems.

Steel roof decking may have a sitecast lightweight concrete or gypsum topping or be covered with some form of insulation board and finished roofing system such as tar and gravel.

Light-gauge steel framing systems use studs, joists, channels, and runners to frame wall and floor systems. Metal studs are used in the same manner as wood studs. They have a metal runner for a top and bottom plate. Metal joists are welded to the top runner and have a perimeter channel for a header. The floor and roof deck can be metal or plywood (see 5-14).

Expanded steel mesh is made by slitting metal sheets and stretching them to form diamond-shaped openings. It is used for many purposes, including gratings, decks, partitions, and as a base for the application of trawled stucco and plaster. Expanded metal mesh is usually designated by the size of the width of the mesh opening and the gauge of the steel sheet. For example, a ¾-16 expanded metal mesh would have diamond openings ¾ in. (18mm) wide on 16 gauge material. The length of the diamond is approximately twice the size of the width or 1½in. (38mm) in this example.

Metal lath is another sheet steel product. It is a form of expanded steel that has a flat or ribbed mesh ⅜ in.(9mm) in height. It is used as a base for plaster. Expanded metal corner beads are available.

152 X 152 — MW9 X MW9   6 X 6 – W1.4 X W1.4

LONGITUDINAL WIRES

TRANSVERSE WIRES

MW9 – METRIC
SMOOTH WIRE.
STEEL AREA
59.3 mm²/m.
CROSS-SECTIONAL
AREA 9mm².

W1.4
(SMOOTH WIRE,
CROSS-SECTIONAL
AREA 0.014 IN.²)
STEEL AREA
0.028 IN.²/FT.)

152 mm

6"

152 mm

152 mm   152 mm   152 mm

6"   6"   6"

5-16 *Welded-wire fabric is specified by giving the wire spacing and the style of wire and the cross-sectional area.*

OTHER PRODUCTS

Welded-wire fabric is used to reinforce concrete slabs. Inch sizes of welded-wire fabric are designated by two numbers and two letter-number combinations. The first two numbers, as 6 × 6, gives the spacing of the wires in inches. The first number is the spacing of the longitudinal wires. The second number gives the spacing of the transverse wires in inches. The first letter-number combination gives the type and size of the longitudinal wire and the second for the transverse wire. The W indicates a smooth wire and the number following it gives the cross-sectional area of the wire (see 5-15). When the W is replaced by a D it indicated a deformed wire was used. The area is given in hundredths of an inch per foot. For example, a W8.0 wire is a smooth wire with a cross-sectional area of 0.08 sq.in.

Longitudinal wires—are spaced 2, 3, 4, 6, 8, and 12 inches.

Transverse wires—are spaced 4, 6, 8, and 12 inches.

Metric-welded wire fabric is specified in the same manner as inch sizes but the sizes are in millimeters. The first two numbers are the wire spacing in millimeters and the last two numbers indicate the type of wire and gives the cross-sectional area in square millimeters (mm²).

Reinforcing bars (rebars) are placed in concrete members to improve the tensile strength of concrete (5-16). They are made to ASTM requirements for yield and tensile strength, cold bend, and elongation. There are three grades of rebar available; structural, intermediate, and hard.

5-16 *This tunnel is to be cast-in-place concrete and is heavily reinforced with steel-reinforcing bars.*

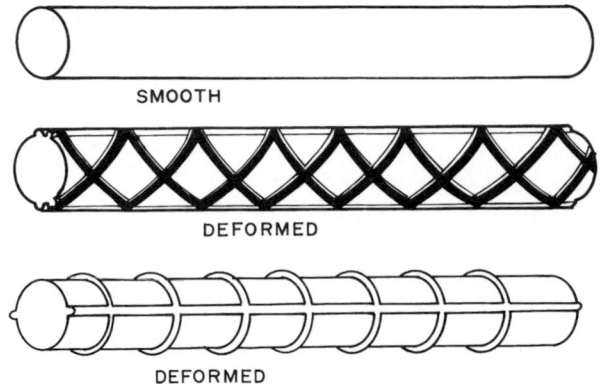

5-17 *Typical types of reinforcing bar (rebar).*

The structural has the lowest yield point and tensile strength while the hard has the highest. Round bars are most commonly used. However, it is also available in squares. The bar may be smooth or deformed (see 5-17). The deformed bar has projections on its surface which, provide a better bond to the concrete. They are available in sizes 3 through 18. The number indicates the diameter in eighths of an inch. For example a No. 5 bar has a diameter of ⅝in. Metric sizes are available.

There are a variety of accessories used to hold reinforcing bars in place until the concrete is poured. These are made from steel wire (see 5-18).

Many fasteners used in construction are made from steel. Commonly used fasteners include bolts, nails, rivets, and screws. **Standard steel bolts** are available in a wide range of sizes and head types. **High-strength steel bolts** are used where tensile strength is important, such as in structural steel framing. There are hundreds of other fastening devices that the constructor will encounter. These include such things as concrete anchors, self-drilling fasteners, eye bolts, and many types of hooks. Other steel products include locks, hinges, and other hardware items.

NONFERROUS METALS

Nonferrous metals are those containing no or very little iron. In other words, metals other than iron and steel. Those commonly found in construction projects include aluminum, copper, zinc, lead, and tin.

Galvanic Corrosion

It should be noted as ferrous and nonferrous metals are studied that they are subject to galvanic corrosion. A metal in contact with a dissimilar metal in the presence of an electrolyte (moisture in the atmosphere) will corrode more rapidly than if similar metals touch. For example, an aluminum gutter secured with copper nails will produce galvanic action at the point of contact. The presence of moisture in the atmosphere sets up an electrolytic action, which will cause the aluminum to corrode. The metal that is higher on the

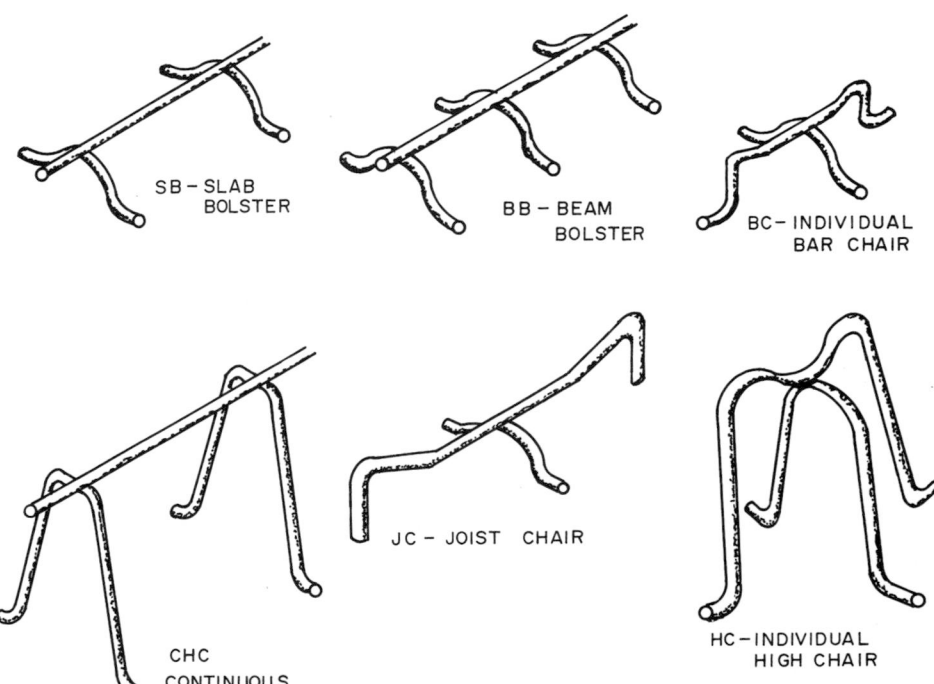

5-18 *Bolsters and chairs used to support reinforcing bars in beams and slabs. Individual chairs support one or possibly two bars while bolsters are continuous and support many bars.*

Table 5-10 THE GALVANIC SERIES

Electrolytic Potential	Metals & Alloys
High potential (anode +)	Magnesium
	Magnesium alloys
	Aluminum (pure & several cast & wrought alloys)
	Zinc
	Cadmium
	Aluminum (wrought 2024 & 356.0 cast)
	Iron or steel
Electric current flows from positive (+) to negative (−)	Cast iron
	Stainless steel
	Lead
	Tin
	Nickel (active)
	Brass
	Copper
	Bronze
Low potential (cathode −)	Chromium stainless steel
	Silver
	Titanium
	Platinum
	Gold

table of electrolytic corrosion potential will be sacrificed (Table 5-10). For example, if aluminum is plated with zinc to protect it from possible corrosion due to electrolytic action, the zinc would corrode protecting the aluminum. Steel is also zinc coated and is called galvanized. When it is not possible to avoid contact between dissimilar metals they can be given a coat of non-load paint or separated with a plastic or other nonconducting material. A joint could be caulked or otherwise sealed to keep out moisture thus eliminating the electrolyte.

Aluminum (Al)

Aluminum is a versatile material widely used in the construction industry. The construction industry is one of the largest consumers of aluminum. Aluminum is light in weight having a specific gravity of only 2.7 times that of water and approximately one-third that of steel. Pure aluminum melts at 1220°F (665°C), which is considerably lower than that of other structural metals. It also is relatively weak as far as mechanical properties are concerned. Aluminum elastically deforms about three times more than steel

under comparable loading. It can be strengthened by alloying, cold working, or strain hardening. Aluminum alloys do not lose ductility or become brittle at cryogenic (low) temperatures.

Aluminum is a good conductor of electricity. When compared with copper wire of the same diameter aluminum is roughly 65 percent of that of copper.

MINING ALUMINUM

Aluminum is found in most rocks and clays. However, concentrations of the aluminum oxide content should be about 45 percent to be economically extracted. Aluminum ores are called bauxites. Most ores are mined by open pit mining. The ore is crushed, washed, screened, ground, and dried.

REFINING THE BAUXITE ORE

Aluminum is refined using a two-step process. The first, the **Bayer process,** produces a very pure alumina (Al_2O_3). Alumina is an oxide of aluminum in crystal form. The second step is to reduce the alumina to a metallic aluminum, which at this point is about 99 percent pure aluminum. This process is referred to as the **Hall-Heroult process.** It is named after the two men who developed the electrolytic method of aluminum production.

The Bayer Process

The ground, dried bauxite is mixed in a digester with soda ash, crushed lime, sodium hydroxide, and hot water. Live steam and mechanical agitators stir the mixture (see 5-19). The mixture is then pumped to digester tanks where, under high pressure and the injection of steam, the mixture is churned. The chemical reaction forms sodium aluminate and the insoluable impurities form a waste material. The mixture flows through a pressure-reducing tank into a settling tank where the waste is removed and the sodium aluminate solution passes through a filter to a cooling tower into the precipitator, where aluminum hydrate is added. The mixture is agitated with compressed air and cooling continues allowing the sodium aluminate to precipitate as aluminum hydrate. This is pumped into filter tanks that separate the aluminum hydrate from the solution. This is calcined in a rotary kiln operating at about 2000°F (1100°C), which produces the alumina used to produce the metallic aluminum in the second step, the Hall-Heroult electrolytic process.

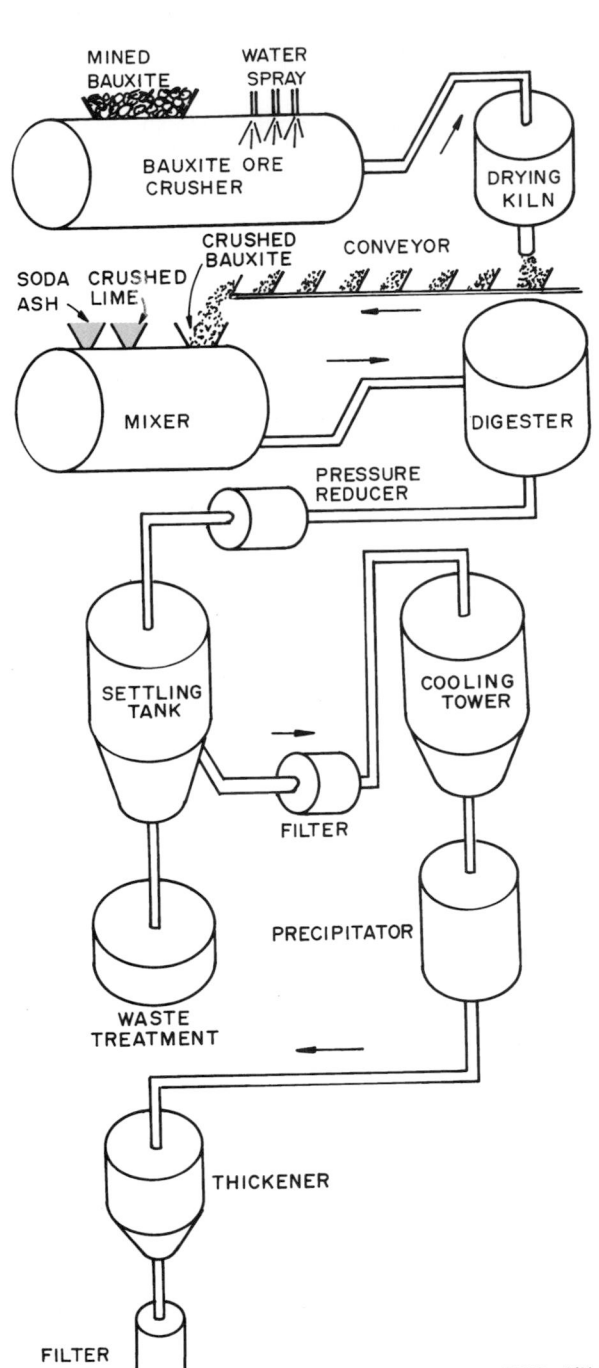

5-19 *The first step in producing aluminum. The Bayer process is to get the alumina from the bauxite ore.*

Table 5-11 CAST ALUMINUM & ALUMINUM ALLOY DESIGNATION SYSTEM

Aluminum 99.00% minimum	1xx.x
Aluminum Alloys Grouped by Major Alloying Elements	
Copper	2xx.x
Silicon, with added copper and/or magnesium	3xx.x
Silicon	4xx.x
Magnesium	5xx.x.
Zinc	7xx.x.
Tin	8xx.x
Other element	9xx.x
Unused series	6xx.x

Courtesy the Aluminum Association

Table 5-12 BREAKDOWN OF 2XXX–9XXX ALUMINUM DESIGNATIONS

X	X	XX
1 commercially pure	0 indicates original alloy developed	in 1xxx series indicates impurity limits
2–9 indicates major alloying element as shown in Table 5-13	2–9 indicate modifications of the original alloy	2XXX through 9XXX series— arbitrary numbers identifying the alloy in the series

The Hall-Heroult Electrolytic Process

Aluminum is produced from the oxide, **alumina**, by a reduction process (see 5-20). **Reduction** is an electrolytic process that uses a carbon-lined vessel containing molten cryolite and alumina. An electric current is passed through this liquid using large carbon anodes suspended in it. When the current is passed through the liquid molten aluminum separates out and settles to the bottom of the vessel where it is siphoned off. Pure aluminum may be cast in molds for use in products requiring these properties. It may also be sent to a holding furnace where alloying elements are added to alter the physical properties.

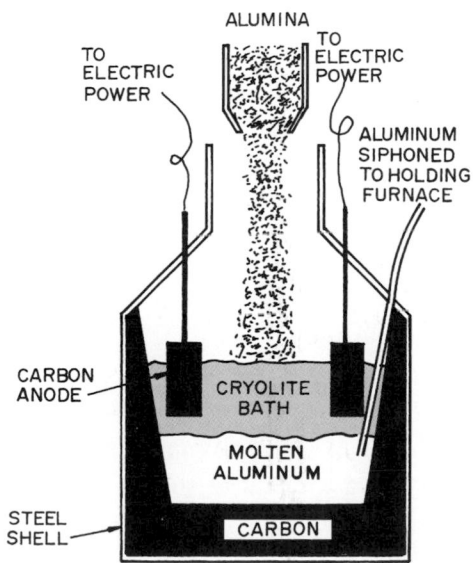

5-20 *The second step in producing aluminum is the reduction of the alumina using the Hall-Heroult process.*

Aluminum Alloys

When aluminum is produced it is between 99.5 and 99.9 percent pure. In this form it is relatively soft, ductile, and has a tensile strength of around 7000 psi (48 258 kPa). To be used for many products it must have alloying elements added to alter its physical properties.

Aluminum alloys are classified as wrought alloys and casting alloys. **Wrought alloys** are those that are mechanically worked by processes such as forging, drawing, extruding, or rolling, such as forming sheet material. **Cast alloys** are those used to produce a product for which the molten metal is cast into the shape of the finished product, such as a grille, in a sand or permanent mold.

WROUGHT ALUMINUM ALLOY CLASSIFICATIONS

The major alloying elements to aluminum are manganese, copper, magnesium, silicon, and zinc. There are a wide variety of aluminum alloys identified by a code system developed by the Aluminum Association, Inc. Each wrought aluminum alloy is specified by a four-digit code number as shown in Table 5-11.

The **first digit** specifies the alloy series and indicates the major alloying element. The **second digit** in the 1xxx series represents a modification of impurity limits and in the 2xxx through 9xxx series represents a modification of the alloy. The last two digits in the 1xxx series indicate an aluminum purity above 99 percent. For example a 1050 specifies an aluminum containing 0.50 percent more aluminum than the minimum 99 percent. In the 2xxx through 9xxx series the second digits are arbitrary numbers indicating a modification of the alloy. A 1150 series has the same aluminum content as the 1050 series but is subject to a modified impurity control. Table 5-12 illustrates this classification system.

Table 5-13 WROUGHT ALUMINUM & ALUMINUM ALLOY DESIGNATION SYSTEM

Aluminum 99.00% Minimum 1xxx	
Aluminum Alloys Grouped by Major Alloying Elements	
Copper	2xx.x
Manganese	3xx.x
Silicon	4xx.x
Magnesium	5xx.x
Magnesium and silicon	6xx.x
Zinc	7xx.x
Other elements	8xx.x
Unused series	9xx x

Courtesy the Aluminum Association

Table 15-14 SELECTED EXAMPLES OF ALUMINUM ALLOY UNS DESIGNATIONS

UNS Designations	Material	Numbers Used by Various Organizations
A02400	Aluminum foundry alloy, casting	AA 240.0, AMS 4227
A91035	Wrought aluminum alloy, nonheat-treatable	AA 1035
A92011	Wrought aluminum alloy, heat-treatable	AA 2011, ASTM B210, FS QQ-A-22513, SAE J454

Courtesy Society of Automotive Engineers

CAST-ALUMINUM ALLOY CLASSIFICATIONS

The **cast alloy** has a three-digit number followed by a decimal (Table 5-13). The first digit identifies the alloy series and the **second** and **third digits** the specific alloy or purity. The decimal indicates whether the alloy composition is for final casting (.0) or for ingot (.1 or .2).

The classification number system also includes a letter on the end of the four-digit number giving information about the temper of the alloys. This is explained in the following paragraphs.

Unified Numbering System

The Unified Numbering System (UNS) is described in detail earlier in this division. It uses a five-digit number

Table 15-15 SUBDIVISION OF H TEMPER: STRAIN HARDENED

First Digit Indicates Specific Treatment:
H1—Strain hardened only
H2—Strain hardened & partially annealed
H3—Strain hardened & stabilized
H4—Strain hardened & lacquered or painted
Digit (0-8) Indicates the Degree of Strain Hardening as Identified by a Minimum Value of the Ultimate Tensile Strength[a]:
0—Annealed
2—Tempers whose ultimate tensile strength is midway between 0 & 4.
4—Tempers whose ultimate tensile strength is midway between 0 & 8.
6—Tempers whose ultimate tensile strength is midway between 4 & 8.
8—Tempers whose ultimate tensile strength exceeds that of 8 by 2 ksi or more.
1, 3, 5, 7—Tempers whose ultimate tensile strength falls between those defined above.

[a]The tensile strength for each number is specified by Aluminum Association tables.

Courtesy Aluminum Association

with a letter prefix to classify and identify various types of aluminum and aluminum alloys that are specified by other organizations such as the Aluminum Association (AA), American Society for Testing and Materials (ASTM), Society of Automotive Engineers/Aerospace Materials Specifications (AMS), Military Specifications (MS), and Federal Specifications (FS). The UNS pulls together those materials with matching specifications from these organizations and gives it a unified number. In Table 5-14 are a few selected examples to show you how this system pulls these together.

Temper Designations

Temper is the degree of hardness and strength imparted to a metal by some process, such as by heat treatment or cold working. Tempering refers to the process of bringing a metal to the proper degree of hardness and elasticity so it can be used for the purpose intended.

Wrought aluminum alloys fall into two classes: heat treatable and nonheat treatable.

Table 5-16   SUBDIVISIONS OF T TEMPER: THERMALLY TREATED

First Digit Indicates Specific Sequence of Treatments:

T1—Naturally aged after cooling from an elevated temperature-shaping process

T2—Cold-worked after cooling from an elevated temperature-shaping process & then naturally aged

T3—Solution heat treated, cold-worked, & naturally aged

T4—Solution heat treated & naturally aged

T5—Artificially aged after cooling from an elevated temperature-shaping process

T6—Solution heat treated & artificially aged

T7—Solution heat treated & stabilized (overaged)

T8—Solution heat treated, cold-worked, & artificially aged

T9—Solution heat treated, artificially aged, & cold-worked

T10—Cold-worked after cooling from an elevated temperature-shaping process & then artificially aged

Second Digit Indicates Variation in Basic Treatment:

Examples:

T42 or T62—Heat treated to temper by user

Additional Digits Indicate Stress Relief:

Examples:

TX51—Stress relieved by stretching after solution heat treating

TX52—Stress relieved by compressing after solution heat treating or cooling

Courtesy Aluminum Association

Heat-treatable alloys are those whose strength characteristics are improved by heat treating. The strength is also improved by adding alloying elements such as copper, zinc, silicon, and magnesium.

Nonheat-treatable alloys have alloying elements added that do not cause an increase in strength when heat treated. The strength of non-heat-treatable alloys depends upon alloying elements such as iron, magnesium, manganese, and silicon. These alloys are strengthened by cold rolling or strain-hardening.

Some temper specifications apply only to cast aluminum while others apply only to wrought aluminum. Alloys specified as F,H, or O can be hardened by cold working and may or may not be nonheat treatable. Heat-treatable aluminum alloys use the T and W specification.

The specification of an aluminum alloy requires the designation of temper or the metallurgical condition.

The following temper designation system for aluminum alloys was developed by the Aluminum Association, Inc. The temper designation consists of letters and numbers. They are placed after the alloy number and separated from it by a dash. The tempers include: "F" represents "as fabricated." No special control over thermal or work hardening conditions is used. "Q" represents "annealed." This indicates wrought products have been heated to effect recrystallation, which produces the lowest strength. Cast products are annealed to improve stability and ductility. "H" represents "strain-hardened." This indicates wrought products are strain-hardened through cold working. The "H" is followed by one or two digits. See Table 5-15. "W" represents "solution heat-treated." This involves heating the alloy to about 1000°F (542°C) and then quenching it. "T" represents "thermally treated." This indicates a product has been heat treated and then strain-hardened. The "T" is followed by one or two digits. See Table 5-16.

Table 15-17 TYPICAL APPLICATIONS OF ALUMINUM ALLOYS

Typical Applications of Nonheat-Treatable Aluminum Alloys

AA Alloy Series	AA Typical Alloys	UNS Designation	Typical Applications
1xxx	1350	A91350	Electrical conductors
	1060	A91060	Chemical equipment, tank cars
	1100	A91100	Sheet metal work, cooking utensils, decorative
3xxx	3003	A93003	Sheet metal work, chemical equipment, storage tanks
4xxx	4043	A4043	Welding electrodes
	4343	A94343	Brazing alloy
5xxx	5005	A95005	Decorative & automotive trim, architectural
	5050	A95050	& anodized, sheet metal work, appliances
	5454	A95454	
	5456	A95456	
5xxx (3% Mg)	5083	A95083	Marine, welded structures, storage tanks, pressure
	5086	A95086	vessels, armor plate, cryogenics
	5454	A95454	
	5456	A95456	

Typical Applications of Heat-Treatable Aluminum Alloys

AA Alloy Series	AA Typical Alloys	UNS Designation	Typical Applications
2xxx (Al-Cu)	2011	A92011	Screw machine products
	2219	A92219	Structural, high temperature
2xxx (Al-Cu-Mg)	2014	A92014	Aircraft structures & engines, truck frames and
	2024	A92024	wheels
	2618	A92618	
4xxx	4032	A94032	Pistons
6xxx	6061	A96061	Marine, truck frames & bodies, structures,
	6063	A96063	architectural furniture
7xxx (Al-Zn-Mg)	7004	A97004	Structural, cryogenic, missile
	7005	A97005	
7xxx (Al-Zn-Mg-Cu)	7001	A97001	
	7075	A97075	High-strength structural & aircraft
	7178	A97178	

Courtesy Aluminum Association

Table 5-17 shows some uses of aluminum in construction.

Aluminum Castings

The conditions involved with the production of aluminum castings influence their physical properties. The metalurgist has to not only consider the characteristics of the alloy—the use to which the product will be put—but also how it will be cast. In general aluminum cast in permanent molds (metal molds) will be stronger than sand-cast products. However, the cost of producing metal molds is high. They can be justified when a large number of products will be cast because they can be reused. A sand-casting mold will produce only one casting.

Heat-treated alloys typically will be stronger and have greater ductility than nonheat-treatable alloys. After casting heat-treatable alloy castings are subject to a very high temperature (but below melting point of the alloy) after which they are quenched cold or hot in oil or water and left to age at room temperatures. This can help control the characteristics of the alloy.

Aluminum Finishes

When exposed to oxygen in the atmosphere aluminum forms a protective oxide coating; therefore under ordinary circumstances, no protective coating need be applied. However, there are various finishes applied to aluminum products to improve appearance and usefulness. The Aluminum Association specifies three categories of finishes: mechanical, chemical, and coatings. Aluminum can also have an anodized finish or be left natural.

The Aluminum Association classifies finishes with a letter and a two digit number. Mechanical finishes are designated by M and chemical finishes by C. Coatings are designated by letters denoting a particular type, and anodic coatings use the letter A. You can see the specifications and designations for these in Table 5-18, 5-19, and 5-20.

NATURAL FINISHES

Natural finishes may be controlled or uncontrolled. **Uncontrolled finishes** are produced on the surfaces of wrought and cast products by the condition of the surfaces of the rollers, molds, or extruding dies used to form them. Hot-rolled products have a brighter surface. **Controlled finishes** are produced by varying the smoothness of the rollers or mold surfaces. The sheet may be rolled with one side having very smooth rollers and the back side uncontrolled. Sheets may be embossed using rollers that have a design on the surface. Aluminum siding that looks like it has a wood grain is an example of an embossed finish.

MECHANICAL FINISHES

Mechanical finishes are used to alter the appearance of the surface of the aluminum product. They are usually performed before the surface is cleaned to receive other finishing processes. The surface may be buffed with an abrasive on a felt or **polished** with very fine abrasive to a high luster. It may be ground with a dry **grinding** wheel. This is often necessary to clean up ridges on castings and produces a rough, scratched finishing. Rotating wire brushes produce a scratched surface. The scratches can vary from very fine to coarse depending upon the size of the wire in the brushes. A matte surface is produced by blasting the surface with sand or steel shot. Round pieces of abrasive cloth or steel wool can be rotated against the surface producing a series of concentric circles. Code designations for mechanical finishes are in Table 5-18.

Table 5-18 MECHANICAL FINISHES ON ALUMINUM

Type of Finish	Designation[a]	Description
As fabricated	M10	Unspecified
	M11	Specular as fabricated
	M12	Nonspecular as fabricated
	M1X	Other
Buffed	M20	Unspecified
	M21	Smooth specular
	M22	Specular
	M2X	Other
Directional textured	M30	Unspecified
	M31	Fine satin
	M32	Medium satin
	M33	Coarse satin
	M34	Hand rubbed
	M35	Brushed
	M3X	Other
Nondirectional textured	M40	Unspecified
	M41	Extra fine matte
	M42	Fine matte
	M43	Medium matte
	M44	Coarse matte
	M45	Fine shot blast
	M46	Medium shot blast
	M47	Coarse shot blast
	M4X	Other

[a]The complete designation must be preceded by AA—signifying Aluminum Association.

Courtesy Aluminum Association

CHEMICAL FINISHES

Chemical finishes are produced by the reaction of the aluminum surface to various chemicals.

Conversion coatings are a major chemical finish. They prepare the surface for the bonding of paints, organic coatings, and laminates. The natural oxide film does not always provide an adequate bonding surface. The surface can be etched with a chemical that produces a frosty surface. Designs can be etched on the surface by protecting all areas except those to be etched. **Chemical oxide films** can be used to produce a surface having greater resistance to corrosion than the natural oxide films. Aluminum can be plated with zinc in a process called zincating. This produces a thin coating of zinc. The zinc coating protects the aluminum from galvanic action and also prepares the surface for electroplating. Aluminum can be used to produce surfaces that are highly reflective of heat and light. These mirror-like surfaces are produced by chemical brightening. Code designations for chemical finishes are in Table 5-19.

ANODIC FINISHES (ELECTROLYTIC FINISHES)

A very important, widely used finish on aluminum is an anodized electrolytic oxide layer. Anodized films can be used as the finish on surfaces to be painted. Anodized coatings are most often used as the finished protecting layer. A film is less than 0.1 mil thick. A coating is a 0.1 mil or thicker.

▼ Following is a typical procedure for the anodizing process:
1. The surface is cleaned using alkaline and/or acid cleaners to remove grease and surface dirt.
2. Next the surface is given a pre-treatment, which may be either etching or brightening. Etching involves producing a matt surface with hot solutions of sodium hydroxide. This removes minor surface imperfections. A thin layer of the aluminum is removed by this process. Brightening involves producing a mirror-like surface with a concentrated solution of phosphoric and nitric acids. These chemically smooth the surface.
3. The third step is the actual anodizing process in which the anodic film is built and combined with the aluminum by passing an electric current through an acid electrolyte bath in which the aluminum is immersed. The coating thickness and finished surface characteristics can be carefully controlled.
4. Coloring the anodized surface can occur in several ways. One method is to combine the coloring with the actual anodizing process (step 3), which simultaneously form and color the oxide cell wall in bronze and black shades. This produces a more abrasive resistant coating than other methods but is more expensive because it requires more electricity.

A second coloring procedure involves a two-step electrolytic coloring. After the aluminum is anodized (step 3) it is immersed in a bath containing an inorganic metal salt and an electric current is passed that deposits the metal salt at the base of the pores. Color is dependent upon the metal salt used (tin, copper, nickel, cobalt). This method provides the greatest variation of colors.

Table 5-19 CHEMICAL FINISHES ON ALUMINUM

Type of Finish	Designation[a]	Description
Nonetched cleaned	C10	Unspecified
	C11	Degreased
	C12	Inhibited chemical cleaned
	C1X	Other
Etched	C20	Unspecified
	C21	Fine matte
	C22	Medium matte
	C23	Coarse matte
	C2X	Other
Brightened	C30	Unspecified
	C31	Highly specular
	C32	Diffuse bright
	C3X	Other
Chemical coatings[b]	C40	Unspecified
	C41	Acid chromate-fluoride
	C42	Acid chromate-fluoride-phosphate
	C43	Alkaline chromate
	C44	Non-chromate
	C45	Non-rinsed chromate
	C4X	Other

[a]The complete designation must be preceded by AA—signifying Aluminum Association.

[b]Includes chemical conversion coatings.

Courtesy Aluminum Association

A third method involves organic dyeing. This produces vibrant colors that are highly weather resistant.

One other coloring process is described as interference coloring. It involves modification of the pore structure produced in sulfuric acid. It produces light-fast colors ranging from blue, green, and yellow to red.

Anodic coatings are classified into four groups—General, Protective and Decorative, Architectural I, and Architectural II.

General coatings are those less than 0.1 mil thick. **Protective and Decorative Coatings** are less than 0.4 mil thick. These two classes are used for general industrial applications. The architectural classes are used on materials to be exposed to weather and wear. Details can be found in Table 5-20.

Architectural Class I coatings are recommended for exterior use where they will receive no regular maintenance. They are also used for interior purposes where extra protection is needed. They must be over 0.7 mil in thickness and weigh more than 27 mg per sq. in.

Architectural Class II coatings are recommended for interior applications not expected to receive heavy wear and for exterior uses where the product will be regularly maintained. They must have a thickness from 0.4 to 0.7 mils and weigh 17 to 27 mg per sq. in. Code designations for architectural classes can be found in Table 5-20.

COATINGS

Many aluminum products are **painted** to provide additional protection or for surface decoration. These products, as aluminum gutters, are painted in a factory and delivered to the job site finished, ready to install. The factory-applied finish is electrostatically sprayed or roller applied with an organic paint and is then passed through an oven where it is baked to a hard, uniform finish. The paint is flexible and will not crack if the product is bent or formed on the job.

Porcelain enamel coatings are produced by using a vitreous inorganic material that is bonded to the metal by fusing it at high temperatures. The coating is very resistant to corrosion, is durable, and is available in a wide range of colors. It is widely used on curtain wall panels as seen on high-rise buildings where it is difficult to provide regular maintenance.

Aluminum can be **electroplated** with chromium, copper, and other materials. If chromium is electroplated the aluminum must first be zincated or plated with copper, brass, or nickel. Chromium plating gives a mirror-like surface and provides resistance to abrasion. Copper plating requires the aluminum to be zincated. It is mainly used where things, as electrical connections, must be soldered to the aluminum product. Tin and brass plating also provides a good soldering surface. Zinc and cadmium plating improves corrosion resistance.

Aluminum surfaces can also have **laminated** finishes. This involves bonding another material, as vinyl or polyvinyl chloride films, to the surface with an adhesive. Usually the aluminum surface must be chemically cleaned to provide the strongest bond.

Aluminum surfaces that are to serve as high-quality mirrors are treated by electropolishing. This is an anodic smoothing of the surface and requires high purity aluminum. The finished surface usually has a final anodic protective coating applied.

Table 5-20 ANODIC COATINGS ON ALUMINUM

Type of Finish	Designation[a]	Description
General	A10	Unspecified
	A11	Preparation for other applied coatings
	A12	Chromic acid anodic coatings
	A13	Hard, wear & abrasion-resistant coatings
	A1X	Other
Protective &	A21	
Decorative	A211	Clear coating
Coatings	A212	Clear coating
less than	A213	Clear coating
10 μm (0.4 mil)	A22	Coating with integral color
thick	A221	Coating with integral color
	A222	Coating with integral color
	A223	Coating with integral color
	A23	Coating with impregnated color
	A231	Coating with impregnated color
	A232	Coating with impregnated color
	A24	Coating with electrolytically deposited color
	A2X	Other
Architectural	A31	Clear coating
Class II[b]	A32	Coating with integral color
10 to 18 μm	A33	Coating with impregnated color
(0.4 to 0.7 mil)		
coating	A34	Coating with electrolytically deposited color
	A3X	Other
Architectural	A41	Clear coating
Class I[b]	A42	Coating with integral color
18 μm (0.7 mil)		
and thicker	A43	Coating with impregnated color
coatings	A44	Coating with electrolytically deposited color
	A4X	Other

[a]The complete designation must be preceded by AA—signifying Aluminum Association.

[b]AA Standards for Anodized Architectural Aluminum.

Courtesy Aluminum Association

Protecting the Finished Aluminum Product

Aluminum products on the construction site need to be protected from damage during delivery, storage, installation, and until the job is finished. Following are several coatings used for this purpose.

The best and most expensive protective coatings are some form of **paper** or **plastic sheet material** bonded to the product with an adhesive that permits them to be easily removed leaving little or no sticky residue behind. If on the job protection is needed masking tape as used by the painters does a good job.

Sometimes the aluminum has a coat of clear lacquer sprayed on the surface. Methacrylate lacquers will chalk off the surface after several years. Protection of aluminum is afforded by waxing the surface. This is a good way to maintain the surface after the job is completed. Any type of automobile wax or polish is good. This will not provide protection from damage by mortar and other abrasive materials during construction.

Joining Aluminum Members

Aluminum members can be joined by any of the standard fastening techniques. **Mechanical** fasteners include screws, bolts, rivets, and a variety of specially designed products. Sheet stock can be joined by **stitching,** which involves sewing the sheets together with aluminum wire. Aluminum can be joined by **welding, brazing,** and **soldering**. This includes gas, arc, resistance, and inert gas shielded arc welding. An ever increasing joining technique is adhesive bonding. There are a variety of adhesives used depending upon the materials and design considerations. Bonding produces a strong joint and increases the design possibilities for aluminum joining and forming laminates.

Aluminum Products

Aluminum is a major construction material. In its many alloys it finds use in all types of construction. Following are examples of some of the most common uses.

One major use for aluminum is the production of maintenance free **windows.** The members used for this purpose are extruded from an aluminum alloy such as 6063.

Sliding glass doors are made with two or more framed glass doors encased in an extruded aluminum frame. One or more of the panels moves. In addition a wide range of interior and exterior aluminum doors are made. These can be panelized, flush, louvered, and have glass lights. Doors and windows often have two or three layers of glass and weatherstripping to prevent air infiltration. In addition the exterior and interior metal parts are separated by plastic strips forming a thermal break. This reduces the heat loss or gain through the aluminum parts.

Specifications developed by the Architectural Aluminum Manufacturers Association (AAMA) have been adopted by the American National Standards Institute (ANSI). These specify minimum frame strength, thickness, corrosion resistance, air infiltration, water resistance, wind-load capacity, and condensation resistance. Units meeting these specifications have the AAMA seal attached.

Aluminum is used for residential and commercial siding and roofing. The specifications are developed by the AAMA. Residential siding is cold rolled to shape and is available in several widths. Horizontal siding is most popular though vertical siding panels are available. The same material is used for soffits, fascia, flashing, gutters, and frieze boards. Panels are secured to the wood framing with aluminum nails.

Aluminum panels for siding and roofing commercial buildings are made in a variety of designs. They are one-third the weight of steel. Aluminum alloy 3004 is often used. Common finishes include mill finish, unpainted, painted, and stucco embossed.

Aluminum curtainwall systems include preformed, insulated wall panels that are integrated with aluminum windows to form a weathertight, maintenance free exterior surface. They are available with anodized or factory-applied baked enamel finishes.

A wide array of **roof accessories,** such as skylights, roof hatches, and smoke and fire vents and louvers, are made with extruded and sheet aluminum. For example, a smoke hatch dome is made with 6063-T-5 aluminum extrusions and an acrylic dome.

Aluminum **structural shapes** are rolled in much the same configurations as discussed for structural steel members. These include shapes such as S and W beams, channels including a number of special shapes, tees, zees, bulb angles, round and square tubes, pipe, plate, and rods of various shapes.

Various fasteners are available including nails, screws, rivets, bolts, and washers.

Aluminum is used for large diameter electrical conductors, various sheet metal applications, and for applications where corrosion exists, as in food processing and chemical plants.

COPPER & COPPER ALLOYS

Copper and copper alloys are used to produce a wide range of products used in construction. While initially more expensive than aluminum they have properties that make them less costly in the long run because of their resistance to corrosion and other damaging conditions.

Copper (Cu)

Copper is a nonmagnetic, reddish brown metal that has excellent electrical and thermal conductivity. It has the highest conductivity properties of all commonly used metals except silver. It is ductile and malleable and easily worked. When alloyed it offers a wide range of properties making it a very valuable and widely used material in construction.

PROPERTIES OF COPPER

The most important properties of copper and copper alloys as used in the construction industry are electrical and thermal conductivity, resistance to corrosion, wear resistance, ductility, and high temperature performance. Copper has a relatively low tensile strength, about 32,000 psi that can be improved by heat treating, cold working, and alloying. High electrical conductivity makes it a good material for electrical wiring and parts in devices that conduct electricity. Its good thermal conductivity makes it useful in heat transfer situations while corrosion resistance properties make it useful for parts exposed to the atmosphere or to corrosive elements, such as chemicals, plumbing pipe, and gas lines. Its ductility properties make it a material easily bent, stretched, stamped, machined, and otherwise formed into useful products. Copper has a melting point of 1981°F (1083°C) and a coefficient of thermal expansion of 0.0000168°F (0.0000093°C).

A unique characteristic of copper is that if exposed unprotected to the atmosphere it will develop a green coating naturally over a period of years. This coating called patina provides a natural protection from additional corrosion. It takes many years of exposure to the atmosphere to build up a fully developed patina coating.

CLASSIFYING COPPER & COPPER ALLOYS

Copper and copper alloys are specified by the Unified Numbering System for Metals and Alloys (UNS). They are identified by a five-digit number code preceded by

the letter "C." Wrought materials are assigned UNS numbers from C1000 to C79999. Cast alloys are numbered from C80000 to C99999. A summary of these can be found in Table 5-21, on the next page.

Coppers are numbered from C10100 to C15999 and are pure or nearly pure having a minimum copper content of 99.3 percent or higher. The coppers in this series are very similar in chemical composition. The numbering system identifies the production or refining processes used to produce the copper. For example, oxygen-free copper is C10200, electrolytic tough pitch copper is C11000, and phosphorus-deoxidized high residual phosphorus copper is series C12200. Coppers

Table 5-21   UNS DESIGNATIONS FOR COPPERS, BRASSES, & BRONZES

Wrought Coppers & Copper Alloys

C10000–C15760	Copper
C16200–C19900	High copper alloys

Cast Coppers & Copper Alloys

C80100–C81200	Copper
C81400–C82800	High copper alloys

Wrought Brasses & Brass Alloys

C21000–C28000	Brasses
C31200–C38500	Copper-zinc-lead alloys
C40400–C48600	Copper-zinc-tin alloys

Cast Brass & Brass Alloys

C83300–C83810	Brasses
C84200–C84800	Copper-tin-zinc & copper-tin-zinc-lead alloys
C85200–C85800	Copper-zinc & copper-zinc-lead alloys
C86100–C86800	Manganese bronze & leaded manganese bronze alloys
C87300–C87800	Copper-silicon alloys

Wrought Bronzes & Bronze Alloys

C50100–C54400	Copper-tin-phosphous alloys
C55180–C55284	Copper-phosphorus and copper-silver-phosphorus alloys
C60800–C64210	Copper-aluminum alloys
C64700–C66100	Copper-silicon alloys
C66400–C69710	Other copper-zinc alloys
C70100–C72950	Copper-nickel alloys
C73500–C79800	Copper-nickel-zinc alloys (nickel silvers)

Cast Bronzes & Bronze Alloys

C90200–C91700	Copper-tin alloys
C92200–C92900	Copper-tin-lead alloys (leaded tin bronzes)
C93100–C94500	Copper-tin-lead alloys (high leaded-tin alloys)
C94700–C94900	Copper-tin-nickel alloys
C95200–C95900	Copper-aluminum-iron and copper-aluminum-iron-nickel alloys
C96200–C96900	Copper-nickel-iron alloys
C97300–C97800	Copper-nickel-zinc alloys (nickel silvers)
C98200–C98840	Copper-lead alloys
C99300–C99750	Special alloys

Courtesy Society of Automotive Engineers

containing less than a total of 0.7 percent of the specified alloying constituents include tellurium-bearing copper is C14500 and zirconium copper is C15000.

The numerical designations indicate the type of copper or copper alloy and the alloying elements and impurities. Detailed handbooks from the Copper Development Association give specific details.

You must identify the copper or copper alloy by its UNS number. Over the years some frequently used types of copper were given trade names as you can see in the tables that follow. The trade name does not indicate the composition of the material and sometimes is used to describe similar materials that have slightly different amounts of various elements.

TYPES OF COPPER

Coppers include those that are pure or nearly pure having a minimum copper content of 99.3 percent. The various coppers available are similar in composition but are identified by the UNS based on the production or refining processes used. For example there is oxygen-free copper (UNS C10200), electrolytic tough pitch copper (UNS C11000), and phosphorus-deoxidized high residual phosphorus copper (UNS C12200). Coppers containing less than a total of 0.7 percent specified alloying elements such as tellurium-bearing copper (UNS C14500) and zironium copper (UNS C15000) are other examples.

Deoxidized copper (UNS C122200) contains 99.90 percent copper and 0.025 percent phosphorus. It has better forming and bending qualities than electrolytic copper and resist embrittlement at high temperature. It is used for water and refrigeration piping, oil burner service, and in sheets and plates where welding is the major joining method. **Electrolytic tough pitch copper** (UNS C11000) is 99.90 percent copper and has high electrical and thermal conductivity. It can be easily formed into useful shapes. Major uses include electrical conductors of all types, such as flashing, gutters, roofing, and forgings.

Copper Alloys

The major copper alloying elements are tin, aluminum, zinc, nickel, silicon, manganese, lead phosphorus, and beryllium. Following is a brief discussion of each. Additional information about brasses and bronzes is given later in this chapter. The UNS designations shown are for wrought products. Designations for cast products are in Table 5-21.

High copper alloys (UNS C16200-C19199) are wrought alloys having specified copper contents from 96 to 99.3 percent copper. Cast high copper alloys have a minimum of 94 percent copper. Wrought high copper alloys have very high electrical and thermal conductivity and are almost as high as pure copper. However, they are much stronger than pure copper, which increases the number of possible uses. They also have good corrosion resistance.

Brasses (UNS C2000-C4999) are copper alloys with zinc being the major alloying element. However, other elements such as lead, phosphorus, nickel, silicon, iron, and aluminum may be added in small specified amounts. They are extremely useful and find application in many products.

Bronzes (UNS C50000-C66399) are copper alloys in which neither nickel or zinc are used as the major alloying element. The various types are used for electrical contacts and corrosion-resistant applications.

Miscellaneous copper-zinc alloys (UNS C66400-C69999) are often referred to as manganese or nickel bronzes. Typically the major alloying element is zinc so they are much like some of the brasses.

Copper-nickel alloys (UNS C70000-C72999) contain from 3 to 33 percent nickel. Other elements may be added to improve corrosion resistance and strength. They are used in marine applications because of their outstanding capacity to resist corrosion. Typical applications include use in heat exchangers, condenser, piping, valve and pump parts, and relay and switch springs. Alloys with more than 50 percent nickel are called monels and form a separate class. **Monels** have high strength at elevated temperature.

Copper-nickel-zinc alloys (UNS C73000-C79999) are referred to as silver nickels because of their silver color. Zinc is the principal alloying element while nickel is secondary. Other elements may be added to alter the properties. They contain no silver. They have good electrical and mechanical properties and have good corrosion resistance. They are used for fasteners and various electrical components.

COPPER ALLOY FINISHES

Copper alloys are available with a variety of finishes. Some are supplied by the mill manufacturing the copper alloy while others are supplied by the company fabricating the copper into a product or into stock shapes. These finishes use the same three classifications as used on aluminum products. These include

Table 5-22 COPPER ALLOY FINISHES

Finish	Copper Development Assn. Finish Designation
Mechanical	
As fabricated	M10 series
Buffed	M20 series
Directional textured	M30 series
Non-directional textured	M40 series
Patterned	M4X (specify)
Chemical	
Cleaned only	C10 series
Matte dipped	
Bright dipped	
Conversion coatings	C50 series
Coatings	
Organic	
Air dry	060 series
Thermo-set	070 series
Chemical cure	080 series
Vitreous	
Laminated	L90 series
Metallic	

Courtesy Copper Development Association

mechanical, chemical, and coatings. You can see a summary in Table 5-22. Notice the Copper Development Association finish designation. An M before identifying digits identifies mechanical finishes, a C designates chemical finishes, and coatings use a three-digit number.

CARE OF COPPER AND COPPER ALLOYS

New copper products, especially sheet stock, have a bright, shiny, light brownish color. If this is to be maintained it must be covered with a protective coating such as a clear lacquer. When left exposed to the atmosphere it will first turn to a darker brown and eventually take on a permanent light green patina.

This is considered highly desirable from an appearance standpoint and can be produced artifically if the natural aging is too slow.

Copper can be washed with liquid soap and water. If scrubbed with an abrasive type cleaner the brown or green color will be damaged and take a period of time to return.

5-21 *Copper and copper alloys are used for a wide range of piping systems, such as this fire protection sprinkler, and it is prized as a roofing material.*
Courtesy Copper Development Association

USES OF COPPER

Copper and copper alloys are excellent for outdoor uses. Some uses in construction include siding, roofing, flashing, guttering, and screen wire. The alloys are extensively used for plumbing pipe in residential and commercial structures as well as in the manufacture of plumbing fittings, such as valves, drains, and faucets (see 5-21). Sewage treatment plants and industrial plants such as chemical processing installations, utilize copper for many purposes, including lining vessels that will contain corrosive materials. Various types of hardware and fasteners, such as nails, screws, and bolts, are made from copper alloys. A major use is in electrical wire, electrical conductors, and parts in electrical appliances that conduct electricity and flows to a concentrator where the **flotation method** is used. Here the pulverized ore is mixed with water, oil, and a foaming agent and is agitated by air. The copper sulfate particles collect in the foam on the surface while other materials settle to the bottom and are removed as waste. The froth flows to a **reverberatory** furnace where it is smelted (roasted) to eliminate sulfur and certain metal impurities by oxidation.

Brasses

Brasses (UNS C2000-C4999) are a copper alloy having zinc as the principal alloying element but variations are produced by adding small quantities of other elements. **Zinc** improves strength, ductility, and produces changes in color. **Lead** is added to improve machinability while tin improves their strength, hardness, workability, and ductility.

Brasses are hardened by cold working. However, hardness is also influenced by the alloy composition. The composition of several wrought brasses and brass alloys used in construction are shown in Table 5-23. Similar information is available for cast brasses and brass alloys. Notice that zinc is the major alloying element with lead and iron in much smaller amounts.

The chemical symbols used to identify the various elements in all metals are shown in Appendix D.

Brasses fall into three general classes. **White brasses** contain less than 55 percent copper and are hard and brittle. They are used for cast products and cannot be hammered or worked without breaking.

Alpha brasses contain 63 to 95 percent copper and are the easiest type to work. They are used to make radiator parts, springs, and grilles.

Alpha-Beta brasses contain 55 to 63 percent copper. They are stronger than alpha brasses and can be worked hot. They are used where strength is important, such as rivets and screws.

In Table 5-21, on page 170, you can find the Unified Numbering System designation for wrought and cast brasses. Brasses are used as much as copper.

COPPER-ZINC ALLOYS (PLAIN BRASSES)

Copper-zinc alloys are sometimes referred to as plain brasses. Several that find use in products related to construction include red brass, commercial bronze, cartridge brass, and muntz metal.

Table 5-23 COMPOSITION OF SELECTED WROUGHT BRASSES & BRASS ALLOYS[a]

UNS Designation	Descriptive Name	Major Alloying Elements in Percent[b]					Other Named Elements
		Cu	Zn	Pb	Fe	Sn	
Copper-Zinc Alloys (Brasses)							
C22000	Commercial bronze	89.0–91.0	REM[c]	0.05 max.	0.05 max.		
C23000	Red brass	84.0–86.0	REM	0.05 max.	0.05 max.		
C26000	Cartridge brass	68.5–71.5	REM	0.07 max.	0.05 max.		
C28000	Muntz metal	59.0–63.0	REM	0.03 max.	0.07 max.		
Copper-Zinc-Lead Alloys (Leaded Brasses)							
C31400	Leaded commercial bronze	87.5–90.5	REM	1.3–2.5	0.10 max.		0.7 Ni
C35000	Medium-leaded brass	60.0–63.0	REM	0.8–2.0	0.15 max.		
C37700	Forging brass	58.0–61.0	REM	1.5–2.5	0.30 max.		
C38500	Architectural bronze	55.0–59.0	REM	2.5–3.5	0.35 max.		
Copper-Zinc-Tin Alloys (Tin Brasses)							
C44300	Admiralty, arsenical	70.0–73.0	REM	0.07 max.	0.06 max.	0.8–1.2	0.02–0.06 As
C46400	Naval brass, uninhibited	59.0–62.0	REM	0.20 max.	0.10 max.	0.5–1.0	
C48500	Naval brass, high lead	59.0–62.0	REM	1.3–2.2	0.10 max.	0.5–1.0	

[a]These are only a few of the many types of brass & brass alloys available.

[b]Cu copper, Zn zinc, Pb lead, Fe iron, Sn tin, Ni nickel, As arsenic

[c]Remainder for the difference between elements specified & 100 percent.

Reproduced with permission from *Standards Handbook*, parts 5 & 6, Copper Development Association.

Red brass contains 85 percent copper, 15 percent zinc, and small amounts of lead and iron. It has excellent resistance to corrosion and has higher ductility and strength than copper. It is used for plumbing pipe, handrails, balusters, stair posts, tubing, and hardware.

Commercial bronze contains 90 percent copper and 10 percent zinc. It has good ductility and good cold-working properties. It is used for screws, forgings, and some types of hardware.

Cartridge brass contains 70 percent copper and 30 percent zinc. It has the best strength and ductility of all the brasses and is easily worked cold. It is widely used where copper products require extensive fabrication, such as stamping or deep drawing. It finds use in the manufacture of electric sockets, reflectors, rivets, and heating units.

Muntz metal contains 60 percent copper and 40 percent zinc, which you will notice is the type with the greatest percent of zinc alloyed with the copper. Muntz metal has low ductility but high strength and is used for things such as sheet stock and exposed architectural features.

COPPER-ZINC-LEAD ALLOYS (LEADED BRASSES)

Lead is added to brass to alter its qualities to make it easier to machine.

Those types used in some products used in construction include architectural bronze, forging brass, and medium-lead brass.

Architectural bronze has the least copper and most lead of the three mentioned above. It contains 55 to 59 percent copper, 41 to 45 percent zinc and 2.5 to 3.5 percent lead. It is widely used for forgings and products produced by machining. On a building it can be found in decorative grilles, handrails, architectural trim, door parts, and other such uses.

Forging brass contains about 60 percent copper, 38 percent zinc, and 2 percent lead. It has great plasticity when hot and is therefore widely used for forgings. Since it has good corrosion resistance it is used for plumbing and hardware items.

Medium-leaded brass contains about 62 percent copper, 34 percent zinc, and 2 percent lead. It is used where good machining properties are required, such as keys, parts of locks, plaques, and various scientific instruments.

COPPER–ZINC–TIN ALLOYS (TIN BRASSES)

When tin is alloyed with copper, zinc, and other elements the alloy has additional properties not present in plain brasses. The two types of tin brasses are admiralty and naval brasses.

Admiralty contains about 71 percent copper, 28 percent zinc, 1 percent tin, and traces of lead, iron, and arsenic. These alloying elements improve strength and ductility and, of greatest importance, increase resistance to corrosion. As such it is widely used in the manufacture of condenser and heat-exchanger plates, various tubes, and in equipment in chemical and electrical power plants as well as in products that must have resistance to seawater.

Naval brasses are used for wrought products. These include uninhibited, arsenical, medium leaded, and high leaded. The copper content for all ranges from 59 to 62 percent. Naval brasses are used in chemical, steam power plant, and marine equipment.

Bronze

Technically bronze has been used to identify a product typically 90 percent copper and 10 percent tin. However, other alloying elements are now added, producing a wider range of materials called bronze having a variation in the properties. Basically bronze refers to alloys of copper having alloying elements of silicon, aluminum, manganese, and other elements. They may or may not have zinc. The phosphor bronzes contain about 89 percent copper. C51800 has 4 to 6 percent tin and 0.10 to 0.35 percent phosphor while C53400 has more lead and is referred to as leaded phosphor bronze. Aluminum bronzes have 6 to 7.5 percent aluminum while the low silicon bronze has no tin or lead but .8 to 2.0 percent silicon. Cast bronzes have similar elements.

HISTORY

Bronze was used by early mankind after copper found widespread use. See the history of copper earlier in this chapter. Bronze is an alloy of copper and tin. Since tin deposits were often mixed with copper ore deposits it is believed that the early discovery or use of bronze was accidental. They probably noted that some ores (those with some tin ore) were harder and more useful for making tools and weapons. Some time through the years it was discovered that mixing tin with copper produced this improved product.

Bronze is used for various cast products and hardware. Screws, washers, nuts and bolts, and weatherstripping are other uses.

In Table 5-21, on page 170, you can find the Unified Numbering System designations for bronzes.

Lead (Pb)

Lead is a soft, heavy metal easily worked, with good corrosion resistance and having a special feature in its ability to resist penetration from radiation.

LEAD PRODUCTION

The major source of lead is the mineral galence or lead sulfide. Other sources are cerussite (lead carbonate) and anglesite (lead sulfate). Lead ores frequently contain zinc and some have gold, silver, and other metals.

The lead-bearing ore is first crushed and ground into fine particles. The metal-bearing material is separated from the rock particles by flotation. Flotation involves mixing water, oils, and chemicals with the ground ore. As air is blown into the mix from the bottom the lead-bearing particles are wetted by the oil and float to the top in a froth of air bubbles. The lead particles are drawn off and the waste material (gangue) settles to the bottom and is removed.

The concentrated ore is roasted in the air, which changes the lead sulfide to lead oxide. Sulfur escapes as the gas sulfer dioxide and is recovered and made into sulfuric acid. The lead oxide is now smelted in a blast furnace in which the lead settles to the bottom. Any gold or silver settles with it. Waste materials float to the top forming a slag and are removed. The lead is now processed to remove the gold and silver.

GRADES OF LEAD

Of the several grades of lead, chemical lead, desilverized lead, and corroding lead are used in some way in construction. Chemical lead and desilverized lead are used for pipes, sheets, and alloys. Corroding lead is used for white lead, red lead, and litarge (used in the manufacture of batteries, pottery, lead glass, and ink).

UNIFIED NUMBERING SYSTEM DESIGNATIONS

Lead and lead alloys are designated by the UNS numbers L50001 through L59999. For example, L50121 is described as a solder alloy containing 98.0 percent lead while L50770 is a battery grid alloy, lead-calcium containing 99.6 percent lead.

PROPERTIES OF LEAD

The advantageous properties for lead are high density and weight, softness and malleability, low melting point, 620°F (327°C), high resistance to corrosion, and good electrical conductivity. Lead is low in strength and elasticity.

Lead Alloys

Lead is alloyed with antimony to improve hardness and strength. However, many other elements are also added. Among these are arsenic, nickel, zinc, copper, iron, and manganese.

USES OF LEAD

Lead pipes and tank liners are used in installations processing highly corrosive materials but are never used for piping to carry drinking water. Since it is a good self-lubricant it is used where high pressure lubricating is necessary. Lead solder is used for electrical connections because it is a good conductor but is not used on water pipe connections. Hard solders have antimony added while high temperature solders are alloyed with silver and are generally referred to as silver solder.

The use of lead in solder for joining copper water pipes has been banned because of the possibility of increasing the lead present in the water, which will cause lead poisoning. It is not used to move drinking water for the same reason. This danger is especially present when soft or distilled water is used. However, lead pipe and lead-lined tanks have high corrosion resistance, and they can find use in industrial production applications such as in a chemical manufacturing industry. Sheet lead is used for roofing, flashing, and spandrels in areas where there is severe industrial air contamination or in areas along the seacoast.

The use of lead in paint has been completely stopped because of the danger of children eating pieces of peeling paint.

Another interesting use for lead, in the form of lead azide that is easily exploded by an electrically heated wire, is in the manufacture of blasting caps, which are used to set off other explosives.

Lead is used in products such as adhesives, caulking, pigments, glazing, pipes and tanks, compounds, and protective coatings over steel and copper. It is an element added to produce products such as brass, bronze, certain asphalt products, glazes, various fusible alloys, glass, porcelain enamel, iron, steel, certain plastics, solder, rust-resistant prime coatings before painting, tin, and as an additive to certain wood preservative preparations.

Architects specify lead for waterproofing, sound proofing, reduction of vibration, and radiation shielding.

Following are more detailed examples of several lead products which will give you a closer look at the elements and properties.

Lead strip and **sheet products** are made from lead that is almost 100 percent pure and from an alloy with about 7 percent antimony, which improves the strength and stiffness. It is used for roofing and flashing.

Red lead is a lead oxide that is a widely used primer applied generally to steel before it receives the final coats of finish paint. It is available in paste, dry, or as a liquid paint. Red lead is available dry in three grades A,B, and C. Grade C has the highest lead content. The red lead paste in an oil is available in grades B and C. Again grade C has a higher percentage of lead.

Solders

Solders are nonferrous metals used to join metals in a waterproof joint and make secure electrical connections. Soldering is a process that many people do without thinking about the composition of the metal. However, the effectiveness of the completed job depends upon using the proper metal composition.

PROPERTIES OF SOLDERS

Solders have low melting temperatures typically around 375°F (192°C) to 595°F (315°C). However, some high temperature solders range up to about 740°F (396°C). The low melting point enables the solder to join metals without raising their temperature to their melting point; therefore, the solder has no alloying action with the metals being joined.

Solders have little shear, tensile, or impact strength and are therefore used where there will be no load on the joint or on a joint that has other fasteners, rivets, bolts, or interlocking seams to carry a load or stress.

TYPES OF SOLDER

Solders are mainly an alloy of lead and tin with small amounts of other elements included. The four major classifications of solder are tin-lead, tin-lead-antimony, silver-lead, and a variety of special alloys.

The **tin-lead solders** are general purposes alloys used for joining metals. The tin-lead composition varies from an alloy with 70 percent lead and 30 percent tin through a series of about 10 combinations with the lead increasing and the amount of tin decreasing. The maximum lead solder has 90 percent lead and 10 percent tin. Tin lead solders have a less than one percent antimony. Typical 50-50 tin-lead solder is most commonly used.

Tin-lead-antimony solders have from 1 to 2 percent antimony and of the five commonly available types all have more lead than tin. Tin content varies from 20 to 40 percent.

Silver-lead solders contain over 97 percent lead, from 0 to 1.25 percent tin, traces of antimony and silver content of 1.5 to 2.5 percent. They have melting points of about 580°F (307°C) and produce fairly strong, corrosion-resistant joints. They are used for soldering copper and brass using a torch-heating method.

Special purpose solders are designed for a specific application, such as soldering copper roofing or where a high-temperature solder is required. They typically contain various amounts of lead, zinc, silver, and cadmium.

FLUXES

If solder is to bond to the metal surface you must be certain it is clean and free of oxides. In addition to mechanically cleaning the surface (washing, buffing, brushing etc.) any oxides on the surface must be removed. Fluxes are materials used to remove these oxides.

Three general types of fluxes are available—neutral, corrosive, and noncorrosive.

Neutral fluxes are mild and used on metals that can be easily soldered, such as copper, lead, brass, and tin plate. They are typically a form of a mild acid that is wiped on the surface. Often they do not need to be wiped off before you begin to solder.

Corrosive fluxes include salt-type and acid-type. They are more effective in cleaning the surface than the neutral fluxes. However, it is important that they along with the oxide residue be removed. Otherwise their corrosive action will continue to attack the base metal being soldered. Corrosive fluxes (often called acid fluxes) cannot be used on electrical connections.

Noncorrosive fluxes tend to be only on metals easily soldered and on electrical connections. Typically they use rosin as the main flux. It is a good flux for you to use when soldering electrical connections because it will not cause the wires to corrode.

Tin (Sn)

Tin is produced from the ore containing the mineral cassiterite, which is a tin oxide. Since there is little cassiterite in North America tin has to be imported from Malaysia, Brazil, Russia, Indonesia, Thailand, China, and Bolivia. The ore usually contains little tin so an extensive refining process is required.

PROPERTIES OF TIN

Tin is a soft metal that is malleable and ductile and is a blue-white color. It is corrosion resistant when exposed to air and moisture. It has a low melting point of 450°F (232°C) and can be cast. It will take a high polish and has properties enabling it to coat other metals.

WORKING CHARACTERISTICS

Tin is soft and malleable; therefore, it can be worked by rolling, spinning, extrusion, and casting.

USES OF TIN

Since tin and tin alloys have high corrosion resistance and excellent coating ability it is used extensively to pro-

---

HISTORY

The use of tin can be traced back to around 2000 B.C. where it was used as an element in producing bronze in Egypt. The Chinese and the peoples of the Malay Peninsula used tin several hundred years B.C. Some tin was mined in France, Spain, and England around 500 B.C. Around 100 A.D. tin was used as an alloying element producing a lead-tin mixture used as a solder to join metals. The Romans coated copper with tin and by about 1500 A.D. tinplate was being used in Europe. Tinplate is iron or steel sheets coated with tin.

vide a protective coating on other metals, especially steel. One example is the coating on cans used to store food for retail sale. It is also used as an alloying element in other metals. Its low melting point makes it useful in some solders. It finds applications in mirrors, hardware, and fusible alloys. Tin compounds are used in the production of glazes, glass, and porcelain enamel.

UNS DESIGNATION

Tin alloys are designated in the Unified Numbering System by the numbers L13001 through L13999. For example, UNS L13630 is a lead-tin solder containing 37 percent lead and 63 percent tin.

Terneplate

Terneplate is a mixture of lead and tin applied to copper-bearing steel sheet or stainless steel sheet to produce a corrosion-resisting coating. Tin is added because lead alone will not alloy with the iron.

CHEMICAL COMPOSITION

Terneplate sheets are available in two types, short terne and long terne.

Short terne used for roofing is available with stainless steel or copper-bearing steel bases. The stainless steel type uses UNS S30400 stainless steel in 26 and 28 gauge thicknesses. The copper-bearing steel terneplate is made in 26, 28, and 30 gauge thicknesses. Both types use a coating on both sides consisting of 75 to 80 percent lead and 20 to 25 percent tin.

Long terne is used for various industrial purposes and is available in 14 to 30 gauge low-carbon steel with a coating on both sides consisting of 75 to 87½ percent lead and 12½ to 25 percent tin.

USES OF TERNEPLATEP

Short terne, often called roofing terne, is used for finished roofing, gutters, downspouts, and flashing. Terne plate roofing is installed with batten and standing seams. Flat-locked and horizontal seams can be used when it is installed over wood sheathing. You can refer back to 16-36 for some examples.

Long terne is used for fireproof doors and frames, other fireproofing items, and roofing. The various types may be available in sheets and rolls. Terne on steel should be painted after installation. Both sides require painting so any pinholes are sealed. If pinholes are allowed to exist in terne-coated steel corrosion will result. Most of the time the mill applies a red iron-

oxide primer. However, a high quality finish coat of paint is required over this. Terne-coated stainless steel does not require a primer or painting.

PRODUCTION OF TERNE SHEETS

The sheets are chemically cleaned usually by passing them through a dilute solution of sulfuric or hydrochloric acid. They then are fluxed by passing through a heated solution of zinc chloride. Finally, the sheets are passed through a molten solution of lead and tin. The terne-coated sheet is then passed between smooth metal rollers to produce the finished surface. After it has cooled it is cleaned to remove any traces of oil and is ready to be primed.

Titanium (Ti)

Titanium is found in large quantities in the earth's surface. It is a very light, strong, ductile, silvery metal and is one of the most common elements.

PROPERTIES OF TITANIUM

Titanium has low electrical conductivity and a low coefficient of thermal expansion. It is also paramagnetic, meaning when it is placed in a magnetic field it possesses magnetization in direct proportion to the field strength. It has a melting point of 3300°F (1820°C) and a coefficient of thermal expansion of 0.0000085/°C.

Two important characteristics of titanium are the high strength-to-weight ratio and its ability to resist corrosion by salt water and the atmosphere. These have an important impact on its use in various products.

Titanium Alloys

The strength of titanium varies depending upon the purity of the metal. The higher the purity the weaker the metal; therefore, various elements are added to produce titanium alloys with greatly increased strength. Typically vanadium, molybdenum, aluminum, iron, chromium, and manganese are used as alloying elements.

WORKING CHARACTERISTICS

Pure titanium is easier to use than titanium alloys; however, in general all types can be fabricated using standard manufacturing processes. These include machining, welding, riveting, drilling, punching, hot and cold rolling, extruding, forging, and drawing.

USES OF TITANIUM

The major use for titanium is in aircraft and aerospace industries and other military applications. Its strength and light weight make it a desirable material for these applications. It can also be formed in sheet, strip, pipe, and tube products and can be forged and cast. Future uses in construction could be as doors, curtain walls, and other exterior applications such as flashing and guttering. Since titanium has a high melting point it will also find use as a structural material.

PRODUCTION OF TITANIUM

Titanium is produced by the Kroll processes named after the person who developed it. The process is chemically driven using magnesium in an inert atmosphere of helium or argon to produce the reduction of the titanium tetrachloride. These elements react releasing magnesium chloride which is distilled off. Left is a sponge metal that is crushed and melted into ingots. The ingots are used (similar to pig iron ingots) to produce titanium and titanium alloys that are processed into various products.

UNS DESIGNATION

Titanium alloys are designated in the Unified Numbering System by the numbers UNS R50001 through R59999. For example R56210 is a titanium alloy containing 90.2 percent titanium.

Nickel (Ni)

Nickel is a silver-colored metal mainly used as an alloying element. It provides increased resistance to atmospheric and chemical corrosion and increases the strength of the alloy.

PRODUCTION OF NICKEL

There are several processes used to produce nickel from the ore. The most recent is the Hybinette process developed in Canada. The ore is processed using the Bessemer process from which the molten metal goes to a cooling chamber. From here it is crushed and ground and magnetically separated. The result is a nickel-copper platinum alloy that is treated by the electrolysis process separating the nickel, copper, and platinum.

PROPERTIES OF NICKEL

Nickel is resistant to strong alkalis and many acids. It has good resistance to corrosion and oxidation, and it is strong and tough. It has a melting point of 2651°F (1455°C) and a coefficient of thermal expansion of 0.000013/°C. It is magnetic up to 680°F (360°C).

WORKING CHARACTERISTICS

Nickel can be fabricated using most of the commonly used processes such as hot and cold rolling, extruding, bending, forging, and spinning. It can be joined by some welding processes, soldered, brazed, or with mechanical fasteners.

UNS DESIGNATIONS

Nickel and nickel alloys are designated in the Unified Numbering System (UNS) by the numbers N02001 through N99999. For example, UNS N02250 is a commercially pure nickel alloy that has 99.0 percent minimum nickel.

Nickel Alloys

A major use for nickel is as an alloying element. The alloying of nickel to other metals provides increased ductility, corrosion resistance, strength, hardness, and toughness. Nickel alloyed to nonferrous metals improves electrical resistance, magnetism, and helps control expansion. Nickel is a commonly used alloying element. It is also widely used in monel metals and aluminum alloys. Monel alloy is about 66 percent nickel and 34 percent copper. Inconel 600® is a special nickel alloy containing about 75 percent nickel, 15 percent chromium, and 7 percent iron. Since it has excellent oxidation resistance it is used in food processing and chemical industries. Other nickel alloys include those used for electrical resistance coils, magnetic and nonmagnetic alloys containing iron, alloys designed to have a high coefficient of thermal expansion and are used in the production of glass, and copper-nickel alloys used for products exposed to marine conditions.

USES OF NICKEL

The major use of nickel is as an alloying element in ferrous and nonferrous metals. It is also an excellent material to use for electroplating and electroless plating. Electroplating is the process of depositing a coating of metal on another metal by electrolysis. Electroless is a method of plating a material by chemical means in which the piece to be plated is immersed in a reducing agent, which when catalyzed by certain materials, changes metal ions to metal forming a deposit on the surface of the piece. Nickel is used in electric heating elements, lamp filaments and plumbing fittings.

Zinc (Zn)

Zinc is a bluish white metal that is brittle and has low strength. It is often referred to as a white metal and is widely used as a protective coating over steel.

PROPERTIES OF ZINC

Zinc has low strength and is brittle. While it can be damaged by alkalis and acids it resists corrosion by water and forms a protective oxide when exposed to air. Zinc has a melting point of 787°F (419°C). It is also subject to creep. The tensile strength can be greatly increased by cold working and alloying.

UNS DESIGNATION

Zinc and zinc alloys are designated in the Unified Numbering System by the numbers Z00001 through Z99999. For example Z13001 is identified by the name, zinc metal, and contains 99.90 percent zinc minimum.

WORKING CHARACTERISTICS

Zinc being a soft material can be hot and cold rolled, drawn, extruded, cast, and machined. It can be joined by welding, soldering, and various mechanical fasteners.

Zinc Alloys

Zinc alloys used for die casting consist of about 95 percent zinc and 4 percent aluminum and magnesium. Some copper may be present.

USES OF ZINC

The major use for zinc is to form a protective coating over steel. This is referred to as galvanizing. Galvanizing involves placing the steel to be coated into a bath of molten zinc, which bonds to the surface. It is important that the coating be free of imperfections such as pin holes which would permit moisture to reach the steel causing it to rust. Both galvanized sheet and strip material are available

Since zinc and zinc alloys have low melting temperatures it is easy to cast and is used for some types of hardware and plumbing items. It is usually die cast and finished by polishing or plating with chromium, brass, or other materials.

Zinc also finds use as an alloying element in brasses. Various zinc compounds find use in the production of paper, plastics, ceramics, rubber, abrasives, paint, and other products. Zinc is also used for specialized products where corrosion resistance is important such as anchors, flashing, screws, nails, expansion joints, and corner beads. Solid zinc strip material is used to produce a wide range of products such as low voltage bus bars, cavity wall ties, electric cable binders, electric motor covers, grading screens, and as roofing and fascia material.

Zinc is high on the galvanic table of electrolytic potentials. This means it can be used to coat a material lower on the table to protect the material if galvanic action does occur. The zinc will be sacrificed thus protecting the coated metal.

ZINC GALVANIZING PROCESSES

Zinc sheet, strip, coils, and wire ore galvanized in a continuous process. The processes in use include electrogalvanizing, hot-dip galvanizing, metallic spraying, and sherardizing.

Electrogalvanizing involves placing cleaned steel or iron in an electrolyte solution of zinc sulfate. The electrolytic action deposits a layer of zinc on the material. This thickness of the coating can be controlled but the thickness is limited. Typical thicknesses range from 0.0001 to 0.0005 in. (0.0025 to 0.0127 mm). Hot-dip galvanizing involves immersing clean steel in a bath of molten zinc. It is a semiautomatic process.

Metallic spraying involves coating the sheet iron or steel by applying a fine spray of molten zinc. It can be applied after an installation is complete thus coating the bolts, rivets, and welds.

Sherardizing is a process in which the cleaned iron or steel is placed in a container filled with zinc dust. The temperature in the container is raised and the objects to be coated are tumbled in the dust. The heated zinc bonds to the metal forming a thin coating. This is usually limited to coating small parts and provides minimum protection.

STEEL-FRAME CONSTRUCTION

Steel-framed buildings are a type of skeleton-frame construction in which the walls, floors and roof are supported by a structural framework of steel beams, columns, girders, and related structural elements (see 5-22). Some designs rely in part on the skeleton-frame for support and utilize other methods such as wall-bearing construction where the walls, as masonry, panelized, etc., form part of the structural system. The interior walls in a skeleton-framed building are non-load-bearing, which permits great freedom for utilizing the interior space. The members of the frame are manufactured to design sizes, transported to the construction site, and erected.

The design of the structural frame requires careful engineering analysis. Design loads include live and dead loads as well as loads due to forces such as earthquake, wind, rain, and snow. Soil and hydrostatic pressures act horizontally below grade and must be considered. The stresses to which steel is subjected in use must be considered as the design process continues. Provisions must be made for temporary stresses that occur during construction, such as supporting a crane. The allowable working stresses are regulated by building codes. Detailed information is available from the American Institute of Steel Construction (AISC). Design information for light-gauge steel structural members can be obtained from the Steel Joist Institute and the American Iron and Steel Institute.

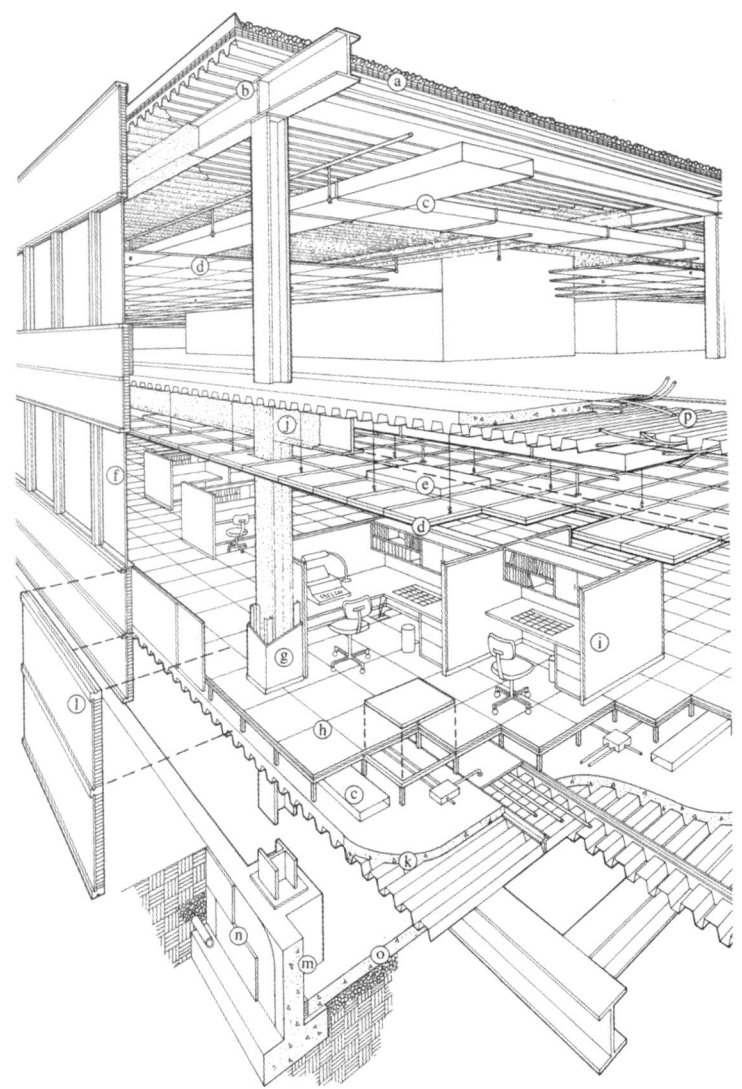

5-22 *This steel-framed building has a cellular steel floor through which electrical wiring can be run. Notice the sprinkler system and suspended ceilings.*

Reproduced with permission from *The Building Systems Integration Handbook,* Richard Rush, editor, Butterworth-Heinmann Publishers, Newton, MA, 1986.

Another design consideration is the economic use of structural steel members. The spacing of columns influences the span of beams and girders. Spans that are too short or too long produce uneconomical use of framing members. Another engineering consideration is the design of connections and the method of securing these. Typically rivets, bolts, or welding is used.

The procedure for the design and construction of structural steel-framed buildings includes the engineering design of the structure, the preparation of shop drawings, the manufacture of the structural members, and their erection on the site.

Structural Steel Drawings

Several types of drawings are developed including engineering design, drawings, shop drawings, and erection plan. The design drawings are prepared by the structural engineer. They indicate the type of construction, and give data on shears, loads, moments and axial forces, which must be resisted by each member and all of the connections. A partial drawing is shown in 5-23. This drawing shows the elevation of the beam as (75'-0") above an established on-site datum. The size of the beams is given and the forces indicated in k (kips). Notes are used to provide additional information. From the information given the structural detailer can prepare shop drawings for the various members.

Shop drawings have all the information needed for the fabrication of the member. It includes the size of the member and the exact location of holes for connections. Provision must be made to allow for clear-

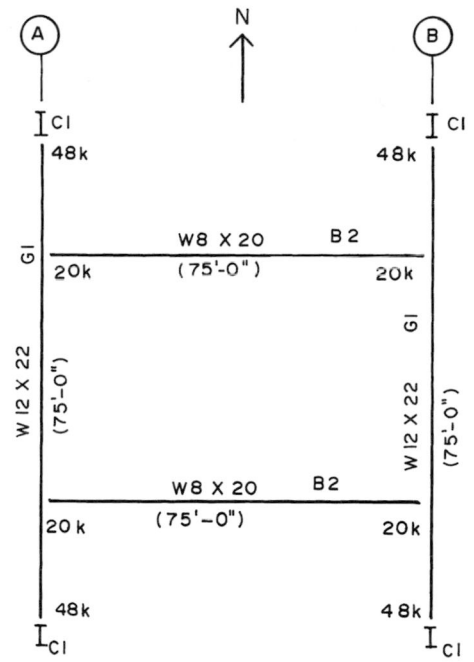

A PARTIAL DESIGN DRAWING

GENERAL NOTES
SPECIFICATIONS: LATEST AISC EDITION
MATERIAL: ASTM A36
FASTENERS: 3/4 Ø A325 IN BEARING TYPE CONNECTIONS
THREADS IN SHEAR PLANES
CONNECTIONS: PER AISC MANUAL AND DEVELOP
INDICATED END REACTIONS

5-23 *A partial engineering design drawing contains the data needed by the structural detailer.*

ances needed so erection can proceed without interference between joining members. You can see a typical example in see 5-24. Notice the use of notes and the mark (B1) identifying the beam.

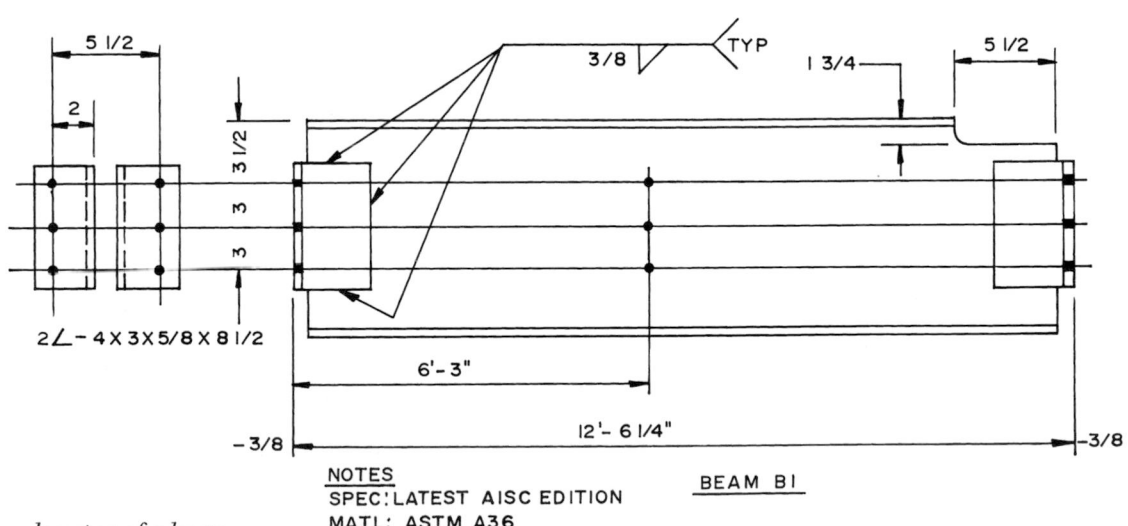

5-24 *A typical shop drawing of a beam that has shop weld connectors that are prepared for bolting on the site.*

NOTES
SPEC: LATEST AISC EDITION
MATL: ASTM A36
OPEN HOLES 13/16" Ø
ONE COAT RED LEAD
WELDS MADE WITH E70XX ELECTRODE

BEAM B1

Erection plans are assembly drawings that are very much like the design drawings. They are assembly-type drawings used on the site to locate each member in its desired location. One example is in 5-25. They show each piece or any subassembly of pieces and their assigned shipping or assembly mark. The erection plan will also include details showing the anchor bolt locations in the foundation. These are used to secure columns to the foundation and are placed in the concrete as the foundation is poured. Therefore their location is critical. The engineering design drawings are basically the same as the erection drawings but have more design detail and in some cases can be used as erection drawings. Typically they are small and crowded with details so erection drawings are usually provided at a larger scale and with only the information needed to identify and locate each member.

Each steel member is identified by an **erection mark,** usually a letter and number giving the part a unique identification. For example, B2 means beam number 2, C2 means column number 2, and G2 means girder number two. All members that are identical will use the same mark. The mark can also include a number indicating the floor where it is to be used. For example, B2(3) means beam B2 is to be erected on the third floor. A column designation as C2(2-4) means that column 2 is in the second tier of the building including the second to fourth floors.

The structural steel members are delivered to the job site with a prime coat and numbered as shown on the erection plan. Connectors are also primed and numbered as required. If a steel member is to be encased in concrete it usually is not primed. This helps the concrete to bond to the member. After erection some field painting is required such as in areas where welding occurred. The welds are chipped free of any slag formed on the surface before they are painted.

The Erection Process

The finished members are delivered to the job site in the order in which they are needed as erection occurs. The erecting company is responsible for setting in place each member of the frame and securing them as required by the design engineer. As the steel arrives on the site it must be placed as near the point of erection as possible and in the order it will be needed.

The erection crew will lift the first level of columns with a ground level crane and place them over anchor bolts cast in the foundation. Columns in multistory buildings are generally two stories high. The columns have a steel baseplate that distributes the load on the column over a larger area of the foundation. The base plates are usually welded to the column when it is manufactured. There are several methods for setting columns. Smaller columns have a steel plate leveled over a bed of grout before the column is erected (see 5-26). Larger columns will have leveling nuts on the baseplate. These are adjusted until the column is plumb. Then grout is worked below the baseplate (see 5-27)

5-25 *An erection drawing of the roof framing, which identifies each structural member by a mark and shows their location.*

The very large baseplates that are required for large, heavy columns may have the baseplate leveled and grouted before the column is welded to it. Large-diameter holes may be specified near the center of the plate so that the grout can be forced below the center of the plate (see 5-28). For additional examples consult the publications of the American Institute of Steel Construction.

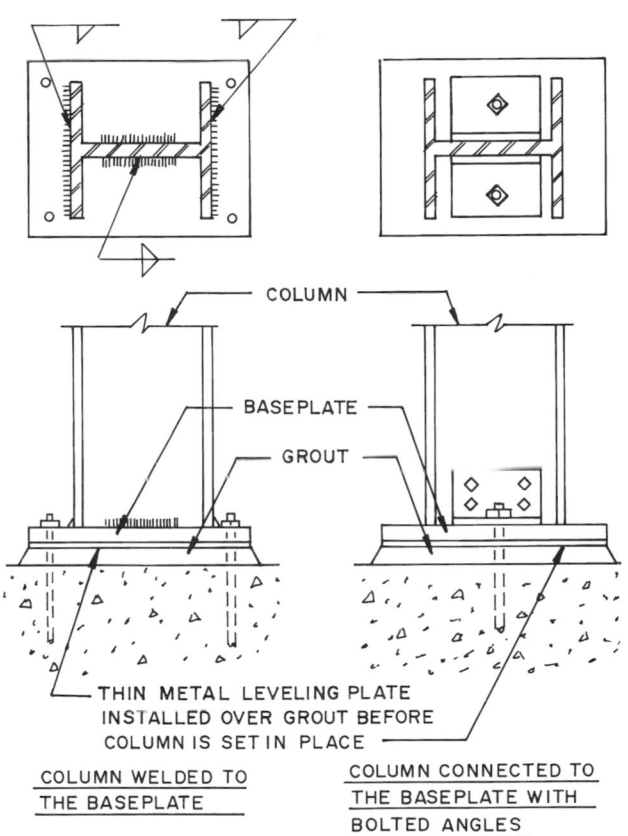

COLUMN

BASEPLATE

GROUT

THIN METAL LEVELING PLATE INSTALLED OVER GROUT BEFORE COLUMN IS SET IN PLACE

COLUMN WELDED TO THE BASEPLATE

COLUMN CONNECTED TO THE BASEPLATE WITH BOLTED ANGLES

5-26 *Typical base connections for small columns.*

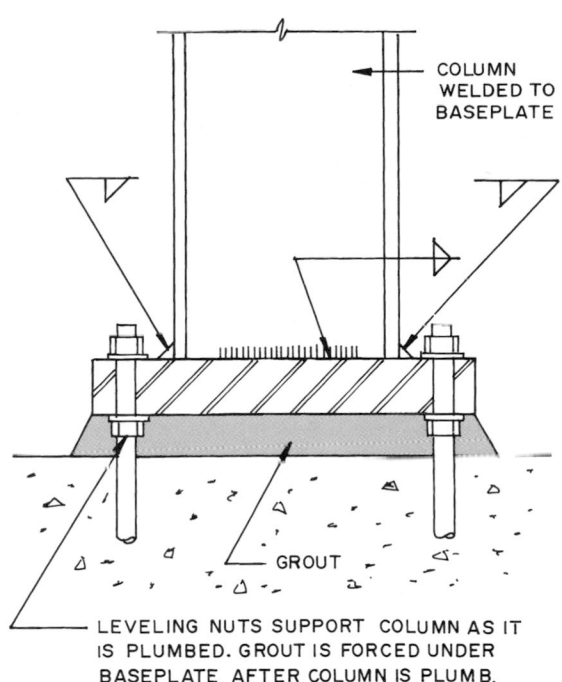

COLUMN WELDED TO BASEPLATE

GROUT

LEVELING NUTS SUPPORT COLUMN AS IT IS PLUMBED. GROUT IS FORCED UNDER BASEPLATE AFTER COLUMN IS PLUMB.

5-27 *Some columns are plumbed with leveling nuts before grouting.*

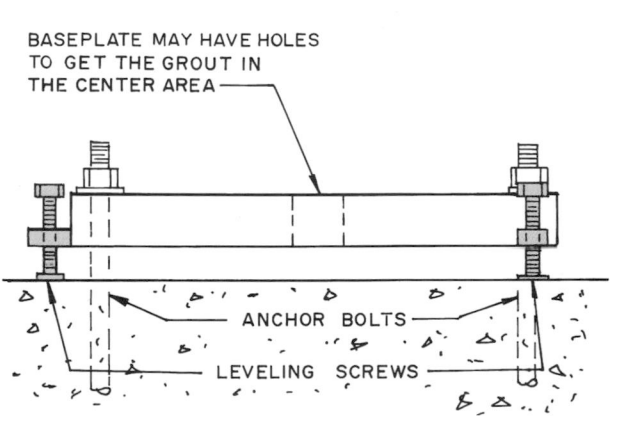

BASEPLATE MAY HAVE HOLES TO GET THE GROUT IN THE CENTER AREA

ANCHOR BOLTS

LEVELING SCREWS

THE BASEPLATE IS SET LEVEL BY ADJUSTING THE LEVELING SCREWS

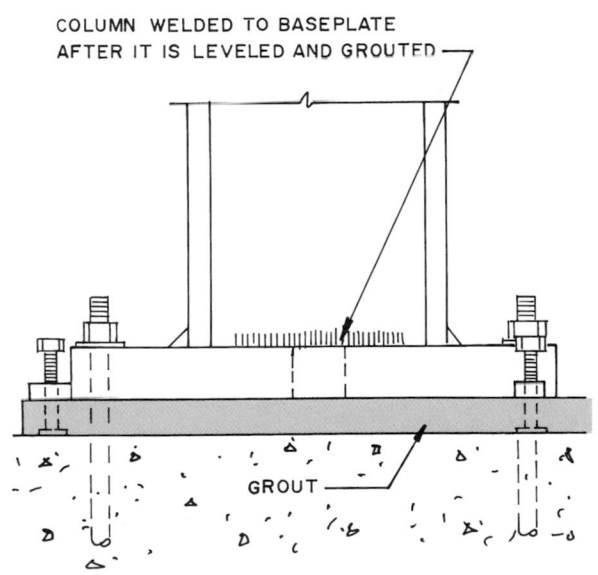

COLUMN WELDED TO BASEPLATE AFTER IT IS LEVELED AND GROUTED

GROUT

BASEPLATE GROUTED AND COLUMN IS WELDED TO IT

5-28 *The baseplate is leveled with the screws before it is grouted or has the column welded to it.*

5-29 *A steel beam is being welded to a steel column.*
Courtesy Bethlehem Steel Corporation

5-31 *These ground-based cranes set the first columns and the connecting beams and purlins.*
Courtesy Bethlehem Steel Corporation

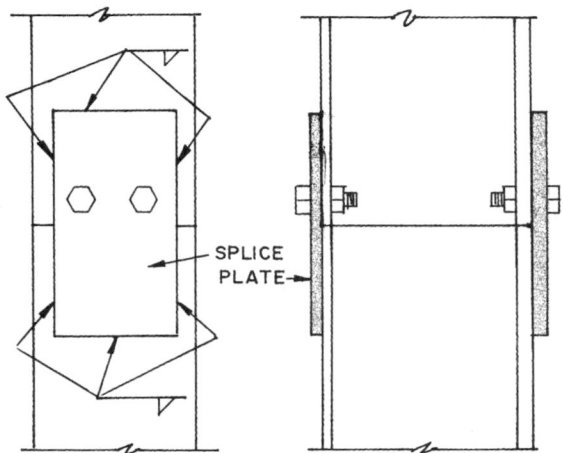

SPLICE PLATE

A TYPICAL WELDED COLUMN SPLICE.

5-32 *Steel decking is welded to the beams and purlins forming the base for the finished floor or roof.*
Courtesy Vulcraft Division of Nucor Corporation

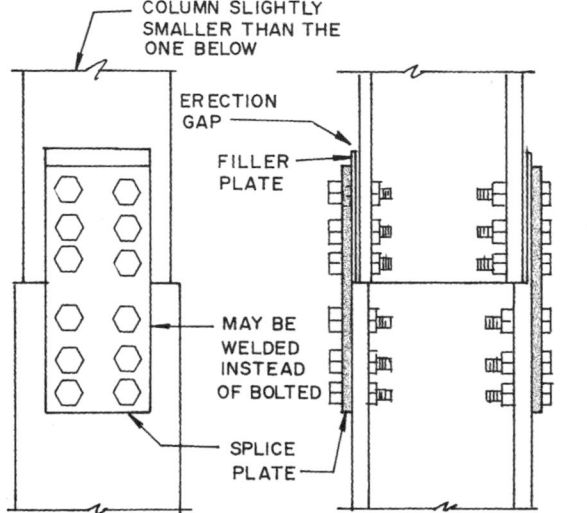

COLUMN SLIGHTLY SMALLER THAN THE ONE BELOW

ERECTION GAP

FILLER PLATE

MAY BE WELDED INSTEAD OF BOLTED

SPLICE PLATE

5-30 *Typical W-column splices.*

5-33 *(Right) Tower cranes are used to reach the upper levels of multistory buildings.*

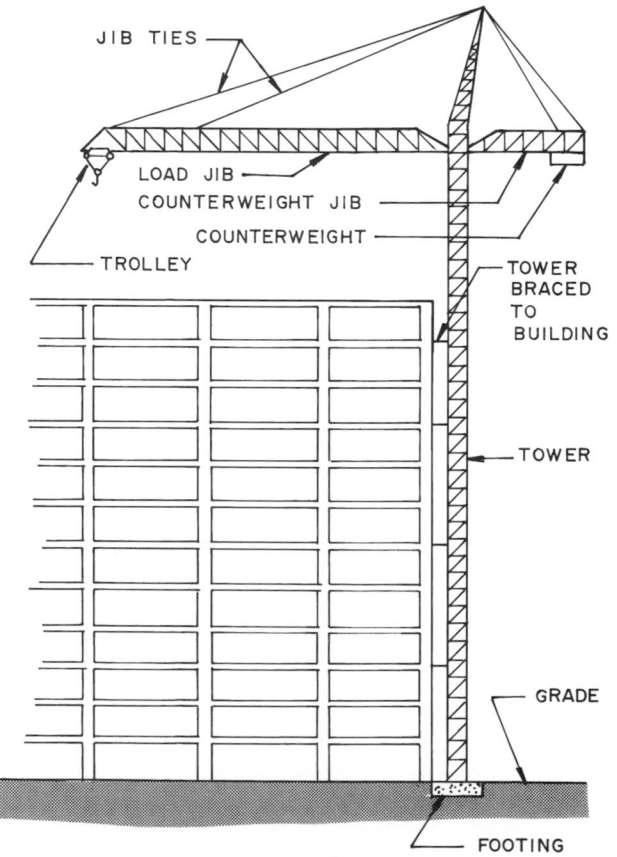

JIB TIES

LOAD JIB
COUNTERWEIGHT JIB
COUNTERWEIGHT
TROLLEY

TOWER BRACED TO BUILDING

TOWER

GRADE

FOOTING

Once the columns are in place the crane lifts the girders and beams and they are secured to the columns with the required connections (see 5-29). Now the next level of columns is erected. The second level of two-story columns is lifted with a crane and connected to the columns below with some type of splice plate (see 5-30). Welded splices are also used. A crane sets the second-level columns, beams, and girders (see 5-31). As the floors are framed the specified decking is installed (see 5-32). In all cases safety regulations are observed, including safety nets, railings, and protection of openings.

As the height of building gets beyond the reach of the ground-based crane a tower crane can be used to reach the upper levels. This might be installed outside the building frame as shown in see 5-33 or be located inside the structure in an elevator shaft opening or a special temporary opening planned for this purpose. This type of crane will have some means of increasing its height as shown in 5-34.

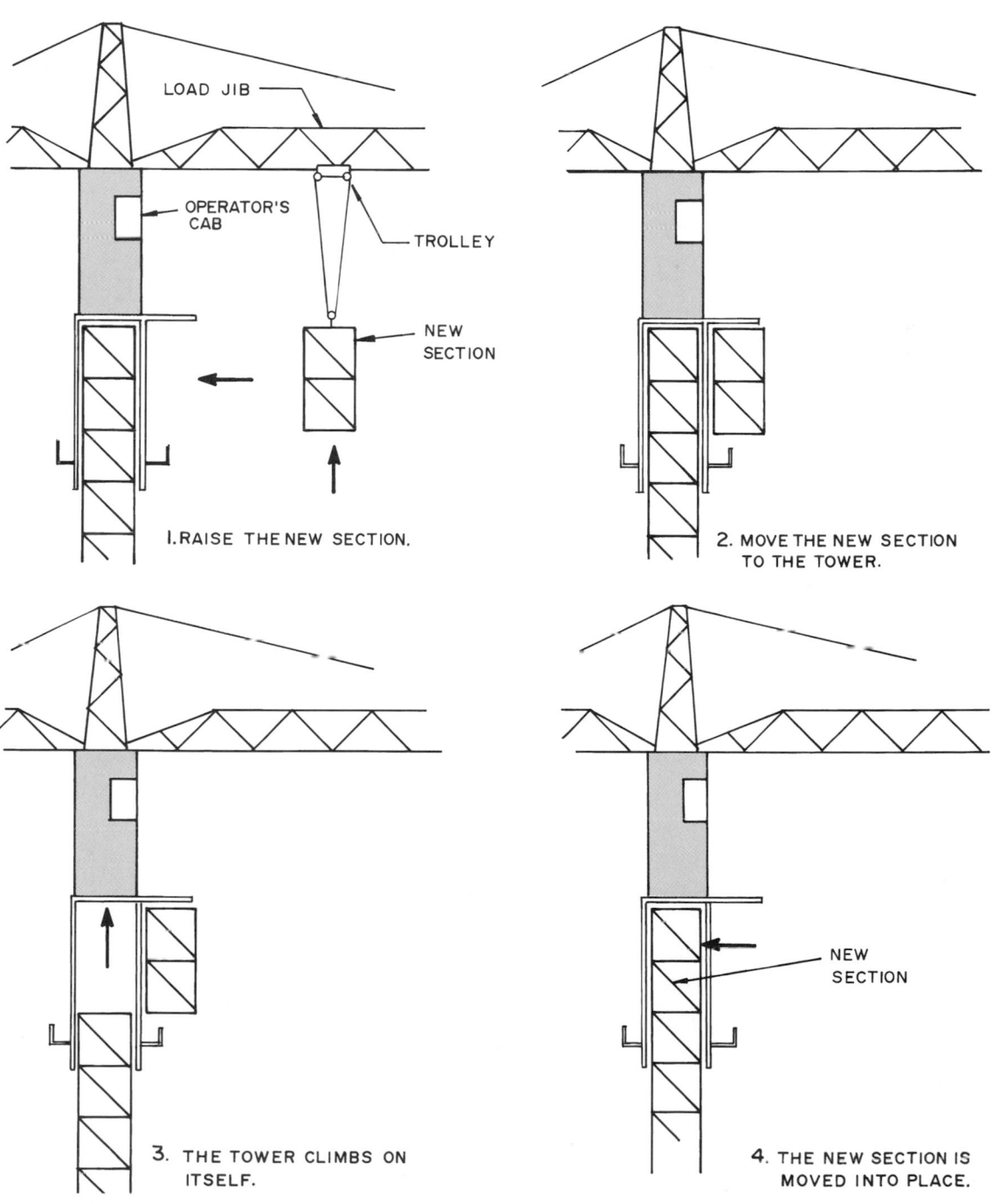

5-34 *The height of the tower crane can be increased by adding additional sections.*

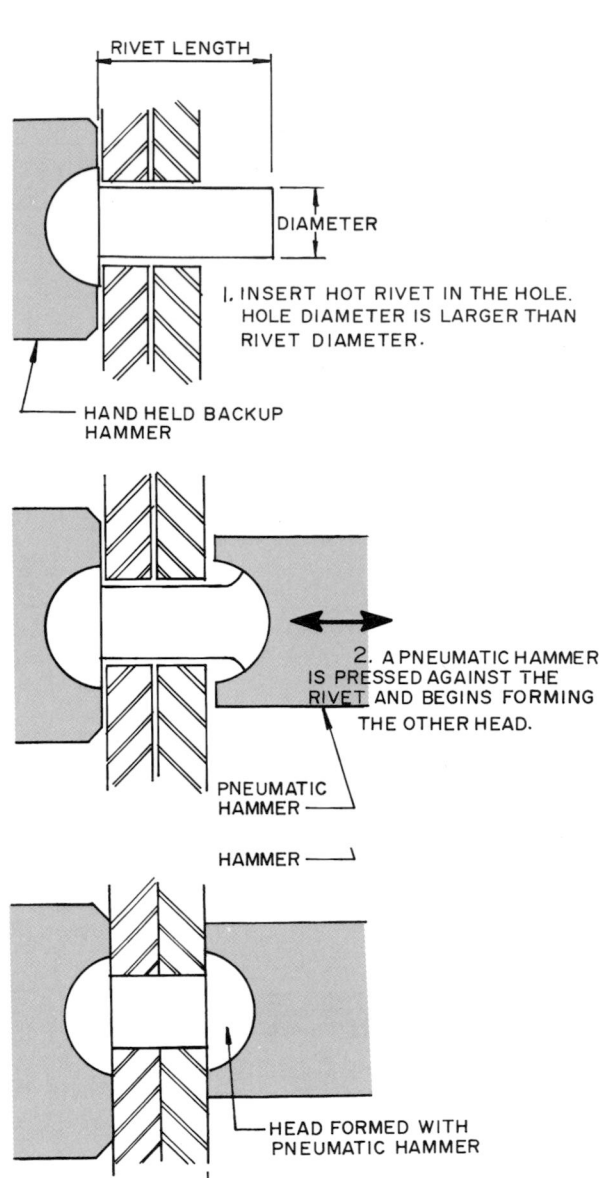

RIVET LENGTH

DIAMETER

1. INSERT HOT RIVET IN THE HOLE. HOLE DIAMETER IS LARGER THAN RIVET DIAMETER.

HAND HELD BACKUP HAMMER

2. A PNEUMATIC HAMMER IS PRESSED AGAINST THE RIVET AND BEGINS FORMING THE OTHER HEAD.

PNEUMATIC HAMMER

HAMMER

HEAD FORMED WITH PNEUMATIC HAMMER

GRIP

3. THE FINISHED HEAD.

5-35 *Structural steel connections can be joined using rivets.*

5-36 *High-strength steel bolts are tightened with a pneumatic impact wrench.*
Courtesy Bethlehem Steel Corporation

Another lifting device for multistory construction is a guy derrick. It consists of a mast, a boom that pivots at the base of the mast, hoisting tackle, and supporting guy lines. The guy derrick can lift itself to floors constructed above it. The hoisting operation is controlled by a power winch located on grade. As the derrick is raised to higher levels the winch remains on the ground. Since the derrick is mounted on the structural frame the loading must be considered as the design is developed. When the frame is completed the derrick is disassembled and lowered to the ground with smaller winches. Additional information can be found in books on rigging.

Fastening Techniques

The steel members, such as beams and columns, in a building frame can be joined with rivets, bolts, or by welding. In some cases a combination of these is used.

Rivets used to connect structural steel members must have properties that enable them to handle the shear, bending, and other stresses developed by the frame. Rivets are made from high-strength, carbon, and alloy steels according to ASTM dimensional and chemical specifications.

Rivets are installed in holes drilled or punched in the framing members. They are heated to a white heat, inserted into the holes of the members to be joined, and a head is formed by a pneumatic hammer producing a head on the other side (see 5-35). When the rivet cools it shrinks, shortens in length, and pulls the members together. The compression and friction developed between the connections being joined assist in resisting shear and tensile stressed in the joint.

Bolts are more commonly used than rivets (see 5-36). Those used to join steel frame members are either carbon steel as specified by ASTM A449 or ASTM A325

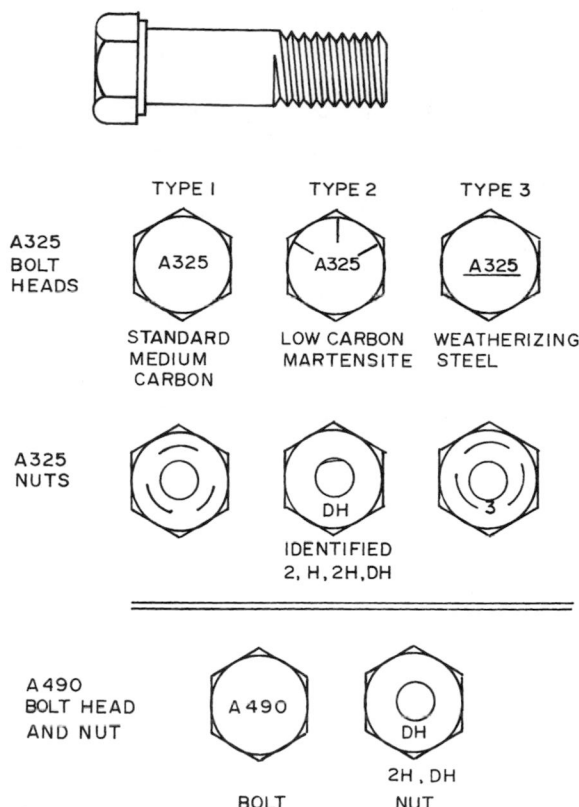

TYPE 1 · TYPE 2 · TYPE 3

A325 BOLT HEADS

A325 — STANDARD MEDIUM CARBON

A325 — LOW CARBON MARTENSITE

A325 — WEATHERIZING STEEL

A325 NUTS

DH — IDENTIFIED 2, H, 2H, DH

A490 BOLT HEAD AND NUT

A 490 — BOLT

DH — 2H, DH — NUT

5-37 *High-strength steel bolts have the ASTM identification markings on the head and nuts.*

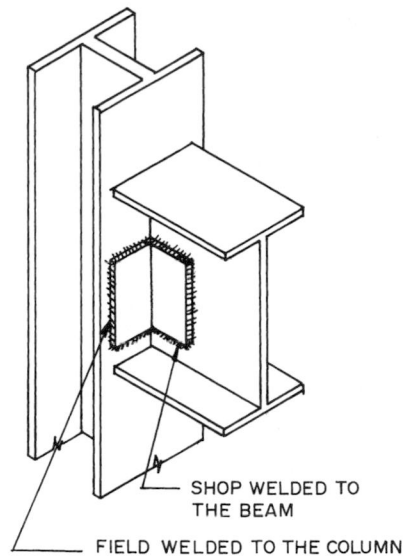

SHOP WELDED TO THE BEAM

FIELD WELDED TO THE COLUMN

5-38 *This angle connector has been welded to the beam during fabrication and to the column during erection.*

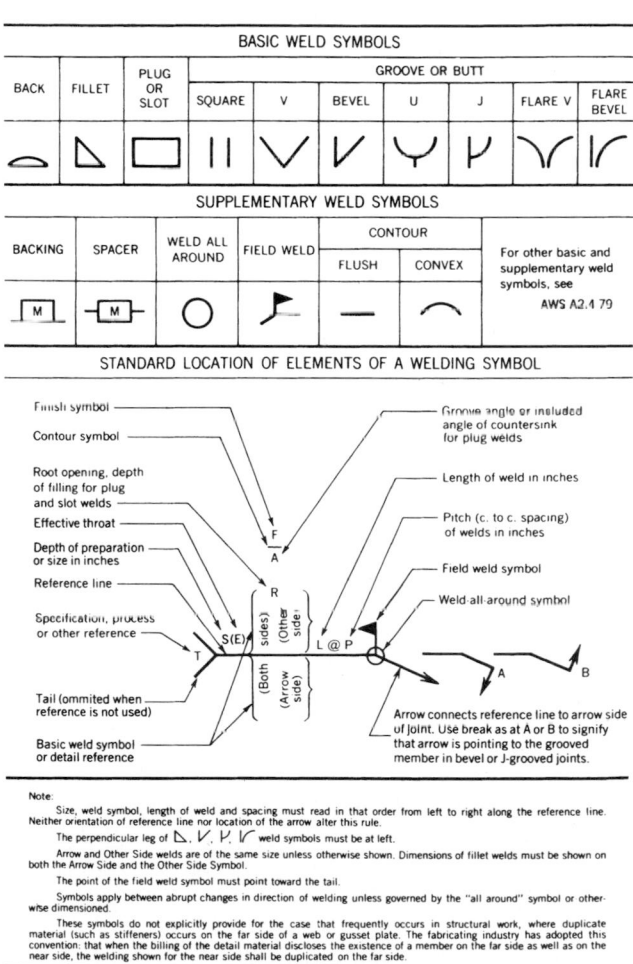

5-39 *Standard symbols for welded joints.*

Courtesy American Institute of Steel Construction

and A490 high-strength bolts. ASTM A449 bolts are used in bearing connections (shear) where their lower strength is adequate. ASTM A325 and A490 high-strength bolts arc used for friction connections. They have a higher shear strength than carbon A449 bolts and high tensile strength. They are identified by markings on the head and nut (see 5-37).

Welding is the third method used to fasten structural steel members. Occasionally welding and bolting are used together on some connections.

Welded-beam connections are of the same general types used for bolted and riveted connections. The connections may be riveted or bolted to the member in the fabrication shop and welded in the shop and the field (see 5-38). When field connections are welded the members are held with several bolts, which hold them together as the field welding occurs. Welds in the fabrication shop are made by clamping the connector to the member and welding it in place. Weld information is shown on structural drawings by symbols. **Weld symbols** are made up of several parts as shown in 5-39.

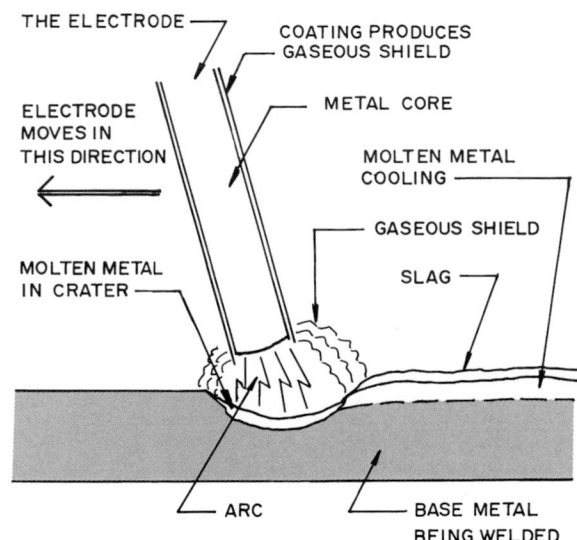

5-40 *The welding arc merges the molten electrode core with the molten base material.*

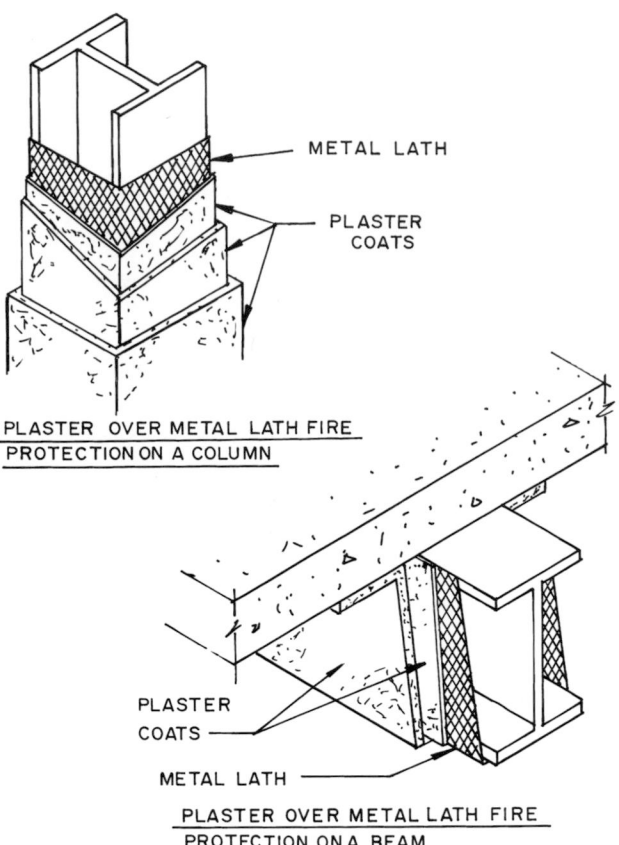

PLASTER OVER METAL LATH FIRE PROTECTION ON A COLUMN

PLASTER OVER METAL LATH FIRE PROTECTION ON A BEAM

5-41 *Lightweight plaster over metal lath provides fire protection for structural steel.*

While there are several types of welding processes **electric arc welding** is generally used for structural steel connections. This involves passing an electric current through a metal electrode (the welding rod) to the piece to be welded. The electrode is kept slightly above the piece causing an arc to jump the gap. This produces considerable heat which melts the electrode and a small area of the members being welded. As the electrode is moved along the line of the weld it leaves a continuous bead of metal that fuses the members together as it cools (see 5-40). Large welds will require several beads be laid on top of each other. The structural engineer determines the size, length, and type of weld to be used. Some welds that are in critical locations (as in a nuclear plant) require that each weld be inspected. There are a number of techniques available for checking welds for flaws.

Fire Protection of the Steel Frame

While structural steel is an incombustible material and will not even melt during a building fire, sustained extreme heat applied to it affects its properties including its strength. This can lead to steel columns and beams failing during a prolonged building fire. To protect the occupants and the structural integrity of the building codes require certain steel frames to be protected with a fire-resistant material. The requirements depend upon the type of construction, building height, floor area, occupancy, the fire-protection system to be installed, and the location of the building. Structural steel may be protected by many materials, including concrete, tile,

brick, stone, gypsum board, gypsum blocks, fire-resistant plasters, sprayed-on mineral fibers, intumescent fire-retarding coatings, liquids, and flame shields.

Unprotected mild steel loses about half its room temperature strength at temperatures exceeding 1000°F (542°C). Fire-resistance ratings are given as the number of hours a material can withstand fire exposure, as specified by standard test procedures. Most standard fire tests on structural steel members are conducted by yhe National Institute of Standards and Technology, Gaithersburg, MD and the Underwriters Laboratories, Northbrook, Il.

Insulating concrete can be used to protect steel members. The amount of protection depends upon the thickness of the cover, the concrete mix, and the method of support. Lightweight concrete (aggregates such as perlite, vermiculite, expanded shale and slag, pumice, and sintered flyash) have greater fire resistance than normal concrete because it has greater resistance to heat transfer and has a higher moisture content. Other insulating materials provide greater protection and are much lighter than concrete.

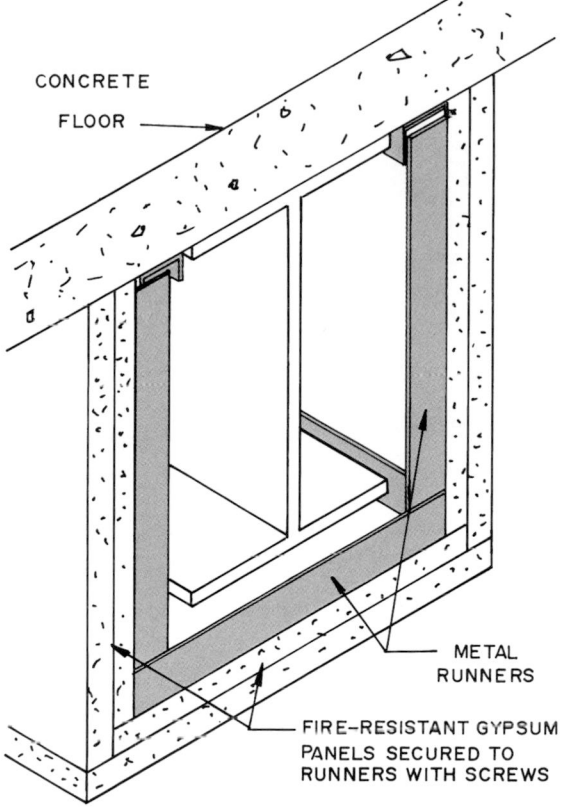

GYPSUM BOARD FIRE RESISTANT BEAM CLADDING

CONCRETE
FLOOR

METAL
RUNNERS

FIRE-RESISTANT GYPSUM
PANELS SECURED TO
RUNNERS WITH SCREWS

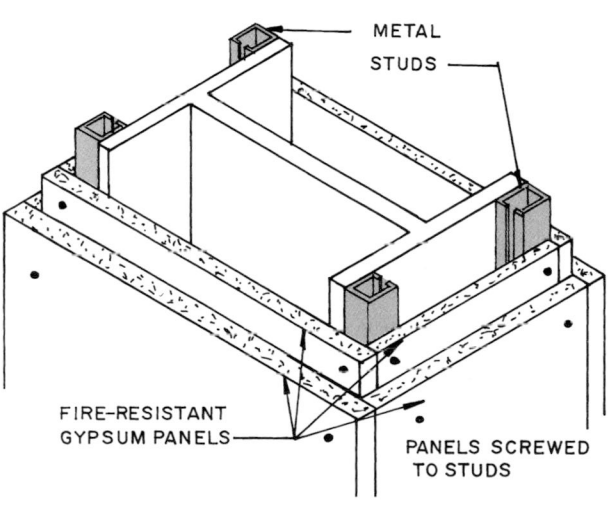

METAL
STUDS

FIRE-RESISTANT
GYPSUM PANELS

PANELS SCREWED
TO STUDS

GYPSUM BOARD FIRE-RESISTANT COLUMN CLADDING

5-42 *Gypsum board products are used to provide fire protection for steel members.*

Masonry units, such as concrete block, brick, gypsum block, and hollow clay tile, are used to enclose structural steel members. In addition to the protection offered by the materials, those with hollow cores can be filled with insulating materials (such as vermiculite) or mortar to increase their fire protection qualities. Masonry protection is heavy and this limits where it is used. Information on the fire resistance of

5-43 *A cementitious material is being sprayed on structural steel members to provide fire protection.*
Courtesy W.R. Grace and Co.

masonry units is available from the National Concrete Masonry Association, Herndon, VA.

Lightweight plaster on metal lath is a lightweight fire resistant protection material. It is also applied over gypsum block (see 5-41). The plaster is troweled over the lath or block. Lightweight plaster using perlite or vermiculite aggregates has good insulating qualities.

Fire-resistive sheet products, such as gypsum board, are secured mechanically to the structural steel members. Those with a hard surface, as gypsum board, can also be painted and serve as the finished surface when used inside a building (see. 5-42).

Sprayed cementitious materials such as gypsum plaster with perlite, vermiculite, or other insulating materials are applied directly to the structural steel (see 5-43). The steel must be clean and primed to receive the coating. Since these coatings can be easily damaged by abrasion they are used where they will not be exposed and are covered with a durable protective material. Thick applications (over 2 in. or 50 mm) require a metal mesh be used.

Mineral fiber slabs are used to enclose structural steel members. Mineral fibers have excellent heat flow retardation properties and can withstand temperature above 1000°F (542°C). The coating is easily damaged and requires a protective covering if exposed to possible impact or the weather.

Intumescent coatings are sprayed on mastic fire-retarding coating. It dries to a hard durable finish much like a paint. When exposed to heat it expands, increasing its thickness and forming an insulation blanket. They are available in colors and can serve as the finished surface.

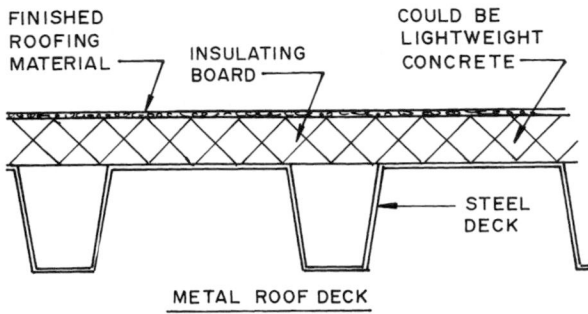

FINISHED ROOFING MATERIAL — INSULATING BOARD — COULD BE LIGHTWEIGHT CONCRETE — STEEL DECK

METAL ROOF DECK

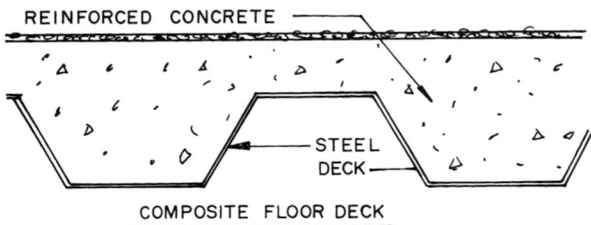

REINFORCED CONCRETE — STEEL DECK

COMPOSITE FLOOR DECK

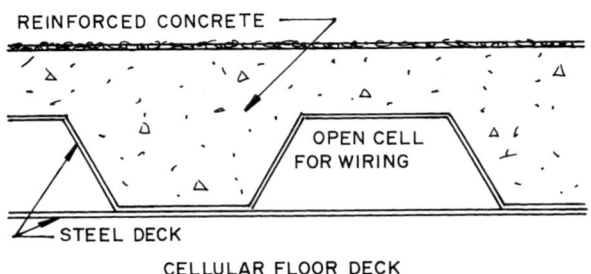

REINFORCED CONCRETE — OPEN CELL FOR WIRING — STEEL DECK

CELLULAR FLOOR DECK

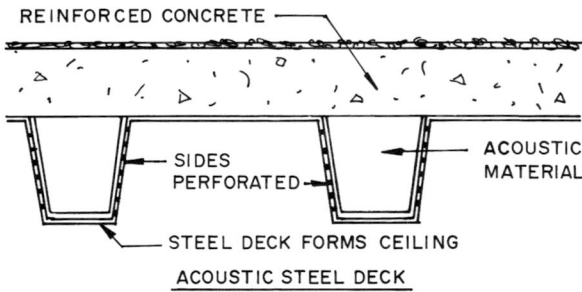

REINFORCED CONCRETE — SIDES PERFORATED — ACOUSTIC MATERIAL — STEEL DECK FORMS CEILING

ACOUSTIC STEEL DECK

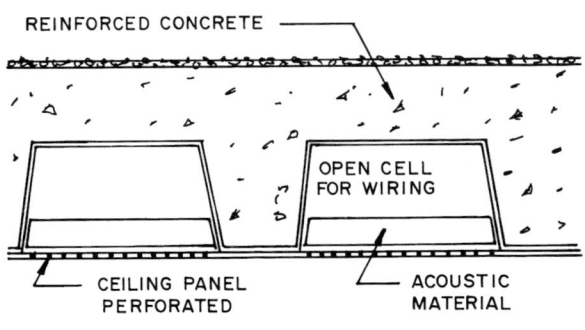

REINFORCED CONCRETE — OPEN CELL FOR WIRING — CEILING PANEL PERFORATED — ACOUSTIC MATERIAL

ACOUSTIC CELLULAR DECK

5-44 *Typical types of metal decking.*

5-45 *Composite metal decking serves as a base for a concrete slab floor or roof.*

Courtesy Vulcraft Division of Nucor Corporation

Liquid-filled columns are used to reduce the heat of the steel. The tube- or box-type columns are filled with water, which is supplied from a central source, such as a water main. This replaces water lost due to heat from the fire. Vent valves release the steam produced. Pumps are sometimes used to maintain circulation within the system. Antifreeze is added if the columns are exposed to freezing temperatures.

Flame shields are metal barriers that deflect the flames and reflect the heat from a fire away from exterior structural steel members. The shield deflects the heat from the flanges of a beam.

The installation of ceilings that provide the needed fire protection for floors and the roof is another procedure used. The design depends upon the situation but plastered ceilings, fire-rated dropped ceilings, acoustic tiles of various types, and drop-in ceiling panels all provide various degrees of fire protection.

Decking

Several types of floor and roof decking are used with structural steel framing. Various types of metal decking and precast concrete units are typically used.

METAL DECKING

Sheet and cellular steel decking are shown earlier in this division. **Sheet decking** is available in various thicknesses of metal and depths of corrugations (see

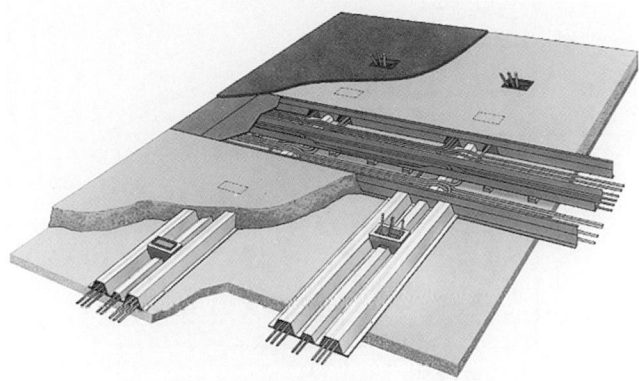

5-46 *A steel cellular floor wire distribution system for installation on a slab-on-grade.*
Courtesy H.H. Robertson

5-47 *These lightweight precast concrete roof deck channels provide a roof deck and can be made with an acoustical treatment on the underside forming an attractive ceiling.*
Courtesy Federal Cement Products.

5-48 *Precast concrete hollow core floor and roof decking can be used with steel or concrete structural frames.*
Courtesy Flexicore Manufacturing Association

5-44). The designer considers the span between supporting members and the loads to be carried as the product is selected. When used for roof decking it is often covered with rigid insulation and the finish roofing material. Under these conditions it must carry all of the imposed loads.

Metal decking provides a base upon which a concrete slab is poured (see 5-45). The slab is reinforced providing most of the structural strength. The metal decking provides tensile reinforcing to the concrete slab because it is bonded to the slab by perforations in the metal or by welded-wire fabric that is tack welded to the decking.

Cellular steel floor raceway systems consist of metal decking that has wiring raceways and a structural concrete slab on top. The cellular raceways provide space to run electrical wiring, telephone, and data cables. These connect to a large, main header duct that has a removable cover for lay-in wiring. Inserts are set that provide access through the concrete slab to the wiring in the raceway (see 5-46).

The cellular system provides a fire-resistant barrier between floors, serves as a form for pouring the concrete deck, and provides tensile reinforcement for the concrete floor slab.

Metal decking is generally plug welded to the steel beams, girders, and joists. The edges are usually joined by welding or self-drilling screws.

CONCRETE DECKING

A number of precast concrete decking units are available for use on structural steel framing. These are shown in Division 3. They include hollow core units and precast slabs and channels of various designs.

Long-span precast concrete roof and floor units are generally prestressed. Short-span members have reinforcement as are cast-in-place members.

Precast slabs and channels are placed on the structural steel framing and form the deck and in some cases the ceiling below (see 5-47). They are secured to beams and joists by welding metal plates cast in them to the steel structure or by using clips.

Precast hollow core decking units are used for long spans and are placed on the steel framing with a crane (see 5-48). Generally the top surface is left rough so the topping placed over it will bond. The topping helps tie the precast units together and adds to the structural integrity of the roof or floor. Reinforcing can be cast in the topping.

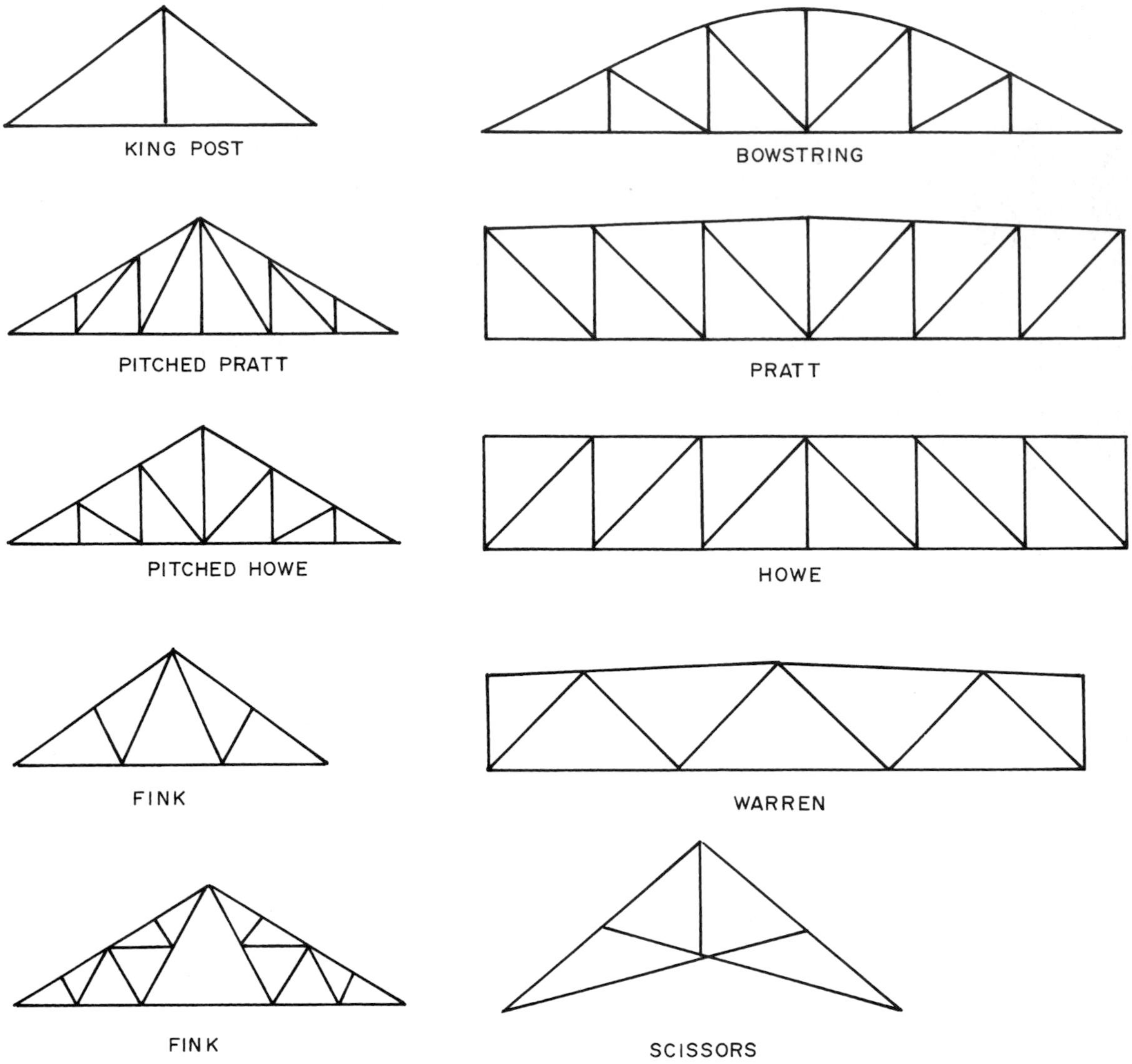

KING POST

BOWSTRING

PITCHED PRATT

PRATT

PITCHED HOWE

HOWE

FINK

WARREN

FINK

SCISSORS

5-49 *Commonly used roof trusses.*

5-50 *Open web joist girders connect the columns and support the long span open web joists. Notice the metal decking being installed.*

Courtesy Vulcraft Division of Nucor Corporation

5-51 *Space frames span long distances and permit the integration of electrical and mechanical systems within the structure.*

Reproduced with permission from *The Building Systems Integration Handbook*, Richard Rush, Editor, Butterworth-Heinmann Publishers, Newton, MA, 1986.

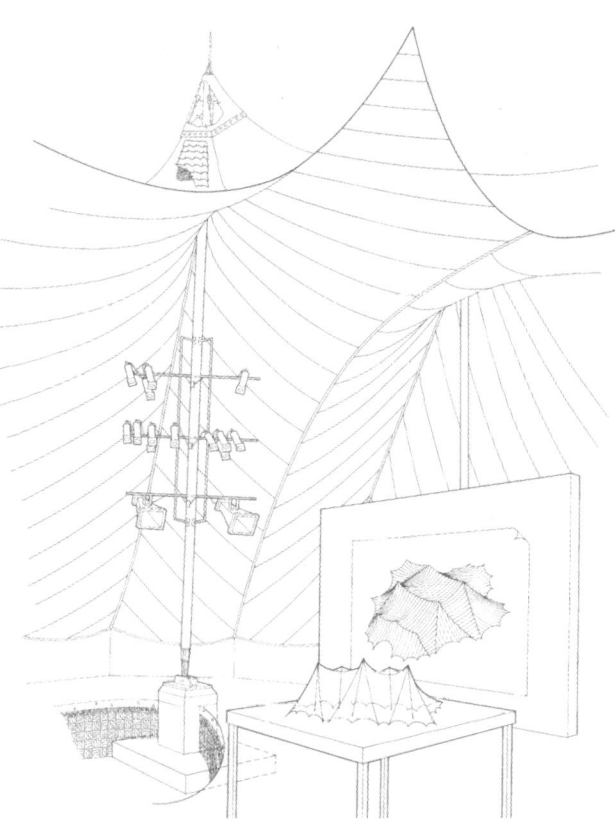

5-52 *Tensile or tensioned fabric structures use a network of steel cables in sleeves in the fabric covering to form the structural support and exterior finish.*

Reproduced with permission from *The Building Systems Integration Handbook*, Richard Rush, Editor, Butterworth-Heinmann Publishers, Newton, MA, 1986.

Steel Trusses

There are a number of steel trusses used for construction purposes. Included are roof trusses, joist girders, open web joists, domes, space frames, and tensile structures. A truss is a coplanar (forces operating in the same plane) assembly of structural members that are joined at their ends. The diagonals form triangles producing a rigid framework. Triangulation is the concept behind the design of a truss. A triangle is the only multisided geometric figure that is rigid and will not move or rock unless a structural member bends or a connection fails.

Roof trusses act like beams and support the roof and all imposed loads. The common types of roof trusses are in 5-49. Parallel chord trusses are used for floors, flat roofs, and as major structural beams and girders.

Open web joists are another type of truss. They are lightweight structural members made from steel angles and bars. They, like other trusses, are end bearing and can span long distances and are used for floor and roof construction. One installation is shown in 5-50.

Space frames are another truss type member. They provide a structural frame that can span in several directions and over large areas with a minimum number of vertical supports. There are various designs used. The space frame is generally made from tubular members (see 5-51). However, wide flange beams and tee members are also used.

The attractiveness of the assembly is lost if it must be covered to meet fire codes. Some codes do not require fire protection if the space frame is over 20 ft. (6.1 m) above the floor.

Tensile Structures

Tensile structures, also referred to as tensioned fabric structures, are framed using high-strength steel cables suspended between supporting members (see 5-52). The steel cables are covered with a tensioned membrane made of a woven electrical grade fiberglass substrate with a Tedflon PTFE coating. These structures can cover large areas. The engineer must design the structure to resist wind uplift and flapping of the membrane.

Light-Gauge Steel Framing

Light-gauge structural steel shapes are formed from flat cold-rolled pieces of carbon steel. The gauge thickness range from No. 12 to No. 20. Some shapes are formed from a single steel sheet while others have several forms shaped and welded together. They are available in nailable and non-nailable types and either galvanized or primed with zing chromate. Some of the commonly available shapes are in 5-53. Load-carrying capacities and spans should be obtained from the manufacturer. Light-gauge metal is widely used in commercial, industrial, and residential construction.

The floors can be steel decking with a concrete topping or precast concrete planks. Roofs can be any decking material available. Stud walls can support lightweight steel or open web joists and various roof trusses or metal rafters. The finished exterior can be any of the materials typically used.

Preengineered Metal Building Systems

Preengineered metal building systems utilize a variety of structural steel and light-gauge steel framing members to construct the building frame. The vertical members are supported on cast-in-place concrete footings. The structural components are designed to carry specifically specified loads. The company manufacturing and selling the systems distributes them through a network of franchised dealers. The dealers work with the customer and architect to produce the structural system required.

An assembly of a preengineered building using rigid frames as the main structure and enclosing materials supported by light-gauge steel members is shown in 5-54. The complete assembly, including the steel frame, girts, purlins, and bracing, provide the strength that is needed.

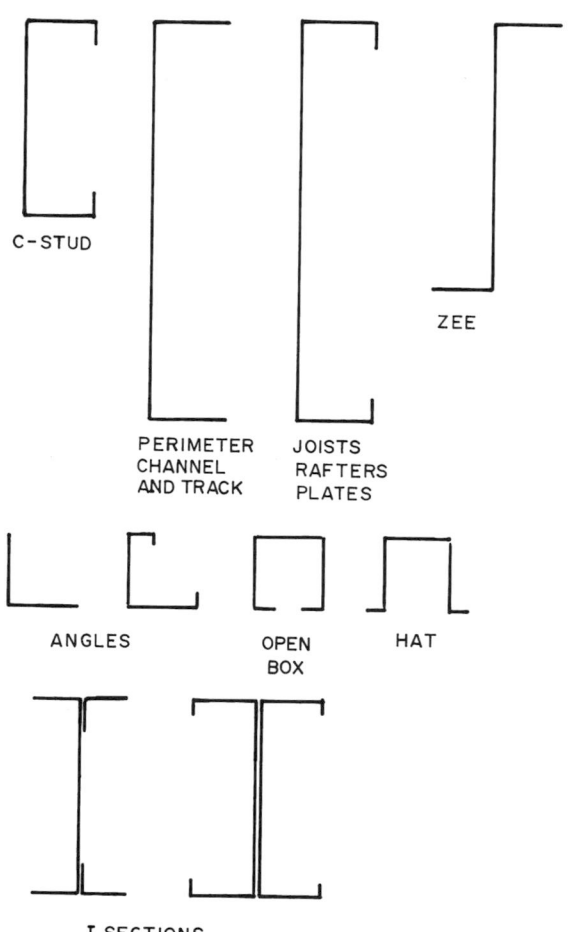

5-53 *Typical cold-formed lightweight structural steel shapes.*

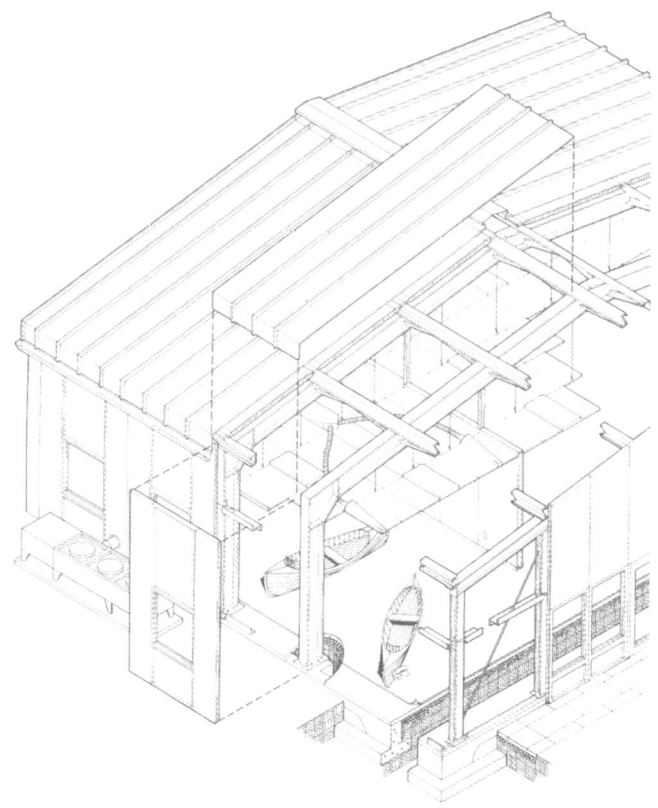

5-54 *This preengineered metal building uses factory-assembled rigid frames for the wall and roof structural members. Z-purlins running perpendicular to the rigid frames support the metal roof.*

Reproduced with permission from *The Building Systems Integration Handbook*, Richard Rush, Editor, Butterworth-Heinmann Publishers, Newton, MA, 1986.

DIVISION 6

WOOD & PLASTICS

CSI MASTERFORMAT™

Courtesy Southern Forest Products Association

WOOD

Wood is a natural organic material. It is unique in that it is a renewable resource. Carefully managed tree farms and natural wild growth provide a continuing source of wood. Wood in the form of lumber and timbers is one of the most familiar construction materials. In addition wood is used to produce a variety of reconstituted wood products such as plywood, particleboard, and hardboard. Wood is also used in the manufacture of paper and cardboard. Wood fibers provide a source of nitrocellulose for the manufacture of explosives.

There are many different species of trees; therefore there is a wide variation in the properties of wood resulting in its being useful for many applications, both structural and decorative. It is important to understand the properties of a species of wood before selecting it for a particular application.

Tree Species

There are several hundred different species of wood growing over the world. Some are in abundance while others are rather rare. Some species, because of their properties, location, and abundance, are used commonly in construction. Woods are divided into two classes, hardwoods and softwoods. This division is based on botanical differences, not on their actual hardness or softness.

SOFTWOODS

Much of the production of wood for commercial use is in the class softwoods. These are used for framing, sheathing, roofing, subflooring, siding, trim and millwork. Softwoods are referred to as **coniferous** because they bear cones and with a few exceptions have needle-like leaves that stay green all year. Refer to 6-10 for examples of commonly used softwoods.

HARDWOODS

Hardwood trees are broadleaved and shed their leaves in the winter. They are called **deciduous**. They are more expensive than softwoods and find use in cabinets, furniture, paneling, interior trim, and flooring.

The Structure of Wood

The root system anchors the tree to the ground. The **tap root** draws minerals and water from the ground. **Side roots** grow out around the tree and **feeder roots** grow off the side roots. They catch soil water and min-

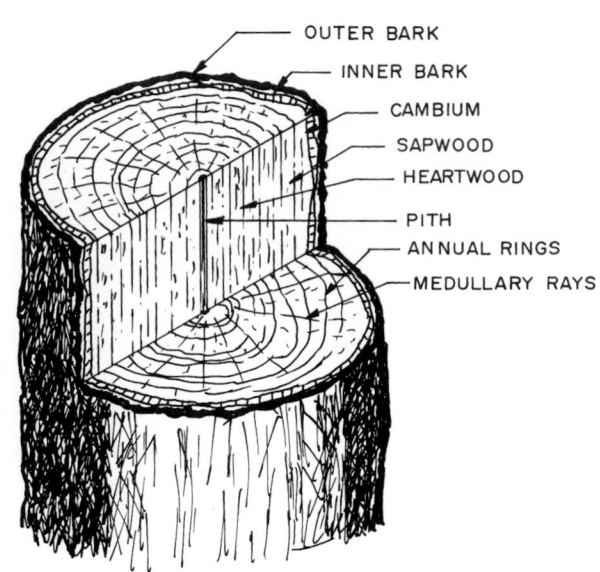

6-1 *The parts of a tree.*

erals in the ground and feed the tree. The water and minerals flow up into the sapwood (xylem). The leaves change water and minerals into food using sunlight and carbon dioxide. This food flows down into the tree through the inner bark (phloem).

Each part of a tree serves a specific purpose in its growth and development (see 6-1). At the center is a small core called the **pith**. During the early years of growth it helps support the stem and feed the tree. As the tree matures it ceases to function. Next to the pith is the **heartwood,** which is hard, mature wood that forms the largest part of the trunk. It serves to strengthen the tree and has the color associated with the particular species. The next layer, **sapwood**, is a living layer and it carries water and food throughout the tree. It is soft, usually light in color, and contains more moisture than heartwood. The heartwood and sapwood form growth rings that can be clearly seen when a tree is cut down. The light, soft ring develops in the spring when the tree grows rapidly and is referred to as **springwood**. The dark, hard ring is formed in the summer when growth is slow and is called **summerwood.** When the tree is cut down its approximate age can be determined by counting the hard summerwood annual rings.

Next to the sapwood is the **cambium layer.** It forms new cells that are either xylem or phloem. Sapwood and heartwood are formed with xylem cells. On the outside of the cambium layer are the phloem cells that

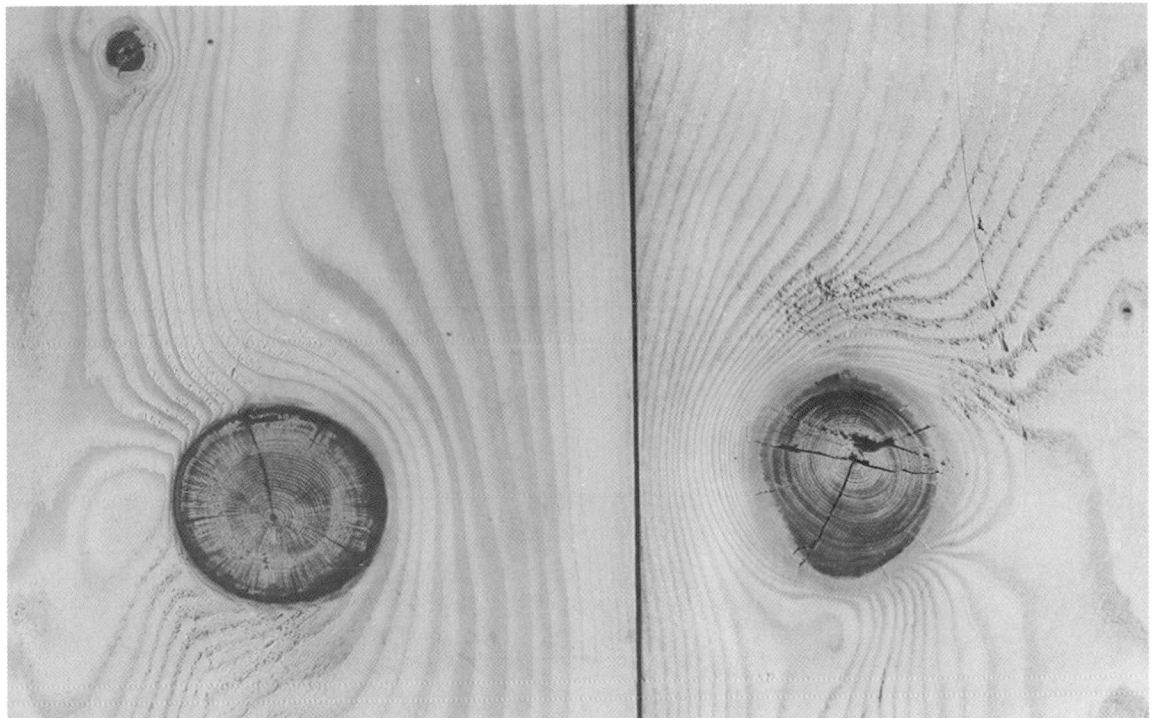

6-2 *Typical knots in softwood lumber. These often are loose and drop out.*
Courtesy Forest Products Laboratory, USDA Forest Service, Madison, Wisconsin.

form the inner bark layer. Phloem cells in the inner bark carry food to the roots. As more phloem cells are formed, the outer layer of the inner bark changes into outer bark. The **outer bark** is the layer that protects the tree.

Within the tree are meduliary rays that run perpendicular to the growth rings. They carry food and water from the cambium layer to the interior of the tree.

WOOD DEFECTS

A piece of wood may have **natural defects** which occur while the tree was growing or have **seasoning defects** produced as the wood is dried for use.

▼ The natural defects include knots, shake, wane, insect holes, and pitch pockets (see 6-2).

Knots—occur when a branch that is imbedded in the trunk as the tree grows is cut. It weakens the wood and is one factor a lumber grader considers as wood is graded.

Pitch—is the accumulation of sap or resin in pockets in the tree. It occurs in softwoods and usually does not cause major problems.

Shake—occurs when a tree is racked or bent as in a wind storm. Shake appears as small cracks running with or across the annual rings.

Wane—refers to the absence of wood or the presence of bark along the edge of a board.

Insect holes—are caused by boring insects that cut their way into the wood.

Seasoning defects include warp, checks, stain, honeycombing, and casehardening. Warp refers to any variation of shape of a board other than a flat, true surface. **Warp** develops as the wood loses moisture and can be controlled by proper seasoning. Some of the common types of warp are cup, bow, crook, and twist (see 6-3).

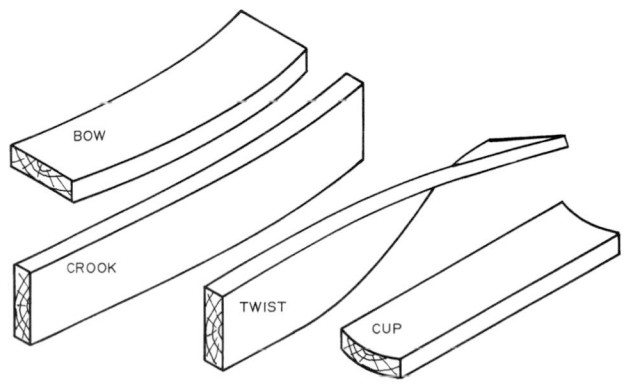

6-3 *Common types of warp.*

Table 6-1 SEASONING REQUIREMENTS FOR SOFTWOODS[a]

Items (Nominal)	Moisture Content Limit	
	Maximum (Dry)	Kiln-Dried (KD or MC15)
D&btr grades		
1" & 1¼"	15%	12% on 90% of pieces 15% on remainder
1½", 1¾" & 2"	18%	15%
Over 2" not over 4"	19%	15%
Over 4"	20%	18%
Paneling		
1"		12%
Boards		
2" and less and dimension 2" to 4"	19%	15%
Decking		
2" thick	19%	15%
3" and 4" thick		15% on 90% of pieces 18% on remainder
Heavy dimension		
Over 2" not over 4"	19%	15%
Timbers		
5" and thicker	23%	20%

[a]As specified by the Southern Pine Inspection Bureau

6-4 *Lumber stacked for air drying before being moved to the kiln.*
Courtesy Southern Forest Products Association

Checks—are small cracks on the surface or ends of the boards that are perpendicular to the growth rings. Wood sometimes develops a **stain** on its surface after it is cut into lumber. It may be brown, green, or blue. This does not influence the strength.

Honeycombing—occurs when cracks occur on the interior of a board. You will notice these when you cut into a board. They occur if the lumber is not seasoned properly.

Casehardening—occurs when the outer surface is drier than the interior of the board and has greater stress than the inner section. Again it is caused by improper seasoning.

Seasoning Lumber

Seasoning is the process of drying wood from the green state to the recommended moisture content desired for finished lumber. The methods of seasoning lumber include air drying, kiln drying, dehumidification, and solar drying. Air drying and kiln drying are most commonly used. Seasoning requirements for softwoods as specified by the Southern Pines Inspection Bureau are in Table 6-1.

AIR DRIED LUMBER

Air drying involves stacking the boards with stickers (wood strips) in between each layer. This permits air to circulate between the layers, aiding the drying process. Air drying is a slow process and does not provide control over how fast the drying occurs. On hot, dry, windy days the wood will season too fast and checking and warping will occur. On cool days with high humidity the wood will dry very little. It is difficult to air dry lumber to the percentage of moisture desired (see 6-4). Some dimension lumber and lower grades of softwood lumber are often air dried. Structural timbers are so large they would take a very long time to air dry and are often shipped green.

The moisture content of air dried lumber for construction purposes should be in the 15% to 19% range in the United States. In some parts of the country, such as the Southwest, it could be somewhat lower than 15%, while in the moist Pacific Northwest it may be closer to 19 to 20%. Lumber grading rules for the various softwoods specify the maximum moisture content for each grade regardless of how it is dried.

Hardwoods are usually air dried for a while, then dried in a kiln. It is difficult to get the moisture content of hardwoods low enough for use in furniture and cabinets by air drying only. Normally a moisture content of 6% to 8% is required for these purposes, so hardwoods have to be kiln dried.

6-5 *Lumber being loaded into a dry kiln.*

Courtesy Southern Forest Products Association

KILN DRIED LUMBER

Kiln drying involves stacking and sticking the lumber as described for air drying. It is stacked in a kiln that is an enclosed building. In the kiln the temperature, humidity, and air circulation are controlled to carefully reduce the moisture content of the wood. Air temperatures in the kiln reach 1800F (820C) with an equally high relative humidity (see 6-5). Since the temperature and humidity are controlled the lumber can be quickly dried to any desired moisture content. Most woods can be dried in less than two weeks in a kiln. Kiln drying reduces the defects in the wood and produces a product that will not expand or contract as much as air dried wood.

DEHUMIDIFICATION AND SOLAR KILNS

These are two relatively new types of kiln. The **dehumidification method** uses electricity to dry the lumber. **Solar kilns** use the energy of the sun to produce the heat needed and are the most economical. Currently those in use can only handle small amounts of wood.

UNSEASONED LUMBER

Most lumber over 2 in. in thickness is air dried. The thicker the stock, the longer the drying time equircd to achieve the desired moisture content of 19% for this type of material. Most stock in these thicknesses is used in a green condition (above 19% moisture content) and continues to dry after it has been used. This means it will shrink in size and sometimes this causes problems. The moisture content can be checked on the job site with a battery-operated moisture meter.

Table 6-2 STANDARD SIZES OF SURFACED DIMENSION LUMBER

Thickness of Stock					
Customary Units (in.)			Metric Units (mm)		
Nominal	Actual Dry	Actual Green	Nomen-clature[a]	Actual Dry	Actual Green
2	1½	1⁹⁄₁₆	38	38.10	30.60
2½	2	2¹⁄₁₆	51	50.80	52.39
3	2½	2⁹⁄₁₆	64	63.50	65.09
3½	3	3¹⁄₁₆	76	76.20	77.79
4	3½	3⁹⁄₁₆	89	88.90	90.49
4½	4	4¹⁄₁₆	102	101.60	103.19
Width of Stock					
2	1½	1⁹⁄₁₆	38	38.10	39.69
3	2½	2⁹⁄₁₆	64	63.50	65.09
4	3½	3⁹⁄₁₆	89	88.90	90.49
5	4½	4⅝	114	114.30	114.47
6	5½	5⅝	140	139.70	142.87
7	6½	6⅝	165	165.10	168.28
8	7¼	7½	184	184.15	190.50
10	9¼	9½	235	234.95	241.30
12	11¼	11½	286	285.75	292.10
14	13¼	13½	337	336.55	342.90
16	15¼	15½	387	387.35	393.70

[a]Nomenclature means the size used to describe the members. Actual size may be slightly larger or smaller.

Courtesy Canadian Wood Council and the Southern Forest Products Association

Lumber Sizes

After the rough sawed wood has been dried it is taken to a planing mill to be surfaced and shaped into products such as boards, dimension stock, timbers, molding, and trim. Softwood lumber is produced in standard sizes. Hardwoods are often dressed to thickness but not to any specific width or length. After the lumber is planed or shaped it must be stored in weatherproof sheds so the moisture content is not increased. Some manufacturers wrap it in plastic sheets.

SOFTWOOD LUMBER

Softwood lumber is sold by its **nominal size**—the size of the board after it has been rough cut at the sawmill. Inch and metric sizes are shown in Tables 6-2, 6-3, and 6-4. The dressed or finished size is the actual size after it has dried and been surfaced—e.g., a typical inch stud has a nominal size of 2 × 4 in. and a finished size of 1½× 3½in. A dry, dressed metric stud is actually 38.10 × 88.90 mm but is referred to as 38 × 89 mm.

Table 6-3 STANDARD SIZES OF SURFACED TIMBERS

Thickness & Width of Stock					
Customary Units (in.)			Metric Units (mm)		
Nominal	Actual Dry	Actual Green	Nomen-clature[a]	Actual Dry	Actual Green
5	-	4½	114	-	114.3
6	-	5½	140	-	139.7
7	-	6½	165	-	165.1
8	-	7½	191	-	190.5
9	-	8½	216	-	215.9
10	-	9½	241	-	241.3
12	-	11½	292	-	292.1
14	-	13½	343	-	342.9
16	-	15½	394	-	393.7
18	-	17½	445	-	444.5
20	-	19½	495	-	495.3

[a]Nomenclature means the size used to describe the members. Actual size may be slightly larger or smaller.

Courtesy Canadian Wood Council and the Southern Forest Products Association

Table 6-4 METRIC LENGTHS FOR SOFTWOOD LUMBER

Nominal Length (ft.)	Metric Length (m)
3	0.91
4	1.22
5	1.52
6	1.83
7	2.13
8	2.44
9	2.74
10	3.05
11	3.35
12	3.66
13	3.96
14	4.27
15	4.57
16	4.88
17	5.18
18	5.49
19	5.79
20	6.10
21	6.40
22	6.71
23	7.01
24	7.32

Courtesy Canadian Wood Council

If stock is surfaced green it is made harder than the dressed size of the dry stock so that when it dries and shrinks it will be the same size as the dried stock. The size of worked stock includes the overall size as shown in 6-6. Worked stock refers to boards that have some machining operation performed on them such as tongue-and-groove flooring. The dressed sizes of green and dry softwood lumber are specified in American Softwood Lumber Standard PS 20-70 and the Canadian National Lumber Grades Authority Standard CSA 0141. American and Canadian standards are coordinated and accepted in both countries. This is based on dry lumber having a moisture content of 19% or less and green lumber over 19%. The stan-

dard sizes for finish lumber, boards, siding and other wood products are in Table 6-5.

Softwood lumber is sold in standard lengths with two-foot multiples ranging from 6 ft. to 24 ft. Some special types, as precut studs, are cut to the exact desired length. Metric lumber lengths are specified in meters (m) and decimal parts of a meter. Metric lengths are shown in Table 6-4.

HARDWOOD LUMBER

Hardwood lumber is sold in random widths and lengths and often is not surfaced to standard thicknesses. Since it is used in the manufacture of cabinets and furniture the boards are cut into many sizes and thicknesses. Cutting to standard widths and lengths would cause great waste. Standard rough thickness in imperial units ranges from ⅜ to 1½ in. in ¼ in. increments. It is also available in 2, 3 and 4 in. thicknesses.

Hardwood lumber in Canada is sized so the metric units and imperial units are within 1% of each other. The standard thicknesses of rough lumber are 15 mm to 60 mm in 5 mm increments and 70 mm to 100 mm

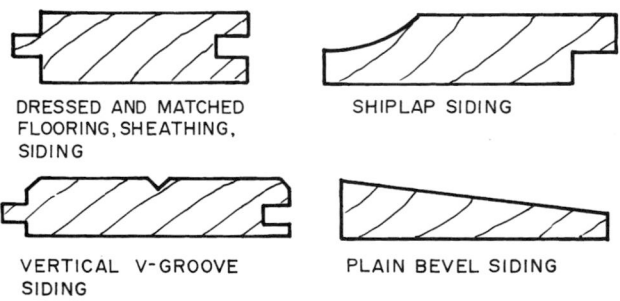

DRESSED AND MATCHED FLOORING, SHEATHING, SIDING

SHIPLAP SIDING

VERTICAL V-GROOVE SIDING

PLAIN BEVEL SIDING

6-6 *Examples of worked stock.*

Table 6-5 STANDARD SIZES FOR SOFTWOOD PRODUCTS

	Thickness (in.)		Width (in.)		
	Nominal	Worked	Nominal	Face	Overall
Bevel siding	½	$^3/_{16} \times ^7/_{16}$	4	3½	3½
	⅝	$^7/_{16} \times ^9/_{16}$	5	4½	4½
	¾	$^3/_{16} \times ^{11}/_{16}$	6	5½	5½
	1	$^3/_{16} \times ¾$	8	7¼	7¼
Drop siding	⅝	$^9/_{16}$	4	3⅛	3⅜
Rustic & drop siding (dressed) and matched)	1	$^{23}/_{32}$	5	4⅛	4⅜
			6	5⅛	5⅜
			8	6⅞	7⅛
			10	8⅞	9⅛
Rustic & drop siding (shiplapped)	⅝	$^9/_{16}$	4	3	3⅜
	1	$^{23}/_{32}$	5	4	4⅜
			6	5	5⅜
			8	6⅝	7⅛
			10	8⅝	9⅛
			12	10⅝	11⅛
Flooring	⅜	$^5/_{16}$	2	1⅛	1⅜
	½	$^7/_{16}$	3	2⅛	2⅜
	⅝	$^9/_{16}$	4	3⅛	3⅜
	1	¾	5	4⅛	4⅜
	1¼	1	6	5⅛	5⅜
	1½	1¼			
Ceiling	⅜	$^5/_{16}$	3	2⅛	2⅜
	½	$^7/_{16}$	4	3⅛	3⅜
	⅝	$^9/_{16}$	5	4⅛	4⅜
	¾	$^{11}/_{16}$	6	5⅛	5⅜
Partition	1	$^{23}/_{32}$	3	2⅛	2⅜
			4	3⅛	3⅜
			5	4⅛	4⅜
			6	5⅛	5⅜
Paneling	1	$^{23}/_{32}$	3	2⅛	2⅜
			4	3⅛	3⅜
			5	4⅛	4⅜
			6	5⅛	5⅜
			8	6⅞	7⅛
			10	8⅞	9⅛
			12	10⅞	11⅛
Shiplap	1	¾	4	3⅛	3½
			6	5⅛	5½
			8	6⅞	7¼
			10	8⅞	9¼
			12	10⅞	11¼
Dressed & matched	1	¾	4	3⅛	3⅜
	1¼	1	5	4⅛	4⅜
	1½	1¼	6	5⅛	5⅜
			8	6⅞	7⅛
			10	8⅞	9⅛
			12	10⅞	11⅛

Courtesy Southern Forest Products Association

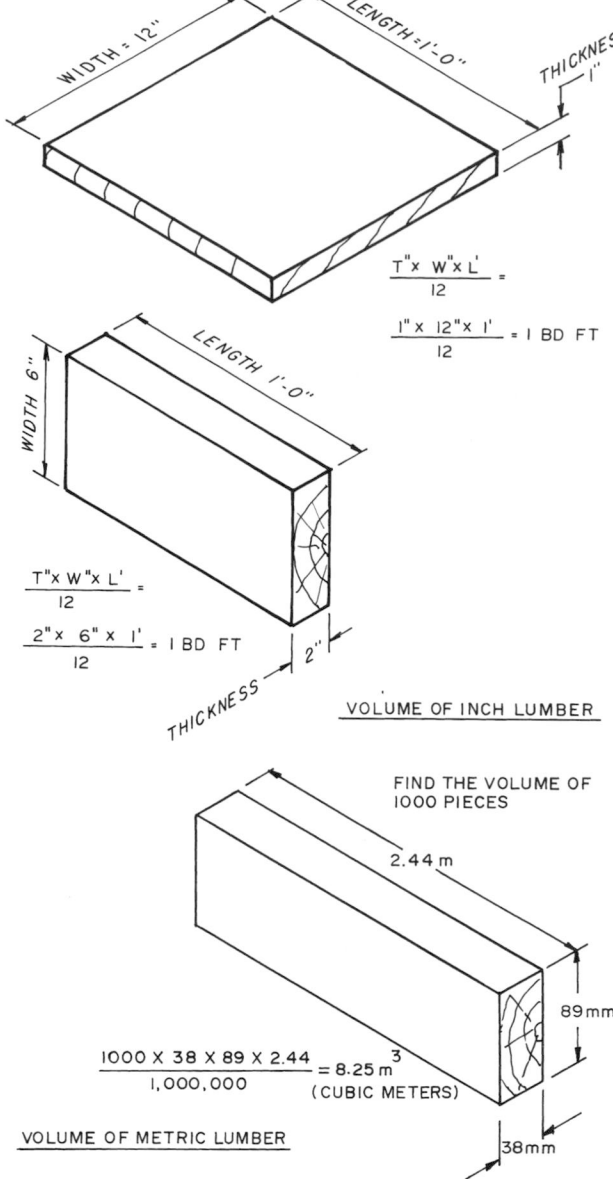

6-7 *Examples representing one board foot and metric volume of lumber.*

6-8 *Southern pine logs being moved to a loading area where they will be put on trucks and moved to a mill.*

Courtesy Southern Forest Products Association

in 10 mm increments. Surfaced lumber thicknesses are 5 mm less than the rough size for green or air dried lumber and 5 to 8 mm for kiln dried lumber. Standard lengths are 1.2 m through 4.8 m in increments of 30 cm. It is sold by volume in cubic meters (m³).

Buying Lumber

Most lumber in the United States is sold by the board foot. A board foot is equal to a piece of lumber with an actual size of 1 in. thick, 12 in. wide, and 1 ft. long. For example, a board 1 in. thick, 12 in. wide, and 8 ft. long contains 8 board feet. These sizes are the nominal sizes. The actual size is ¾ in. × 11¼ in. × 8 ft. Usually stock under 1 in. thick is figured as 1 in. To calculate board feet multiply the thickness in inches by the width in inches by the length in feet and divide by 12— which converts the width to feet (see 6-7).

Metric lumber is sold by the cubic meter (m³). The volume is based on the actual size. This gives the actual volume of the piece in cubic meters. It is computed by the formula:

Volume = thickness (mm) × width (mm) × length (m)

Thickness and width are given in millimeters (mm) and the length is in meters (m). An example is shown in 6-7. To convert to board feet requires the use of tables available from the Canadian Wood Council.

Other materials, such as molding and trim, are sold by the lineal foot. Posts and pilings are sold by the piece. Wood shingles are sold by the square (enough to cover 100 square feet).

Lumber Manufacturing

The production of lumber of all kinds begins in the forest where trees are felled. After the branches are cut away the log is moved to a loading site. It may be skidded to the site by a tractor or hauled down a steep hillside on a long cable called a choker line. Once it reaches the loading site it is generally loaded on trucks (see 6-8). The trucks operate on dirt roads built into the forest especially for the logging operation. At the mill the logs are stacked into large piles and often sprayed with water to keep them from drying out and splitting. Most mills use a mechanical peeler to grind off the bark. After the log has been pealed it is ready to be cut into lumber of various types.

A log is hauled onto the carriage of a large machine called a **heading**. This machine has a saw blade that is most often a band type. Small mills use circular saw blades 3 to 4 feet in diameter. The carriage moves the log into the blade, cutting off slabs of wood. The saw operator, called the sawyer, judges how best to get the most marketable wood from each log and adjusts the machinery to rotate and advance the log into the saw. Some mills use computer controls to assist with this process.

The slabs cut from the log fall on a conveyor and move to an **edger,** where the slabs are cut to desired widths and edges are cut square. Then the slab goes to a **trim saw,** which cuts the rough, square-edged slabs to standard lengths. Hardwoods are not edged or cut to length.

Most lumber used for construction purposes is **plain sawed.** This method produces the maximum yield from the log and the widest stock (see 6-9). **Plain sawed** lumber produces boards that have a broad grain running the length of the board. Another common sawing method, **quarter sawing,** produces boards with the edges of the annual rings showing on the face. This is often done with flooring because the annual rings are hard and withstand wear better than the wide areas of springwood exposed on plain sawed boards. Examples of commonly used softwood showing plain sawed, quarter sawed, and end grain are in 6-10 (on this page and the following page).

Next it passes on a conveyor past a lumber **grader.** The grader usually grades each board and marks it as it passes from the trim saw. The mark indicates the grade. This is a very technical job because of the extensive specifications for each of the many grades and types of lumber. Finally the green lumber goes to an air drying area where it is stacked with sticks between each layer. After a number of weeks the lumber is usually moved, still stacked, to a kiln for the final reduction of moisture.

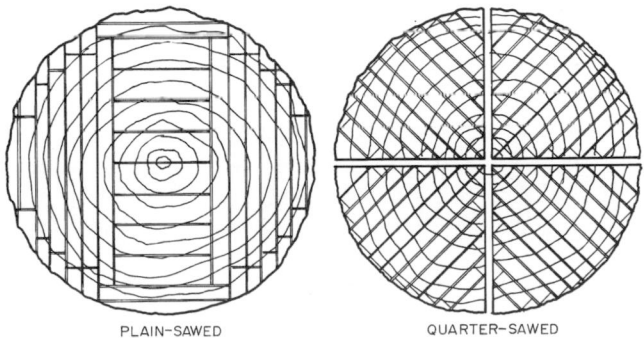

6-9 *Lumber is produced by plain sawing and quarter sawing the logs.*

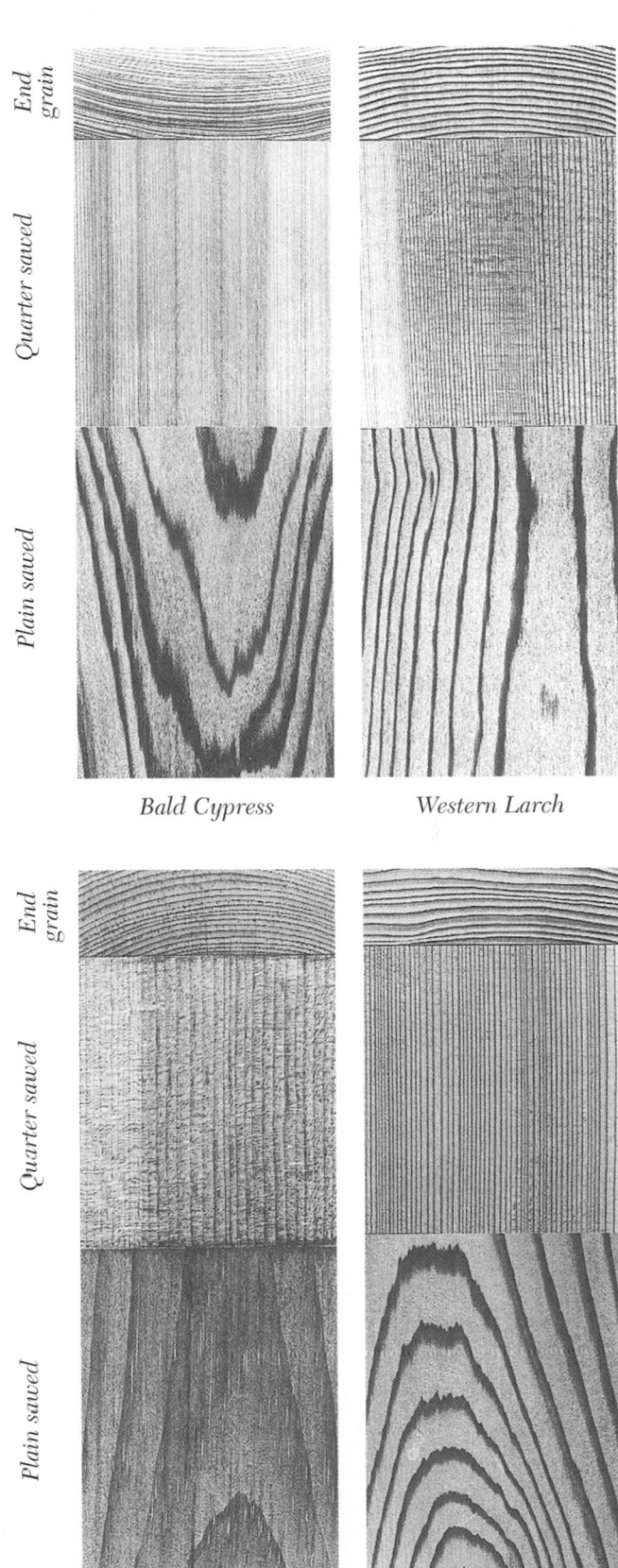

Bald Cypress   *Western Larch*

Western White Pine   *Western Red Cedar*

6-10 *Some of the woods used for construction products. (Continued on the following page.)*

Courtesy Forest Products Laboratory, USDA Forest Service, Madison, Wisconsin

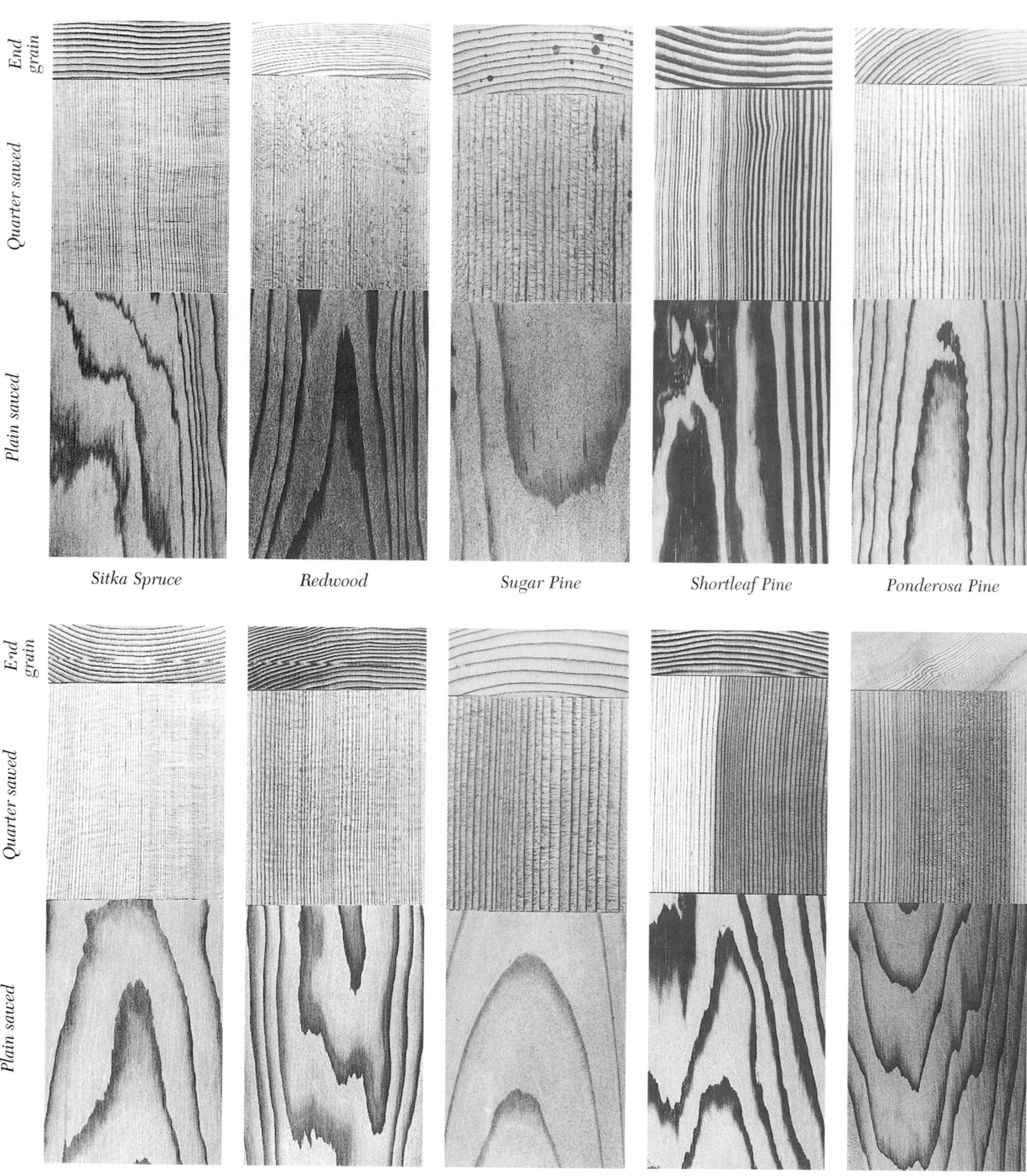

End grain

Quarter sawed

Plain sawed

Sitka Spruce   Redwood   Sugar Pine   Shortleaf Pine   Ponderosa Pine

End grain

Quarter sawed

Plain sawed

Engelmann Spruce   Western Hemlock   White Spruce   Douglas Fir   Incense Cedar

6-10 *(Continued from the previous page.)*
Some of the woods used for construction products.

Courtesy Forest Products Laboratory, USDA Forest Service, Madison, Wisconsin

Table 6-6 U.S. & CANADIAN SOFTWOOD GRADING AUTHORITIES BY REGION

Softwood Region	Species	Lumber Association	Grading Authority
		United States	
Western wood region	Western red cedar Ponderosa pine Douglas fir White fir Western hemlock Englemann spruce Western larch Sitka spruce Lodgepole pine Idaho white pine Sugar pine	Western Wood Products Association	Western Wood Products Association West Coast Lumber Inspection Bureau
Southern pine region	Shortleaf pine Longleaf pine	Southern Forest Products Association	Southern Pine Inspection Bureau
Redwood region	Redwood Douglas fir	California Redwood Association	Redwood Inspection Service
		Canada	
	Spruce Pine, several types Fir, several types Larch Douglas fir Hemlock Western red cedar Aspen Poplar	Canadian Wood Council and 12 lumber grading authorities	Canadian National Lumber Grades Authority

After the lumber has been dried it is dressed to its finished size in a planing mill. Under the American Softwood Lumber Standard, PS 20-70, and Canadian CSA 0141, the lumber can be surfaced green or after it has been dried. When surfaced green it is made slightly larger than the dry size because wood shrinks as it dries. All kiln dried dimensional lumber (up to but not including 5 in. (114 mm) must be marked S-Dry (surfaced dry); however, if it was surfaced green (above 19%) it must be grade stamped S-GRN. If the lumber is surfaced at 15% moisture content it is often stamped MC15. Hardwood lumber is generally shipped dried but unsurfaced to manufacturers using it for furniture, cabinets, and other products.

Lumber Grades

The grades of softwood lumber are based on the American Softwood Lumber Standard, PS 20-70, published by the U.S. Department of Commerce. Under the provisions of this standard, a National Grading Rule Committee (NGRC) was established to develop uniform national wide-grade requirements for dimensional lumber. The specifications developed are known as the National Grading Rule for Dimensional Lumber (NGRDL). They form a part of the grading rules of all wood association grading rules such as those of the Western Wood Products Association, the Southern Pine Inspection Bureau, the West Coast Lumber Inspection Bureau, and the Redwood Inspection Service. The grading rules of these individual associations pertain to those species produced in their geographic area and contain the National Grading Rules for Dimensional Lumber specifications plus additional specifications unique to that association (Table 6-6).

Grades of Canadian lumber are identical with those used in the United States, or the same as the requirements of American Softwood Lumber Standard PS-20-70, and are specified in the National Lumber Grades Authority (NLGA) Publication, *NLGA Standard Grading Rules for Canadian Lumber.* The grades are published by the Canadian National Lumber Grades Authority and appear in the publication *Standard Grading Rules for Canadian Lumber* (Table 6-6).

The mills that manufacture lumber according to the standards of the United States and Canada place a grade stump on each piece (see 6-11).

A description of the most commonly used lumber grades for softwoods as used by the Southern Pine Inspection Bureau is shown in Table 6-7. Other U.S. regional and Canadian associations have established similar grades for the species harvested in their region. The higher the grade the higher the allowable stresses. However, the lowest grades of lumber cost less.

TYPICAL GRADE MARKS USED IN THE UNITED STATES

WEST COAST LUMBER INSPECTION BUREAU

No. 2 grade structural framing, Douglas fir species, maximum 19% moisture content, Mill No. 10.

WESTERN WOOD PRODUCTS ASSOCIATION

Select grade decking incense cedar species, maximum 15% moisture content, Mill No. 12.

SOUTHERN PINE INSPECTION BUREAU

No. 3 light framing, maximum 15% moisture content, Mill No. 7.

REDWOOD INSPECTION SERVICE

Construction grade, redwood species, moisture content over 19 %, Mill No. 50.

TYPICAL GRADE MARKS USED IN CANADA

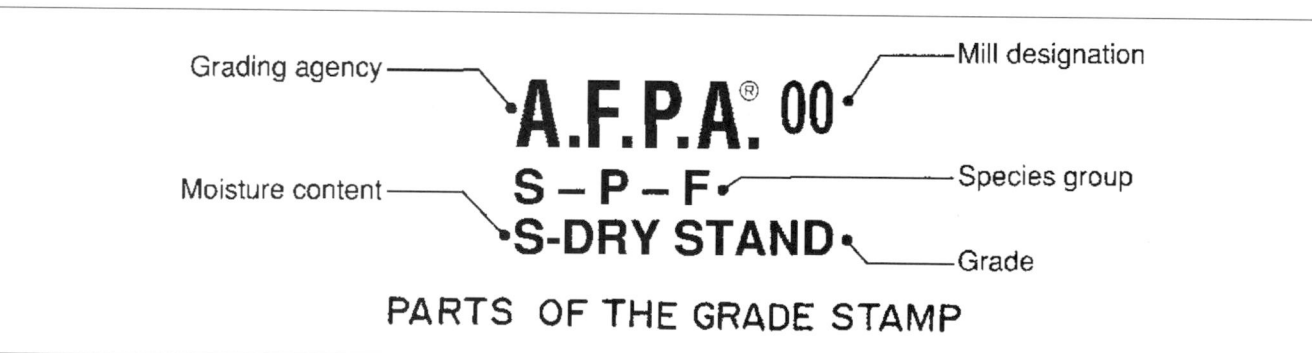

PARTS OF THE GRADE STAMP

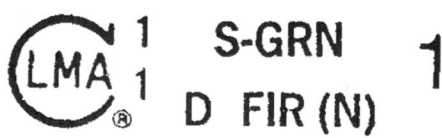

CARIBOO LUMBER MANUFACTURERS ASSOCIATION

205-197 North 2nd Ave
Williams Lake BC V2G 1Z5

6-11 *Typical grade stamps for dimension lumber used by various grading authorities in the United States and Canada.*

Table 6-7 SOUTHERN PINE SOFTWOOD LUMBER GRADES

Product	Grade	Character of Grade and Typical Uses
Finish	B&B	Highest recognized grade of finish. Generally clear, although a limited number of pin knots permitted. Finest quality for natural or stain finish.
	C	Excellent for painted or natural finish where requirements are less exacting. Reasonably clear but permits limited number of surface checks and small tight knots.
	C&Btr	Combination of B&B and C grades; satisfies requirements for high-quality finish.
	D	Economical, serviceable grade for natural or painted finish.
Boards S4S	No. 1	High quality with good appearance characteristics. Generally sound and tight-knotted. Largest hole permitted is ¹⁄₁₆". A superior product suitable for wide range of uses, including shelving, form, and crating lumber.
	No. 2	High-quality sheathing material, characterized by tight knots. Generally free of holes.
	No. 3	Good, serviceable sheathing, usable for many applications without waste.
	No. 4	Admit pieces below No. 3 which can be used without waste or contain usable portions at least 24" in length.
Dimension Structural light framing 2" to 4" thick 2" to 4" wide	Select Structural Dense Select Structural	High quality, relatively free of characteristics that impair strength or stiffness. Recommended for uses requiring high strength, stiffness, and good appearance.
	No. 1 No. 1 Dense	Provide high strength; recommended for general utility and construction purposes. Good appearance, especially suitable where exposed because of the knot limitations.
	No. 2 No. 2 Dense	Although less restricted than No. 1, suitable for all types of construction. Tight knots.
	No. 3	Assigned design values meet wide range of design requirements. Recommended for general construction purposes where appearance is not a controlling factor. Many pieces included in this grade would qualify as No. 2 except for single limiting characteristic. Provides high-quality, low-cost construction.
Studs 2" to 4" thick 2" to 6" wide 10¢ and shorter	Stud	Stringent requirements as to straightness, strength, and stiffness adapt this grade to all stud uses, including load-bearing walls. Crook restricted in 2" × 4"-8' to ¼", with wane restricted to ⅓ of thickness.
Structural joists and planks 2" to 4" thick 5" and wider	Select Structural Dense Select Structural	High quality, relatively free of characteristics that impair strength or stiffness. Recommended for uses where high strength, stiffness, and good appearance are required.
	No. 1 No. 1 Dense	Provide high strength; recommended for general utility and construction purposes. Good appearance; especially suitable where exposed because of the knot limitations.
	No. 2 No. 2 Dense	Although less restricted than No. 1, suitable for all types of construction. Tight knots.
	No. 3 No. 3 Dense	Assigned stress values meet wide range of design requirements. Recommended for general construction purposes where appearance is not a controlling factor. Many pieces included in this grade would qualify as No. 2 except for single limiting characteristic. Provides high-quality, low-cost construction.
Light framing 2" to 4" thick 2" to 4" wide	Construction	Recommended for general framing purposes. Good appearance, strong, and serviceable.
	Standard	Recommended for same uses as Construction grade, but allows larger defects.
	Utility	Recommended where combination of strength and economy is desired. Excellent for blocking, plates, and bracing.
	Economy	Usable lengths suitable for bracing, blocking, bulkheading, and other utility purposes.
Appearance framing 2" to 4" thick 2" and wider	Appearance	Designed for uses such as exposed-beam roof systems. Combines strength characteristics of No. 1 with appearance of C&Btr.
Timbers 5" × 5" and larger	No. 1 SR No. 1 Dense SR No. 2 SR No. 2 Dense SR	No. 1 and No. 2 are similar in appearance to corresponding grades of 2" dimension. Recommended for general construction uses. SR in grade name indicates Stress Rated.
Structural lumber	Dense Str. 86 Dense Str. 72 Dense Str. 65	Premier structural grades from 2" through and including timber sizes. Provides some of the highest design values in any softwood species with good appearance.

Courtesy Southern Forest Products Association

Table 6-8 CLASSIFICATIONS, SIZES, AND GRADES OF SOFTWOOD LUMBER

| Classification | Standard Units | | |
	Thickness Nominal (in.)	Width Nominal (in.)	Grades
Finish	⅜–4	2–16	B&B, C, C&Btr, D
Boards	1½	2–12+	No. 1, No. 2, No. 3, No. 4
Structural light framing	2–4	2–4	Select Structural, No. 1, No. 2, No. 3
Light framing	2–4	2–4	Construction, Standard, Utility
Studs	2–4	2–6	Stud
Structural joists and planks	2–4	5 and wider	Select Structural, No. 1, No. 2, No. 3
Appearance framing	2–4	2 and wider	Appearance
Timbers, nonstress	5 and larger	5 and larger	Square–edge and sound, No. 1, No. 2, No. 3
Timbers, stress-rated	5 and larger	5 and larger	Sel, Str, SR, DNS, Sel Str SR, No. 1 SR, No. 1 DNS SR, No. 2 SR, No. 2 DNS SR

| Classification | Metric Units | | |
	Thickness Nomenclature[a] (mm)	Width Nomenclature[a] (mm)	Grades
Finish	8–64	38–286	B&B, C, C&Btr, D
Boards	17–32	38–387	No. 1, No. 2, No. 3, No. 4
Structural light framing	38–102	38–387	Sel Str, No. 1, No. 2, No. 3
Light framing	38–102	38–387	Construction, Standard, Utility
Studs	38–89	38–140	Stud
Structural joists and planks	38–89	114–387	Sel Str, No. 1, No. 2, No. 3
Appearance framing	38–89	38 and wider	Appearance
Timbers, nonstress	114 and larger	114 and larger	Square edge and sound, No. 1, No. 2, No. 3
Timbers, stress-rated	114 and larger	114 and larger	Sel Str SR, DNS Sel Str SR, No. 1 SR, No. 2 DNS SR, No. 3 DNS SR

[a]Nomenclature means the size used to describe the members. Actual size may be slightly larger or smaller.

Courtesy Southern Forest Products Association (standard) and Canadian Wood Council (metric)

LUMBER CLASSIFICATIONS AS TO MANUFACTURE

Lumber can be classified by the manner of its manufacture.

Rough lumber has not been dressed and shows saw marks on all four surfaces.

Planked lumber is dressed to a size larger than standard dressed sizes but smaller than the nominal size. It may be surfaced on one surface, S1S, one edge, S1E, onto surfacing the four sides, S4S.

Dressed lumber has one or more surfaces smoothed in any combination of surfaces and edges such as S1E, S2E, S1S1E, S1S2E, S4S, etc. For example, S1E means surface one edge, S4S means surface four sides, and S1S1E means surface one side and one edge.

Worked lumber has been dressed and been matched, shiplapped or patterned. **Matched lumber** has a tongue on one edge and a groove on the other, providing a T and G joint. **Shiplapped lumber** has been worked or rabbeted on both edges of each piece providing a lapped joint when the pieces fit together. **Patterned lumber** has been worked to a shaped or molded form in addition to being dressed, matched, or shiplapped.

Resawn lumber is produced by resawing any thickness of lumber into thinner lumber.

Ripped lumber is produced by sawing any width lumber in narrower pieces.

SIZE CLASSIFICATIONS

Following are size classifications used by the Southern Pine Inspection Bureau. A summary is in Table 6-8.

Boards are less than 2 in. in nominal thickness and in widths from 2 in. (50 mm) to 12 in. (305 mm) wide and sometimes up to 16 in. (406 mm).

Dimension lumber is from 2 in. to but not including 5 in. thick and 2 in. or more in width. It is subdivided into five classes, Structural Light Framing, Light Framing, Structural Joists and Planks, Appearance Framing, and Stud grades.

Timbers are 5 in. or more in their least dimension and are subdivided in classes such as beams, posts, girders, etc.

STRESS-RATED LUMBER

Each piece is stress-rated at the mill. The lumber is evaluated by mechanical stress rating equipment that subjects each piece to bending stress. The modulus of elasticity (E), which is a measure of stiffness, is measured by instruments on the machine. The stress grade is electronically calculated and includes consideration of the effect of the slope of the grain, knits, growth rate, moisture content, and density. The machine then puts the grade stamp on the piece.

The grade stamp on the machine stress-rated lumber indicates the stress rating system used to make the test meets certification requirements. The grade stamp shows the agency trademark, the mill name or number, the phrase "Machine Rated," the species and the "E" rating in millions of pounds per square inch (see 6-12).

The Southern Pine Inspection Bureau has established fifteen categories of stress-graded lumber 2 in. or less in thickness. These can be used for most structural purposes such as for trussed rafters. Five classes with lower allowable bending stresses in relation to the modulus of elasticity are used for members where lower bending stress can be used. These are used for floor joists.

The Southern Pine Inspection Bureau also has six grades of stress-rated timbers, 5" × 5" and larger. The grades Select Structural SR, Dense Select Structural

Table 6-9 WESTERN LUMBER SPECIES GROUPS

Douglas Fir-Larch
  Douglas fir
  Western larch
	Spruce-Pine-Fir (South)
	  Engelmann spruce
	  Sitka spruce
	  Lodgepole pine

Douglas Fir-South
  Douglas fir grown in AZ,
  CO, NV, NM, & UT
	Western Woods
	  Ponderosa pine

Hem-Fir
  Western hemlock
  Noble fir
  California red fir
  Grand fir
  Pacific silver fir
  White fir
	Sugar pine
	  Idaho white pine
	  Mountain hemlock
	Western Cedars
	  Incense cedar
	  Western red cedar
	  Port Orford cedar
	  Alaska cedar

SR are No. 1 SR, No. 1 Dense SR, No. 2 SR, and No. 2 Dense SR. There is also a series of grades for stress rated industrial lumber. Other regional associations in the United States and Canada have similar grading procedures.

In-Grade Testing Program

The In-Grade Testing Program was a 12-year research program conducted by the U.S. Department of Agriculture Forest Products Laboratory in cooperation with the United States and Canadian Lumber Industry associations. The program was initiated to verify the softwood lumber design values for visually graded lumber and to provide a scientific basis for wood engineering similar to those used for steel and concrete. Thousands of lumber specimens of many species, grades, and sizes were tested and design values were applied.

The In-Grade Testing Program developed new **species groupings.** Six species groups were developed for Western lumber species and Eastern lumber species (Table 6-9).

```
MACHINE  RATED
(W)®  12   /D\
 WP   S-DRY  \FIR/
1650 Fb 1.5E
```

WESTERN WOOD PRODUCTS ASSOCIATION

Douglas fir species, machine-rated to 2650 psi extreme fiber stress in bending, modulus of elasticity 1,5000,000 psi, maximum moisture content 19%, Mill No. 12.

6-12 *Grade stamp for machine stress-rated lumber.*
Courtesy Western Wood Products Association

The Canadian Species groupings are in Table 6-10. Southern pine is a separate grouping for woods in the southern and southeastern United States. The design values are available in the Supplement to the National Design Specifications for Wood Construction. The various lumber associations also have publications pertaining to the species in their area.

The design date are presented in the form of **Base Values** for the various specie groupings. The Base Values can be adjusted for a particular application in which the structural lumber is to be used. The Southern Forest Products Association refers to these as empirical values. Base Values are assigned to six **Basic Properties** of wood. The six Basic Properties are (1) extreme fiber stress in bending (F_b) (bending strength); (2) tension parallel to the grain (F_t); (3) horizontal shear (F_v); (4) compression parallel to the grain ($F_c//$); (5) compression perpendicular to the grain (F_c) (side-grain crushing), and (6) the modulus of elasticity (E_c or MOE) (stiffness).

▼ These Base Values are adjusted for various Conditions of Use. The seven **Conditions of Use** include:

Size Factors (C_f)—Applied to dimension Base Values

Repetitive Member Factors (C_r)—Applied to size-adjusted Fb (bending stress).

Duration of Load Adjustment (C_d)—Applied to size-adjusted values.

Horizontal Shear Adjustments (C_h)—Applied to Fv (horizontal sheer) values.

Flat Use Factors (C_{fu})—Applied to size-adjusted Fb (bending stress).

Adjustments for Compression Perpendicular to Grain (C_c) (compression values).

Wet Use Factors (C_m)—Applied to size-adjusted values.

The various lumber associations have published new span tables for members, as joists and rafters. These tables include some applications of Conditions of Use. Before using the span tables it is necessary to note which Conditions of Use have been applied.

Physical Properties of Wood

Since wood is a naturally occurring material, it has considerable variation in its properties. For example, it varies in color, density, weight, and strength. The physical and chemical composition of wood determine its

Table 6-10 CANADIAN LUMBER SPECIES GROUPS

Douglas Fir-Larch	Lodgepole pine
Douglas fir	Jack pine
Western larch	Alpine fir
Hem-Fir	Balsam fir
Western hemlock	Northern Species
Amabilis fir	All species graded in
Spruce-Pine-Fir	accordance with the
White spruce	NLGA standard
Red spruce	grading rules for
Black spruce	Canadian lumber
Engelmann spruce	

properties and therefore its uses. Following is a discussion of the properties of wood.

Chemical Composition

Wood is a cellular material. The cells are made of fibers that are mainly **cellulose** and **hemicellulose** that are bonded together with an organic substance called lignin. Cellulose and hemicellulose are complex glucose compounds. Since glucose is a sugar it is useful to fungi and insects as a food. While the exact amounts of each varies with different species of wood, the general composition in kiln dried woods is shown in Table 6-11. These elements are what give wood its strength, susceptibility to decay, and hygroscopic properties. Hygroscopic refers to the property of wood that permits it to absorb and retain moisture. Cellulose provides strength in tension, toughness, and elasticity. Lignin bonds the fibers together in fiber bundles, giving wood its compressive strength. The remaining substances in wood do not contribute to its structure but give each species its unique color, odor, taste, density, resistance to decay, and flammability.

Hygroscopic Properties

Wood expands when it absorbs moisture and shrinks when it loses moisture. Therefore, wood is **hygroscopic**. During the life of a tree it contains considerable water. In this condition it is called green. To be useful in construction, furniture, and other products, it must be in a dry condition. The moisture content of wood must be reduced to a level acceptable for the purpose it is to serve.

Table 6-11 COMPOSITION OF KILN-DRIED WOOD

Material	Softwoods %	Hardwoods %
Cellulose	40–50	40–50
Hemicellulose	20	15–35
Lignin	23–33	16–25
Extraneous materials	5–10	5–10

MOISTURE CONTENT

The moisture content of wood is the weight of the water in the wood divided by the weight of the wood when oven-dryed expressed as a percent.

The moisture content of lumber from newly cut trees varies by species but ranges from 30% to as much as 200%. The moisture content of sapwood is higher than that in heartwood.

FIBER SATURATION POINT

Moisture in the living tree or in freshly cut wood (green) is present in the **cell fibers** as absorbed water and in the **cell cavities** as free water. When the wood begins to dry, the water in the cell cavities begins to disappear. Once all free water in the cell cavities is gone the cell fibers making up the cell walls contain any remaining water. This is the **fiber saturation point.** Most softwoods have a moisture content of about 30% at this point. As moisture is removed from the cell walls they begin to shrink. Therefore the fiber saturation point is the point at which shrinkage begins. The reduction mentioned above, 30 percent to 15 percent, is about half the possible shrinkage.

EQUILIBRIUM MOISTURE CONTENT

Green lumber gives off moisture to the air and if left long enough will lose moisture until it has the same moisture content as the surrounding air. It will continue to take on and give off moisture as the moisture content of the air varies. When this point is reached the wood has reached its **equilibrium moisture content.**

As the wood seeks equilibrium it is gradually shrinking or expanding. Excessive expansion or contraction can cause problems in a building. Doors and windows stick if expansion occurs. Cracks occur in the interior wall finish if studs and plates have excessive shrinkage. Therefore, to minimize the amount of expansion and contraction, wood should be installed at a moisture content as close as possible to the equilibrium moisture content it will have when installed.

HOW MOISTURE AFFECTS WOOD PROPERTIES

The moisture content of wood affects its size, dimensional stability, strength, stiffness, decay resistance, glue bonding, and paintability.

Size and **dimensional stability** are directly affected by the moisture content of wood. Shrinkage and swelling occur only after the moisture content falls below the fiber saturation point. Most woods shrink and swell very little **parallel with the grain** (longitudinally). This shrinkage has very little influence on construction uses. Wood shrinks and swells a great deal in thickness and width **across the grain.** Shrinkage is greatest in the direction parallel to the annual growth rings (tangential) and is about twice as much as the shrinkage across the rings (radial) (see 6-13).

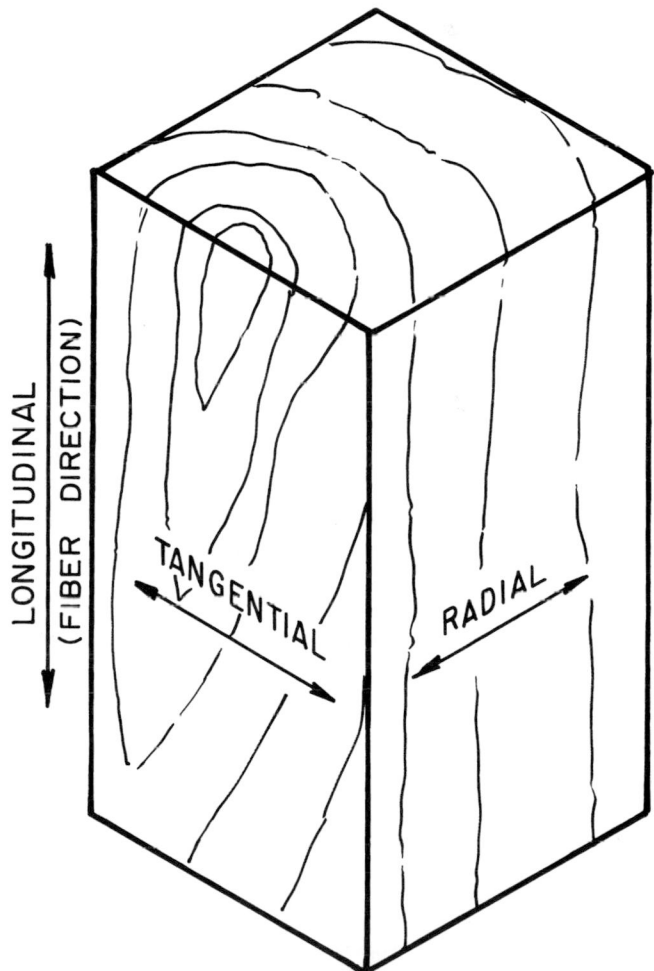

6-13 *Wood shrinks in three directions in relationship with the direction of the grain and growth rings.*

Most hardwoods shrink more than softwoods. Heavier species shrink more than lighter species. Stock with a large cross-sectional area, as 6 × 6 timbers, do not shrink as much proportionately because the inside does not dry at the same rate as the outside. The outer layers dry faster, become set, and keep the inner area from shrinking normally as it dries.

Softwood lumber shrinks about ⅟32 in. per inch of face width while drying from a green condition to about 19% moisture content. This means a rough cut 2 × 10 in. board will shrink about ¼ in. in width as it dries to 19 percent moisture content and about 5/16 to 3/8 in. when it reaches 15 percent moisture content.

As the moisture content of wood decreases its **strength** increases. This is caused by the stiffening and strengthening of the fibers in the cell walls and the fact that the wood shrinks and becomes a denser material.

Crushing strength and bending strength increase a great deal in dry woods. Stiffness increases moderately. Shock resistance depends upon the pliability of the material and is less for dry wood than the original green stock. Green wood will also bend further before it breaks than dry wood.

Wood that maintains a moisture content below 20% will be free from **decay**. This is why it is important to use wood where it is possible to control the moisture content. This can be any of several things including paint and treatment with wood-preserving chemicals.

POROSITY OF WOOD

Wood is a cellular material. Softwood cellular structure contains large longitudinal cells called **trachelds** and smaller radial cells called **rays**. These cells store and transfer nutrients. In addition the annual rings are cellular. The structure of hardwoods is more complex. It has two different types of longitudinal cells, small-diameter **fibers** and larger-diameter **vessels** or **pores**, which transport the sap of the tree. It also has a higher percentage of rays than found in softwoods.

The surface area of these cells is very large and gives wood several important properties. First, it makes it possible for wood, when dry, to absorb toxic chemicals needed to prevent decay and insect attack. Second, it can absorb moisture repellents to minimize moisture exchange and therefore control shrinkage. Third, the cells are air pockets providing insulating qualities. Fourth, it enables wood to shrink and swell as moisture content varies. Fifth, it contributes to the ease of adherence of paint, adhesives, and other synthetic resins used on wood surfaces.

Structural Properties of Wood

Wood is an **orthotropic** material. Orthotropic pertains to or exhibits a mode of growth that is more or less vertical. Wood has unique and independent mechanical properties in the directions of three mutually perpendicular axes: longitudinal, radial, and tangential. The longitudinal axis is parallel to the fibers (grain). The radial axis is normal to the growth rings (perpendicular to the grain in the radial direction). The tangential axis is perpendicular to the grain but tangent to the growth rings. Refer back to 6-13.

Wood is a **fibrous** material. The fibers are bonded together with ligin, forming the walls of the cells making up the material. The fibers in hardwoods are about ⅟25 in. long and from ⅛ to ⅓ in. in softwoods. The strength of wood does not depend upon the length of the fibers but upon the thickness of the cell walls and the direction of the fibers in relation to applied loads.

Structural members when under an **external load,** such as wind, furniture or people, produce internal forces called stresses in a member to resist these external forces. **Tensile** stresses result when an external force tends to stretch a member. **Compressive stresses** occur when a member is under a squeezing force.

BENDING

Wood beams when under load deflect, and **bending stresses** are produced in the fibers. Wood has high fiber strengths in bending. However, if the beam is loaded enough to produce stresses greater than the fiber strength of the wood, the beam will break.

SHEAR

A beam is subject to vertical and horizontal shear. **Vertical shear**, V, refers to the tendency for one part of a beam to move vertically in relationship to an adjacent part, allowing the beam to slip down between supports. **Horizontal shear** refers to the tendency of the sapwood fibers to move horizontally in relationship to the bottom fibers.

MODULUS OF ELASTICITY

The modulus of elasticity, E, is a measure of a beam's resistance to deflection or its stiffness. It is important when selecting columns, beams, rafters, joists, and other structural members.

TENSION

Wood under tension has good tensile strength **parallel** with the grain. It is weak in tension **perpendicular** to the grain.

COMPRESSION

The unit of stress in compression for wood parallel to the grain is several times greater than that perpendicular to the grain.

GLUING PROPERTIES

The glue bond between wood pieces is improved as moisture content decreases with 10 to 12 percent moisture effective for exterior products. In addition, some species of wood bond better than others (Table 6-12). A satisfactory glue joint is one that is as strong as the wood itself. Properly made joints are often stronger than the wood and the wood will break before the joint.

Other Properties of Wood

Wood has other properties, some unique to it as a material, that need to be considered. These include thermal properties, decorative features, decay resistance and insect damage.

THERMAL PROPERTIES

Softwoods have a **thermal conductivity** of approximately 1 Btu/in. of thickness. Therefore, wood is a fairly good insulator but not as good as other materials used specifically for insulation. As the moisture content of the wood increases it becomes less of an insulator because the moisture increases thermal conductivity. The lighter (less dense) woods are better insulators because of their larger cell structure.

Wood also experiences **thermal expansion**. It expands when heated and contracts when cooled. For most construction purposes this is not a factor, because the amount of change is so small. Expansion due to moisture increases is much greater and is the factor to consider.

DECORATIVE FEATURES

Wood has unique decorative features that make it a prized material. Various species have unique **colors** including the mature heartwood and the new growth sapwood. They vary in density, producing wood with **smooth**, closed grain to **rough**, very open grain producing beautiful decorative appearances. Wood can also be stained, bleached to alter the color, and a paste filler can be worked into the grain.

Table 6-12 GLUING PROPERTIES OF SELECTED WOODS

Bond easily	Alder
	Aspen
	Fir: white, grand, noble, pacific
	Pine: eastern, white, western white
	Red cedar, western
	Redwood
	Spruce, Sitka
Bond well	Elm, American rock
	Maple, soft
	Sycamore
	Walnut, black
	Yellow poplar
	Douglas fir
	Larch, western
	Pine: sugar, ponderosa
	Red Cedar, eastern
	Mahogany: African, American
Bond satisfactorily	Ash, white
	Birch: sweet, yellow
	Cherry
	Hickory: pecan, true
	Oak: red, white
	Pine, southern
Bond with difficulty	Persimmon
	Teak
	Rosewood

Courtesy Forest Products Laboratory

Wood also takes on decorative colors and textures when it is permitted to **weather** with no protective coating. Some woods, such as cedar and cypress, are decay resistant and weather to a silver gray. Some woods develop a rough, checked surface when weathered while others do not. All woods being weathered tend to warp, some more than others. Decay-resistant woods, such as redwood, cedar, and cypress, warp less than most. Narrow boards, such as 1 × 6 or 1 × 8 siding, warp less than wider boards.

DECAY RESISTANCE

Some species of wood have **natural decay-resistant** substances in their cell structure that enable the heartwood to resist attack by decay or insects, such as termites. Decay-resistant woods include cedars, redwood, cypress, black walnut, and black locust. Insect-resistant woods include cypress, some cedars and redwood. These species may be used for all aboveground

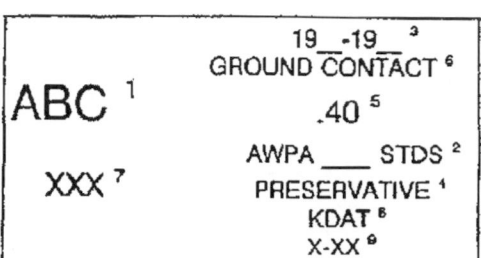

1 - The identifying symbol, logo or name of the accredited agency.
2 - The applicable American Wood Preservers' Association (AWPA) commodity standard.
3 - The year of treatment if required by AWPA standard.
4 - The preservative used, which may be abbreviated.
5 - The preservative retention.
6 - The exposure category (e.g. Above Ground, Ground Contact, etc.).
7 - The plant name and location; or plant name and number; or plant number.
8 - If applicable, moisture content after treatment.
9 - If applicable, length, and/or class.

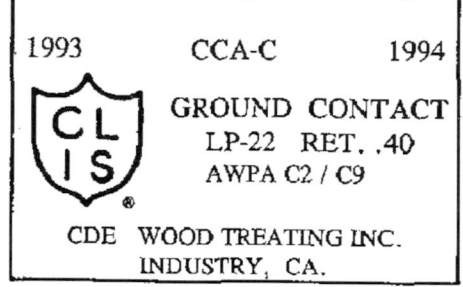

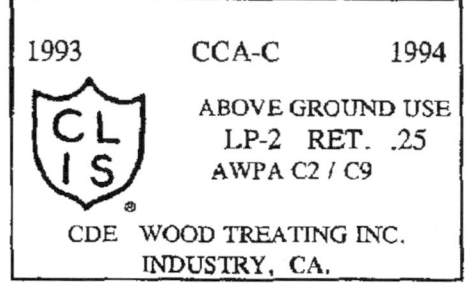

6-14 *The quality mark indicates that the wood has been pressure treated following industry standards.*

Courtesy American Lumber Standard Committee and the California Lumber Inspection Service

uses where they are exposed to the weather. To be most effective they should be 100 percent heartwood. Wood used in the ground as foundations for permanent structures should be pressure treated with a wood preservative.

Wood can be attacked by **fungi** (microscopic plants) causing decay, molds, and stains. fungi develop in wood when the moisture content is above 20 percent, when temperatures are mild (400 to 1000F), there is sufficient oxygen, and the wood provides an adequate food supply.

INSECT DAMAGE

Insects bore holes in living trees and cut lumber. Some holes are very small and are called **pinholes.** Others are larger and are **grub holes.** Lumber graders watch for this damage and take it into account because it can influence the strength of the stock. When structural integrity is important do not use insect-damaged lumber.

Completed buildings are also subject to attack by insects. The termite is the major offender. Most parts of the country have termites, but they occur in larger numbers in the milder climates. Most termites are the **subterranean type**. They live in nests in the ground

and build tunnels through the earth to get to material containing cellulose, which is their food. This type of termite must have a moist atmosphere to exist. When they attack the wood in a house they must build mud tunnels up the foundation to reach the floor joists.

Wood Preservatives

Wood preservatives are used to protect wood from delay and insect attack. Preservative chemicals are forced into the wood to provide needed protection and are a must for wood that is to be in contact with the ground. Nonpressure treatments, such as dipping, brushing, or soaking, are used only on exterior wood that will not be in contact with the ground.

The major types of wood preservative are pentachlorophenol, inorganic arsenic, creosote, and ACQ Preserve®. Pentachlorophenol and creosote are oil borne whereas inorganic arsenic is water borne. Other types finding some applications include copper naphthenate, zinc naphthenate, and chromated zinc chloride.

The American Wood Preservers Association (AWPA) is a professional society responsible for establishing consensus standards for the wood preserving industry. The American Lumber Standards Committee certifies wood

6-15 *The seal of the National Wood Window and Door Association certifies that millwork has been treated with a nonpressure preservative following NWWDA standards.*
Courtesy National Wood Window and Door Association

preservation inspection agencies to provide quality control services to individual wood preserving plants. The accrediting agencies permit the use of their quality stamp indicating to the consumer that the product meets established standards (see 6-14).

The National Wood Window and Door Association (NWWDA) has testing, plant inspection, and certification programs for nonpressure treated products. They provide standards for water-repellent preservatives, wood window units, wood flush doors, wood sliding patio doors, wood skylight and roof windows, and wood swinging doors. A typical NWWDA quality certification seal is shown in 6-15.

Fire-Retardant Treatments

There are many applications where wood must be treated with fire-retardant chemicals. The two general methods for applying fire-retardant chemicals are **pressure impregnating** the wood with water-borne or organic solvent-borne chemicals or applying fire-retardant **chemical coatings** to the wood surface. Pressure treating chemicals include inorganic salts and more complex chemicals. Salts are most commonly used. They react to temperatures below the ignition point of wood, causing the combustible vapors generated in the wood to break down into non-flammable water and carbon dioxide. After treatment the wood should be dried to its original required moisture content.

CHEMICAL COATINGS

Coatings have low surface flammability and when exposed to fire form an expanded low-density film. The film insulates the surface from high temperatures.

FIRE-RETARDANT CHEMICALS

Most fire-retardant chemicals do not resist exposure to weather so it is necessary to use leach-resistant types for exterior use, such as on wood shakes. Fire-retardant treatment results in some slight reduction in the strength properties of wood so design values for allowable stresses are reduced. Most chemicals cause fasteners to corrode. Designers must select a combination of chemicals and metal fasteners that can coexist without corrosion. Crystal salts in wood have an abrasive effect on cutting tools. Carbide-tipped cutting tools should be used when working with fire-retardant wood. Gluing is also a problem. Special resorcinol-resin adhesives have proven acceptable. Fire-retardant wood can be painted if the moisture content has been reduced sufficiently.

PLYWOOD

Plywood is a structural wood panel made by bonding three or more thin layers or plies of wood. It is a strong, dimensionally stable product that resists splintering and splitting. It has less warp and twist than solid wood panels.

Some grades of veneered plywood panels are manufactured under specifications or performance testing standards of U.S. Product Standard PS 1-83 Construction and Industrial Plywood. This is a manufacturing specification developed cooperatively by members of the plywood industry and the Office of Product Standards Policy of the National Bureau of Standards. Other veneered panels including a number of **Performance Rated** composite and nonveneered panels are manufactured under provisions of the performance standards of APA-The Engineered Wood Association. APA rated panels meeting PS 1-83 requirements have the designation "PS 1-83" in the APA trademark.

▼ In Canada there are three standards for softwood plywood, all of which are in metric terms.
　　CSA 0121-M Douglas Fir Plywood
　　CSA 0151-M Canadian Softwood Plywood
　　CSA 0153-M Poplar Plywood

Plywood Panel Construction

Plywood panels are made by bonding together thin layers or plies of wood. The grain in each ply is perpendicular to the ply below it. Panels always have an odd number of plies, such as three, five or seven. Each ply may be a single thickness veneer or two veneers glued together to form a thicker ply.

The plywood constructions commonly used in building construction are the veener-core, lumber-core, particleboard-core, and medium-density fiberboard core (6-16). The **veneer-core panel** has three to nine plies. The **lumber-core panel** has strips of solid wood glued together as do the core and two veneers glued to both sides. Particleboard core and medium-density fiberboard-core panels have a single sheet of one of these materials on the core and a single ply glued on each side. Notice that each type has the same number of plies on each side of the core. This provides **balanced construction.**

Construction & Industrial Plywood

Construction and industrial plywood are manufactured according to the specifications in U.S. Product Standard PS 1-83. Canadian plywood is made following the standards detailed earlier in this chapter.

The outer veneers of construction and industrial plywood are classified in five appearance groups: N, A, B, C, and D. N is the best and D is the poorest grade (6-17).

Construction and industrial plywood is made using about 70 species of wood. These may be mixed within a panel and may be hardwoods and softwoods. The species used are divided into five groups depending upon their strength and stiffness. Group 1 includes the strongest and stiffest species. Group 5 includes those with the lowest properties (Table 6-13). The inner plies in Groups 1, 2, 3, and 4 may be any of the species in Groups 1, 2, 3, and 4. Inner plies in Group 5 may be any of the species listed. The front and back plies are of the same species and must be from the group indicated by the group number.

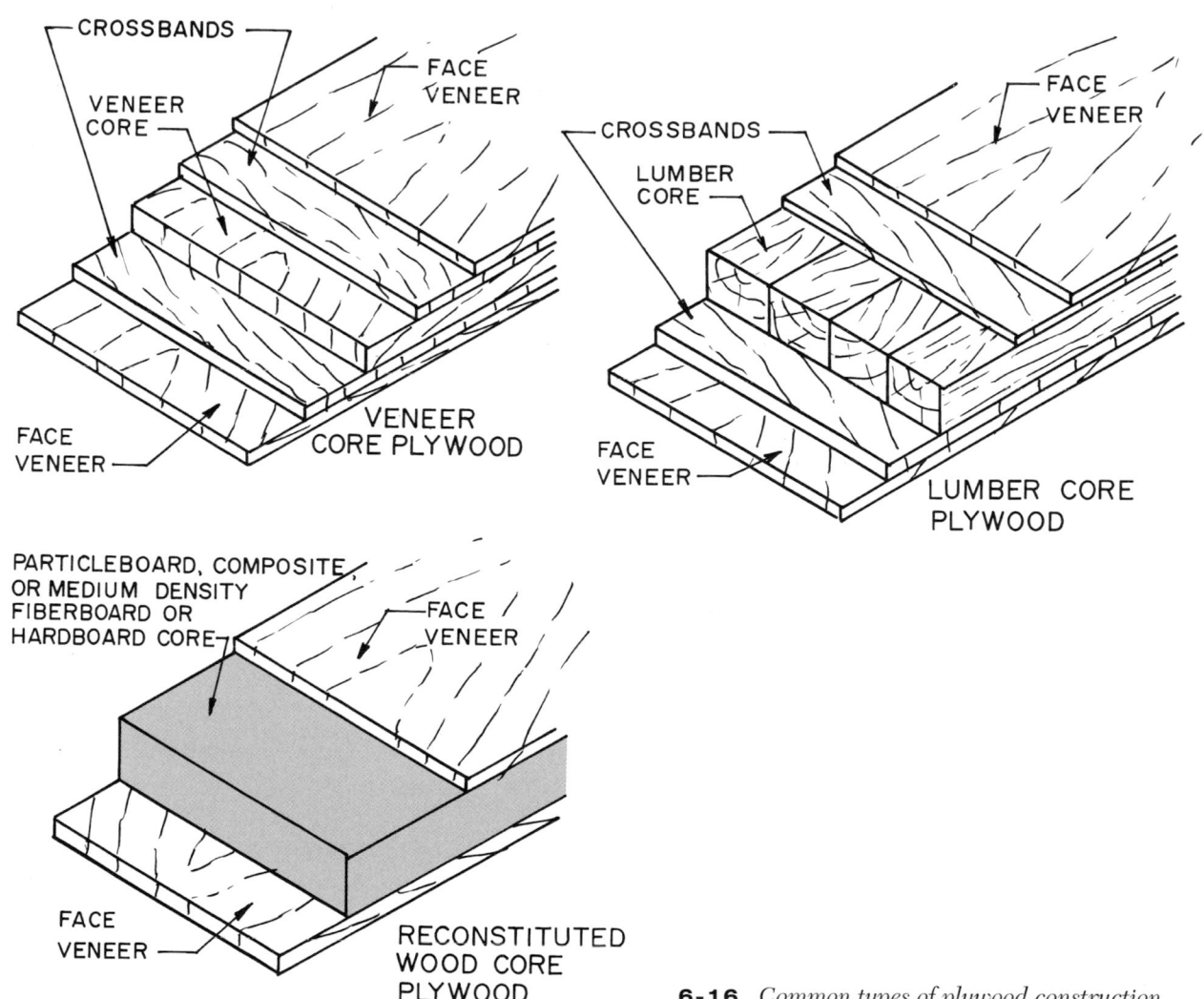

6-16 *Common types of plywood construction.*

Table 6-13 PLYWOOD OUTER VENEER GROUP NUMBERS

Group 1	Group 2		Group 3	Group 4	Group 5
Apitong	Cedar, Port Orford	Maple, black	Alder, red	Aspen	Basswood
Beech, American	Cypress	Mengkulang	Birch, paper	Bigtooth	Poplar, balsam
Birch	Douglas fir 2	Meranti, red	Cedar, Alaska	Quaking	
Sweet	Fir	Mersawa	Fir, subalpine	Cativo	
Yellow	California red	Pine	Hemlock, eastern	Cedar	
Douglas fir 1	Grand	Pond	Maple, bigleaf	Incense	
Kapur	Noble	Red	Pine	Western red	
Keruing	Pacific silver	Virginia	Jack	Cottonwood	
Larch, western	White	Western white	Lodgepole	Eastern	
Maple, sugar	Hemlock, western	Spruce	Ponderosa	Black (western	
Pine	Lauan	Red	Spruce	poplar)	
Caribbean	Almon	Sitka	Redwood	Pine	
Ocote	Bagtikan	Sweetgum	Spruce	Eastern white	
Pine, southern	Mayapis	Tamarack	Engelmann	Sugar	
Loblolly	Red	Yellow poplar	White		
Longleaf	Tangile				
Shortleaf	White				
Slash					
Tanoak					

A *Smooth, paintable. Not more than 18 neatly made repairs, boat, sled, or router type, and parallel to grain, permitted. May be used for natural finish in less demanding applications. Synthetic repairs permitted*

B *Solid surface. Shims, circular repair plugs, and tight knots 1 inch across grain permitted. Some minor splits permitted. Synthetic repairs permitted*

C Plugged *Improved C veneer with splits limited to ¼ inch width and knotholes and bored holes limited to ¼ x ½ inch. Admits some broken grain. Synthetic repairs permitted*

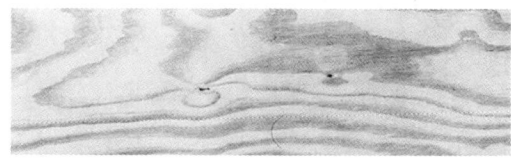

C *Tight knots to 1½ inch. knotholes to 1 inch across grain and some to 1½ inch if total width of knots and knotholes is within specified limits. Synthetic or wood repairs. Discoloration and sanding defects that do not impair strength permitted. Limited splits allowed. Stitching permitted*

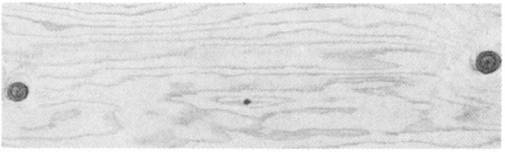

D *Knots and knotholes to 2½-inch width across grain and ½ inch larger within specified limits. Limited splits allowed. Stitching permitted. Limited to Interior, Exposure 1 and Exposure 2 panels*

6-17 *Plywood veneer grades.*
Courtesy APA-The Engineered Wood Association

Table 6-14 PLYWOOD PANEL THICKNESSES & DIMENSIONS

Nominal Thickness	
In.	**mm[a]**
¼	6.4
5/16	7.9
11/32	8.7
3/8	9.5
7/16	11.1
15/32	11.9
½	12.7
19/32	15.1
5/8	15.9
23/32	18.3
¾	19.1
7/8	22.2
1	25.4
1 3/32	27.8
1 1/8	28.6

Nominal Dimensions (width × length)		
ft.	**mm**	**m[a]**
4 × 8	1219 × 2438	1.22 × 2.44
4 × 9	1219 × 2743	1.22 × 2.74
4 × 10	1219 × 3048	1.22 × 3.05

[a]Soft converted metric sizes

Courtesy APA-The Engineered Wood Association

SIZES OF PANELS

Standard nominal imperial thicknesses of sanded construction and industrial plywood panels are ⅛ through 1¼ in. in ⅛ in. increments and in unsanded panels are 5/6 in. to 1¼ in. in increments for panels over ⅜ in. thick. Panel widths are 36, 48, and 60 in. and lengths from 60 to 140 in. in 12 in. increments (Table 6-14). Soft converted metric thicknesses are shown in Table 6-14. Panels sized 1200 × 2400 mm are designed for use in hard metric designed buildings, while panels 1219 × 2438 are soft conversions and can be used in buildings that are designed using imperial units.

Performance-Rated Panels

Performance-Rated Panels are manufactured under provisions of the American Plywood Association performance standards that establish performance criteria for designated construction applications. The four panels specified by APA are plywood, oriented strand board, composite (Com-Ply®), and APA-Rated Siding (see 6-18). Waferboard and particleboard are also available but are not manufactured under APA specifications.

Plywood
All-veneer panels consisting of an odd number of cross-laminated layers, each layer consisting of one or more plies. Many such panels meet all of the prescriptive or performance provisions of U.S. Product Standard PS 1-83/ANSI A199.1 for Construction and Industrial Plywood

APA-Rated Siding
All-veneer panels constructed in the same manner as plywood panels. Available in a variety of panel sizes and thicknesses. Various surface textures and designs available

Composite (Comply®)
Panels of reconstituted wood cores bonded between veneers face and back plies

Oriented Strand Board
Panels of compressed strand-like particles arranged in layers (usually three to five) oriented at right angles to one another

6-18 *APA-The Engineered Wood Association (APA) trademark Performance Rated Panels.*

Courtesy APA-The Engineered Wood Association

Plywood panels are made using all veneer plies producing the strongest type of plywood. Oriented strand board is manufactured from strands or wafers oriented in general in one direction. These layers of oriented strands are bonded together at right angles to each other. Usually three to five layers are bonded to form a panel. Most OSB panels are textured on one side to produce a non-slick surface.

Composite or **Com-Ply**® panels have a core of reconstituted wood bonded between solid wood veneers. This produces a panel that allows for efficient use of wood materials yet has a wood grain surface on the front and back.

APA-rated siding is made by bonding wood veneers in the same manner as plywood. It is available in a variety of surface textures and designs. Panels are available in 4 × 8, 4 × 9, and 4 × 10 foot dimensions. A guide to APA Performance-Rated Panels is in 6-19.

Waferboard panels are made from large, waferlike wood chips bonded together, usually in random positions, but occasionally some panels are made with directionally oriented wafers. Canada has two standards pertaining to the manufacture of waferboard.

Particleboard is made from wood chips, water, and a synthetic resin binder. It is manufactured to standards developed by the National Particleboard Associ-

---

Guide to APA Performance-Rated Panels [1][2]
For application recommendations, see following pages.

APA RATED SHEATHING Typical Trademark	APA RATED SHEATHING 24/16 7/16 INCH SIZED FOR SPACING EXPOSURE 1 000 NER-QA397 PRP-108		Specially designed for subflooring and wall and roof sheathing. Also good for a broad range of other construction and industrial applications. Can be manufactured as conventional veneered plywood, as a composite, or as a nonveneer panel. EXPOSURE DURABILITY, CLASSIFICATIONS: Exterior, Exposure 1, Exposure 2. COMMON THICKNESSES: 5/16, 3/8, 7/16, 15/32, 1/2, 19/32, 5/8, 23/32, 3/4.
APA STRUCTURAL I RATED SHEATHING[3] Typical Trademark	APA RATED SHEATHING STRUCTURAL I 32/16 15/32 INCH SIZED FOR SPACING EXPOSURE 1 000 PS 1-83 C-D NER-QA397 PRP-108	APA RATED SHEATHING 32/16 15/32 INCH SIZED FOR SPACING EXPOSURE 1 000 STRUCTURAL I RATED DIAPHRAGMS-SHEAR WALLS PANELIZED ROOFS NER-QA397 PRP-108	Unsanded grade for use where shear and cross-panel strength properties are of maximum importance, such as panelized roofs and diaphragms. Can be manufactured as conventional veneered plywood, as composite, or as a nonveneer panel. EXPOSURE DURABILITY, CLASSIFICATIONS: Exterior, Exposure 1. COMMON THICKNESSES: 5/16, 3/8, 7/16, 15/32, 1/2, 19/32, 5/8, 23/32, 3/4.
APA RATED STURD-I-FLOOR Typical Trademark	APA RATED STURD-I-FLOOR 24 OC 23/32 INCH SIZED FOR SPACING T&G NET WIDTH 47-1/2 EXPOSURE 1 000 NER-QA397 PRP-108		Specially designed as combination subfloor-underlayment. Provides smooth surface for application of carpet and pad and possesses high concentrated and impact load resistance. Can be manufactured as conventional veneered plywood, as a composite, or as a nonveneer panel. Available square edge or tongue-and-groove. EXPOSURE DURABILITY, CLASSIFICATIONS: Exterior, Exposure 1, Exposure 2. COMMON THICKNESSES: 19/32, 5/8, 23/32, 3/4, 1-1/8.
APA RATED SIDING Typical Trademark	APA RATED SIDING 24 OC 15/32 INCH SIZED FOR SPACING EXTERIOR 000 NER-QA397 PRP-108	APA RATED SIDING 303-18-S/W 16 OC 11/32 INCH GROUP 1 SIZED FOR SPACING EXTERIOR 000 PS 1-83 FHA-UM-64 NER-QA397 PRP-108	For exterior siding, fencing, etc. Can be manufactured as conventional veneered plywood, as a composite, or as a nonveneer siding. Both panel and lap siding available. Special surface treatment such as V-groove, channel groove, deep groove (such as APA Texture 1-11), brushed, rough sawn, and texture-embossed (MDO). Span Rating (stud spacing for siding qualified for APA Sturd-I-Wall applications) and face grade classification (for veneer-faced siding) indicated in trademark. EXPOSURE DURABILITY, CLASSIFICATIONS: Exterior. COMMON THICKNESSES: 11/32, 3/8, 7/16, 15/32, 1/2, 19/32, 5/8.

(1) Specific grades, thicknesses, and exposure durability classifications may be in limited supply in some areas. Check with your supplier before specifying.

(2) Specify Performance Rated Panels by thickness and Span Rating. Span Ratings are based on panel strength and stiffness. Since these properties are a function of panel composition and configuration as well as thickness, the same Span Rating may appear on panels of different thickness. Conversely, panels of the same thickness may be marked with different Span Ratings.

(3) All plies in Structural I plywood panels are special improved grades and panels marked PS 1 are limited to Group 1 species. Other panels marked Structural I Rated quality through special performance testing.

Structural II plywood panels are also provided for, but rarely manufactured. Application recommendations for Structural II plywood are identical to those for RATED SHEATHING plywood.

6-19 *APA-The Engineered Wood Association trademarks and uses of APA Performance-Rated Panels.*

Courtesy APA-The Engineered Wood Association

ation. It is mainly used for furniture and cabinet construction (see 6-20). Canada has four metric standards for various types of particleboard and waferboard.

APA rated sheathing is used for subfloor, wall, and roof sheathing. It is used where strength and stiffness are required. Nominal panel thicknesses range from $5/16$ to $3/4$ inch. **Structural 1** is a type of rated sheathing used where increased resistance to wracking and cross-panel strength is needed. It is used on structural diaphragms and panelized roofs. **APA-Rated Stur-1-Floor®** is a single-layer flooring for use under carpets. It may avoid the need for installing an underlayment. Panels are available with square and tongue-and-grooved edges. It is available in thickness from $19/32$ to $1 1/8$ inch.

EXPOSURE DURABILITY CLASSIFICATIONS

The APA performance rated panels are manufactured in four exposure durability classifications, Exterior, Exposure 1, Exposure 2, and Interior. These classifications pertain to the ability of the bonding agent used to join the veneers to resist exposure to moisture.

Exterior panels use a waterproof adhesive and are designed for use on applications subject to permanent exposure to moisture and weather.

Exposure 1 panels use the same waterproof adhesive as Exterior panels and are designed to be used where they will be exposed to moisture or the weather for long periods before they are finally protected. They are also used where the panels may be subjected occasionally to moisture after they are in service. They differ from exterior panels in some compositional aspects and therefore are not recommended for permanent exposure to the weather.

Exposure 2 panels are used in applications where during construction they will be briefly exposed to the weather before being permanently protected. They are actually an interior type panel with an intermediate adhesive.

Interior panels are manufactured with an interior glue and intended for interior use only. They are identified by abbreviation INT-APA and the omission of glueline information on their trademark.

Veneer grades used on Performance Rated Panels are A, B, C, C plugged, and D.

PRODUCT IDENTIFICATION

American Plywood Association trademarks used on the various types of plywood used in construction are in 6-21. The **sanded** panels with B-grade or better veneer panels are used for various construction applications. The **specialty grades** include panels designed for a specific use such as Plyform for concrete forms and underlayment. Most of the information on the trademark is self explanatory except for span rating. A span rating of $3/16$ means the panel can be used as roof sheathing with rafters spaced up to 32 in. on-center and as subflooring with floor joists spaced up to 16 in. on-center.

Waferboard
Panels of compressed wafer-like particles or flakes randomly or directionally oriented

Oriented Strand Board
Panels of compressed strand-like particles arranged in layers (usually three to five) oriented at right angles to one another

Structural Particleboard
Panels comprised of small particles usually arranged in layers by particle size, but usually not oriented

6-20 *Other panels composed of reconstituted wood that are designed to carry known loads over known distances.*

Specialty Plywoods

Many plywood manufacturers make various plywood products that are not included in the national standards for plywood manufacture. These are classified as **specialty plywoods**. The user must refer to information supplied by the manufacturer when selecting these products, because they meet no standard. Several of the most frequently used are overlaid plywood, siding panels, and interior paneling. APA-The Engineered Wood Association does have standards and trademarks for several specialty panels.

Overlaid plywood is a high-grade, exterior type panel that has bonded to one or both sides a resin-impregnated fiber ply. It is made in two types, high density and medium density. **High-density panels** have a hard, smooth, chemically resistant surface. It is made in several colors and requires no additional finish. **Medium-density panels** have a smooth, opaque, non-glossy surface that hides the grain of the veneers below it. It is used where a high-quality paint finish is needed.

Siding panels are used as finished exterior siding and are available in a variety of surface finishes such as rough sawed, V-grooved, and reverse batten grooves.

Paneling is used on interior walls as the finish material. It is usually prefinished and only requires installation. It is available in a wide variety of wood species and surface fixtures.

Hardwood Plywood

Hardwood plywood is manufactured following the standards in ANSI/HPMA HP 1983, American National Standards for Hardwood and Decorative Plywood. These standards are published by the Hardwood Plywood Manufacturers Association, P.O. Box 2789, Reston, VA 22090. Canadian hardwood plywoods are made according to metric standard CSA 0115-M, Hardwood and Decorative Plywood.

SPECIES

Species used in hardwood plywood are divided into four categories. These categories reflect the modulus of elasticity (stiffness) of each specie and its specific gravity. Category A is the stiffest and D has the lowest rating.

GRADES OF VENEERS

Hardwood plywood offers six grades of hardwood veneer. They also have specific requirements for softwoods used in hardwood plywood panels.

A grade (A) is the best. The face is made of hardwood veneers carefully matched as to color and grain.

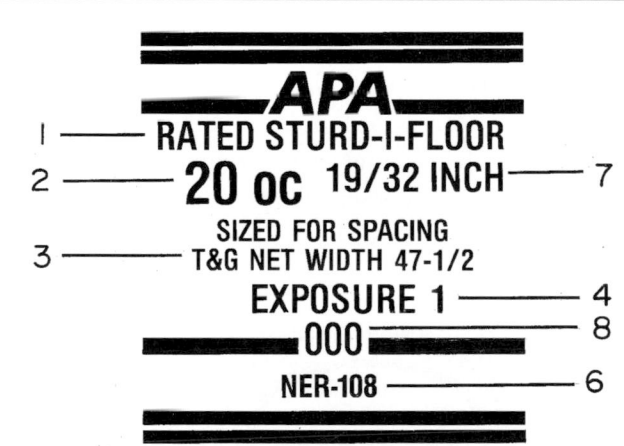

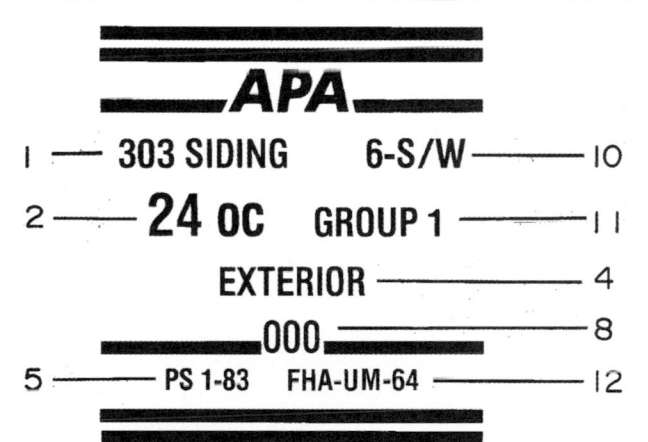

1. Panel grade
2. Span rating
3. Tongue and groove
4. Exposure durability classification
5. Product standard
6. Code of recognition of APA as a quality assurance agency
7. Thickness
8. Mill number
9. *(Not shown)* National Research Board report number
10. Siding face grade
11. Species group number
12. FHA recogntion

6-21 *A few of the trademarks of the APA-The Engineered Wood Association*

Courtesy APA-The Engineered Wood Association

Table 6-15 TYPES OF HARDWOOD PLYWOOD[a]	
Technical	(Exterior use)
Type I	(Exterior use)
Type II	(Interior use)
Type III	(Interior use)

[a]Based on water-resistance capacity

Table 6-16 CLASSIFICATIONS OF HARDBOARD PANELS	
Class 1	Tempered
Class 2	Standard
Class 3	Service-tempered
Class 4	Service
Class 5	Industrialite

B grade (B) is suitable for a natural finish, but the face veneers are not as carefully matched as on A grade.

Sound grade (2) provides a face that is smooth. All defects have been repaired. It is used as a smooth base for a paint finish.

Industrial grade (3) face veneers can have surface defects. This grade permits knotholes up to 1 in. (25 mm) in diameter, small open joints, and small areas of rough grain.

Backing grade (4) uses unselected veneers having knotholes up to 3 in. in diameter and certain types of splits. Any defect permitted does not affect the strength of the panel.

Specialty grade (SP) includes veneers having characteristics unlike any of those in the other grades. The characteristics are agreed upon between the manufacturer and the purchaser. For example, species such as wormy chestnut or bird's-eye maple are considered Specialty Grade.

TYPES OF HARDWOOD PLYWOOD

There are four types of hardwood plywood rated according to their water-resistance capacity. The Technical type is the most water-resistant while Type III is the least (Table 6-15).

CONSTRUCTION OF HARDWOOD PLYWOOD

The construction of hardwood plywood is described by identifying the core. Following are the commonly available constructions:

Hardwood veneer core—has an odd number of plies, such as 3-ply, 5-ply, and so on.

Softwood veneer core—has an odd number of plies, such as 3-ply, 5-ply, and so on.

Hardwood lumber core—used in 3-ply, 5-ply, and 7-ply constructions.

Softwood lumber core—used in 3-ply, 5-ply, and 7-ply constructions.

Particleboard core—used in 3-ply, and 5-ply constructions.

Medium-density fiberboard core—used in 3-ply construction.

Hardboard core—used in 3-ply construction.

Special cores—used in 3-ply construction. Special cores are those made of any other material than those listed above.

SIZES AND THICKNESSES OF PANELS

Hardwood plywood is available in panels 48 in. wide and 96 and 120 in. long. Standard thicknesses are ¼, ⅜, ½, and ¾ in. Lumber-core and particleboard-core panels are available in only the ¾ in. thickness.

Canadian hardwood plywood is available in incremental thicknesses from 4.8 to 32 mm.

PRODUCT IDENTIFICATION

The trademark stamp is placed on the back of the panel (see 6-22). Most of the information shown is self explanatory. The flame spread rating shows the rating given the panel following tests. Flame spread ratings are set by ASTM E162.

6-22 *An HPMA trademark stamp for hardwood plywood.*
Courtesy Hardwood Plywood and Veneer Association

Table 6-17 GRADES & USES OF PARTICLEBOARD

Type	Grade	Use
High density	H-1, H-2, H-3	High-density industrial
	H-1, H-2, H-3 Exterior Glue	High-density exterior industrial
Medium density	M-1	Commercial
	M-2, M-3	Industrial
	M-1, M-2, M-3 Exterior Glue	Exterior construction
		Exterior industrial
Medium density—specialty grade	M-S	Commercial
Low density	LD-1, LD-2	Door core
Underlayment	PBU	Underlayment
Manufactured home decking	D-2, D-3	Flooring in manufactured homes

Courtesy National Particleboard Association

RECONSTITUTED WOOD PRODUCTS

In addition to the performance-rated panels discussed earlier there are a number of reconstituted wood panels used primarily in cabinet and furniture construction. These include hardboard, particleboard, fiberboard, and waferboard.

Hardboard is made from wood chips converted into fibers and bonded into panels under heat and pressure. It is manufactured following standards developed by the American Hardboard Association. It is available in thicknesses from ¹⁄₁₂ to 1⅛ in. The most used panel size is 4 × 8 ft. However, other sizes are available on special order.

Metric hardboard thicknesses are 1.9, 2.7, 4.0, 5.4, 6.9, and 8.3 minimum. Panels are 1200 × 2400 mm.

Hardboard is available in five classes (Table 6-16). Hardboard is manufactured in the United States according to the following standards: ANSI/AHA A135.4-1988, Basic Hardboard; ANSI/AHA A135.5-1988, PF Hardboard Paneling; and ANSI/AHA A135.6-1990, Hardboard Siding. These are promoted by the American Hardboard Association. Panels manufactured to this standard have an AHA grade stamp.

Canadian hardboard products are manufactured following CGSB 11-GP-3M, Hardboard and CGSB 11-GP-5M, Hardboard for Exterior Cladding.

Particleboard is a generic term for a panel composed of cellulosic materials such as wood particles combined with a synthetic binder. The particles are bonded with heat and pressure. Particleboard is manufactured following standards developed by the National Particleboard Association. These are detailed in ANSI A208.1-1993, Particleboard. The general uses and grades are in Table 6-17.

▼ The grades are identified by a letter designation followed by a hyphen and a digit or letter. The meaning of the first letter designations are:

H	High density
M	Medium density
LD	Low density
D	Manufactured home decking
PBU	Underlayment

The second digit or letter designation indicates the grade identification within a particular density or product description such as **M-2.** It indicates a medium density particleboard, Grade 2.

If there is a third designation it indicates a special characteristic such as panel **M-3 Exterior Glue.** This is a medium-density panel, Grade 3 with exterior glue.

Waferboard is made by bonding large wood flakes 1½ in. or longer into panels having the same thicknesses and panel sizes as particleboard.

STRUCTURAL BUILDING COMPONENTS

Trusses are used for floor and roof construction. They are engineered to carry known loads over specified distances. Most trusses are made from 2 × 4 and 2 × 6 in. lumber joined with wood or metal gusset plates (see 6-23).

Glued laminated wood members are made by bonding ¾ in. wood strips for curved members and 1½ in. thick wood strips for straight members to form a beam or column. Laminations in Canadian products are 19 and 38 mm. These are commonly called "glulam" members, as a short way of saying glued laminated members. Members of any size can be made but manufacturers produce a wide range of standard sizes (see 6-24).

Wood can be laminated into various shapes, as curved, and be stronger than solid wood the same size. They have the advantage of being able to be made to a high level of quality because wood defects can be cut away.

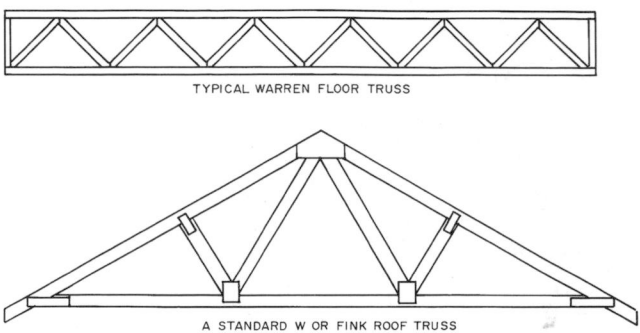

TYPICAL WARREN FLOOR TRUSS

A STANDARD W OR FINK ROOF TRUSS

6-23 *Two types of wood truss.*

Glued laminated members are manufactured in the United States according to ANSI/AITC A190.1-1983, Structural Glued Laminated Timber. In Canada the standards include CSA 0177-M, Qualification Code for Manufacturers of Structural Glued-Laminated Timber; CSA 0122, Structural Glued-Laminated Timber; and CSA3-086-M, Code for Engineering Design in Wood.

Laminated veneer lumber is a manufactured structural member made by bonding wood veneers with an exterior adhesive. It has almost no shrinking, checking, twisting, or splitting. It has more load-bearing capacity per pound than sawed lumber. It is available in depths from 9¼ to 18 in. (235 to 457 mm) and thicknesses of 1½ and 1¾ in. (38 to 44 mm) (see 6-25).

Plywood-lumber (box) beams are made according to carefully engineered specifications as to the size of members and nailing requirements. They are glued and built in factories under controlled conditions. One example is shown in 6-26.

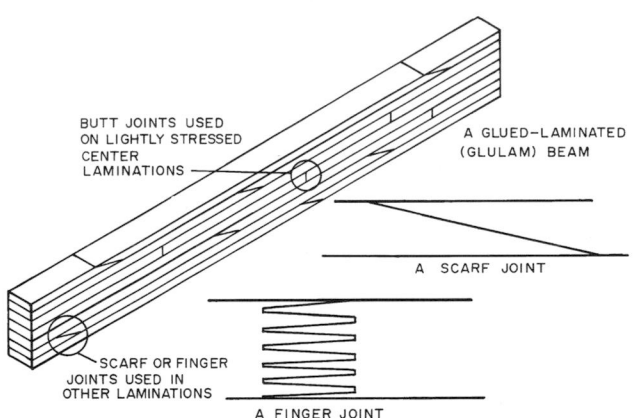

BUTT JOINTS USED ON LIGHTLY STRESSED CENTER LAMINATIONS

A GLUED–LAMINATED (GLULAM) BEAM

A SCARF JOINT

SCARF OR FINGER JOINTS USED IN OTHER LAMINATIONS

A FINGER JOINT

6-24 *Glued laminated beams can span long distances and carry heavy loads.*

Courtesy Southern Forest Products Association

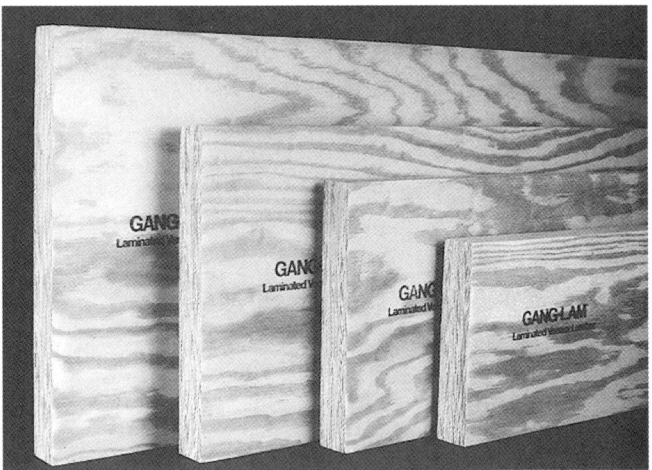

6-25 *Laminated veneer lumber is made by bonding together layers of wood veneer with an exterior adhesive.*

Courtesy Louisiana-Pacific

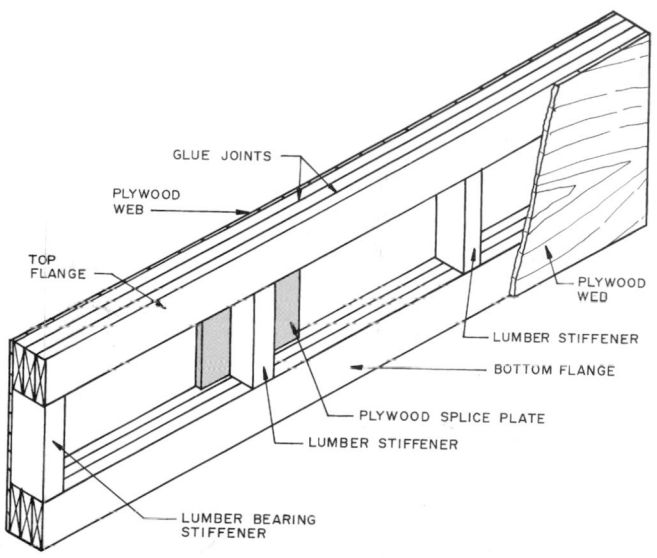

6-26 *A typical plywood-lumber beam (box beam) with plywood webs and solid wood flanges and stiffeners.*

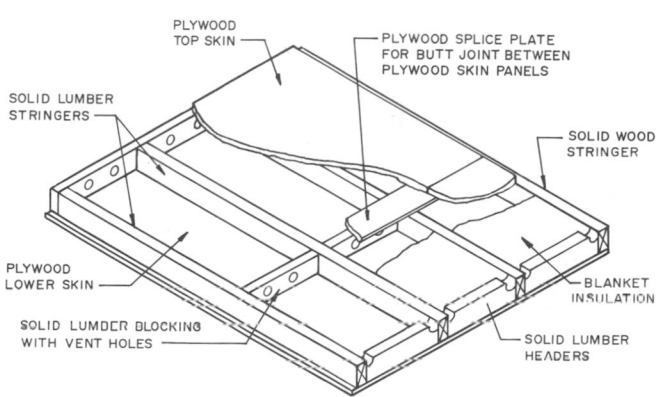

6-27 *A stressed skin panel.*

Stressed skin panels are prefabricated panels using solid lumber stringers and headers and plywood skins. They are used for floors, walls, and roof. They enable a constructor to rapidly erect a building because they cover a large area and span fairly long distances. For example, on a floor the joists would be replaced with beams, possibly box beams, spaced 4 ft. on-center. (see 6-27).

A proprietary type of joist is an **I-joist** made of soft-wood veneers bonded together to make the top and bottom flanges with a composite panel core (see 6-28). Another type of manufactured structural members is made from **parallel strand lumber** (PSL). These are manufactured under the registered name Parallam®. Parallam structural products include beams, headers, posts, and columns. They are made from Douglas fir or southern pine. The logs are peeled to produce a veneer that is dried and screened to remove strength reducing defects. Then the sheets of veneer are clipped into strands up to 8 ft. in length and ¹⁄₁₀ and ⅛ in. thick. Small defects are removed and the strands are coated with a waterproof adhesive. The long oriented strands are fed into a rotary belt press and cured under pressure using microwave energy. This produces a PSL billet that can be cut to standard sizes. It is available in lengths up to 66 ft. (6-29).

6-28 *This I-joist is made with flanges made of solid lumber and a composite wood core. Some are made with laminated veneer flanges.*

Courtesy Louisiana-Pacific

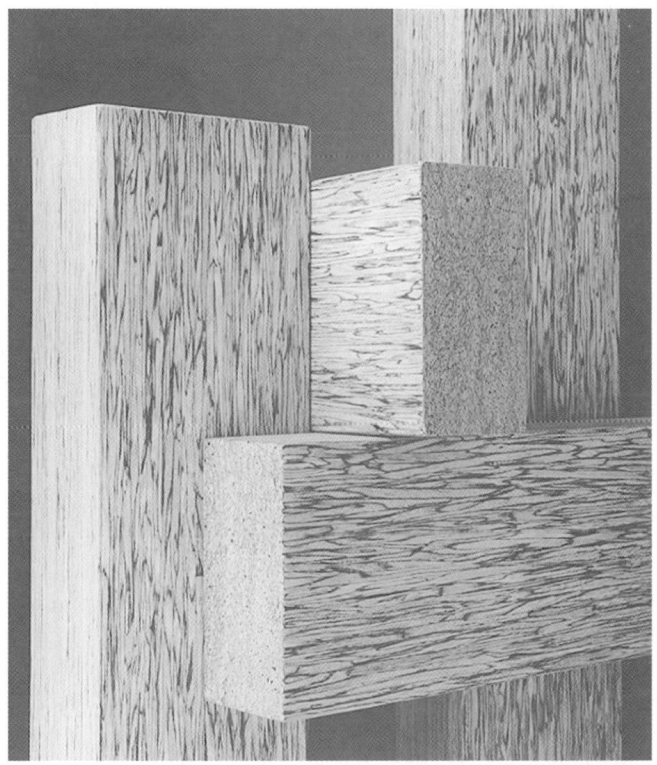

6-29 *Parallam® parallel strand lumber (PSL) is made by bonding thin wood strands into structurally sound members.*

Courtesy MacMillan Bloedel Limited

6-30 *This wall panel is made by bonding wood composition panels to a rigid foam core.*

Courtesy AFM Corporation

(1) Indicates structural use:
    B - Simple span bending member.
    C - Compression member.
    T - Tension member.
    CB - Continuous or cantilevered span bending member.
(2) Mill number.
(3) Identification of ANSI Standard A190.1, Structural Glued Laminated Timber.
(4) Applicable laminating specification.
(5) Applicable combination number.
(6) Species of lumber used.
(7) Designates appearance grade, INDUSTRIAL, ARCHITECTURAL, PREMIUM.

6-31 *A typical APA-EWS trademark. An explanation of several of the markings follows: 24F—the allowable bending stress in fiber is 24,000 psi; V4—the lamination lay-up for Douglas Fir; 117-93—the manufacturing standards and specification for glulam beams published by the American Institute of Timber Construction.*

Courtesy APA-The Engineered Wood Association

There are a wide variety of **prefabricated wall panels** available from manufacturers. One of these is shown in 6-30. The framed panel is much like the conventional framed wall, floor, or roof. The stressed skin panel has the skin adhesive bonded to interior wood members, forming a stronger panel. The sandwich panel has facing material, such as plywood, bonded to an insulating foam core. This is highly energy efficient.

American Wood Systems

American Wood Systems (AWS) was created by APA - The Engineered Wood Association to serve the needs of the engineered wood systems industry. The AWS created a new trademark, APA-EWS (Engineered Wood Systems). This trademark appears on glued laminated beams and other engineered products such as wood I-beams, structural composite lumber, and other engineered wood products. These products receive the same technical and promotional services APA provides to the manufacturers of structural wood panels. A typical APA-EWS trademark is shown in 6-31.

Wood Shingles & Shakes

Wood shingles and shakes are popular as a finished roof covering and are also used as exterior siding on wood-framed buildings (see 6-32). Red cedar **shingles** are smooth sawn on both faces and are uniform in thickness at the butt end. They are tapered from the thick butt end to the thin top edge. They are available in grades No. 1 (blue label), No. 2 (red label), No. 3 (black label), and in an undercoursing grade (green label) and in lengths of 16, 18 and 24 inches. Widths range from a 3 in. minimum to a 14 in. maximum.

Red cedar **shakes** are available in two types, Certi-Sawn® and Certi-Split®. Certi-Sawn shakes are sawed on both faces as is a shingle, but are not as precisely manufactured. They have a rough, split face and a smooth back.

Cedar shakes and shingles are available impregnated with fire-retardant polymers that penetrate the innermost cells of the wood.

Southern pine taper-sawn shakes are available in grade No. 1 (the best) and No. 2. Special hip and ridge units are made from No. 1 shakes.

Doors & Windows

There are many types and styles of wood doors and windows manufacturered. These are discussed in Division 8.

SHAKE

SHINGLE

6-32 *Cedar shingles and handsplit shakes are widely used for finished roof coverings.*
Courtesy Cedar Shake and Shingle Bureau

Wood Flooring

Both hardwoods and softwoods are used for finish flooring. They are popular because of the beauty of the grain and color. The most commonly used hardwoods are oak, beech, birch, pecan and maple. Softwoods include southern pine, western larch, bald cypress, eastern Engelmann, eastern spruce, red pine, ponderosa pine, eastern hemlock, and Douglas fir.

Grading rules for wood flooring vary with the species and sections of the country. They specify requirements for kiln drying, grading, control of moisture, and establish standard sizes. Many manufacturers produce additional sizes for special applications such as very thick flooring for industrial use.

Wood flooring is available quarter sawed and plain sawed. It is produced in four basic types: strips, planks, parquet (thin wood blocks), and solid end grain blocks (see 6-33). The construction of these products varies widely and manufacturers should be consulted. Additional information on flooring is in Division 9.

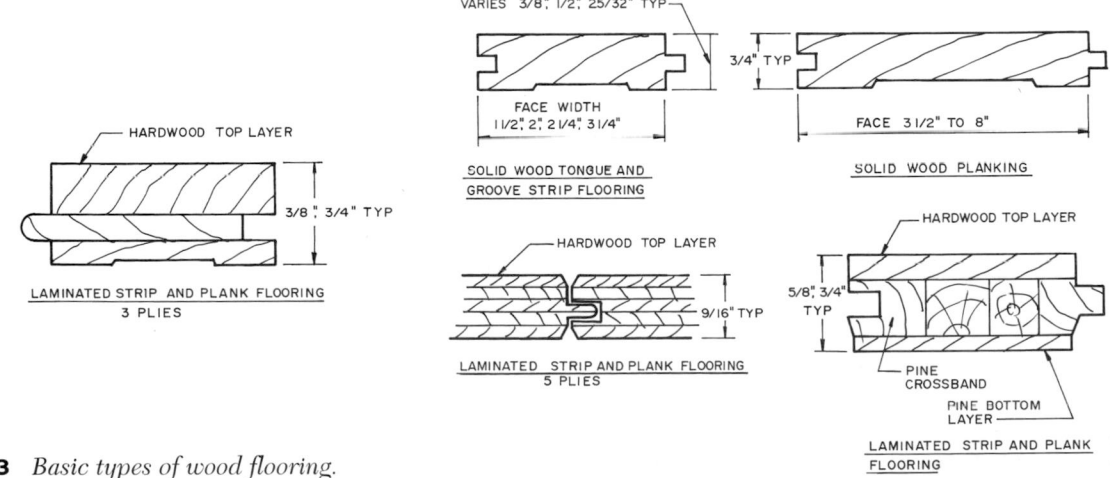

HARDWOOD TOP LAYER

3/8" 3/4" TYP

LAMINATED STRIP AND PLANK FLOORING
3 PLIES

VARIES 3/8", 1/2", 25/32" TYP

FACE WIDTH
1 1/2", 2", 2 1/4", 3 1/4"

SOLID WOOD TONGUE AND
GROOVE STRIP FLOORING

3/4" TYP

FACE 3 1/2" TO 8"

SOLID WOOD PLANKING

HARDWOOD TOP LAYER

9/16" TYP

LAMINATED STRIP AND PLANK FLOORING
5 PLIES

HARDWOOD TOP LAYER

5/8", 3/4" TYP

PINE CROSSBAND

PINE BOTTOM LAYER

LAMINATED STRIP AND PLANK FLOORING

6-33 *Basic types of wood flooring.*

LIGHT-FRAME CONSTRUCTION

Wood light frame is the most widely used system for the construction of residences and small apartments in the United States and Canada. Wood, a renewable resource, is in abundance in these countries, and is easily harvested, dried, and processed into structural members. The exterior finish can be wood, stucco, masonry, and other available siding products.

Wood light-frame construction is a very flexible system permitting an almost unlimited range of design possibilities. The architect can produce designs that are contemporary or classic or simple or complex, low cost or very expensive, and that can accommodate almost any electrical, heating, air conditioning, plumbing, and security system desired. With the vast array of products on the market the building can be insulated, sealed, waterproofed, producing a building with long life and low maintenance. It can be built in almost any climate and on any site that will accept an adequate foundation.

The system has evolved from basically a solid wood structure to the use of a variety of reconstituted wood products. Considerable effort has been made to utilize factory-assembled panels and modules, thus reducing the on-site labor costs associated with what is often referred to as "stick built" construction.

Types of Foundation

The footing is cast-in-place reinforced concrete. The foundation wall may be concrete block, brick, cast-in-place concrete, or pressure-treated wood. The building may have a basement, crawl space, or a concrete slab floor. Additional information on foundations is in Division 2. Here we will review those commonly used when building wood light framed buildings. These include buildings with a basement, crawl space and concrete slab-on-grade floor.

Basements are used in some parts of the country, typically in the colder climates They provide considerable living space for a small cost per square foot, especially

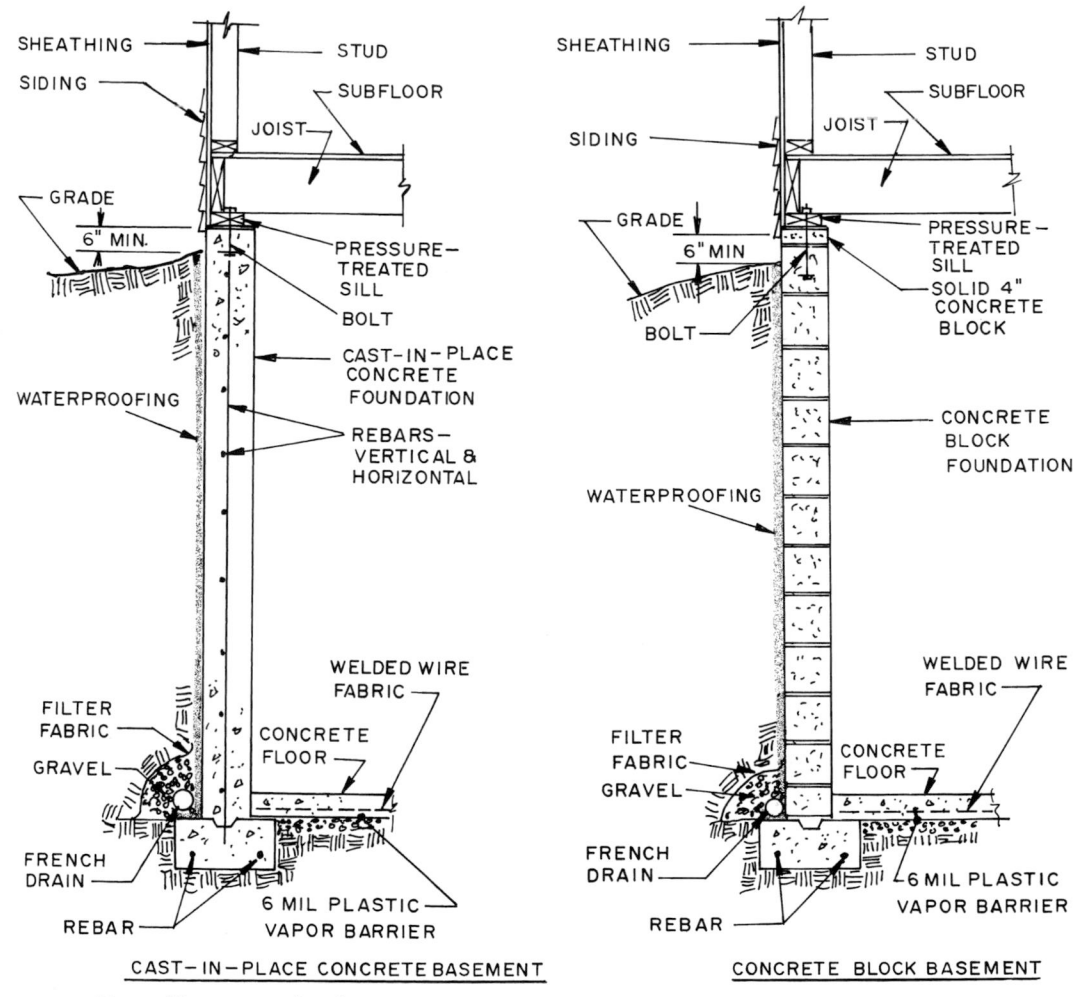

SHEATHING — STUD
SIDING — SUBFLOOR
JOIST
GRADE
6" MIN.
PRESSURE-TREATED SILL
BOLT
CAST-IN-PLACE CONCRETE FOUNDATION
WATERPROOFING
REBARS — VERTICAL & HORIZONTAL
FILTER FABRIC
GRAVEL
CONCRETE FLOOR
WELDED WIRE FABRIC
FRENCH DRAIN
REBAR
6 MIL PLASTIC VAPOR BARRIER

CAST-IN-PLACE CONCRETE BASEMENT

SHEATHING — STUD
SIDING — SUBFLOOR
JOIST
GRADE
6" MIN
PRESSURE-TREATED SILL
BOLT
SOLID 4" CONCRETE BLOCK
CONCRETE BLOCK FOUNDATION
WATERPROOFING
FILTER FABRIC
GRAVEL
CONCRETE FLOOR
WELDED WIRE FABRIC
FRENCH DRAIN
REBAR
6 MIL PLASTIC VAPOR BARRIER

CONCRETE BLOCK BASEMENT

6-34 *Typical basement details.*

when you consider that a crawl space requires a footing and some foundation wall and provides no living space. The building needs a foundation and footing below the frost line, so in many areas it does not take much more to get a full eight-foot (2.4m) deep basement.

A typical basement detail can be found in 6-34. The walls can be cast-in-place reinforced concrete or concrete block.

The foundation in 6-35 is typical for a building with a crawl space and wood exterior siding. When a wood framed building is to have a brick veneer, the detail would be as shown in 6-35. It requires that the foundation have a brick ledge that is usually near the grade.

Several foundations for houses with concrete **slab-on-grade floors** are illustrated in 6-36. The monolithic poured foundation and floor is used in areas where freezing does not occur. The others are used where the footing has to go below grade. The anchor bolts for the bottom plate are set in the concrete before it hardens.

WOOD FOUNDATIONS

Wood foundations can be used for buildings with basements or crawl spaces. They use pressure-treated plywood and pressure-treated solid wood members. The materials are stress-graded so it is known that they will withstand lateral soil and subsurface water pressures. The wood materials must have the stamp of the American Wood Preservers Bureau (AWPB).

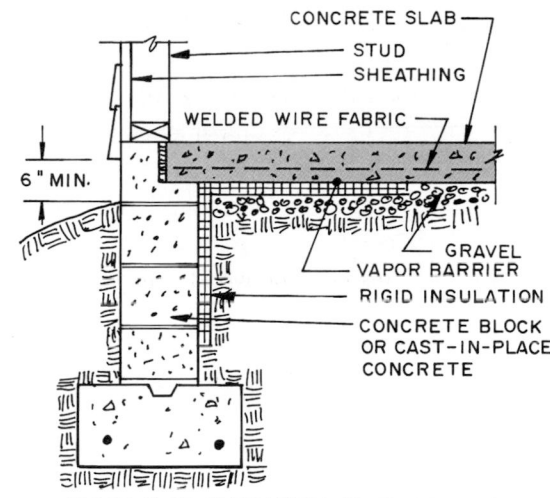

FLOOR SLAB SUPPORTED ON FOUNDATION

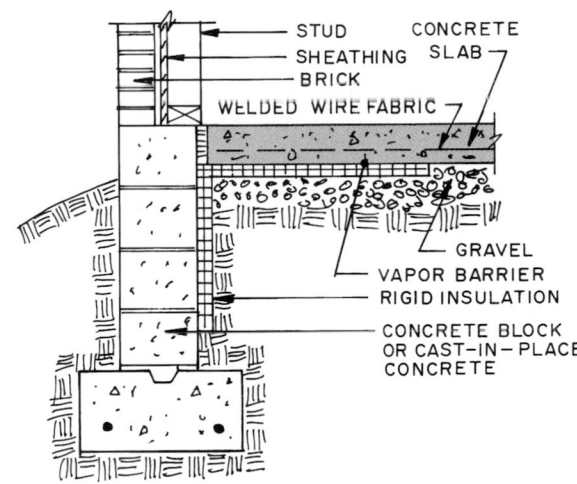

A GROUND SUPPORTED CONCRETE SLAB

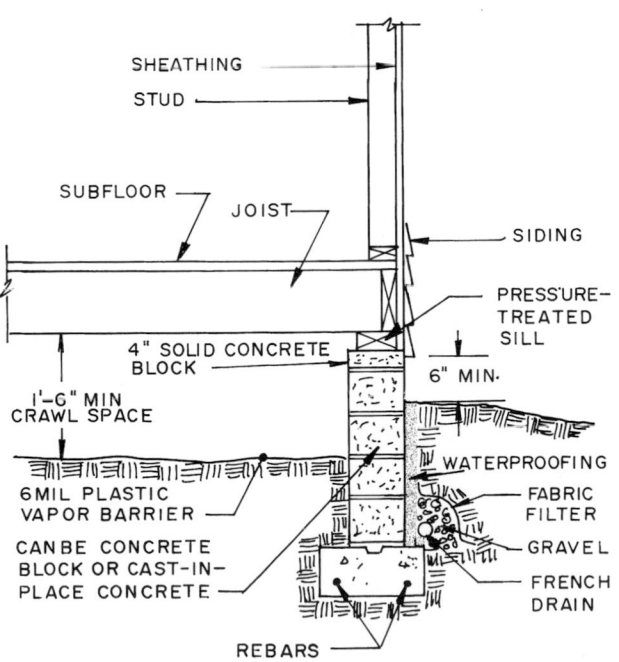

6-35 *This foundation is typical of those used on buildings having a crawl space.*

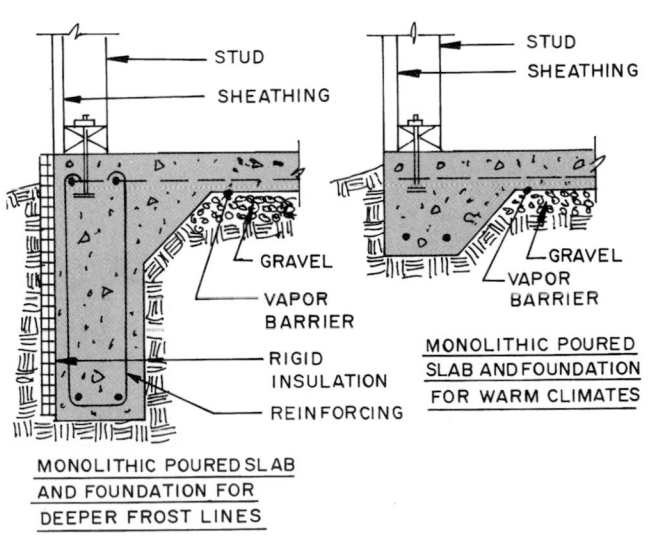

6-36 *Several types of concrete slab floor construction used with wood frame construction.*

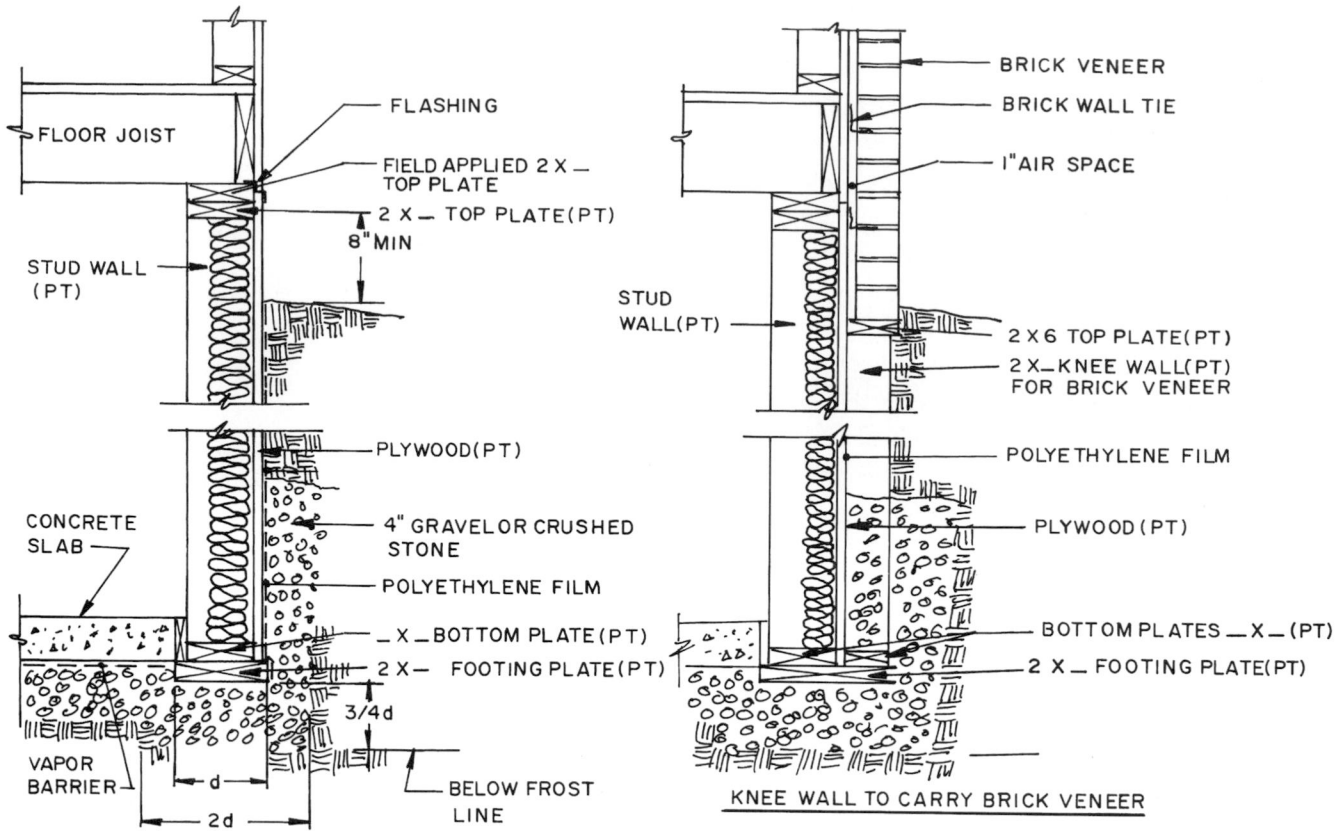

6-37 *Typical details for a permanent wood foundation basement wall.*

Courtesy American Forest and Paper Associations

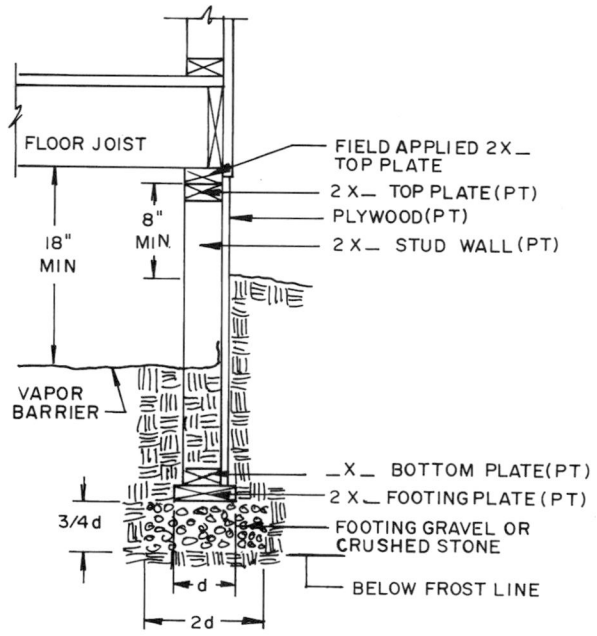

6-38 *The permanent wood foundation can be used on buildings having a crawl space.*

Courtesy American Forest and Paper Associations

The panels are often built in a shop and shipped to the site. This makes their assembly easier and more accurate since jigs can be used. The studs forming the panel face the inside of the building and can be insulated just like you would an exterior above-grade wall.

A typical basement foundation panel is shown in 6-37. The panels must be carefully assembled using bronze, copper, silicon, or stainless steel nails. They must follow the designs engineered by The American Forest and Paper Associations. Notice the bottom plate is set on a bed of gravel rather than a concrete footing. The floor joists rest on the double top plate. A detail for a foundation for a building with a crawl space is in 6-38. Notice in 6-37 the wood foundation can be used to support a brick veneer exterior wall.

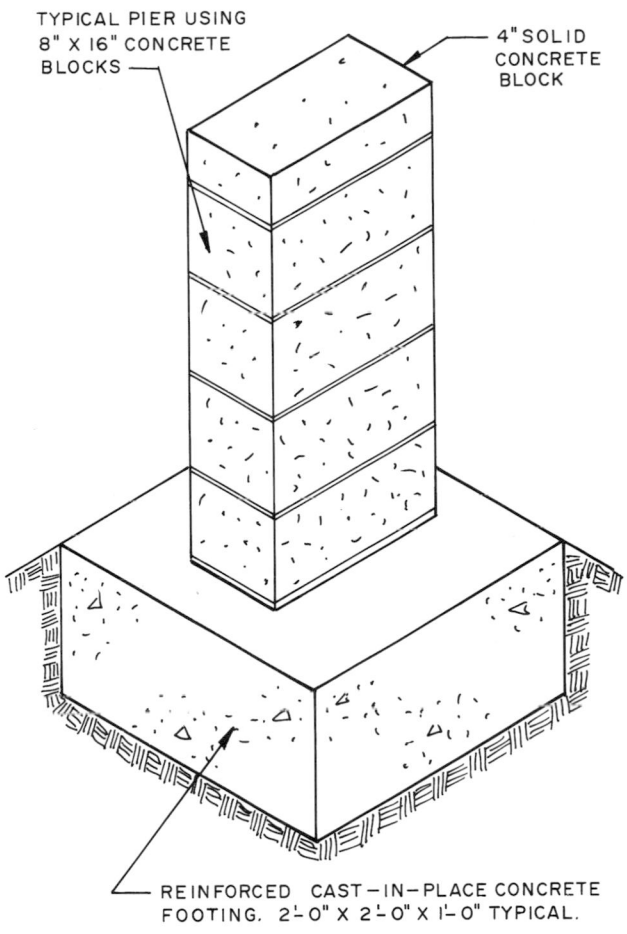

TYPICAL PIER USING 8" X 16" CONCRETE BLOCKS

4" SOLID CONCRETE BLOCK

REINFORCED CAST−IN−PLACE CONCRETE FOOTING. 2'−0" X 2'−0" X 1'−0" TYPICAL.

6-39 *A typical concrete block pier frequently used in residential construction.*

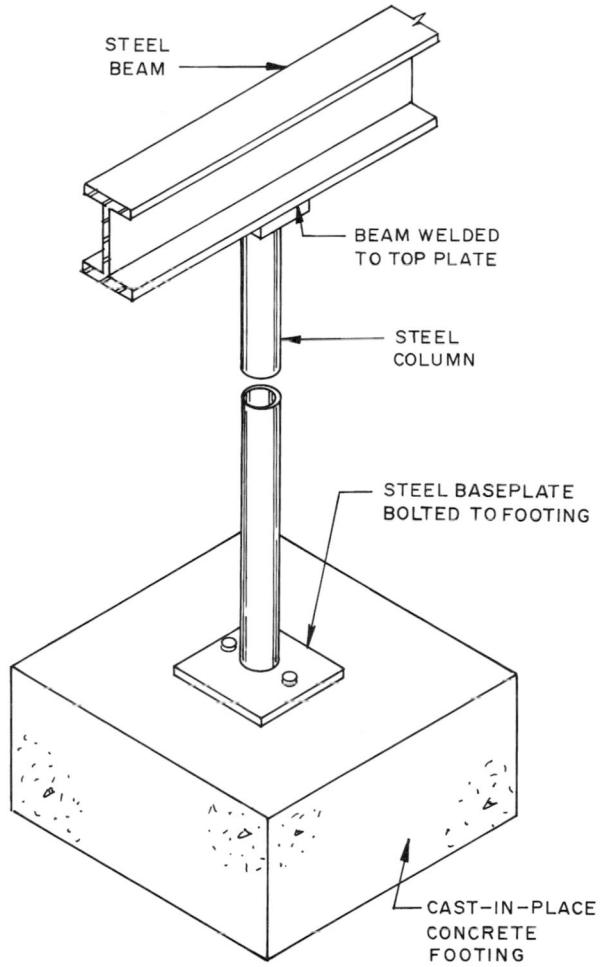

STEEL BEAM

BEAM WELDED TO TOP PLATE

STEEL COLUMN

STEEL BASEPLATE BOLTED TO FOOTING

CAST−IN−PLACE CONCRETE FOOTING

6-40 *A typical steel column supporting a beam that will carry the wood floor joists.*

PIERS AND COLUMNS

In addition to constructing the foundation walls, the contractor must pour footings for piers or columns to support the beams that will carry the floor joists. A typical pier used on wood light-frame buildings with crawl space is often built with concrete blocks, as you can see in 6-39. The size depends upon the loads to be carried.

If the building has a basement, steel columns are used. One example is shown in 6-40. Again, the size of the footing and type of column depend upon the load to be carried.

When the foundation is ready for the carpenters to install the first floor platform, it will look like the one shown in 6-41.

6-41 *A completed foundation for a wood light-framed building.*

The floor joists are assembled on the foundation.

The floor joists are covered with subflooring, forming a platform upon which the interior and exterior walls are assembled and raised.

6-42 *Building the first-floor platform.*

The walls are assembled on the platform.

The assembled wall is raised and nailed in place. It must be braced with wood members to the floor.

6-43 *Assembling the walls.*

Courtesy APA-The Engineered Wood Association

Platform Framing

The most frequently used method of light wood framing is the platform method. Platform framing involves building the first-floor deck on top of the foundation (see 6-42).

Upon this the walls for the first floor are assembled, erected, and braced (see 6-43). The joists for the second floor are laid on top of the double plate of the exterior first-floor walls and supported by load-bearing interior walls.

The second-floor joists are covered with subflooring, forming the second-floor platform and the walls for the second floor are assembled, erected, and braced (see 6-44).

The ceiling joists are laid on the double plate of the second-floor exterior walls and supported by interior load-bearing walls. Finally the carpenters install the rafters which rest on the double plate of the second-floor walls (see 6-45).

After the walls are sheathed the roof sheathing is installed. This greatly strengthens the roof structure and the rigidity of the entire structure (see 6-46).

6-44 *This structure has the second-floor platform and walls in place. Notice the diagonal bracing let into the studs.*
Courtesy Southern Forest Products Association

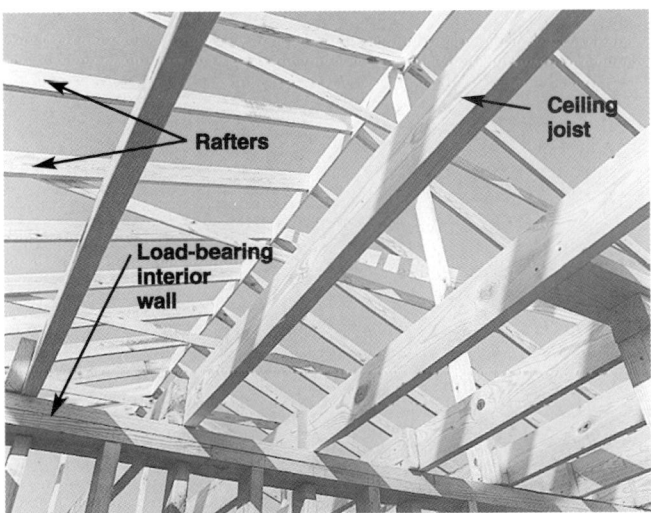

6-45 *The carpenters have installed the ceiling joists, shown resting on an interior load-bearing wall. The stick-built roof rafters are also in place.*
Courtesy Southern Forest Products Association

6-46 *The joints between the plywood sheathing panels are staggered as the panels are laid.*
Courtesy APA-The Engineered Wood Association

The completely assembled two-story wood light-framed building is shown in 6-47. This is a typical example of platform framing.

Wood Roof Trusses

Wood roof trusses are widely used for framing wood light-framed buildings. A roof truss is a structural unit, usually in a triangular form, made by assembling structural wood members into a rigid frame. A typical example is seen in 6-48.

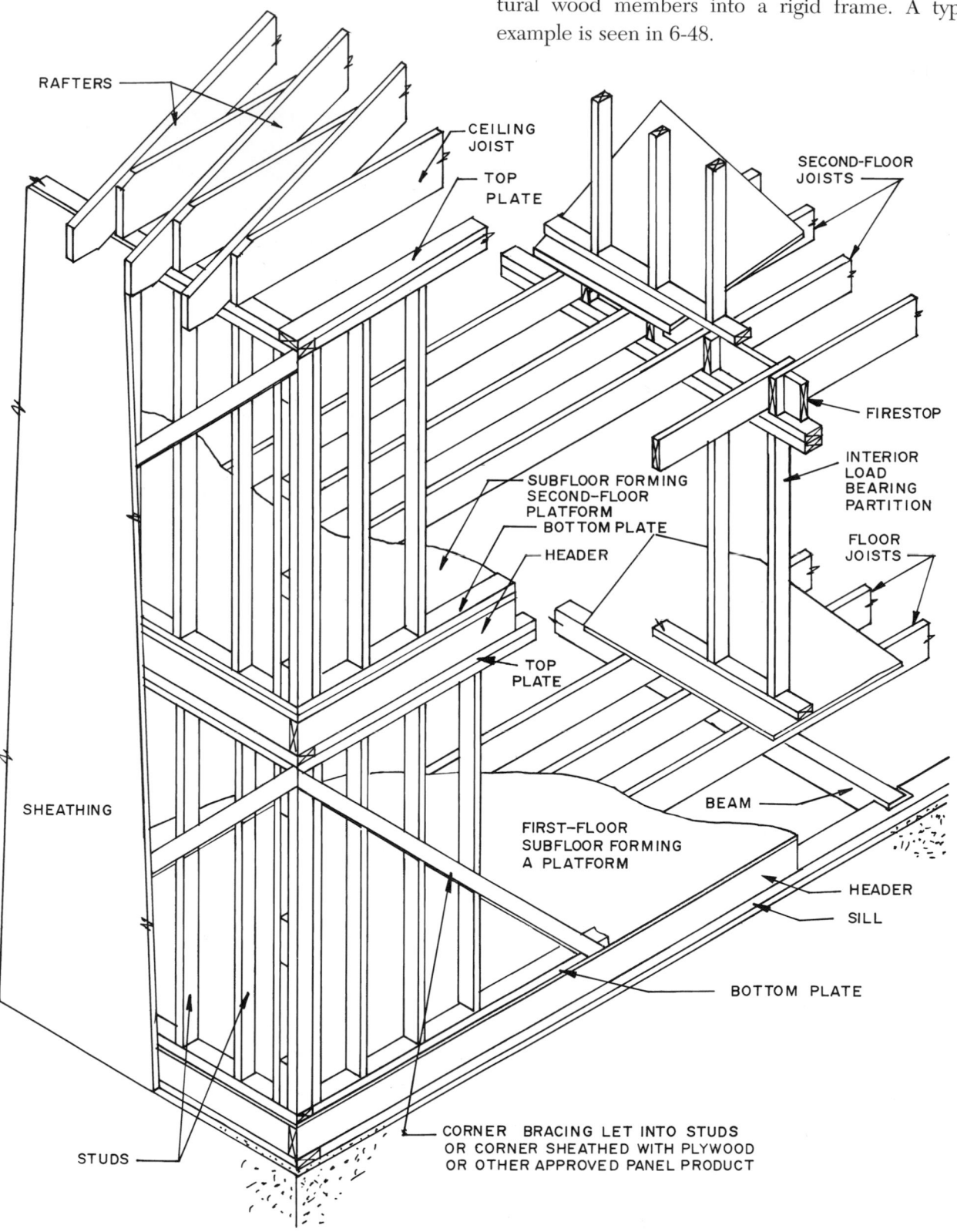

6-47 *Typical platform framing for a two-story building.*

The **advantages** of wood trusses are that they speed up on-site erection time, can span the width of most buildings without interior load-bearing walls, and since they are carefully engineered and factory built have a consistent, reliable quality. One **disadvantage** is they make it difficult to use the attic for storage. However, trusses can be designed to allow the center of the building to be open for storage or second-floor rooms. A few of the many types of wood trusses available are shown in 6-49. Many other designs are available.

RIDGE GUSSET
TOP CHORD
WEB GUSSET
WEBS
HEEL GUSSET
WEB GUSSETS
BOTTOM CHORD
FINK OR W TRUSS

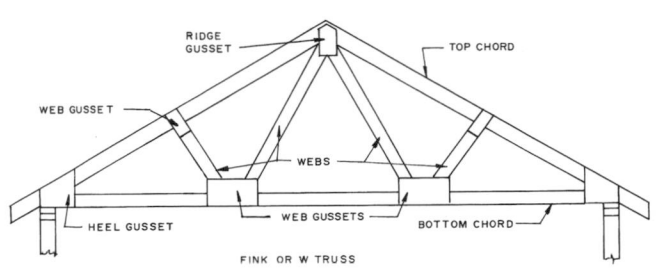

6-48 *The parts of a typical wood framed truss.*

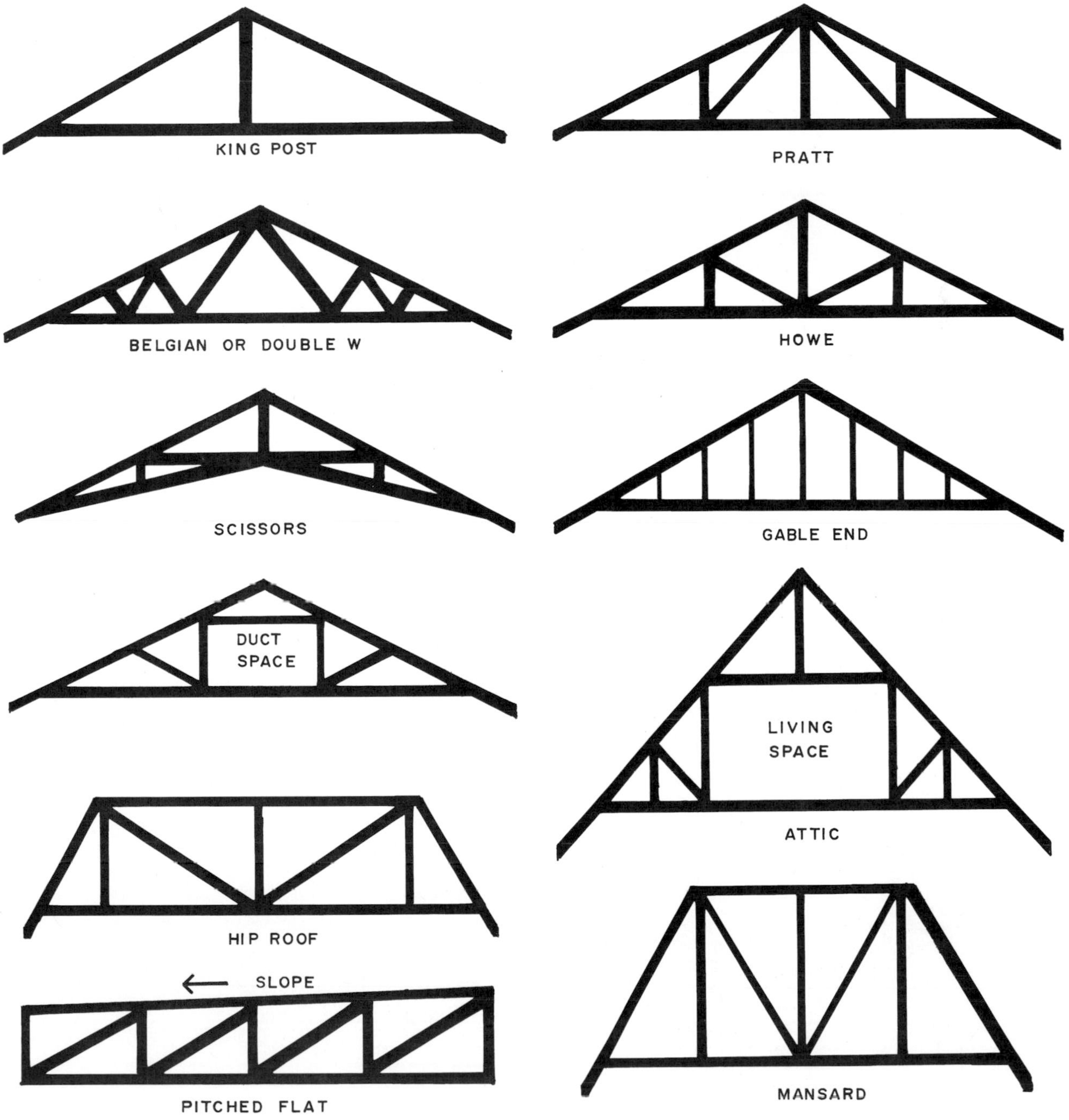

KING POST
PRATT
BELGIAN OR DOUBLE W
HOWE
SCISSORS
GABLE END
DUCT SPACE
LIVING SPACE
ATTIC
HIP ROOF
SLOPE
PITCHED FLAT
MANSARD

6-49 *Some of the commonly used wood roof trusses.*

Small trusses can be raised by workers standing on the platform using wood poles to spin the point into an upright position. Workers on the top plate set it in position, secure the truss to the top plate and brace it. Generally, trusses are set by a crane which raises them over the building and lowers them to the top plate where workers on the plate secure and brace them as shown in 6-50.

6-50 *Large roof trusses are lifted into place with a crane.*

Wood Floor Trusses

Wood floor trusses are frequently used instead of solid wood joists. They provide long spans and often eliminate the need for one or more beams to support the floor in long spans. Since they have open webs it is easy to run electrical, mechanical, and plumbing systems through them. An all wood floor truss is shown in 6-51. They are also made with metal webs. Wood floor trusses must be installed following the manufacturer's instructions.

Using Wood I-Joists

Wood I-joists are structural members made in a factory using plywood or oriented strandboard webs and laminated veneer lumber for flanges. They are widely used for floor and ceiling joists and sometimes for rafters. They must be installed following the manufacturer's directions. They must not be installed using only the techniques common for solid wood joists. Some typical floor installation details are in 6-52. They are also used as rafters.

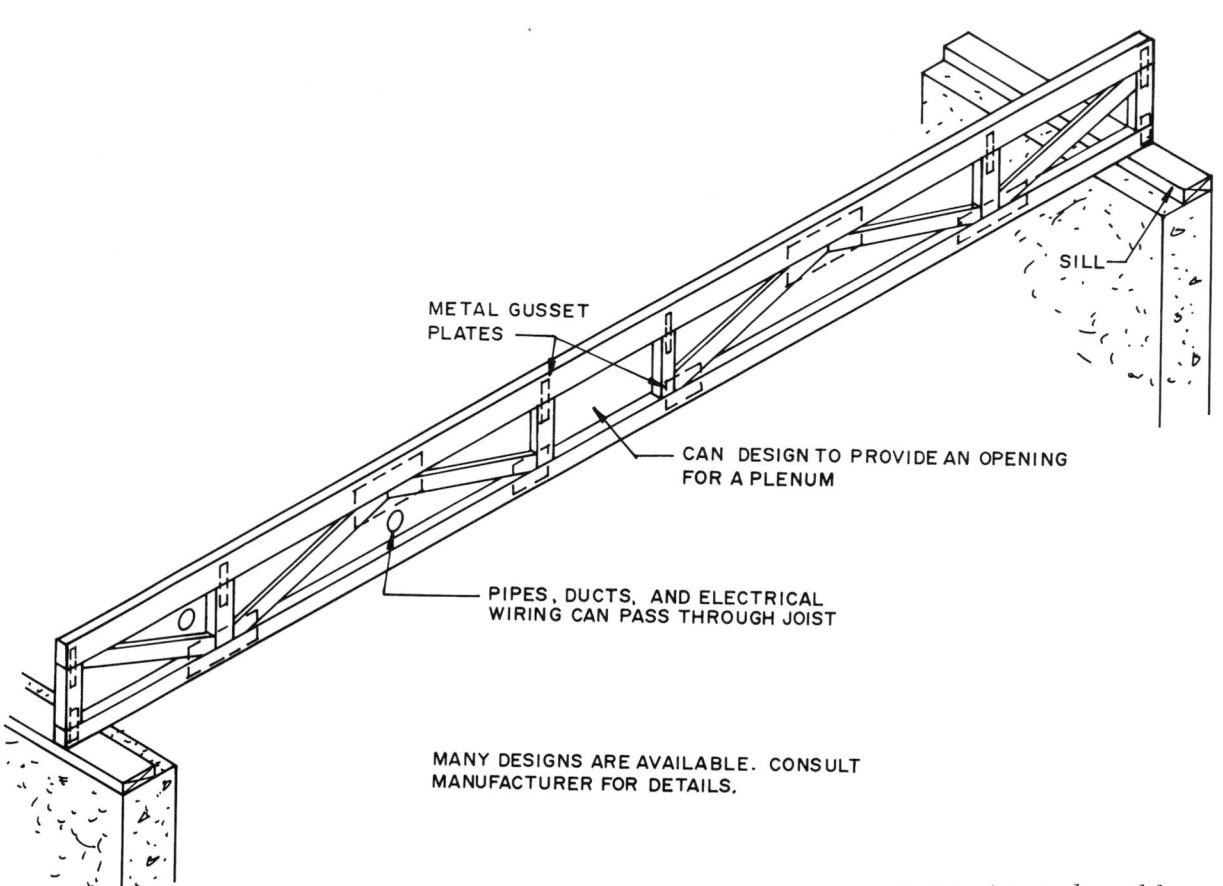

METAL GUSSET PLATES

CAN DESIGN TO PROVIDE AN OPENING FOR A PLENUM

SILL

PIPES, DUCTS, AND ELECTRICAL WIRING CAN PASS THROUGH JOIST

MANY DESIGNS ARE AVAILABLE. CONSULT MANUFACTURER FOR DETAILS.

6-51 *A typical wood frame truss joist is used for floor and ceiling construction.*

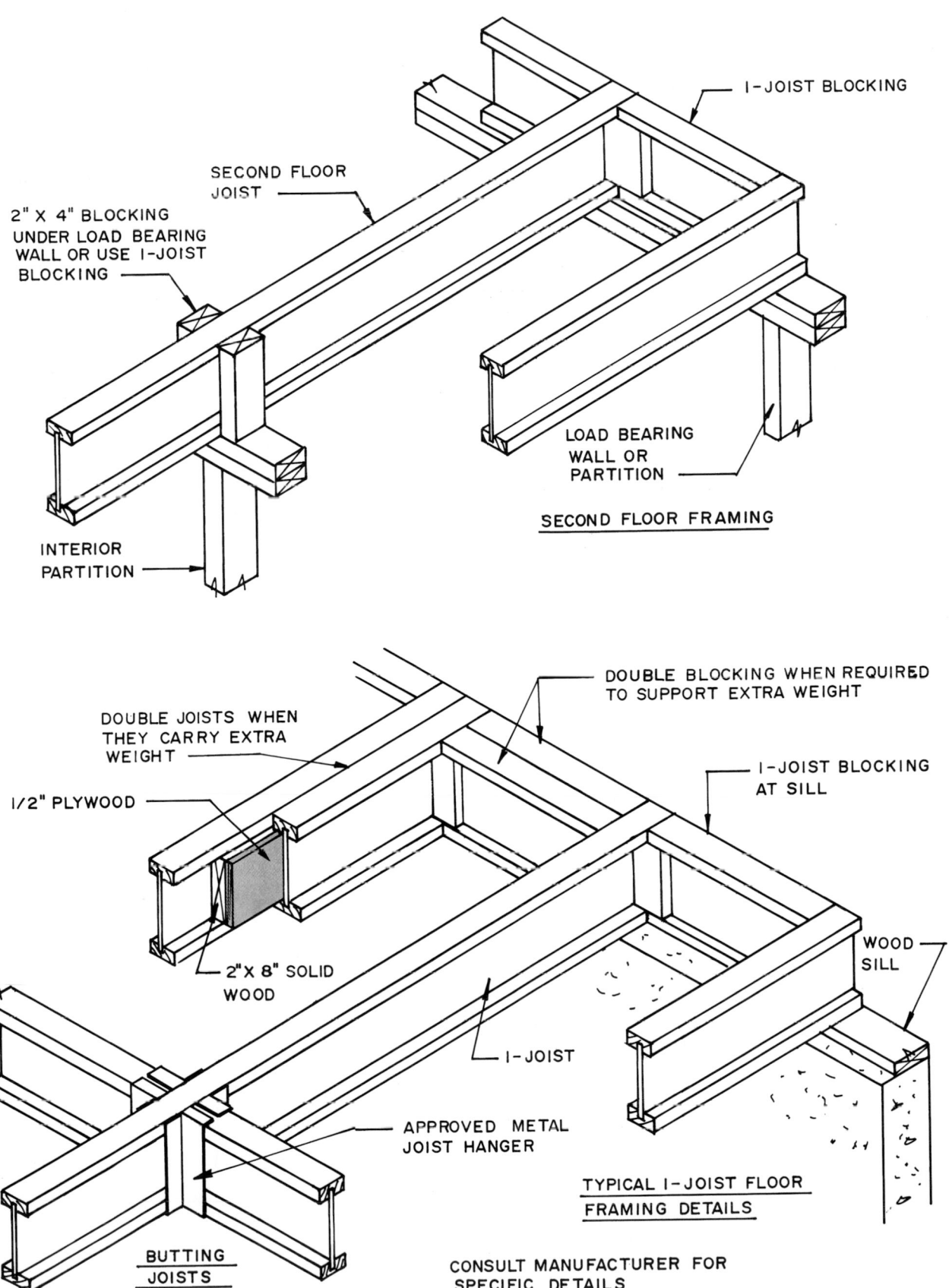

I-JOIST BLOCKING

SECOND FLOOR JOIST

2" X 4" BLOCKING UNDER LOAD BEARING WALL OR USE I-JOIST BLOCKING

LOAD BEARING WALL OR PARTITION

SECOND FLOOR FRAMING

INTERIOR PARTITION

DOUBLE BLOCKING WHEN REQUIRED TO SUPPORT EXTRA WEIGHT

DOUBLE JOISTS WHEN THEY CARRY EXTRA WEIGHT

I-JOIST BLOCKING AT SILL

1/2" PLYWOOD

WOOD SILL

2" X 8" SOLID WOOD

I-JOIST

APPROVED METAL JOIST HANGER

TYPICAL I-JOIST FLOOR FRAMING DETAILS

BUTTING JOISTS

CONSULT MANUFACTURER FOR SPECIFIC DETAILS

6-52 *Typical framing details for use with manufactured wood I-joists.*

HEAVY-TIMBER CONSTRUCTION

Heavy-timber construction is used for both residential and commercial construction. The wood products commonly used include timber and dimension lumber, glulams, parallel strand lumber, laminated veneer lumber, and prefabricated wood I-joists. The selection of the members will depend upon the load-bearing capacity, appearance of the member, and availability of the product. Various types of metal connector are available. Sheathing and decking can also be thick wood members capable of spanning the distances between the wood structural members (see 6-53).

When properly designed, heavy timber construction will meet established engineering standards and building code, fire safety, and energy efficiency requirements.

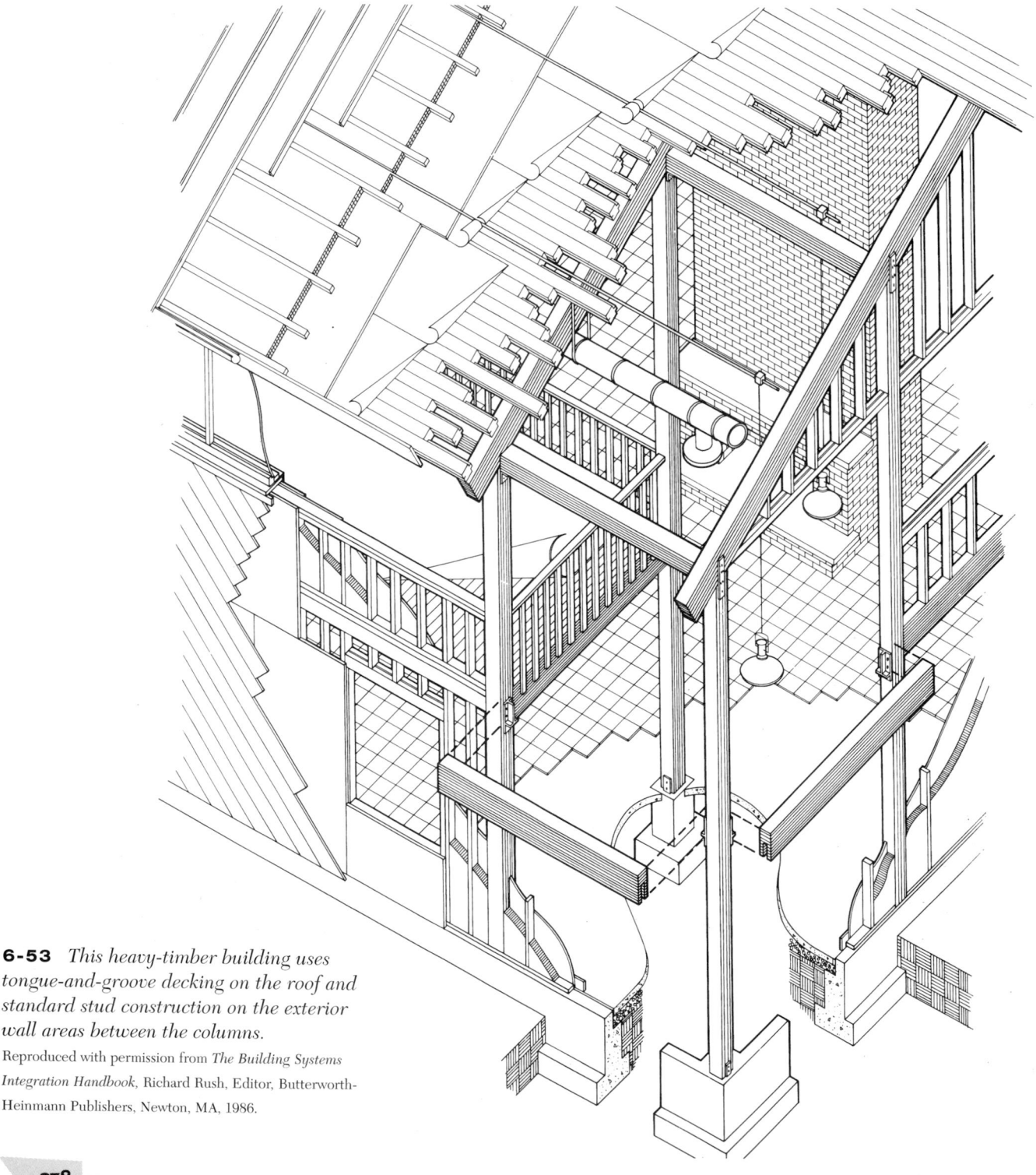

6-53 *This heavy-timber building uses tongue-and-groove decking on the roof and standard stud construction on the exterior wall areas between the columns.*

Reproduced with permission from *The Building Systems Integration Handbook*, Richard Rush, Editor, Butterworth-Heinmann Publishers, Newton, MA, 1986.

Table 6-18 ACTUAL SIZES OF COMMONLY AVAILABLE RECTANGULAR GLUED-LAMINATED MEMBERS

Laminations 1½ in. (38 mm) Thick				Laminations 1⅜ in. (35 mm) Thick			
width (in.)	depth (in.)	width (mm)	depth (mm)	width (in.)	depth (in.)	width (mm)	depth (mm)
3	7½	79.3	190.5	3	6⅞	76.2	174.6
5	6	130.1	152.4	5	6⅞	127	174.6
5	9	130.1	228.6	5	8¼	127	209.5
5	10½	130.1	266.7	5	11	127	279.4
6	9	171.4	228.6	6¾	8¼	171.4	209.5

Building Codes

The designer must observe the requirements of the building code when designing any building including those with heavy timber construction. Heavy timber construction requirements are specified in the various building codes. They typically specify the size and type of approved structural members such as columns, floor framing, floors, and roof decks. The approved types of connection are detailed. The fire resistance rating of the structural elements is specified. Codes also require seismic analysis in areas where earthquakes are a known factor. This includes such things as diaphragms, shear panels, and selection and spacing of fasteners. The building codes also limit the height of heavy timber buildings depending upon the design and occupancy.

FIRE RESISTANCE

Heavy-timber framed structural systems are able to absorb heat and char when exposed to fire. They retain much of their structural strength until the cross-sectional area becomes so small it will not carry the load. The char actually tends to protect the wood in the member from sustained fire damage.

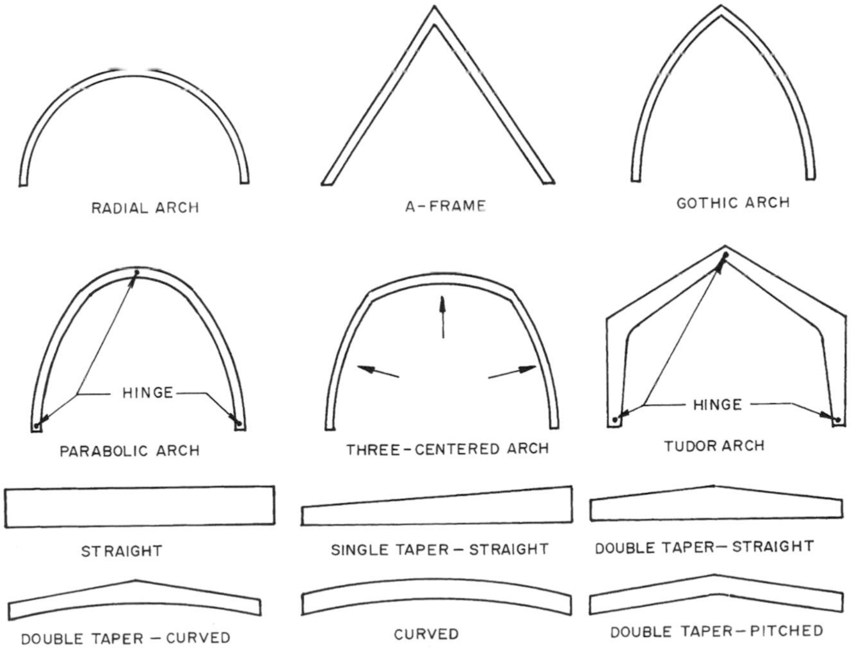

Glued-Laminated Construction

Columns, beams, joists, rigid frames, arches, domes and decking are made from glued-laminated wood. Information about these products is available from the American Institute of Timber Construction (AITC).

Glued-laminated members (called glulams) are engineered, stress-rated members made by laminating wood strips, usually 1⅜ in. (35 mm) and 1½ in. (38 mm) thick for straight members and ¾ in. (19 mm) for curved members such as arches. Commonly available glulam timber members are shown in 6-54. The actual sizes of commonly available rectangular glued-laminated members are shown in Table 6-18 above. In addition rectangular, tapered, and spaced columns are manufactured.

6-54 *Some of the glued-laminated structural timber members available from various manufacturers.*

RADIAL ARCH
A-FRAME
GOTHIC ARCH
HINGE
PARABOLIC ARCH
THREE-CENTERED ARCH
HINGE
TUDOR ARCH
STRAIGHT
SINGLE TAPER – STRAIGHT
DOUBLE TAPER – STRAIGHT
DOUBLE TAPER – CURVED
CURVED
DOUBLE TAPER – PITCHED

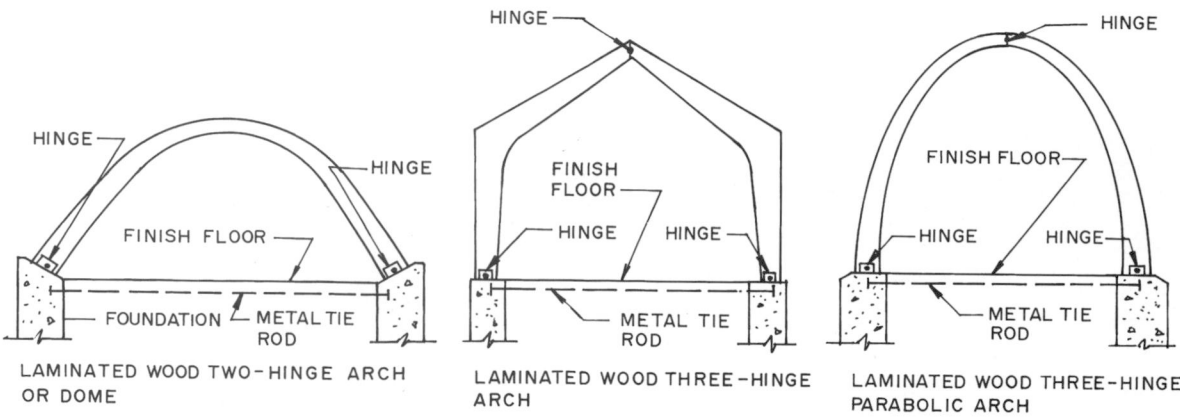

HINGE

HINGE

HINGE

HINGE

HINGE

HINGE

HINGE

FINISH FLOOR

FINISH FLOOR

FINISH FLOOR

HINGE

HINGE

HINGE

HINGE

FOUNDATION — METAL TIE ROD

METAL TIE ROD

METAL TIE ROD

LAMINATED WOOD TWO-HINGE ARCH OR DOME

LAMINATED WOOD THREE-HINGE ARCH

LAMINATED WOOD THREE-HINGE PARABOLIC ARCH

6-55 *Arches and domes may be two-hinged or three-hinged.*

Arches & Domes

Generally, arches may be two-hinged with hinges at each base or three-hinged with a hinge at the crown. A hinged joint is any joint which permits action similar to a hinge and in which there is no appreciable separation of adjacent members.

Glued laminated arches and domes may be two-hinged with hinges at each base or three-hinged with an additional hinge at the crown (see 6-55). These members produce considerable horizontal thrust at their base and they are controlled by the foundation and tie rods.

Laminated arches span long distances, some spanning 70 ft. (21.4 mm) or more. The span depends upon the type of arch, loads, roof pitch, and species of wood. They are usually left exposed inside the building, forming an attractive ceiling (see 6-56).

Typical construction details for arches are shown in 6-57 and 6-58.

6-56 *These parabolic glued-laminated arches provide a lofty atmosphere and a beautiful ceiling.*

Courtesy Canadian Wood Council

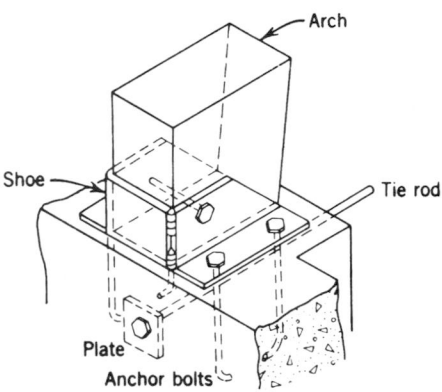

Arch

Shoe

Tie rod

Plate

Anchor bolts

TIE ROD IN CONCRETE.
Thrust is taken by anchor bolts in shear into the concrete foundation and tie rod.

6-57 *Base connection for glued-laminated arches use some type of steel shoe. Notice the tie rods shown.*

Courtesy American Institute of Timber Construction

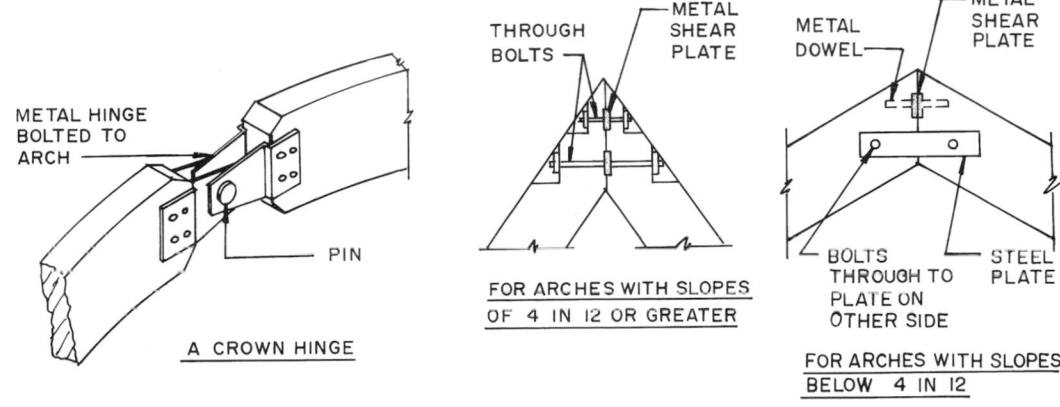

6-58 *Typical connections for the crown of arches.*

Connections for wood three-hinged arches are shown in 6-58. Long span arches may require a section be added on the site because the assembled arch would be too large to transport.

Glue-laminated wood domes are designed as a radial arch or a triangulated system. The triangulated arch can span greater distances than the radial arch. A triangulated glue-laminated wood dome is shown in 6-59, spanning an entire football field. Arched laminated wood structural members are also used for a variety of other projects such as pedestrian bridges over a freeway, ice skating and hockey rinks, and churches.

6-59 *This triangulated glue-laminated wood dome spans a large area.*
Courtesy American Institute of Timber Construction, 7012 S. Revere Parkway, Suite 140, Englewood, CO

Several ways that are used to secure glued-laminated beams and columns to foundations are shown in 6-60. The heavy steel clips and bolts resist vertical and horizontal forces. It is recommended that the beam rest on a metal plate.

Steel connectors are used to secure beams and girders to the columns (see 6-61). The design engineer must indicate the type of fastener to be used and the size and number of bolts. Several beam to beam and beam to girder connections are shown in 6-62.

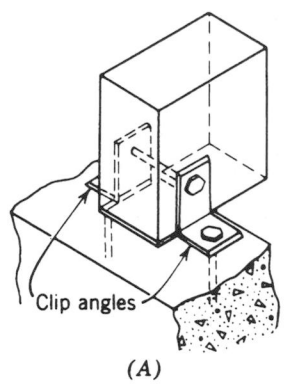

Clip angles

(A)

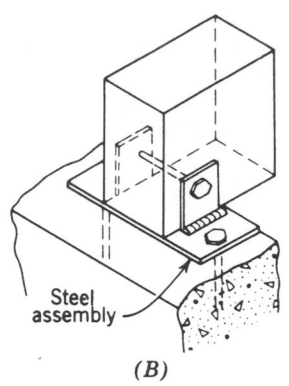

Steel assembly

(B)

BEAM ANCHORAGES

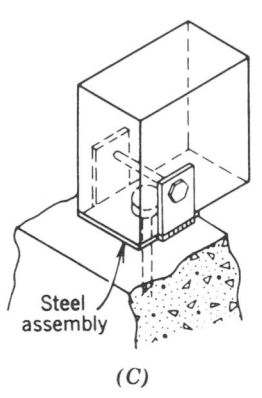

Steel assembly

(C)

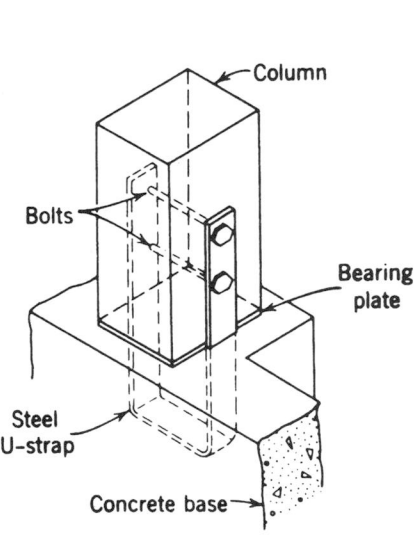

Column

Bolts

Bearing plate

Steel U-strap

Concrete base

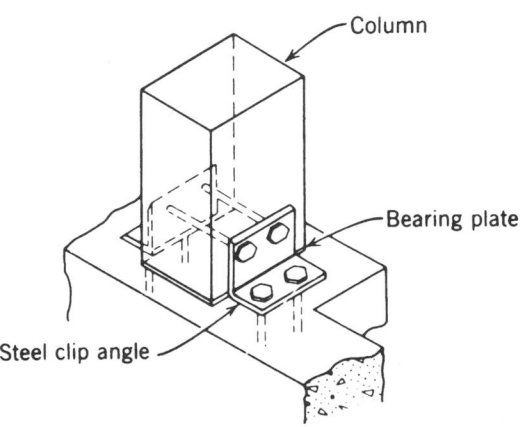

Column

Bearing plate

Steel clip angle

6-60 *Typical glued-laminated connections used to secure a beam and columns to the foundation.*

Courtesy American Institute of Timber Construction, 7012 S. Revere Parkway, Suite 140, Englewood, CO

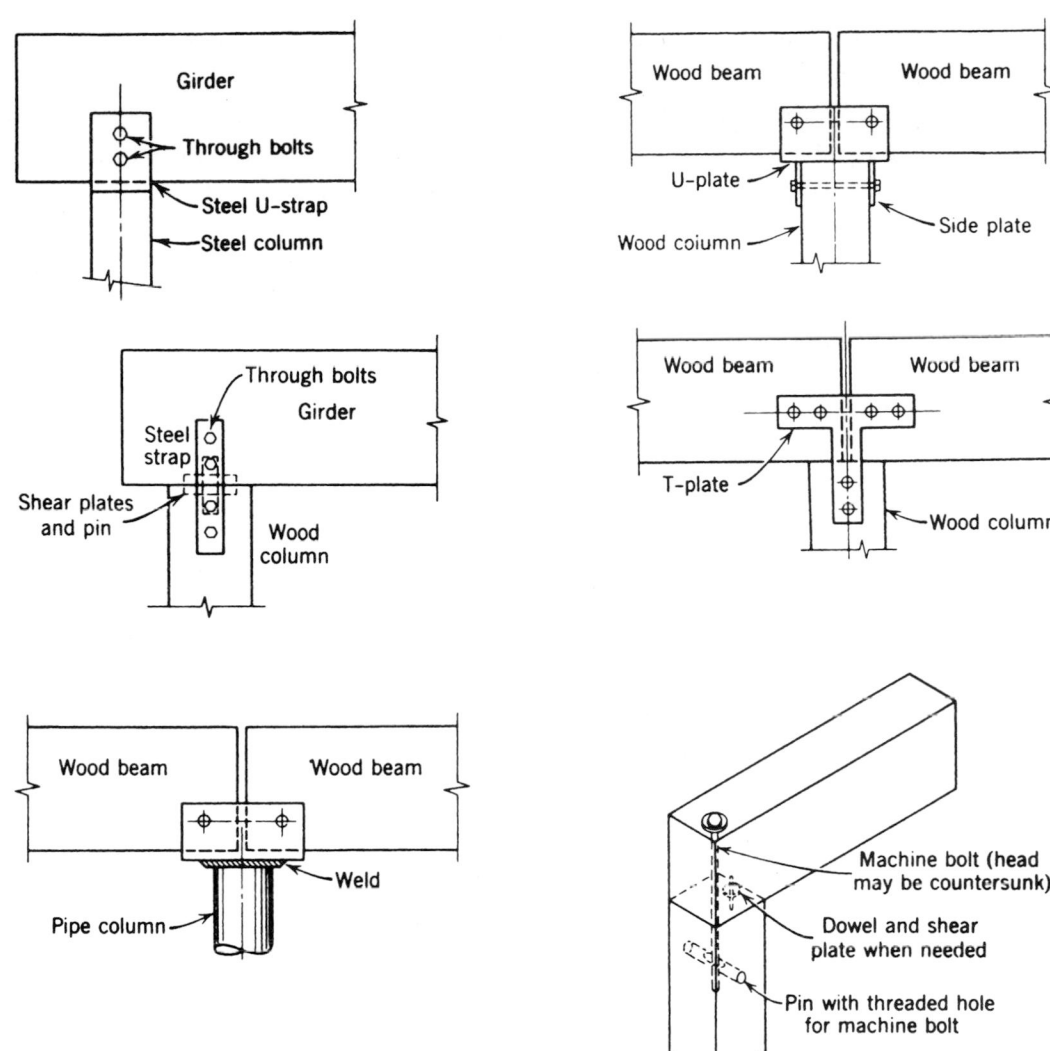

6-61 *Typical beam and girder to column connections.*

Courtesy American Institute of Timber Construction, 7012 S. Revere Parkway, Suite 140, Englewood, CO

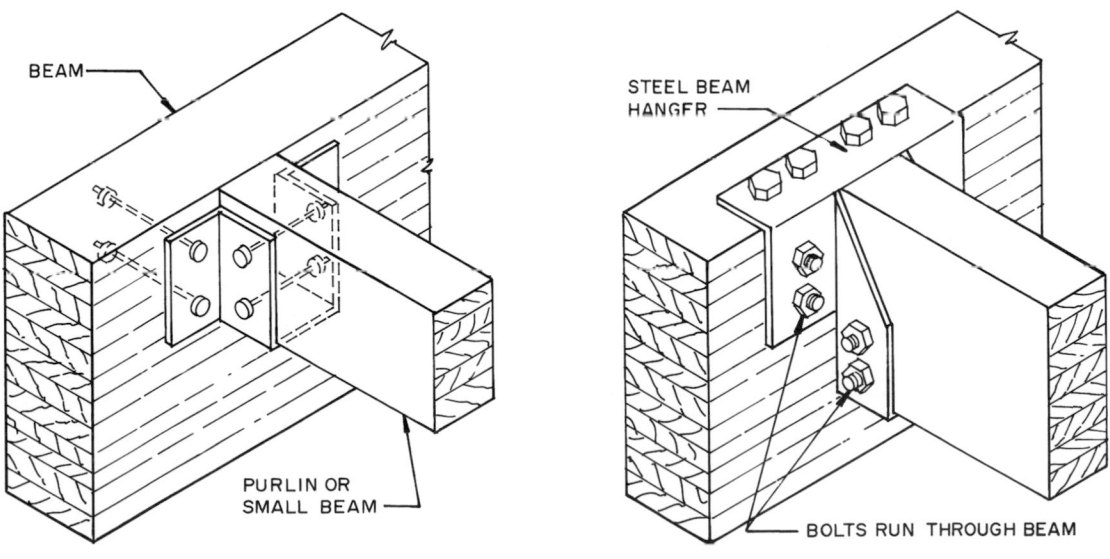

6-62 *Typical beam to girder connections.*

Decking systems include laminated and solid wood decking, stressed skin panels, and 1⅛ in. (29 mm) plywood panels as shown in 6-63. The sizes of laminated decking vary depending upon the species of wood used. Typical actual sizes include 2⅞ × 5⅜ in. and 3 × 7⅛ in. while solid wood sizes include 1½ × 5¼ in., 2½ × 5¼ in., and 3½ × 5¼ in. Several ways for installing laminated and solid wood decking are shown in 6-64. Notice that the single span and continuous two span have the end joints meet on a beam or purlin. Manufacturer's directions should be followed when installing decking.

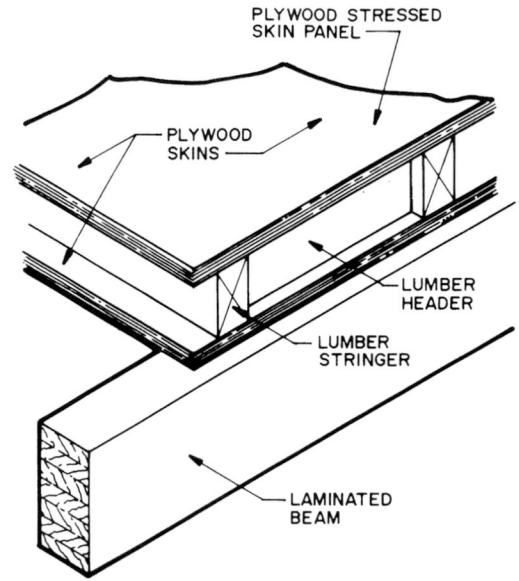

STRESSED SKIN PANELS ON LAMINATED BEAM SYSTEM.

Stressed skin panels, which have a practical span range of 32 ft, are fastened directly to the main laminated timber beams by lag screws or gutter spikes.

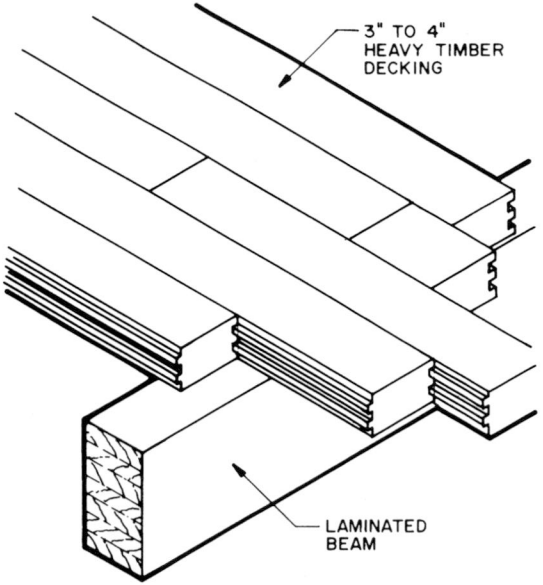

HEAVY TIMBER DECKING ON A LAMINATED BEAM SYSTEM.

Heavy timber decking either laminated or solid 3 or 4 in. nominal thickness is nailed directly to the main laminated beams. The economical span range for the heavy timber decking is 8 to 20 ft depending upon the thickness and loading conditions.

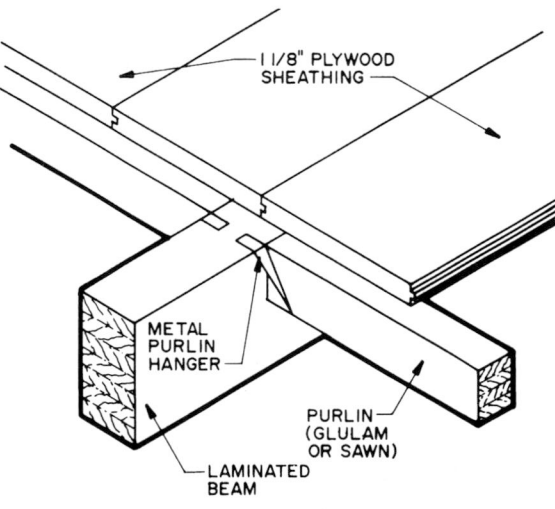

ONE AND ONE-EIGHTH-INCH PLYWOOD ON A LAMINATED BEAM AND PURLIN SYSTEM.

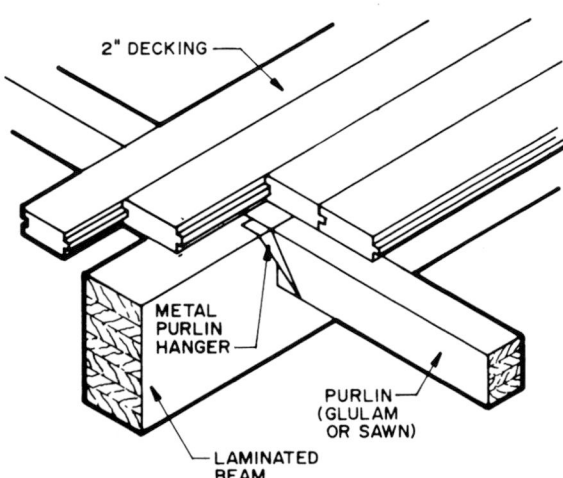

TWO-INCH DECKING ON A LAMINATED BEAM AND PURLIN SYSTEM.

Two-inch nominal thickness decking with an economical span range of 6 to 12 ft is nailed directly to glulam or sawed wood roof purlins, typically on 8 ft centers. Purlins are connected to the main laminated timber beams by metal purlin hangers.

6-63 *Various types of decking are used with glued-laminated framing.*

Courtesy American Institute of Timber Construction, 7012 S. Revere Parkway, Englewood, CO

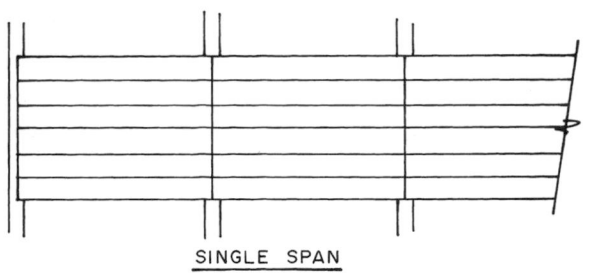

SINGLE SPAN

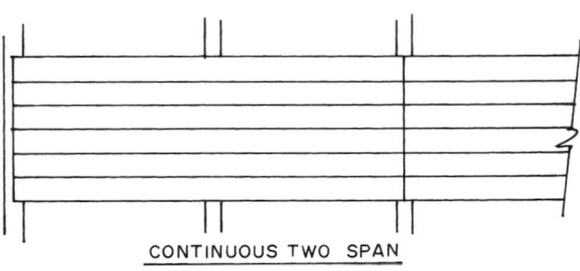

CONTINUOUS TWO SPAN

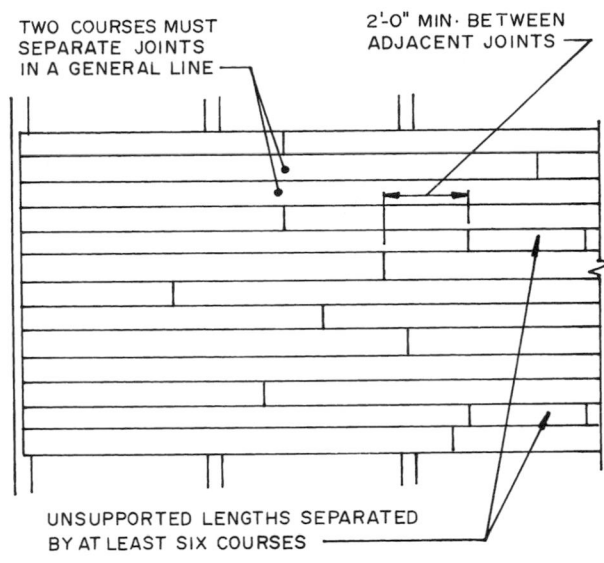

TWO COURSES MUST SEPARATE JOINTS IN A GENERAL LINE

2'-0" MIN· BETWEEN ADJACENT JOINTS

UNSUPPORTED LENGTHS SEPARATED BY AT LEAST SIX COURSES

CONTROLLED RANDOM PATTERN

6-64 *Typical ways glued laminated wood decking may be installed.*

Solid Heavy-Timber Construction

A wide range of techniques are used when constructing heavy-timber buildings. In 6-65 is a building using specially designed roof truss construction supported by columns.

Some of the connections are made by mortising the timber to receive a tenon. This type of construction makes it possible to erect a building with minimum intrusion on the surroundings.

6-65 *This heavy-timber framing uses mortised connections producing an exposed interior framing free from metal connectors. Heavy-timber framing can be erected with a minimum of disturbance of the surrounding area.*

Courtesy The Beamery

FINISHING THE EXTERIOR & INTERIOR

After the building is framed and sheathed the exterior finish work can begin. The contractor tries to close up the building from the weather as rapidly as possible so interior work may begin. After the exterior is complete the drive and landscape work can proceed. However, some delay this until the interior work is complete because it could sustain considerable damage from the trades doing the interior work.

Exterior Finish Work

The exterior finish work includes installing the windows and exterior doors, finishing the eaves, shingling the roof, installing the siding, gutters and doing exterior painting. With the installation of the windows and finishing the roof, the building is weather-tight.

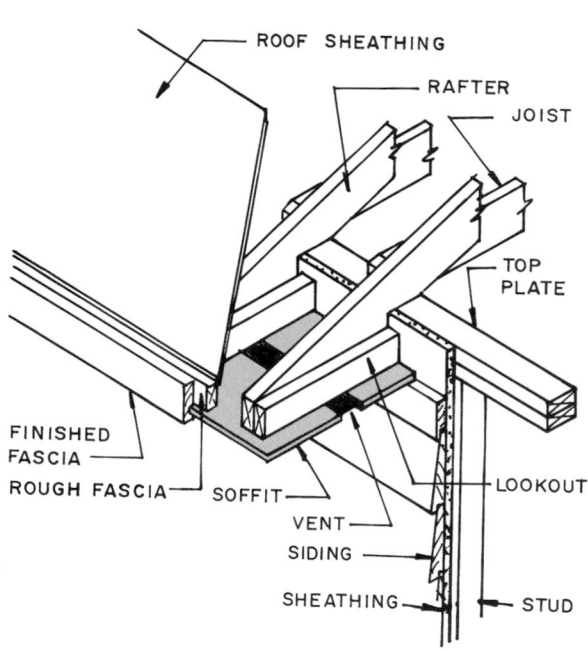

6-66 *A ontinuous attic vent strip in a wide boxed cornice.*

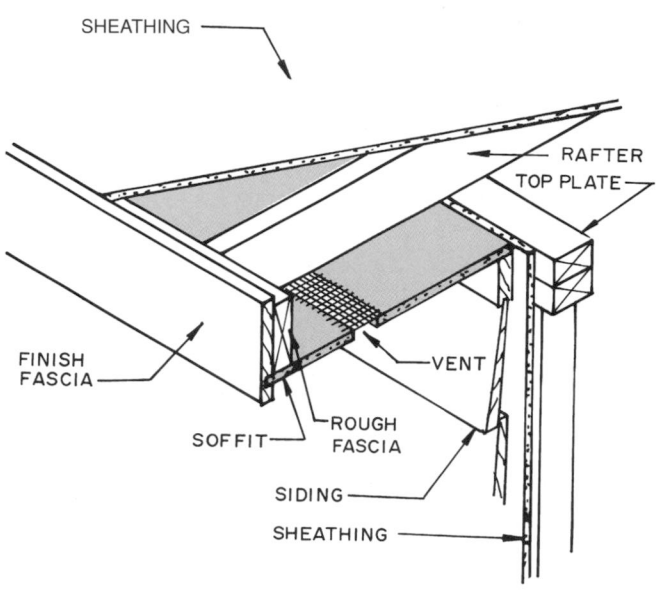

6-67 *This is one way to frame an overhanging cornice on a flat roof.*

6-68 *This vinyl soffit is perforated to provide attic ventilation. The fascia is also vinyl clad.*
Courtesy of Alcan Building Products

FRAMING THE EAVES & RAKE

The eaves and rake are framed after the roof is sheathed. There are many ways these can be constructed. The carpenter will find the design required on the working drawings. The architect designs the eave and rake to suit the architectural style of the house. A typical design for sloped roofs is in 6-66 and a flat roof design is in 6-67. The rough fascia is 1½ in. (38 mm) materials. It provides spacing support for the ends of the rafters and a means of securing gutters and the finished fascia. The finished fascia can be painted wood or wood covered with aluminum or vinyl. The soffit can be plywood, hardboard, or other exterior reconstituted wood product, vinyl, or aluminum. The vinyl and aluminum soffits are perforated, providing a flow of air into the attic (see 6-68). Solid soffits require that metal vent strips be installed.

The rakes are finished similar to the eave. The rake fascia may be set flush over the siding or extend out from the gable end as shown in 6-69.

VENTILATING THE ATTIC

Attic ventilation is required to remove high temperature air in the summer and the accumulation of moisture in summer and winter. Moisture in the attic in the winter will freeze on the rafters and sheathing. As the air temperature gets above freezing it will melt and drip on the ceiling insulation, damaging it. Attic ventilation is also required to keep the air temperature in the winter about the same as the outside air. If the ceiling is poorly insulated or ventilation is inadequate, the air temperatures will rise, causing snow on the roof to melt. As it runs to the uninsulated eave it will again freeze. This causes an ice dam on the eave and moisture will back up under the shingles and possibly leak through into the interior wall and on the ceiling insulation. In cold climates extra layers of builder's felt are laid from the eave up to the roof to provide additional waterproofing. To get adequate ventilation, the soffit must have some form of vent allowing air to flow up into the attic.

It is necessary to install some type of baffle at the exterior wall so the ceiling insulation does not block the flow of air from the soffit into the attic (see 6-70).

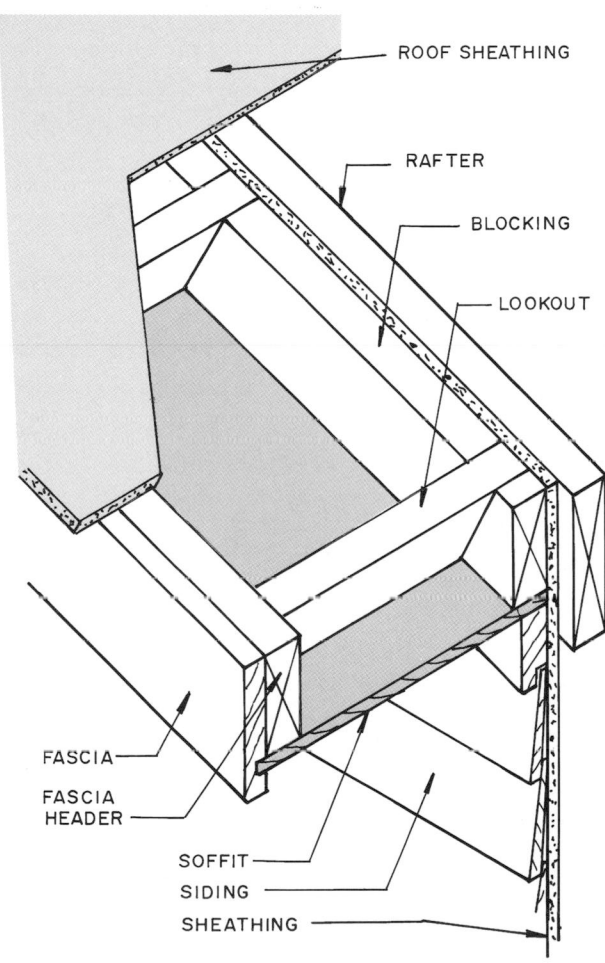

6-69 *Rake fascias are generally extended beyond the end wall and have a soffit returning to the wall.*

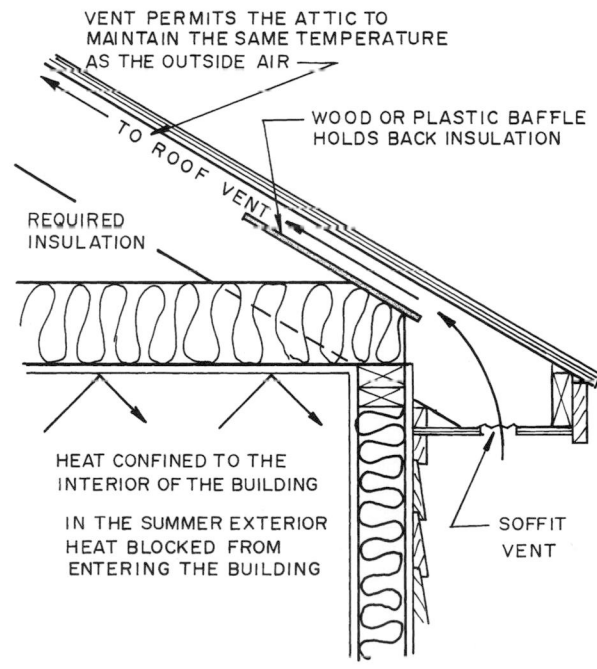

6-70 *Baffles are used to provide for passage of air at the eave into the attic.*

Vents near the ridge are used to allow the flow of air to exit the attic. Power fans, ridge vents, and gable end vents are commonly used (see 6-71 and 6-72). Typical venting details for flat roofs are in 6-73. They use soffit vents and require that a space be left between the sheathing and insulation.

THE FINISHED ROOF MATERIAL

After the roof sheathing is covered with builder's felt, the finished roofing may be installed. These materials are covered in Division 7. A commonly used roofing material is composed of a fiberglass base saturated with asphalt and having mineral granules embedded in the surface to form a protective coating (see 6-74).

Wood shingles and shakes, clay, perlite, and concrete tile and metal roofing materials are also used. Flat roofs will have a built-up or single-ply sheet membrane. Details can be found in Division 7.

Shingles are usually installed by a roofing contractor. The crews are trained to properly install flashing and the shingles. It is important they be installed according to the manufacturer's recommendations. The fire resistance of roofing materials is an important consideration and they are classified by their ability to resist exposure to fire.

After the shingles are in place the gutters may be installed. However, this is often delayed until the exterior is finished in order to keep them from being damaged. The gutters are either aluminum or vinyl. The roof shingles extend about one inch over the edge of the fascia and direct the water into the gutter. The gutters also serve as a decorative finish on the fascia.

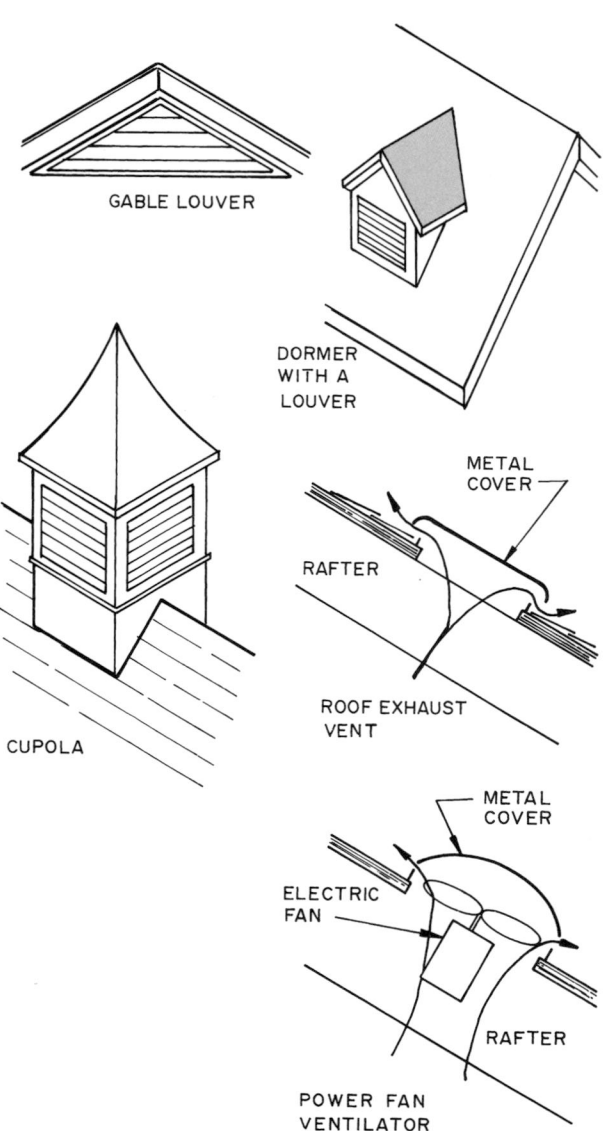

6-71 *Some ways to ventilate an attic.*

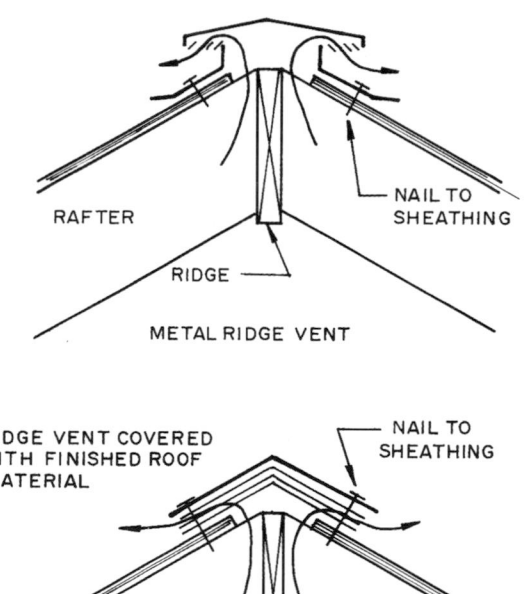

6-72 *Ridge vents are widely used for attic ventilation*

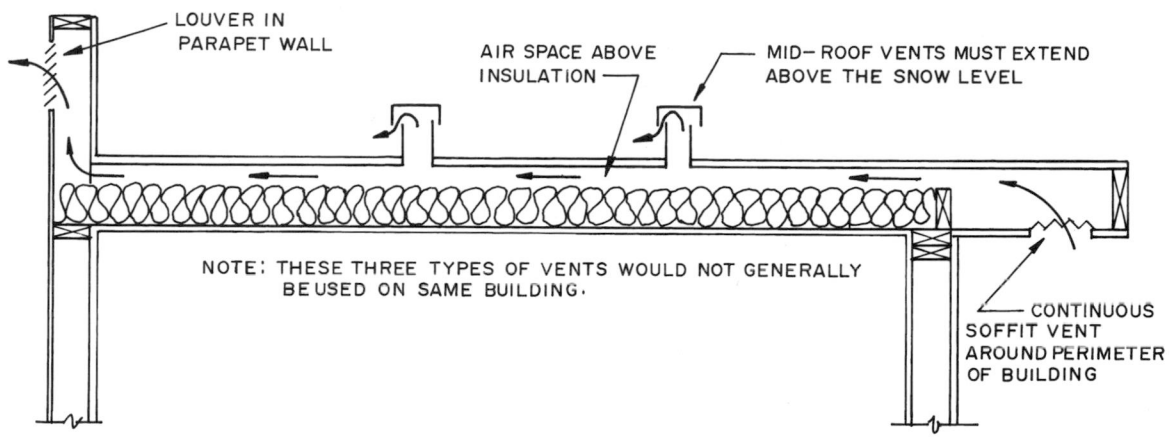

LOUVER IN PARAPET WALL

AIR SPACE ABOVE INSULATION

MID–ROOF VENTS MUST EXTEND ABOVE THE SNOW LEVEL

NOTE: THESE THREE TYPES OF VENTS WOULD NOT GENERALLY BE USED ON SAME BUILDING.

CONTINUOUS SOFFIT VENT AROUND PERIMETER OF BUILDING

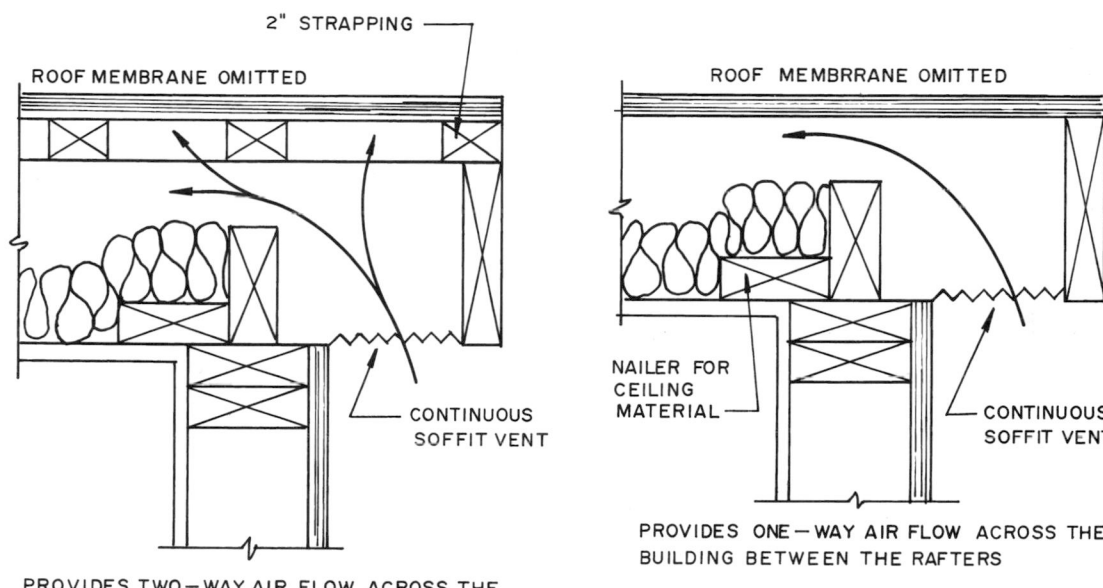

2" STRAPPING

ROOF MEMBRANE OMITTED

ROOF MEMBRANE OMITTED

CONTINUOUS SOFFIT VENT

NAILER FOR CEILING MATERIAL

CONTINUOUS SOFFIT VENT

PROVIDES TWO–WAY AIR FLOW ACROSS THE LENGTH AND WIDTH OF THE BUILDING

PROVIDES ONE–WAY AIR FLOW ACROSS THE BUILDING BETWEEN THE RAFTERS

6-73 *Techniques for ventilating flat roofs.*

6-74 *Fiberglass asphalt shingles are widely used as a finish roofing material.*

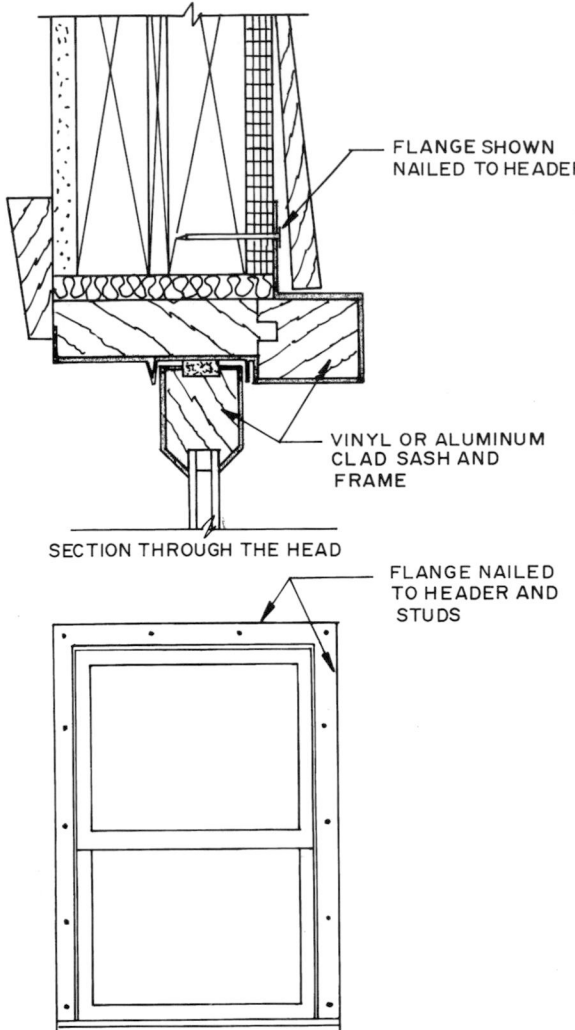

FLANGE SHOWN NAILED TO HEADER

VINYL OR ALUMINUM CLAD SASH AND FRAME

SECTION THROUGH THE HEAD

FLANGE NAILED TO HEADER AND STUDS

CASING NAILED TO HEADER

WINDOW HEAD JAMB

WINDOW SASH

SECTION THROUGH THE HEAD

CASING NAILED TO HEADER AND STUDS

6-75 *Windows made with flanges on the frame are installed by nailing through the flanges into the sheathing and studs.*

6-76 *Windows made with brick casing attached are installed by nailing through the casing into the sheathing and framing.*

INSTALLING WINDOWS

There are many types and designs of windows used in light wood-frame construction. These are described in Division 8. The recommended way to install windows varies somewhat with the manufacturer's. Following is a general description of two frequently used types.

Most windows are installed from the outside of the building after the sheathing is in place. One type has a metal or plastic flange secured to the unit. The window is placed in the opening, checked for plumb, and nailed to the studs and header through the flange. A simplified illustration is in 6-75. Another installation is shown in 6-76. Here the window unit is nailed to the studs and header through the casing (sometimes called the brick molding).

The installation procedure begins by checking the opening to see that the sill is level. Usually the sides of the rough opening are covered with builder's felt or plastic sheeting. Techniques used to install the unit are in 6-77. The wood, aluminum, and vinyl siding or brick veneer is installed after the windows are securely in place.

EXTERIOR DOOR INSTALLATION

A variety of exterior doors and frames are available. These are described in Division 8. The doors may be wood, metal, or fiberglass.

The door and frame come in an assembled unit. It is important that the frame be kept square.

1. Lay a bead of caulking on the sheathing around the window opening.

2. Set the window unit in the opening.

3. Check for levelness and plumb. Then nail through one corner. Recheck for levelness and plumb, and nail the other corner.

4. Check the diagonals to be certain that the unit is square before nailing the rest of the flange. Also check to see that the windows open easily. Then nail the flange on all sides.

6-77   Procedures for installing windows getting the unit level and plumb before final nailing.

Courtesy Pella Corporation

To install an exterior door, cover the edges of the rough opening with builder's felt or sheet plastic. Set the door frame in the opening and check to be certain the sill is level. Shim it level if necessary. Check the sides for plumb.

Place wedge-shaped blocking between the studs and the door jamb. Drive one casing nail through the casing into the studs on both sides. Adjust the blocking, keeping the jambs plumb, and nail through the blocking into the studs.

INSTALLING THE SIDING

Exterior siding may be wood, plywood, wood shingles, hardboard, plastic, vinyl, stucco, brick, or stone. The methods for installing wood siding are shown in 6-78 and 6-79. Notice that the nails in the lower end of the horizonal siding do not penetrate the piece of siding directly below. This allows each piece to expand and contract independently of the other. Hardboard siding is available in lap siding and panel siding. Plywood siding is available in lap siding and panel siding. It is made with a variety of surface textures. Panel siding is available in 4 ft. (1220 mm) widths and 8, 9, and 10 ft.

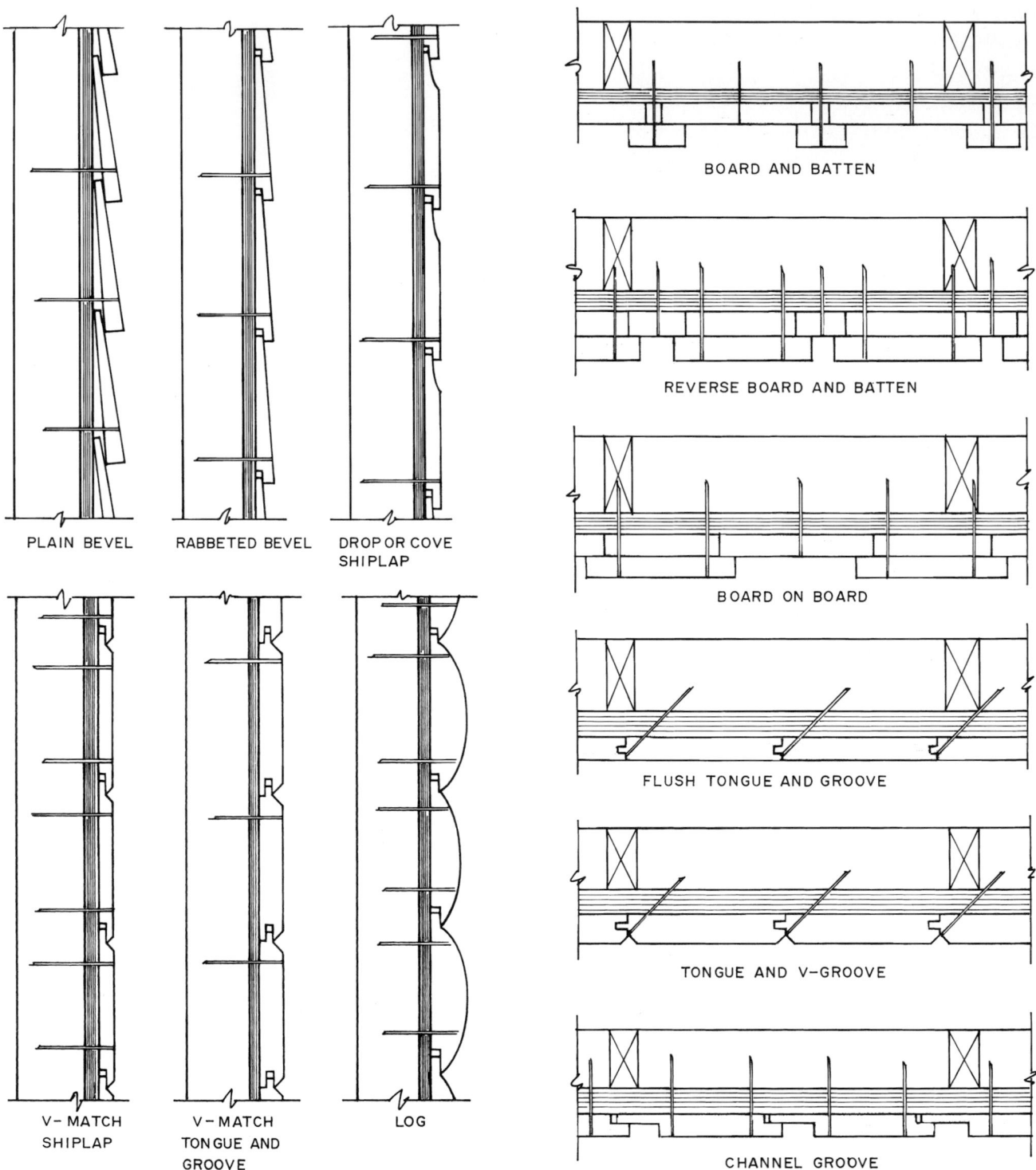

PLAIN BEVEL

RABBETED BEVEL

DROP OR COVE SHIPLAP

V-MATCH SHIPLAP

V-MATCH TONGUE AND GROOVE

LOG

BOARD AND BATTEN

REVERSE BOARD AND BATTEN

BOARD ON BOARD

FLUSH TONGUE AND GROOVE

TONGUE AND V-GROOVE

CHANNEL GROOVE

6-78 *Types of horizontal wood siding with nailing patterns.*

6-79 *Types of vertical wood siding with nailing patterns.*

(2440, 2745, 3050 mm) lengths and in a variety of panel patterns. Sheathing or bracing is not required because the panel provides the required structural strength.

Aluminum and vinyl siding are available in a variety of colors and surface textures. They must be nailed to nailable sheathing or directly into the studs because they have no structural strength. Specially designed channels are placed at internal and external corners and where the siding butts a surface such as the wood frame of a door or window. A typical installation detail is shown in 6-80.

Wood shingles used as siding are applied over nailable sheathing, such as plywood, and may be installed as a single or double course (see 6-81 and 6-82).

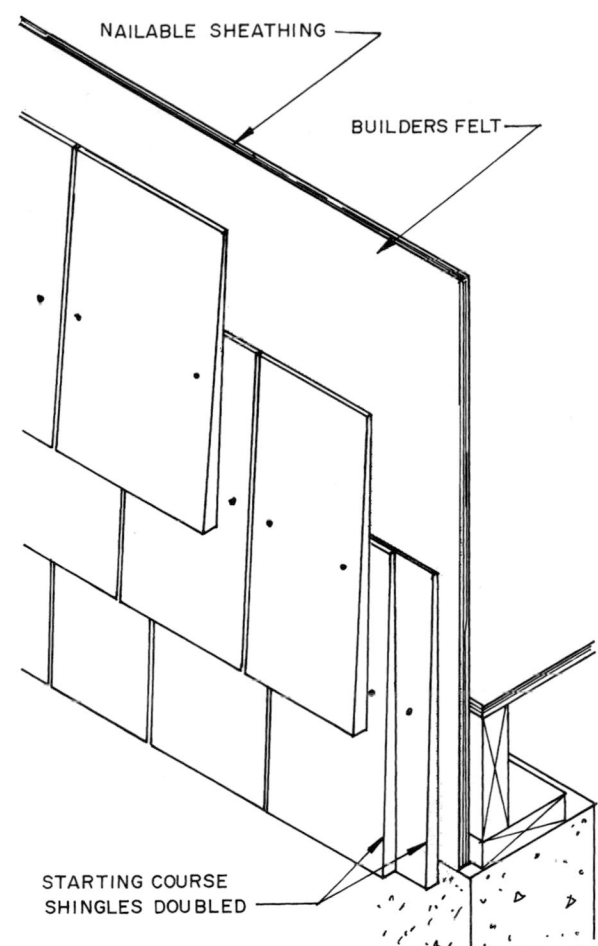

6-81 *These wood shingles are applied in a single course.*

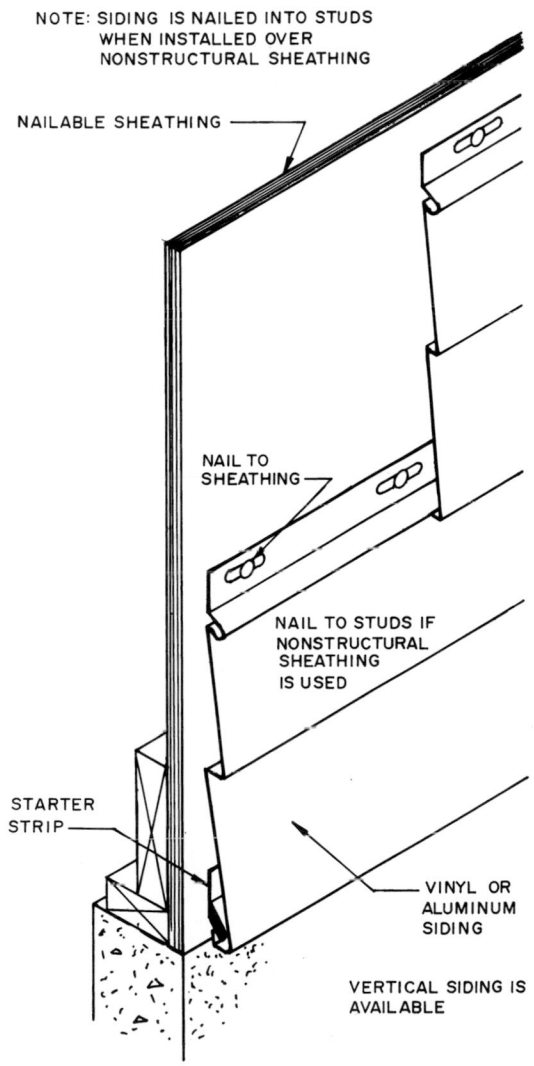

6-80 *A typical installation design for vinyl and aluminum siding. The connection system will vary depending upon the manufacturer of the siding.*

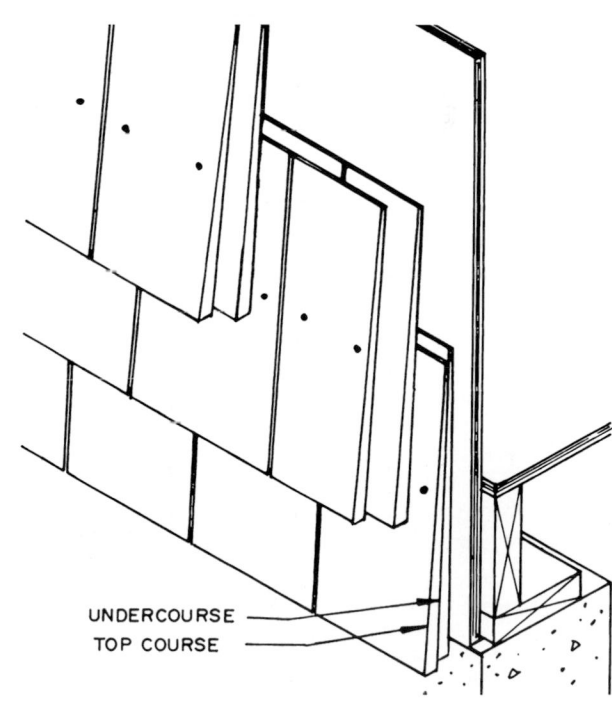

6-82 *Wood shingles and shakes can be used as siding. These are applied in a double course.*

Stucco exterior surfaces may be constructed with the finish coat being a portland cement, lime, and sand mixture (see 6-83). It is troweled over a wire mesh that has been nailed to the sheathing (see 6-84).

Another exterior finish similar to stucco is finding increasing use. It is composed of a glass fiber-reinforced portland cement mixture containing special bonding agents. It is applied over a glass fiber mesh bonded to the sheathing. In 6-85 a wall assembly is flashed and permits water to weep from the wall. This is the USG Corporation Water-Management Finish System.

Brick and stone veneers are also used as exterior siding on light-frame construction. They rest on a brick ledge constructed as part of the foundation. Metal ties are nailed to the studs and inserted in the mortar joints between the masonry units. Wall ties are generally spaced every 16 in. (406 mm) vertically (see 6-86). Masonry and stone materials are covered in Division 4.

Before the siding is installed the sheathing is usually covered with asphalt-impregnated builder's paper in a water-tight, vapor-permeable plastic sheet. These provide a waterproof layer permitting water vapor in the wall cavity to pass through to the outside, thus prevent-ing it from being trapped inside the wall cavity. Some types of sheathing have a cover of perforated aluminum foil that serves the same purpose.

Interior Finish Work

After the building is weather-tight the electricians, plumbers, and air-conditioning–heating contractor may begin their work. The electricians run the wires and mount the required boxes (see 6-87). The plumbers run the water and waste disposal lines.

The heating contractor crew places the ducts or runs the hot water pipes to the locations of registers. After they have completed this, the local building inspector checks the work for compliance. The electrician installs the light fixtures, switches, and outlets after the interior walls and ceilings are completed. The plumber sets the fixtures, except the tub or shower which are installed as the interior partitions are built. The heating contractor covers the duct openings until the carpet is installed (see 6-88) or the hardwood floors are finished. Then they are uncovered, vacuumed carefully, and the registers are set in place. Information on flooring and carpet can be found in Division 9.

6-83 *Stucco coats are troweled over the base.*

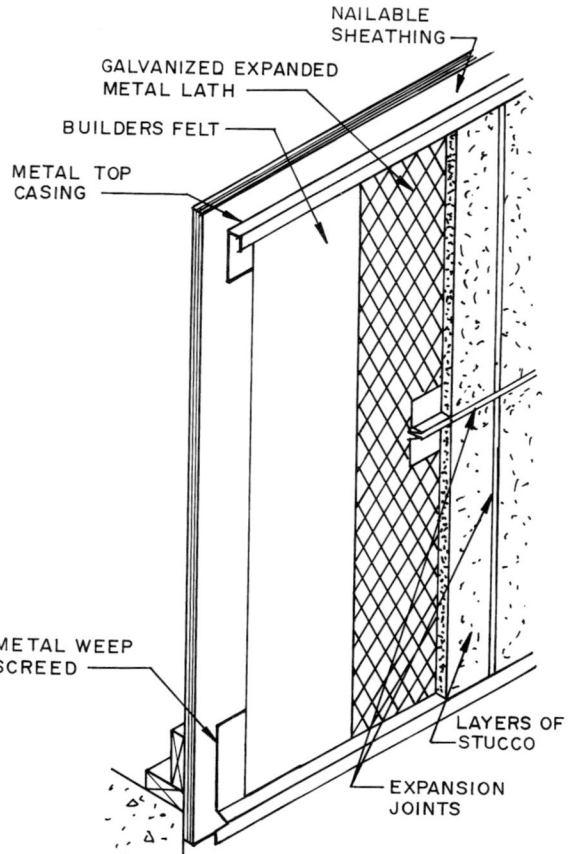

6-84 *Typical construction for a portland cement stucco installation.*

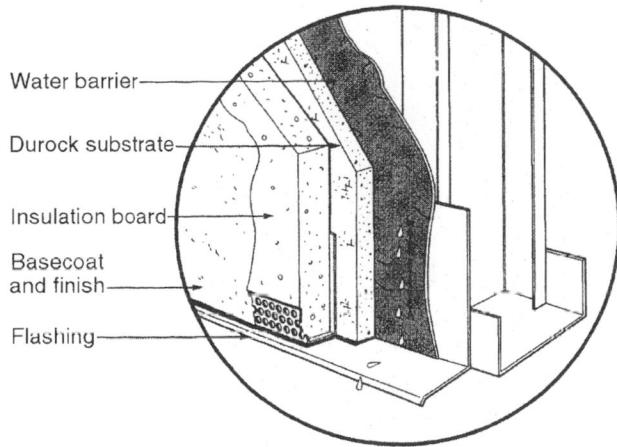

Water barrier

Durock substrate

Insulation board

Basecoat and finish

Flashing

Insulscreen 2100 Water-Managed Exterior Finish System

6-85 *This Water-Management Finishing System uses a cement board (Durock®) as the substrate for the insulation over which the synthetic stucco is applied. Note the use of flashing and the provision for the wall to weep at the bottom.*

Courtesy United States Gypsum

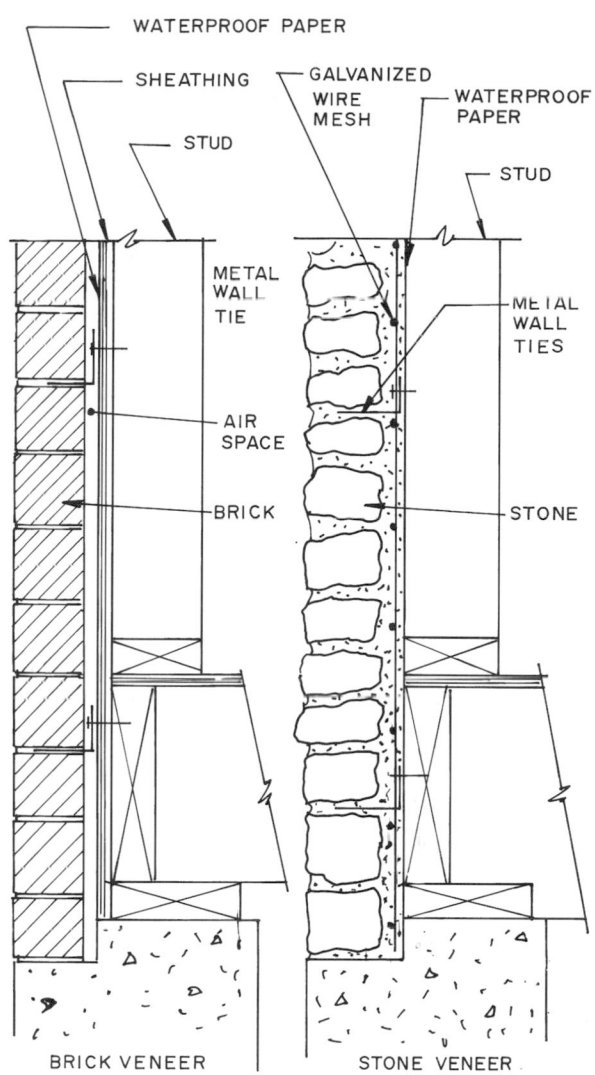

WATERPROOF PAPER

SHEATHING

GALVANIZED WIRE MESH

STUD

WATERPROOF PAPER

STUD

METAL WALL TIE

METAL WALL TIES

AIR SPACE

BRICK

STONE

BRICK VENEER

STONE VENEER

6-86 *Typical brick and stone veneer over frame construction.*

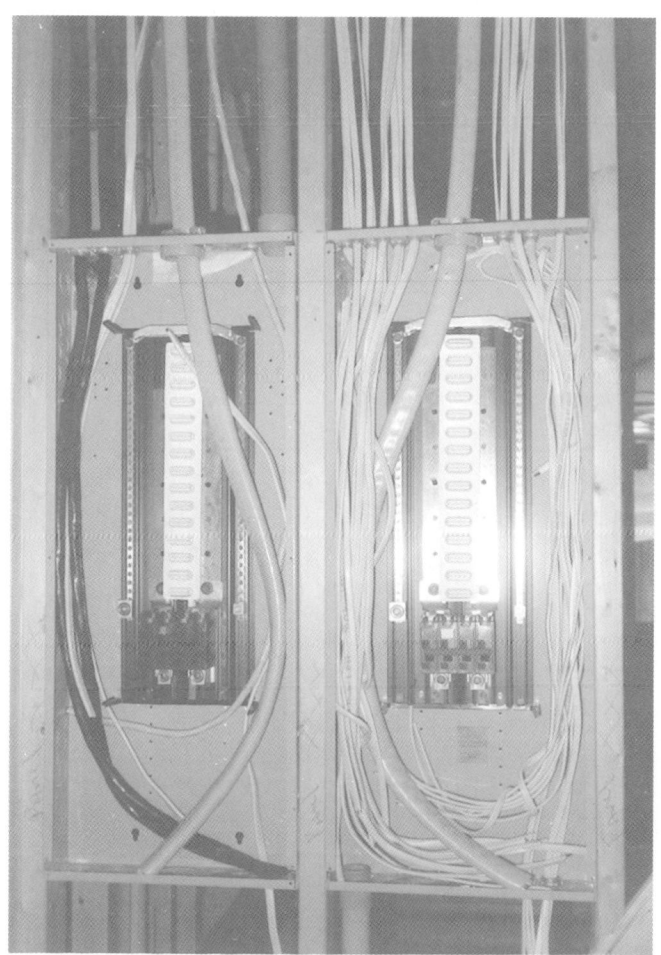

6-87 *These main electric panels are installed after the building is weather-tight.*

6-88 *The carpet is installed after the walls are finished, the base is in place, and the painting is complete.*

Courtesy NO-MUV Corporation, Inc.

INSTALLING INSULATION

Thermal insulation and vapor barriers are discussed in Division 9. Once the interior is protected from the weather and the electrical and plumbing systems are installed, the insulation can be installed in the walls, floors, and ceilings. Blankets and batts may have a water-resistant paper covering that forms a vapor barrier. It is placed facing the interior of the building. The paper provides tabs used to staple the insulation to the studs (see 6-89). When unfaced blankets are used, the entire wall is covered with plastic sheets, forming a vapor barrier. The blankets should fit snugly against the top and bottom plates. They are cut to fit tightly around plumbing and electrical boxes. Another form of insulation is a rigid batt. It is pressed between the studs and is held in place by friction. A plastic sheet vapor barrier is stapled over these batts. Insulation is placed wherever the interior is exposed to exterior temperatures. The ceiling can be insulated after the drywall has been installed by laying the blankets on top of the gypsum. Floor insulation is often friction fit. Blankets can be supported with wires placed between the joists (6-90). Ceilings are also insulated by spraying in loose fiber insulation.

The insulation properties of the wall can be improved by using a rigid plastic foam insulation type sheathing. This can be run down over the foundation into the ground.

This must be protected by a hard material where it is exposed to the weather. Some sheets are made with a protective coating already applied.

Insulation may also be used to reduce the passage of sound through walls, floors, and ceilings. Various types of sound control batts are available. More information on acoustical treatments is in Division 9.

INTERIOR WALL & CEILING FINISHES

Interior walls and ceilings are generally covered with gypsum board after the insulation has been installed. Other finishes include plaster, plywood, and hardboard paneling, and solid wood paneling. Gypsum and plaster products are discussed in Division 9 and wood products in Division 6.

Generally the gypsum panels are installed by subcontractors specializing in this work (see 6-91). Plaster finishes require the services of highly skilled plasterers and these craftsmen are not generally available in some areas. Wood, plywood, and hardboard paneling is usually installed by finish carpenters. Usually wood panel-

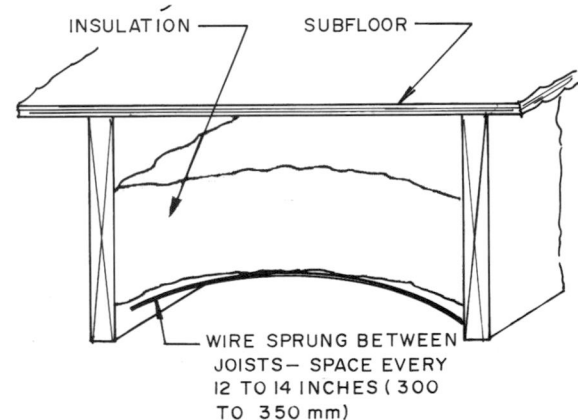

6-90 *Insulation blankets between joists can be supported with wires sprung between the joists.*

6-89 *These insulation blankets have a water-resistant paper vapor barrier that must face the side of the wall.*
Courtesy Gold Bond Building Products

6-91 *Gypsum wallboard is a widely used material for finishing interior walls.*
Courtesy United States Gypsum Corporation

ing is applied over a substrate of gypsum wallboard. Masonry partitions are used in many commercial projects but seldom in light-frame construction, because they require a footing. They can be conveniently used on buildings with concrete slab floors.

Ceilings are typically finished with gypsum board, plaster, some form of composition tile, or panels set in a suspended metal frame.

Interior Finish Carpentry

Interior finish carpentry includes projects such as installing interior molding, wainscoting, cabinets, bulletin, shelves and other units and stairs. It requires the efforts of the most skilled craftsmen and cabinetmakers.

INSTALLING INTERIOR DOORS

One task performed by the finish carpenter is to install interior doors and jambs. There are many varieties and types of interior door available. Manufacturers often supply installation instructions with each unit. Additional information about doors is in Division 8.

Most interior swinging doors arrive prehung on a one-piece door jamb. The carpenter places the jamb into the rough opening, gets the sides plumb by driving wood wedges between the frame and the wall stud, and nails through the side jamb and wedges into the stud. A long level is used to check for plumb (see 6-92). The door is closed and the spacing between it and the jambs is checked to be certain it is uniform and that the door does not rub on the jamb. There are several types of split door jamb available. They are made with the casing attached to each half. The frames are slid into the rough opening from opposite sides and nailed to the studs.

INSTALLING CASINGS & BASE MOLDING

After the door jambs are in place and the gypsum wallboard is installed, the finish carpenter can install the casings and baseboards. The casings are available in a wide range of sizes and patterns. The architect has specified the type to use. A typical cased interior door opening is shown in 6-93. This particular example shows mitered corners, but other types of corners may be used.

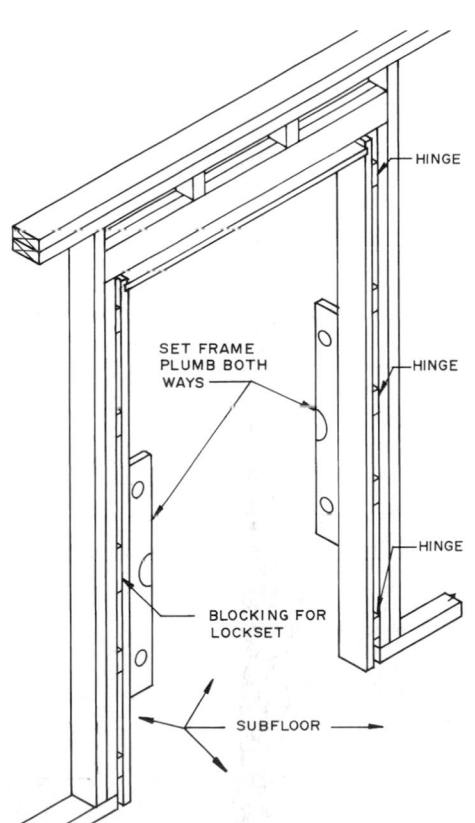

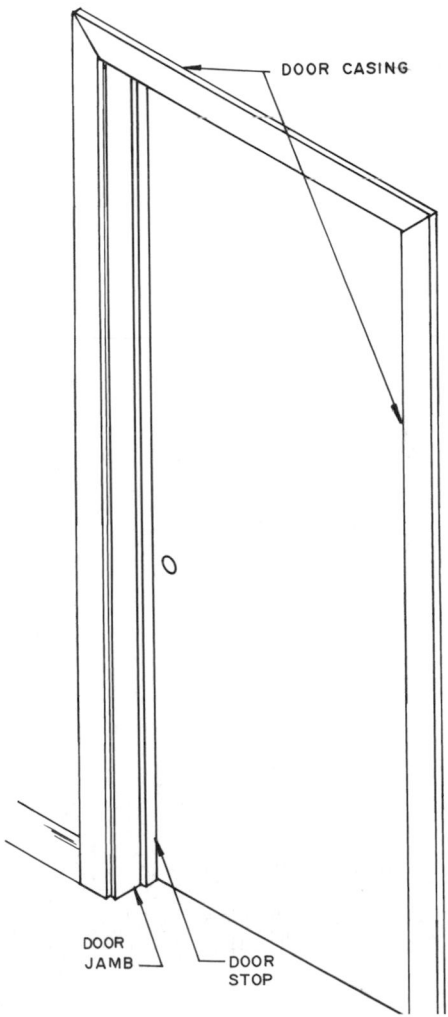

6-92 *Interior door frames are set plumb using wood wedges to position them in the rough opening.*

6-93 *Interior door frames are trimmed on both sides with a casing after the finish wall material is installed.*

Window casings are usually like those shown in 6-94 and 6-95. One type shows the casing on all sides of the opening. The other has a stool with an apron below.

Base molding is used where the wall meets the floor. It is usually mitered at outside corners and coped at inside corners. Coping refers to cutting the ends of the base molding to the shape of the piece it will butt (see 6-96). If the floor is to be carpeted the base is usually placed a little above the subfloor. If hardwood flooring is to be used, the base is installed after the flooring is in place and a small molding covers the crack between the flooring and the base molding.

Other types of moldings are often shown on the architectural drawings. For example, some form of crown molding is often used where the wall and ceiling meet (see 6-97). Chair rails are used to protect the wall from damage, especially in formal dining rooms, and to add a decorative feature. Often the wall below the chair rail is paneled or at least finished in a different way than the wall above it. Moldings are often mounted on the wall to form decorative panels.

INSTALLING CABINETS

Cabinets are available in several styles and types of construction. Standard cabinet sizes and construction details are shown in Division 6. Most cabinets used today are made in a factory and shipped to the site in strong cardboard boxes. The finish carpenter installs them following the architectural drawings. Many times the local company supplying the cabinets will also do the installation.

Some installers prefer to install the wall cabinets first. This enables them to stand directly below the cabinet while they are working. Others set the base cabinets first and use them as a support to hold the wall cabinets as they are installed.

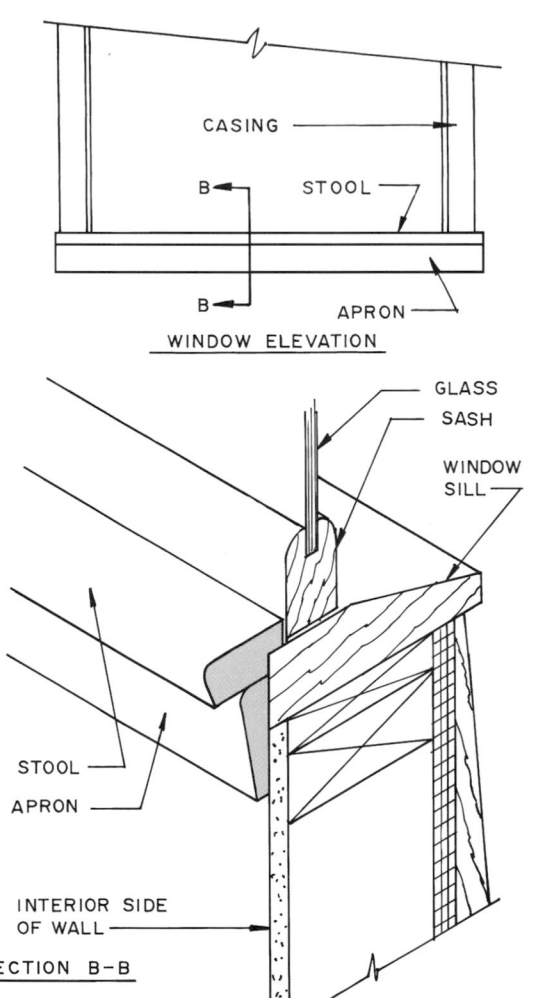

6-94 *This window is cased on three sides and has a stool and apron on the bottom.*

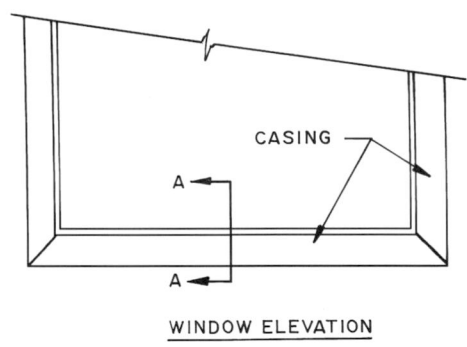

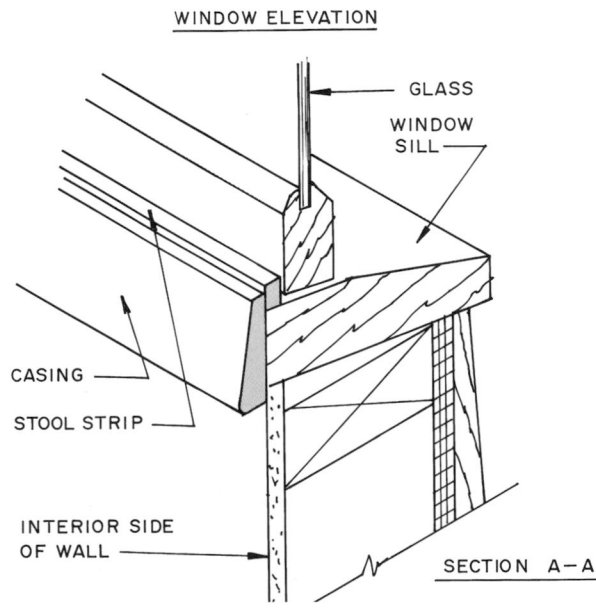

6-95 *The casing on this window is on all four sides and the stool is omitted.*

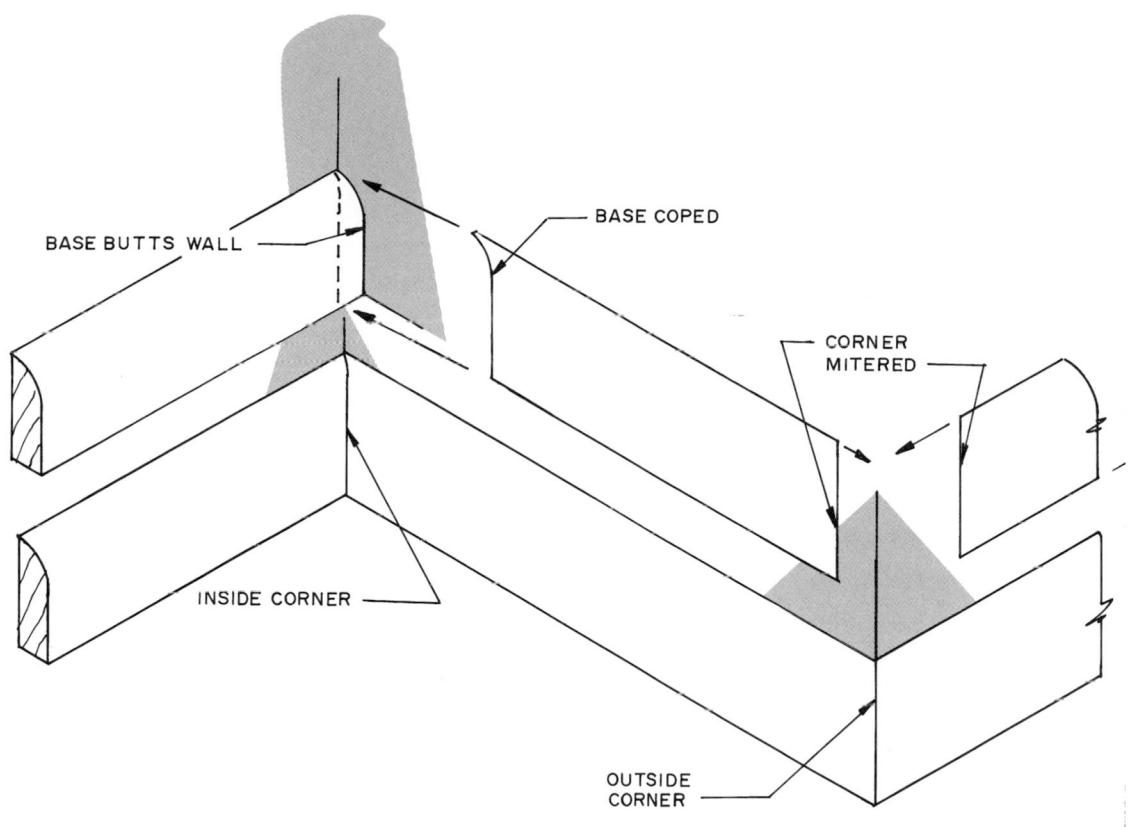

6-96 *The base is coped at the inside corners and mitered at the outside corners.*

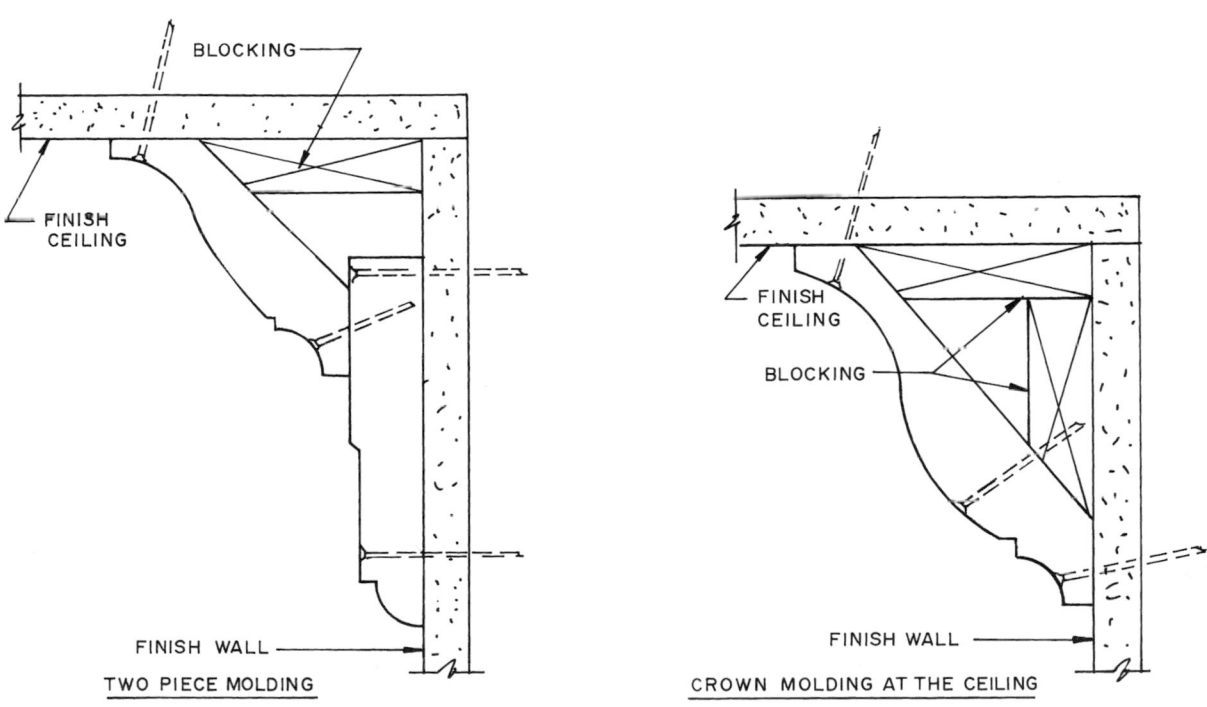

TWO PIECE MOLDING

CROWN MOLDING AT THE CEILING

6-97 *Moldings are often used at the intersection of the wall and ceiling.*

The base cabinets are placed on the subfloor and leveled by shimming them at the floor. They are screwed through the back rail to the wall studs (see 6-98). Some types have metal legs that can be adjusted in length to level the cabinet. The wall must be checked for straightness. If it has a bow, the base cabinets must be shimmed out so they are straight (see 6-99).

The wall cabinets are installed the specified distance above the base cabinets. They are often supported on the base cabinets and adjusted until they are level. Then they are screwed through the back rails into the studs (6-100). They are shimmed to get them plumb.

The counter top is usually plywood or oriented standboard with a plastic laminate top. Ceramic tile is also used over the wood base. The carpenter cuts the openings needed for the sinks and lavatories. The plumber installs these later on as the building nears completion.

INSTALLING STAIRS

The stair design is shown on the architectural drawings. The type and quality of the treads, handrail, newel posts, balusters and other parts is specified. Some stairs are built by the carpenter on the site. Usually the carpenter builds the stringers and buys manufactured parts. Others buy completely manufactured stairs and assem-

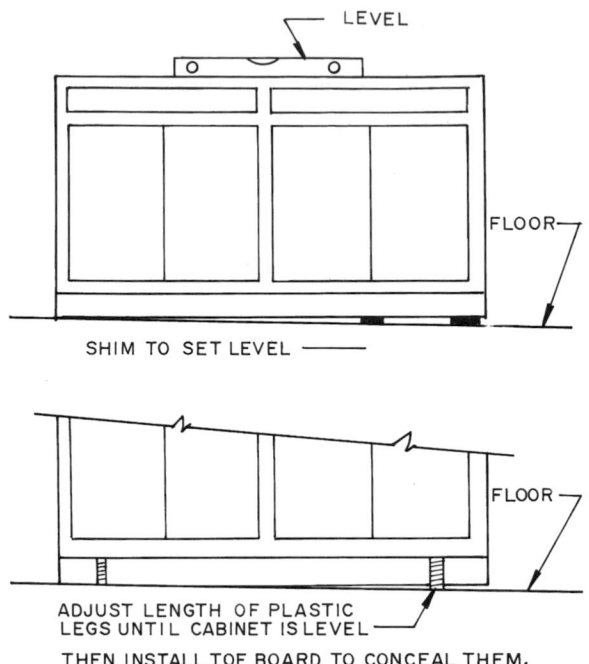

6-98 *The base cabinet can be leveled with wood shims or by adjusting metal legs if they were used.*

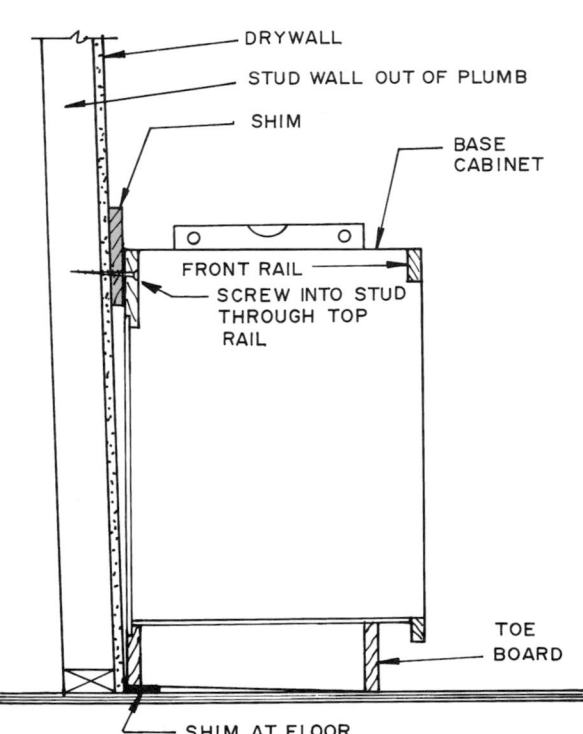

6-99 *Shim the base cabinet against the wall when the wall is not plumb.*

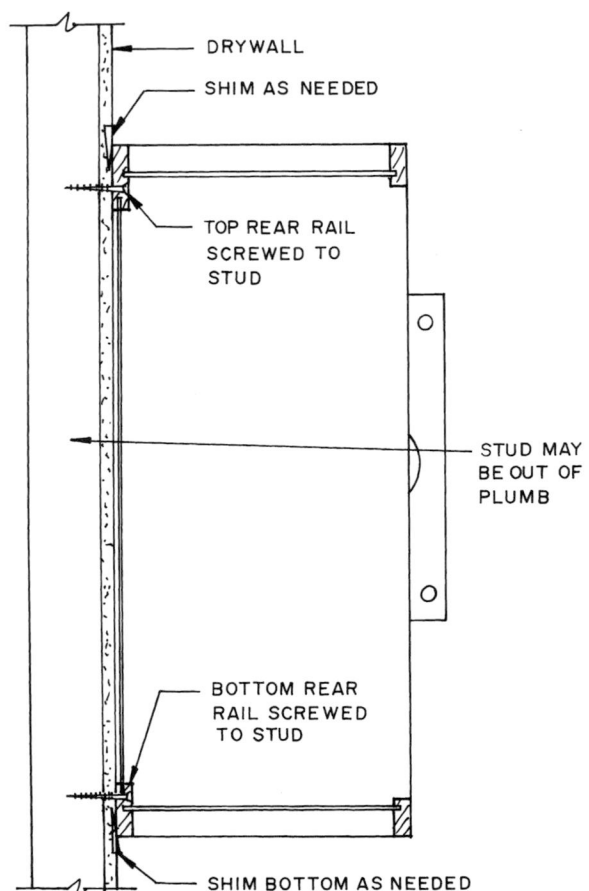

6-100 *Wall cabinets are plumbed and leveled and screwed to the studs.*

ble them on the site. As the stair was designed the architect had to observe the requirements in the local building codes that regulate riser height, tread width,, stair width and handrail requirements.

The framing for a typical wood-framed stair is seen in 6-101. The load is carried by the stringer. Generally a stair will have three or more stringers. The riser is a board covering the vertical distance between treads. There are many factory-manufactured stairs that provide an interesting and attractive appearance (see 6-102).

OTHER FINISH ITEMS

There are many things the finish carpenter has to do as the building nears completion, including installing locks, door stops, bath accessories, and fireplace mantels. Finally, the building is thoroughly cleaned, subfloor scraped clean and sanded if necessary, and the carpet or other floor covering is installed and finished as required.

Other trades have to phase in their work at the proper time. This includes painting and installing wallpaper and other wall coverings. Window shades and curtains are installed after the carpet is down and all interior finish is complete.

6-102 *This stair was built using commercially available parts.*
Courtesy L.J. Smith Stair Systems

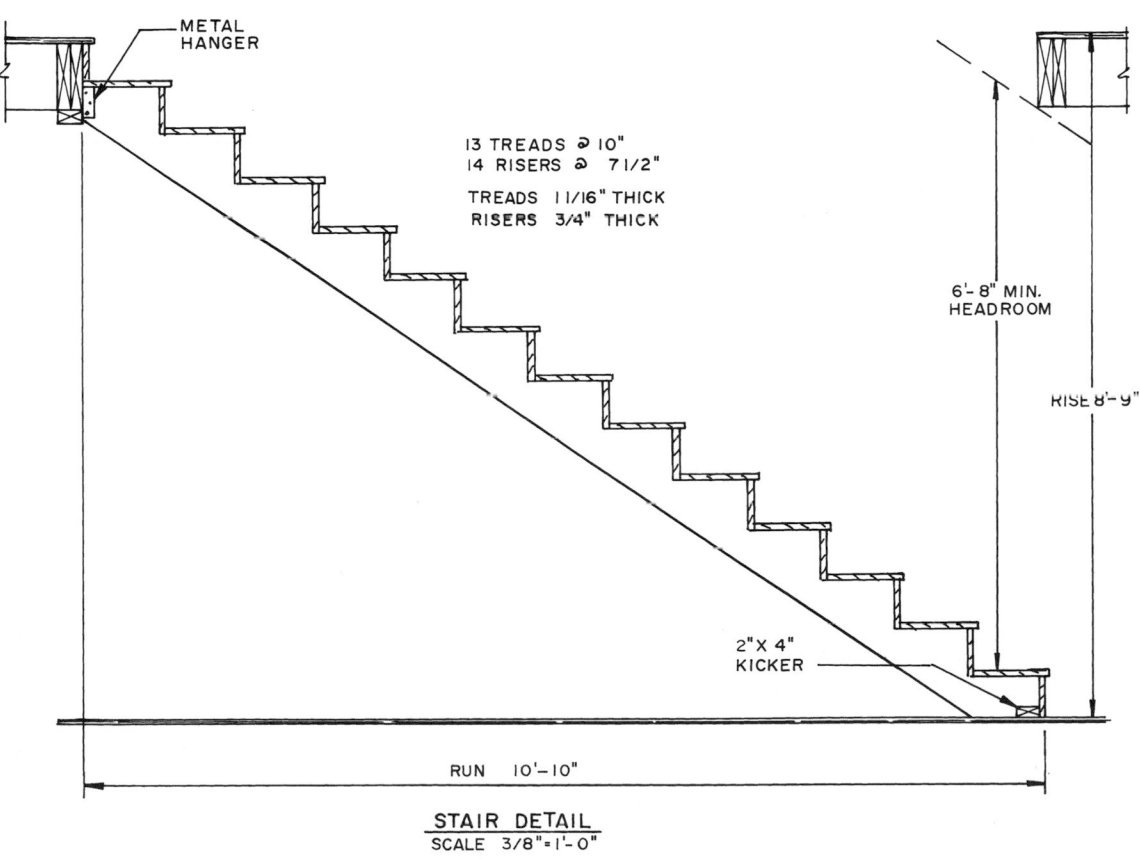

METAL HANGER

13 TREADS @ 10"
14 RISERS @ 7 1/2"

TREADS 1 1/16" THICK
RISERS 3/4" THICK

6'- 8" MIN. HEADROOM

RISE 8'- 9"

2" X 4" KICKER

RUN 10'- 10"

STAIR DETAIL
SCALE 3/8"=1'-0"

6-101 *Framing for a stair with a landing.*

PLASTICS

While the use of plastics is relatively new in the construction field, it has rapidly become a major material and is being used for structural and nonstructural purposes (see 6-103). It is replacing conventional materials, such as wood and metal, in many parts of a building. This is occurring because of extensive research in plastic development, resulting in a large variety of plastic materials that have properties not available in conventional materials. In general, plastics resist corrosion and moisture, are tough, light in weight, and are easily formed into useful products. Since plastics are chemically derived, a wide range of plastics with special properties can be developed. This makes them particularly useful as a construction material because of the extensive range of applications that exist.

Vinyl siding is available in a variety of colors, surface textures, and sizes, and provides a tough, maintenance-free exterior. Wood windows are clad in vinyl, eliminating the need for painting, and some windows are framed entirely with solid plastic extruded parts. Some types of plastics are replacing glass glazing because the plastic is lighter and more resistant to shattering. Plastic lavatories, showers, and bathtubs have largely replaced those of ceramic coated metal and cast-iron fixtures. A high percentage of items used in electrical systems, such as boxes and wiring insulation, are plastic. Plastic film is used for vapor barriers and to reduce air infiltration, while plastic foams are used for insulation.

Since there are so many different kinds of plastics and these have varying properties, the designer and constructor must be careful to use a plastic for the purpose for which it was designed. For example, using plastic pipe designed to carry cold water for hot water lines will eventually cause problems. There are many families of plastic materials and within each the properties can be varied widely. Plastics from several families may be able to serve the same purpose, while others exhibit unique, special characteristics.

The term **plastics** is used today to describe manmade **polymers** that contain carbon atoms covalently bonded with other elements. Plastic is obtained by breaking down materials found in nature into other elements, such as petroleum, coal and natural gas (see 6-104). Plastics are **synthetic materials** resulting from chemical manipulations of natural materials. In other words, they are man-made and not found in nature. In their formulation they are soft, can be formed into the shapes desired, and exhibit **plastic behavior.** Plastics are **organic materials.** They are made up of molecules built around a **carbon atom.** An exception to this is a group of materials considered a plastic that is built around the **silicon atom.**

This chapter will discuss plastic materials commonly used in construction and describe the products for which they are intended.

Molecular Structure of Plastics

The way that the properties of plastics can be made to vary is understood by examining the molecular structure of the material. Most materials accepted in the plastics family are based on the carbon atom. The carbon atom bonds with other atoms through covalent bonding. **Covalent bonding** is the process in which small numbers of atoms are bound into molecules. A single molecule is known as a **monomer.** When several monomers link together to form a chain, they form a

6-103 *Plastic products are widely used in the construction industry.*

polymer. This chain formation is called **polymerization**. Plastic polymers are called **macromolecules** because they are composed of many smaller molecules (macro means large). Since the small molecules are joined together in a chainlike condition, they are often called chains. Plastics are made up of these polymers (chains). Note that many plastics are described by the prefix "poly" such as polyvinyl and polystyrene. Plastics is therefore a polymer. Polymers are moldable materials sold in the form of granules, powder, flakes, liquids, or pellets for processing into useful products.

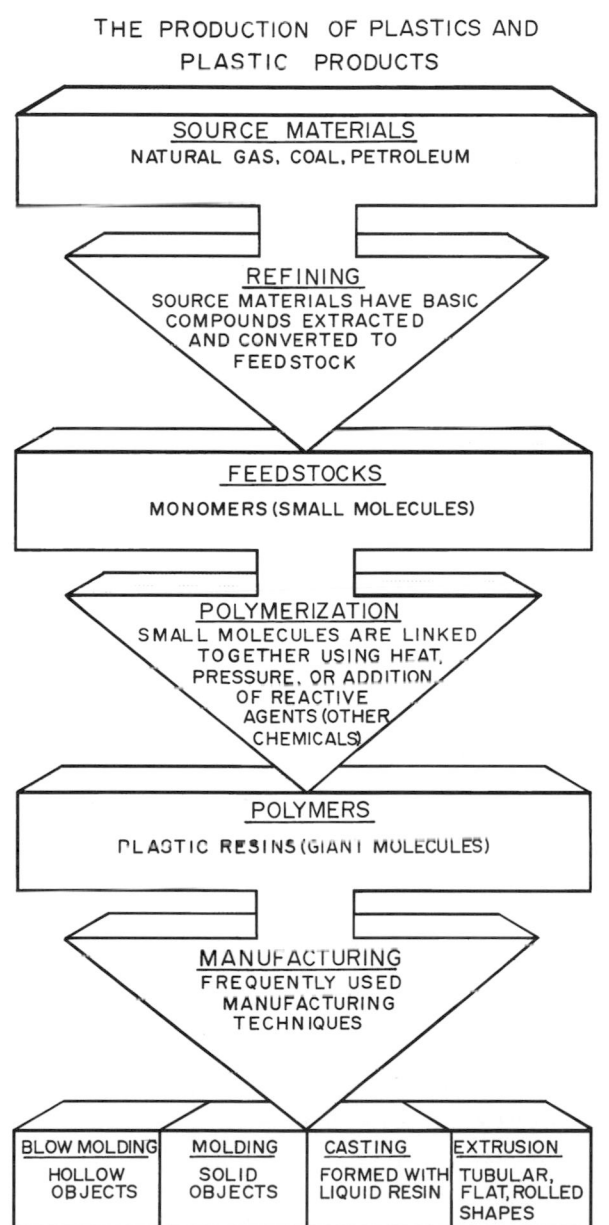

6-104 *The production of plastics and plastic products.*
Courtesy The Society of the Plastics Industry, Inc.

The molecular bonding process can be shown by the following example.

The carbon atom has a **valence** of four, which means there are four points at which other atoms can attach themselves to the carbon atom, forming covalent bonds. Other elements, such as oxygen, have different valences. For example, oxygen atoms have a valence of two. Two oxygen atoms can bond to the four carbon valence points, creating carbon dioxide (CO_2) (one carbon atom and two oxygen atoms).

Classifications

Plastics can be divided into two basic classifications, thermoplastic and thermosetting materials.

THERMOPLASTICS

Thermoplastics are plastic materials that can be softened or remelted by application of heat and reformed. Some thermoplastic materials will experience contamination and chemical degradation if reheated frequently.

Thermoplastic materials are composed of long chainlike molecules that are unattached to each other. These molecules can slide past each other and change shape. At normal temperatures, 70°F (21°C), the material retains its shape because the movement of molecules is slight. However, when heat is applied the bonding between molecules weakens and the material expands and becomes soft. At high temperatures (which vary according to the composition), the plastic will melt and have sufficient flow to permit it to be molded. As the molded part cools, the plastic holds the new shape.

THERMOSETTING PLASTICS

Thermosetting plastics (also called thermosets) are those that once they have been softened, formed and cured cannot be reheated and reformed. They are formed by a chemical process that produces a strong bond between the molecules and they cannot slide by each other. The final form of the material is irreversible.

Additives

Most plastic resins in pure form do not have the properties needed for particular applications. Therefore additives are mixed with the resin to modify the properties. Frequently more than one additive is needed to secure the properties desired. The amount added also has great influence on the final result. The commonly used additives include plasticizers, fillers, stabilizers, reinforcing agents, and colorants.

PLASTICIZERS

Plasticizers are added to the plastic resin to reduce brittleness and increase flexibility, resiliency, moldability, and in some cases to improve impact resistance.

FILLERS

Some fillers are used to provide bulk, thus reducing the cost of the plastic. Others are used to improve a specific property. **Fillers** to increase **bulk** or **ease of molding** include finely ground hardwood or nut shells.

Hardness is improved by adding mineral oxides and mineral powders.

Heat resistance is improved by adding inorganic fillers such as clay, silica, ground limestone, or asbestos.

Electrical resistance is increased by adding quartz or mica.

Toughness is improved by adding fibers such as hemp, cotton, sisal, rayon, polyester, or nylon.

Mechanical properties are strengthened by adding glass fibers or metal fibers.

Stabilizers, or lead compounds, are used to stabilize the plastic by helping it resist heat, loss of strength, and resist the effect of radiation on the bonds between the chains.

Colorants added to the resin include **organic dyes** and **inorganic pigments** (metal-based).

Properties of Plastics

The properties of plastics that are important for consideration of construction purposes include mechanical, electrical, thermal, flammability, chemical, density, and optical. To make proper use of plastic materials the properties of the specific formulation of the resin and additives must be known.

MECHANICAL PROPERTIES

The mechanical properties that are generally most important to consider for plastics used in construction are tensile strength, stiffness, toughness (or impact strength), hardness, and creep.

ELECTRICAL PROPERTIES

Plastics have excellent electrical insulating properties and have enabled the development of greatly improved electrical equipment. In addition, they have good heat resistance, which is necessary for the use of a material in many electrical devices.

THERMAL PROPERTIES

Two major factors to consider when examining thermal properties are expansion and contraction of the material and the influence of temperature on its strength. Many plastics lose strength at comparatively low temperatures. The **service temperature** (maximum temperature at which the plastic can be used without affecting its properties) of plastics is low when compared with other construction materials.

Some plastics will soften below 200°F (94°C) and therefore will soften if placed in boiling water. In general, thermosets have higher service temperatures than thermoplastic materials.

All building materials **expand** and **contract** and this must be taken into consideration as a designer selects materials and prepares the design drawings. Most plastics have a higher **coefficient of thermal expansion** than other construction materials. In general, thermosets have a lower rate of expansion than thermoplastics.

Fiberglass-reinforced plastics and plastic laminates have a low coefficient of thermal expansion, approaching that of wood. Plastics are poor **conductors** of heat. Unmodified plastics compare favorably with brick, concrete, and glass.

Foamed plastics have the lowest thermal conductivity of construction materials and are therefore **excellent insulators** and are widely used for all types of insulation. When comparing plastics in general with other materials, it can be seen that plastics have the lowest coefficient of thermal conductivity.

FLAMMABILITY

Flammability is the ability of a material to resist burning. Since plastics are an organic material, they will burn. The range of flammability, however, is great. Cellulosics are highly flammable and are often banned by building codes. Polyurethane and polyvinyls give off toxic fumes when burning and building codes specify protective measures for many uses. A widely used product, such as vinyl siding, begins to burn at 698°F (370°C), while wood will ignite at about 400°F (206°C). In general, thermoplastics are more flammable than thermosets, though these properties can be changed with additives.

While plastics are no more combustible than wood, they can produce toxic fumes that are more likely to cause death in a fire than the flames. Most building codes have a separate chapter devoted to the use of plastics in building construction.

CHEMICAL PROPERTIES

While plastics do not corrode like metals, they can deteriorate and be damaged by chemical attack.

Plastics deterioration is caused by gaining weight due to the attacking chemicals, combining with the plastic resin. This causes swelling, crazing, and discoloration. In addition, there is loss of impact, as well as flexural and tensile strengths.

The determination of the **weatherability** of a plastic involves consideration of moisture, ultraviolet light, heat, and chemicals found in the air as ozone and hydrochloric acid. High-density polyethylene (HDPE) has great resistance to acids, water absorption, and weathering. It is used for electrical wire insulation. Acrylics also have great resistance to weathering (and excellent optical qualities) which makes them widely used for glazing.

DENSITY

The density of plastic materials is in general lower than other commonly used construction materials. Glass reinforced plastics (GRP) are lighter than steel and aluminum.

OPTICAL PROPERTIES

Some plastics have optical properties equal to those of glass. Acrylics are as transparent as fine optical glass, having a light transmission of 92%. Polystyrene, polypropylene, and polycarbonates are also 90% or better.

PLASTIC CONSTRUCTION MATERIALS

The following discussion gives a brief description of the plastics most commonly used in construction and some of their major users. They are divided into thermoplastics and thermosets.

Since plastic materials are frequently identified in technical literature and professional magazines by "approved" abbreviations, the designer and constructor must be familiar with them. A list of these abbreviations as compiled by the American Society for Testing and Materials is in Table 6-19.

Thermoplastics

More products used in construction are made from thermoplastic materials than from thermoset. Following are commonly used thermoplastic materials and products using them.

Table 6-19 ASTM[a] ABBREVIATIONS FOR PLASTICS

Type	Abbreviation
Thermoplastics	
Acrylonitrile-butadiene-styrene	ABS
Acrylic:	
Polymethyl methacrylate	PMMA
Cellulosics:	
Cellulose acetate	CA
Cellulose acetate-butyrate	CAB
Cellulose acetate-propionate	CAP
Cellulose nitrate	CN
Ethyl cellulose	EC
Fluorocarbons:	
Polytetrafluoroethylene	PTFE
Nylons:	
Polyamide	PA
Polyethylene	PE
Polypropylene	PP
Polycarbonates	PC
Styrene:	
Polystyrene	PS
Styrene-acrylonitrile	SAN
Styrene-butadiene plastics	SBP
Vinyl:	
Polyvinyl acetate	PVAC
Polyvinyl butyral	PVB
Polyvinyl chloride	PVC
Thermoset Plastics	
Epoxy, epoxide	EP
Fiberglass-reinforced plastics	FRP
Melamine-formaldehyde	MF
Phenolic:	
Phenol-formaldehyde	PF
Polyester	—
Polyurethane:	
Urethane plastics	UP
Silicone plastics	SI
Urea-formaldehyde	UF

[a]American Society for Testing and Materials

ACRYLONITRILE-BUTADIENE-STYRENE (ABS)

ABS plastics are a combination of high-impact polystyrene that is very tough and acrylonitrile which improves its rigidity, tensile strength, and chemical resistance. ABS plastics are widely used for pipe and pipe fittings for water lines with water up to 180°F (83°C) (non-pressure), gas supply lines, waste, drain, and sewage vent systems. It also is used for hardware such as handles and knobs.

ACRYLICS

The most widely used acrylic is polymethyl methacrylate (PMMA). It has excellent optical clarity and is used for glazing. Typical uses include door and window lights and roof domes and skylights. It is also used to make lighting fixtures but will soften at 200°F (94°C). It finds use in outdoor signs, corrugated roofing, and molded hardware.

Many acrylics are also dispersed as fine particles in a liquid, producing a latex that is used to make **latex** paints.

CELLULOSICS

Two common plastics based on the cellulose molecule are cellulose acetate (CA), and cellulose acetate-butyrate (CAB). Cellulose acetate is not used for construction purposes but CAB can be made resistant to weathering and is used in coating compounds and adhesives.

FLUORCARBONS

The most widely used fluorocarbon is polytetrafluoroethylene (PTFE). It has high resistance to chemical degradation and has a service temperature from -450°F (-234°C) to +500°F (262°C). It has a low coefficient of friction and is marketed under the trade name Teflon. It is used for pipe that must handle corrosive liquids at high temperatures and for parts that require easy-sliding surfaces.

NYLONS

Polyamides (PA), known also as nylon, are tough, high strength, and have good chemical resistance. Since they resist abrasion well, they are used for molded parts, such as locks, rollers, gears, and cams. They do not weather well.

POLYCARBONATES (PC)

Polycarbonates have high impact strength, good heat resistance, are dimensionally stable, and are transparent. They are used for light fixtures, molded parts, and signs. They are used in place of glass in areas where damage is likely, as in skylights.

Polycarbonate plastic sheets such as Lucite® or Lexan® have high impact strength, reaching up to 250 times the strength of glass and 30 times that of acrylic. They are suitable for glazing openings in areas where high security is needed. They are laminated to produce bullet-resistant panels.

Double-skinned units with internal ribs and dead air spaces are available for use on vertical and sloped glazing.

POLYETHYLENE (PE)

Polyethylene is light, strong, and flexible even at low temperatures. It has good water resistance and low vapor transmission. Its major use in construction is as a vapor barrier on walls, floors, and ceilings. It is also used on basement walls as part of the waterproofing application.

POLYPROPYLENE (PP)

Polypropylene is much like polyethylene but is more heat-resistant and stiffer. It is used for pipes to carry hot water and waste disposal systems. It is used to make strong fibers for carpeting.

POLYSTYRENE (PS)

Polystyrene is a water-resistant, dimensionally stable, transparent plastic that maintains its properties at low temperatures but begins to soften around 200°F (94°C). It is brittle and has poor weathering qualities. When produced in a foamed condition it is widely used as an insulation. The foam is also used as the core for insulated doors and sandwich panels, as well as roof insulation.

VINYLS

The term vinyl describes a large group of plastics developed from the ethylene molecule. Those used in construction are polyvinyl chloride (PVC) and polyvinyl butyral (PVB).

Polyvinyl chrloride (PVC) is the most widely used vinyl in the manufacture of products used in construction. It has high impact resistance, abrasion resistance, good aging qualities, and is dimensionally stable. A long list of products are made from PVC including siding, gutters, floor tile, pipe, and window frames. It is bonded to other materials, as plywood panels to give a protective skin. PVC is available as a rigid or flexible foam and used as a core material in panel construction. It is copolymerized with other plastics to produce a binder for terrazzo floors and a number of adhesives.

Polyvinyl butyral (PVB) is used as an inner layer in safety glass and as a protective coating on fabrics.

Another type of vinyl, **polyvinyl acetate** (PVAC) is used in mortars, paints, and adhesives.

Thermosets

Thermoset materials are used for products requiring greater heat resistance and stiffness than afforded by thermoplastics; however, they find limited use because they are brittle and harder to form.

EPOXIES

Epoxies have good chemical and moisture resistance but are mainly used because of their excellent adhesive qualities. They are widely used in coating compounds and as adhesives. They bond to almost any material and are used in the assembly of panels, bonding of veneers and overlays, and as protective coatings. They are used as mortars for bonding concrete block and in patching material for damaged concrete. Epoxies are also used to produce some types of fiberglass reinforced plastic.

FORMALDEHYDE

Formaldehyde plastics are incapable of plastic deformation. These plastics are hard, strong, heat resistant, and brittle. There are three types that find some use in construction.

Phenol-formaldehyde, generally referred to as **phenolics** (PF), has fillers, such as glass fibers, added to improve impact resistance and strength. It has good electrical and thermal properties. Phenolic plastics are the most widely used of the thermosets. They are used to mold electric parts, as switches, boxes, and circuit breakers, and hardware.

An important use is the coating of Kraft paper that forms the base of high-pressure plastic laminates used for counter tops and other surfaces. The cardboard interior structure of hollow-core doors and panels is impregnated with phenolic resin. It is also used to make some types of adhesive, protective coating material, and foamed insulation.

Melamine-formaldehyde, generally referred to as melamines (MF), are a hard, scratch-resistant plastic that withstands chemical attack. It is used in the production of high-pressure plastic laminates as used for counter tops. It is used as an adhesive in the production of plywood. It is used to mold hardware and electrical fixtures.

Urea-formaldehyde (UF) has the same uses as melamine-formaldehyde. Urea-formaldehyde is not as hard or have as good heat resisting properties as melamine-formaldehyde.

POLYESTERS

There are a large number of plastics that fall under the generic names **polyester**. These are a group of plastics that include the mylars from which drafting film is made, alkyds used for paints and enamels, and fiberglass-reinforced plastics. Saturated and unsaturated polyesters are produced. Major uses include molded bathtubs, showers, sheets for roofing and partitions, curtain wall exterior laminated, and window frames and sash.

POLYURETHANES (UP)

Polyurethanes are used to produce low-density foams that can be varied from soft, open-cell types which are flexible to a tough, closed-cell, rigid material. These foams have a very low thermal conductivity and make excellent insulation for wall, ceiling, and floor cavities by spraying them as a creamy foam on the surface or in the cavity after which the foam solidifies. They are also used to insulate pipes, ducts, and wall panels. Rigid insulation sheets are widely used for many purposes such as insulating flat roofs over which a hot tar and gravel roof covering is applied.

SILICONES

Silicone plastics are not based on the carbon atom as all those previously discussed, but on the silicone atom. They have good corrosion resistance and are good electrical insulators. Heat resistance properties enable them to have a service temperature ranging from -80°F (-27°C) to 500°F (260°C). An important property is that silicone plastics withstand exposure to the elements.

Since silicons are very water repellent, they (in liquid form) are applied to exterior masonry materials to provide a water-resistant coating. This stops water from penetrating a wall, such as a brick exterior wall, yet has the permeability to allow internally developed moisture to pass through as a vapor.

Silicone rubber is soft, heat resistant, and does not harden at low temperature. It is used as sealants and gaskets where watertightness is desired.

Manufacturing Processes

Common manufacturing processes include blow molding, calendaring, compression molding, casting, extrusion, expandable bead molding, form molding, injection molding, laminating, rotational casting, transfer molding, and thermoforming.

Blow molding involves placing a heated, reformed plastic tube called a paison in a forming die. Air pressure is raised inside the plastic tube, forcing it to conform to the shape of the die. When it is cool, the die is opened and the part removed.

Calendaring involves moving a plastic material in a liquid state through a series of rollers that form a thin plastic film as the material solidifies. Vapor barrier and floor-covering materials are typical products.

Compression molding involves placing plastic resin in powder form in a heated mold where heat and pressure are applied. The resin melts and fills the mold.

Casting involves pouring plastic in a liquid state into a cavitiy in a mold where it fills the cavity and hardens.

Extrusion is a process in which a semiliquid plastic is forced under pressure through an opening in a die. The shape of the die opening determines the shape of the extruded member. This is much the same as squeezing toothpaste out of a tube.

Expandable bead molding is a process in which small granules of resin, such as polystyrene, are mixed with an expanding agent and placed in a steam heated rolling drum. When the granules have expanded they are cooled and are transferred to a mold where they are heated until they fuse together.

Form molding requires that an expanding agent be mixed with plastic granules or powder and injected into the mold where it is heated, melting the resin and forming a gas that expands the resin, filling the mold cavity.

Injection molding uses granules or powder resin. This is fed into the heated cylinder of the injection molding machine and forced by a ram into a cold mold where it solidifies.

Laminating is a process in which several layers of material are bonded together to form a single sheet. An example would be laminating a colored plastic veneer to a plywood backing for use as wall paneling. The materials to be laminated are impregnated with the plastic resin, placed together, and bonded by the application of heat and pressure.

Rotational molding forms hollow one-piece items from polyethylene powders. The resin is placed inside the mold, which is heated as it rotates about two axes. The resin is distributed to the surfaces of the mold by centrifugal force and is fused by the heat.

Transfer molding is a combination of compression and injection molding. The resin is made liquid in a chamber outside the mold and is then injected into the mold where it fills the cavity and solidifies.

Thermoforming involves two commonly used procedures: vacuum forming and pressure forming. Vacuum forming involves placing a heated sheet of plastic over a mold cavity and pulling a vacuum below it. Atmospheric pressure forces the sheet to the shape of the mold.

Pressure forming involves placing a heated sheet of plastic over a mold cavity and increasing the pressure behind the sheet, forcing it into the cavity of the mold.

DIVISION 7

THERMAL & MOISTURE PROTECTION

CSI MASTERFORMAT™

Courtesy Gerard Roofing Technologies

A wide variety of materials are in use as thermal insulation. These include wood, plastics, and metal products. Considerable attention is now given to the design and construction of energy efficient buildings. Manufacturers have responded with an array of products for use in almost any environment or situation. A key to energy efficient design is understanding how heat is transferred.

THERMAL INSULATION

Thermal insulation is manufactured from a variety of materials. **Metallic** insulation is in the form of aluminum or copper and other metallic foil or as an organic insulation material with a metallic laminate. **Organic fibrous** insulation materials include cane, cotton, wood, cellulose, and synthetic fibers. **Organic cellular** materials include polyurethane, polystyrene, cork, and foamed rubber. **Mineral cellular** insulation materials include perlite, vermiculite, and foamed glass. Mineral fibrous materials include rock, glass, slag, and asbestos melted and spun into fibers.

Thermal insulation is available as **loose fill,** which is granular or fibrous, **flexible** in the form of wool-like blankets and batts, **rigid** in the form of sheet material,

liquid spray using a mineral fiber or insulating concrete, **cast-in-place** using insulating concrete, and **foamed-in-place** as with polyurethane foams.

Methods of Heat Transfer

Heat is transferred through materials from the warm side to the cool side. In order to avoid a loss of heat in a building in the winter, insulation is used to break this transfer. Likewise, in the summer high exterior temperatures cause heat to pass into a building unless the transfer is retarded by insulation.

The amount of heat that can be transferred depends upon the characteristics of the material and its thickness. Porous and fibrous materials, as wood, which have many air pockets, transfer less heat than a solid material, as brick. Thicker materials will transfer less heat than a thin section of the same material.

Heat is transferred by radiation, convection, or conduction. **Radiation** involves the transmission of heat by electromagnetic waves. The heat energy passes through the air between the source and the body to be heated without heating the intervening air. Some materials accept radiant energy while others reject it. Shiny materials reject radiant energy and dark materials absorb it. A white shingle roof will reflect more radiant

Table 7.1   COEFFICIENTS OF HEAT TRANSMISSION

Material	Conductance (C)	Resistance (R)
Building Panels		
½-inch Gypsum Board & Plaster	2.22	0.45
1-inch Plywood	25.00	0.04
½-inch Fiberboard	0.76	1.32
Flooring		
Carpet with Rubber Pad	0.81	1.23
Hardwood Floor	1.41	0.71
Insulation		
Mineral Wool Insulation (3 to 4-inches)	0.09	11.0
1-inch Glass Fiber Insulation	0.25	4.00
1-inch Expanded Polystyrene	0.25	4.00
1-inch Vermiculite	0.44	2.27
Masonry		
1-inch Stucco	5.00	0.20
1-inch Common Brick	5.00	0.20
4-inch Concrete Block	1.40	0.71

energy than a brown or black roof, thus increasing the energy efficiency of the building.

Convection heat transfer involves the transfer of heat by the circulation of movement of heated liquids or gases. The heat is moved by natural or forced (fan) means by currents of air that absorb heat brought to the space, as by a hot water heating system convector. As warm air passes over a cool surface it transfers some of its heat to the cool surface. This convection current is a means for heat transfer and occurs in every building.

Conduction heat transfer involves the transmission of heat by its passing through a solid or a liquid. For example, the sun can heat a brick wall and the heat is transferred to the other side through the brick by conduction.

Designating Thermal Properties

Heat is the result of the movement of molecules of a substance. Heat quantity is measured in **British thermal units** (BTU). A BTU is the amount of heat needed to raise the temperature of 1 lb. of water 1°F. In the metric system heat quantity is measured in calories. One **calorie** is the heat required to raise the temperature of 1 gram of water 1°C. Calories can be converted to joules (J) by multiplying calories by 4.18. A **joule** is a derived metric unit used to describe the thermal properties of materials include thermal conductivity (k), thermal resistance (R), thermal conductance (C), and thermal transmittance (U).

Thermal conductivity, k, is a measure used to indicate the amount of heat that will be conducted through a square foot of area of a material per a specified unit of thickness. The lower the k value the better the insulating qualities.

In the U.S. customary system of measurement, thermal conductivity is indicated at Btu / in./ft.^{2}/ h /°F or Btu per inch of thickness per square foot per hour per degree Fahrenheit. In the metric system it is w/m · K or watts per meter-kelvin.

Thermal resistance, R, is a measure used to indicate the ability of a material to resist the flow of heat through it. The larger the R-value the greater the resistance and therefore the better the insulating value. Thermal resistance, R, is the reciprocal of conductance, C, or R = 1/C. In the U.S. customary system of measurement thermal resistance R = (h / ft.^{2} /°F)/Btu (hours per square foot per degree Fahrenheit per Btu). In the metric system thermal

TABLE 7-2 FORMS OF BUILDING INSULATION

Materials
Metallic, organic fibrous, organic cellular, mineral cellular, mineral fibrous

Forms
Loose fill, flexible blankets, rigid sheets, liquid spray, cast in-place, foamed-in-place

resistance is specified as R SI. In this system R SI = (K · m^2)/W or kelvin-meter squared per watt.

Total thermal resistance, R$_t$, is the resistance to heat flow through an assembly of materials. R$_t$ is the reciprocal of C or R$_t$ = 1/C.

Thermal conductance, C, is a measure used to indicate the amount of heat that will pass through a specified thickness of material. It is the reciprocal of thermal resistance or C = 1/R$_t$. In the U.S. customary system of measurement thermal conductance C = Btu (h /ft.^{2} /°F) or Btu per hour per square foot per degree Fahrenheit. In the metric system thermal conductance is W/m^2·°C or watts per square meter-degree Celsius. Examples of the resistance and conductance of a few materials can be seen in Table 7-1.

Thermal transmittance, U, is a measure of the amount of heat that would pass through an assembly of various materials, such as an exterior wall. It is the reciprocal of the total resistance, Rc, of the assembly, U = 1/R. The smaller the U-value the greater its resistance to the transmission of heat.

Insulation Materials

There are a wide array of insulating materials available in Table 7-2. The decision of which to use depends upon their location in the building, the environment to which they will be exposed, their effectiveness in resisting heat transfer, and the cost relative to the savings expected from improving the energy efficiency of the building. The major classifications of insulation include loose fill, batts and blankets, rigid, reflective, foamed-in-place, and sprayed. Insulation materials are rated by their ability to resist heat flow, which is indicated by their R-value. Typical R-values are in Table 7-3 on the following page.

Table 7-3   R-VALUES OF INSULATION MATERIALS[a]

Thickness (in.)	Thickness (mm)	R-Value (customary units)[b]	R = Value (SI units)[c]
Fiberglass Board Panels			
1	25	4.30	30
1.5	38	6.50	45
2	50	8.70	60
2.5	63	10.90	75
3	75	13.00	90
4	100	16.50	114
Extruded Polystyrene Rigid Panels			
0.75	18	3.80	26
1	25	5.00	35
1.5	38	7.50	52
2	50	10.00	69
2.6	66	16.67	115
3.1	78	20.00	138
Polyisocyanurate Rigid Panels			
1.2	30	7.14	49
1.6	40	10.00	69
2.0	50	12.50	86
2.6	66	16.67	115
3.1	79	20.00	138
Polyurethane Rigid Panels			
1.0	25	6.25	43
1.5	38	10.00	69
2.0	50	14.30	98
2.7	68	20.00	138
Wood Fiber Rigid Panels			
0.5	12	1.40	10
0.75	18	2.10	14
1	25	2.80	19
1.5	38	4.20	29
2	50	5.60	38
Granular Insulation			
Vermiculite per inch	per 25 mm	2.1 to 2.3	14 to 16
Perlite per inch	per 25 mm	2.6 to 3.5	18 to 24
Aluminum Reflective Insulation			
Multilayer foil batts with 4 three-quarter inch air spaces	18 mm air space	summer 17 / winter 13	117 / 89
Multilayer foil batts with 6 one-inch air spaces	25 mm air space	summer 30 / winter 18	207 / 124

[a]Data obtained from various manufacturer catalogs. Consult manufacturer's specification sheets for specific data.

[b]R-value (customary units) hr. $\times$ ft.^{2} $\times$ °F/Btu

[c]R SI (metric units) m^2 $\times$ K/W

LOOSE FILL INSULATION

Loose fill insulation is available as a granular or loose fibrous material. Granular insulation is poured and fibrous is machine-blown into the areas to be insulated. **Granular insulation** includes perlite (expanded volcanic rock), vermiculite (expanded mica), cork, or expanded polystyrene. These are available in different densities, which influences the R-value (Table 7-4). Granular insulation materials are also used to produce a lightweight concrete roofing base material that is pumped and leveled on the roof deck. Expanded polystyrene roof insulation boards are placed on this and are bonded to the deck. The roof base material is flowed over the insulation board forming a base for the application of the finished roofing material (see 7-1).

Fibrous loose insulation is made by blowing a jet of air through molten glass, slag, or rock forming thin fibers that when gathered form a wool-like substance. One type is made from cellulosic fibers obtained from wood chips, newsprint, and other organic fibers. Cellulosic fibers must be fireproofed. This treatment is noted by the Underwriters Laboratory rating on the label. When filling vertical cavities, such as cores in concrete block and wall cavities, use granular insulation (see 7-2). Fibrous loose insulation is difficult to work into such places. It is widely used in large areas, such as insulating ceilings, where it can be sprayed. It rapidly covers the area with a uniform layer.

Table 7-4 PROPERTIES OF LOOSE-FILL INSULATION

Material	Density (lb./ft.^{3})	R-Value (hr. · ft.^{2}·°F/Btu)	Water Vapor Permeability
Cellulose	2.2–3.0	3.1–3.7	High
Expanded polystyrene	0.9–1.8	3.8–5.0	High
Fiberglass	0.6–1.0	2.9–3.7	High
Mineral wool	1.5–2.5	2.9–3.7	High
Perlite	2–11	2.7–2.9	High
Vermiculite	4–10	2.1–2.3	High

BATTS AND BLANKETS

Batts and blankets are flexible insulation matts made from fiberglass, mineral wool, cotton fibers, or wood fibers. Batts are 48 in. (1219 mm) long, while blankets are available in rolls up to 8 ft. long (2438 mm). Some thinner blankets are available in longer rolls. Common thicknesses are 3½, 6½, 9½, and 12 in. (89, 158, 241, 305 mm) and widths 15 and 23 in. (380 and 584 mm). Other sizes are available for special applications such as noise barrier insulation, 2¾ and 4 in. (70 and 100 mm), and furred masonry walls, ¾ in. (18 mm). Batts and blankets are available for a variety of applications such as light density insulation as used in residential construction, types with extra low flame spread qualities used in commercial construction where the insulation will remain exposed, special batts for use over ceiling panels in suspended ceilings, and one type that forms a noise barrier increasing the sound transmission class performance of the wall, floor, or ceiling.

7-1 *Casting an insulating roof basecoat using Zonolite®, which uses granular vermiculite aggregate in a Type I or Type III portland cement.*

Courtesy W.R. Grace and Co.

7-2 *Granular insulation is a good choice for filling cavities.*

Courtesy W.R. Grace and Co.

Batts and blankets are available unfaced, faced on one side with moisture resistant kraft paper forming a vapor barrier, and faced with aluminum foil forming a fire-resistant facing. Some types have a facing on both sides and are used on vertical applications such as walls, and horizontal applications as floors and ceilings in commercial buildings. They can be used to wrap around items needing to be insulated, such as a water heater.

Some types are designed to be placed between studs or joists and held by friction. Other types are stapled to the face or side of the stud. The vapor barrier side faces the inside of the building (see 7-3).

RIGID INSULATION

Rigid insulation board is made using organic fibers, such as wood or cane, mineral wool fibers, glass fibers, corkboard, several forms of expanded plastics such as expanded and extruded polystyrene and polyisocyanurate foam, and some forms of cellular hard rubber. These products are used in all parts of a building, including roof, wall, and floor insulation (see 7-4).

Wood and **cane fiberboard** are commonly used for exterior sheathing and shingle backer boards and are asphalt impregnated. They are also used for roof insulation. Rigid insulation sheets made from **granulated cork** are used for roof, wall, and floor insulation and are available in thicknesses from 2 to 12 in. (50 to 305 mm) and sheet sizes 24 × 36 in. (610 × 915 mm). **Mineral wool** wall panels have the insulation matt bonded to a rigid back sheet and are commonly used for roof insulation.

Expanded and **extruded plastic** rigid sheet insulation is used on walls, floors, roofs, and foundations. Some are made tapered so a flat roof will have some slope. Usually a built-up room membrane is placed over the rigid insulation panel. One type is bonded to a particleboard sheet providing a nailing surface for shingles (see 7-5). Rigid polystyrene insulation panels are also used on roof decks to be covered with insulating concrete. The variation in panel thicknesses gives the roof slope to drainage sources. The insulating concrete has a cellular structure that reduces weight and increases the insulation properties.

Flat panels containing a fiber insulation core may be unfaced or faced with an aluminum foil vapor barrier. Common thicknesses are 1, 1½, 2, 2½, and 3 in. (25, 38, 50, 63, 75 mm). Standard fiber core panel sizes are 24 × 48 in. (610 × 1 220 mm). Extruded and expanded plastic panels typically are 16 and 24 in. (406 and 610 mm) in width and 48 to 96 in. (1 220 to 2 438 mm) in length. Common thicknesses are 1, 1½, 2, 2½, 3, and 4 in. (25, 38, 50, 63, 75, 100 mm). Other sizes are available. Manufacturers should be consulted for sizes and R-values.

Most rigid insulation panels are manufactured from flammable materials. Building codes require they be covered with a fire-resistant material such as gypsum board. Polystyrene products deteriorate when exposed to ultraviolet rays from the sun and must be covered to avoid exposure. Molded polystyrene and expanded perlite board must be protected from water. There are many other forms of insulating board products available for use primarily on commercial buildings.

TABS STAPLED INSIDE STUDS — KRAFT PAPER OR ALUMINUM FOIL

TABS STAPLED TO FACE OF STUD

FRICTION FIT FIBER INSULATION

7-3 *Insulation batts and blankets are installed between studs and joists in several ways.*

7-4 *Rigid insulation board is used to insulate concrete roof decks.*
Courtesy U.S. Industries, Inc.

7-5 *This roof decking is made by laminating rigid foamed insulation sheets to particleboard providing a nailing surface.*
Courtesy Tectum, Inc.

7-6 *Reflective insulation is available in rolls.*
Courtesy Advanced Foil Systems

REFLECTIVE INSULATION

Reflective insulation is usually aluminum or copper foil in sheets or rolls (see 7-6). Rolls are usually 24 and 48 in. wide and 500 ft. long. It is available in single thickness layers or in a multilayer batt that has dead air spaces between the layers. The reflective foil utilizes the reflective properties to reject the passage of heat plus the effectiveness of the dead air spaces. The foil may be bonded onto a heavy kraft paper, insulation board, or gypsum lath. Some forms of fiber insulation batts have reflective foil bonded to one side where it also serves as a vapor barrier. Reflective insulation is used in residential and commercial construction in walls, floors, ceilings, and roofs. It can reduce heat flow by as much as 25 percent reducing the amount of conventional insulation needed and in some cases reducing the size of the air-conditioning units required. It is easily cut with a knife and installed by stapling.

FOAMED-IN-PLACE AND SPRAYED INSULATION

Foamed-in-place insulations are generally polyurethane or phenol-based compounds that provide excellent insulation. When mixed they are pumped through hoses into cavities, such as wall cavities, and sprayed in layers on flat and sloping surfaces, such as roof decks. The ingredients are carefully measured and mixed by special equipment. The equipment meters the isocyante and polyol at a one-to-one ratio. These are pumped as separate materials into the proportioning unit, which heats each and pumps them into separate hoses that are heated. The two components are mixed in a spray gun and sprayed upon the substrate. It can also be applied by power or hand rollers. Since the material expands as it hardens, the amount injected into a cavity must be carefully measured. If a residential wall cavity is overfilled the pressure could break the gypsum. Polyurethane will burn and must be covered with a fire-resistant finish.

After the sprayed polyurethane foam insulation is in place is must be protected from exposure to moisture and ultraviolet radiation. Commonly used protective coatings include acrylics, butyls, chlorinated synthetic rubber, modified asphalts, silicones, and urethanes.

Sprayed-on insulations are used on ceilings, walls, tanks, and other items. While polyurethane is widely used, vermiculite or perlite aggregate mixed with a gypsum or portland cement binder and sometimes wood or asbestos fibers are mixed with an inorganic binder and used as insulation. Severe restrictions are placed on insulation with asbestos fibers. Sprayed-on insulation is also used to increase the fire resistance, moisture resistance, and improve acoustical properties. They have the advantage of being able to bond to irregular shaped and sloped surfaces. Consult manufacturers for R-value, fire resistance, flame spread, smoke developed, compressive, and impact strength.

VAPOR BARRIERS

Another component of the insulation of a building is the installation of vapor barriers. A vapor barrier is used to keep water vapor generated inside a building, such as by cooking, from penetrating the wall and condensing as moisture on the building insulation. Where circumstances warrant it can be used to reduce the penetration of moisture from outside sources into the building as shown in 7-7.

Various types of insulation are made with a vapor barrier as part of the sheet or blanket. **Kraft® paper** coated with wax or asphalt is commonly used on one side of fibrous insulation batts and blankets. Aluminum foil is a more expensive coating and has heat-reflective values as well. **Polyethylene film** is sometimes stapled to the studs on the side facing inside the building and placed under the subfloor and on the ceiling. It is also used to cover the ground in a crawl space to retard moisture from leaving the ground into the air in the crawl space and is laid below welded-wire fabric before concrete slabs are poured. It is an excellent vapor barrier and also assists in reducing air infiltration. It is available in thicknesses of 2, 3, 4, and 6 mils and rolls 3 to 20 ft. (0.9 to 6m) wide. Air infiltration from the exterior of a building is also reduced by applying a plastic barrier on the exterior over the sheathing. This material is a spunbound olefin formed into a sheet of very fine high density polyethylene fibers. It resists tearing, puncture, and will not rot. It has a perm of 94 so moisture vapors that get into the wall from inside the building can pass through it to the exterior. In all cases vapor barriers and air infiltration wrap must be carefully installed, overlapped at joints, and sealed to reduce penetration.

There are various liquid materials that also serve as vapor barriers. Enamels, various primers, latex, and oil based paints are used.

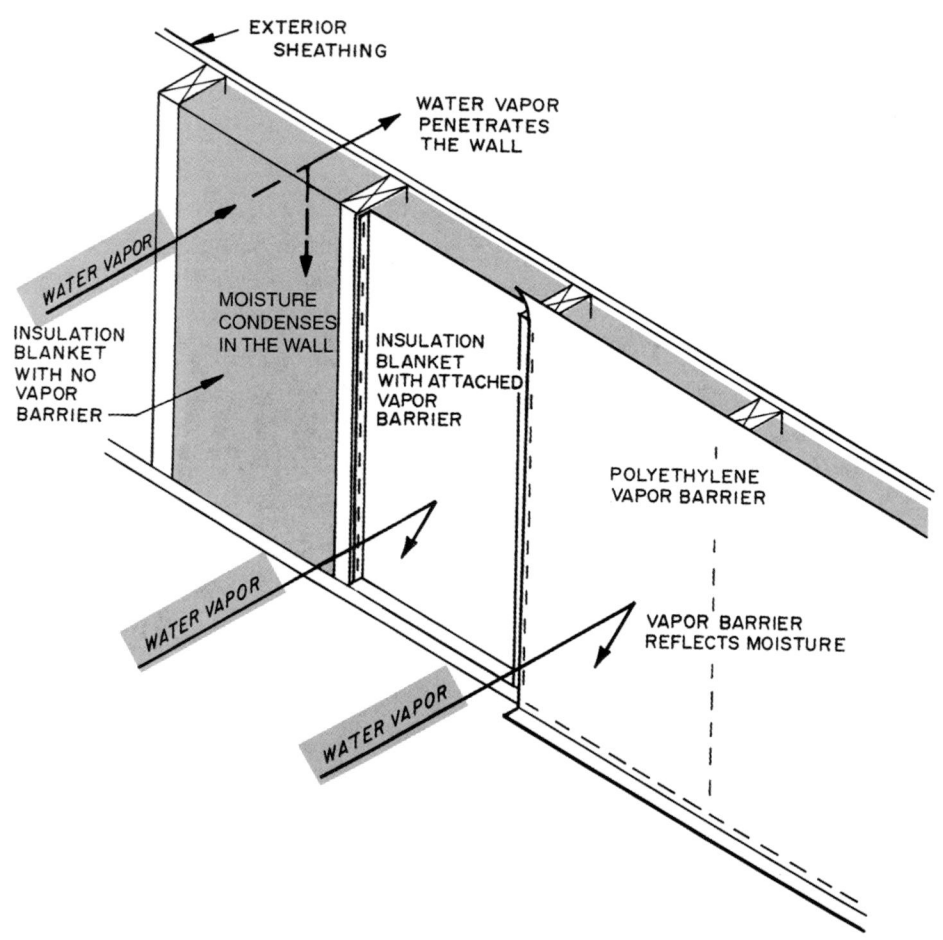

7-7 *Polyethylene vapor barriers reflect moisture generated inside a building, thus keeping it from penetrating the insulation.*

Table 7-5 COMMONLY USED ADHESIVES & THEIR APPLICATIONS

Adhesive	Bonded Material	Typical Uses
Acrylic	Plastics to metal, plastics to plastics, rubber to metal	Curtain walls
Casein	Wood to paper, wood to wood	All interior wood-joining needs
Cyanoacrylate (anaerobic)	Acrylics, phenolic, rubber, glass, polycarbonates, ceramics, steel, copper, aluminum	Any use (known as "Super Glue"); electronic and electrical devices
Epoxy	Almost any material except a few plastics and silicones	Interior and exterior uses, panels, glass to metal, curtain walls
Melamine formaldehyde	Paper, textiles, hardwood, interior plywood	Interior uses, plywood manufacturing
Natural rubber	Leather, paper, cork, foam rubber	Pressure-sensitive tape
Neoprene rubber (contact cement)	Many plastics, ceramics, aluminum	Plastic laminates, other interior uses
Nitrile rubber	Many plastics, ceramics, glass, aluminum	General uses
Phenol	Wood, cardboard, cork	Exterior plywood, any exterior use
Polyvinyl acetate	Porous materials (paper and textiles)	Various interior applications
Resorcinol	Rubber, paper, cork, asbestos, wood	Furniture, wood beams, columns
Silicone	Glass, ceramics, aluminum, polyester, acrylics, phenolic, rubber, steel, textiles	Sealant, gasket material
Polyurethane	Many plastics, glass, copper, aluminum, ceramics	Bonding dissimilar materials (e.g., on steel and glass sun roofs)
Urea	Many plastics, glass, copper, aluminum, ceramics	Particleboard, furniture, cabinets
Vinyl butyral	Glass	Laminating glass

BONDING AGENTS

A **bonding agent** is a compound that will hold materials together by bonding to the surfaces to be joined. There are many parts of a building and many products where bonding agents are used to permanently fasten things together, such as wood doors and plywood veneers. Bonding agents commonly used in the fabrication of construction products and in on-site applications typically join materials by **mechanical action** or **specific adhesion.** Those that join by **mechanical action** are best represented by the bonding of wood. The bonding agent used enters into the pores of the wood, hardens, and forms a mechanical link. Bonding agents that join by **specific adhesions** are used to bond dense materials without pores such as glass and metal.

Bond agents can have their properties varied for specific conditions or use on a particular material. Some are a combination of two or more types, such as a phenol and resorcinal resin combination or urea resin blended with a melamine resin. They are available as powders, solids, liquids, and pastes. Some require the addition of a catalyst.

The commonly used adhesives, the materials they join, and the agent's uses are summarized in Table 7-5.

Types of Bonding Agent

Bonding agents may be divided into three major classes: adhesives, cements, and glues. Following are some of the more commonly used types. A summary of some of the properties of these bonding agents is in Table 7-6.

ADHESIVES

Adhesive bonding agents are made from synthetic materials. They fall into two types, thermoplastic and thermosetting.

Thermoplastic adhesives are moisture resistant but are not used where they will be exposed to moist conditions.

Thermoset adhesives have good bonding strength and some types, such as resorcinal resins, produce a waterproof joint. Others are water resistant but do not produce a waterproof bond. It is important to examine the manufacturers directions when choosing a bonding agent.

GLUES

Glues are bonding agents made from animal and vegetable products such as bones, hides, fish, and milk. Those finding some use are animal, blood albumin, vegetable, and fish glues. One widely used glue is liquid hide glue, which is a ready-mixed ready-to-use form of animal glue. It is not moisture resistant and is used only on interior products, such as furniture, paper, and textiles.

Another glue used on construction products is casein glue. It is in powdered form and is mixed with water. It is water resistant and used for interior wood and on exterior products that will be sheltered and not be directly subjected to the weather, such as laminated timbers.

CEMENTS

Cements are made from synthetic rubber, such as neoprene, nitrile, and polysulfide, suspended in a liquid.

Cellulose cements such as cellulose acetate, cellulose nitrate, and ethyl cellulose are used for interior purposes and mainly for bonding plastics, glass, and porous materials, such as wood and paper.

Contact cements are neoprene cements that stick immediately upon contact and require no clamping time. They are used to bond plastic laminates on countertops, plastics to plastics, plastics to wood, and wood to wood. While they have good resistance to moisture, they are generally used for interior purposes. They have a very high bonding strength.

Mastic is a elastomeric construction cement. It is sold in tubes and is applied with a caulking gun. Typical uses include bonding plywood subfloors to joists, wall paneling to studs, laminating gypsum wallboard, styrene, and similar materials and assembling wall, floor, and roof panels. Mastic is water resistant.

Buna N resins are a form of acrylonitrile butadiene rubber. They are liquid cements that have good moisture resistance and strength. They are good general purpose interior cements.

SEALERS FOR EXTERIOR MATERIALS

Sealing materials are applied to the surface of a material to seal it against penetration of water or to keep water from passing through the material. Brick walls are often sealed to keep moisture from penetrating the mortar joints and into the inside of the wall. Sealers have adhesive properties and are closely related to the bonding agents just discussed. In addition to bonding to the surface they must form an unbroken film over it and fill any minute pores, cracks, or other minute openings. Large cracks or defects must first be filled with some form of sealant before a sealer is applied because the sealer will not span most cracks.

ACRYLIC SEALER

Acrylic sealer is a high solid content clear acrylic sealer. It maintains the original appearance of masonry or concrete while protecting it from moisture, airborne dirt, and other pollutants. It is used on products such as exposed aggregate panels, brick, and stone. Acrylic sealer is virtually unaffected by prolonged exposure to moisture, common acids, ultraviolet, oils, and aliphatic solvents. It reduces damage from the freeze-thaw cycle, efflorescence, and stains. It is available in a number of variations for different applications such as methyle methacrylate acrylic polymer.

ASPHALT DRIVEWAY SEALER

Asphalt driveway sealer is a think quick-drying sealer that gives a black protective coating. It is applied by brush or roller.

Table 7-6 CHARACTERISTICS OF BONDING AGENTS

Thermoplastic Adhesives

	Form	Moisture Resistance	Application Temperature	Clamp Time
Aliphatic	Liquid	Low	45°F (7°C)	1 hr.
Alpha-cyanoacrylate	Liquid	Low	75°F (21°C)	Minutes
Hot melts	Solid	Low	Electronic welder 300°F (150°C)	None
Polyvinyl acetate	Liquid	Low	60°F (16°C)	1 hr.

Thermosetting Adhesives

	Form	Moisture Resistance	Application Temperature	Clamp Time
Epoxy	Liquid	Water resistant	60°F (16°C)	None
Melamine resin	Liquid	Water resistant	Hot press 250°F (122°C)	Minutes
Phenol resin	Liquid and powder	Waterproof	300°F (150°C)	Minutes
Resorcinol resin	Liquid	Waterproof	70°F (21°C)	16 hr.
Urea resin	Powder and liquid	Water resistant	70°F (21°C)	1 to 3 hr., or seconds with high-frequency glue curing machine

Glues

	Form	Moisture Resistance	Application Temperature	Clamp Time
Animal	Flakes and powder	Low	70°F (21°C)	2 to 3 hr.
Liquid hide	Liquid	Low	70°F (21°C)	2 to 3 hr.
Blood albumin	Liquid	Moderate	70°F (21°C)	2 hr.
Vegetable or fish	Liquid	Low	70°F (21°C)	2 hr.
Casein	Powder	Water resistant	32°F (0°C)	2 hr.

Cements

	Form	Moisture Resistance	Application Temperature	Clamp Time
Contact cement	Liquid to thick	Water resistant	70°F (21°C)	None
Mastic	Paste in tubes	Water resistant	70°F (21°C)	15 min.
Cellulose	Liquid	Low	70°F (21°C)	Minutes
Buna N	Liquid	Water resistant	70°F (21°C)	Minutes

COAL-TAR COATINGS

Coal-tar coatings are used to seal surfaces and provide protection against corrosive conditions encountered in industrial plants, water and sewage systems, chemical plants, refineries, and other such industries. They resist moisture and most acids, alkalies, corrosive vapors, and atmospheric corrosion. They can be applied to metal, concrete, and masonry surfaces.

SILICON SEALERS

Silicon sealers are in liquid form and are used on brick, concrete block, stucco, cement plaster, and concrete. They produce a water repellent coating that protects the surface but does not prevent water vapor from escaping. They seal the surface, minimize efflorescence, and protect from absorbing staining materials, such as dirt and soot. Since protection against moisture penetration is provided, damage due to freeze-thaw cycles is reduced.

EPOXIES

Epoxies are used to provide a waterproof coating. They are often used on floors in areas where moisture and chemicals exist, such as food, meat processing, and other industrial plants. They are also used on floors around swimming pools, decks, showers, restrooms, loading docks, and on stair treads. When mixed with an aggregate, as emery, it provides a slip-resistant surface. These are available in a variety of formulations.

POLYURETHANE SEALERS

Polyurethane sealers are used to minimize concrete problems, such as scaling, spalling, chloride penetration, and damage due to the freeze-thaw cycle. Some types also have an ultra-violet stabilizer added. They do not inhibit the transmission of vapor out of the concrete.

SEALANTS

A **sealant** is a material used to seal joints between members in construction materials and protect materials against the penetration of moisture, air, corrosive substances, and foreign objects. Examples include expansion joints in large masonry walls, spaces between glass and frames, and openings between exterior siding and door and window units. The sealing of the joints between these and other parts of the building is essential to assure the integrity of the entire structure.

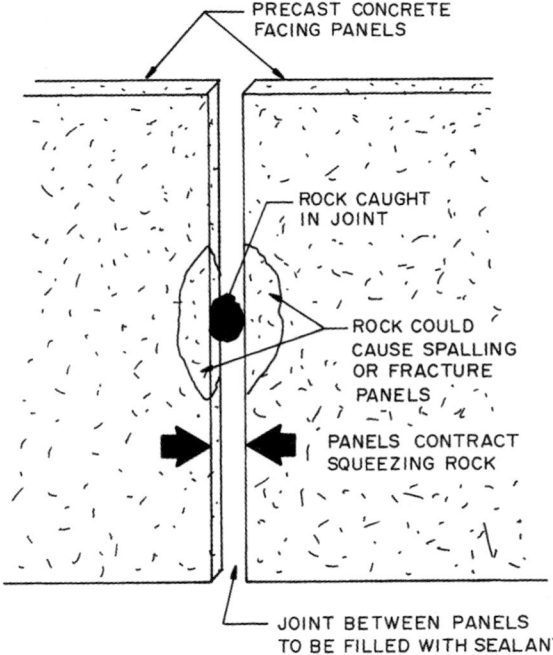

7-8 *Joints between masonry materials to be sealed must be free of foreign objects.*

Joints in the exterior walls are needed to allow for expansion and contraction of the materials. These are called **working joints.** The joint must allow for this movement. If a foreign object, such as a rock, gets into the joint, contraction may be blocked and the rock may cause the materials on each side to spall or crack (see 7-8).

The two basic methods for protecting joints are with **sealants** and **prefabricated covers.** Sealants are a flexible, adhesive material that is worked into the joint, bonds to the sides, and sets into a firm but rubbery, stretchable material. Prefabricated covers are usually metal and are made so they can allow for the movement. Both must maintain the integrity of the assembly of materials but allow for the movement between them.

Moldable sealants may be deformable or elastic/elastomeric. The **deformed sealant** is installed in its natural stretched shape, stretches as the joint widens, and deforms as the joint width is reduced. The **elastic/elastomeric sealant** stretches as the joint widens and shrinks back to its normal size as the joint width is reduced (see 7-9).

Types of Sealant

Sealants are of two basic types, as (1) a solid, flexible preformed shape, and (2) as a flexible, moldable, adhesive compound.

PREFORMED SEALANTS

Preformed sealants are available in a variety of materials and shapes and materials. Examples of several are in 7-10.

7-9 *Elastic/elastomeric and deformable sealants allow for expansion and contraction at each joint.*

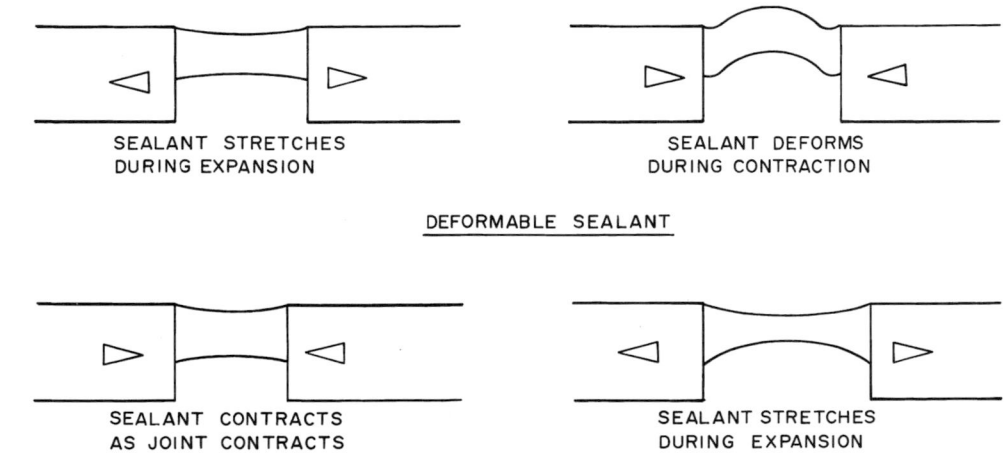

Moldable sealants are manufactured in three performance levels. **Low-performance sealants** are less costly, have a short life (4 to 7 years), and are used where limited joint movement is expected. **Intermediate performance sealants** are more expensive but last 7 to 14 years. They are used where joint movement is greater than those for low-performance sealants. **High-performance sealants** are the most expensive but have a life expectancy of 20 to 30 years. They are used in joints when the movement is the greatest. Manufacturers recommendations on joint size and the allowable percent of elongation must be observed.

Moldable bulk sealants are available in pourable, knife grade (as glazing compounds), gunable in manual or pneumatic caulking guns, and as preformed tapes that may be cured or not cured.

Sealant Performance Considerations

A key factor is the **percent of elongation** the sealant can safely stretch and still give expected protection. Some high performance sealants have a 50 percent elongation while intermediate types are usually rated up to 25 percent. The sealant must have excellent **adhesiveness** to the material to which it is expected to bond. It must be **flexible** and have minimum internal shrinkage through the years of use, must resist **stain-ing** material around it, and have a tough **nontacky** elastic surface skin so it does not have dirt and solid objects stick to it or penetrate it.

Commonly used sealants include polyurethane, epoxy, silicone rubber, latex, polysulfide polymers, and urethane bitumen.

Sealants & Joint Design

The width of the joint between two members, as precast concrete facing panels, must be carefully determined. For example, one type of sealant available tolerates a joint movement of +100 percent to -50 percent. The manufacturer requires the joint width be two times the expected joint movement and at least ¼ in. (6 mm) wide. A backup rod is used to control the depth as shown in 7-11.

Caulking & Glazing

Caulking is a procedure for sealing joints, cracks, or other small openings with caulking compound (sealer). Caulking compound is a resilient mastic material.

Glazing compounds are a form of sealer used to set glass in place in frames. They serve to seal out water and air and form a cushion allowing for expansion and contraction of the glass and frames.

Caulking and glazing compounds are commonly silicone, acrylic, butyl, polysulfide, or polyurethane. These five types are "gunable," meaning they are of a consistency to be applied by a gun-like device. There are other types referred to as "knife" grade, meaning they are applied with a putty knife.

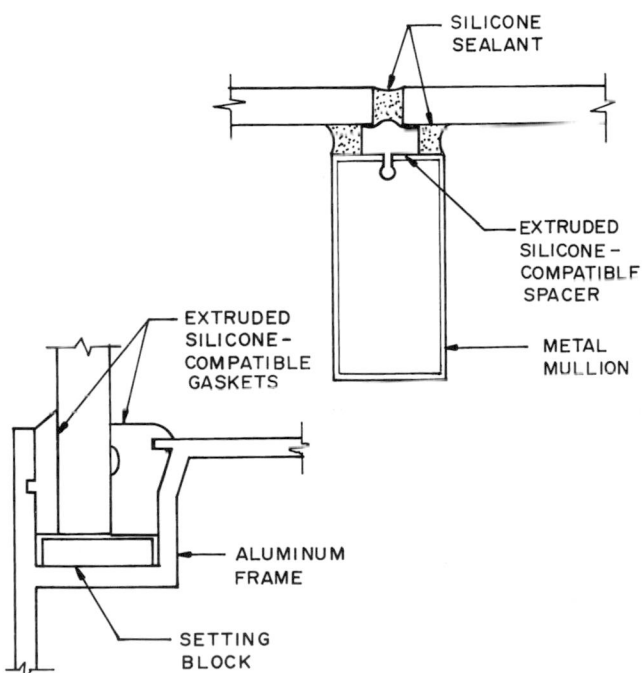

7-10 *Preformed sealants are designed for specific applications.*

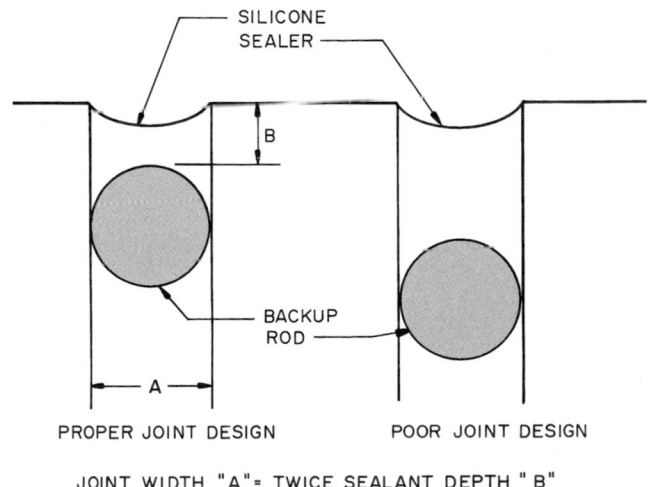

7-11 *A backup rod is used to control the depth of the sealant.*

Table 7-7 TYPES OF WATERPROOFING

Sheet membranes	Composite membranes	Built-up Membranes		Liquid membrane	Applied coating
		Hot applied	Cold applied		
Butyl	Elastomeric, backed	Asphalt, type I, II, III	Bitumen emulsion	Butyl	Acrylic, silicone
Ethyene propylene	Polyethylene and	Coal tar pitch, type B	Bitumen, fiberated	Urethane	Asphalt emulsions,
Neoprene	rubberized bitumen	Felts, saturated and	cement	Polychlorene	cut backs
Polyethylene	Polyvinyl chloride backed	coated	Felts, coated	(neoprene)	Cementitious with
Polyvinyl chloride	Saturated felts and		Bentonite clay	Polyurethane,	admixtures
	bitumen coated		Fabric, saturated	coal tar	Epoxy, bitumen
			Glass fiber mesh,		Urethane, bitumen
			saturated		Bitumen,
			Cementitious		rubberized
			membrane		

WATERPROOFING MEMBRANES & COATINGS

Waterproofing involves applying a material on the surface of an assembly of materials, such as a foundation, to make it impervious to water. Most common among these are foundation walls, roofs, exterior wood or masonry walls, and exposed structural components including steel. A summary of the most frequently used types of waterproofing are in Table 7-7.

Waterproofing must be adequate to resist the forces that tend to force water through the assembly of materials.

▼ These forces include:
Gravity that forces water through horizontal areas as a roof or deck.
Hydrostatic pressure on one side of a horizontal or vertical assembly (most commonly due to subsoil water).
Difference in air pressure on one side of a horizontal or vertical assembly.

For surfaces not subjected to hydrostatic gravity or air pressure differences waterproofing can be a light duty coating, such as silicone or coal tar pitch. This provides a moisture-resistant membrane that **dampproofs** the assembly. Surfaces under pressure require a heavy duty membrane, such as tar and felt or a synthetic membrane. Waterproofing is most effective if applied on the surface directly facing the source of moisture.

▼ Waterproofing can be accomplished by:
Adding waterproofing admixtures to the concrete as it is mixed.
Applying a thin film or coating to the exterior of the wall, such as liquid silicone or coal tar pitch.
Applying a heavy coating such as portland cement plaster or a trowelable asphalt.
Applying a dry coating that will emulsify in place such as bentonite clay.
Applying a built-up bituminous membrane of felt and hot or cold tar pitch.
Bond an elastomeric membrane to the wall.

Waterproofing coatings and membranes are not self-supporting, but must be bonded to the surface to be treated. In addition, waterproofing must be able to adjust to the stresses caused by movements of the assembly and any cracks or deterioration without losing the waterproofing capabilities. It should be noted that waterproofing coatings and membranes can be damaged during installation and construction. One example is damage to roof membranes by someone walking on them. Protective pads are available for installation on top of the roof membrane to protect the areas where workers may walk. Another frequently occurring damage occurs during backfilling a foundation. A protective material can be placed over the waterproofing to keep rocks from piercing it.

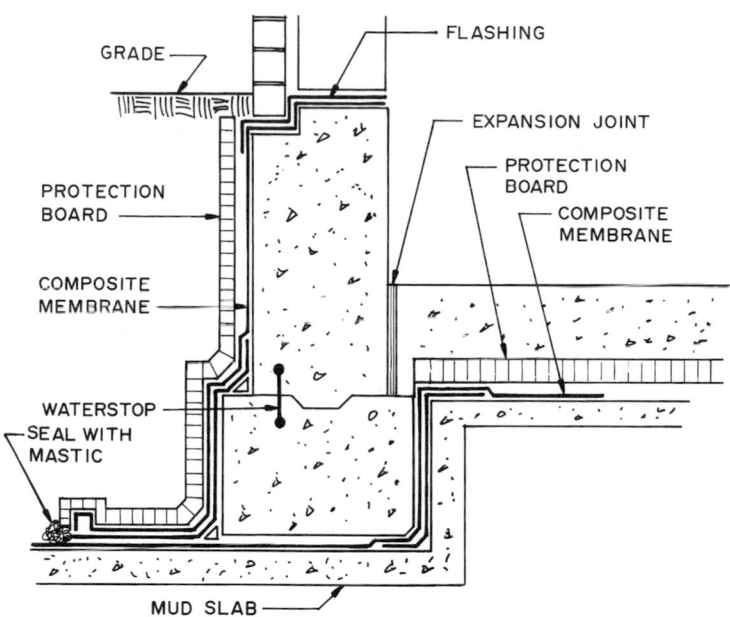

GRADE

FLASHING

EXPANSION JOINT

PROTECTION BOARD

COMPOSITE MEMBRANE

PROTECTION BOARD

COMPOSITE MEMBRANE

WATERSTOP

SEAL WITH MASTIC

MUD SLAB

7-12 *Composite membranes provide high-quality waterproofing.*

Bituminous Coatings

Hot applied and emulsified coal tar pitch, hot-applied and cold-applied asphalt, and emulsified asphalt can be applied to the foundation by brush, roller, or spray. These are effective only for situations where hydrostatic pressure is not a factor.

A waterproofing system that will resist hydrostatic pressure consists of alternate layers of hot mopped coal tar pitch, asphalt, or cold mopped emulsified asphalt over layers of mineral or glass fiber felts in much the same way as laying a built-up roof. The number of layers of felt and asphalt depends upon the hydrostatic conditions. Manufacturers of these systems have established specifications for various conditions.

Synthetic Sheet Membranes

Synthetic membranes are available made from neoprene, polyvinyl chloride, polyethylene, butyl, and ethylene propylene. These sheet materials are bonded to the foundation wall using adhesives recommended by the system manufacturer. Since they are flexible membranes they tend to adjust to settling and compaction of the soil and are not likely to rupture. They are available as single materials or composite membranes. Following are examples of a few of these products.

COMPOSITE MEMBRANES

Composite membranes are sheet products made by laminating two or more waterproofing materials. One frequently used composite membrane is made of a rubberized asphalt layer with a polyethylene film bonded to the outer surface. It remains stable below ground and below water. When the sheets are lapped the rubberized asphalt back bonds to the polyethylene face of the sheet beside it. Under most conditions it will bridge gaps up to ¼ in. (6 mm). It is available in rolls 48 in. (121 mm) wide and 60 ft. (18.3 m.) long. Since the back surface is very tacky, it is covered with a peelable paper covering (7-12).

Another composite membrane is laminated chlorinated polyethylene (CPE) film to a nonwoven polyester fabric. This composite is bonded with waterbased acrylic adhesives, avoiding problems that occur from fumes of solvent-based adhesives. It can be installed on surfaces that are damp at the time of installation.

SHEET MEMBRANES

Sheet membranes are waterproofing products composed basically of one major waterproofing material. One such product is made from **polyvinyl chloride** alloyed with high density **polymer resins**. It is not affected by aging, mildew, or corrosion. It remains flexible at low temperatures and has good abrasion and tear resistance. It is bonded and laps are sealed with a special adhesive. The seams can be welded with hot air or bonded with an adhesive specified for that purpose.

Another type of sheet membrane is a **chlorinated polyethylene** (CPE) product. It is available in a range of thicknesses with and without integral reinforcement. It is suitable for use above and below grade and on horizontal and vertical surfaces. Laps can be joined by chemical or thermal fusion.

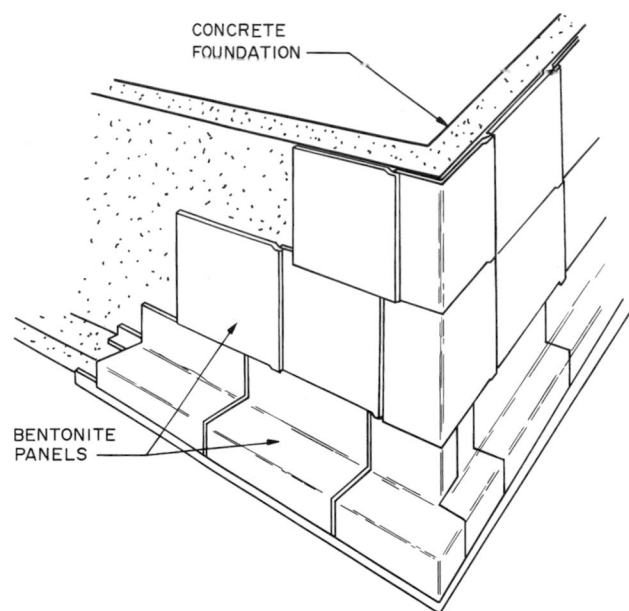

7-13 *Bentonite panels can be nailed to concrete foundations and are overlapped at the joints between panels.*

Asphalt Waterproofing Materials

There are a variety of solvent-based asphalt dampproofing compounds. They are available as a thick mastic, a semisolid mastic, and a spray mastic. They give dampproofing properties to interior and exterior above-grade and below-grade surfaces. They are used on metal to prevent corrosion.

Dampproofing & Waterproofing Coatings

An **acrylic copolymer** waterproof coating is available in a variety of colors. It may be obtained with fillers and texturing aggregates that are fused on to the concrete or masonry surface. This offers waterproofing protection and an attractive finish coating. Important uses are on above-grade exterior concrete walls, columns, and spandrels. It is also used on portland cement plaster and stucco walls giving a textured, sandlike finish in color, and will minimize surface defects. It is applied by brush, roller, or spray.

Clear silicone dampproof liquids are widely used on exterior concrete, masonry, and wood surfaces. They are not a surface coating but penetrate into the surface carrying solids into the pores. They do not color the wall but do reduce efflorescence. They can be applied with a brush, roller, or spray.

A liquid waterproofing material that is self-curing is available in a **polyurethane rubber** with a coal tar additive. It is applied to decks, roof slabs, and floors with a brush, roller, or squeegee. It will cover small cracks and is flexible enough to cover surfaces of irregular shapes. It can be covered with a concrete slab, hardboard, or mineral surface roll roofing.

Another form of liquid waterproofing uses a **silicate liquid gel,** which, when applied to concrete, reacts with the soluble calciums in the concrete, forming a glass gel in the microscopic pores. The gel will penetrate 1.5 to 2.0 in. (38 to 50 mm), providing a strong moisture-proof barrier that also reduces efflorescence.

Cement-Based Waterproofing

There are a number of cement-based heavy duty waterproof coatings available. These have carefully graded aggregates that produce a high density and high strength waterproof coating. Some types are applied with a trowel while others require special procedures. They are used to waterproof concrete masonry and stone in interior and exterior locations. They are used on water reservoirs, swimming pools, basements, parapet walls and other heavily exposed surfaces.

Lead Waterproofing

Lead waterproofing sheets are frequently used for projects where waterproofing must be of uncompromised security. Examples include areas of grass, fountains, and reflecting pools that may be built over underground facilities such as parking or stores or built on the roof of a building. While the gauge of the lead sheet varies for different purposes, it must be at least 6 lb. lead (3/32 in. or 2.3 mm) thick to allow for the burning (welding) necessary to join the sheets. If lead is in contact with cementitious materials it must be protected with a bituminous coating.

Benonite-Clay Waterproofing

Bentonite is a clay formed from decomposed volcanic ash, with a high content of the mineral montmorillonite. It has the capability of absorbing large amounts of water that causes it to swell many times its original volume, forming a waterproof barrier. The dried, finely ground particles are usually applied as a waterproofing membrane in three ways.

Bentonite panels, usually 4 × 4 ft. (1200 × 1200 mm) are made consisting of a biodegradable paper covering over bentonite clay particles. One type is 3/16 in. (2.5 mm) thick with a corrugated kraftboard core. It is used on vertical walls and under structural slabs. Another type is 5/8 in. (16 mm) thick composed of the layers of corrugated kraftboard with the center layer holding the bentonite clay. The hollow outer layers allow space for the expansion of the bentonite clay, reducing upward pressure against a thin nonstructural slab.

The panels are ready to apply when received on the job and can be applied at all temperatures and over moist substrates. They can be nailed to green concrete walls. A hydrated sodium bentonite gel is used to fill gaps around pipes and fittings. If the backfill contains rocks that may pierce the panel, cover it with a protective material (see 7-13).

A bentonite clay mixed with a modified asphalt, which serves as an adhesive to bond the clay to the surface, is applied by spraying. A ⅜ in. (10 mm) membrane is built up for normal applications but thicker layers can be used for situations having severe hydrostatic pressure.

Bentonite clay mixed with sand is used to produce a waterproof barrier below concrete slabs. The mix is carefully measured and spread over the area to be covered with concrete. It is then covered with a polyethylene sheet to protect it from the moisture in the concrete. The reinforcing is placed on top and the slab is poured. If too much bentonite is used the slab could be cracked by the forces of expansion.

BITUMINOUS MATERIALS

Bitumen is a mixture of complex hydrocarbons that occur naturally or are heat produced from materials such as coal and wood. Bitumens may be in a gaseous, liquid, semisolid, or solid state.

Asphalt, tar, and coal-tar pitch are those most commonly used on construction. Asphalt is found in natural deposits but most is produced from petroleum. Tar is produced by the distillation of wood and coal. Fractional distillation of tar produces coal-tar pitch.

Properties of Bitumens

The properties of bitumens are determined by a series of standardized tests. These include penetration, softening point, ductility, viscosity flash point, thin-film oven, and solubility tests.

Possibly the most important property is **water resistance.** The water resistance of bitumens is excellent. While this will vary with the different products produced, overall it is a good waterproof material.

Bitumens **adhere well** to dry solid surfaces. Bitumens bond well because they are in a semifluid state needed by adhesives to bond. They will not bond to wet surfaces.

Bitumens are **flammable** and will ignite when heated to their **flash point.** This influences the temperatures used as it is heated for various applications.

Another property to consider is the **softening point.** While this varies with the composition of the product, it becomes a very important property when the temperatures to be experienced by various applications are considered. For example, the softening point for roofing asphalts ranges from 200° to about 220°F (95 to 104°C), while coating grade asphalts are used for waterproofing, have a softening point of 50° to 55°F (11 to 13°C).

Bitumens exhibit **cold flow** properties. This means they tend to flow or spread or lose their shape. This is more pronounced when temperatures rise. It does offer the advantage of causing a check or crack on a roof membrane to seal or heal when the sun heats up the asphalt.

The **viscosity** of asphalt is an important property when considering applications. Viscosity is its ability to stay in place when subjected to heat. Asphalts have good viscosity properties up to 135°F (58°C) and some can withstand temperatures up to 275°F (136°C).

Asphalts rate high in their **ductility properties.** Their molecules hold fast even when they are extended by heat and pressure. They can expand and still remain bonded to the materials upon which they have been placed.

Materials Made with Bituminous Products

A wide range of products are manufactured using bituminous materials as the major components. Most common among these are coal-tar pitch, asphalt, felts, stabilizers, and surfacing materials. Various solvents and fillers are also used.

COAL-TAR

Coal-tar pitch is a dark brown to black hydrocarbon obtained by the distillation of coke-oven tar. coal-tar pitch is available in several grades and is used as the basis for a number of paints, roofing, and waterproofing materials.

Coal-tar enamel is made from coal-tar pitch and is used to protect pipe in pipeline work. Cold-applied coal-tar products have a solvent added to liquify them. Hot applied coal-tar coatings give better protection than cold coal-tar coatings.

Table 7-8 GRADES OF ASPHALT USED FOR PAINTS, COATINGS, ADHESIVES & CEMENTS

Grades of Liquefaction
0 (Thinnest)
1
3
4
5 (Thickest)

Table 7-9 TYPES OF ASPHALT USED FOR PAINTS, COATINGS, ADHESIVES & CEMENTS

Type	
Steam Refined with Petroleum Solvent	**Emulsified in Chemically Treated Water**
Slow curing (SC)	Slow setting (SS)
Medium curing (MC)	Medium setting (MS)
Rapid curing (RC)	Rapid setting (RS

ASPHALT

Asphalt is a dark brown to black cementitious material in semisolid or solid form made up of bitumens found in deposits of natural asphalt. A similar product is a residue from the distillation of petroleum. Petroleum provides most of the asphalts used today.

Asphalt is used in many products used in construction. Most apparent is its use in roofing and siding and as a paving material. It is used in some paints, adhesives, acid- and alkali-resistant coatings, and dampproofing and waterproofing solutions. It is used to coat organic fiber and fibrous plastic panels, coat papers of various types, in adhesives and cementitious materials, and the manufacture of drain and sewer pipe. Some types, as waterproof coatings, are applied on the site. Others, such as coating fibrous panels, are applied in the factory.

The grades and types of asphalt used for coatings, cements, paints, and adhesives are in Table 7-8 and Table 7-9. The grade refers to the liquification of the asphalt. Grade O is the thinnest and grade 5 has the thickest consistency, which is much like a thick paste. The types are slow curing (SC), medium curing (MC), and rapid curing (RC) (Table 7-10). These refer to steam-refined or oxidized asphalt having a petroleum solvent. Asphalt emulsified in chemically treated water has grades. These are slow setting (SS), medium setting (MS), and rapid setting (RS).

▼ Roofing asphalt is available in four types.

Type I	Dead level
Type II	Flat
Type III	Steep
Type IV	Special steep

Physical characteristics for these four types of roofing asphalts are in Table 7-11.

Asphalt cements are binders used to produce high-quality asphalt pavements. They are a highly viscous asphalt that is made in several grades based on consistency and are semisolid at normal ambient temperatures.

Asphalt cements are blended with aggregates graded into a range of sizes. Typical aggregates would include crushed stone, gravel and sand. Aggregates compose about 90 percent of the weight of the paving mix. The aggregates give the mix its strength while the asphalt is the binder. In some cases tars are used because they increase the resistance to damage from gasoline spilled on the paved surface (see7-14). Asphalt cement paving is used for paving road, drives, and parking lots.

Cutback asphalt is a broad classification of residual asphalt materials left after the petroleum has been processed producing gasoline, kerosene, diesel oil, and lubricating oils. The **residual asphalt** is then blended with various solvents producing cutback asphalt. There are three classifications, rapid curing (RC), medium curing (MC), and slow curing (SC).

7-14 *Asphalt paving cement is used to pave roads, driveways, and parking lots.*
Courtesy The Asphalt Institute

Table 7-10 LIQUID ASPHALT PRODUCTS & THEIR SOLVENTS

Classification	Solvent
Rapid curing (RC)	Gasoline or naphtha
Medium curing (MC)	Kerosene
Slow curing (SC)	Slowly volatile or nonvolatile oils
Asphalt emulsions	Water and emulsifiers

Table 7-11 PHYSICAL CHARACTERISTICS OF ROOFING ASPHALTS

Type	Softening Point	Flash Point
Type I, Dead level	135–151°F	475°F
	58–67°C	248°C
Type II, Flat	158–176°F	475°F
	70–80°C	248°C
Type III, Steep	185–205°F	475°F
	85–97°C	248°C
Type IV, Special steep	210–225°F	475°F
	100–108°C	248°C

They are often mixed with granular soil to stabilize the road bed before paving or serve as the binder for a finished road surface for light traffic, such as on secondary roads.

The **emulsion group** of asphalts consists of emulsified asphalt cement mixed with water and is used in road construction. The emulsion is made by adding heated, fluid asphaltic cement into water to which an emulsifying agent, such as soap or bentonite clay, has been added. A stabilization agent, protein, is added to prevent the particles from blending together within the mix. The suspended particles of asphalt blend with the aggregate or soil particles as the water drains away and evaporates.

Plastic asphalt cements are made using asbestos fibers and asphalts that have good plasticity and elastic properties. This black cement is used to bond flashing and make roof repairs.

Quick setting asphalt cement is much like plastic asphalt cement but has greater adhesive properties and sets up rapidly. It is used to cement down the free tabs on shingles and the laps between layers of roll roofing.

Asphalt roofing tape is a porous fabric strip saturated with asphalt. It is available in rolls 4 to 36 in. (100 to 914 mm) wide and 50 yds. (45.5 m) long. It is used with plastic asphalt cements to patch holes in roofs and patching seams or sealing flashing.

Felts

Felt is a sheet material made from cellulose fibers from organic materials such as wood, paper, and rags or glass fibers or asbestos.

Saturated felts are made with an organic mat saturated with coal-tar pitch or asphalt and coated with a surface coat of thin asphalt. This is sometimes called tar paper. It is used as an underlayment below shingles, as sheathing paper and as laminations in built-up roof construction. It is also used to produce roll roofing and shingles in addition to its use as felt underlayment.

Fiberglass Sheet Material

Fiberglass mats are impregnated with asphalt. However, they are not saturated because the glass fibers will not absorb the asphalt. The asphalt forms a coating on the surface and fills the spaces between the fibers.

Fireproofing Paper

Fireproofing paper is made using asbestos fibers either in a pressed mat-like felt or in woven sheets.

The various sheet-type products are used as underlayment under finished roofing materials, as vapor barriers in walls and floors, and other such applications.

Roof Coverings

Building codes classify roof covering requirements. They vary with the type of building, size, occupancy, and fire district. The following pages discuss the various types of bituminous roof-covering materials in common use. It should be noted that coal-tar pitch and asphalt are not compatible and are not used where they will be in physical contact (see 7-15).

7-15 *There are a variety of roof-covering materials using asphalt.*

7-16 *Manufacturers of shingles whose products meet the Safety Standards of Underwriters Laboratories Inc. (UL) may be authorized by UL to use these labels, which include the UL mark on those products.*

FIRE RATINGS

As asphalt roof coverings are considered, one factor is the fire-rating classification of the products available. The following fire ratings of roofing materials are based on specifications of the Underwriters Laboratory, Inc. Underwriters Laboratory, Inc. labels indicating the fire rating appear on packaged asphalt roofing products (see 7-16).

Class A: Highest rating—The covering is effective against severe fire exposure. Roof covering is not readily flammable and offers a high degree of fire protection for the roof deck.

Class B: Moderate protection—Against fire exposure. Roof covering is not readily flammable and offers a moderate degree of fire protection for the roof deck.

Class C: Minimal protection—Covering provides protection against light fire exposure. Covering is not readily flammable and offers a measurable degree of fire protection for the roof deck.

ROLL ROOFING

Roll roofing uses either organic felt or fiberglass mats as the base. Upon this base a viscous bituminous coating is applied forming the exposed surface. It is made in four types, smooth surfaced, mineral surfaced, mineral surfaced selvage -edged, and pattern edged. **Smooth-surfaced roll roofing** has both sides covered with a fine talc or mica to keep the surfaces from sticking as it is made into rolls. **Mineral-surfaced roll roofing** has mineral granules in a wide range of colors rolled into the surface, giving a surface that is attractive and protecting the bitumen from damage from the ultraviolet rays of the sun.

Mineral-surfaced selvage-edge roofing is of the same construction as mineral-surfaced roll roofing except only 17 inches of the 36-inch wide surface is covered with granules. The 19-inch selvage edge is used for lapping with the next layer forming a two-ply covering. **Pattern-edged roll roofing** is a mineral-surfaced product that has a 4-in. uncoated band in the center and the roll is semicut along this strip to form

7-17 *Tar can be applied by hot mopping.*

two 18-in. wide patterned roofing strips which are thin lapped 2 in. over the layer below.

CONVENTIONAL HOT BUILT-UP ROOF MEMBRANES

A built-up roof consists of installing alternate plies of organic or fiberglass roofing felt with a hot bitumen coating mopped over each layer (see 7-17). The design of the roof varies with the situation but generally it consists of three or more layers of felt with a bitumen layer over each and a bitumen top coat with some form of aggregate rolled on top. A conventional hot built-up roof system is in 7-18. The felts give the needed reinforcements to keep the bitumens in each layer from alligatoring. Alligatoring refers to surface cracking due to oxidation and shrinkage stresses giving a repetitive mounding of the asphalt surface similar to those on an alligator hide. The aggregate surface is usually gravel, marble chips, or slag. Gravel and marble chips are usually applied 400 pounds per 100 sq. ft. and slag 300 pounds per 100 sq. ft. Conventional built-up roofs can be designed for slopes up to 3 in. (76 mm) per 12 in. (305 mm). Slopes above ½ in. (12 mm) per 12 in. (305 mm) usually require the use of Steep Asphalt.

MODIFIED ASPHALT ROOFING SYSTEMS

Modified asphalt roll roofing is composed of polymer modified bitumen (modified asphalt) reinforced with one or more plies of fabric such as polyester glass fiber (see 7-19). These membranes are of uniform thickness and have consistent physical properties throughout the membrane. Modified membranes may be used below

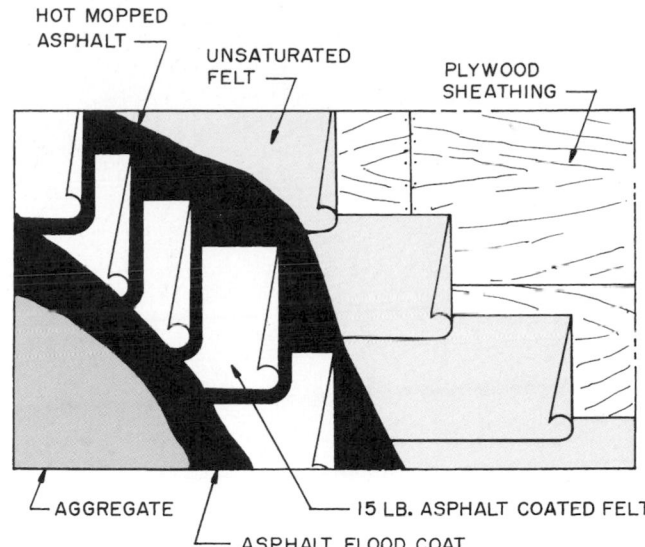

7-18 *A conventional hot asphalt built-up roof system on a nailable deck.*

grade for waterproofing canals, water reservoirs, landfills, and various roofing applications.

COLD-APPLIED ASPHALT ROOFING SYSTEMS

Cold-applied systems use some form of coated base sheet, fabric, or similar reinforcement over which the principal waterproofing agent, which is a liquid, is applied at ambient temperatures. The selection of the cold process application depends upon the level of maintenance, repair, service expected, and the compatibility between the cold-applied materials and any existing substrate.

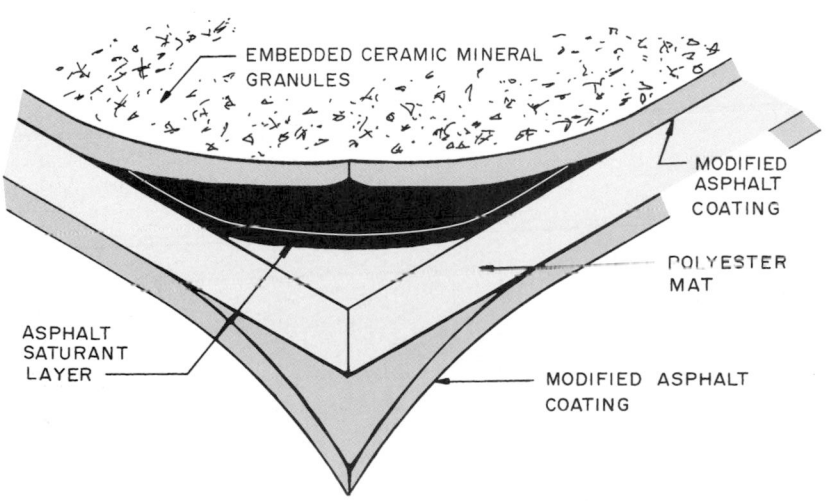

7-19 *A form of a modified asphalt roll roofing installation.*
Courtesy Asphalt Roofing Manufacturers Association

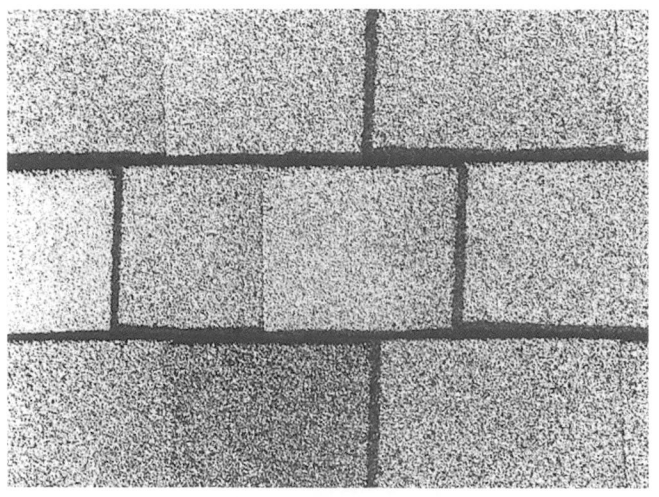

7-20 *Two styles of the many asphalt/fiberglass shingles available.*

ASPHALT & FIBERGLASS SHINGLES

Asphalt shingles are made using an organic felt base saturated with asphalt and coated with a mineral stabilized coating of asphalt on both sides. The top side to be exposed to the weather is coated with mineral granules. The are available in a wide range of colors. The bottom is relatively smooth and is from sticking to the shingle below it as they are bundled for shipping (see 7-20).

Fiberglass shingles are made using a fiberglass mat as the base. The mat does not have to be saturated with asphalt before it receives the asphalt top coating. Mineral aggregates are rolled into the top coating. These shingles are lighter than asphalt shingles (see 7-21).

Shingles are made in a variety of styles. Most commonly they are in strips 12 × 36 in. (305 × 915 mm) with the exposed surface cut to resemble three smaller shingles. They are available in weights of 100 to 300 lbs. per square (45 to 135 kg per square). Commonly available forms of asphalt and fiberglass shingles are in 7-22. Fiberglass shingles are more commonly used today than asphalt shingles.

Asphalt Paving

Asphalt is widely used for paving because it is cementitious and binds other ingredients, such as aggregates. Since it is flexible, plastic, and ductile, it experiences a minimum of cracking. It withstands exposure to the weather and has resistance to salt, acids, and alkalis.

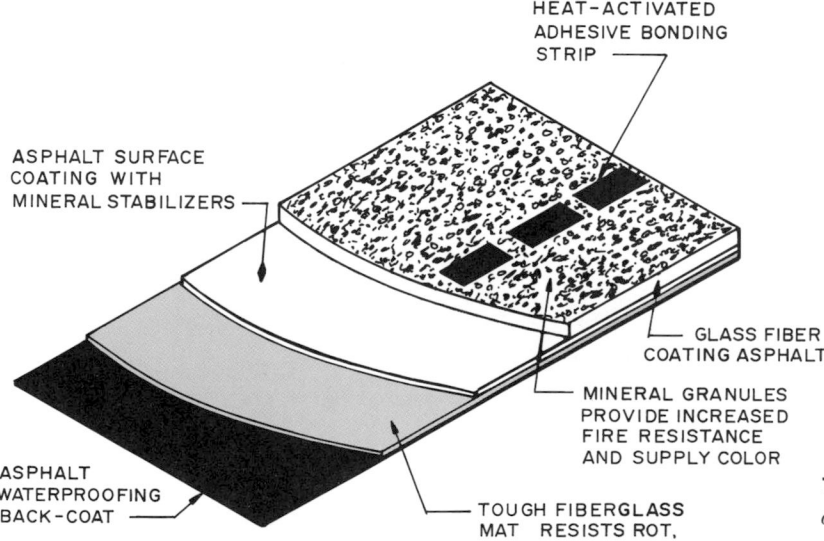

HEAT-ACTIVATED
ADHESIVE BONDING
STRIP

ASPHALT SURFACE
COATING WITH
MINERAL STABILIZERS

GLASS FIBER
COATING ASPHALT

MINERAL GRANULES
PROVIDE INCREASED
FIRE RESISTANCE
AND SUPPLY COLOR

ASPHALT
WATERPROOFING
BACK-COAT

TOUGH FIBERGLASS
MAT RESISTS ROT,
MILDEW, AND MOLD

7-21 *The layers forming the construction of a fiberglass shingle.*

A bituminous binder is used to hold the asphalt paving together and protect the mineral aggregate used on paving surfaces.

Asphalt cements are used as binders for quality paving. **Cutback asphalt** is used to stabilize the road bed before paving and as a binder on the surface of light duty roads. **Emulsion asphalts** may be applied cold to damp aggregates. Detailed information on each is given earlier in this division.

Asphalt paving mix designs are tested by a variety of methods. The purpose of these tests is to get a blend of aggregates of various sizes and the asphalt to give a product that will withstand the traffic and still be able to be easily placed.

ROOFING SYSTEMS

A roof system consists of a number of interacting materials, which, when properly combined, provide a weather resistant top surface to a building. In some systems they also insulate the building.

There are many types of decks, insulation, and roof-covering materials available. Manufacturers of these materials provide technical information and installation instructions. The roofing system can be designed to serve a variety of purposes, such as resistance to fire, water penetration, and wind. In addition they can provide acoustical and thermal insulation and contribute to energy conservation.

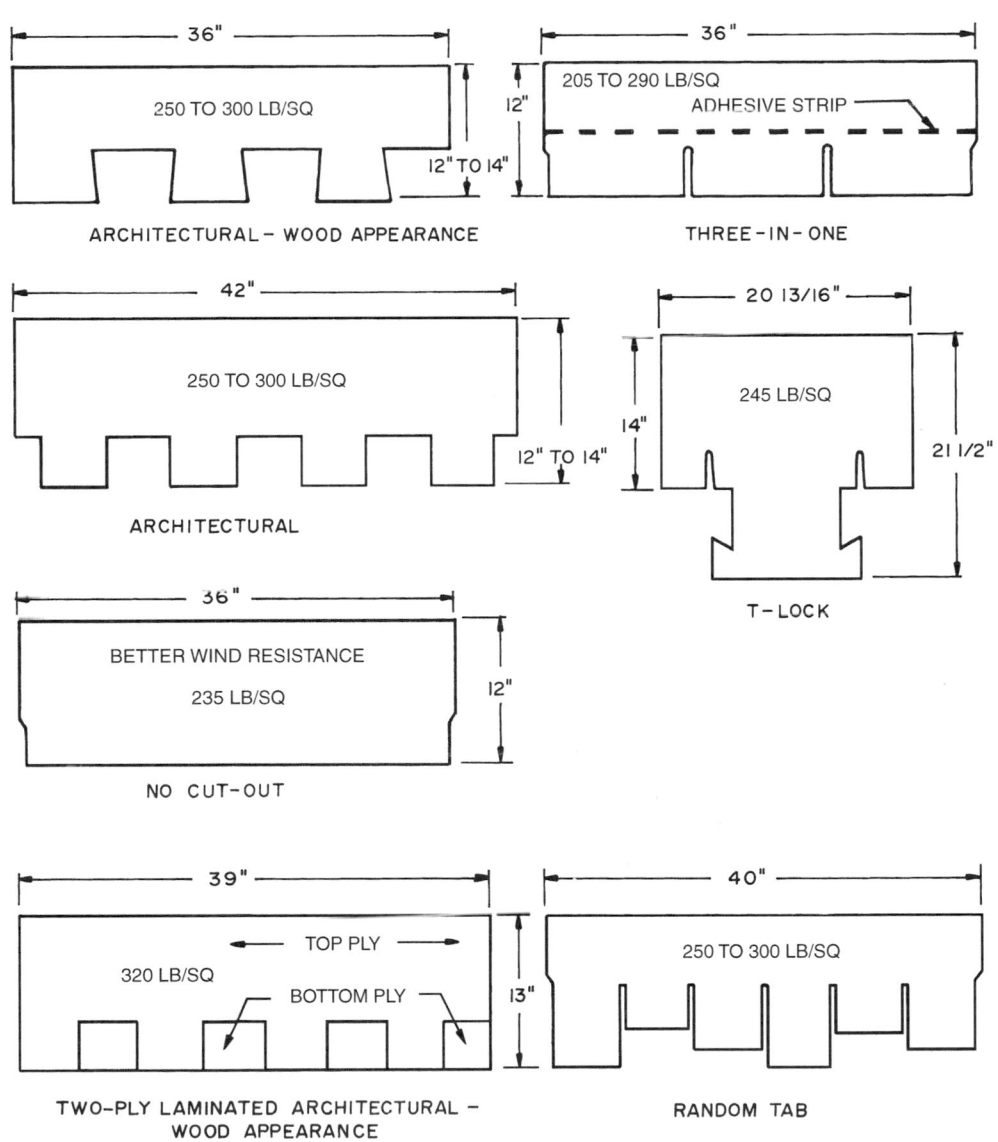

7-22 *Commonly available forms of asphalt and fiberglass shingles.*

Roof systems are generally divided into two categories, low-slope and steep-slope.

Low-slope roofs are flat or nearly flat and provide little runoff of rain.

Steep-slope roofs permit the rapid runoff of rain, which reduces the likelihood of penetration through the water-resistant surface.

LOW-SLOPE ROOFS

Low-slope roofs are generally cheaper to build than are steep-sloped roofs and the low-slope roof can be extended over large buildings whereas steep-sloped roofs are limited by the spans possible by the roof structural materials. Water tends to pond on flat roofs even though drains are provided. As the temperatures change or if there is movement in the deck, the finished membrane may crack or pull loose from parapets and drains. Expansion joints are used to try to diminish this problem. If water vapor gets below the finished membrane, solar heating will cause the roof to get hot and this condition may form blisters in some roofing materials.

STEEP-SLOPED ROOFS

Steep-sloped roofs can be finished with materials such as metal or fiberglass shingles, clay tiles, or standing seam metal roofing.

Steep-sloped roofs may have an attic space below, which, when properly ventilated, reduces possible damage due to vapor or high temperatures below the roofing material. They also provide additional living space. The finished roofing material is highly visible and forms an important part of the overall design of the building.

Building Codes

Building codes govern the materials, design, construction, and quality of the roof and roof covering. This includes the ability to resist rain and wind, durability specifications, compatibility of materials with each other, physical characteristics, and fire protection. Fire protection classifications are specified by ASTM E108.

▼ The four classifications are:

Class A Roofing Coverings—These are effective against severe fire test exposure. They include clay tile, mineral fiber, concrete, slate, some fiberglass asphalt shingles, and other materials that have been so certified by an approved testing agency. They may be used on buildings of all types.

Class B Roofing Coverings—These are effective against moderate fire test exposure. They include metal sheets and shingles, some composition shingles, and other materials certified by an approved testing agency.

Class C Roof Coverings—These are effective against light fire test exposure. They include materials certified as Type C by an approved testing agency.

Nonclassified Roof Coverings—Nonclassified roof coverings are not permitted on any buildings covered by most codes. In some cases these may be approved on various types of storage buildings.

TYPES OF ROOF DECK

Materials commonly used for roof decks are detailed in other chapters related to materials and types of construction. A properly functioning roof depends upon a structurally sound deck that is compatible with the roofing system. Following are the frequently used decking materials.

Cement-wood-fiber panels made by bonding treated wood fibers with portland cement or some other binder that are compressed into structural panels.

Concrete decks made by using cast-in-place techniques and precast concrete structural members.

Gypsum concrete decks are produced by mixing gypsum, wood fibers, or mineral aggregates with water, and casting it on formboards. Precast gypsum planks are also used.

Lightweight insulating concrete is made by mixing insulating aggregates, such as perlite, portland cement, and water, and casting it on top of metal or bulb-tee and formboard decking systems.

Reconstituted wood panels are produced by bonding wood veneers or wood chips, such as plywood and oriented strand board, two common materials.

Steel decks are produced by cold roll forming sheet steel into various structural shapes.

Wood planks are solid wood decking or glued laminated members made by bonding solid dimensional lumber into a structural decking member.

VAPOR RETARDERS

It is important to stop the flow of water vapor from inside the building up into the roofing system. This flow will cause some decking to deteriorate and damage some types of finished roofing surfaces. The moisture will also damage insulation in the exposed roofing assembly. Changes in temperature and humidity cause a difference in water vapor pressure between the inside and outside of the building, causing water vapor migration into the insulation (see 7-23).

The vapor retarder must resist the passage of water vapor. Any tears or holes in the material must be sealed. The protected roof membrane also serves as a vapor retarder. Vapor retarders should usually be applied to the warm side of the insulation in roof decks in cold climates. In warm humid climates winter condensation in the roof assembly is not a problem and vapor retarders are often not used. Air-conditioning in warm climates may actually cause a reverse vapor migration. Vapor retarders must be used on top of concrete and poured gypsum decks.

The performance of a vapor retarder is indicated by its Water-Vapor-Transmission Rate (WVTR), which is specified in perms. The perm is a measure of the porosity of the material to the passage of water vapor. Materials must have a perm rating of 0.00 to 0.50 perms. A perm rating indicates the number of grains of water vapor that will pass through one square foot of a material per hour when the vapor-pressure differential between the two sides equals 1 in. of mercury (0.49 psi). The metric equivalent is expressed in terms of grams/m^2 hr mm of mercury pressure difference between the two faces.

▼ The following are some frequently used vapor retarders.

Bituminous materials such as applying layers of asphalt roofing felt covered with hot asphalt.

Kraft® paper layers bonded with asphaltic adhesive and a glass-fiber reinforcement. They are bonded to the decking with cold-applied asphalt adhesive.

Polyethylene sheet material may be laid loose on the deck or attached with mechanical fasteners. The sheets are overlapped and taped or joined with an adhesive.

Aluminum foil is used on the face of insulation batts and rigid insulation. It provides a vapor barrier and serves as a reflective insulation. The joints between insulation panels should be covered with aluminum tape.

Other vapor retarders contain a combination of materials. Examples of products available include combinations such as a polypropylene scrim, kraft paper sheet, vinyl scrim, aluminum foil, or polypropylene scrim and aluminum foil. The scrim used is a fiberglass fabric. Scrim is a fabric of open weave.

ROOF INSULATION

The location of roof insulation varies with the design of the roof. In a typical steep-slope roof with an attic no roof insulation is required. Instead the ceiling is insulated. A low-slope roof requires some form of insulation and ventilation to be part of the assembly. All of the various types of insulation can find some use in the design of the roof system.

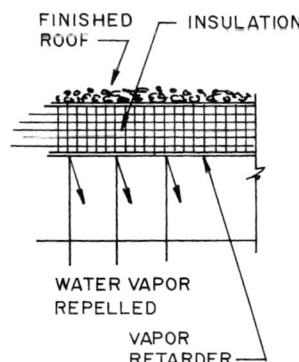
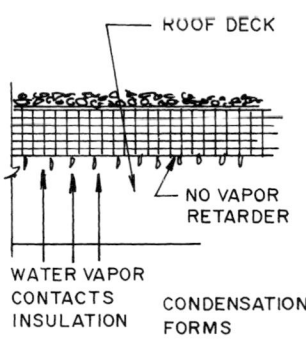

7-23 *The vapor retarder is placed on the warm side of the roof assembly below the insulation.*

Table 7-12 MATERIALS USED FOR FINISHED ROOFING

Material	Type of Roof	Descriptive Factors	Weight per Square (lb./100 ft.^{2})	Weight per Square Meter (kg/m^2)
Aluminum (sheet, shingles)	Steep-slope	Fire resistant, long life, range of colors	5–90	2.44–4.39
Asphalt (built-up)	Steep-slope, low-slope	Granular topping applied influences fire class, life 20–30 years	100–600	4.88–29.3
Asphalt shingles (fiberglass, asphalt)	Steep-slope life 20–30 years	Fire resistance varies with product, range of colors,	235–325	11.47–15.86
Cement-fiber tile	Steep-slope	Fire resistance, long life, heavy, use in warm climates	950	46.4
Clay tile	Steep-slope	Fire resistant, long life, heavy	800–1600	39–78
Copper (sheet)	Steep-slope, low-slope	Fire resistant, long life, can be soldered	0.019" thick 160 0.040" thick 320	7.8 15.6
Lead, copper coated	Steep-slope, low-slope	Fire resistant, long life	1/32" thick 200 1/16" thick 400	9.76 19.52
Monel (Ni-Cu)	Steep-slope	Fire resistant, long life	22 gauge 1424 26 gauge 827	69.5 40.4
Perlite-portland cement	Steep-slope	High fire rating, lightweight, long life	900–1000	43.9–48.8
Plastic (single-ply membrane)	Low-slope	Long life, requires careful installation, limited fire classification, several types available	Loose laid Ballasted, 1000–1200 Fully adhered, 30–55	48.8–58.6 1.5–2.7
Plastic (liquid applied)	Low-slope	Limited fire classification follow manufacturerÕs directions	20–50	0.98–2.4
Slate	Steep-slope	Fire resistant, heavy, long life	3/8" thick 800 1/4" thick 900 3/8" thick 1100 1/2" thick 1700 3/4" thick 2600	39.0 43.9 53.7 83.0 126.9
Stainless steel, terne coated	Steep-slope	Fire resistant, long life	90	3.89
Steel (sheet, shingles)	Steep-slope, low-slope	Fire resistant, long life, durable colors	Copper coated 130 Galvanized 130	6.3 6.3
Wood (shingles, shakes)	Steep-slope	No fire resistance unless treated, limited life	200–450	9.8–22.0
Zinc	Steep-slope	Fire resistant, long life, can be painted	9 gauge 670 12 gauge 1050	32.7 51.2
Terneplate copper-bearing sheet	Steep-slope	Fire resistant, long life	30 gauge 540 26 gauge 780	26.4 38.1
Mineral-surfaced cap sheet	Low-slope	Limited fire resistance, limited life	55–60	2.68–2.9
Modified bitumen	Low-slope	Fire resistance, 10 years or more	100	4.9

MATERIALS USED FOR THE FINISHED ROOF

The finished roofing is exposed to the weather. A wide range of products are available providing the designer with considerable choice. The specific properties of the products provided by the various manufacturers must be known and their installation recommendations followed. Table 7-12 contains many of the finished roofing materials available.

Low-Slope Roof Assemblies

The following discussion and illustrations will detail some of the construction used for roofing on low-slope roofs. A low-slope roof system is one with a slight slope and upon which water may collect and then slowly flow to drainage outlets. The actual slope permitted will vary with the type of roofing system used. The manufacturers recommendations must be observed. Typically these will fall in a range of ¼ in. rise per 12 in. run (6.4 mm per 305 mm) in the range of 2 inches rise per 12 inch of run (50.8 mm per 305 mm). Special provisions can be made for roofs with a 3 in. per 12 in. (76.2 in per 305 mm) slope.

Low-slope roofing systems include built-up, thermoset single-ply coverings, thermoplastic single-ply coverings, modified bitumen coverings, spray-applied polyurethane-foam coverings, liquid-applied coatings, and metal-sheet coverings. An important feature in the design of low-slope roofs is adequate insulation.

ROOF INSULATION FOR LOW-SLOPE ROOFS

Roof insulation may be applied to the top side of the decking. The insulation must meet building code and insurance bureau requirements. Roofs insulated this way increase the possibility of condensation occurring within the roof system. Therefore a vapor retarder is required.

7-24 *Rigid insulation is placed over hot asphalt on a concrete deck.*

Courtesy National Roofing Contractors Association

Since the insulation increases the roof temperature on hot days the roof materials age faster and the expansion and contraction stresses on the membrane are greater.

Frequently used rigid roof insulation includes cellular glass, composite boards, glass fiber, perlitic boards, polyisocyanurate foam boards, polystyrene boards, polyurethane foam boards, and wood fiber boards.

Roof insulation may be placed as a single or double layer. The double layer eliminates any leakage of heating or cooling energy that may occur between the joints in the first layer. The first layer may be mechanically joined to the deck. The second layer is bonded with hot bitumen or an adhesive. The first layer on concrete decks is bonded to the deck with hot bitumen (see 7-24).

BUILT-UP ROOFING

The traditional built-up roofing system used on low-slope roofs consists of bitumen (asphalt or coal tar) usually applied hot over felts, which may be glass-fiber, organic, or polyester, and a finished top surface, such as an aggregate (gravel or slag) or a cap sheet. Various manufacturers offer a variety of materials, including the composition of the felts and roofing asphalts to be used. Their recommended methods and application systems also vary. The following examples are typical of those in use.

Typical built-up roof construction on uninsulated nailable roof decks is shown in 7-25.

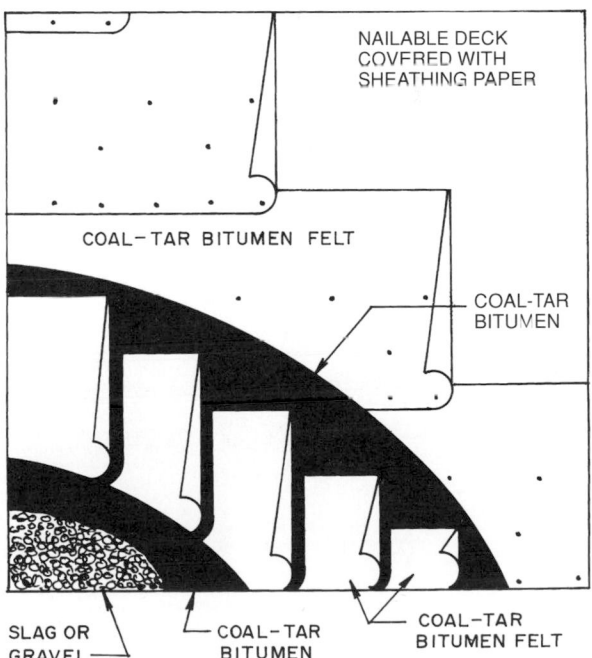

7-25 *One of several ways built-up roof constructions are applied over a nailable roof deck.*

The deck of the built-up roof is covered with one ply of sheathing paper (a vapor retarder) nailed to the deck. Nailable decks include wood, plywood, structural wood fiber panels, lightweight insulating concrete, and precast and poured gypsum. Next, three to five layers of an asphalt-coated base felt are applied bonded with coatings of a hot mopped bitumen. The top coating is covered with roofing asphalt and gravel or slag. Usually 400 lb. of gravel or 300 lb. of slag are applied per 100 sq. ft. of surface area. Manufacturers recommend that dead flat roofs be given a slight slope such as ½ in. per lineal foot.

Non-nailable decks, as steel, precast concrete and poured concrete, have the insulation bonded with hot bitumen or an approved adhesive. This is followed by layers of asphalt saturated roofing felt and hot roofing asphalt. The layers of felt are laid in a full bed of hot asphalt and broomed in place.

The roofing asphalt is brought to the site in a tank truck and heated in an asphalt kettle. The heated asphalt is pumped to a tank on the roof and moved to the area where it is needed (see 7-26).

Roof penetrations have to be flashed, roof drains flashed, and parapets and other places where the roof butts against a wall are flashed (see 7-27). Finally the gravel forming the top protective coating is lifted to the roof and spread in a bed of hot roofing asphalt. A typical construction detail is in 7-28.

Another type of built-up roofing is an assembly using an asphalt glass fiber roof membrane covered with a mineral-surfaced inorganic cap sheet. **A mineral-surfaced roof material** consists of a base felt that is coated on one or both sides with asphalt and surfaced with mineral granules. A typical assembly is shown in 7-29. An asphalt glass fiber membrane is nailed to the nailable deck and additional layers are bonded with hot asphalt. The mineral surfaced inorganic cap sheet is bonded to the asphalt glass fiber base with hot asphalt. It is recommended that this system be used on roofs with a slope of ¼:12 in. (6.25:305 mm) or greater.

MODIFIED BITUMEN MEMBRANES

Modified bitumen membranes combine polymer-modified asphalt and a polyester or fiberglass mat resulting in a product of exceptional strength. The two membranes available are SBS (styrene-butadiene-styrene) and APP (atactic polypropylene). SBS sheets have a reinforcement mat coated with an elastomeric blend of asphalt and SBS rubber. APP membranes have a reinforcement mat coated with a blend of asphalt and APP plastic. They are available from various manufacturers with several reinforcements. However, fiberglass reinforcement is most frequently used.

The major difference between SBS and APP products is the blended asphalt used. The blends create a product that has greater elongation, strength, and flexibility than traditional roofing asphalts.

SBS products are generally installed using hot asphalt as the bonding material. They are applied as cap sheets over a base of hot asphalt and roofing felts as shown in 7-30. The cap sheet (SBS membrane) may have a ceramic granule surfacing protecting it from ultraviolet light or be unsurfaced. The unsurfaced type must be coated with asphalt and gravel to give it ultraviolet protection.

7-26 *The hot roofing asphalt is pumped up to a tank on the roof.*

Courtesy National Roofing Contractors Association

7-27 *Layers of asphalt saturated organic felt are mopped with hot roofing asphalt to build up a multilayer membrane along the edge of the roof.*

Courtesy National Roofing Contractors Association

APA products are applied by a method called "torching." This is made possible by the unique properties of the modified bitumen. The back coating of modified asphalt is heated with a propane torch to the point at which it becomes able to bond the sheet to the substrate. APP products cannot be installed with hot mopped asphalt.

SINGLE-PLY ROOFING SYSTEMS

Single-ply roofing systems can be applied over almost any commonly used roof decking and over existing asphalt or built-up roofs when reroofing a building. Roofing insulation is required over the roof deck except very smooth concrete, plywood, and splinter-free solid wood decks. The roof must drain freely and have sufficient drain outlets to carry away the water.

Single-ply roofing membranes are available in a number of materials and manufacturers specifications and installation instructions should be carefully observed. Basically single-ply membranes are either thermoset or thermoplastic materials.

7-30 *These modified SBS bitumen membranes can be laid in hot roofing asphalt.*
Courtesy Schuller International, Inc.

7-28 *A construction detail for an insulated low-slope roof deck with built-up roofing.*

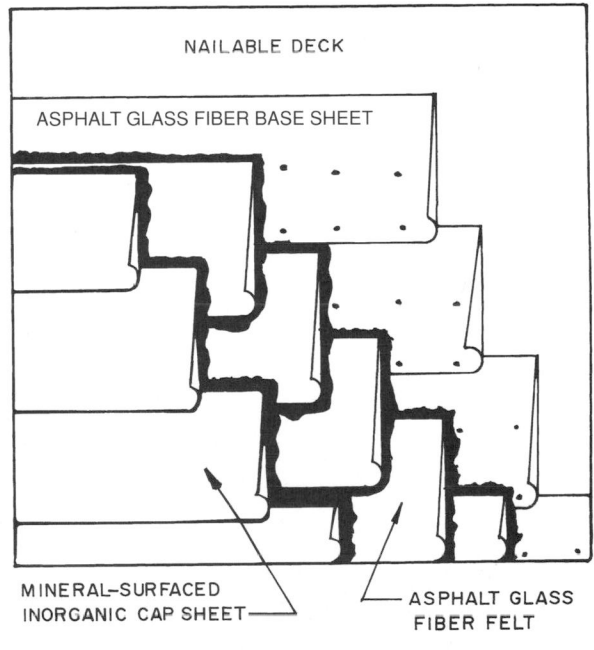

7-29 *One installation detail for using mineral surfaced cap sheets.*

▼ The following are some of the commonly used single-ply roofing membranes:

Chlorosulfated Polyethylene (CSPE) is a therm0set membrane that completes its cure after it is installed. It has excellent weathering qualities and is resistant to ozone, sunlight, and most chemicals.

Ethylene Propylene Diene Monomer (EPDM) is a thermoset elastomeric compound produced from propylene, ethylene, and diene monomer. It has good resistance to weathering, ultraviolet, abrasion, and ozone (see 7-31 and 7-32).

Polyvinyl Chloride (PVC) is a thermoplastic membrane produced by the polymerization of vinyl chloride monomer and stabilizers and plasticizers. PVC membranes are resistant to weather and chemical atmospheres and have good fire resistance and are easy to bond. They do not work well with bituminous materials.

Styrene-Butadiene-Styrene (SBS) is a thermoplastic membrane made by blending SBS with a high-quality asphalt over a fiberglass mat. It has good fire resistance and can be applied with hot or cold asphalt or be torched. The method used depends upon the specific composition of the membrane.

SPRAY-APPLIED ROOF COATINGS

Spray-applied roof coatings consist of a polyurethane foam insulation layer applied to the deck and then topped with a protective coating. The protective coating is usually acrylic, polyurethane, or silicone. Sometimes mineral granules or aggregates applied to the wet top coating are used to provide additional protection.

LIQUID-APPLIED ROOF SYSTEMS

Liquid-applied coverings are available from various manufacturers as one or two-component elastomeric materials. They are applied by spraying, brushing, or rolling over the roof decking, which is typically plywood or concrete. Any joints require special attention as recommended by the manufacturer.

METAL ROOF ASSEMBLIES

Various manufacturers have metal roofing systems designed for low-slope roof assemblies (see 7-33). They have developed panel profiles and seaming methods to be used to produce a watertight roof. The panels are available in a wide range of colors and may be galvanized or Galvolume steel, aluminum, copper or terne-coated stainless steel. Galvolume is a trade name for a patented steel sheet coated with a corrosion resistant aluminum-zinc alloy applied by a continuous hot-dip

7-31 *The EPDM single-ply membrane is unrolled on the deck.*
Courtesy Schuller International, Inc.

7-32 *The EPDM single-ply membrane is unfolded and laid over the surface for the deck.*
Courtesy Schuller International, Inc.

process. Terne-coated stainless steel panels are nickle-chrome stainless steel coated on both sides with an alloy of 80 percent lead and 20 percent tin. They are recommended for use in severe chemical and marine environments. Metal roofing panels are available embossed to create a stucco-like finish. Various types of other coatings are available, such as siliconized polyester and epoxy-based coatings.

Low-slope standing seam metal roofing systems require some slope. Typically a slope of ¼: 12 is minimum. This type of roofing is generally held to the structure with some form of metal clip. The clip is joined to the metal frame as a bar joist or "Z" purlin with metal fasteners. One such design is shown in 7-34.

ROOFTOP WALKWAYS

Rooftop walkways are placed over standing seam metal roofs and other flat roof membranes to provide maintenance workers access to mechanical units on the roof and other parts of the roof where access is denied. One type consists of lightweight perforated steel planks designed to fit over the standing seams. They are anchored with clips and supports attached directly to the standing seat. One walkway is shown in 7-35.

Walkways used on aggregate-ballasted roof systems use concrete pavers. The pavers may be set on plastic or concrete blocks or a bed of gravel is laid over the membrane and the concrete pavers or slabs are laid on it with cracks left open between each piece. In both types drainage below and between the pavers is provided. In no situation should the roof membrane be pierced to support a walkway.

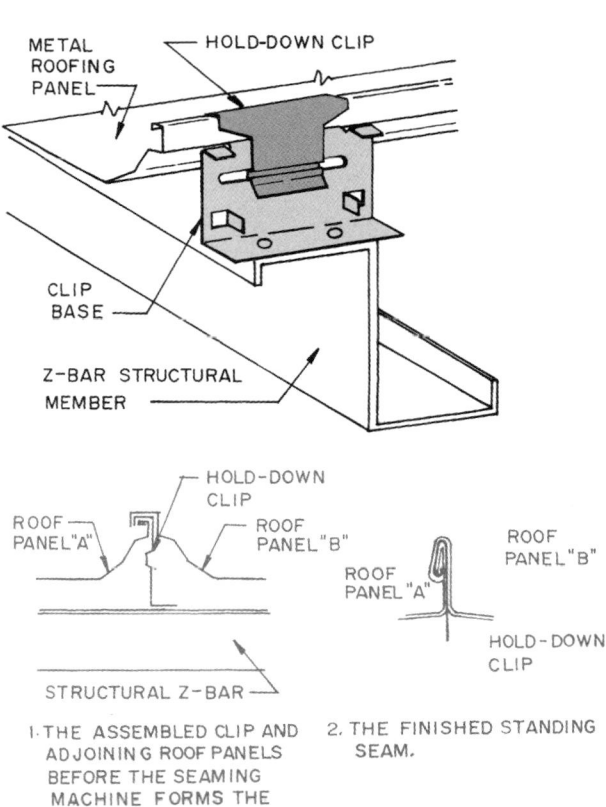

7-34 *This metal roofing panel is held to the structural frame with a metal clip that is rolled into a standing seam.*
Courtesy CECO Building Systems

7-33 *This metal roofing system is designed for use on low-slope roofs.*
Courtesy CECO Building Systems

7-35 *These lightweight perforated steel planks are anchored to the standing seam.*
Courtesy Unistrut Corporation

FLASHING

Flashing is a thin material impervious to penetration by water that is placed where needed to prevent water penetration and provide for water drainage. Typical places in roof construction include joints between the roof and a wall, at expansion joints, and around objects, such as pipes, that penetrate the roof membrane.

Materials used for flashing include copper, galvanized steel, lead, aluminum, stainless steel, bituminous sheet material, and plastics. In some cases a combination of these are used, such as galvanized steel covered with bitumen that prevents corrosion of the steel when in contact with mortar.

As the size of the roof increases the danger of tears from expansion and contraction of the membrane also increases. The designer locates expansion joints to provide allowances for this movement. Different designs are available and one for a built-up roof is shown in 7-36.

Steep-Slope Roof Assemblies

Steep-slope roof assembles have sufficient slope to permit water to drain and flow to the ground. Usually gutters and downspouts are used to control the flow. There are a variety of roofing materials available for use on steep slopes. The choice depends upon the architectural appearance and the slope. Minimum slopes are recommended for the various products available.

For example, if a sloped roof is below the minimum acceptable slope for various types of shingles it must be roofed with a built-up membrane or metal standing seam roof as described for low-slope roofs. Generally accepted slopes for steep-sloped roofs are in Table 7-13.

FIBERGLASS/ASPHALT SHINGLES

Asphalt shingles may have an organic or fiberglass base. Organic asphalt shingles have a wood-fiber base

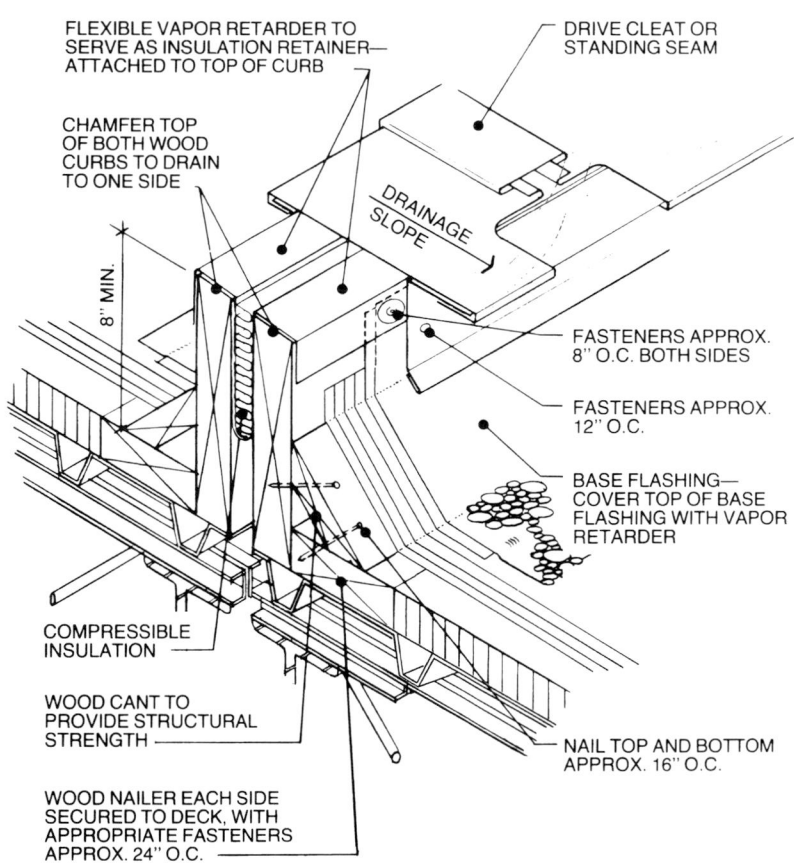

7-36 *The expansion joints allow for expansion and contraction in the roof.*
Courtesy National Roofing Contractors Association

Table 7-13 TYPICAL MINIMUM SLOPES FOR ROOFING ON STEEP-SLOPE ROOFS[a]

Material	Allowable Slope
Asphalt shingles	4:12 if one layer of asphalt-saturated felt underlayment is used. 2:12 if two layers of asphalt-saturated felt underlayment are used.
Clay & concrete tile	4:12 if interlocking tile are used with one layer of 4 lb. cap sheet underlayment. Less than 3:12 when noninterlocking tile are used with two layers of No. 40 asphalt-saturated felt underlayment.
Slate	4:12 with one layer of 30 lb. asphalt-saturated felt underlayment. 2:12 with double underlayment.
Wood shingles & shakes	3:12 for shingles & 4:12 for shakes with 30 lb. felt underlayment & interlayment with shakes or shingles.
Metal shingles & shakes	3:12 & 4:12 depending on the material & style. Requires 30 lb. asphalt-saturated felt underlayment.

[a]Consult local building codes & observe the manufacturer's recommendations on acceptable slope and installation details.

that is saturated with asphalt and coated with colored mineral granules. Fiberglass asphalt shingles have a fiberglass mat that has top and bottom layers of asphalt. It is also coated with mineral granules. The shingles are die cut from the mat and are available in a variety of tabs as shown earlier in 7-22.

A typical installation starts with the underlayment. The material is doubled at the eave as shown in 7-37. In cold climates this double layer should extend up the roof until it is 24 in. (610 mm) inside the exterior wall.

A metal drip edge is placed along the edge of the roof at the eave and rake. The underlayment goes below the drip edge on the rake and on top at the eave. The placement of shingles can be done in several ways. One way is to keep the cutouts centered over the tabs as shown in 7-38.

The nails or staples are placed just above the notch in each shingle. Nails should be 11 or 12 gauge hot-dipped galvanized roofing nails having shanks 7/8 to 1 in. (25 mm) long. Codes specify staple sizes.

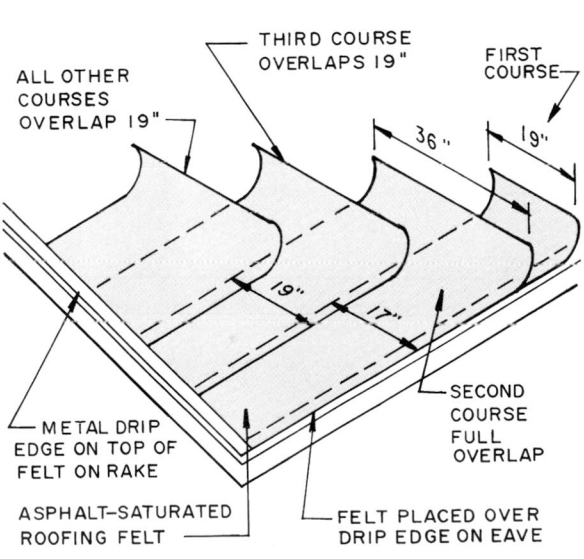

7-37 *This illustrates a double layer asphalt saturated roofing felt underlayment with drip edge flashing.*

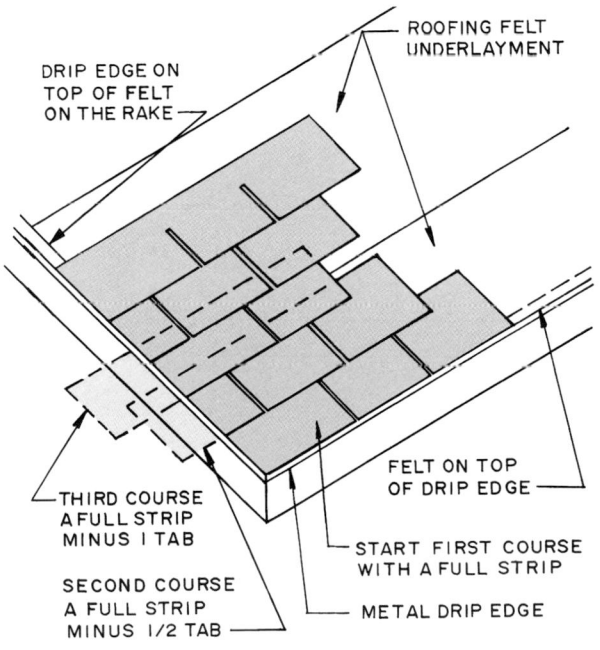

7-38 *Asphalt shingles are spaced so the slots between the tabs on overlapping shingles do not line up.*

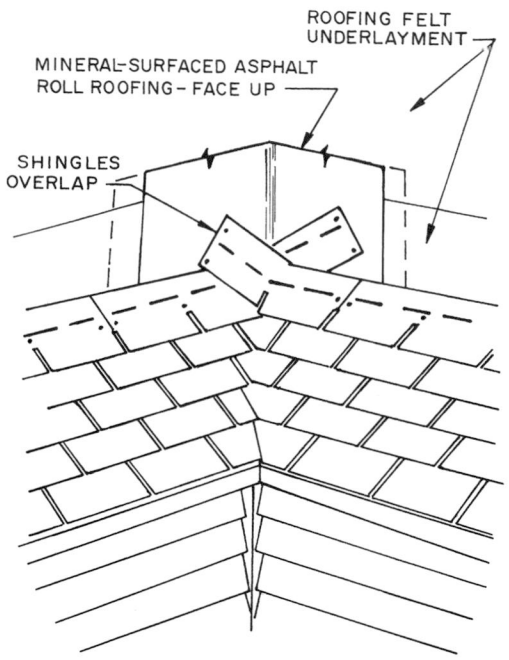

ROOFING FELT UNDERLAYMENT

MINERAL-SURFACED ASPHALT ROLL ROOFING – FACE UP

SHINGLES OVERLAP

7-39 *The shingles are woven over roll roofing valley flashing forming the valley.*

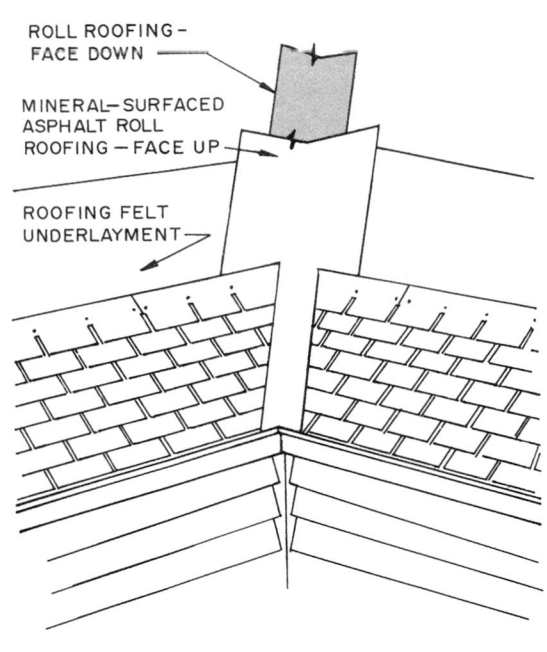

ROLL ROOFING – FACE DOWN

MINERAL-SURFACED ASPHALT ROLL ROOFING – FACE UP

ROOFING FELT UNDERLAYMENT

7-40 *This valley is flashed with roll roofing material.*

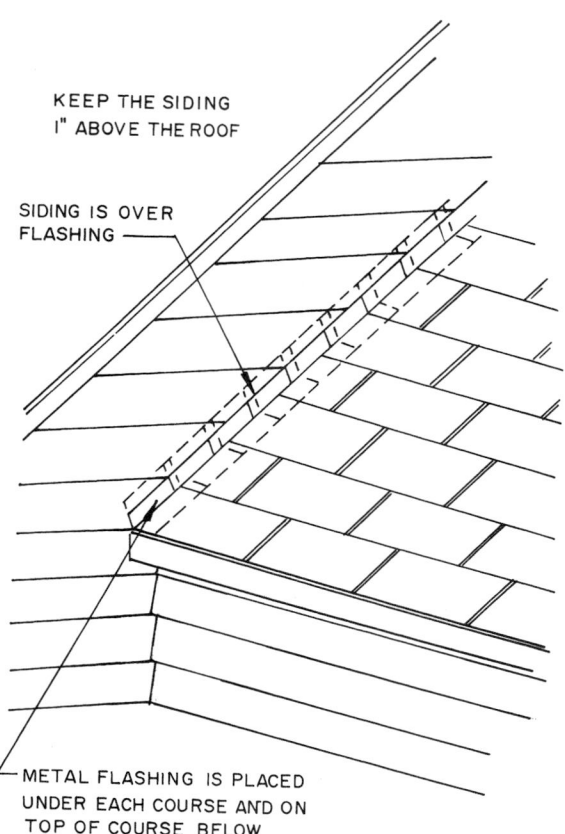

KEEP THE SIDING 1" ABOVE THE ROOF

SIDING IS OVER FLASHING

METAL FLASHING IS PLACED UNDER EACH COURSE AND ON TOP OF COURSE BELOW

7-41 *The siding is placed over the flashing and kept 1 in. above the surface of the shingles.*

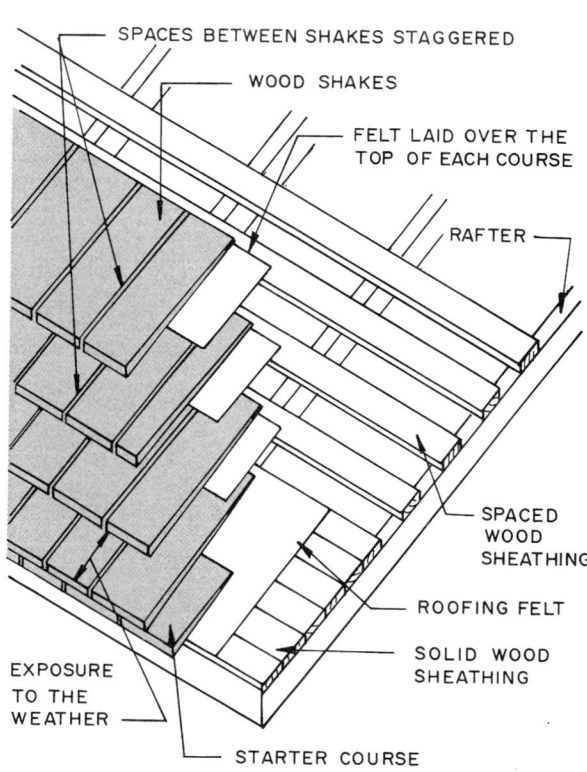

SPACES BETWEEN SHAKES STAGGERED

WOOD SHAKES

FELT LAID OVER THE TOP OF EACH COURSE

RAFTER

SPACED WOOD SHEATHING

ROOFING FELT

SOLID WOOD SHEATHING

STARTER COURSE

EXPOSURE TO THE WEATHER

7-42 *Typical installation detail for wood shakes.*

The valleys can be flashed as shown in 7-39 and 7-40. When a roof meets a wall step flashing is used. Each piece is nailed to the wall and rests on top of each shingle. The wood siding is laid over the flashing (see 7-41). When the roof meets a brick wall the flashing is set into the mortar joint. Finally, hips and ridges are capped with single tabs overlapping each other.

WOOD SHINGLES AND SHAKES

Wood shingles and shakes are made from cedar, redwood, southern pine, and other woods. Shakes are hand hewn and shingles are sawed. Building codes in many areas require they be treated with fire-retardant chemicals.

Usually wood shingles are applied over spaced 1 × 4 in. or 1 × 6 in. wood sheathing boards. The size used depends upon the amount of shingle to be exposed to the weather. Wood shakes are placed over 1 × 6 spaced wood sheathing wood. Unspaced sheathing is used at the eave and runs up until it is 24 in. (610 mm) inside the exterior wall. It is covered with roofing felt (see 7-42).

Shingles are doubled or tripled at the eave while shakes are usually doubled. A layer of felt is laid over the top row of shakes. The hips and ridges are capped with shingles. Valleys use metal flashing. Shakes can be applied on roofs with slopes below 4:12 if special construction is used.

SLATE SHINGLES

Slate shingles are available in a variety of colors depending upon the rock from which it was quarried. They are split to thickness and trimmed to size from larger slabs of slate. Then holes for fasteners are drilled or punched. Slate is a heavy material and the roof structure must be designed to carry the weight. It is fire resistant and has a long life but is more costly than most roofing materials. It produces a roof that enhances the architectural beauty of the building.

▼ Slate roofing is divided into two classifications: textural or random and standard commercial.

Textural slate—is delivered to the job in a range of thicknesses and sizes and must be sorted on the job by the slaters.

Standard commercial slate—is graded at the quarry by thickness, length, and width. Thickness is usually about ¼ in. (6 mm).

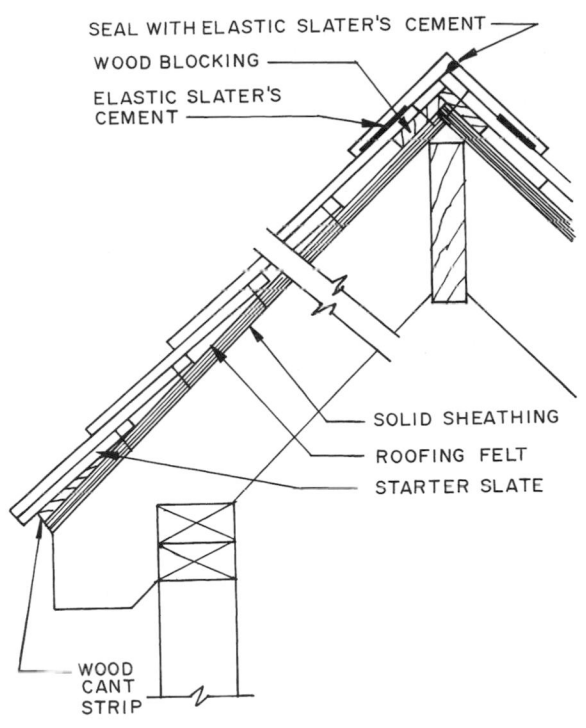

7-43 *Slate shingle installation is started with a wood cant strip at the eave. Wood blocking is used at the ridge.*

ASTM C406 Standard Specification for Slate Roofing addresses material characteristics, physical requirements, and sampling procedures for the selection of roofing slate. Roofing slate is in three grades, which predict the service life of the material. Grade S1 has an expected life of 75 years, S2 40 up to 75 years, and S3 20 to 40 years.

Slate installation details are shown in 7-43. They are applied over solid sheathing covered with asphalt-saturated roofing felt. Codes require double underlayment for roofs below certain specified slopes. Each course of slate covers the joints in the course below. Plastic cement is applied over the joint to be covered. Large head copper slaters nails are used. Each slate must have at least two fasteners.

CEMENT-FIBER SHINGLES

Cement-fiber shingles are made by combining portland cement and various fiberous material. One such product is a composition of portland cement, organic and inorganic fibers, perlite, and iron-oxide pigments for color.

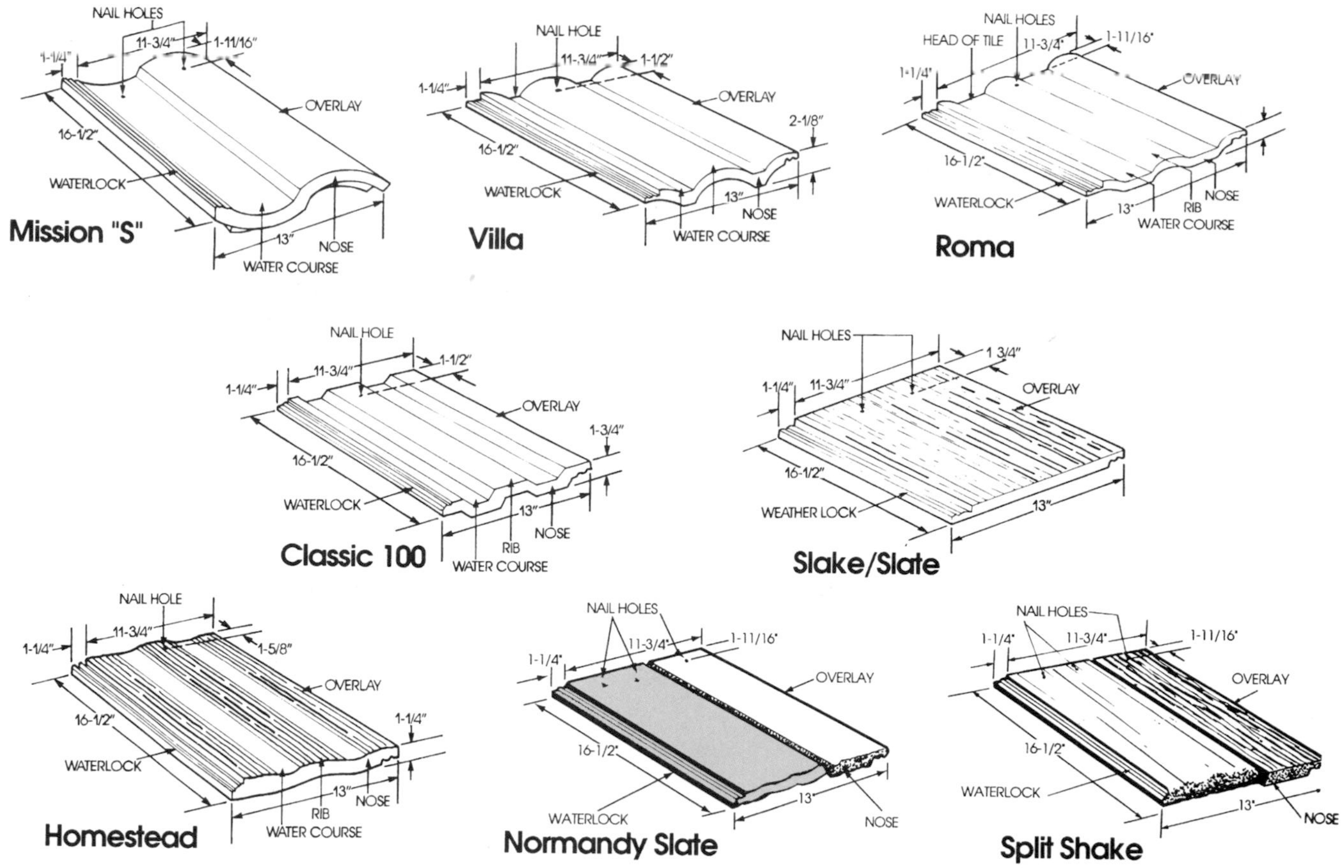

The following figure shows roofing tile shapes with labels: Mission "S", Villa, Roma, Classic 100, Slake/Slate, Homestead, Normandy Slate, Split Shake, each with dimensions including NAIL HOLES, OVERLAY, WATERLOCK, NOSE, WATER COURSE, RIB, and measurements such as 1-1/4", 11-3/4", 1-11/16", 16-1/2", 13", etc.

7-44 *The shapes of concrete roofing tile available from one manufacturer.*
Courtesy Monier

CLAY & CONCRETE ROOFING TILE

Tile is made from concrete or clay. It is durable, fire resistant, heavy, and expensive. Some forms of concrete tile are made using lightweight aggregates. It is important that the aggregates be carefully screened so the sizes are controlled. They are manufactured in the same basic types as clay tile. The colors can be varied by adding iron oxides (see 7-44).

Clay roofing tile may be unglazed or glazed. Unglazed tile range from an orange-yellow to a dark red depending upon the clay. Glazed tile are available in a wide range of solid colors. Interlocking type tiles are used on roofs with slopes of 3:12 or higher. Shingle tiles, often called Norman tiles, require of 5:12 slope. Tile roofing is heavy and fire resistant.

The specific installation details for clay and concrete tile vary by the tile manufacturer. Products should be installed following the instructions of the manufacturer.

In general, the tiles are installed much like slate shingles. Spaced sheathing can be used for roofs over 4:12 slope. Roofing felt is laid between each layer (see 7-45).

There are a number of circular profile tiles available. These tiles use a gable rake tile and a curved ridge and hip tiles. In addition to being nailed to the sheathing, they are set in a cement mortar. Valleys are flashed as described for other roofing products.

METAL ROOF SYSTEMS

Some commonly used seams on metal roofing on steep-slope roofs are the standing, capped, and batten. Some metal roofing is installed over decking while others have sufficient strength to span between properly spaced purlins.

7-45 *Procedures for installing concrete roofing tiles.*
Courtesy Monier

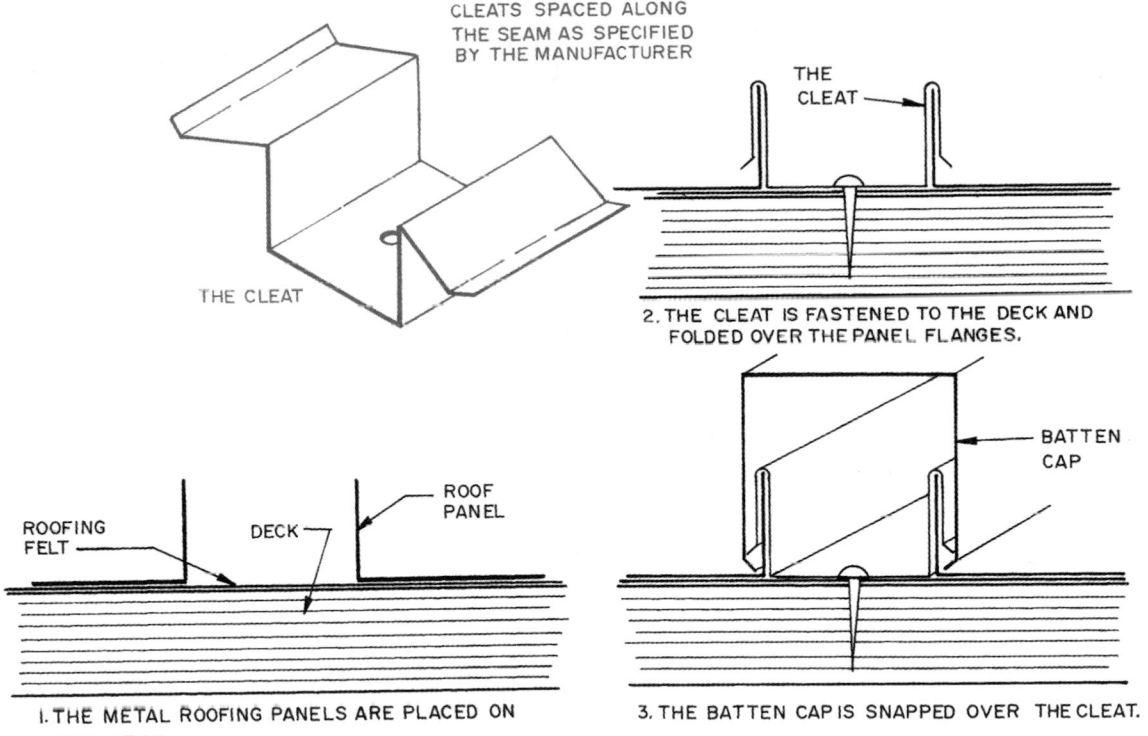

CLEATS SPACED ALONG
THE SEAM AS SPECIFIED
BY THE MANUFACTURER

THE CLEAT

2. THE CLEAT IS FASTENED TO THE DECK AND
FOLDED OVER THE PANEL FLANGES.

ROOFING FELT

DECK

ROOF PANEL

THE CLEAT

BATTEN CAP

I. THE METAL ROOFING PANELS ARE PLACED ON
THE DECK.

3. THE BATTEN CAP IS SNAPPED OVER THE CLEAT.

7-46 *A batten seam is constructed by snapping a metal batten cap over the cleats that secure the edge of the metal roofing to the deck.*

A typical batten seam is shown in 7-46. The roofing panels have an L-flange that butts a cleat. The cleat is fastened to the deck and folded over the flanges. A metal batten cap is snapped over this assembly.

A capped seam is shown in 7-47. A T-shaped cleat is fastened to the deck and the roofing flanges butt it. The cleat is folded over the flanges and a metal cap is snapped over the assembly.

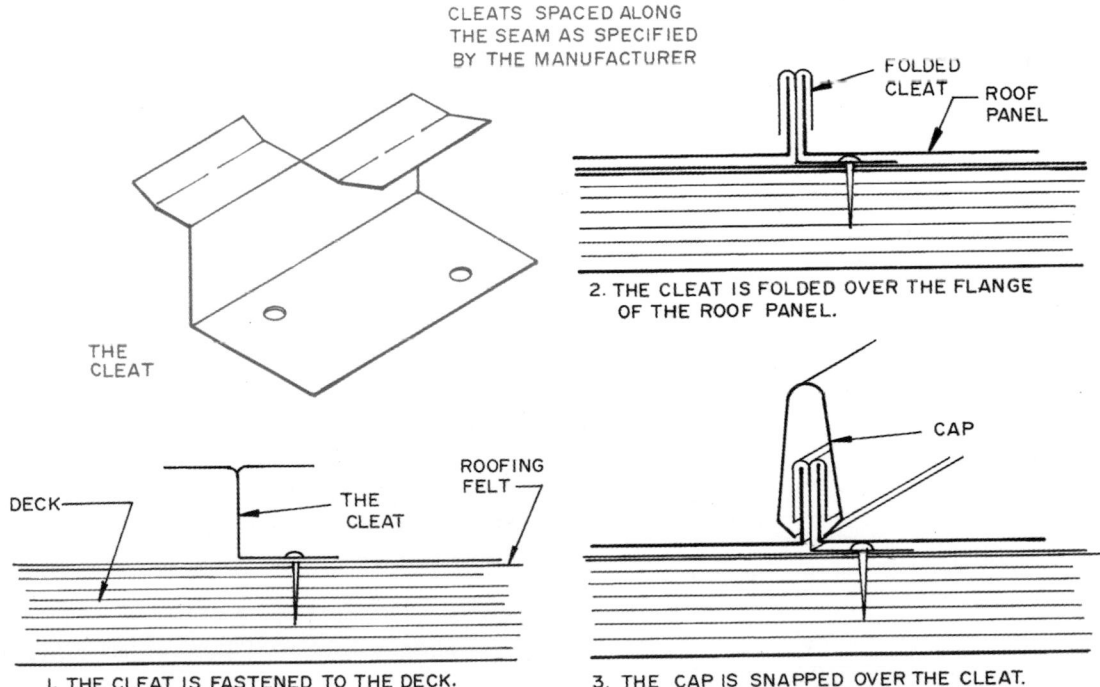

CLEATS SPACED ALONG
THE SEAM AS SPECIFIED
BY THE MANUFACTURER

FOLDED CLEAT

ROOF PANEL

THE CLEAT

2. THE CLEAT IS FOLDED OVER THE FLANGE
OF THE ROOF PANEL.

DECK

THE CLEAT

ROOFING FELT

CAP

I. THE CLEAT IS FASTENED TO THE DECK.

3. THE CAP IS SNAPPED OVER THE CLEAT.

7-47 *A capped seam provides a watertight seal over the union of metal roofing panels.*

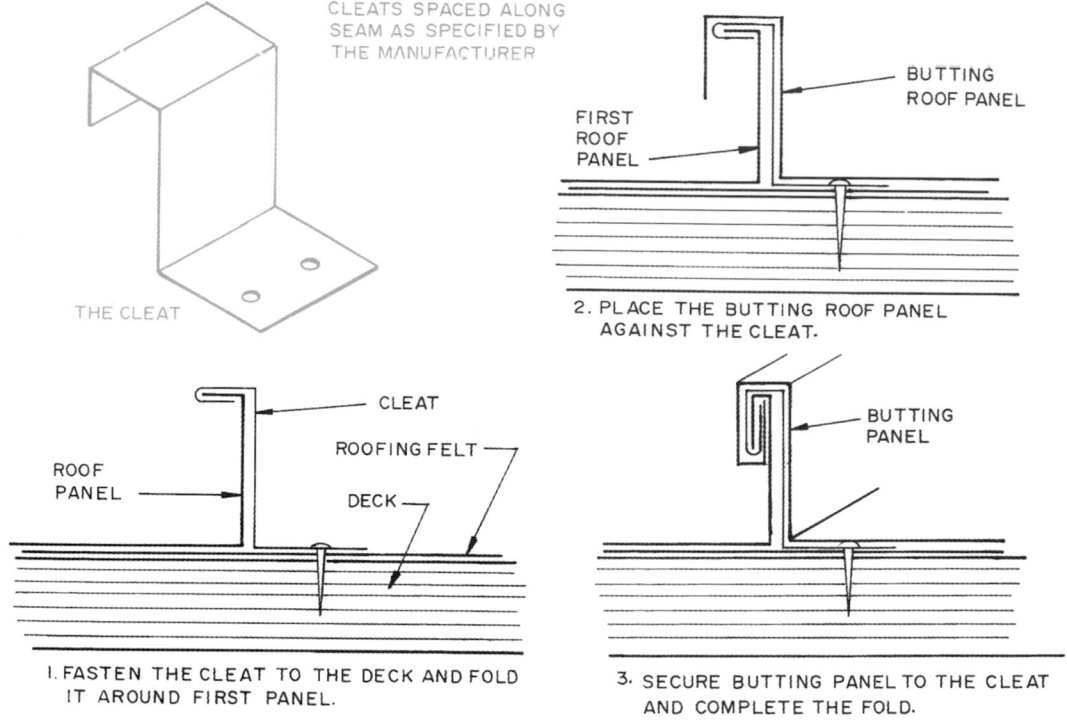

CLEATS SPACED ALONG
SEAM AS SPECIFIED BY
THE MANUFACTURER

THE CLEAT

FIRST
ROOF
PANEL

BUTTING
ROOF PANEL

2. PLACE THE BUTTING ROOF PANEL
AGAINST THE CLEAT.

CLEAT

ROOFING FELT

ROOF
PANEL

DECK

1. FASTEN THE CLEAT TO THE DECK AND FOLD
IT AROUND FIRST PANEL.

BUTTING
PANEL

3. SECURE BUTTING PANEL TO THE CLEAT
AND COMPLETE THE FOLD.

7-48 *This standing seam is formed by rolling over the edges of each metal roof panel and the cleat that secures them to the deck.*

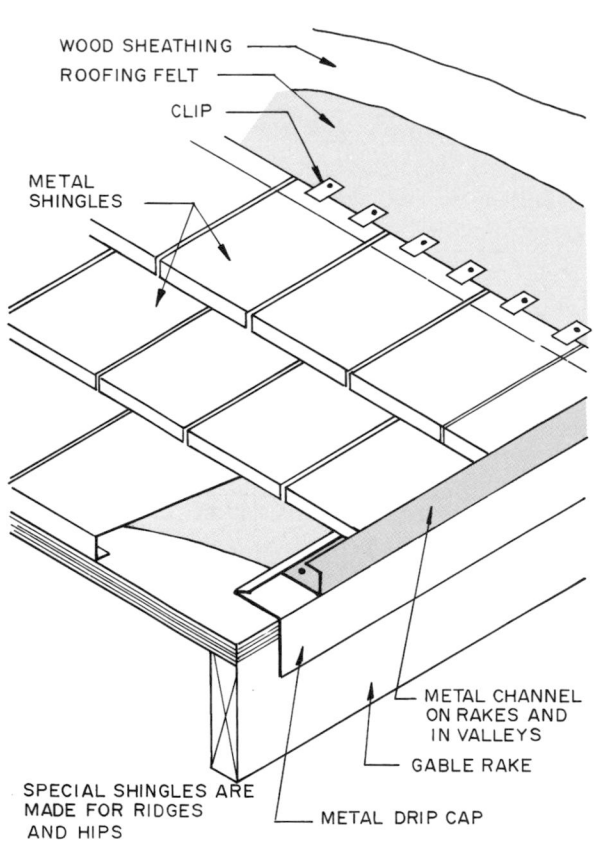

WOOD SHEATHING
ROOFING FELT
CLIP

METAL
SHINGLES

METAL CHANNEL
ON RAKES AND
IN VALLEYS
GABLE RAKE
SPECIAL SHINGLES ARE
MADE FOR RIDGES
AND HIPS
METAL DRIP CAP

7-49 *A typical construction detail showing the installation of metal shingles and shakes.*

One type of standing seam is shown in 7-48. A cleat is fastened to the deck. One roof panel flange butts it and the cleat is folded around it. The butting panel is folded around it. The butting panel is placed next to the cleat and folded over the assembly and the entire assembly is folded over again. In all cases it is important that the cleats and fasteners provided by the manufacturer of the roofing panels be used.

Metal shingles and shakes are available that have the surface embossed with a range of patterns and textures. A typical construction detail is in 7-49. This shows the shingles locked together and secured to the deck with metal clips. The fastening system used will vary with the manufacturer of the shingles. A close-up view of a finished metal shake roof is seen in 7-50. Notice the surface texture stamped into the metal.

7-50 *Detail of a finished metal shake roof.*
Courtesy Berridge Manufacturing Co.

DIVISION 8

DOORS & WINDOWS

CSI MasterFormat™

Courtesy H.H. Robertson

GLASS

Glass is an inorganic mixture that has been fused at a high temperature and cooled without crystallization. It has an unusual internal structure because mechanically it is rigid and has the characteristics of a solid—yet the atoms in glass are arranged in a random order similar to those in a liquid. It is technically supercooled liquid.

▼ There are six basic types of glass. They are classified by their ability to resist heat (see Table 8-1).

Soda-lime-silica glass—is the type commonly used for door and window glazing and bottles. About 90 percent of all glass produced is this type. Its general composition includes 74 percent silica, 15 percent soda, 10 percent lime, and 1 percent alumina. It is easy to form and cut. It has fair chemical resistance and does not resist high temperatures or rapid thermal changes. It is easily shattered into small, sharp pieces.

Lead-alkali-silica glass—is more expensive than soda-lime and has the same properties.

Fused-silica glass—is composed of about 99 percent silicon dioxide and is the most expensive. It has the highest resistance to heat handling temperatures ranging from 1650° to 2190°F (900° to 1200°C). It also has the highest corrosion resistance and affords excellent transmission of ultraviolet rays.

Ninety-six percent silica glass—is used where high thermal hardness and the ability to withstand thermal shock (going from hot to cold rapidly) is required.

Table 8-1 THERMAL PROPERTIES OF COMMONLY USED GLASS

Categories of Glass	Thermal Expansion	Heat Resistance
Soda-lime-silica	High	Low
Lead-alkali-silica	High	Low
Fused silica	Low	High
96% silica	Low	High
Borosilicate (Pyrex®)	Medium	Medium
Aluminosilicate	Medium	Medium

Borosilicate glass—is a thermally hard glass that has been used for many years. It is best recognized by its Corning trade name, Pyrex®.

Aluminosilicate glass—is more costly than borosilicate but can withstand high service temperatures and is similar in its ability to withstand thermal shock.

The following discussion is devoted to the uses of soda-lime-silica glass because it is the most widely used glass for construction applications.

Manufacture of Glass

Although several glassmaking processes have been used for many years, most glass produced today is produced by the **float process** (see 8-1). The first production of glass by this method was in 1959 by the English firm, Pilkington Brothers, Ltd. This process is now used worldwide.

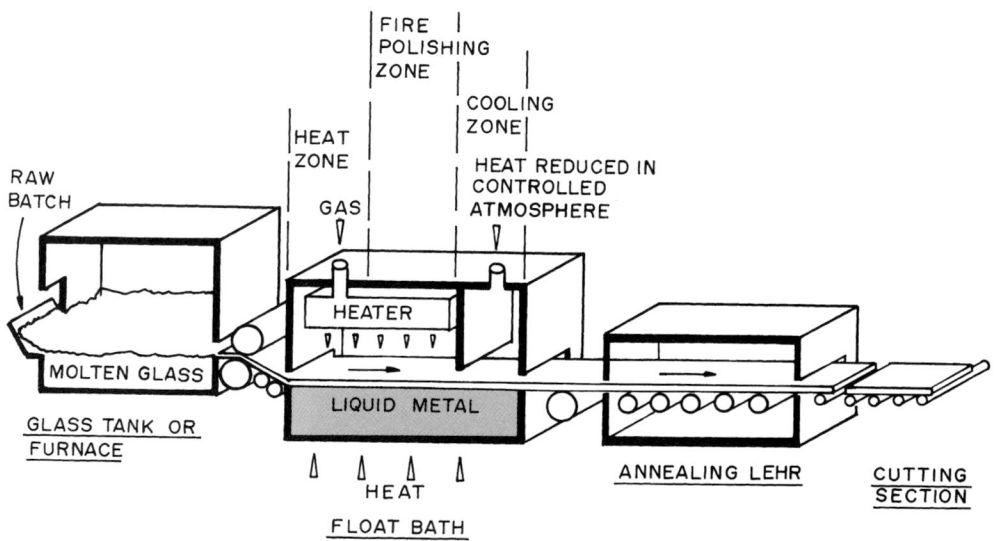

8-1 *The process for making float glass.*

Table 8-2 THICKNESSES OF COMMONLY USED FLOAT GLASS SHEETS[a]

Thickness in.	mm
3/32	2.5
1/8	3.0
5/32	4.0
3/16	5.0
1/4	6.0
3/8	10.0
1/2	12.0

[a]Sheet size depends on the type of glass and design pressures.

FLOAT GLASS

The **float process** involves producing the molten glass in a furnace from which it is conveyed to a float bath. Here the molten glass is floated across a bath of molten tin (8-1). The molten tin gives a very flat surface which supports the glass as it is polished by the application of heat from above. The heat melts out irregularities in the glass. The ribbon of glass moves on to a cooling zone where heat is reduced permitting the glass to solidify enough to be conveyed on to the annealing lehr. After annealing the glass is moved to a section where it is cut into lengths, inspected, and packed. The sheets of glass produced by this method have parallel surfaces, a smooth, clear finish, and have high optical clarity.

Float glass is a flat glass that is available as **regular float glass** or **heavy float glass.** Thicknesses range from 3/32 to 1/2 in. (2.5 to 12mm). Regular float glass is made in three types: **silvering,** which is selected high quality pieces for optical uses and mirrors; **mirror-glazing,** which is for general purpose mirrors; and **glazing,** which is for door and window glazing. Heavy float glass is available in glazing type. Float glass is used where clarity and visual transparency with a minimum of distortion are desired. Float glass is used for many products, such as reflective glass, mirrors, tinted glass, laminated glass, and insulating glass. Selected data are in Table 8-2.

SHEET GLASS

Sheet glass is a type of flat glass that is less expensive than float glass. It is made by older methods, which involve drawing a ribbon of molten glass along a series of rollers where its thickness is established. It is annealed, cooled, and cut to size. It has more distortion than float glass and is not as widely used as in the past. It is available in single strength, 3/32 in. (2.3mm) thick, double strength, 1/8 in. (3.1mm) thick, and heavy sheet, 3/16 in. (4.7mm) and 7/32 in. (5.6mm) thick. Picture glass is a thinner version, 3/64, 1/16, and 5/64 in. (1.2, 1.6, and 2.0 mm) thick, used for covering pictures, charts, and other such purposes where strength is not a factor. Sheet glass is available in three grades, AA (best), A (good), and B (general glazing) and as clear, tinted, reflective, tempered, or heat-treated products.

Properties of Soda-Lime-Silica Glass

The following discussion presents general properties of soda-lime-silica glass. These properties can be varied by altering the composition of the ingredients.

MECHANICAL PROPERTIES

A major consideration is the ability of glass to withstand breakage. Glasses are brittle yet remain elastic up to their ultimate tensile strength. This means that glass can be bent up to the breaking point and if released will return to its original position. Glass breaks by bending or stretching, therefore, **tensile strength** is a major mechanical property to be considered.

Glass does not have a clearly defined tensile strength because the actual strength depends upon the condition of the surface of the glass. A small scratch or nick is sufficient to cause tensile stresses to concentrate at that weaker point. Therefore the actual tensile strength is less than the theoretical tensile strength. The mechanical strength can be increased by chemical and heat treatments.

THERMAL PROPERTIES

Heat can be transferred by conduction, convection, and radiation. The thermal conductivity of glass is high. However, it is lower than most metals.

The heat transfer of the inner surface of glazing can be reduced a great deal by adding low emmittance metallic films (referred to as Low-E) to the glass. This also reduces the U-value for glazing having airspaces. A typical double glazed opening with suspended Low-E film has a U-value of 0.31 to 0.32.

Other glass products, such as tinted glass, reflective glass, and insulating glass, reduce heat gain and loss.

The thermal expansion of glass must be considered as units containing glass are designed. The coefficient of expansion of soda-lime glass is about 4.5×10^{-6} while the coefficient for aluminum is about 13×10^{-6}. An aluminum window frame contracting under low temperatures imposes stress on the edge of the glass unless suf-

Table 8-3 THERMAL PROPERTIES OF FLAT GLASS

Glass	Thickness (in.)	Winter Nightime U-value/R-value	Summer Daytime U-value/R-value
Clear single	¼	1.3/0.78	1.04/0.96
Clear double	⅛	0.49/2.04	0.52/1.92
Clear double	¼	0.49/2.04	0.56/1.79
Clear double with Low-E film	¼	0.31/3.32	0.32/3.03
Light brown single	¼	1.13/0.88	1.10/0.91
Dark brown single	¼	0.89/0.88	0.89/1.88
Double glass light brown/clear	¼	0.31/2.04	0.33/1.75
Double glass dark brown/clear	¼	0.41/2.04	0.47/1.72
Light green single	¼	1.13/0.88	1.10/0.91
Dark green single	¼	0.95/1.14	0.98/1.12
Double light green/clear	¼	0.50/2.04	0.59/1.75
Double dark green/clear	¼	0.42/2.50	0.50/2.13
Clear insulating glass with suspended Low-E film	¼	0.23/4.30	0.37/2.70

ficient clearance is allowed for the difference in thermal movement. Typical thermal properties for selected flat glass products are in Table 8-3.

CHEMICAL PROPERTIES

Glass used in typical applications in building construction is a very durable material and more resistant to corrosion than most other materials. Since it is not porous it will not absorb moisture or chemical elements in the ground or atmosphere. There are few exceptions that may occur in isolated industrial applications. Hot concentrated alkali solutions and superheated water can cause soda-lime-silica glass to dissolve. Hydrofluoric acids will cause corrosion.

ELECTRICAL PROPERTIES

Glass is a good electrical insulator. It is widely used for applications where this property plus its other properties make it useful. Light fixtures often use glass and the envelope on a light bulb in glass because it is strong, heat resistant, and an electrical insulator.

OPTICAL PROPERTIES

When light falls on glass some is absorbed and some is reflected. Clear sheet glass permits the passage of about 86 to 89 percent of visible light. Double glazed lights permit about 80 to 82 percent light passage. Typical daylight transmittance figures are in Table 8-4.

As glass is treated to reflect or absorb light the percent of light transmitted is greatly affected. Tinted glass will frequently transmit less than 50 percent of the daylight striking it. Those with reflective coatings will transmit 6 to 50 percent of the visible light.

The optical quality is also influenced by the lack of distortion. If the front and back surfaces of a sheet of glass are not parallel the image will, from some angles, appear wavy or distorted. With the increased use of float, glass distortion has been reduced and optical properties improved.

Heat-Treating Glass

The properties of glass can be improved by various heat-treating methods. These include annealing, tempering, and heat strengthening.

ANNEALING

Annealing occurs in the normal production of glass. After the glass ribbon has been formed it is passed through an annealing lehr where the temperatures are carefully controlled. The temperature is raised high enough to relieve strains developed during the forming in the float bath. Then the temperature is slowly lowered allowing all parts of the glass to cool uniformly.

TEMPERING

Tempering is used to increase the strength of glass. Tempered glass is used where additional strength beyond that of standard annealed glass is required.

Tempering involves raising the temperature of the glass near the softening point and then blowing jets of cold air on both sides suddenly chilling it. The surfaces harden and shrink and the interior is still fluid. As the interior cools and shrinks the exterior remains unchanged in size. This causes the surfaces and edges to be in compression and the interior in tension. The opposing compression and tension forces balance each other resulting in a stronger glass. Tempered glass is 3 to 5 times more resistant to damage than annealed glass. When the thin tempered skin on the glass is broken the entire sheet will disintegrate into small pebblelike particles instead of sharp slivers. Tempered glass is made in accordance with Federal Specification DD-G-1403B, which requires a minimum compression of 10,000 pounds per square inch (surface) or 9,7000 pounds per square inch (edge). Fully tempered glass can meet safety standards required by ANSI Z97.1-l975 and Federal Standard CPSC 16 CFR 1201, *Safety Standard for Architectural Glazing Materials.*

HEAT STRENGTHENING

Heat strengthening glass is heated and cooled much like that described for tempered glass. Heat strengthened glass has a compression range of 3,000 to 10,000 pounds per square inch (surface) or 5,500 to 9,700 pounds per square inch (edge). It is about twice as strong as annealed glass.

Chemical Strengthening

Glass can also be strengthened by chemical treatment involving immersing the glass in a molten salt bath causing larger potassium ions in the salt to replace the smaller sodium ions already in the glass. This crowds the surface causing compressive stresses.

Finishes

The finish on glass depends upon its end use. While most glass produced is clear and transparent various finishes make it useful for other applications.

FIRE FINISH

As glass leaves the annealing lehr it has a smooth, transparent finish. This is referred to as a **fire finish.**

ETCHING

Etching produces a surface that can range from almost opaque to rather smooth but translucent. The degree of opacity depends upon the length of time the glass is exposed to hydrofluoric acid or some other etching compound. The glass may be dipped or sprayed with the etching fluid. The finished surface, while opaque, is rather smooth to the touch.

Table 8-4 DAYLIGHT TRANSMITTANCE FOR SELECTED GLASS UNITS

Glass	Percent Daylight Transmittance
Clear single	86–89
Clear double	80–82
Clear double with	
Low-E film	72
Clear triple	72–74
Clear insulating with	
Low-E film	37–69
Light brown single	52
Light green single	75
Light gray single	41

SANDBLASTING

Glass is sandblasted by bombarding the surface with coarse-grained sand particles blown by compressed air. This produces a translucent finish that is generally a bit rougher than etched glass. Etching and sandblasting abrade the surface which greatly lowers the strength of the sheet.

PATTERNED FINISH

Patterned glass has texture or pattern rolled into the surface as the glass is drawn through a furnace. It may be colorless or tinted and is available in a wide variety of patterns. It may be relatively transparent from one side and translucent from the other.

SILVERING

Mirrors are produced by spraying silver nitrate and tin chloride on the surface of the glass as it moves on a conveyor. For additional protection this layer can be covered with shellac, varnish, paint, or by electroplating a layer of copper over it.

Float glass mirrors are the highest quality and are available in **silvering** quality, mirror-glazing quality, and glazing quality. Silvering quality is the very best. Sheet glass can be used for low-quality mirrors.

CERAMIC FRIT

Glass surfaces are sometimes coated with a ceramic frit that is then fired at a high temperature. **Ceramic frit** is composed of small powdered particles produced by quenching a molten glossy material. This produces a colored layer—translucent or opaque—on the surface that is highly weather resistant. A major use is a finished exterior surface on all spandrels.

TINTING

Another way to infuse color in glass is by tinting. Tinting is done by adding color producing ingredients to the glass both at the beginning of the glass production process. Therefore the color is not a surface finish but is infused in the glass.

This type of glass is used for the reduction of glare. The blue-green glass gives a moderate reduction in glare and brightness. The bronze tint is used primarily for heat reduction. The grey tint is produced by giving a wide range of light transmittance. Translucent glass diffuses most of the direct sunlight yet permits transmission of considerable diffused light.

Flat-Glass Products

Following are examples of the more commonly used flat glass products. Specific specifications will vary with the manufacturer.

LOW-E GLASS

Low-E (low emissivity) glass reduces energy costs by creating a heat barrier that helps keep heat outside in the summer and inside in the winter. It is used on residential and commercial buildings, greenhouses, and buildings using passive solar design. It is suitable for use in all climates and is available in double and triple insulating glass units. If the outer pane is tinted the energy efficiency is increased.

Low-E glass insulating units have a thin metallic coating on the inside glass surface so the coating is fully protected. This coating selectively reflects the ultraviolet and infrared wavelengths of the energy spectrum. Low-E coatings permit the use of natural daylighting techniques because they are virtually invisible and have a high visible light transmission (see 8-2).

LAMINATED GLASS

Laminated glass is used in areas where the glazing is subject to possible breakage or where security of glazed openings is required. Typical applications from a safety standpoint include residential and commercial door glazing or windows in a gymnasium. Security applications occur in prisons and banks.

Laminated glass is made by bonding layers of float glass with interlayers or plasticized polyvinyl butyral (PVB) resin or polycarbonate (PC) resin. The glass is chemically strengthened by immersing it in a molten salt bath. Laminated glass is made to meet ANSI Z97.1-1975 and Consumer Products Safety Commission Regulation 16 CFR1201 Categories 1 and 11.

There are many variations and thicknesses available from various manufacturers. Laminated glass is available with ultraviolet filtering laminates and wire glass laminates. Laminated mirrors are also available. The thicknesses of the various products vary with the design with thickness from ¼ to 3 in. (6 to 75mm) common. Typical designs are shown in 8-3.

Another type of laminated glass is **acoustical glass.** It is a construction of sound absorbing plastic between two or more pieces of glass. The sound reduction is achieved because this soft interlayer allows the glass unit to bend in response to pressure from the sound waves. It is also available in insulated constructions thus reducing heat loss and gain as well as reducing sound transmission. Common uses are in office partitions or radio, TV, and recording studios.

INSULATING GLASS

A wide range of insulating glass systems are designed for use on commercial and residential buildings. Examples of these are shown in 8-4.

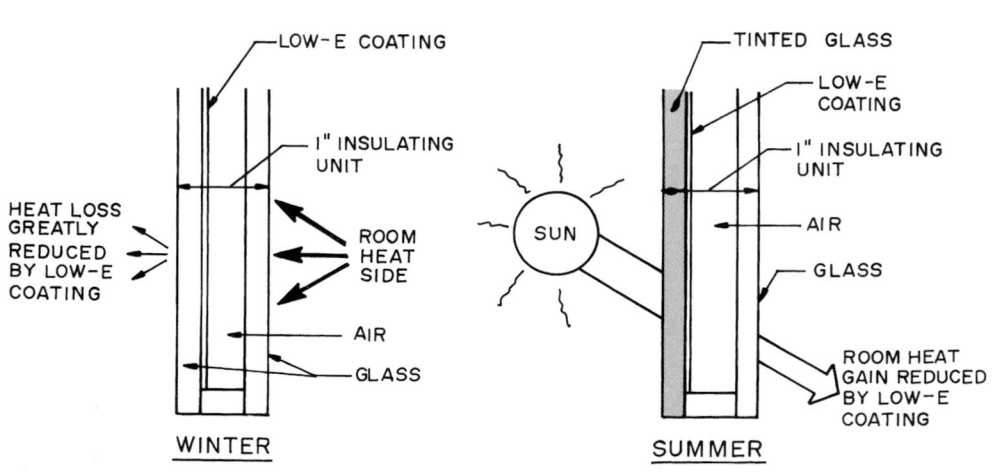

8-2 *Low-E glass helps keep heat in the building in the winter and out of the building in the summer.*

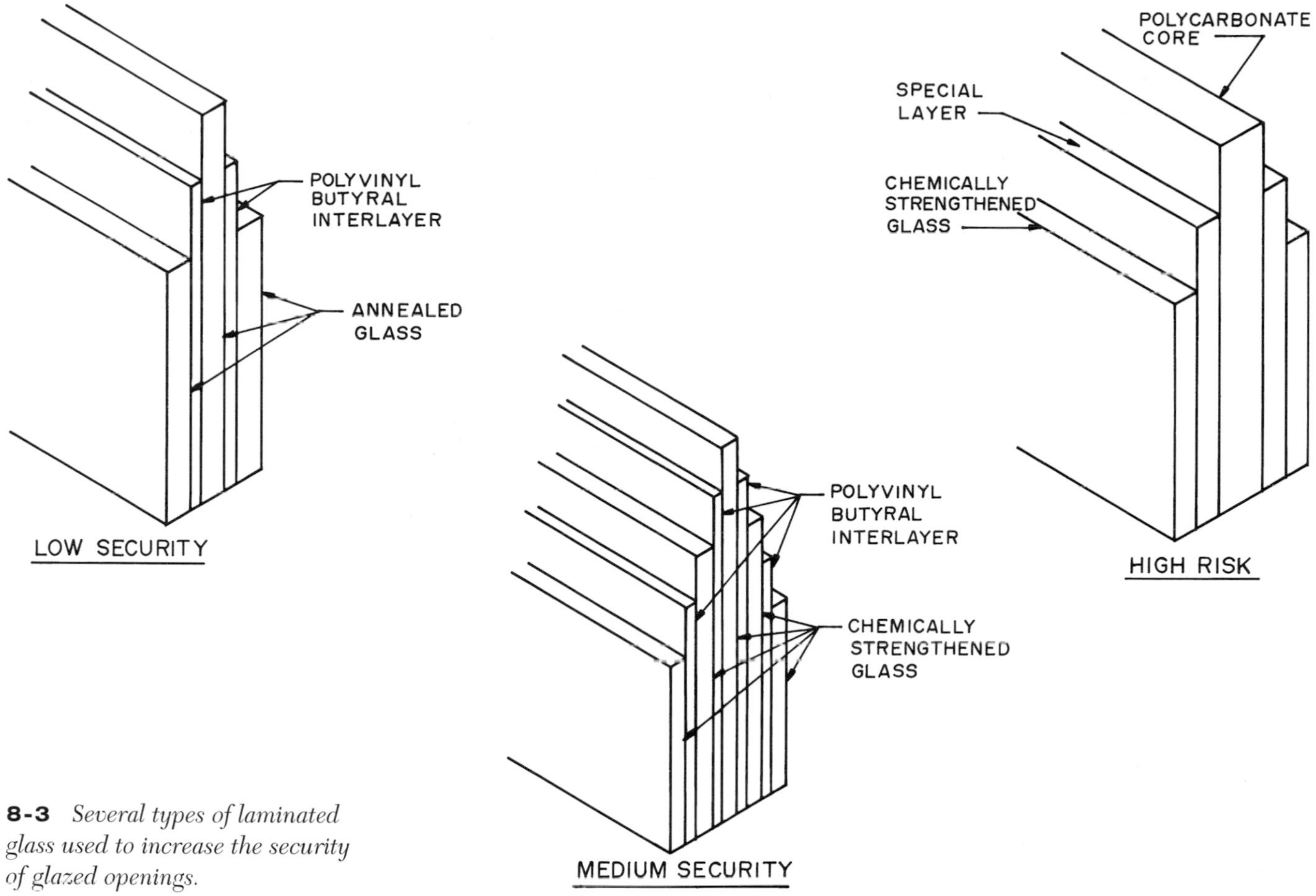

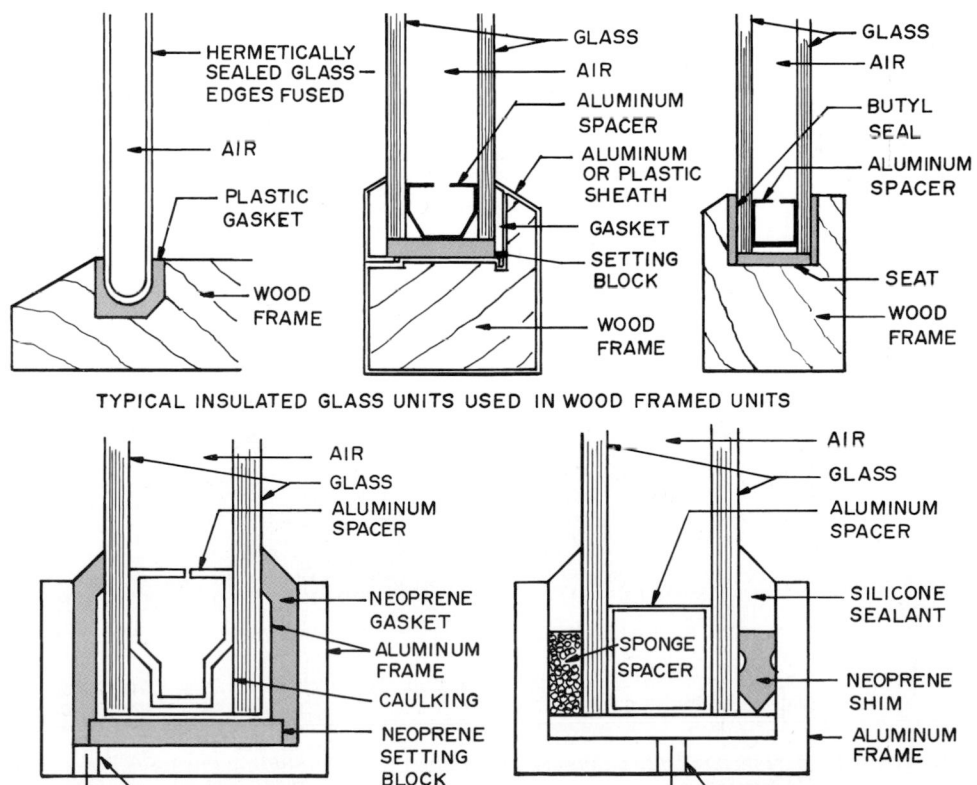

POLYVINYL
BUTYRAL
INTERLAYER

ANNEALED
GLASS

LOW SECURITY

POLYCARBONATE
CORE

SPECIAL
LAYER

CHEMICALLY
STRENGTHENED
GLASS

HIGH RISK

POLYVINYL
BUTYRAL
INTERLAYER

CHEMICALLY
STRENGTHENED
GLASS

MEDIUM SECURITY

8-3 *Several types of laminated glass used to increase the security of glazed openings.*

HERMETICALLY
SEALED GLASS –
EDGES FUSED

AIR

PLASTIC
GASKET

WOOD
FRAME

GLASS
AIR
ALUMINUM
SPACER
ALUMINUM
OR PLASTIC
SHEATH
GASKET
SETTING
BLOCK
WOOD
FRAME

GLASS
AIR
BUTYL
SEAL
ALUMINUM
SPACER
SEAT
WOOD
FRAME

TYPICAL INSULATED GLASS UNITS USED IN WOOD FRAMED UNITS

AIR
GLASS
ALUMINUM
SPACER
NEOPRENE
GASKET
ALUMINUM
FRAME
CAULKING
NEOPRENE
SETTING
BLOCK
WEEP HOLE

AIR
GLASS
ALUMINUM
SPACER
SILICONE
SEALANT
SPONGE
SPACER
NEOPRENE
SHIM
ALUMINUM
FRAME
WEEP HOLE

8-4 *Some of the types of insulating glass units available.*

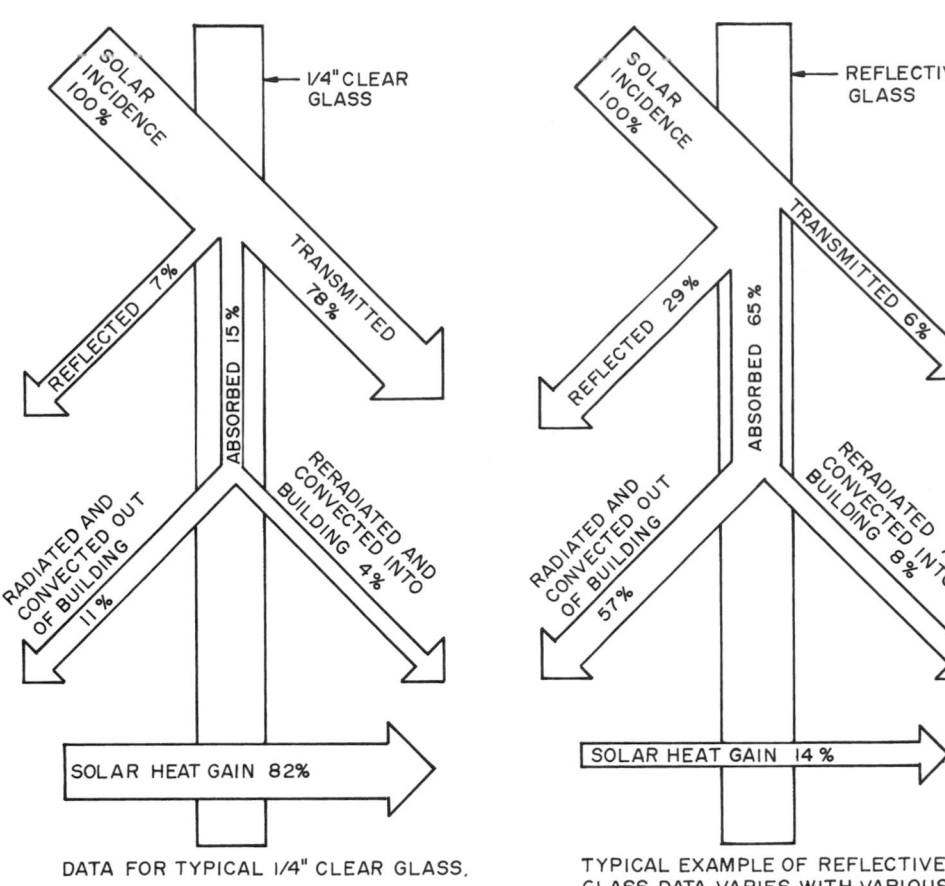

SOLAR INCIDENCE 100%

1/4" CLEAR GLASS

REFLECTED 7%

ABSORBED 15%

TRANSMITTED 78%

RADIATED AND CONVECTED OUT OF BUILDING 11%

RERADIATED AND CONVECTED INTO BUILDING 4%

SOLAR HEAT GAIN 82%

DATA FOR TYPICAL 1/4" CLEAR GLASS.

SOLAR INCIDENCE 100%

REFLECTIVE GLASS

REFLECTED 29%

ABSORBED 65%

TRANSMITTED 6%

RADIATED AND CONVECTED OUT OF BUILDING 57%

RERADIATED AND CONVECTED INTO BUILDING 8%

SOLAR HEAT GAIN 14%

TYPICAL EXAMPLE OF REFLECTIVE GLASS. DATA VARIES WITH VARIOUS MANUFACTURED PRODUCTS.

8-5 *Reflective glass reduces the amount of solar energy transmitted through a glazed opening.*

8-6 *High-performance reflective glass lets daylight enter the building but reflects a large percentage of the sun's rays.*
Courtesy PPG Industries, Inc.

Insulating glass is a manufactured glazing unit composed of two layers of glass with an airtight, dehydrated air space between them. They are made with hermatically sealed glazing with edges fused together, or set in gaskets. Insulating glass lowers heating and cooling costs by reducing air-to-air heat transfer.

REFLECTIVE GLASS

To understand how reflective glass reduces the amount of solar energy transmitted through a glass opening consider the following explanation.

When solar energy strikes a glass surface it can be (1) transmitted through the glass, (2) reflected back, or (3) absorbed in the glass. Of that portion that is absorbed in the glass, part of the energy will be reradiated and convected inward and part of it will be reradiated and convected outward. The solar heat gain consists of that portion of the absorbed energy that is reradiated and convected to the inside. The solar heat rejected is that portion of the original solar incidence that is returned to the exterior. This includes the energy reflected back from the glass surface plus any absorbed energy that is reradiated and convected outward.

Another source of energy transfer through a glass window is due to convection. The total heat gain of a window is the sum of solar heat gain and convective heat gain (see 8-5). It is these factors designers consider as they choose the type of glazing opening. Reflective glasses have one surface covered with thin, transparent layers of metallic film. Several metals and mineral oxides are used and produce a variety of colors and heat reflection ratings. On single sheet glazing the metallic film is on the surface facing the inside of the building. On insulating units it is on the outside face of the glass facing the inside of the building.

Reflective glass is available in a variety of colors, such as silver, green, blue, bronze, copper, and gold. Variations of these are available from various manufacturers (8-6).

Visible light transmittance available varies with the coating but range in general from 8 to 45 percent. Coatings are available on clear or tinted, heat-absorbing glass. Most reflective glasses are ¼ in. (6mm) thick but thicker sheets are available.

WIRED GLASS

Wired glass is a type of safety glass that has a mesh of small diameter wires rolled into a sheet of molten glass. When wired glass breaks the wires hold the glass shards together. It also maintains its integrity as a fire

8-7 *Bent, laminated glass provides a strong transparent exterior partition.*
Courtesy Laminated Glass Corporation

barrier and is used in fire doors and window glazing. Tempered and laminated glasses have replaced wire glass in many applications.

SPANDREL GLASS

Ceramic coated spandrel glass is an effective way to clad exterior wall areas. The spandrel panel eliminates the problem of corrosion greatly reducing deterioration of the building. Glass spandrel panels use ¼ in. (6mm) heat-strengthened glass with an opaque ceramic frit fired on the surface. Some fire the frit on the exposed exterior surface while other companies fire it on the protected interior surface. The spandrel is insulated usually with 1 to 2 in. (25 to 50mm) fiberglass or other acceptable insulation with a ½ in. (12mm) air space and an aluminum foil vapor barrier facing the interior side of the panel. Panel sizes vary with the design.

BENT GLASS

Bent glass provides the designer with the opportunity to produce a building with a unique and dramatic look (see 8-7).

Bent glass is available in clear float, tinted float, obscure, wire, pyrex, patterned, and other types of glass. Some types of reflective glass can be bent. Laminated safety glass and double glazed thermal insulating units can be fabricated. Thickness and size specifications must be developed with the help of the manufacturer.

ARCHITECTURAL BEVELED GLASS

Architectural beveled glass provides a unique transition from outside to inside and outstanding architectural detail. The glass is made in a wide range of standard sizes and designs as well as custom designs. The pieces of beveled glass are assembled into panels with lead caming. Energy efficient triple glazed panels are available.

GLASS BLOCKS

Glass blocks are made as solid and cavity units and are laid in a mortar bed similar to masonry units. Various designs permit light to be diffused, reduced, or reflected. The translucence or transparency is varied by design (see 8-8).

PLASTIC GLAZING MATERIALS

Several plastic sheet products find use as glazing materials replacing glass. Two commonly used plastics are acrylics and polycarbonates. These are discussed in Division 6.

8-8 *Glass blocks are available in a wide range of sizes and styles.*

Courtesy Pittsburgh Corning Corporation

Acrylic sheets are available in clear form having a high transmittance value. Colored panels have a greatly reduced light transmission value. They withstand considerably more impact than window glass and even more than normal tempered glass of the same thickness.

Polycarbonate sheets have greater resistance to impact than acrylic panels. They do have a short life but provide great safety in an application where impact is likely.

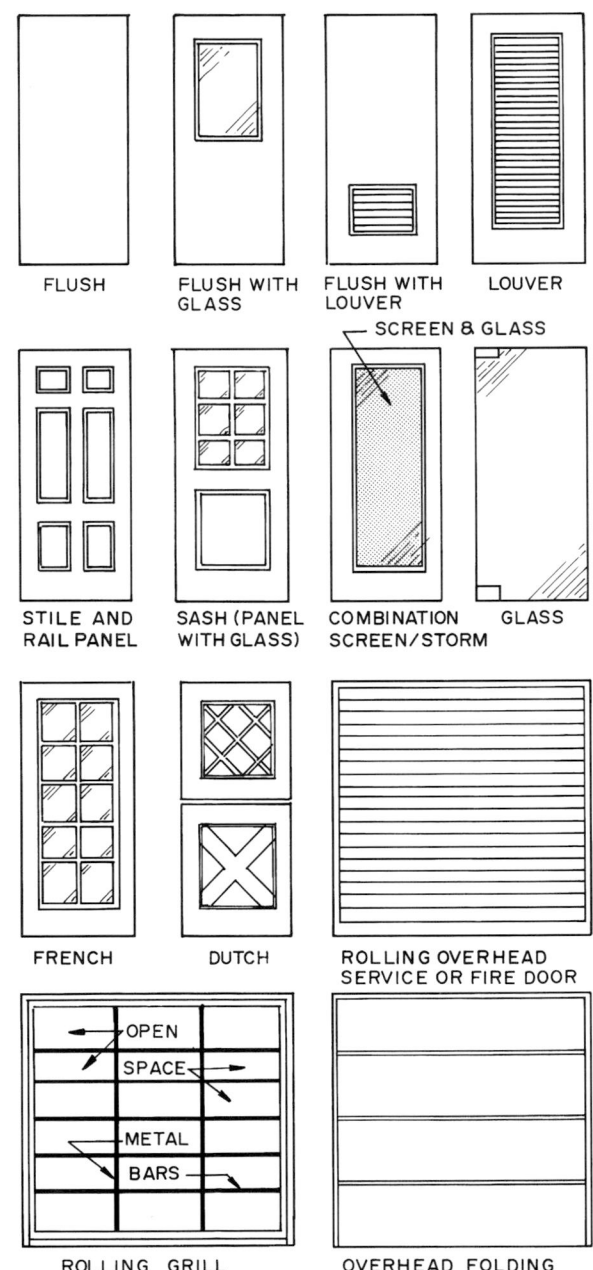

8-9 *Some of the doors available for residential and commercial applications.*

Plastic glazing tends to scratch easily and great care must be exercised when cleaning it. It also has a high coefficient of thermal expansion, which must be considered for any application.

DOORS

Doors are used to provide security, privacy, and fire protection to access openings in interior and exterior walls. The choice of a door depends upon its appearance and the amount of traffic passing through the opening. In residential work interior doors provide privacy and some degree of security while exterior doors provide protection from the weather, security, and often are an important part of the exterior design. In commercial buildings doors must provide access to permit ingress and egress as required by building codes. They are subject to considerable use and must be strong and have hardware to withstand constant use. Other doors provide access for large equipment as trucks and aircraft. They are usually mechanically opened and closed because they are large and heavy. Doors are also used to divide large interior spaces into smaller rooms.

Doors are available manufactured from wood, hardboard, plastic, metal, and glass. The typical types are shown in 8-9 and 8-10. These types are used in residential and commercial construction.

Hollow-Core Metal Doors

Hollow-core metal doors are available in steel and aluminum. They are used on interior and exterior openings and are available in a wide range of designs (8-11). The exterior surface may be flush or paneled, have windows or louvers, and be finished with a baked enamel finish on a smooth or textured surface or be covered with plastic veneers that may be wood grain or a variety of colors and patterns.

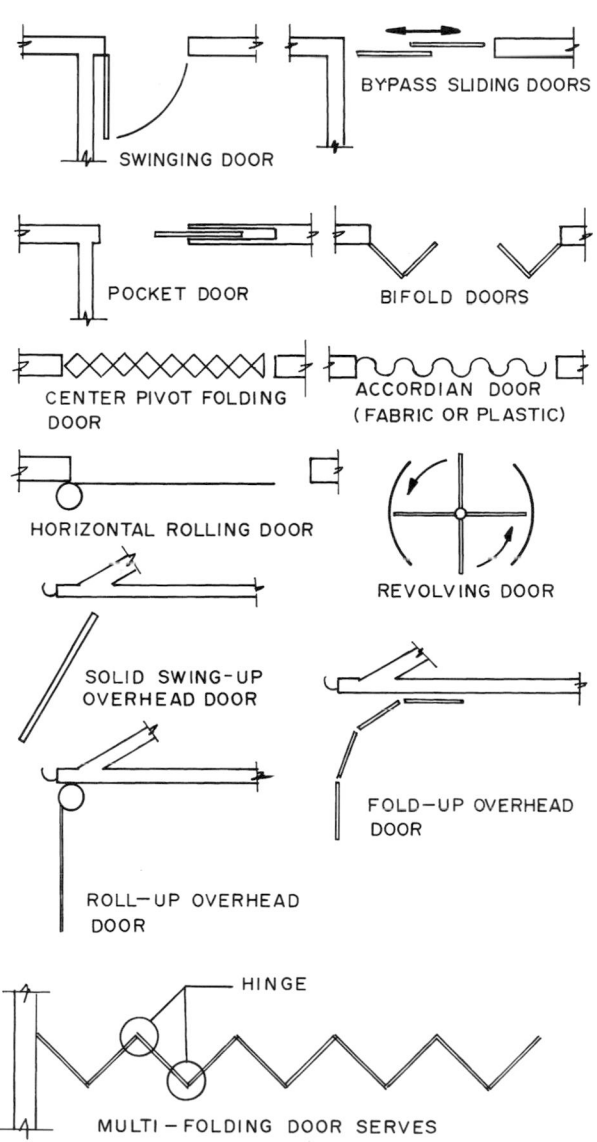

8-10 *Common types of door operations.*

8-11 *A typical hollow core metal door.*

Table 8-5 GRADES & MODELS OF STANDARD STEEL DOORS

Grade	Model
Grade I—Standard duty, 1⅜" and 1¾" thick	Model 1—Full flush design, hollow metal and composite
	Model 2—Seamless design, hollow metal and composite
Grade II—Heavy duty, 1¾" thick	Model 1—Full flush design, hollow metal and composite
	Model 2—Seamless design, hollow metal and composite
Grade III—Extra heavy duty, 1¾" thick	Model 1 and 1A[a]—Full flush design, hollow metal and composite
	Model 2 and 2A[a]—Full flush design, hollow metal and composite
	Model 3—Stile and rail, flush panel

[a]1A and 2A are made with a heavier gauge metal than models 1 and 2.

Courtesy Steel Door Institute

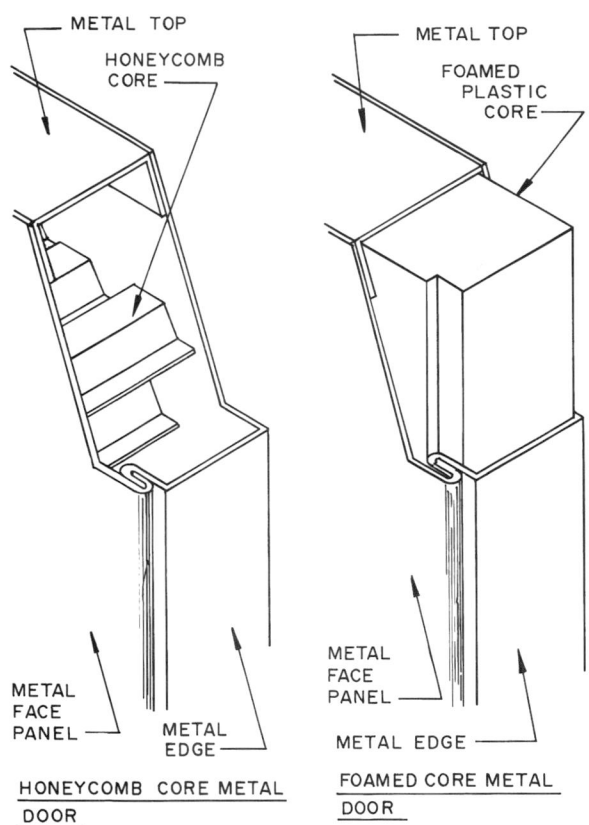

8-12 *Typical metal door construction*

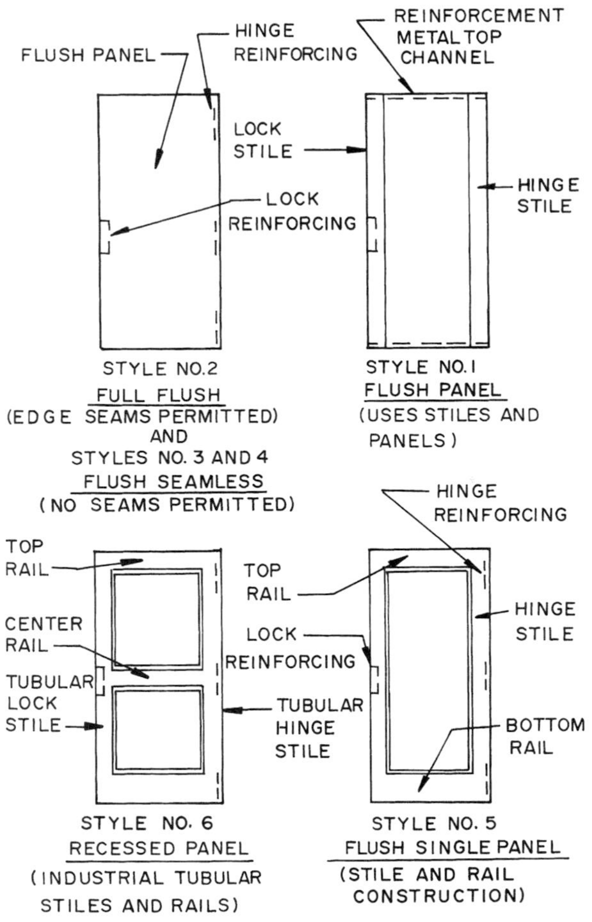

8-13 *Standard types of steel door.*

While the exact construction of hollow metal doors varies with the manufacturer two types of construction are common. One uses a tubular framework over which the face panels are welded. Insulation and sound deadening material is placed in the hollow core. Another type uses some type of core material such as Kraft paper honeycomb, polyurethane, polystyrene, a steel grid, mineral fiberboard, or vertical steel stiffeners over which metal face panels are bonded (see 8-12). The door may be seamless, full-flush, flush-panel, or industrial-tube construction (see 8-13). Seamless doors have no seams showing on the face or edges. The edge seams are hidden by welding them and grinding the weld smooth. Full-flush doors have visible seams on the edges but no seams on the face. Flush-panel doors use either a stile and panel or a stile and rail. The face panel may be slightly recessed. Industrial-tube doors are made using tubular steel stiles and rails and have recessed panels. Metal doors are classified by grades and models as shown in Table 8-5. The higher the grade the thicker the metal specified. The nomenclature for hollow steel door designs is shown in 8-14. The letter symbols are used as the basis for the description of steel doors. Some typical construction details for the top and bottom edges of the doors are shown in 8-15. Some of the commonly used meeting stiles for hollow metal doors are shown in 8-16.

NOMENCLATURE LETTER SYMBOLS

F — Flush
L* — Louvered (bottom)
TL* — Louvered (top)
LL* — Louvered (top and bottom)
V — Vision Lite
NL* — Narrow Lite
G — Half Glass (options G2, G3, G4, and G6)
VL* — Vision Lite and Louvered
N — Narrow Lite
NL* — Narrow Lite and Louvered
GL* — Half Glass and Louvered (options G2L* and G3L*)
FG — Full Glass (option FG3)
FL* — Full Louver
D — Dutch Door

* Louvered door designs : specify design, louver size and/or free area requirements
ADD SUFFIX I TO INDICATE INSERTED LOUVER
ADD SUFFIX P TO INDICATE PUNCHED LOUVER
ADD SUFFIX A TO INDICATE AIR CONDITIONING GRILLE

8-14 *Nomenclature for steel doors and frames.*
Courtesy Steel Door Institute

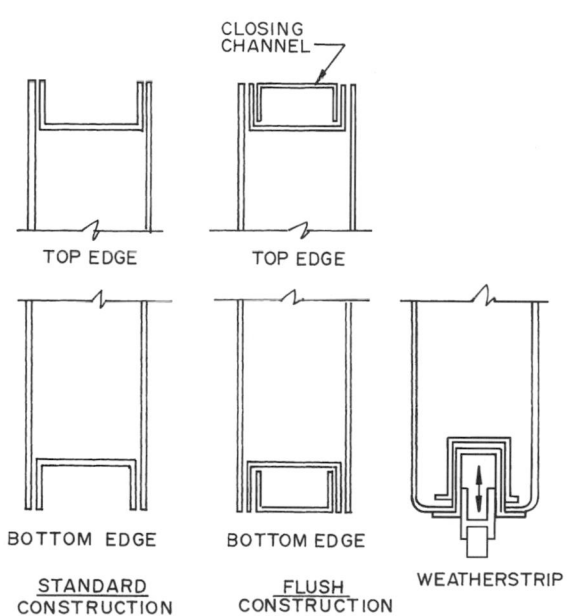

8-15 *Some commonly used top and bottom construction details for hollow metal doors.*

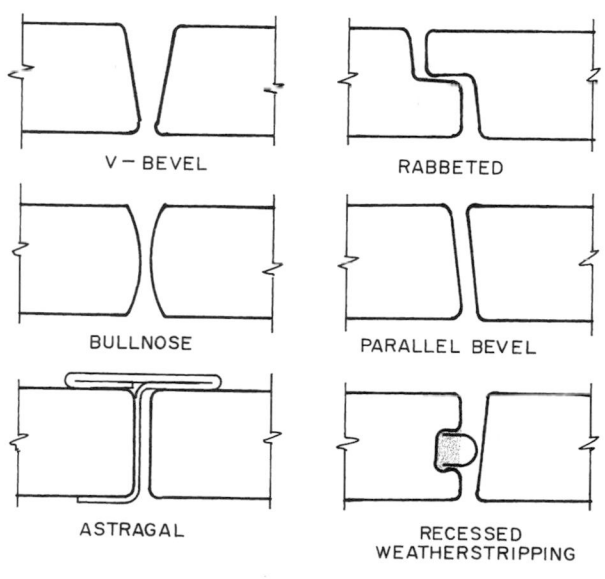

8-16 *Typical meeting stile profiles used on metal doors.*

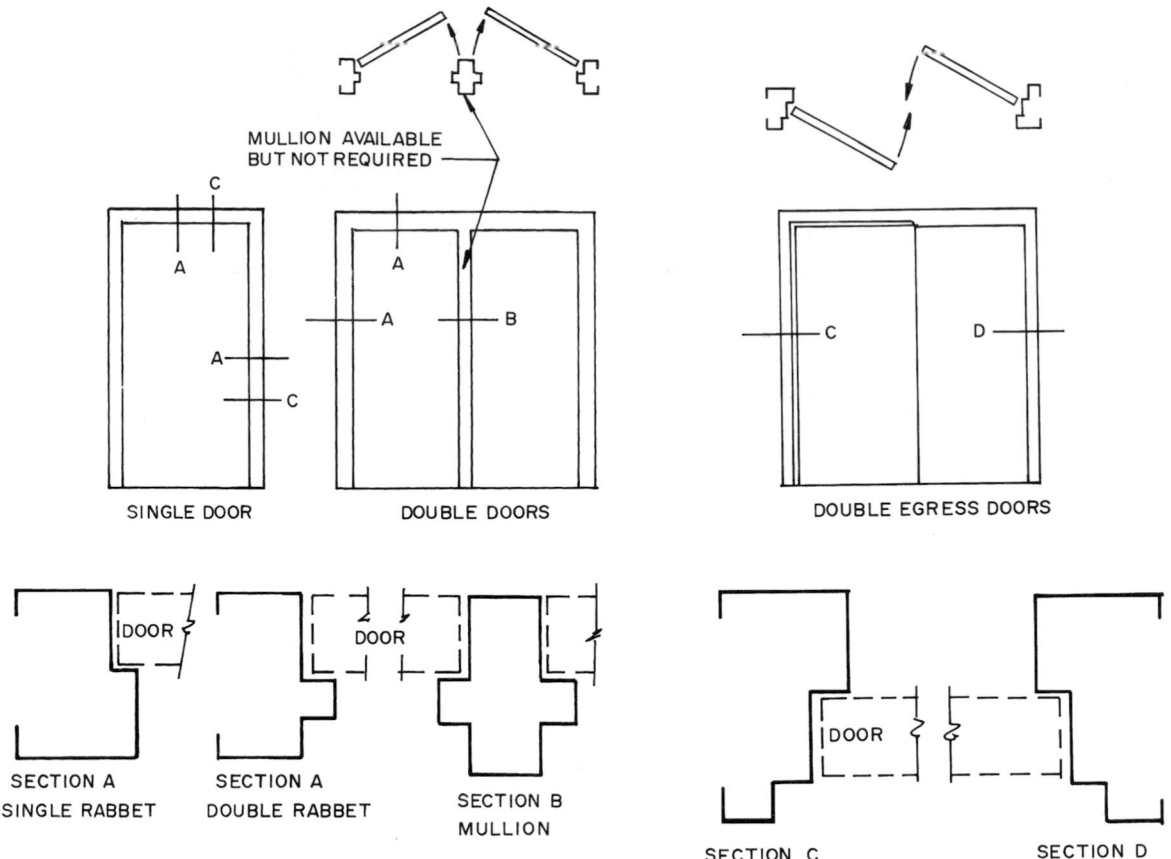

MULLION AVAILABLE
BUT NOT REQUIRED

SINGLE DOOR

DOUBLE DOORS

DOUBLE EGRESS DOORS

SECTION A
SINGLE RABBET

SECTION A
DOUBLE RABBET

SECTION B
MULLION

SECTION C

SECTION D

8-17 *Some typical metal door frame details.*

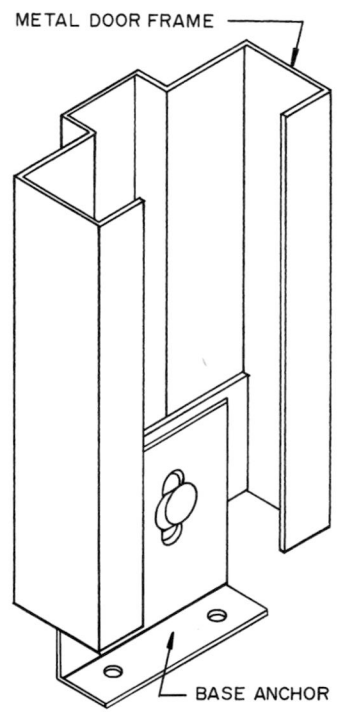

METAL DOOR FRAME

BASE ANCHOR

8-18 *Metal door frames are anchored to the floor with some form of metal angle.*

Hollow Metal Frames

Hollow metal frames are available in steel or aluminum. They are supplied knocked down or pre-assembled. A wide range of designs are available providing for many types of interior and exterior wall construction (see 8-17). They are anchored to masonry walls with some type of strap or wire loop. Usually three anchors are required on each jamb. Several strap designs are used to secure the hollow metal door frames to wood or metal studs. Brackets are used to secure the base to the floor (see 8-18). The installation of some of these frames is detailed in 8-19.

The frame and the door are prepared to receive the lockset and strike plate. The holes, recesses, and screw holes are cut. Assembled door frames frequently have the door installed before the unit is shipped to the job (see 8-20).

The required steel gauge for door frames is related in the standards to the door grade and model. The higher the door grade the thicker the steel used for the door frame.

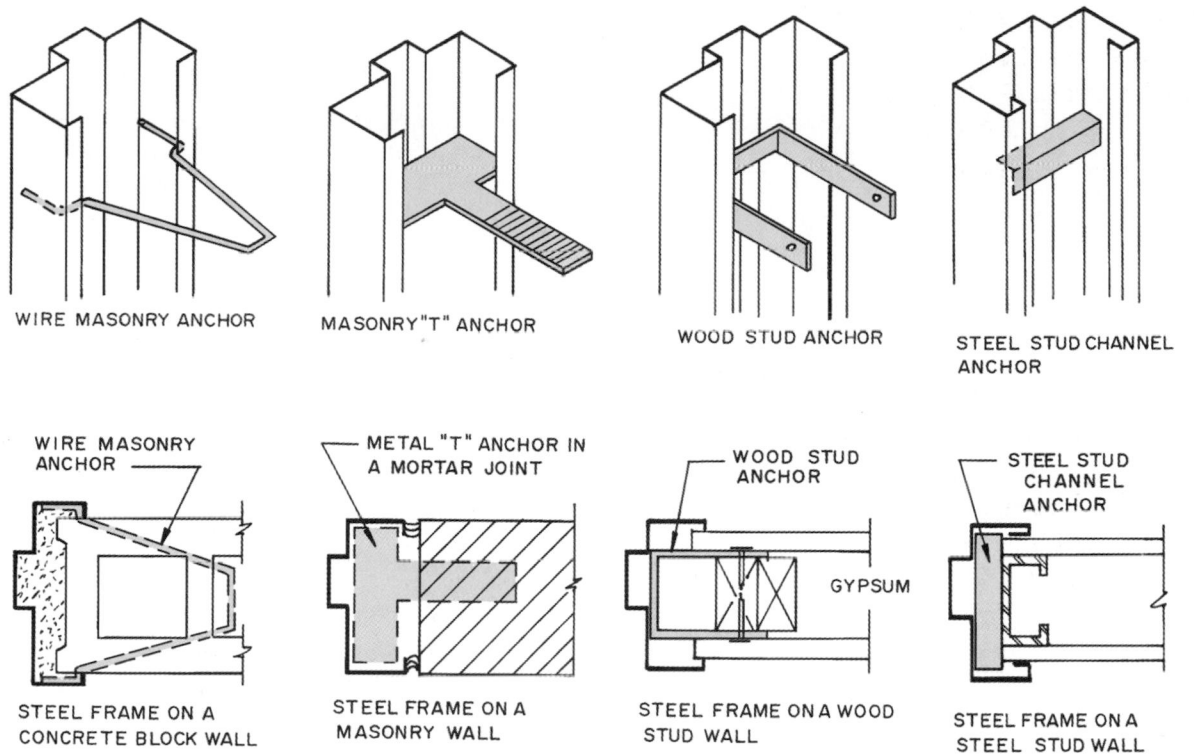

WIRE MASONRY ANCHOR

MASONRY "T" ANCHOR

WOOD STUD ANCHOR

STEEL STUD CHANNEL ANCHOR

WIRE MASONRY ANCHOR

METAL "T" ANCHOR IN A MORTAR JOINT

WOOD STUD ANCHOR

STEEL STUD CHANNEL ANCHOR

GYPSUM

STEEL FRAME ON A CONCRETE BLOCK WALL

STEEL FRAME ON A MASONRY WALL

STEEL FRAME ON A WOOD STUD WALL

STEEL FRAME ON A STEEL STUD WALL

8-19 *Some of the types of anchors used to secure steel door frames to the wall.*

Metal Fire Doors & Frames

Fire doors are an assembly of a door and frame manufactured and installed to give protection against the passage of fire through an opening in a wall.

The design, location, and installation are controlled by national standards and local building codes. Approved fire door assemblies are required to meet the test requirements specified in ASTM E152, *Methods of Fire Tests of Door Assemblies.* Standards for fire doors and windows are in the publication ANS/NFPA 80. This publication was prepared by a committee of the National Fire Protection Association and has been approved by the American National Standards Institute.

All fire door assemblies must bear the label of an approved agency and include the name of the manufacturer, the fire protection rating, and the maximum transmitted temperature end point. Some assemblies are automatic and keep the door (as on a long hall) open and will close automatically if a fire raises the temperature or they are exposed to smoke.

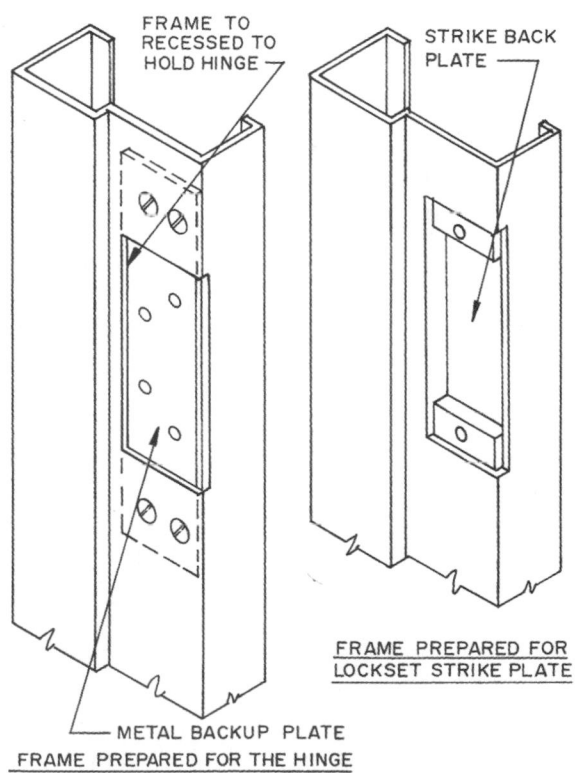

FRAME TO RECESSED TO HOLD HINGE

STRIKE BACK PLATE

FRAME PREPARED FOR LOCKSET STRIKE PLATE

METAL BACKUP PLATE
FRAME PREPARED FOR THE HINGE

8-20 *The metal door frame has recessed hinge and lockset back plates to receive the hinges and strike plate.*

Table 8-6 FIRE DOOR RATINGS

Class	Hour	Glazing	Location
A	3	No glazing	In fire wall and walls that divide a building into fire areas
B	1½	100 sq.in. of glazing	In enclosures of vertical communication through buildings
C	¾	1296 sq.in. of glazing per light	Openings in walls requiring a fire resistance rating of 1 hr. or less
D	1½	No glazing	Openings in exterior walls subject to fire exposure from the outside of the building
E	¾	720 sq.in. glazing	Openings in exterior walls subject to moderate or light fire exposure from outside the building

▼ Fire doors may be classified by the following:
  1. Hourly rating designation.
  2. Alphabetical letter designation.
  3. A combination of hourly and a letter designation
  4. Horizontal access doors use a special listing indicating the fire-rated floor, floor-ceiling, or roof-ceiling assembly for which the door will be used.

The hourly designation indicates the duration of the fire test exposure in hours and is called the fire protection rating.

▼ The alphabetical letter designation in use follows:
  Class A—Openings in fire walls and in walls dividing a single building into fire areas.
  Class B—Openings in enclosures of vertical communications through buildings and in 2-hour rated partitions providing horizontal fire separations.
  Class C—Openings in walls or partitions between rooms and corridors having a fire resistance rating of 1 hour or less.
  Class D—Openings in exterior walls subject to severe fire exposure from outside of the building.
  Class E— Openings in exterior walls subject to moderate or light fire exposure from outside of the building.

Special listings are tested in accordance with NFPA 251, *Standard Methods of Fire Tests of Building Construction and Materials*. It indicates the fire rated assembly and its hourly rating.

Fire door rating details are shown in Table 8-6. Labels on the doors specify the class of the door which indicates the time interval the door will meet. For example, a Class A door has a 3-hour time interval.

▼ The typical types of construction for fire doors include:
  Wood-core fire doors with wood, hardboard, or plastic laminate faces bonded to solid wood core or a wood particleboard core.
  Hollow-metal fire doors may be of flush or panel design with the steel face being 20 gauge or thicker.
  Metal-clad fire doors have wood cores or stiles and rails with insulated panels. The face is 24 gauge steel or lighter.
  Sheet-metal fire doors are made using 22 gauge or lighter steel.
  Tin-clad fire doors have a solid wood core covered with a 24 to 30 gauge terne plate or galvanized steel facing.
  Composite fire doors consist of some combination of wood, steel, or plastic laminate bonded to a solid core material.
  Rolling-steel fire doors are steel doors that move on an overhead barrel that is enclosed in a hood. It may have an automatic closing mechanism (see 8-21).
  Curtain-type fire doors have interlocking steel blades forming a steel curtain in a steel frame.
  Wood-core-type fire doors have a solid wood or particleboard core covered with wood, plastic, or hardboard face sheets.

Special-purpose fire doors include a fire door and frame assembly. These include acoustical fire doors, security fire doors, armored attack-resistant fire doors, radiation shielding fire doors, and pressure-resistant fire doors. Also a variety of automatic fire vents are manufactured (see 8-22).

8-21 *A rolling steel fire door.*
Courtesy Cornell Iron Works, Inc.

8-22 *This automatic fire door opens during a fire to let the rising smoke and gases clear from the building.*
Courtesy The Bilco Company

Metal door frames have fire ratings of ¾, 1½, and 3 hours. The hardware used must be fire rated to maintain the fire rating of the door and frame.

Wood & Plastic Doors

Flush doors have wood veneers, hardboard, or fiberglass face panels bonded to a core of solid wood strips (called a solid core door) or a core made of wood or kraft paper forming a hollow, honeycomb interior—called a hollow core door (see 8-23). They can have openings cut for windows and louvers.

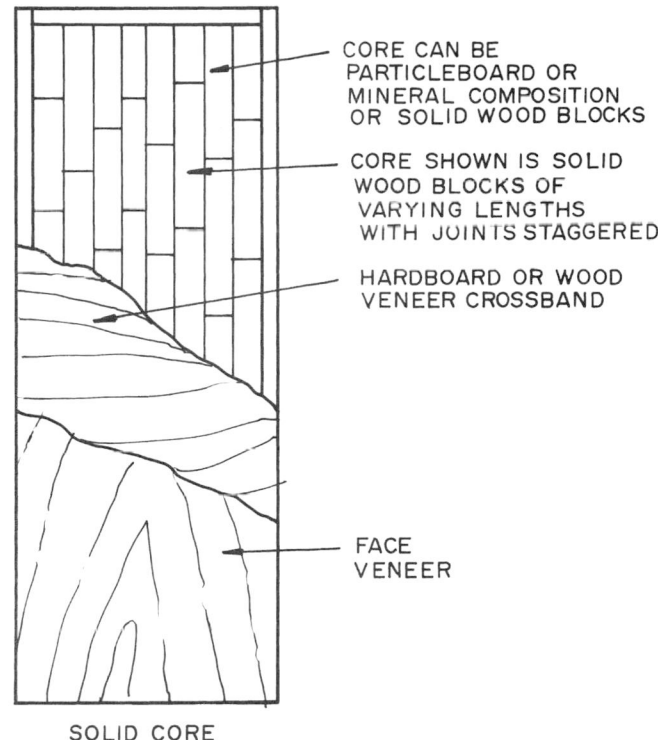

CORE CAN BE PARTICLEBOARD OR MINERAL COMPOSITION OR SOLID WOOD BLOCKS

CORE SHOWN IS SOLID WOOD BLOCKS OF VARYING LENGTHS WITH JOINTS STAGGERED

HARDBOARD OR WOOD VENEER CROSSBAND

FACE VENEER

SOLID CORE

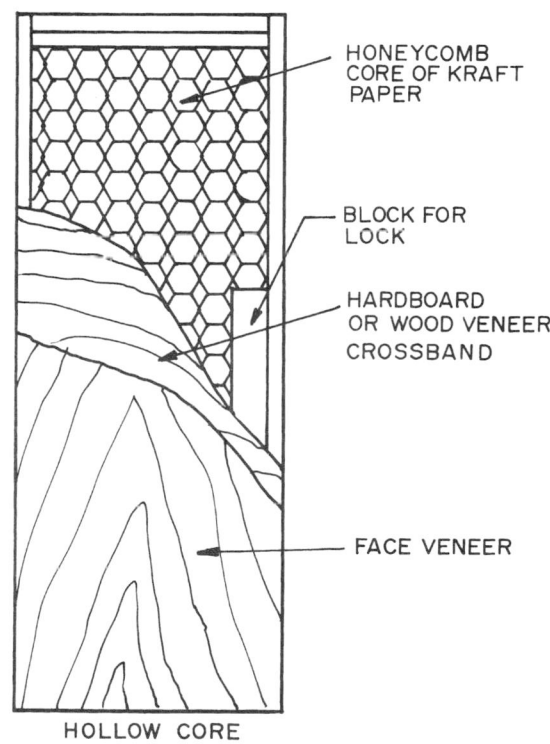

HONEYCOMB CORE OF KRAFT PAPER

BLOCK FOR LOCK

HARDBOARD OR WOOD VENEER CROSSBAND

FACE VENEER

HOLLOW CORE

8-23 *Construction of hollow-core and solid-core flush doors.*

8-24 *A panel door made of molded hardboard.*

8-25 *A wood-framed door with fiberglass outer skin.*

QUALITY ɲ̩̩ɯɯɒ̩ CERTIFIED

WOOD FLUSH DOOR   CONFORMS TO ɲ̩̩ɯɯɒ̩ I.S.-1

8-26 *Quality certification seal for wood doors.*
Courtesy National Wood Window and Door Association

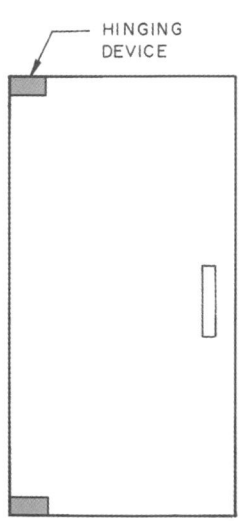

HINGING DEVICE

FRAMELESS TEMPERED
GLASS DOOR

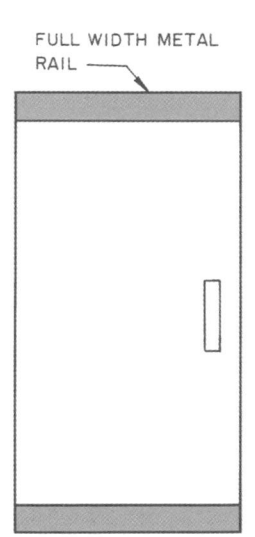

FULL WIDTH METAL
RAIL

FRAMELESS TEMPERED
GLASS DOOR WITH FULL
WIDTH METAL TOP AND
BOTTOM RAILS

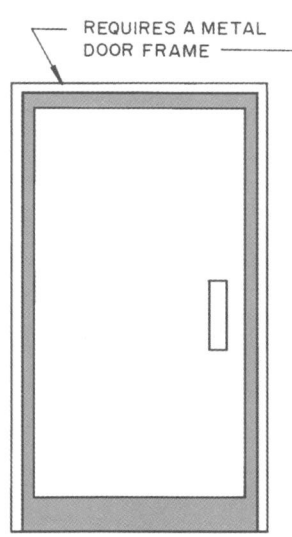

REQUIRES A METAL
DOOR FRAME

NARROW METAL FRAME
DOOR WITH TEMPERED
GLASS

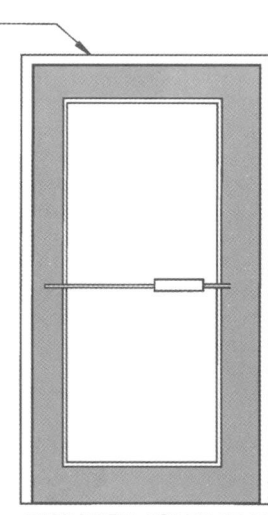

WIDE METAL FRAME DOOR
WITH TEMPERED GLASS

8-27 *Some of the glass doors used in commercial construction.*

8-28 *A roof scuttle provides safe and easy access to the roof.*
Courtesy Bilco Company

Stile-and-rail doors have solid wood vertical and horizontal members enclosing panels made of wood, glass, or louvered sections. Some are made from pressed wood fibers forming a paneled hardboard door (see 8-24) or a fiberglass face veneer (see 8-25). Fire rated doors use a mineral composition core. Acoustical doors use sound dampening cores or sheets of lead over the core. Solid wood core doors reduce sound transmission and retard fire better than hollow core doors.

Many companies manufacture wood doors to the specifications of the National Wood Window and Door Association. These standards establish minimum requirements for material, design, construction, and pressure treatment. Those doors meeting these standards are recognized by the NWWDA seal (see 8-26).

Glass Doors

Glass doors may be frameless and have metal channels to hold the pivot type hinging apparatus or have a very narrow metal frame on all sides or a wider frame that provides a stronger door to be used in areas where heavy traffic is expected (see 8-27). The glass must meet building code safety requirements.

8-29 *These steel rolling service doors provide a fire-resistant closure between areas separated by this interior wall in a manufacturing plant.*
Courtesy Cornell Iron Works, Inc.

8-30 *During the hours this business is open it is totally exposed to the public and rolling grille doors provide security during closed hours yet permit the interior to be visible.*
Courtesy Overhead Door Company

Special Doors

There are many doors manufactured to meet special needs other than providing human access within a building. These include units providing access to an area such as a roof (see 8-28), various types of sliding doors, doors to resist explosions or to provide an airtight or a watertight closure. Special doors and glazing are used for security purposes. Various types of folding and accordion doors are widely used in residential and commercial construction. There are many variations of rolling doors, such as those used on aircraft hangers and openings in manufacturing plants. Many are power activated (see 8-29). Various types of rolling grille doors are used to provide security to contents of stores opening onto a public mall (see 8-30).

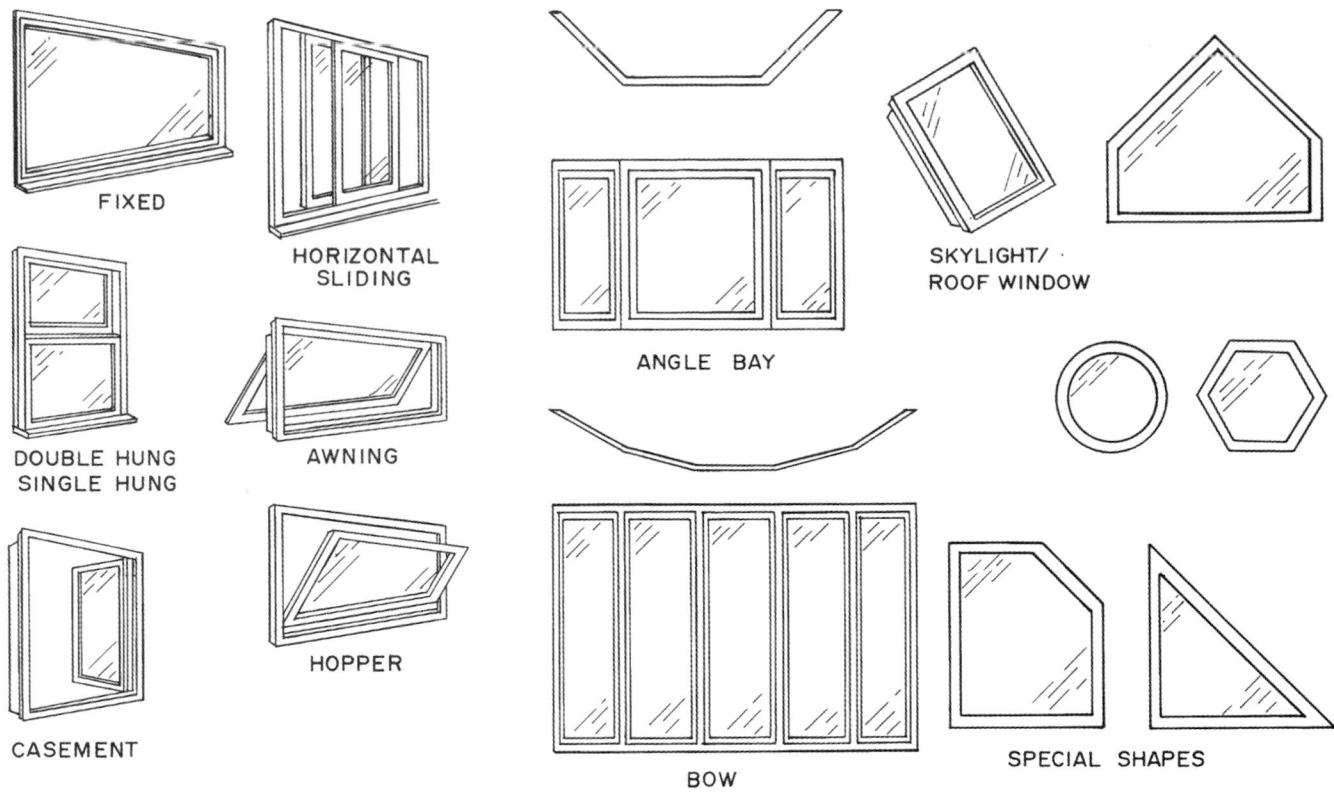

8-31 *Wood windows in common use.*

8-32 *A generic example of the certification label of fenestration products rated by the National Fenestration Rating Council. The blank space is used to indicate solar heat gain, air infiltration, long-term energy performance, and other energy performance attributes.*

Courtesy National Fenestration Rating Council

WINDOWS

Windows are available made from solid wood, wood clad with plastic or aluminum, solid plastic, steel, stainless steel, aluminum, bronze, and composite materials. Those commonly used in residential construction are shown in 8-31.

A major consideration in the selection of windows is energy savings. Various types of energy-efficient glazing are available, such as energy-efficient glass, double glazing, and glass with louver blinds set between the panes of glass. Some double-glazed units have a gas inserted in the space between the panes. Windows must have airtight unions between both fixed and moving parts to reduce air infiltration. Metal units transmit heat and cold rapidly through the material so the design must provide insulation between touching interior and exterior parts. These are called **thermal breaks.** Windows are also a major source of light to the inside space. From a design standpoint they are a major feature of the appearance of the exterior. They used to be a major source of ventilation but this need has been reduced to some degree by year-round interior temperature control and air filtering systems.

Quality windows are tested and certified if they meet standards for air leakage, water infiltration, uniform wind load structural requirements, and uniform load deflection. These tests are made following ASTM standards.

The **energy performance** of **fenestration products** is rated by the National Fenestration Rating Council (NFRC). Fenestration refers to the design of windows and other exterior openings of a building. The NFRC combines U-value, solar heat gain factors, optical properties, air infiltration, condensation resistance, and other characteristics into a uniform rating system to reflect annual energy performance. Products are certified and labeled providing a means for comparing the products manufactured by various companies. A generic example of the NFRC certification label is shown in 8-32.

A wide range of **locking** and **operating devices** are available and should be considered as window units are selected. These include various locking devices, manual and electric operating devices, hinges and other hardware, and safety devices such as a keyed lock or safety bar that limits the amount a window can be opened.

Local fire codes also influence the choice of windows. These codes typically control the size of windows, height of sill above the floor, and the ability to open the window from the inside without the need of a special tool. Windows used in firewalls must be fire rated and labeled to restrict the spread of fire and smoke. Security windows are tested to be certain they meet the standards required to provide resistance to forced entry. Finally, be certain the material and finish fit in with the surrounding area and that it can withstand any corrosive elements to which it will be exposed.

Wood Windows

Wood-framed windows are available with fixed and operable sash. They may be clad with plastic or aluminum or left natural and are painted. Wood windows provide better insulation than metal and plastic windows. However, they do swell and shrink as the moisture content changes. The cladding of wood window exteriors has reduced this considerably. Wood windows are installed in wood-framed walls by nailing through a metal or plastic flange through the sheathing into the wood studs and headers in the rough opening (8-33).

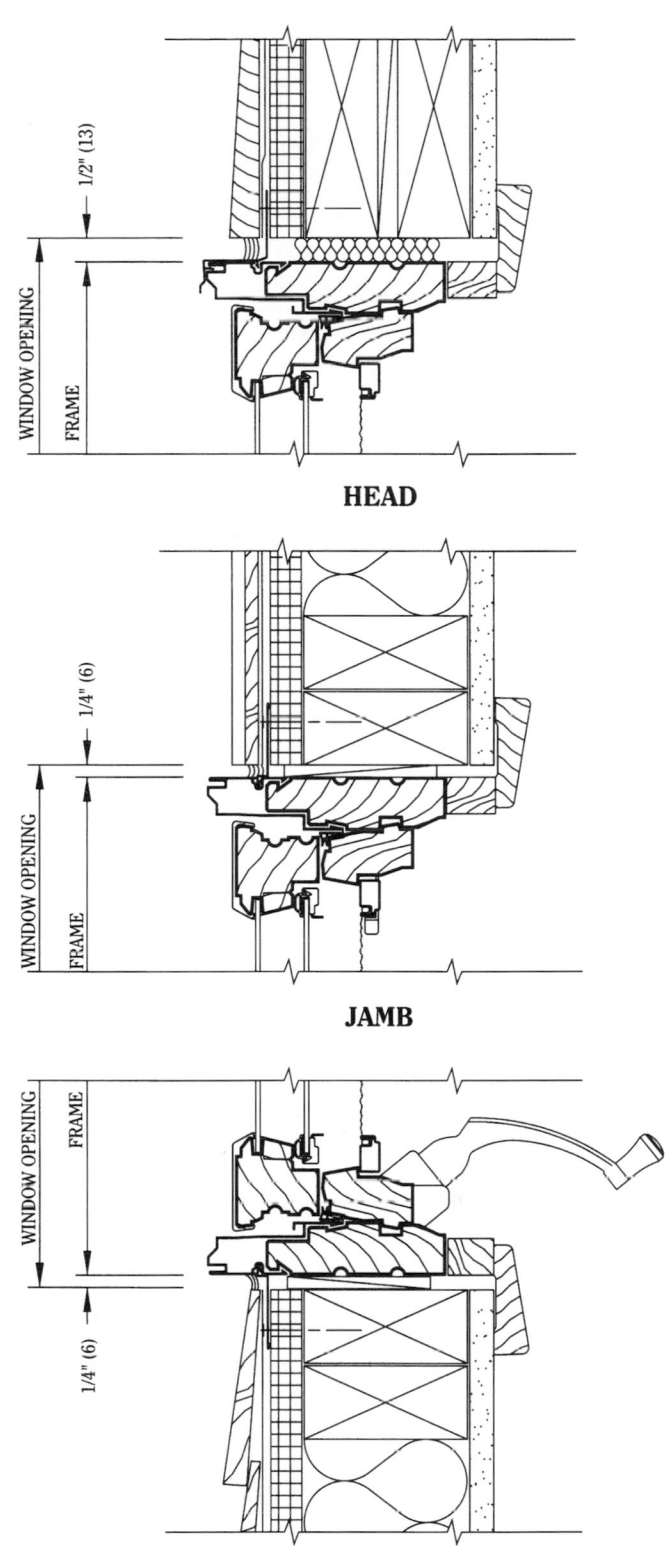

8-33 *An installation detail for a clad casement window secured in a wood-framed wall by nailing through the plastic flange.*
Courtesy Pella Windows and Doors.

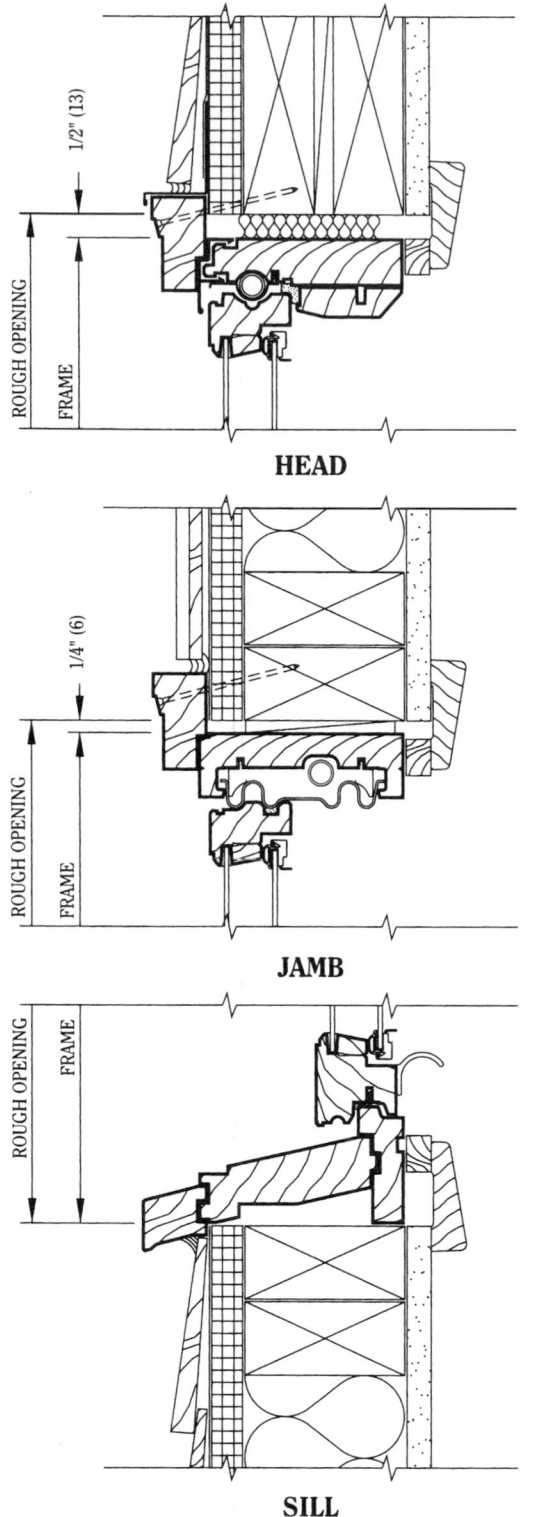

HEAD

JAMB

SILL

NOTE: THESE DETAILS ARE FOR TYPICAL SINGLE PUNCH OPENINGS.

8-34 *An installation detail for a wood-framed double hung window in a wood-framed wall installed by nailing through the exterior molding.*

Courtesy Pella Windows and Doors.

Some wood-framed windows are designed with a wood molding that is nailed to the sides of the wood-framed rough opening (8-34). An example of another installation method is the use of metal clips to secure the window in a wall framed with metal studs as shown in 8-35. Windows come with various types of glazing and insect screens.

The National Wood Window and Door Association standards IS-2 and IS-3 established specifications for wood windows and sliding glass doors. Their seal on a product certifies it meets these standards (refer to 8-26). NWWDA also has standard IS-4, which pertains to water-repellent preservative treatment.

Plastic Windows

Plastic windows are available in the same basic types as described for wood windows. The structural frame and casement frame are made from polyvinyl chloride (PVC) and glass fibers and a polyester resin. The material will not rust, swell, pit, peel or corrode and never needs painting. It is a fairly good insulator, and the frame is made with dead air pockets that increase the insulation value. PVC material also has excellent sound insulation value and members can be produced with the accuracy needed to provide airtight fits (see 8-36).

8-36 *This vinyl window has low air infiltration, is maintenance free, and is assembled by fusion welding.*

Courtesy Georgia-Pacific Corporation

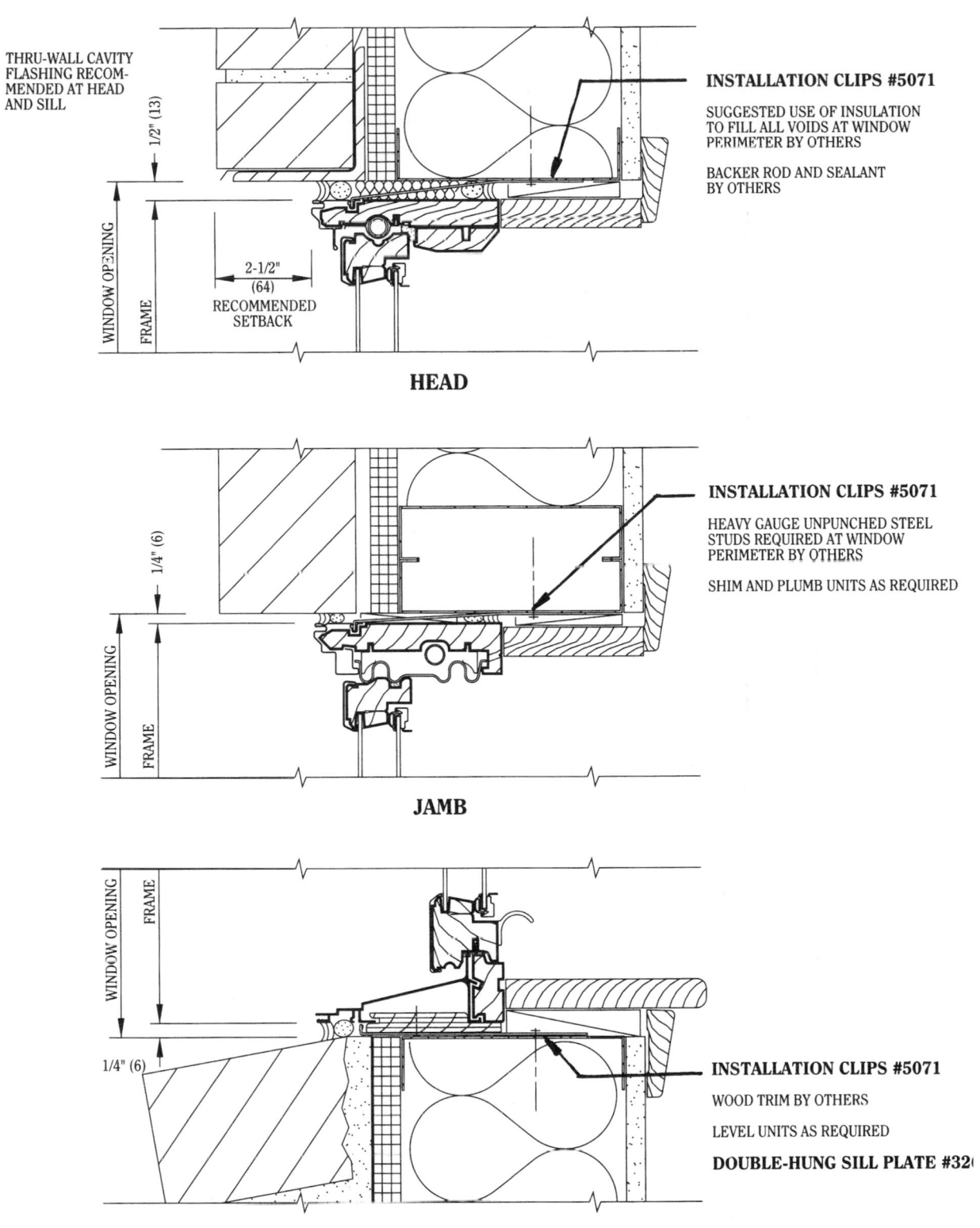

THRU-WALL CAVITY FLASHING RECOMMENDED AT HEAD AND SILL

1/2" (13)

WINDOW OPENING

FRAME

2-1/2" (64) RECOMMENDED SETBACK

HEAD

INSTALLATION CLIPS #5071

SUGGESTED USE OF INSULATION TO FILL ALL VOIDS AT WINDOW PERIMETER BY OTHERS

BACKER ROD AND SEALANT BY OTHERS

1/4" (6)

WINDOW OPENING

FRAME

JAMB

INSTALLATION CLIPS #5071

HEAVY GAUGE UNPUNCHED STEEL STUDS REQUIRED AT WINDOW PERIMETER BY OTHERS

SHIM AND PLUMB UNITS AS REQUIRED

WINDOW OPENING

FRAME

1/4" (6)

SILL

INSTALLATION CLIPS #5071

WOOD TRIM BY OTHERS

LEVEL UNITS AS REQUIRED

DOUBLE-HUNG SILL PLATE #32

8-35 *An installation detail for a clad double hung window installed in a steel stud-framed wall with clips that are secured to the metal studs.*

Courtesy Pella Windows and Doors

PVC windows are available in a range of colors. Some color the exterior members and leave the interior exposed surfaces an off white. A section through a plastic window unit is in 8-37. Windows are fastened to the wall using steel wall anchors with zinc coat screws applied through slots in the frame as specified by the manufacturer. Units are available with various types of glazing and insect screens.

Metal Windows

Metal windows used in residential construction are most often aluminum. While some types of steel windows are designed for use on residential building they mainly are used for commercial, industrial, and monumental buildings. Monumental buildings include structures such as schools, public buildings, hospitals, and churches.

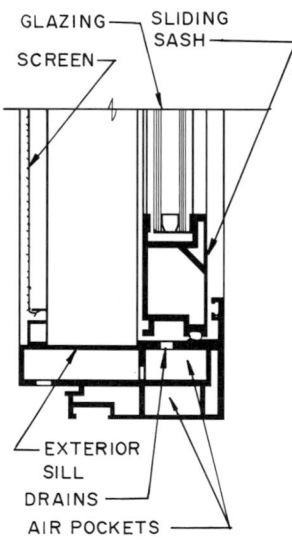

8-37 *A section through the sill of a typical plastic double hung window.*

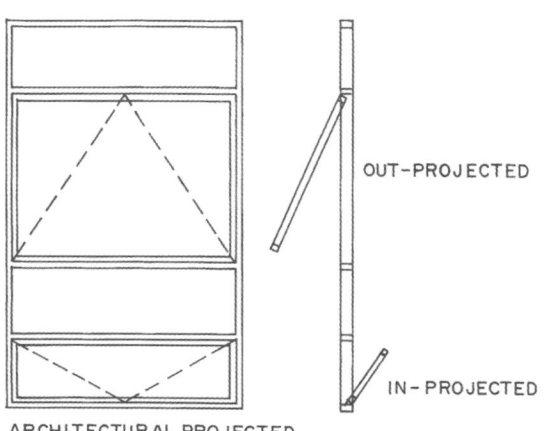

8-38 *Typical types of steel windows available. Other designs are available.*

Steel Windows

Steel windows are made from structural grade steel. Residential grade windows are made from lighter gauge steel than windows designed for use in monumental, industrial, and commercial buildings and are classified as intermediate grade. Heavy-duty intermediate grade windows are used where the size of the window requires additional strength or it must meet other conditions, such as resisting high winds. Manufacturers' catalogs show the gauges and recommended uses for their products. Steel window units are rigid and strong and require narrower members than other materials producing a smaller sight line. They are available primed or galvanized and ready for painting. Some types have a polyvinyl chloride or urethane plastic coating which gives permanent protection. The units are weatherstripped and available with single and double glazing. Manufacturers produce a range of stock sizes but will produce custom designed units.

Steel windows used in residential construction are typically casement type. In commercial, industrial, and monumental type buildings other designs are available as shown in 8-38. A section through a typical steel window is in 8-39.

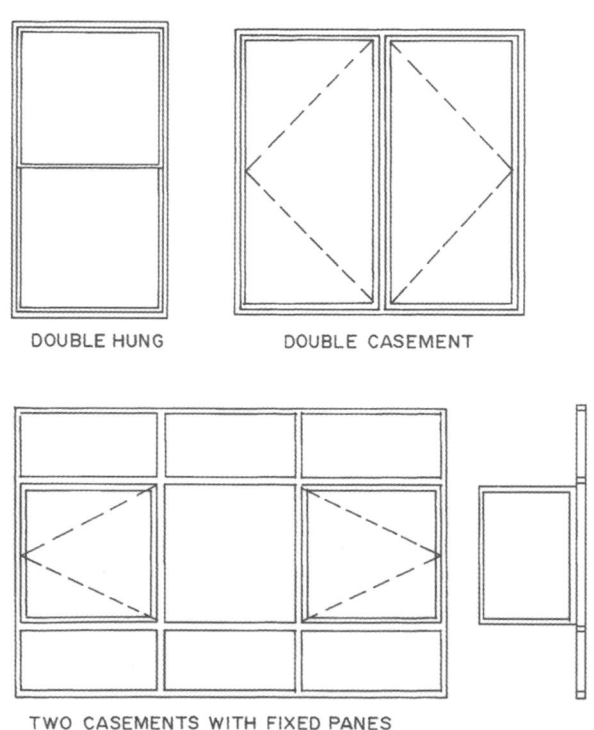

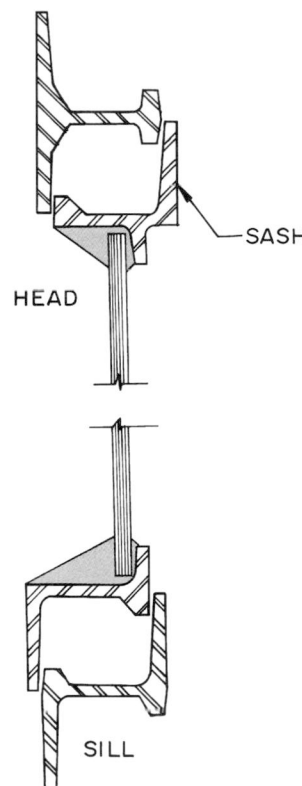

8-39 *A section through a typical steel window.*

8-40 *These windows are made from a composite material composed of wood by-products and a cellular PVC material.*

Courtesy Jeld-Wen, Inc.

Stainless Steel Windows

Stainless steel windows are usually more expensive than steel or aluminum. However they have a long life and may save money over a period of years. They are made from various types of stainless steel formed into structural members and welded forming the unit. The sizes and types of window vary by the manufacturer and many are custom built to the architects design. Typical types available include folding, awning, casement, and various hinged types. Some are like those shown for commercial and industrial steel windows.

Aluminum Windows

Aluminum windows are made from extruded aluminum structural members. Since aluminum is easy to work complex designs on these members are readily produced. Aluminum has good structural properties, and it will resist corrosion and damage from weather. If used in an atmosphere where damage may occur it can be given various organic protective coatings. It can also be colored using anodized finishes.

Aluminum windows are assembled using both mechanical connections and welding. Any connectors, fasteners, hardware, or anchors must be aluminum or a material, such as stainless steel, that is compatible with aluminum so corrosion does not occur because of galvanic action.

Several aluminum alloys are used in the manufacture of aluminum windows. These affect the properties of the product and vary with the purpose for which the window will be used. Standards for the quality of aluminum windows are established by the Architectural Aluminum Manufacturers Association. These vary with the type of window and its intended use. The windows are of three types—residential, commercial, and monumental. They are available in the same types as described for plastic and steel windows. Some manufacturers will have other special designs and will produce custom-design units for special applications.

Composite Materials

Increasing use is being made of composite materials in construction. Composites are materials made by combining several different materials. One such product is shown in 8-40. It is a window made entirely of composites (a cellular PVC material) and wood by-products producing a product that, while looking like wood, will not split, warp, or rot. It has twice the insulation value of wood and is available in several colors.

8-41 *These skylights are installed as a string to enhance the lighting of the interior of this room.*
Courtesy Eagle Window and Door, Inc.

8-42 *This barrel skylight provides considerable light into the interior.*
Courtesy O'Keeffe's Inc.

Special Windows

Special windows are those designed for a special application such as skylights, security windows, pass windows, and storm windows.

Skylights and roof windows perform the same basic function as windows except they are in the roof. Skylights have fixed glazing while the sash on roof windows can be opened to provide ventilation. In addition they can be used to vent smoke and fumes developed during a fire. They may be installed as single units, as is common in residential construction or in strings or clusters admitting considerable light (see 8-41). They are glazed with glass or plastic lights. Some operable units are opened manually with a crank while other types are electrically opened. They may be designed to blow out the glazing if an explosion occurs, or open automatically if the temperature increases as a fire vent.

Most skylights used in commercial, industrial, and monumental work come with aluminum frames. The small units used on residential work are wood framed. Small domed and barrel skylights may have single or double glazing (see 8-42).

Skylights are available for enclosing large spaces, such as on a mall, reception area, swimming pool, or restaurant. They are also used to enclose walkways protecting the interior from the weather and providing abundant light. They are glazed with glass and plastic materials. One example is in 8-43.

8-43 *Walkway covers protect those moving between buildings.*
Courtesy Lin-El, Inc.

Security windows (also called detention windows) are tested to ascertain their resistance to forced entry. This includes testing the locking device, impact resistance of the sash and frame, resistance of security bars, and the resistance of the glazing.

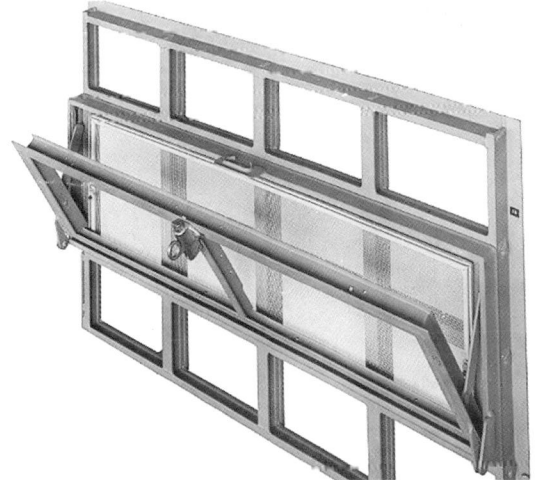

Protected out-awning security window

Medium-security guard window

Top-pivoted awning security window

8-44 *Detention windows are available with fixed or operating glazing.*
Courtesy William Bayley Company

▼ The degree of security is specified as one of four classes.

Class 1 minimal
Class 2 moderate
Class 3 medium
Class 4 relatively high

In some cases window security is specified in building codes requiring a high grade window if located in certain areas of the wall. If windows are protected by devices such as bars, screens, or shutters provisions for meeting exit requirements in the fire code must be met. Manufacturers can indicate if the windows meet ANSI/ASTM security standards. In addition to a secure window, the window frame, lock, hinges, glazing, are designed to prohibit insertion of tools through the unit is considered. Several types are shown in 8-44.

ENTRANCES & STORE FRONTS

Entrances and store fronts are made using flat architectural glass products, such as tempered and heat-strengthened glass, laminated glass, and insulated glass. Typically tempered glass doors are available with door closers, top and bottom pivots, locks, and push-pulls. The unit may be manually or automatically opened. The metal frame, typically aluminum, is usually anodized and a variety of colors are offered. Stainless steel frames are also available. The glazing is available in clear, obscure, and in several colors. Some door units are frameless (8-45).

8-45 *This entrance to a multistory building uses frameless glass doors.*
Courtesy EFCO Corporation

8-46 *These heavy-duty entrance doors and frames are designed to provide years of service under severe conditions.*

Courtesy Kawneer Company, Inc.

Entrances may be custom built for a particular opening using stock door sizes and custom built frames. The frames are mechanically joined and then welded. In 8-46 is a high traffic entrance system for shopping malls, schools, concert, and convention centers. The doors, frames, and hardware are all reinforced, heavy-duty assemblies. The door walls are heavier and the hinges and pivots are directly mounted into the reinforced frame. In 8-47 the entrance shelters the walkway and is the focal point of the building. It is painted red for emphasis. The unique circular entrance in 8-48 also contains a stair providing access to the second level of the building.

Store fronts use various types of entrance units plus glazing and solid panels of various types. These large glass units are backed by a metal rolling grill door to provide security. The grill can be raised during business hours to permit a clear view of the store interior. When designing a glazed store front it is essential to include wind load calculations. Many window manufacturers have this information for their products. In 8-49 is a store front typical of those used on many small retail establishments. It provides access to the store and an area in which retailers can display their merchandise. These stores open onto the street.

On malls the stores face onto a wide pedestrian walkway. These stores use a number of entrances that not only provide access but a clear view of the products inside the store (8-50).

8-47 *This striking entrance provides a focal point for the building, shelter from the weather, allows natural light to enter the walkway, and is reflected by the designer in the construction of the curtain wall mullions.*

Courtesy Kawneer Company, Inc.

8-48 *This custom-designed entrance contains stairs to provide access to a second level and includes a short covered walkway to the right. It shelters the area from the weather and allows natural light to enter the area.*

Courtesy Lin-El, Inc.

8-49 *A typical store front used on many small retail shops providing access and a merchandise display area.*

8-50 *This glass wall and entrance provides those passing by on the mall a view of the interior of the store or office and those inside a degree of security.*
Courtesy EFCO Corporation

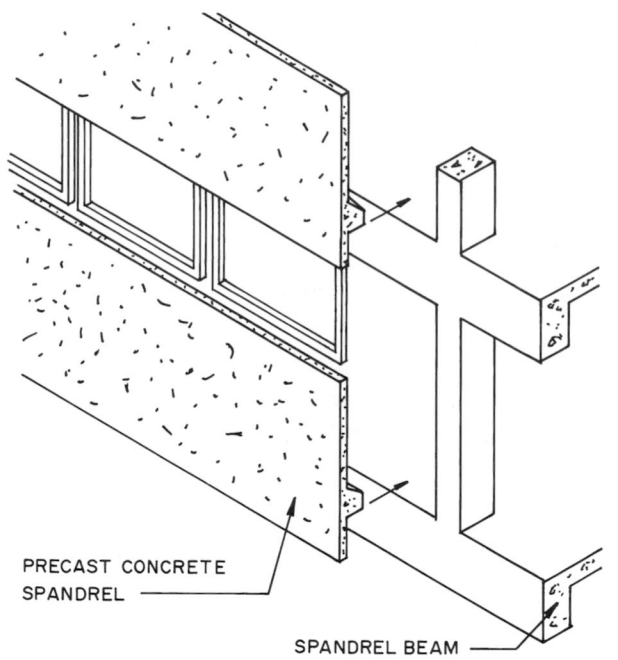

PRECAST CONCRETE
SPANDREL

SPANDREL BEAM

PRECAST CONCRETE CURTAIN WALL

RIBBON
WINDOWS

BRICK OR STONE
SPANDREL

ANGLE IRON

SPANDREL BEAM

MASONRY CURTAIN WALL

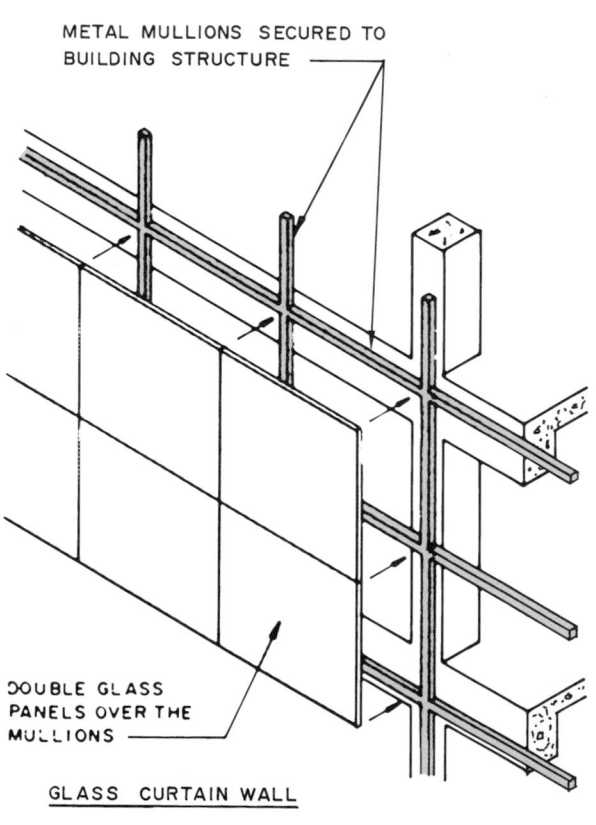

METAL MULLIONS SECURED TO
BUILDING STRUCTURE

DOUBLE GLASS
PANELS OVER THE
MULLIONS

GLASS CURTAIN WALL

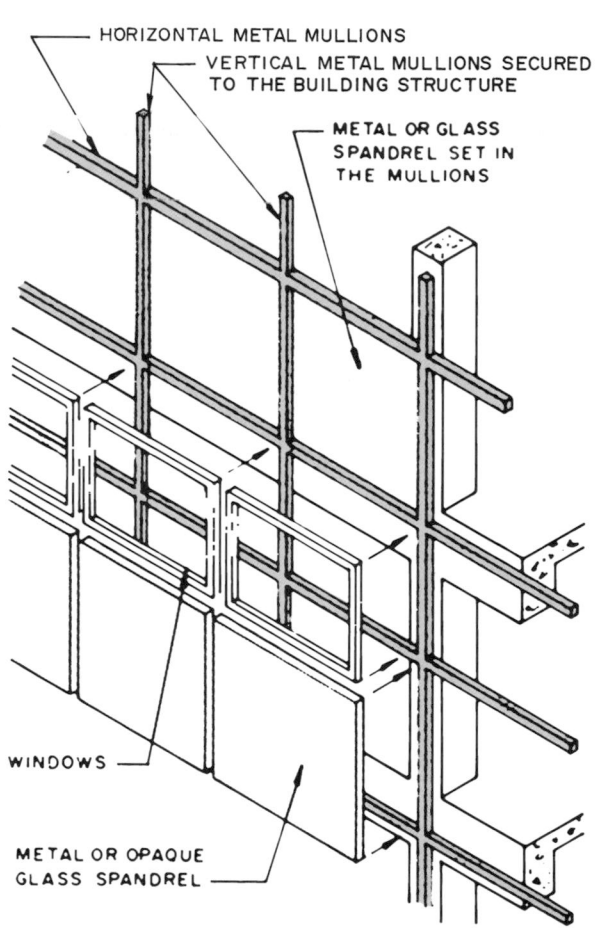

HORIZONTAL METAL MULLIONS

VERTICAL METAL MULLIONS SECURED
TO THE BUILDING STRUCTURE

METAL OR GLASS
SPANDREL SET IN
THE MULLIONS

WINDOWS

METAL OR OPAQUE
GLASS SPANDREL

METAL AND GLASS CURTAIN WALL

8-51 *Some of the commonly used cladding systems.*

Building Codes

Entrances and store fronts are subject to a variety of requirements in local building codes. These include considerations of fire resistance, wind and snow loads, hazards such as human impact loads, the means for supporting the doors and related glazing, hardware, locking arrangements, requirements for power operated doors, and size and number required.

CLADDING SYSTEMS

Cladding is a nonload-bearing exterior wall enclosing a building. It may be brick, aluminum, steel, bronze, plastic, glass, stone, or other acceptable material. Cladding is exposed to the weather and must resist the forces generated by the wind, rain, earthquakes, and temperature changes. It must control penetration by rain, condensation, high winds, transfer of heat into and out of the building, and meet building code requirements, including weather resistance, structural requirements, and fire resistance. Cladding therefore enables the indoor environment to be controlled all year round so it maintains conditions required by the occupancy of the building. A building may be clad with precast concrete panels, masonry panels, or some type of lightweight curtain wall (8-51).

Cladding Design Considerations

The design details of cladding systems require careful engineering analysis and the consideration of a wide range of factors. The various elements and the final assembled materials must be tested to assure they will meet the design requirements and building codes. Following are some of the factors to be considered when selecting and designing cladding systems.

STRUCTURAL PERFORMANCE

Panels can run continuously one or more stories in height and may be made up of spandrels and vision glass or one solid continuous panel. They must carry their own weight, have connectors to secure the panel to the structure, resist wind forces, provide for movement after they are installed, and meet fire codes.

The structural design of the panel must enable it to be transported to the site, be lifted into place, and support its own weight after installation. Stiffness requirements are also a major structural factor enabling the panel to resist wind loads.

ADEQUATE CONNECTIONS

The connections securing the panels to the structural frame must support the weight of the panel and resist other forces that may act upon it, such as wind loads. The connections are generally designed and tested by the manufacturer of the panels and must meet structural requirements. In 8-52 and 8-53 are shown connectors designed for installing a crystallized glass curtain wall system. One shows a connection to a concrete structural frame and the other to a steel-framed building. Examples of other connections are shown later in this section.

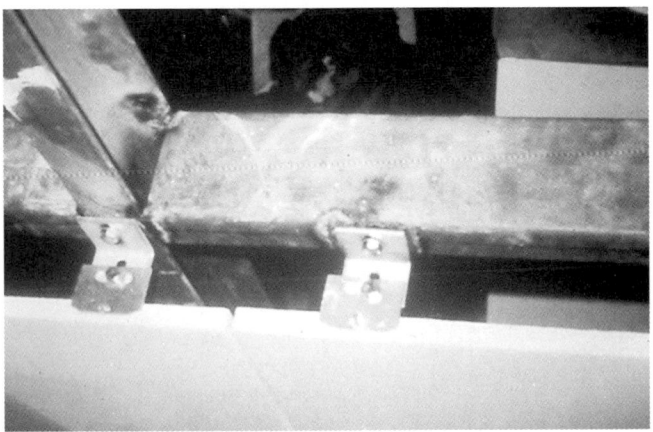

8-52 *These connections were designed to secure crystallized glass curtain wall panels to a structural steel framework.*
Courtesy N.E.G. America, Inc.

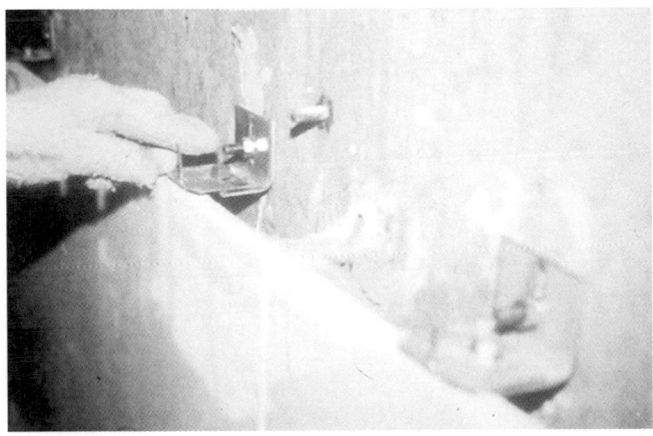

8-53 *These connections were designed to secure crystallized glass curtain wall panels to a structural concrete framework.*
Courtesy N.E.G. America, Inc.

WIND FORCES

The cladding must have sufficient structural strength to resist forces produced by winds. These forces may be positive pressure against the panel that may cause deflection or in some situations create a negative pressure (suction) that tends to blow the panel from the structure. Wind blowing by a building is likely to create a vacuum (negative pressure). On high-rise buildings these negative pressures are usually greatest near the corners of the building. The indoor air pressure, due to the heating and cooling system, may be higher than the outdoor air pressure thus putting a stress on the curtain wall from the inside of the building. If the panel lacks adequate stiffness and/or connectors it could blow off (8-54).

Wind forces are more severe on the upper floors high rise buildings because they are subject to winds of greater velocity than lower floors or low rise buildings. In an area with surrounding multistory buildings the wind pressures are influenced by factors such as the direction of the prevailing winds, the shape of the building and neighboring buildings, the positioning of the building on the site, and the topography of the surrounding area. Detailed information is available from the America Architectural Manufacturers Association.

PROVIDING FOR MOVEMENT

Since a building is constantly subject to various forces the cladding system must be designed to allow for the movement within the structural system. For example, wind forces and earthquakes may cause the structural frame to deflect or twist, putting the cladding panels, windows, and connections under stress (8-55).

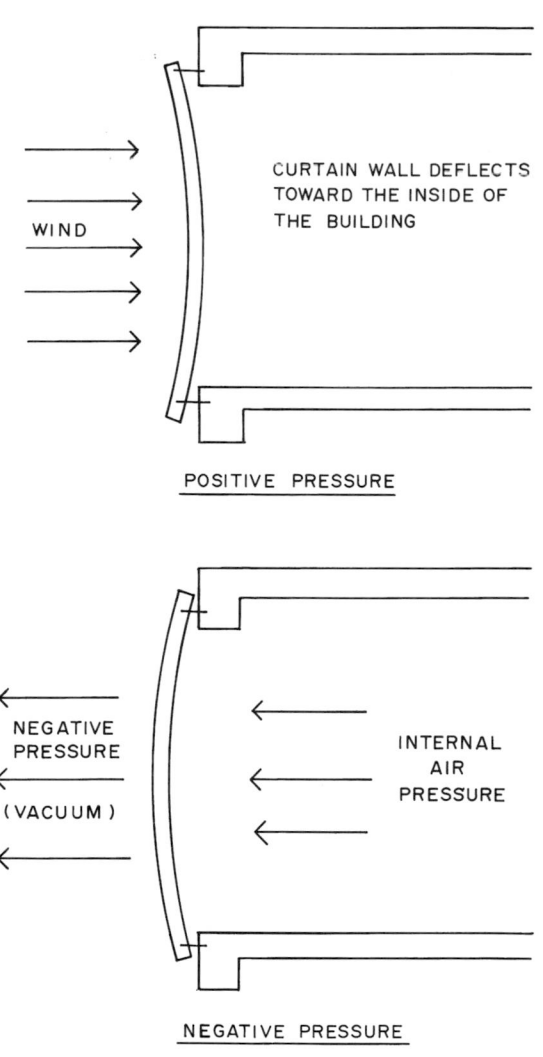

8-54 *Wind forces subject the curtain wall to positive and negative forces.*

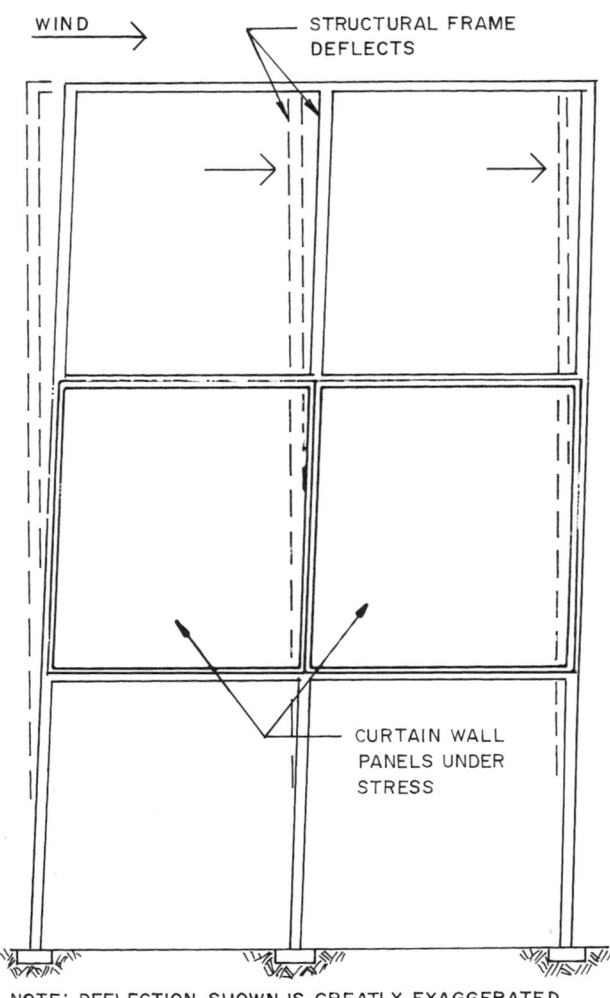

NOTE: DEFLECTION SHOWN IS GREATLY EXAGGERATED.

8-55 *Earthquakes cause wracking and twisting of the structural frame. Wind causes movement in the frame. This puts stress on the curtain wall panels, windows, and connections, possibly leading to a failure.*

In addition the cladding may be subject to forces created by gravity, thermal expansion and contraction, and moisture penetration and condensation. The weight of construction materials and differential (uneven) settling or heaving (lifting) of a foundation can cause beams to deflect. Creep that may occur over a long period of time can cause beams, columns, and girders to produce stress on the cladding. Moisture on the curtain wall can cause movement because of swelling and drying. The designer must produce a product that will serve to shelter the interior of the building and remain structurally intact (8-56).

Cladding panels must have expansion joints that permit some type of sliding overlap or other method of accommodating movement. Large glass panes must be glazed into frames that provide a watertight seal yet permit movement between the glass and the frame. Should the design prove inadequate windows could be broken, cladding attachments pulled loose, panels ruptured, and parts could even fall off the building.

CONTROL AIR INFILTRATION

The control of air infiltration is closely related to the methods that are used for controlling water penetration. Air leaks permit unconditioned air to enter and require energy to heat or cool the unwanted air. Water vapor can enter the cladding panel through air leaks and condense inside the panel. They also permit pollution to enter and, if extensive, permit the entrance of outside noise.

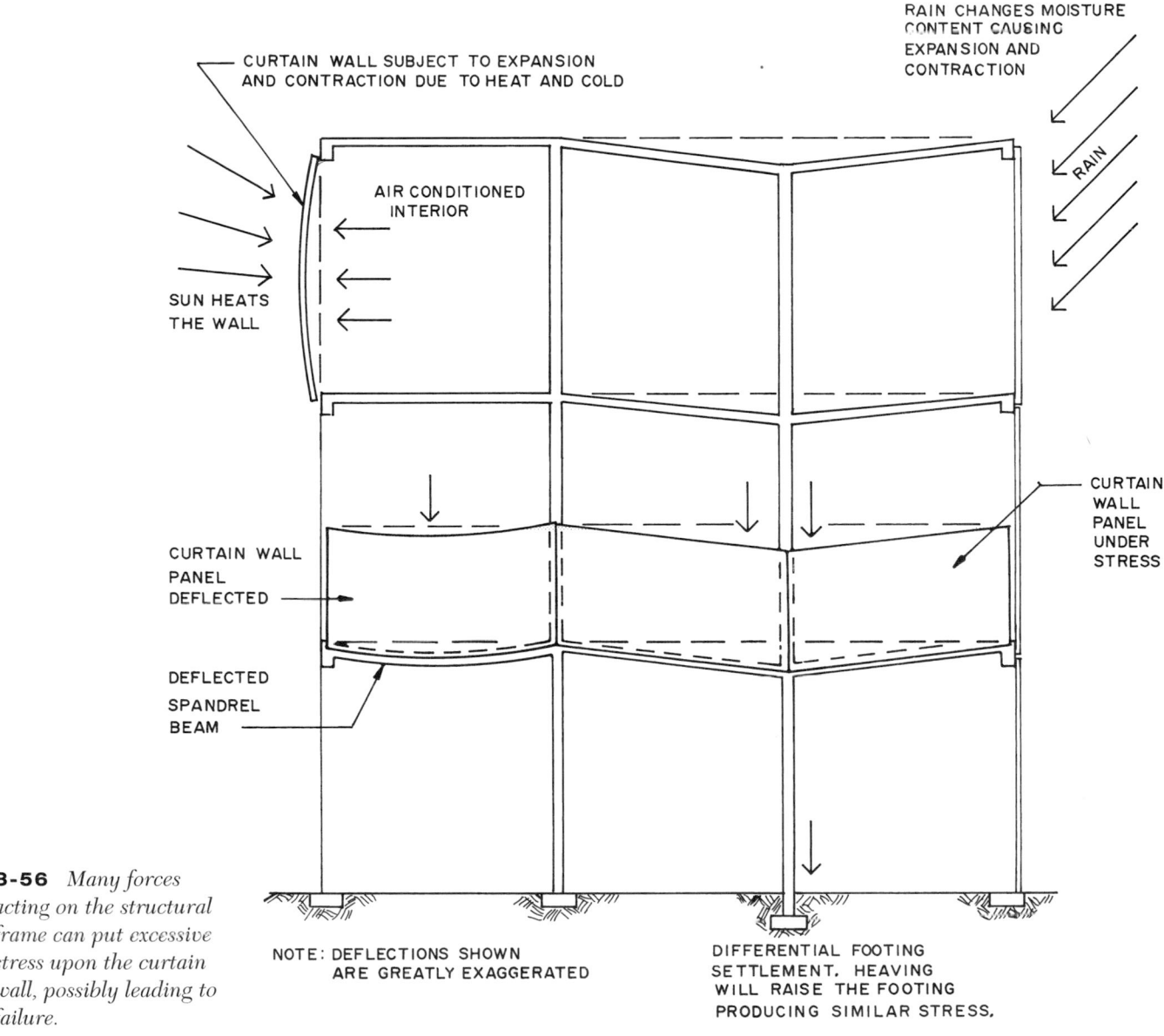

8-56 *Many forces acting on the structural frame can put excessive stress upon the curtain wall, possibly leading to failure.*

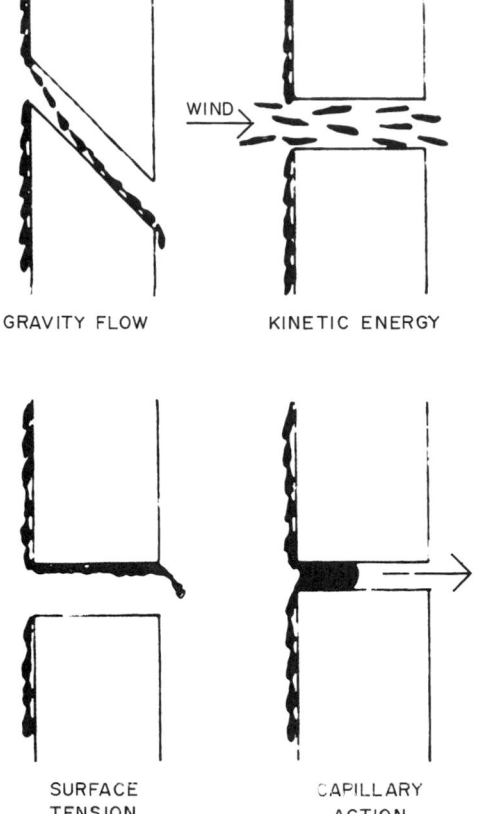

GRAVITY FLOW   KINETIC ENERGY

SURFACE TENSION   CAPILLARY ACTION

8-57 *These forces can cause water penetration in joints and other openings in exterior walls.*

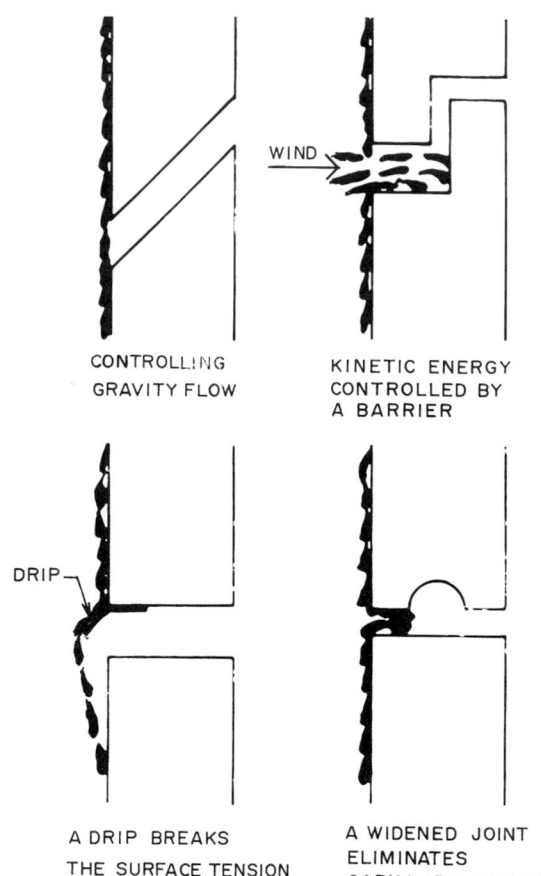

CONTROLLING GRAVITY FLOW   KINETIC ENERGY CONTROLLED BY A BARRIER

DRIP

A DRIP BREAKS THE SURFACE TENSION   A WIDENED JOINT ELIMINATES CAPILLARY ACTION

8-58 *Techniques for controlling water penetration at joints in exterior walls.*

Courtesy American Architectural Manufacturers Association, from *Curtain Wall Manual CW 1-9,* "The Rain Screen Principle"

RESISTING WATER PENETRATION

A typical curtain wall is an assembly of different materials and has a number of joints between similar and dissimilar materials. The watertightness of the exterior cladding is critical to the success of the installation. The sealing material must bond to the surfaces of the joint, withstand temperature changes, rain, wind, and the stress due to expansion, contraction, and wind loads. All of these directly influence the watertightness of the cladding.

Since the cladding is directly exposed to the weather it is subjected to wind-driven rain, snow, and hail. The upper levels of high-rise buildings will be subject to greater possible penetration because of higher wind velocities at these levels. In some areas water penetration seals must withstand chemical attack by polluted air. Variations in temperatures also attack seals.

In multistory buildings large amounts of wind-driven rain cascade down the face of the building. Provisions must be made to carry away this water and maintain the integrity of the water penetration seals. The major factor in controlling water penetration is the

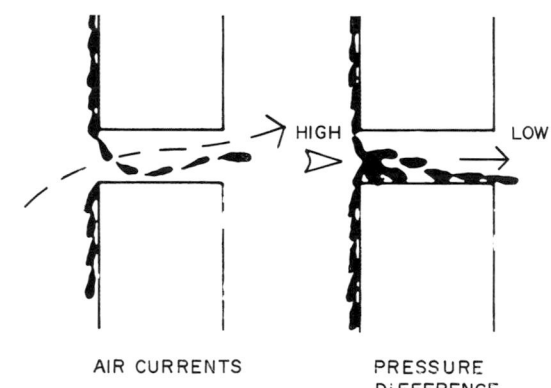

AIR CURRENTS   PRESSURE DIFFERENCE

8-59 *Wind currents can drive water through openings in the wall and cause a difference in the air pressure between the outside face and the air space inside the panel.*

design of the joints and the proper installation of the seals. In some types of panels, such as aluminum-faced panels, provision is made for internal drainage of water that may have penetrated a joint. This consists of flashing, drainage channels, and weep holes.

Water penetration of cladding may be caused by wind-driven rain and other forces as shown in 8-57. Gravity works constantly to force water through improperly designed joints. Proper joint design can reduce this type of penetration. Wind-driven rain has sufficient kinetic energy to carry the water through openings in joints and wall surfaces. This can be reduced by covering the joints with some type of batten or internal baffle. Water also can penetrate poorly designed joints due to the surface tension of the water, which allows it to cling to and flow along soffit areas. This can be prevented by adding a drip on the outer edge. Another force to consider when designing joints is capillary action. This can occur when the butting surfaces of a joint provide a very small opening. One way to control this is to have an air gap in the joint, which breaks the capillary path. These four forces can be easily controlled by proper joint design (8-58).

Wind forces may also cause a difference in pressure on the outside face and inside the cladding panel. When the air pressure is greater on the outside than the inside, water may be forced through the joint into the panel (8-59). These penetrations can be controlled by the **rain screen principle**.

RAIN SCREEN PRINCIPLE

The **r**ain screen principle is a pressure equalization method. The rain screen is the exposed outer surface of the wall. It is backed by an air space. The joints are designed to prevent water from penetrating and are not sealed. This permits the air pressure on the face of the panel to equalize with the air pressure inside the panel.

The air space behind the rain screen must be limited in size by subdividing it into relatively small subdivisions and each area should have one opening to the exterior. It is necessary to provide a structural air barrier on the interior side of the panel. This is necessary to prevent leakage to the interior side of the panel. This is necessary to prevent leakage to the interior of the building. The use of the rain screen principle with aluminum curtain wall panels is shown in 8-60 and with precast concrete curtain wall panels in 8-61.

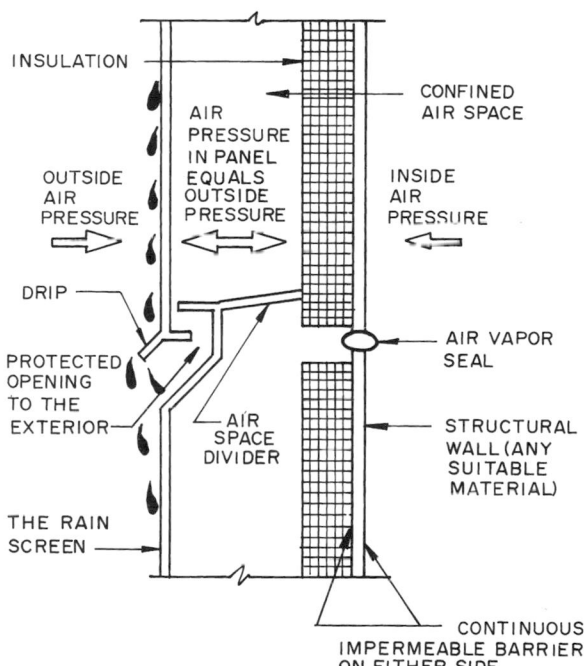

8-60 *This illustrates the rain screen principle. This aluminum curtain wall panel has divided air spaces that are vented with protected openings to the exterior, which equalizes the air pressure in the panel with the outside air.*

Courtesy American Architectural Manufacturers Association, from *Curtain Wall Manual CW 1-9,* "The Rain Screen Principle"

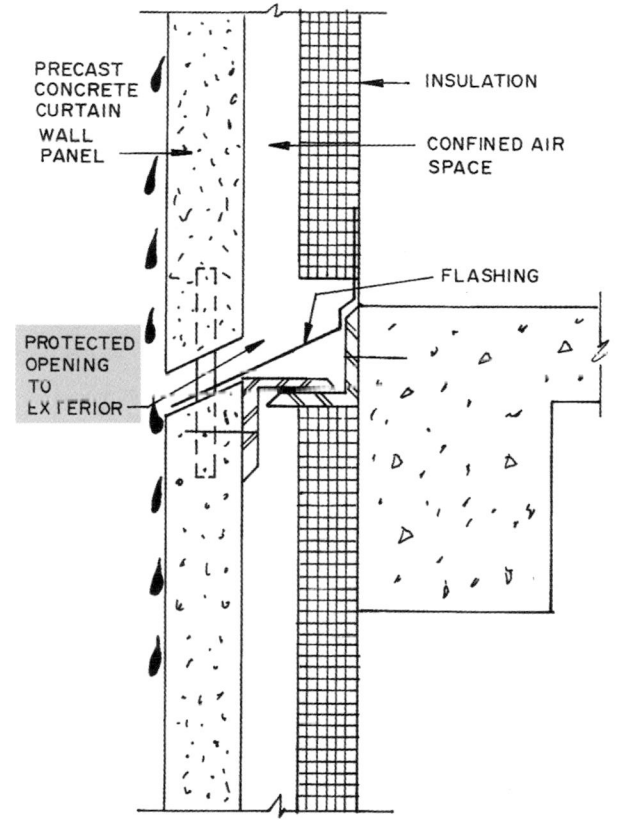

8-61 *A generalized example of how the rain screen principle can be used with precast concrete curtain wall construction.*

Courtesy American Architectural Manufacturers Association from *Curtain Wall Manual CW 1-9,* "The Rain Screen Principle"

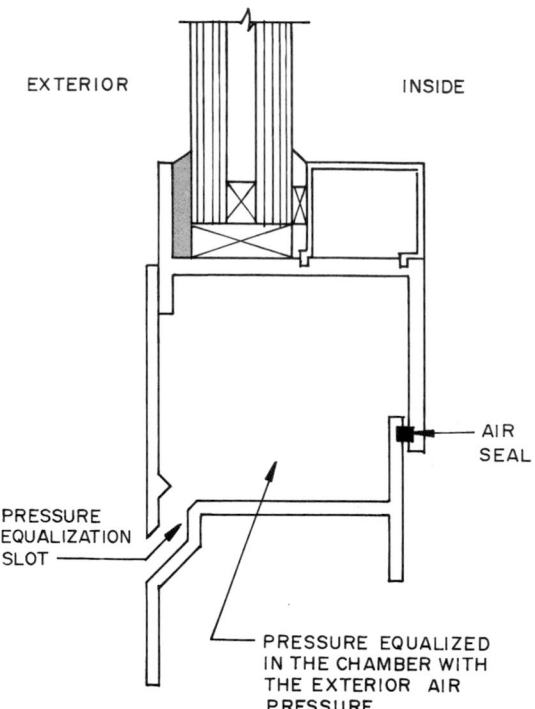

EXTERIOR   INSIDE

AIR
SEAL

PRESSURE
EQUALIZATION
SLOT

PRESSURE EQUALIZED
IN THE CHAMBER WITH
THE EXTERIOR AIR
PRESSURE

8-62 *A typical detail for using the rain screen principle with glazed units.*
Courtesy American Architectural Manufacturers Association, from *Curtain Wall Manual CW 1-9*, "The Rain Screen Principle"

Pressure equalization can also be used with glazing details. The units have a water-resistant seal, a pressure equalization chamber along the edge of the glass, and an inner vapor seal (8-62). The rain screen principle can also be used to resist water penetration by air pressure differences in traditional wood and masonry walls.

ASTM tests for water penetration and wind infiltration are made on curtain wall units with glazing sealants or gaskets installed. The actual conditions can vary during a storm and may under certain circumstances exceed those normally specified.

CONDENSATION CONTROL

As is true with any type of wall construction the cladding must have a vapor barrier on the inside wall facing to prevent water vapor from passing into the wall assembly. If this happens the water vapor will condense and run down the surface of masonry and concrete cladding. If it penetrates into a hollow assembly, such as an aluminum panel assembly, it will condense, reduce the value of the insulation, and cause damage due to freezing and thawing. Some form of drainage is provided to remove any condensation that may occur. Provision for the escape of mois-

ture vapors to the exterior of the building can be provided. Interior surfaces of the panel should be insulated so they are warmer than the dew point of the air within the panel, thus reducing condensation.

DURABILITY & MAINTENANCE

Cladding should last over the expected life of the building. Materials used should have a proven record of durability, including the connections, seals, and joints. If a reduced durability is accepted then maintenance over the life of the building will be greater. The decision may involve choosing a more expensive but durable material that has higher initial cost but a lower maintenance cost or a less costly but less durable material providing a building at lower initial cost but one that will have increasing maintenance costs over the years.

Generally masonry coverings require little maintenance and tend to show weather streaking less than impervious skins. The exterior of a building can be routinely cleaned using vertical ladders or a scaffold or cradle hanging from the gantry arms on a trolley (8-63). Some windows are designed so the outside can be cleaned by someone from inside the building. However, codes may prohibit operating windows. In many buildings the windows are all fixed and must be cleaned from the exterior.

RADIATION & CONDUCTION CONTROL

The cladding must control the influx of radiant heat from the sun into the building that could cause physical discomfort and large air-conditioning costs. Radiant heat is actually radiant energy and not heat. Heat is induced when the radiant energy strikes something. The energy moves from a hot body to a cold body. Therefore a person sitting near a cold wall will radiate heat to the wall making the person feel cold. Radiant energy from the sun will pass through windows and heat up people and materials it strikes inside the building.

Heat is also transferred by thermal conduction. The cladding must prevent the conduction of heat out of or into the building through the wall. Thermal conduction is the process of heat transfer through a material. The cladding must use insulation and thermal breaks in materials that are good thermal conductors. A thermal break is an insulating material placed between thermal conducting materials that touch retarding the passage of heat or cold.

SOUND & POLLUTION CONTROL

The importance of controlling the influx of exterior sounds and air pollution into the building varies with the type of occupancy. For example, a hospital would require greater control than many other occupancies. Likewise, in some cases it is necessary to keep the noise or pollution inside the building, such as with some type of manufacturing or chemical processing operation. The cladding system used would have to be airtight, well insulated, and have in many cases sound deadening material as part of the assembly. Generally the glazed wall areas admit more sound than through spandrel panels.

NATURAL LIGHT

Natural light is used to provide some illumination and relieve the closed-in feeling. In actual practice artificial lighting is on all the time regardless of the amount of natural light available. Sunlight must be controlled to prevent distracting glare and unwanted radiant energy (heat). While some control can be had using various types of solar screening and energy efficient glass the amount of glazing used is a major factor.

WINDOWS

Windows are a part of the exterior cladding systems. They are exposed to the same factors as the actual panels, such as wind load, thermal movement, air infiltration, and water penetration. Their design must enable them to meet the structural requirements as well as have adequate fastening systems and durability.

FIRE RESISTANCE

Cladding is subject to the requirements of the building code. Of great importance is its ability to meet the code in terms of fire resistance. The fire resistance ratings of cladding are determined by tests of the assembly and framing following ASTM test procedures. These involve the combustibility of materials in the cladding, the fire-resistance ratings of the assembled panels and spandrels, and the nearless of other buildings. The required fire resistance of nonload-bearing cladding varies with the occupancy of the building and the **fire separation distance.** The fire separation distance is the distance between buildings and is used to reduce the risk of fire spreading from one building to another. The distances specified in the code typically might be measured in feet between the buildings and the closest lot line, the centerline of the street, or a halfway distance between two buildings on the same site.

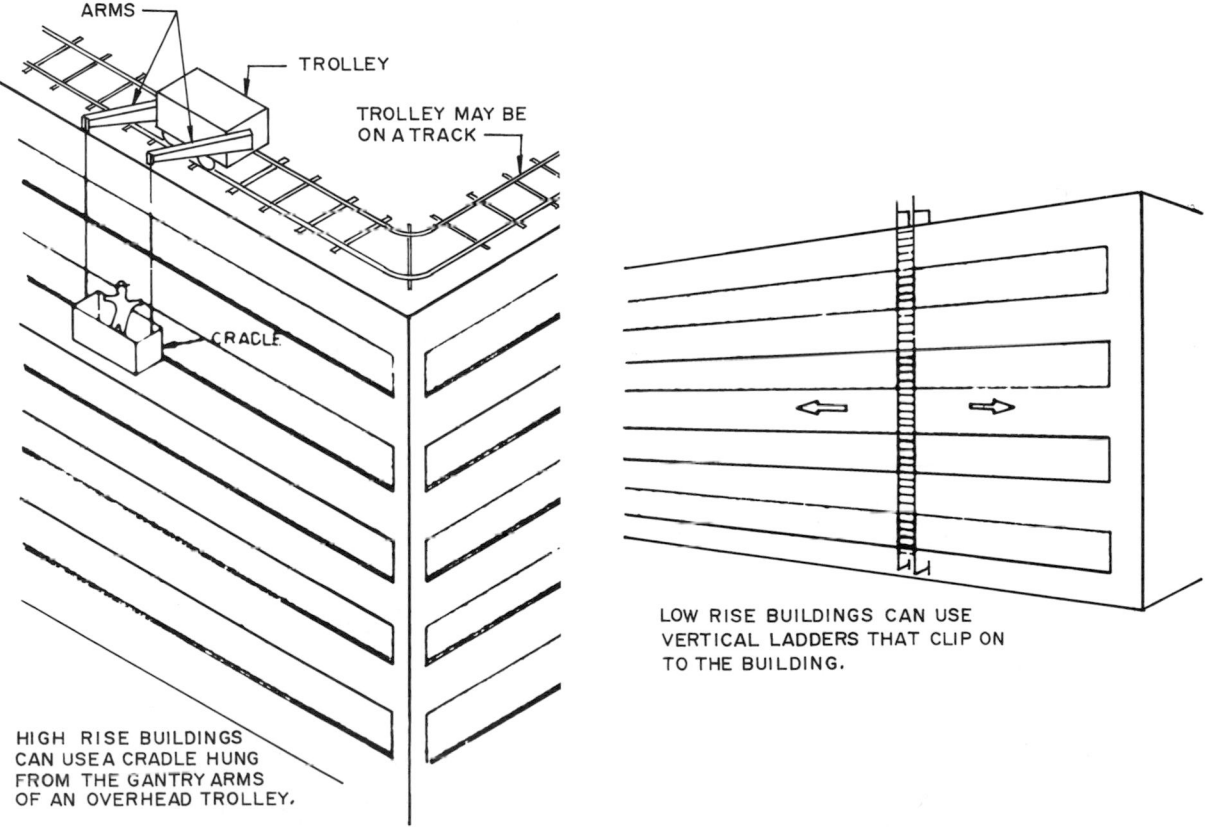

8-63 *Two methods for performing maintenance on the exterior of multistory buildings.*

The connectors should hold the cladding in place during a fire for at least the time specified for the fire-resistance rating of the assembly. Exterior trim and other wall finish materials are also subject to established fire-resistance ratings. Codes also contain requirements for the fire protection of openings in the wall. Another requirement is to install firestopping. Firestopping uses approved materials to prevent the movement of flame and gases through small vertical and horizontal concealed openings in building components (see 8-64). Materials commonly approved for use as fire stops include masonry set in mortar, concrete, mortar or plaster on metal lath, plasterboard, sheet metal of approved gauge, asbestos-cement board, mineral, slag or rock wool if solidly compacted into a confined space. These materials must be permanently fastened in place. They are commonly referred to as **safing.**

8-65 *This building is clad with several types of curtain walls. On the left is an all-glass wall with exposed mullions and rails. On the right it has ribbon windows with a solid material forming the spandrel below.*
Courtesy H.H. Robertson

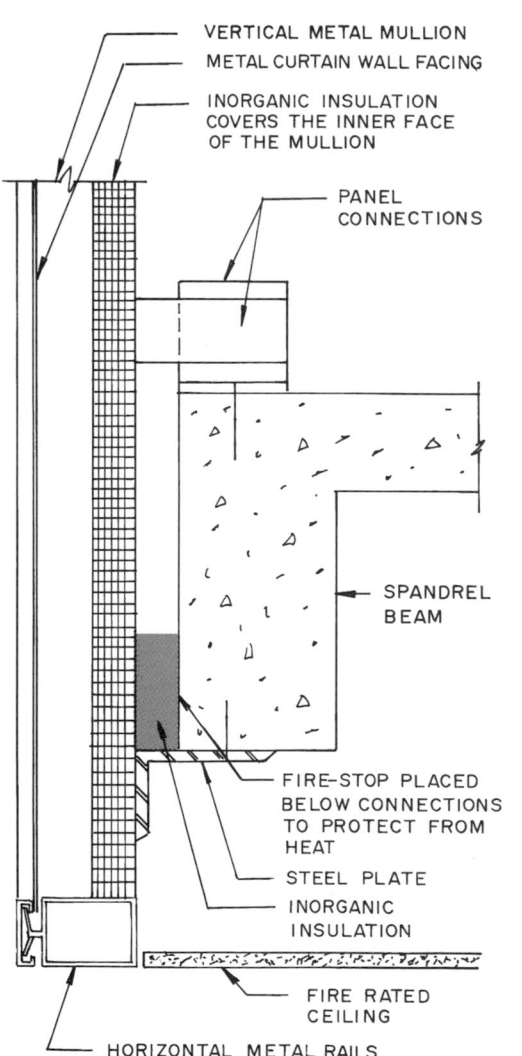

8-64 *One example of firestopping a curtain wall at the spandrel beam.*

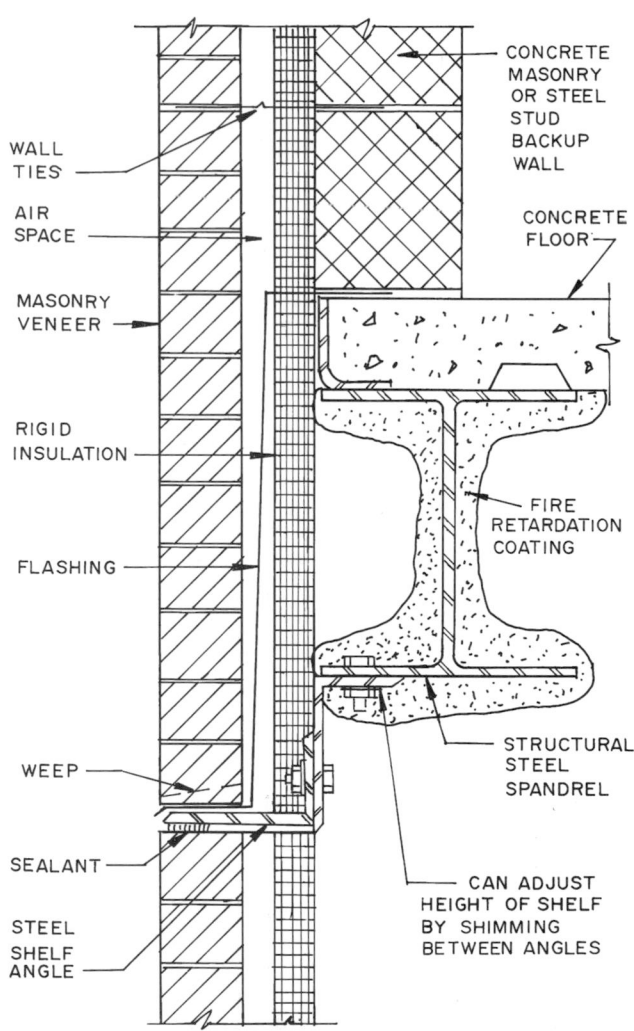

8-66 *A typical detail for a brick masonry curtain wall placed over a steel-framed building. The brick is mounted on a steel shelf and is tied to a backup wall.*

CURTAIN-WALL CLADDING

A **curtain wall** is defined as a non-load-bearing exterior building wall supported by the building framework. A **metal curtain wall** is an exterior non-load-bearing building wall consisting of metal, glass, or other surfacing materials that are supported by a metal framework. A **window wall** is a type of metal curtain wall composed of metal framing members containing operable sash, fixed lights, ventilators, or opaque glass panels. All of these systems are the result of years of engineering development and testing. Various manufacturers supply completely designed systems engineered for a wide range of applications (8-65).

A curtain wall system typically includes panels forming the exposed exterior walls, glazing, mullions, connections, gaskets, and sealants. Curtain walls are lightweight, non-load-bearing, and are anchored to columns, spandrel beams, and floors, and they are generally supported at their bottom edge. Therefore the materials making up the panel are not under tension. The facing panels are made from metal, glass, modified stucco, molded glass fiber reinforced polyester or other code approved materials. Since the curtain wall panels are nonload-bearing

they can be rather thin and lightweight thus reducing the loads on the foundation. This can be a significant saving in weight and construction costs on high-rise buildings. Curtain walls could be a simple single thickness material, such as metal siding, to a multilayer sandwich panel containing insulation. The choice depends upon the desired appearance and the requirements for the interior environment. Following are general examples of details for commonly used curtain walls.

Masonry Curtain Walls

Masonry curtain walls that cover the structural frame are usually laid on a steel shelf angle secured to the frame. The shelf may be anchored to the concrete spandrel beam or bolted or welded to a structural steel spandrel beam (8-66). The wall may be laid up a brick at a time as is done with low-rise masonry construction. Provisions are made for horizontal and vertical expansion joints to permit the panel to expand and contract as temperatures vary. Masonry curtain wall panels can be prefebricated on the ground and the assembly lifted into place with a crane. High-bond mortars are used because they have excellent bonding characteristics and high compressive and tensile strength (8-67).

8-67 *A preassembled curtain wall panel is hoisted into place.*
Courtesy American Brick Company

The masonry curtain wall has a backup wall, typically concrete masonry units or steel studs. It is non-load-bearing and supports sheathing, insulation, air and vapor barriers and interior finish materials. The curtain wall is tied to the backup wall with ties. While the ties are stiff they can flex enough to accommodate differential movements between the veneer and the backup wall.

Most single-width masonry walls will experience some penetration by driving rain. This moisture is handled by the rain screen principle which provides a cavity behind the masonry and weep holes to drain the moisture. The cavity acts as a pressure equalization chamber and provides some degree of pressure equalization between the exterior and interior surfaces of the masonry.

A typical detail showing a ribbon window installation on top of a masonry curtain wall is in 8-68. The aluminum frame has a sill extending into the room and a slot to hold the top of the interior wall finish material.

Stone Curtain Walls

Stone is a widely used material for facing low-rise, mid-rise, and high-rise buildings. This facing material is secured to the structural frame of the building. Commonly used stones include granite, slate, marble, and limestone. The material is cut into panels of various thicknesses ranging generally from 1¼ in. to 4 in. (32 to 102mm) depending upon the material and the size of the panel.

Installation details are recommended by the manufacturer of the panels. They often supply or recommend engineering approved connections. A typical way to support stone panels is with some type of metal subframe (see 8-69).

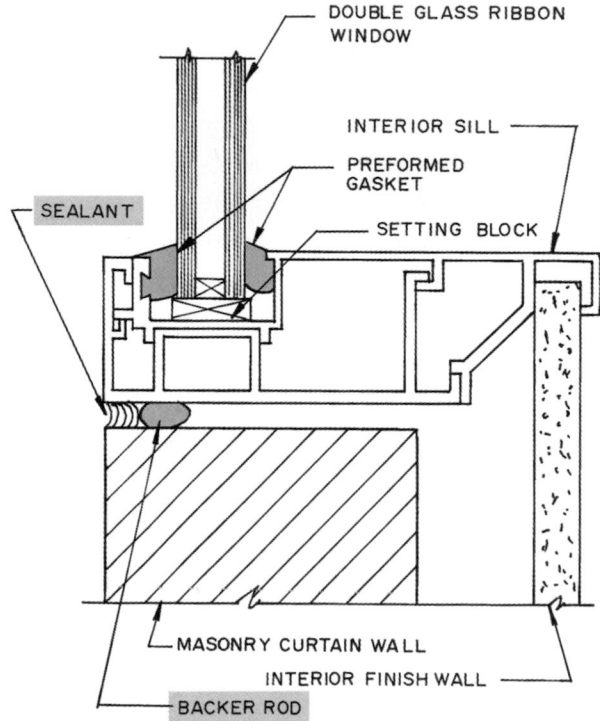

8-68 *A typical installation of a metal window on a masonry curtain wall panel.*

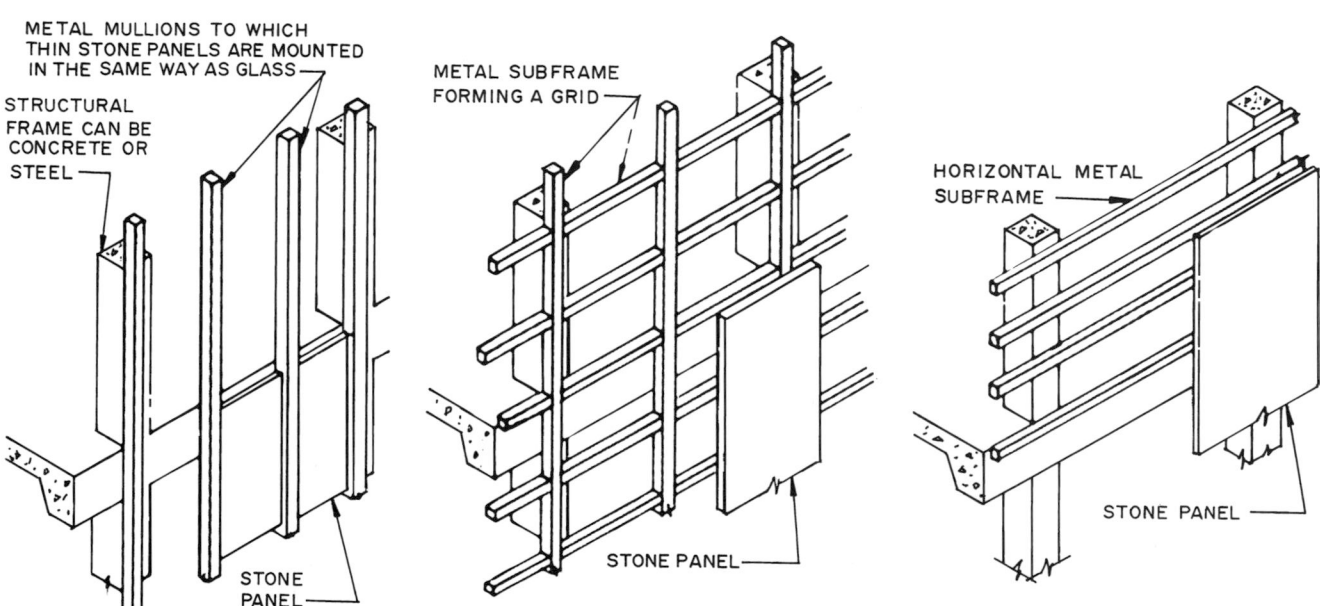

8-69 *Typical types of steel curtain wall subframes used with stone panels. The subframes may be secured to steel on concrete structural systems.*

They can be mounted to the structural concrete frame or to steel or masonry backup walls. One connection system is shown in 8-70.

Precast Concrete Curtain Walls

Precast concrete curtain wall panels are available with a variety of surface finishes and thicknesses. The surface finish may be produced by the surface texture of the casting form or after casting prior to hardening, such as broomed, stippled, floated, troweled, or exposed aggregate. The surface may be recessed or panelized. The design of the reinforced precast concrete curtain wall panels can vary depending upon the exterior appearance specified by the architect and the structural design of the engineer. Several typical panels are in 8-71. They may be made using conventional reinforcing or as pre-

stressed units. The use of glass fiber reinforcing along with steel reinforcement enables lighter and thinner panels to be produced than those using only steel reinforcement. The glass fibers reduce the need for the secondary steel reinforcing used to resist cracking and help control thermal expansion and contraction.

Precast panels are installed basically in the same manner as stone panels. Typically anchors are cast into the panel and angles are bolted to the panel and the structural frame. The panels can be insulated by bonding rigid insulation to the inside surface of the panel or sandwiching rigid insulation in the center of the panel. Allowances between panels must be made to allow for expansion and contraction. An adequate sealant is required to provide a waterproof wall.

Modified Stucco Curtain Walls

Modified stucco curtain wall panels consist of an insulation board base covered with one or more fiberglass reinforcing mesh layers bonded to it and fully embedded in a basecoat. Over this is trowelled a synthetic plaster finish coat made of an acrylic polymer and Type 1 portland cement.

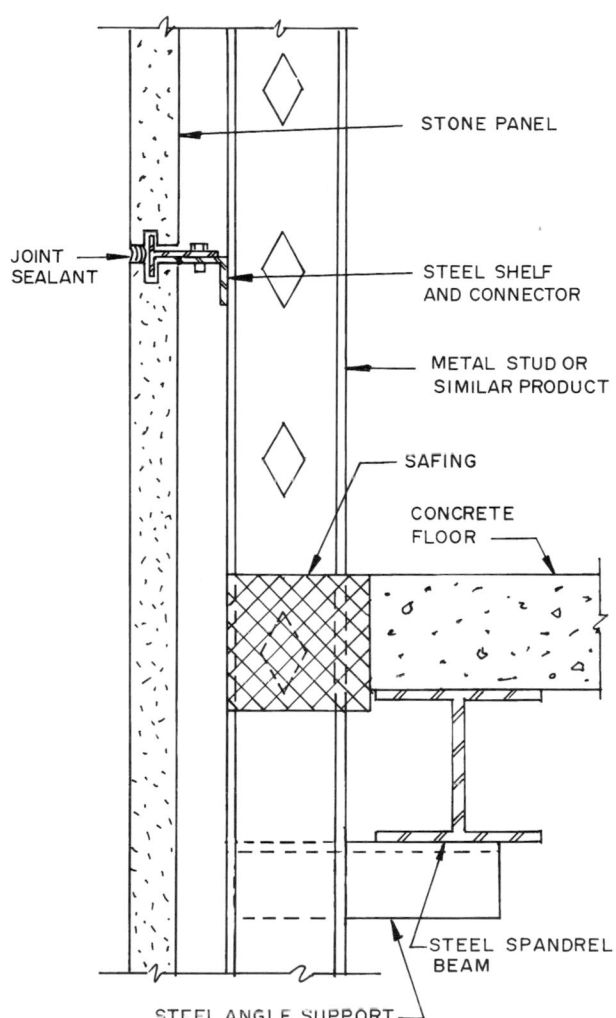

STEEL STUD WALL WILL CONTAIN INSULATION

STONE PANEL

JOINT SEALANT

STEEL SHELF AND CONNECTOR

METAL STUD OR SIMILAR PRODUCT

SAFING

CONCRETE FLOOR

STEEL SPANDREL BEAM

STEEL ANGLE SUPPORT

8-70 *A stone curtain wall panel mounted over a steel structural system and steel studs.*

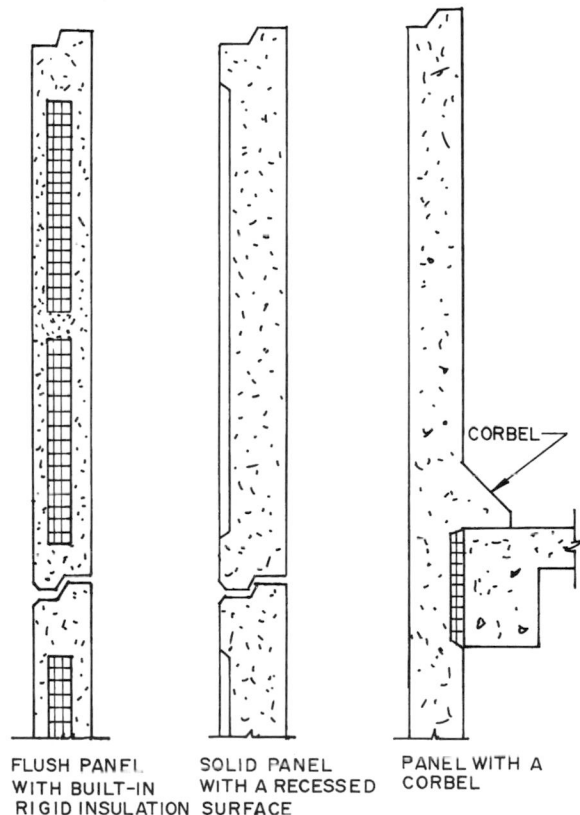

FLUSH PANEL WITH BUILT-IN RIGID INSULATION

SOLID PANEL WITH A RECESSED SURFACE

CORBEL

PANEL WITH A CORBEL

8-71 *Some of the types of precast concrete curtain wall panels available.*

The finish coat is available in many colors and applied in various surface textures. This system is used for residential and commercial cladding. When used for curtain wall construction on low-, mid- and high-rise building the assembly may be secured to a backup wall such as steel studs or masonry as shown in 8-72. An approved substrate is joined to the steel studs and the panel is bonded to it with a special adhesive specified by the manufacturer. Approved substrates are those recommended by the manufacturer. Typically they include unpainted brick, unit masonry, concrete, gypsum sheathing, stucco brown coat, diamond mesh metal lath, and certain cementitious and wood sheathings.

Curtain wall panels one or more stories high are constructed by securing the assembly to a light gauge steel frame covered with exterior grade gypsum sheathing. The gypsum sheathing is screwed to the lightweight steel frame. The panels are bonded to the gypsum substrate with a special adhesive. When finished the assembled panels are lifted into place and bolted or welded to connectors on the structural frame of the building as described for other types of curtain wall construction (8-73).

Metal & Glass Curtain Wall Systems

Glass and metal curtain walls are widely used because they provide a lightweight exterior wall and are rapidly erected (8-74). Various types of glass panels are used. The metal framing is usually aluminum. The glass panels are typically insulating glass, laminated glass, heat reflective glass, and spandrel glass. Spandrel glass is a heat-strengthened glass with a ceramic frit fired on it. Frit is a molten glassy material. It is recommended that rigid insulation or some other rigid backing material be bonded to the back of the glass spandrel panel. This helps hold the glass in place should a panel fracture due to high internal temperatures or storm damage.

Metal and glass curtain wall systems are available in two basic types, custom and standard. **Custom walls** are those designed especially for a particular building and are built by the manufacturer for that application. **Standard walls** are those that are standardized by the manufacturer and can be assembled from stock parts. Custom and standard curtain wall systems can be classified by the method of installation. The following classifications are detailed in the publication *Aluminum Curtain Wall Design Guide Manual* published by the American Architectural Manufacturers Association.

The **stick system** mullions are installed followed by horizontal rails into which glazing and spandrel panels are placed (see 8-75). A typical aluminum-framed stick system with vision glazing and ceramic tile-faced spandrel panels is shown in 8-76.

The **unit system** uses large preassembled wall panels. The vertical edges of the units are joined, serving as a mullion, and the bottom of one panel joins the top of the one below forming a horizontal rail (8-77).

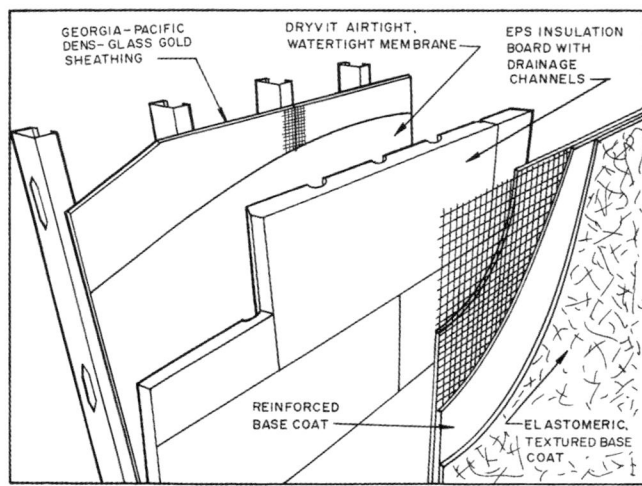

8-72 *A typical installation detail showing construction of a modified stucco cladding curtain wall as recommended for commercial construction.*

Courtesy Dryvit® Systems, Inc.

8-73 *Preassembled modified stucco curtain wall panels can be several stories high.*

Courtesy Dryvit® Systems, Inc.

8-74 *A glass-to-the-front, stick fabrication framing system has the advantage of fast field erection and significant savings. Used effectively in ribbon-type applications, this multipurpose framing system may be inside or outside glazed and has an option for structural silicone glazing. It accepts infills of ¼ inch or 1 inch with a minimum front mullion projection.*

Courtesy Kawneer Company, Inc.

8-76 *A typical stick system curtain wall with vision glazing and ceramic tile-faced spandrels.*

Courtesy United States Ceramic Tile Company

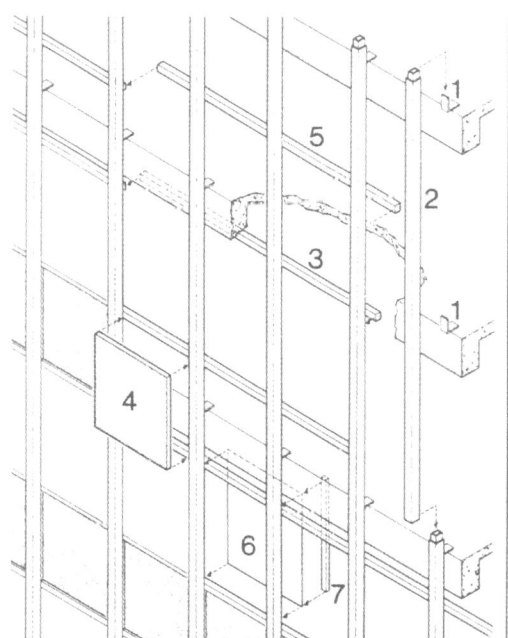

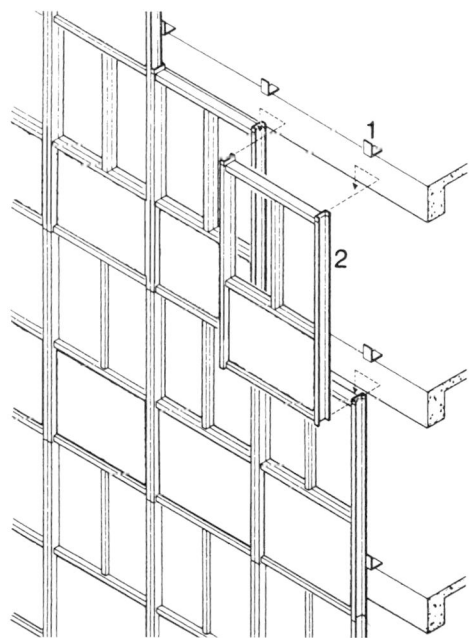

8-75 *The stick curtain wall system. 1. anchor 2. mullion 3. horizontal rail 4. spandrel panel 5. horizontal rail 6. vision glass*

Courtesy American Architectural Manufacturers Association, from *Aluminum Curtain Wall Design Guide Manual*

8-77 *The unit curtain wall system. 1. anchor 2. preassembled framed curtain wall unit.*

Courtesy American Architectural Manufacturers Association, from *Aluminum Curtain Wall Design Guide Manual*

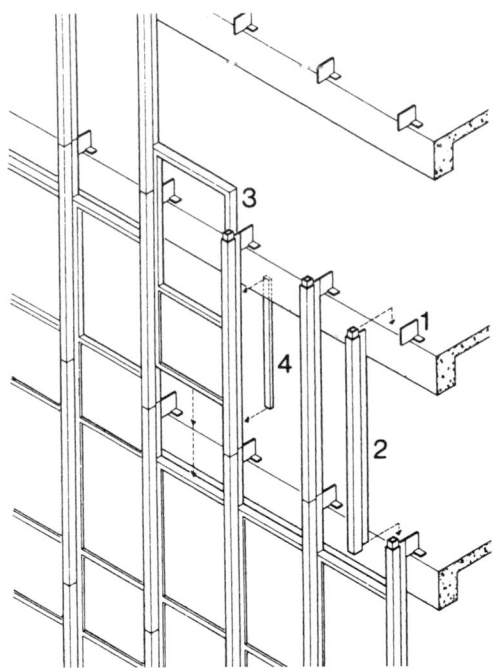

8-78 *The unit-and-mullion curtain wall system. 1. anchors 2. mullion 3. preassembled curtain wall unit lowered into place behind the mullion from the floor above 4. interior mullion trim.*

Courtesy American Architectural Manufacturers Association, from *Aluminum Curtain Wall Design Guide Manual*

The **unit-and-mullion system** uses mullions secured to the building structure and preassembled wall panels are installed between the mullions (8-78).

The **panel system** uses homogenous wall panels that can be precast concrete, some form of molded plastic, or stamped from sheet metal. They may or may not have openings for windows. They are connected to the structural frame as shown in 8-79.

The **column-cover-and-spandrel system** uses long spandrel panels that span between the column covers. Glazing may be installed above the spandrel (8-80).

Many other system designs are being developed, including some with little or no exposed framing.

There are many factors to consider when designing glass curtain walls and glass spandrels. For example, when glass is exposed to sunlight the glass temperatures rise. If the glass panels are glazed directly to a material that easily absorbs heat, such as concrete, the edge of the glass is cooler than the center of the panel. The center being hotter expands more than the edges causing increased stresses at the edges and may cause the panel to fracture. The design must allow for air circulation. This is also a consideration when interior shades are

8-79 *A panel curtain wall system. 1. anchor 2. preassembled curtain wall panel.*

Courtesy American Architectural Manufacturers Association, from *Aluminum Curtain Wall Design Guide Manual*

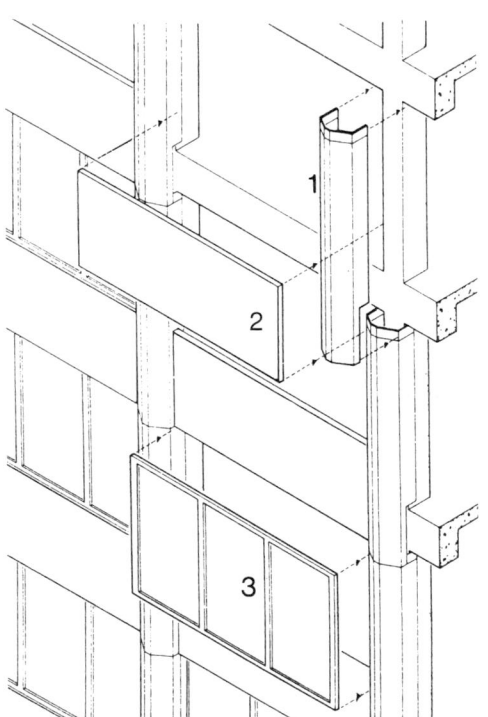

8-80 *The column cover and spandrel curtain wall system. 1. column cover 2. spandrel panel 3. glazing unit.*

Courtesy American Architectural Manufacturers Association, from *Aluminum Curtain Wall Design Guide Manual*

used. They must be hung several inches from the glass curtain wall to permit air to circulate between the shade and the glass. Exterior sun shading of glass curtain walls also can produce uneven temperatures in the panel because often only a part of the glass is shaded, producing differences in temperature. If most of the panel is shaded the stresses are minimal.

The large glass panels used in curtain wall construction are heavy. They must have the strength to span a distance equal to the width of the panel and to resist wind loads. They must be isolated from the frame so expansion, contraction, and possible bending from wind loads will not fracture the glass. The curtain wall system is designed so the frame provides the amount of grip around the edge of the glass to hold it in place when under stress. The frame and mullions are designed to withstand the combined stresses of weight and wind loads.

Glass manufacturers frequently specify the minimum compressive pressure of the gasket on the glass surface. Typically a pressure of at least 4 pounds per linear inch (700 N/m) is needed to provide support for the glass edge and a watertight seal. Excessive pressure, usually over 10 pounds per linear inch (1750 N/m), can increase mechanical stress and may contribute to glass breakage.

Curtain Wall Joint Sealants

The choice of joint sealant depends upon the materials to be sealed, the design of the joints, and possible changes that may occur in the joints after the sealant has been installed. Joints may be **working joints,** which are designed to allow movement, or **nonworking joints,** which are joined by a fastener so they do not move (see 8-81). Sealants may be flowable compounds or solid materials.

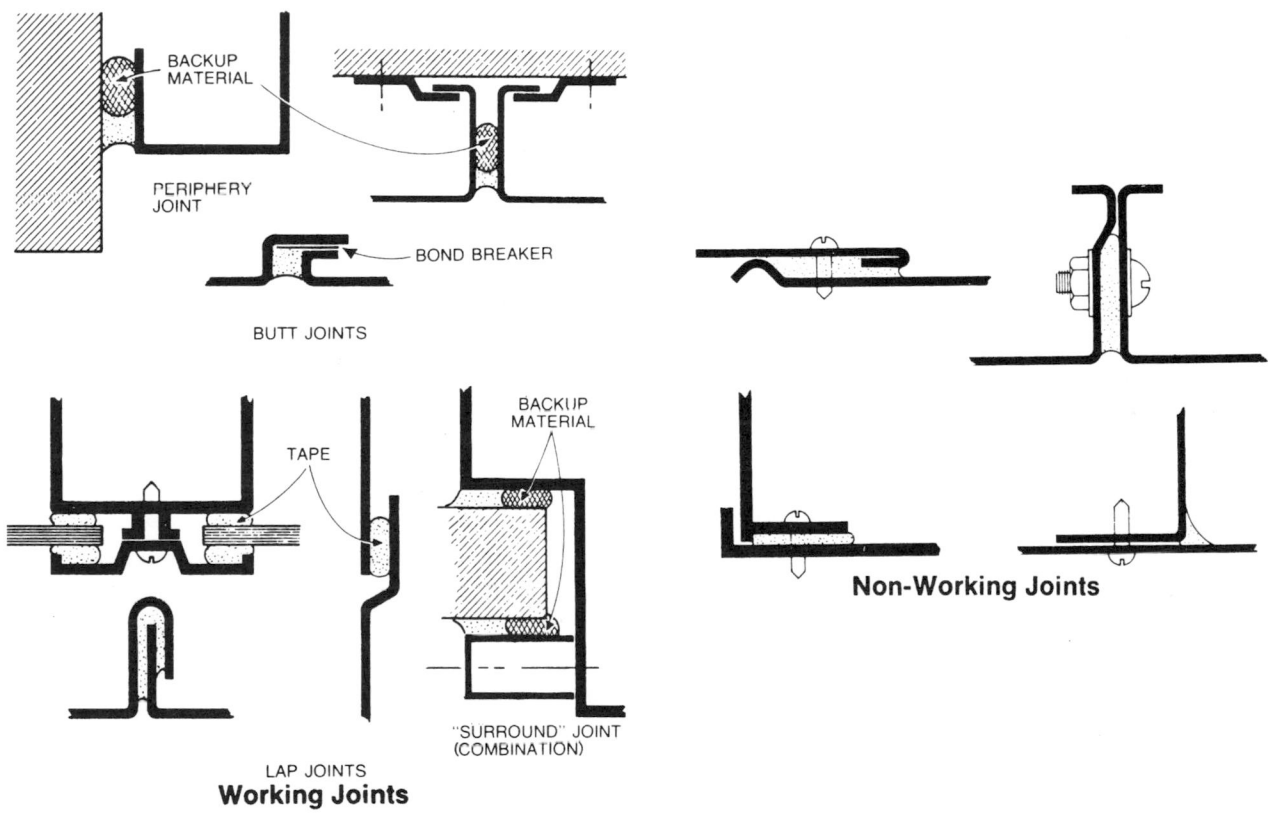

8-81 *Some of the frequently found shapes used as edge joints of joining members. Notice the use of sealants.*
Courtesy American Architectural Manufacturers Association, from AAMA *Joint Sealants Manual*

FLOWABLE SEALANT MATERIALS

Flowable sealant materials have adhesive qualities and are applied either by a gun or a knife. Some types are in an extruded preformed strip. They adhere to the surfaces upon which they are applied and cure into a rubbery material sealing the joint. They allow for expansion and contraction in the joint because they can stretch and contract without fracturing.

▼ Flowable sealant materials are grouped into three classes depending upon the amount of change in joint size they can accommodate.

Low-performance sealants have minimum movement capacity and are used in stable joints. Typical materials include oil-and-resin-based compounds, bituminous based caulks and mastics, and polybutrene compounds.

Medium-performance sealants can accommodate elongations in the range of +5% to +12.5%. Typical materials in this group include acrylics, butyls, and neoprene.

High-performance sealants are used on situations where the cyclic movement is expected to be between +12.5% to +25%. Typical materials in this group include polymercaptans, polysulfides, polyurethanes, silicones, and some solvents release acrylics.

SOLID SEALANT MATERIALS

Solid sealant materials include tapes and gaskets.

Preformed solid tapes are made from polybutene or polyisobutylene and have adhesive on one or both sides. They are available in rectangular, square, circular, and wedge shapes. They are completely cured when manufactured and seal by being compressed. They are mainly used on lap joints.

Preformed cellular tapes are made from polyurethane, neoprene, vinyl nitrile and polyvinyl chloride. They contain a chemical solvent and are delivered in a tightly compressed condition. When installed they expand, filling the joint and bond to the contacting surfaces as they cure.

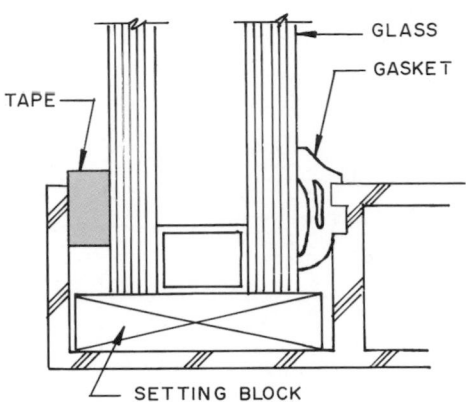

8-82 *Typical gasket and tape sealant applications on a glass unit set in an aluminum frame.*

Gaskets are preformed solid elastomeric materials designed to fit the configuration of materials and panels to be sealed. They are compressed and forced into the joint forming a watertight seal (see 8-82).

Building Code Requirements

Building code requirements consider factors such as fire resistance, structural characteristics, and connections. Fire requirements specify the ability of the cladding to resist fire and prevent the spread of fire to or from neighboring buildings. Critical is the inclusion of firestops between the cladding and each floor. Firestops are also required on other vertical openings such as covers on columns. Firestops are usually mineral wool safing, gypsum on metal lath, or steel plates with grout. All firestops (safing) must be securely fastened to the edge of the floor or other structural member.

Structural requirements include factors as the strength of the cladding to support its own weight over the distance it spans between connectors, accommodating forces of cyclical expansion and contraction, resisting wind loads, and meeting seismic requirements. The method of connecting the cladding to the structural frame is also subject to analysis. One consideration is whether the designer should rely on local and national standards pertaining to wind loads or conduct wine tunnel tests of the building. Some cities require a wind analysis of buildings above a specified height.

DIVISION 9

FINISHES

CSI MasterFormat™

Courtesy Harris Tarkett, Inc.

9-1 *These exposed glued laminated wood arches and roof purlins provide a major architectural detail to the interior of the room. The ceiling is exposed finished wood decking.*
Courtesy American Institute of Timber Construction

INTERIOR FINISHES

Interior finishes refer to the wall, ceiling, and floor finishes, and other material applied to them such as paneling, wainscoting, or those used for acoustical treatment, insulation, decoration, and other features. Their installation is carefully regulated by the various building codes, which vary depending upon the proposed occupancy of the building. Interior finishes include a wide range of materials as shown by the Master-Format™ outline on the previous page introducing this division. These systems and materials are described in other chapters.

The selection of interior finish materials, in addition to meeting code requirements, is made considering factors such as cost, durability, appearance, coatings, and acoustical and fire requirements. The architect and the owner work to select materials that meet the requirements of the building and the building code.

Interior finishing begins after the building has been enclosed protecting the interior from the weather. The roof is finished, cladding installed, and doors and windows set. The electrical and mechanical trades can begin to install the electrical, communications, telephone, and computer systems as well as waste and potable water lines, automatic sprinkler system, and the heating, cooling, and ventilation system including the air handlers, boilers, pumps, chillers, fans, cooling towers, and related ductwork. Installation of transportation systems such as elevators, escalators, and moving walks and ramps can also be started.

Cost

The cost of interior finish materials involves consideration of the price of the material plus the labor to install it, and the contractor's overhead and profit. Other cost factors include the expected life of the installation before replacement is required, regular maintenance, minor repairs, and the possibility of increased replacement years later because of inflation and increased labor costs.

Calculations can reveal if it would be less costly over the long run to use higher quality, more expensive materials. This will frequently be the case if the owner plans to retain title to the building over a long time period.

Appearance

The interior finish generally hides the mechanical and electrical systems and provides an attractive interior surface. In some cases portions of the structural system are left exposed and become part of the interior design (see 9-1).

The materials used are directly influenced by the desired appearance of the interior spaces and their proposed use. For example, a hotel lobby may have darker textured walls, carpeted floors and stairs, and elaborate wall hangings.

The lighting system can provide soft general illumination yet spotlight features as paneled walls and paintings. The interior of a health and fitness center will be in sharp contrast to this by having light durable wall surfaces, acoustical ceilings, considerable natural light, and floor coverings to suit the activity in the area. The halls and lobby may have a durable clay tile floor, aerobics room a soft resilient floor covered with carpet, and a basketball area with a composition floor.

The color of the interior finish establishes the mood and reactions to the area. Warm colors such as those ranging from reds through orange and yellow create a feeling of friendliness and warmth. Cool colors such as those ranging from greens through blues and purples create a feeling of coldness. Lighter colors increase the level of illumination while darker colors absorb some light lowering the amount of reflective light. Generally interior finish uses a range of both warm and soft colors. The texture of the finished wall and ceiling surfaces also influences the appearance and other qualities such as acoustical properties and light reflection.

COATINGS

Protective and decorative coatings are applied over many types of interior walls, ceilings, and floors. Coatings are layers of material in liquid form applied to a surface to decorate, preserve, protect, seal, or smooth it. When the coat solidifies it may leave a flat, semi-gloss, or gloss finish. The coatings may be transparent, semi-transparent, or opaque. They are made from a wide range of materials that determine the color, hardness, and opacity of the coating. The type of coating used is regulated by building codes. Fire resistance and the support of combustion are important. Codes will specify the flame spread, smoke, and fuel contributed ratings of the coatings.

Other surface coverings may be paper, plastic, fabric, or thin wood veneers. They are subject to the same code restrictions as liquid coatings.

Durability

The durability of interior finishes includes the ability of the exposed material and any protective or decorative coating to resist damage due to bumps, abrasion, or soil. Hard materials or hard-surfaced materials will provide considerable protection and are used in areas where wear and tear are to be expected. In some areas, such as a commercial kitchen or restrooms, the ability to withstand water and high humidity is important. Floor coverings are especially vulnerable to wear and damage. A lobby of a busy building will require very durable floor covering while other areas may be able to use something less expensive. The ability to easily clean the materials with minimum damage to them is also a factor. Finally, the expected life of the product and replacement costs must be a factor when considering the degree of durability.

Acoustical Considerations

The acoustical properties of the finish materials on walls, ceilings, and floors are another major consideration. The type of material, texture of the surface, and any coatings will influence the acoustical properties. Under consideration is the ability of a finish to control the sound created within an area as well as restrict its flow through walls, ceiling, and floor to adjoining areas. Sound transmitted through materials may be due to vibrations in the material caused by actions such as someone walking on the floor above or by machinery operated on the floor. Assemblies can be tested to ascertain their **impact noise rating.** This can be reduced by insulating the floors and covering them with soft materials such as a soft underlayment, carpet padding, and carpet and must be considered as the finish is planned.

Sound transmission of airborne noise through a material or assembly of materials is indicated by its **sound transmission class (STC).** This is a single number that indicates the effectiveness of the material to reduce the transmission of airborne sound through the material into the next area. The higher the STC rating the more efficient is the material in blocking the transmission of sound. Building codes have regulations for the transmission of sound between areas for some types of construction such as noise from a hall passing into apartments. The manufacturers of various finishing materials have them tested and can provide the needed information.

The amount of airborne sound energy absorbed by a material is indicative of its **noise reduction coefficient** (NRC). It is a single number rating used to calculate the quantity of sound absorbing material required. Various materials can be compared using this rating.

Sound can also be transmitted between areas by such things as poorly fitting doors, openings around electrical outlets and piping, partitions that are not sealed at the floor or ceiling, through air ducts, and ceilings that have inadequate sound proofing.

Fire Considerations

Interior finish and trim are subject to a wide range of building code requirements pertaining to fire.

The combustibility of interior finish materials are rated by testing the flame spread of the surface of the material. **Flame spread** is the rate at which combustion will spread across a material. The **flame spread rating** is the measurement of the ability of a material to resist flaming combustion over its surface designated by a single number. The rate of flame travel is measured under ASTM testing procedures. Noncombustible cement-asbestos board has a rating of O. An untreated specified species of wood has a designated rating of 100. The lower the rating the slower the flame will spread across its surface.

▼ Flame spread ratings are:
   Class 1—flame spread, 0-25
   Class 11—flame spread, 26-75
   Class 111—flame spread, 76-200

Table 9-1 MAXIMUM FLAME-SPREAD CLASS[a]

Occupancy Group[b]	Enclosed Vertical Exitways	Other Exitways[c]	Rooms or Areas
A	I	II	II[d]
E	I	II	III
I	I	I[e]	II[f]
H	I	II	III[g]
B, F, M & S	I	II	III
R–1	I	II	III
R–3	III	III	III[h]
U	No restrictions		

[a]Foam plastics shall comply with the requirements specified in Section 1001.5. Carpeting on ceilings and textile wall coverings shall comply with the requirements specified in Sections 804.2 and 805, respectively.

[b]A—Assembly building

B—Factories, office buildings, retail stores

E—Educational buildings

H—Hazardous materials

I—Nursing homes

M—Private garages, agricultural buildings

R-1—Hotels, apartment houses

R-3—Residences, lodging houses

[c]Finish classification is not applicable to interior walls and ceilings of exterior exit balconies.

[d]In Group A, Divisions 3 and 4 Occupancies, Class III may be used.

[e]In Group I, Divisions 2 and 3 Occupancies, Class II may be used or Class III when the Division 2 or 3 is sprinklered.

[f]In rooms in which personal liberties of inmates are forcibly restrained, Class I material only shall be used.

[g]Over two stories shall be of Class II.

[h]Flame-spread provisions are not applicable to kitchens and bathrooms of Group R, Division 3 Occupancies.

The acceptable flame spread ratings for various types of buildings and occupancies are specified by the building code. See Table 9-1.

Smoke development ratings classify materials by the amount of smoke it will give off as it burns. Most codes prohibit materials having a rating of 450 or more to be used inside a building. Materials are tested using ASTM test procedures. Class 1, Class 11, and Class 111 ratings permit smoke developed to range from 0 to 450.

Fire-resistance ratings indicate the capacity of a material to withstand fire for a specified time and under conditions of standard intensity so it will not fail structurally and will not permit the side away from the fire to become hotter than a specified temperature. The ratings are given in hours. Codes will also specify regulations pertaining to flame resistance materials. Any material such as curtains, draperies, or other decorative materials are required by code to be flame resistant.

Floor finish materials, such as wood, composition tile, and carpet are also regulated by building codes. Doors and windows are also controlled by code.

PROTECTIVE & DECORATIVE COATINGS

Coatings are layers of materials in liquid form applied to a surface to decorate, preserve, protect, seal, or smooth it. When the liquid coat solidified it leaves a thin layer over the surface (substrate) to which it was applied. Coatings are used to protect a material from heat, soiling, abrasion, solar radiation, moisture, chemicals in the air or in a solution, and corrosion. They also are used to provide a decorative surface.

A coating consists of a primer coat, an intermediate coat sometimes called an undercoat, and a finish coat also referred to as a topcoat.

Primer coat—is applied directly to the substrate (such as steel, wood). It must have good adhesion to the surface, have appropriate flexibility, permit the intermediate coat to bond to it, retard corrosion of the substrate, and resist weathering long enough to permit the application of the intermediate and finish coats.

Intermediate coat—is applied over the primer coat and must provide an adequate film thickness and structural strength, bond to the primer coat and permit the finish coat to bond to it, and be an excellent barrier to chemicals and other environmental contaminates.

Finish coat—(topcoat) is applied over the intermediate coat and provides the initial barrier to action of the environment, gives a pleasing appearance, and may be required to serve other functions such as providing a nonskid surface or resist fouling agents as found in marine situations.

It can be seen therefore that the total coating system must be one that is compatible with the substrate and each coating must be compatible with the next one to be applied. If this does not occur the coating will deteriorate and have to be completely removed before the substrate can be recoated.

Air Quality Regulations

The 1970 **U.S. Clean Air Act** mandates National Air Quality Standards requiring that health-based limits on ozone and carbon monoxide be observed. This includes regulating emissions from automobile exhausts, **construction, manufacturing,** and other areas including all products that emit volatile organic materials (VOM), which includes paints and other coatings. Emissions from solvents in coatings are one of the largest sources of air pollution.

Volatile organic materials (VOM) include any solvent, propellant, or other substance, except water, which evaporates from a coating as it dries. VOM is calculated as the combined weight of **all solvents** in a given volume of coating, excluding certain exempted materials. VOM is specified in **grams per liter (g/l),** which can be converted to pounds per gallon (lb/gal) by dividing g/l by 119.8.

LEAD IN COATINGS

Lead is now prohibited from being used in paints. An ingestion of loose paint dust or chips can cause lead poisoning. Typical hazards include harm to workers on remodeling projects where paint containing lead was used in previous years and for children living in older houses and apartments who may eat loose paint chips pulled from the trim and cabinets.

Architectural coatings are limited to not more than 250 g/l (2.09 lb/gal) and **industrial maintenance** primers and topcoats to no more than 420 g/l (3.5 lb/gal). Architectural coatings are used on surfaces exposed to mild environments. Industrial maintenance coatings are used to protect surfaces against chemical corrosion, high temperatures, abrasion, immersion in water, and solvents.

Some currently used water-borne coatings available meet VOC requirements. However just because they are a water-based product does not mean they meet the requirements. Some water-based coatings contain sufficient co-solvent to raise the VOM above acceptable limits. Many water-borne coatings cannot meet the severe conditions faced by the industrial maintenance environment. Currently some coating manufacturers have solvent- and water-based products that meet VOC requirements and provide the required protection. For example, the Rust-Oleum Corporation has a water-based two-compound amine epoxy coating having a VOM level of 250 g/l that meets emission requirements. Contractors must observe the VOM ratings of the coatings they use or see if an exempt product is available. Manufacturers are placing VOM ratings of their products in their sales literature.

The implications of the Clean Air Act as it is enforced by various states means that many of the coatings mentioned in this chapter may cease to be used as newly formulated coatings that meet the limitations are developed. New coatings will have less solvents, be water-based with little or no co-solvent, or be comprised of 100% solids. This will also influence how these new materials will be applied. Brush application will possibly become less viable and new application equipment, including newly designed spray systems, will be developed. It is possible that more coatings will be factory applied and less on-the-job application will occur.

The constructor must be constantly aware of these changes as they take place.

Types of Coating

Coatings may be clear, semi-transparent, or opaque.

Clear coatings—let most of the natural color and texture of the substrate show through. They are used when it is desired to show the natural appearance, such as the color and grain of wood or the color and texture of colored concrete or the color of the exposed concrete aggregate.

Semitransparent coatings—allow some of the color and texture to show but obscure much of it. They are used when the appearance of the substrate is to be changed, such as adding a bit of color, yet permitting the material, as the grain in wood, to show.

Opaque coatings—completely obscure the color and much of the texture of the substrate. They are used when a solid, uniform colored surface is desired. The material composing the substrate is often not identifiable.

Composition of Coatings

There are many different kinds of coating available and their composition varies widely. However, they all have the same basic composition. This includes a binder, solvent, and pigments.

Binder—is a nonvolatile film that forms the base of the hardened coating.

Solvent—is the volatile part of the coating in which the binder is dispersed.

Vehicle—of the coating is formed from the mixture of the binder in the solvent.

Pigments—are insoluble particles that are suspended in the vehicle and give it color and opacity.

Solids—of the coating are made up of the binder plus the pigments and produce the layer remaining after the solvent has evaporated.

Clear coatings are made of binder and solvent but contain no pigments. **Semitransparent coatings** have a small amount of pigment providing some opacity. **Opaque coatings** contain considerable pigment and totally obscure the face of the substrate. The binder bonds itself to the substrate. It also must have the properties needed to meet the requirements of the coating. For example, it must have plasticizers so it is flexible, stabilizers to enable it to resist solar radiation or sources of heat, and driers to control the rate of curing.

Table 9-2 WET FILM THICKNESS BASED ON THE SPREADING RATE

Spreading Rate (sq. ft./gal.)	Wet Film Thickness (mils)
1600	1.0
1000	1.6
700	2.3
400	4.0
200	8.0
160	10.0

Coatings are either solvent-based or water-based. **Solvent-based coatings** are volatile while **water-based coatings** have the binders and pigments dissolved in water. Solvent-based coatings require the use of a solvent, as paint thinner, to thin the mix and clean brushes and spray guns. The solvent used depends upon the composition of the coating. Water-based coatings are thinned and tools cleaned with water.

Coatings are specified for specific uses. Manufacturer's directions clearly tell whether they can be used for exterior or interior purposes or in some cases both. They specify the substrate (wood, metal, masonry, etc.) upon which the coating can be applied and if it will resist attack by chemicals, heat, and ultra-violet rays. In addition the final appearance is indicated. This includes clear, semitransparent, and opaque coatings and the amount of gloss as high gloss, satin gloss, or flat (no gloss).

Application Specifications

Coating specifications may include wet film thickness and/or dry film thickness.

To ensure expected performance the proper thickness of paint must be applied. This is specified by giving the **minimum wet film (MWF)** thickness. This is stated "applied at the rate of xx mils per coat minimum wet film thickness" (MWF). A mil is a unit of measure equal to.001 in. (0.025mm).

WET FILM THICKNESS

The wet film thickness varies depending upon the spreading rate per gallon. The dry film thickness of a coating can be determined by measuring the wet film thickness with special gauges designed for the purpose or by using an industry-developed formula that relies on the spreading rate. Table 9-2 shows some typical wet film thicknesses found using the formula. The dry film thickness is found by multiplying the wet film thickness in mils by the percent of solids in the paint. This gives a theoretical dry film thickness. The percent of solids in a coating is determined by the manufacturer of the material.

The minimum wet film thickness for various coatings can also be found by noting the recommended spreading rates indicated by the manufacturer of the coating material.

Factors Affecting Coatings

After coatings have been applied to a substrate and have hardened they are subject to many external factors. These must be considered as a choice of coatings is considered.

Water is one of the most common factors. It may rain on external surfaces, which may cause a thermal shock if the coating is very hot. It may penetrate the coating through checks and cracks and freeze or become very hot and cause blisters. **Water vapor** may penetrate the coating from the outside resulting in damage to the substrate, which causes the coating to peel and crack. Water vapor may penetrate the substrate from behind the coating again causing blisters and peeling. This is one reason why vapor barriers are used on the interior side of exterior walls. The substrate, such as wood, may contain moisture before the coating is applied. The moisture content of the substrate must be within specified limits for successful coating. Excess moisture in the substrate will cause poor adhesion of the coating.

Solar radiation bombards the coating with ultraviolet radiation, which could cause fading of the pigment and some chemical reactions in the solvents and binders causing coating deterioration. This permits ultraviolet radiation to reach the substrate causing possible damage to it. Solar radiation causes an increase in the **temperature** of the coating and the substrate, which causes both to expand. The coating must be flexible enough to expand and contract without cracking, including expansion of the substrate that may exceed the normal expansion of the coating at the same temperature. Likewise, freezing temperatures cause damage, especially if moisture has penetrated the coating.

Coatings are subject to **dust** and **dirt.** Their location will indicate the extent of possible damage. A ceiling, for example usually has minimum exposure while a baseboard or door has greater exposure. Location also dictates exposure to **abrasion.** Impact due to natural causes, such as hail, or normal wear as on doors or vandalism must be considered.

In some situations coatings are affected by **chemical fumes, solutions,** and **reactions.** Chemical fumes are generated by many sources, including power plants and automobile emissions. Seawater, oils, solvents, and other chemical solutions impact heavily on coatings. Soluble alkaline salts in mortar and concrete can dissolve and crystalize on the surface damaging the coating. This is called **efflorescence.** Sealants may react unfavorably while rust may streak the coating surface. Wood knots heavy with resin may bleed through the coating. These and other possible reactions to coatings used must be carefully considered.

The **absorption** of the surface of the substrate will affect the coating. Some surfaces are hard and smooth and absorb none of the coating. Such surfaces may require roughing by sanding, sandblasting, or etching. Various porous surfaces have different rates of absorption providing different levels of adhesion. This may cause some cracking of the coating.

New coatings applied over an **old coating** must be compatible. The surface of the old coating must be stable and have good adhesion to the substrate or the new coating will fail. Old coatings **chalk** as they age. If this is the only deterioration a new coating can be applied over it. If there are cracks and loose peeling sections the old coating must be removed before recoating.

Following is a discussion of commonly used field-applied coatings. There are many products finished in the factory and they use mass production application and drying techniques. One example is polyvinyl chloride film used for factory finished metal wall panels.

Clear Coatings

In addition to the clear coatings that have been used for years, such as shellac, varnishes from natural resins, and lacquer, a number of products using synthetic resins are available. Clear coatings used in exterior locations in general do not have the durability of opaque coatings because they lack the pigments that protect against ultraviolet damage from the sun. They let the natural color and texture of the substrate show as well as protect it from moisture, abrasion, and other forms of damage.

Varnishes are one of the oldest finishes used to coat wood surfaces. There are several types currently available and their properties and composition vary considerably. Varnishes fall into two broad classifications: natural-resin varnishes and synthetic varnishes.

There are three basic types of natural resin varnishes—linseed oil varnishes, tung oil varnishes, and spirit varnishes. Natural varnishes are made from resins obtained from a variety of trees in tropical countries or fossil resins. The vehicle is some form of drying oil into which the resin is dissolved. The oil-to-resin ratio determines the classification of the varnish and is expressed as the number of gallons of oil that are mixed per 100 lb. (45 kg) of resin. Varnishes made with natural resins and oil are called oleoresinous varnishes.

Turpentine—is the normally used solvent for varnish although mineral spirits, naphtha, and benzene can be used. The solvent evaporates causing the varnish to harden. The drying oils cure by oxidation and polymerization following the loss of the solvent. This is why varnish is a slow drying coating.

Linseed oil varnishes—are available in three types—long-oil, medium-oil, and short-oil.

  Long-oil varnish—is sold under the name, spar varnish. It is used on exterior surfaces where moisture may be present but not constantly wetted. Some of the synthetic resin varnishes are better for very moist conditions.

  Medium-oil varnishes—dry faster than long-oil types and have a harder film, but are not as water resistant. Synthetic binders, acrylic and alkyd, are used to produce a modified spar varnish.

  Short-oil varnishes—contain the least oil, dry rapidly, are rather brittle, and do not resist abrasion very well.

Tung oil varnishes—are used in areas where heavy use is expected, such as school furniture. It is usually a factory applied product.

Spirit varnish—also known as **shellac**—uses a resin obtained from the exudation of the lac insect, which is found in Southeast Asia and India. The resin is dissolved in denatured alcohol and is orange in color (referred to as orange shellac). It can be bleached producing white shellac.

  Shellac is available in various grades depending upon the amount of resin dissolved in a gallon of denatured alcohol. The grades are referred to as cuts, such as a 4 lb.(1.8 kg) cut which is 4 lbs. of resin dissolved in one gallon of denatured alcohol.

  Shellac dries rapidly but does not resist moisture. It's main use in construction is to seal knots and other resinous places in wood over which water resistant coatings are applied. It will not withstand exposure to sunlight.

SYNTHETIC RESIN VARNISHES

Synthetic resins are plastic materials suspended in a solvent. The most commonly used resins are alkyds, polyurethane, silicone, epoxy, acrylics, and phenolics. The vehicle may be the same drying oils used for natural resins though synthetic materials have been developed. A listing of some clear synthetic resin varnishes and their solvents are in Table 9-3.

Acrylic resin—varnishes produce a thermoplastic film that is resistant to yellowing with age, ultraviolet rays, and oxidation. They have good gloss retention and are almost colorless. They are used on metal, wood, plastics, textiles, and paper. Acrylin resin coatings are made with a solvent base or a water base. The solvent-based formulation has a high gloss and is impermeable to water vapor. The water-based formulation has a semigloss finish and is permeable to water.

Table 9-3 CLEAR FINISHES

Binder	Base	Uses
Acrylic	Solvent or water	Waterproofing, sealing surface against dirt. Used on concrete, masonry, stucco
Alkyd (spar varnish)	Solvent	Used on interior and exterior protected surfaces
Phenolic (spar varnish)	Solvent	Exterior wood exposed to moisture, marine applications
Silcone	Solvent	Waterproofing, sealing surface against dirt. Used on concrete, masonry, stucco, wood
Polyurethane (one part)	Solvent	Resists chemical attack, abrasion, heavy foot traffic

Table 9-4 POLYURETHANE RESIN COATINGS

ASTM Type Designation	Curing Agent	Chemical Resistance
One Component		
Type 1	Oxygen	Good
Type 2	Humidity in air	Very good
Type 3	Heat	Excellent
Two Components		
Type 4	Amine	Excellent
Type 5	Polyester	Excellent

Alkyd resin—varnishes are made by a chemical reaction between an alcohol and an organic acid. They are used for waterproofing and reducing dirt retention on concrete, masonry, and stucco walls. Some types are used on exterior metal surfaces.

Phenolic resin varnishes—include phenol formaldehyde and modified phenolic types. Phenol formaldehyde varnishes are excellent for materials exposed to the weather and can be used on wood marine products. They are resistant to caustic substances and acids. They tend to yellow and darken with age and lose gloss with exposure to the sun. Modified phenolic varnishes do not resist weathering as well but have an abrasion resistant quality that makes them useful for interior applications such as furniture and floors.

Epoxy varnishes—have excellent resistance to caustic materials, excellent adhesion qualities, and excellent resistance to abrasion. There are a variety of epoxy coatings each with special characteristics, such as being highly flexible. They can be used on a variety of materials, such as concrete, metal, and wood. They are often used as a primer. Some are formulated to resist solvents, heat, chemicals, and salt water.

Polyurethane resin varnishes—provide better abrasion resistance than natural varnish and offer resistance to solvents, chemicals, and oxidation. They are fast curing and are used on concrete floors and walls and places where a hard gloss surface is needed. There are three types that are one component coatings and two that are two component. See Table 9-4. Type 1 cures by exposure to

oxygen in the air. Type 2 requires 30 percent humidity in the air to cure. Types 1 and 2 can be applied on the job site. Type 3 cures with heat. Types 3, 4, and 5 are factory applied

Silicone varnishes—have excellent heat resistance, water resistance, and resistance to corrosive atmospheric conditions. A silicone-acrylic coating has a high gloss and high resistance to blistering and crazing. Silicones are also used to modify alkyds improving their durability. Silicone-polyester coatings will resist damage by heat up to 550°F (288°C) if the mix contains 75 percent silicone.

Silicones bond well to wood, but steel must be cleaned and bonderized to get proper adhesion. Silicone solutions are formulated to be applied to concrete and masonry materials to provide a water repellent coating by penetrating into the pores leaving film on the surface.

LACQUER

Lacquer is made from synthetic materials and is quick drying due to the rapid evaporation of the solvent. Lacquer has a nitrocellulose base used in combination with various resins and plasticizers. Drying oils are added to improve adhesion and elasticity. The **natural** and **synthetic resins** improve adhesion, hardness, and give the desired gloss. There are many different resins added. Those used depend upon the end use of the lacquer. Commonly used are alkyd resins, epoxy resins, acrylic resins, and cellulose acetate.

Solvents commonly used in lacquer include diethylene glycol, acetone, and amyl, ethyl, butyl, and isopropyl acetate. In formulating lacquer it takes several different solvents to dissolve the synthetic and natural materials used. The blending of solvents used affects the gloss, ease of flow, setting time, and amount of bubbling as it is applied.

Thinners are sometimes added to lacquer before it is applied. This is especially important when using spraying lacquers. They adjust the consistency and rate of drying. In general lacquers dry to the touch in 5 to 10 minutes and form a firm film in 30 minutes to 3 or 4 hours depending upon the formulation. Some can have a second coat applied just 15 to 20 minutes after the previous coat was sprayed. Commonly used lacquer thinners include toluol, xylol, benzol, and ethyl, amyl, butyl, and isopropyl alcohol. Lacquer thinner will dissolve a hardened lacquer finish if it is spilled on it.

Table 9-5 COMMONLY USED PRIMERS AND TOPCOATS

Material	Primer	topcoat	Remarks
Aluminum	Vinyl red lead	Vinyl	Exposure to weather
	Zinc chromate	Chlorinated rubber	Exposure to rain, salt water spray
	Zinc chromate	Alkyd or acrylic	Used on trim, flashing
	Self-priming	Epoxy ester	Exposure to fumes
Ferrous metal	Self-priming	Phenolic	Exposure to weather, high humidity
	Zinc silicate	Silicate, alkyd	Exposure to weather
	Zinc silicate	Silicone, aluminum pigmented	Exposure to weather
	Self-priming		
	Red lead	Vinyl	Exposure to rain, salt-water spray
	Self-priming	Acrylic	Will not resist abrasion
	Self-priming	Urethane	Exposure to corrosion, chemicals, abrasion
	Self-priming	Epoxy	Exposure to acids, alkalis, chemicals
	Chlorinated rubber with red lead or zinc chromate	Coal tar	Apply hot to metal to be below ground
		Chlorinated rubber	
	Red lead	Oil-based paints	Exposure to rain, salt-water spray, chemicals
	Red lead and alkyd-based		Normal exterior conditions
		Alkyd	Not abrasion resistant, exposure to severe weather
	red lead		
	Zinc-polystyrene	Polystyrene	Chemical fumes, exposure to fresh and salt water
Ferrous metal (galvanized)	Zinc dust or zinc chromate-zinc dust	Alkyd	Does not require topcoat
	Zinc dust or zinc oxide	Chlorinated rubber	Exposure to rain, salt-water spray
Ferrous metal in ground	Self-priming	Coal-tar-epoxy	Used on pipelines, buried structural steel
Gypsum wallboard	Vinyl	Alkyd	Light duty
	Self-priming	Acrylic	Heavy duty
Gypsum plaster	Self-priming	Acrylic	Plaster must be dry
Concrete and concrete masonry (dry), brick masonry	Self-priming	Acrylic	Interior locations, scrubbable
	Self-priming	Vinyl	Dry locations
	Self-priming	Epoxy esters	Exterior use, resists fumes, scrubbing
	Self-priming	Polychloroprene	Resists water, solvents, impact, exterior uses
	Self-priming	Urethane	Washable, interior locations
Concrete floors, no moisture exposure	Self-priming	Urethane	Light to moderate traffic
	Self-priming	Epoxy	Moderate to high traffic
Concrete, heavy moisture	Self-priming	Alkali-resistant chlorinated rubber	Water reservoirs, swimming pools
Portland cement plaster	Self-priming	Vinyl	Dry locations
	Styrene-butadiene	Alkyd	Dry locations
Wood, interior	Self-priming	Vinyl	Walls and floors
	Self-priming	Alkyd	Doors, paneling, trim light-duty floors
	Self-priming	Urethane	Surfaces subject to impact, scrubbing, heavy-duty floors
Wood, exterior	Self-priming	Acrylic	Surfaces subject to impact, scrubbing
	Self-priming	Urethane	Porch decking, exterior stairs
	Self-priming	Alkyd	Siding, plywood, cedar shakes, trim
	Self-priming	Acrylic	Siding, plywood
	Self-priming	Phenolic	Siding, plywood, trim
	Oil-based primer	Oil-based vehicle	Wood siding, exterior trim, plywood siding
	Self-priming	Epoxy	Any exterior wood

Most lacquers are applied by spraying. A **brushing lacquer** is available and can be applied with a pure bristle brush. It is slower drying than spray lacquers. Wood surfaces to be lacquered need a washcoat of shellac or lacquer sealer applied to them before applying the finish coats. It takes several finish coats because the lacquer film is very thin.

Lacquer is used mainly on interior products and is widely used on cabinets and furniture. It can have pigments added giving color and opacity.

Primers for Opaque Coatings

Opaque coatings are used to hide the color of the previous coating or substrate and obscure much of the grain or texture of the substrate. Before the opaque coating is applied the surface must be coated with a primer. A primer is a coating applied to the surface to seal it and prepare it to receive the finish topcoat. It helps hide discoloration, stains, and seals areas that may cause a bleed through the finish topcoat. Resin in wood is a common bleeding problem. Primers are also used to smooth porous surfaces as exist on concrete blocks. The use of the proper primer is vital to the successful use of the finish topcoat, and therefore the manufacturers recommendations must always be followed. In Table 9-5 are examples of some commonly recommended primers and associated topcoats. To coat a surface it must first be cleaned and prepared to receive a primer if required.

CLEANING METAL SURFACES FOR PRIMING

The preparation of **steel** surfaces depends upon the primer and finish topcoat. The directions of the manufacturer shoulder be followed. In some cases cleaning with a detergent or solvent is adequate. Wire brushes, sandpaper, and sandblasting also clean the surface and roughen it. Galvanized metal can be cleaned with a dilute acetic acid while steel can be cleaned with phosphoric acid. After washing and drying the surface the proper primer can be applied. Some prefer to let it weather for 6 to 9 months before trying to prime it.

On-site **aluminum** to be painted should be allowed to weather for at least a month. Then it should be wiped with mineral spirits or any solution recommended by the coating manufacturer. Most aluminum products arrive on the site with a factory applied finish. See Division 5 for more information.

Copper, bronze, and their **alloys** can be cleaned by wiping with mineral spirits or a dilute solution of hydrochloric or acetic acid, which must be completely washed away. Corrosion and material stuck on the surface can be removed by brushing or light sanding.

An etching primer is available for ferrous and nonferrous metals and some alloys. It cleans and chemically etches the surface, leaving a very thin protective film.

PREPARING WOOD SURFACES FOR PRIMING

Exterior wood to be painted should have a moisture content of 12 to 15 percent. Interior wood should have a moisture content of about 6 percent. Knots that might bleed through should have a coat of shellac or other sealer applied before the wood is primed. Mold, fungus, and other stains should be removed.

PREPARING CONCRETE SURFACES FOR PRIMING

It is advisable to let the concrete completely cure before priming. Concrete aged less than 30 days is generally considered not suitable for painting. Paint manufacturers recommendations on curing time should be observed. The surface should be free of dirt, oil, and other substances. Loose material can be removed by brushing or sanding.

PRIMING COATS

A primer is a first coat applied to the substrate. It seals and fills the pores of the surface, inhibits rust in ferrous metals, and improves adhesion of subsequent coats of paint. Following are some of the frequently used primers.

Gypsum plaster can be primed with latex (acrylic), alkyd, or oil-based primer after a 30-day waiting period. Latex flat interior finish may be used as a primer if the topcoat is a gloss or semigloss coating. Damp, new plaster requires an alkali-resistant primer.

Gypsum wallboard and other paper-covered products are primed with polyvinyl acetate if an alkyd topcoating is used. If a latex (acrylic) topcoating is used it is self-priming. **Portland cement plaster** often is painted with a vinyl topcoating that is self-priming.

A wide range of primers and topcoatings are used on concrete and **concrete and brick masonry.** Most of the topcoats are self-priming. The surface of concrete and masonry is usually rough and porous. They can be coated with a latex portland cement grout or latex blockfiller to produce a smoother surface. This results in a better, watertight coverage because the topcoat will not always bridge the pores in the surface. A thinned coat of catalyzed-epoxy coating material can also be used. Concrete walls can be primed with a latex primer-sealer after which a topcoat of latex or alkyd paint can be applied.

Aluminum is often primed with zinc chromate. Bare **ferrous metals** use a variety of primers depending upon the topcoat. Since ferrous metals rust rapidly they must be coated with a rust inhibiting primer. Zinc silicate and zinc chromate are used when there will be considerable exposure to the weather. Coatings designed for special purposes will often have a special primer designed to be used with that product. Etching primers which clean and seal the surface also are effective. **Galvanized metal** can be chemically etched before painting to provide needed adhesion. This can be omitted if a latex metal primer designed for galvanized metal is used. Another primer is a varnished-based material pigmented with zinc dust, zinc oxide, or portland cement.

Opaque Coatings

Opaque coatings obscure the natural color of the surface, give it protection, and add a decorative feature by providing a wide range of colors. After priming the required number of finish coats are applied. Manufacturers specifications must be followed. The type of coating material used depends upon the material in the surface to be coated and its intended use. Exposure to the weather, the ground, chemical fumes, salt water, and other conditions must be considered as the type of coating is considered. Some of the frequently used coatings for various materials are in Table 9-5. The following discussion gives basic details about these.

ALKYD COATINGS

Alkyd coatings are currently a major type of organic coating. However, because of requirements relating to the amount of volatile organic materials (VOM) they release to the environment, their use is becoming limited. Water-based products are replacing alkyd coatings.

Alkyd emulsions are formulated using a synthetic alkyd resin. There are many types of alkyds used in coating formulation. Alkyd coatings have only mild alkali resistance but have excellent water resistance and weather well. This makes them useful for exterior applications such as enamels for porches. Alkyd coatings retain their color well and permit the formulation of a range of light colors. With some reformulation alkyd emulsions are used in making baking enamels as are used on kitchen appliances. Alkyd resins are added to other coatings to improve adhesion and durability. For example, when modified with phenolic resins water resistance and alkali resistance is improved and they will penetrate rusted surfaces. When alkyd resins are formu-

lated with rust inhibiting pigments, such as red lead, iron oxide, or zinc chromate, they are used as rust inhibiting primers. A gloss enamel and resistance to chalking is produced by adding vinyl chloride acetate to alkyd resins. Silicone aids in color and gloss retention.

CHLORINATED RUBBER

Chlorinated rubber base coatings are solvent thinned and have excellent resistance to alkalis, acids, and some resistance to salt water and salt-air exposure. They have excellent water and water vapor resistance and are used on swimming pools and basement walls. They have good abrasion resistance and can be used on masonry, plastic, concrete, and metal surfaces. Chlorinated rubber coatings are not recommended for use on wood because they are permeable and blistering could occur, but they do resist attack by microorganisms. They do bond to metal and are used to stop corrosion due to galvanic action by separating dissimilar metals.

ENAMEL COATINGS

Enamels are a form of pigmented coating using varnish as the vehicle. It forms a gloss or semigloss film that is hard and durable. Oil and resin base paints also form similar coatings and can be classified as enamels.

Baked enamels are formulated to be applied in a factory by spraying. They are generally thermosetting materials that cure to a hard coating at 200° to 300°F (93° to 149°C). The coating becomes insoluble in the solvent used in its formulation. Enamels are available in a wide range of colors, are hard, washable, and resist alkalis and acids.

EPOXY

A variety of epoxy coatings are available.

Epoxy-ester—is an epoxy resin reacted with a drying oil. It has properties similar to phenolic varnishes and alkyd resins. However, it offers better resistance to chemical fumes and exposure to water.

Epoxy-polyamide—offers excellent resistance to chemical fumes, oils, atmospheric acids, and alkalies, resists abrasion, adheres to concrete, metal, or wood, and will cure even if wet.

Epoxy-bitumens—may be formulated with coal tar or asphalt. They are used on items buried in the soil such as tanks and piping.

Epoxy-polyester—coatings are heavy-bodied two-part systems used for protection of masonry and concrete. They have a high solids vinyl filler, which is applied to the concrete surface followed by a high solids epoxy-polyester pigmented topcoat.

Epoxy resins are chemically setting and solventless. They do not harden by the evaporation of solvent but by the application of heat. When exposed to the weather they may fade and chalk but remain undamaged.

LATEX COATINGS

Latex is a term applied to emulsions containing synthetic resins that are thinned with water. Since they contain no flammable solvent they present no fire hazard when in storage or when being applied, and entail minimum release of volatile organic materials into the atmosphere. Latex emulsion coatings dry rapidly. The common types available are acrylic, polyvinyl acetate, and styrene-butadiene. They are used for coating interior and exterior vertical surfaces.

Acrylic Coatings

Acrylic coatings are thermoplastic resins that have a range of properties varying from a rather hard coating to a softer finish. They are water based and available in clear and pigmented coatings. They offer excellent protection to concrete against weathering. They are also used as factory applied coatings on aluminum and steel wall panels because they have good durability and resistance to salt spray and chemicals. Acrylic coatings remain flexible and do not lose their color as they age. Some acrylic coatings are solvent based and are therefore not latex emulsions. They use solvents such as xylol or toluol.

One-part acrylic emulsions (latex) are used for interior and exterior vertical surfaces such as wood, masonry, gypsum wallboard, plaster, and metal. They are permeable to water vapor.

Two-part epoxy modified acrylic coatings are water based and used for interior and exterior vertical surfaces. It is a tough coating that resists stains and will withstand scrubbing.

STYRENE-BUTADIENE COATINGS

This is a water-based coating using styrene and butadiene, producing a rubber-like film. It has good resistance to alkali and is permeable to water vapor. It has some exterior use as a filler over porous concrete surfaces. Some formulations are used on interior masonry, plaster, and gypsum wallboard, producing a film that resists abrasion and is washable. They are generally not used on wood.

VINYL COATINGS

Vinyl and **polyvinyl acetate emulsions** are available as water-based coatings. One formulation is used for interior application while another is used for exterior use. **Polyvinyl chloride** copolymerized with polyvinyl acetate is an opaque, solvent-based coating that has poor adhering properties and requires a special primer. It has good durability and resistance to oils, alkalies, acids, and salt water.

Oil-based coatings are formulated by combining the body, pigment if needed, and a vehicle (drying oil, a thinner, and a drier). The paint **body** is a solid, fine material that provides the hiding power. Lithopone, titanium white, and zinc oxide are used for bodies of white oil-based paint. **Pigments** are added to give the coating color. Both natural and synthetic pigments are used. Natural pigments are obtained from minerals, animal, and vegetable products. For example, red lead is a typical red pigment. Synthetic pigments are generally derived from coal tar.

The **vehicle** is a nonvolatile fluid that suspends the body particles in solution. It is composed primarily of drying oil with small amounts of thinner and dryer. **Thinners** are volatile and evaporate. They are used to regulate the flow of the paint. Turpentine, a product formed by distilling gum from pine trees, is a high-quality thinner. Naptha and benzene are used in some formulations. **Driers** accelerate the oxidation and hardening process. Organic salts of iron, zinc, cobalt, and manganese are commonly used driers.

Oil-based paints are permeable to water vapor and thus minimize blistering over porous surfaces such as wood. They should not be used in corrosive or alkaline conditions. Since they have excellent wetting properties they are widely used as primers.

PHENOLIC COATINGS

Phenolic coatings are made by polymerization of phenol and a formaldehyde reactant and are solvent based. They dry by the evaporation of the solvent leaving a strong flexible coating. They are used on exterior concrete, plaster, wood, metal, and gypsum wallboard. Phenolic coatings are used where resistance to acids, alkalies, and some solvents is required as well as immersion in hot distilled water.

Two part phenolic coatings have a catalyst added at the site and harden due to a chemical reaction. This coating is useful to resist very harsh conditions such as chemical fumes.

Table 9-6 FIRE RETARDANT COATINGS[a]
SURFACE BURNING CHARACTERISTICS (BASED ON 100 FOR UNTREATED RED OAK)

Coating Type[b] Surface	DS - Clear Douglas Fir	PR - Clear Douglas Fir	PR - White Douglas Fir	PR - White Cellulose Board	"DS 11" - Clear Douglas Fir	"DS 11" - Clear Cellulose Board
Flame spread	5	5	5	5	10	10
Smoke developed	0	0	0	0	30	20
Number of preliminary coats	None	None	None	None	None	None
Number of fire-retardant coats	2	2	2	2	2	2
Rate per coat (sq. ft./gal.)	200	200	200	200	200	200
Number of overcoats	None	None	None	None	None	None

[a]Tests conducted in accordance with ASTM E/84 (UL 723 and ULC-S-102)
[b]Coatings tested are Exolit Fire Retardant Coatings manufactured by American Vamag Company, Inc.
Courtesy American Vamag Company, Inc.

URETHANE OR POLYURETHANE COATINGS

Urethane resin coatings are available as one-part and two-part formulations. One-part coatings are moisture-cured (reaction to atmospheric moisture) and are generally clear (no pigment). They have better abrasion resistance than alkyd enamels. An oil-modified one part urethane coating is available. While it has better gloss retention it has lower resistance to chemical attack.

Two-part formulations range from a hard to a rubber-like surface film. Adhesion to steel and concrete is poor and the surfaces must be carefully prepared. It has good resistance to abrasion, water, and solvents.

Both one- and two-part coatings are used for heavy-duty wall coatings and surfaces subjected to heavy traffic, such as gymnasium floors. They resist scrubbing, abrasion, and impact. They have better abrasion resistance than regular varnishes.

Special-Purpose Coatings

Special-purpose coatings are formulated to meet a specific need and are therefore not suitable for most coating situations. Several widely used types include bituminous, asphalt, reflective, and fire-retardant coatings.

Bituminous coatings are formulated by dissolving natural bitumens, as coal tar or asphaltic products, in an organic solvent. They are most effective when used in below ground applications because they do not react well when exposed to sunlight. One special weather-resistant type is a solvent-based asphalt roof repair coating. It has fillers to give it body and prevent it from running. Another type has **coal tar** emulsified in water, which enables it to be used on damp roofs. It has good resistance to sunlight. **Coal-tar paints** are made with a coal-tar produced solvent, as coal-tar naptha or xylol. They are often mixed with other synthetic resins to produce high-quality water resistant coatings for metal.

Asphalt coatings are produced from petroleum and are found in a variety of emulsions, cold applied paints, and enamels. They have good moisture resistance.

Reflective coatings absorb the ultraviolet band of solar radiation and reflect it as visible light. The life expectancy of reflective coatings when exposed to sunlight is about one year.

There are a variety of **pigmented fire-retardant coatings** available from coating manufacturers. Their products are rated for surface burning characteristics on combustible and noncombustible substrates. Class 1 flamespread (0-25) is required by codes for many applications. Pigmented fire-retardant paints will meet this requirements. While these coatings retard the spread of flame on a surface, they do not protect the substrate from fire or heat. If substrate protection is required an **intumescent coating** must be used.

Intumescent fire-retardant coatings develop a thick, rigid foam protective layer that insulates the substrate and prevents the spread of fire. They are applied to wood, hardboard, cellulose board, and other wood-based products. The coating is noncombustible and does not produce toxic fumes. One type uses a urea-formaldehyde resin with an intumescent agent. It is a water-based material and dries rapidly by evaporation of the water. When exposed to fire or heat of about 350°F (178°C) it expands and develops a thick, insulating mat hundreds of times thicker than the original paint film. Usually one or two coats are sufficient. The important factor is not the number of coats but the amount of coat

Table 9-7   WOOD STAINS COMMONLY USED FOR INTERIOR APPLICATION

Type of Stain	Vehicle Solvent	Staining Action	Remarks
Alcohol	Alcohol	Penetrating	Dries quickly but fades easily Sold in powder form Will raise grain
Gelled wood stain	Mineral spirits or turpentine	Pigmenting	Slow drying and does not fade Sold in gelled form
Latex stain	Water	Pigmenting	Slow drying and does not fade Sold in liquid form
Non-grain-raising stain	Alcohol Glycol	Penetrating	Dries quickly and does not fade or raise the grain Sold in liquid form
Oil stain (penetrating)	Mineral spirits or turpentine	Penetrating	Dries quickly and fades easily but does not raise the grain Sold in liquid form
Oil stain (pigmenting)	Mineral spirits or turpentine	Pigmenting	Slow drying and does not fade Sold in liquid form
Penetrating resin stain	Mineral spirits	Penetrating	Sold in liquid form Contains stain and protective coating in one coating
Water	Water	Penetrating	Dries quickly but fades easily Sold in powder form Will raise the grain

ing applied per square foot. The insulating layer delays contact between the flames and the combustible material below it. This impedes flame spread on the surface, holds down smoke, and, since it retards heat transfer, delays the ignition of the substrate giving occupants additional time to evacuate the building (Table 9-6). These coatings should not be applied over other coating materials and are generally not suited for exterior use because they are water sensitive.

Stains

Stains used on exterior wood color and protect it from the weather. Those designed for use on interior wood provide color while protection is given by a transparent topcoat such as varnish or lacquer.

EXTERIOR STAINS

Exterior stains are blends of oil, driers, resins, a coloring pigment, a wood preservative such as creosote or pentachlorophenol, a mildewcide, and a water repellant. They are low in film formation. Low pigmented stains are penetrating types made to soak into the wood. Heavily pigmented stains have the same formulation but contain more pigment. Heavy pigmented stains are frequently used on cedar shakes and shingles. They usually do not have the durability of paints. Both types are used on any type of exterior wood, such as siding, plywood, fencing, decks and trim.

Exterior wood stains are available in solid and semitransparent types. Solid stains hide the color of the wood but allow the texture of the wood to be seen. Semitransparent stains allow the natural wood color to be seen.

Stains that are oil- and water-based are available for application on wood. Oil-based alkyd stains are solvent thinned and are available in opaque and semitransparent types. Acrylic latex stains are water soluble.

INTERIOR STAINS

Interior stains are used on doors, trim, cabinets, and other wood products requiring a quality furniture-like finish. There are many types of stains available for quality interior wood finishing. A summary of these is in Table 9-7. Notice the various vehicles used and that some are penetrating stains while others are pigmented stains. Penetrating stains are made with dyes and do not obscure the grain. Pigments in pigmented stains stay on the surface of the wood and obscure some of the grain.

Woods fall into two general groups: close-grained, as maple, and open-grained, as oak or walnut. When finishing these woods a filler is applied to get a smoother surface. Close-grained woods are rather smooth and use a liquid filler to seal the pores on the surface. Some form of varnish is often used as a liquid filler. Open-grained woods have visible open pores that have to be filled with a paste wood filler to get a smooth surface. Paste wood fillers contain a vehicle such as linseed oil, a solvent such as mineral spirits, silex (ground quartz), and a drier. The silex mixed with the linseed oil makes a paste. It is forced into the pores and the excess is wiped off before it hardens leaving the pores full. The silex is a neutral colored material; therefore, color pigment can be added so the filled pores match the color of the wood.

Flame-Spread-, Smoke- & Fuel-Contributed Ratings

The speed at which flames spread over a surface is indicated by the **flame-spread rating** of the material. Coatings manufacturers have their products tested and provide these data to the consumer. Flame spread is generally specified for different situations by building codes.

Flame spread is measured by following ASTM tests which give noncombustible cement-asbestos board a value of 0 and untreated red oak a value of 100.

Flame spread ratings are specified as Class I (0-25), Class II (26-75), and Class III (76-200).

Other considerations when selecting a coating is the **fuel-contributed rating,** which is an indication of the amount of combustible material in the coating and the **smoke-developed rating,** which classifies the coating by the amount of smoke produced as it burns. This rating is on a scale of 0 to 100. The nearer the rating is to 0 the less smoke is developed. Coatings manufacturers can supply these data. Fuel-contributed and smoke-developed ratings use the same scale as flame spread.

GYPSUM, LIME & PLASTER

Gypsum is a mineral found in rock formations and is a hydrated calcium sulfate. It is identified by the chemical formula $CaSo4.2H20$, which means it is a compound of lime, sulphur, and water. The mined rock is crushed, ground, and calcined (heated). Calcining drives off most of the chemically combined water. The calcined gypsum called plaster of paris, is then ground into a fine powder, which is used to produce products such as wallboard and gypsum plaster. Various materials are added to produce the properties needed for the different gypsum products.

When calcined, gypsum (plaster of paris) is mixed with water the plaster absorbs the water and hardens into a solid state. This bonds the gypsum and any additives into a solid, hardened mass.

Gypsum Plaster Products

A plaster finish consists of a supporting base, as gypsum or metal lath, and one or more coats of plaster. The main ingredient in plaster is usually gypsum but some types use portland cement. Gypsum plasters meet the requirements of ASTM C28, *Standard Specification for Gypsum Plasters,* or ASTM C587, *Standard Specification for Gypsum Veneer Plaster.*

Gypsum plasters include gauging, wood-fiber, neat, and ready-mixed plasters. Following are plaster products related to interior finishing.

Plaster of Paris—is a calcined gypsum mixed with water to form a thick paste-like mixture. It sets in 15 to 20 minutes and is used in construction for ornamental plastering and repairing plaster walls. If mixed with lime putty it can be used for a finish coat on plaster walls. It hardens rapidly and has little shrinkage.

Unfibered gypsum—is a neat gypsum that has no aggregate or filler added and is used to form the plaster scratch coat (first), brown coat (second coat), and any required leveling coats. There are three kinds. Regular is used with sand aggregate and is for hand application. LW is used with lightweight aggregates for hand application. **Machine application** gypsum is used with either sand or lightweight aggregates.

Gypsum neat plaster—is calcined gypsum plaster mixed at the mill with other ingredients to control working quality and setting time. It may be fibered or unfibered. Aggregates as sand, perlite, or vermiculite are used.

Fibered gypsum—is a neat gypsum with cattle hair or organic fibers, as sisal, added. The fibers hold the plaster together. It is not generally used for machine application. It is used for the base coats.

Wood fibered gypsum—is a factory-made product requiring the addition of water on the job. There are two types: **regular** for use over gypsum lath and **masonry** for masonry surfaces.

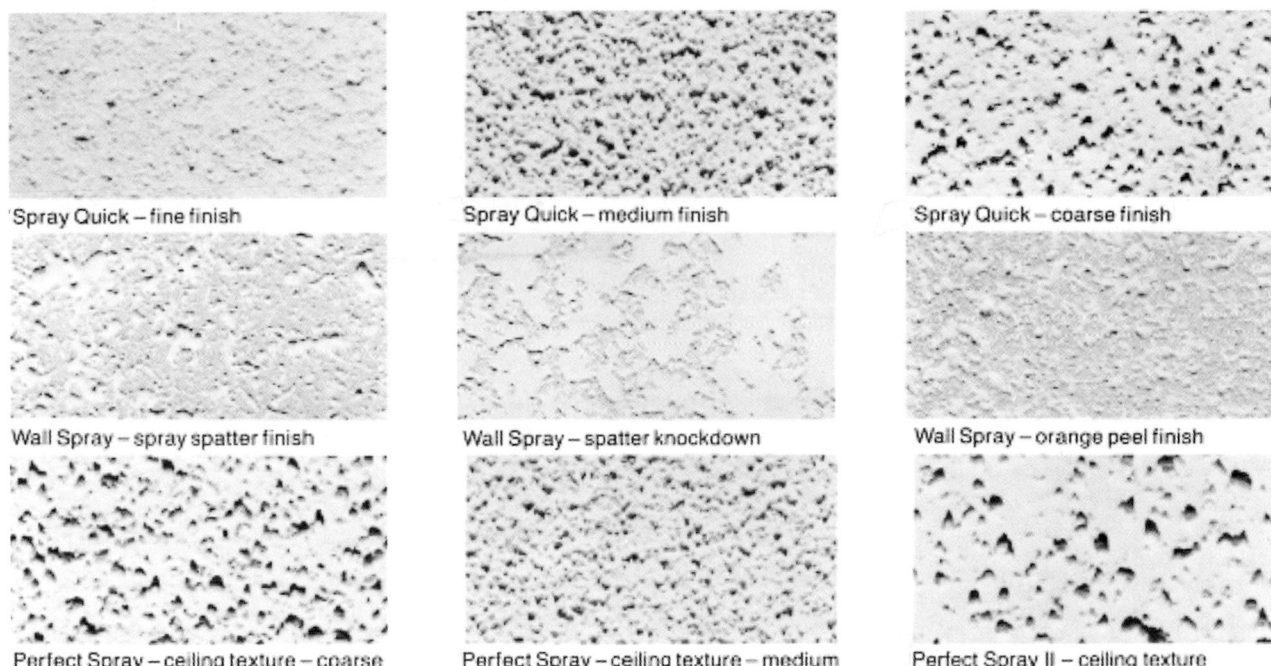

9-2 *Textures typically used to finish interior drywall ceilings and walls. The gypsum product can also be used on plaster and monolithic concrete walls.*

Courtesy Gold Bond Building Products, National Gypsum Company

Bond coat gypsum—is prepared for use as a base coat on monolithic concrete surfaces that are smooth and dense and do not have sufficient suction for standard plaster base coats. Finish plaster is applied over this coat.

Gauging plaster—is a coarsely ground gypsum plaster composed of screened particles that regulate the set time at definite time periods. It is available in quick set (30 to 40 min.) and slow set (50 to 70 min.) mixtures. It is mixed with slaked lime to make a finishing coat.

Keenes or Keene's cement—is a double calcined gypsum (almost all water removed). The two types are regular (slow setting) and quicksetting. It has a high strength and is highly resistant to moisture. It is used with lime and sand for a float or sprayed finish.

Casting plaster—is made from plaster of Paris that has been ground to a finer powder than the regular plaster of paris. It is used for molding ornamental plaster, such as cornices. Lime putty is added to increase plasticity. It sets up slower than regular plaster of paris giving more time to form the ornamental work.

Finish plaster—is used to cover the brown (second) coat with a hard, smooth finish or putty coat that can be painted or papered. It is mixed with hy-drated lime putty and water in proportions of 1 part plaster to 3 parts lime putty. **Prepared finish plaster** has no lime and only requires the addition of water. It does not dry to as white a color as finish plaster but does dry faster.

Acoustical plaster—is a calcined gypsum mixed with a light-weight mineral aggregate. Several compositions of acoustical plaster are in use. These use either gypsum, lime, or Keene's cement with aggregates such as rockwool, pumice, perlite, or vermiculite. Some include a foaming agent, which forms small voids producing a porous plaster.

Texture plaster—is a type of finish plaster used to produce a rough, textured finished surface. It is applied in two coats over the plaster base coat or gypsum lath. The final coat is applied by sponge, brush, or trowel depending upon the desired texture.

Texture spray—is a gypsum-based material that is mixed with water and applied with a spray machine as the final coat. The texture of the finished surface can be varied by adjusting the air pressure or spray orifice size, or by varying the amount of water thus changing the consistency of the mix by the addition of aggregate (see 9-2). It is used to put a textured finish on gypsum wallboard plaster and monolithic concrete walls and ceilings.

Veneer plasters—are specially formulated high-strength gypsum plaster that may be a thin monolithic basecoat plaster over which a finish is applied or may be formulated for application as a finish plaster. Setting time and compressive strength are controlled by the composition prepared by the manufacturer. They may be applied over gypsum plaster base, masonry, or concrete surfaces.

Fire-proofing plasters—contain asbestos (where permitted), inorganic binders, or lightweight aggregates. They are sprayed directly on to bare steel shapes providing fire protection as specified by codes. The finished coating is easily damaged and not intended to be the exposed finish coating.

Joint compound—is used to cover the heads of nails and joints between sheets of gypsum wallboard. It is sold ready mixed for immediate use (see 9-3). It is available with various setting times ranging from 30 to 360 minutes. The rapid chemical hardening and low shrinkage permit some day finishing and next day decorating.

Gypsum Board Products

Gypum board is the generic name for a series of panel products having a noncombustible core of calcined gypsum with paper surfacing on the face, back, and long edges. It is often called drywall or wallboard. These products are manufactured to ASTM specifications.

ADVANTAGES OF GYPSUM BOARD

Gypsum board is highly **fire resistant** and is the major material used on walls, ceilings, floors, and other parts of the building. Type X is formulated to meet fire code requirements. As it is exposed to fire the chemically combined water is slowly released retarding the transfer of heat through the panel. In addition it has a low **flame spread index** and **smoke density index**. Type X panels meet the requirements of ASTM C36 and tests made by the Underwriters Laboratories.

Gypsum board is also an effective barrier to the **transfer of sound,** as through a partition or floor-ceiling construction. It is in addition a durable material producing high-quality walls and ceilings and maintains excellent dimensional stability. Gypsum products are inexpensive, easy to install, and accept a wide range of finishes.

9-3 *Drywall joint compound is available ready mixed for immediate use.*

MANUFACTURING GYPSUM WALLBOARD

Gypsum board products are made by mixing calcined gypsum with water and additives forming a slurry. This is fed between continuous layers of paper on a board machine. The board moves down a conveyer during which the calcium sulfate recrystalizes or rehydrates forming a solid core. The gypsum bonds to the paper. Finally the board is cut to size and passed through dryers to remove excess moisture.

TYPES OF GYPSUM BOARD PRODUCTS

Regular gypsum wallboard—is used as the finished surface of walls and ceilings. It is highly fire-resistant because it has a gypsum core. It has an off-white paper on the finished side and a gray paper on the back side of the gypsum core. It is available with an aluminum foil covering on the back acting as a vapor barrier. The standard sheet is 4 × 8 ft. Metric sizes will be approximately 1200 × 2400 mm. Sheets 10, 12, and 14 feet long are available. It is available in thicknesses of ¼, 5/16, ⅜, ½, and ⅝ in. (6.4, 8, 9.5, 12.7 and 16 mm). The edges manufactured are tapered, square, beveled, rounded, or tongue and grooved (see 9-4).

Fire-resistant gypsum board—(Type X) has improved fire-resistance qualities due to the addition of fire-resistant materials in the core. It is used in assemblies that must meet fire code ratings. It is available in ½ in. (12 mm) and ⅝ in. (16 mm) thicknesses. Some types have an aluminum foil back layer that serves as a vapor barrier.

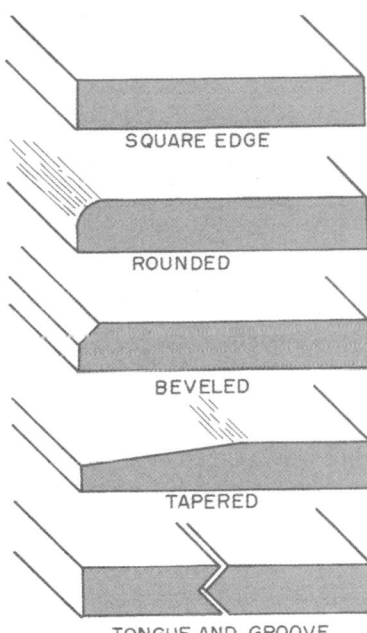

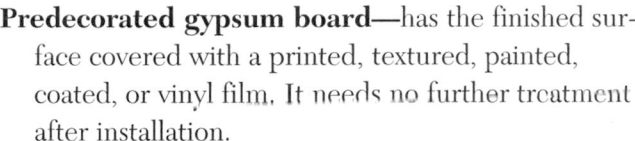

9-4 *Regular gypsum wallboard is available with a variety of edge shapes.*

9-5 *Gypsum panels are used on suspended ceilings*
Courtesy Chicago Metallic Corporation

Predecorated gypsum board—has the finished surface covered with a printed, textured, painted, coated, or vinyl film. It needs no further treatment after installation.

Water-resistant gypsum board—has a water-resistant gypsum core and a green water-repellent paper. It is used as a base for the installation of tile and plastic panels in baths, showers, kitchen, and laundry areas. It is available with a regular or fire-resistant core and in ½ in. (12 mm) and ⅝ in. (16 mm) thickness. It is not recommended for use on ceilings with joist spacing greater than 12 in. because the weight of the tile ceiling could exceed the strength of the panel.

Backer board—is used as the base ply or plys in assemblies where more than one layer of gypsum board is required. It is also used as a backing behind acoustical tile, plywood paneling, and other decorative wall paneling. It is available in regular and fire-resistant panels.

Gypsum coreboard—is a 1 in. (25 mm) thick panel used in shaft walls and laminated gypsum panels.

Gypsum liner board—has a special fire-resistant core enclosed in a moisture-resistant paper. It is used as a liner panel on shaft walls, stairwells, chaseways, corridor ceilings, and area separation walls. It is in ¾ in. (18 mm) and 1 in. (25 mm) thicknesses and 4 in. (610 mm) and 48 in. (1220 mm) widths.

Exterior gypsum soffit board—is available in regular and fire-resistant core in ½ in. (12 mm) and ⅝ in. (16 mm) thicknesses. It is used in exterior areas as soffits, canopies, and carport ceilings where it has indirect exposure to the weather.

Gypsum sheathing—have a fire-resistant and water-resistant gypsum core faced with specially treated water-repellent paper on both faces and the long edges. It is used on wood and steel framing in residential and commercial construction. Panels are available with beveled, square- and V-shaped tongue-and-grooved edges. It is available in ½ in. (12 mm) and ⅝ in. (16 mm) thicknesses.

Sound-deadening board—is a ¼ in. (6 mm) gypsum panel used in connection with fire-resistant gypsum panels to meet the requirements for sound and fire resistance.

Lay-in ceiling panels—are made from ½ in. (12 mm) fire-resistant gypsum wallboard. They are cut into 2 × 2 ft. and (610 × 610 mm) 2 × 4 ft. (610 × 1220 mm) panels for easy installation in suspended ceiling grids (see 9-5).

Gypsum lath—is used as a base to receive the layers of hand or machine laid plaster. It is available in 16 in. (406 mm) and 24 in. (610 mm) panels 48 in. (1220 mm) long. It is made in ⅜ in. (9 mm) and ½ in. (12 mm) thicknesses. It may have an aluminum foil backing providing heat reflective qualities.

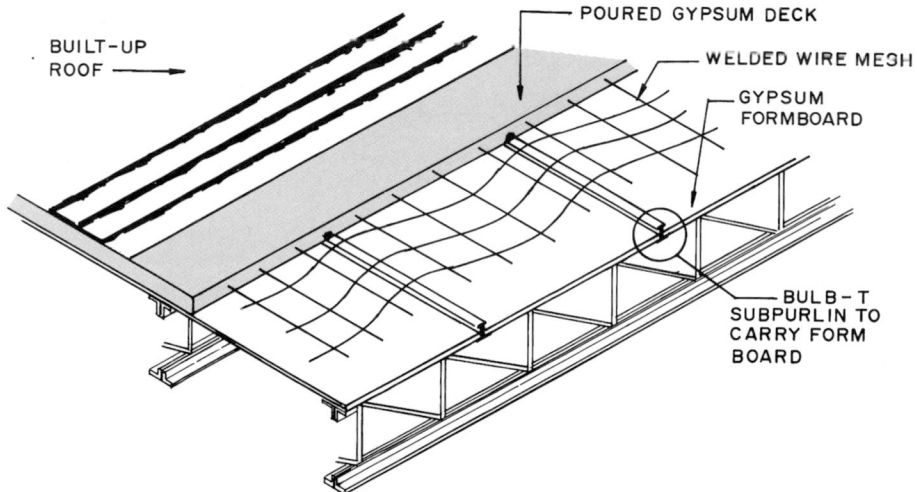

BUILT-UP ROOF

POURED GYPSUM DECK

WELDED WIRE MESH

GYPSUM FORMBOARD

BULB-T SUBPURLIN TO CARRY FORM BOARD

9-6 *Gypsum formboard is supported on the structural members and provides a floor for pouring the gypsum roof deck. The formboard remains in place after the roof deck has hardened.*

Gypsum fiberboard—is a composite material of gypsum, cellulose fibers, fibers from recycled newspapers, and sometimes includes perlite to reduce the weight of the panel. The cellulose fiber reinforces the gypsum that surrounds it making a product that works more like particleboard than the typical gypsum wallboard. It is a solid material and does not have the paper facing used on conventional wallboard. This increases its fire resistance. It can be installed with a pneumatic stapler, nail gun, or drywall screws in wood stud applications. It is strong enough to permit mirrors and towel racks to be anchored to it rather than to studs. It is more stable than conventional gypsum wallboard, and it is moisture, fire, impact, mildew, and sound resistant.

The seams are not taped as are conventional gypsum wallboard but are filled and smoothed with two layers of a specially formulated joint compound.

The panels are available in light weight and a heavier, high-density panel. They are available 4 feet wide and in stock lengths of 8, 10, and 12 ft. (2440, 3050, and 3660 mm). Widths to 8 ft. (2440mm) and lengths to 20 ft. (6100mm) are available by special order. Panels are made in thicknesses of ⅜, ⁷⁄₁₆, ½, and ⅝ in. (10, 11, 13, and 16mm). Panels are available with four square edges or two or four tapered edges. It is available for use as interior wall finish, floor underlayment, tile backing, and as gypsum sheathing.

Poured gypsum roof decks—are formed by pouring gypsum concrete over permanently installed **gypsum formboards**. The formboards are supported by structural steel roof framing. The gypsum is in a thick, creamy condition and is pumped through a hose to the deck. A construction detail for a poured gypsum roof deck is in 9-6.

Applications

Various plaster products are used for finish wall and ceiling construction in residential and commercial construction and for cast ornamental features. Plaster is used for sound attenuation and fire protection of structural members such as steel beams and columns (see 9-7). It is applied over gypsum plaster base panels and metal and wire lath that has been secured to the structural framing.

Gypsum wallboard products are used for finishing interior walls and ceilings (see 9-8 and 9-9) providing sound attenuation, fire resistance on walls and ceiling, and fire protection of structural members such as steel beams and columns. Some types have water resistive coverings and are used on walls to be tiled. Other types are used for exterior sheathing. Gypsum wallboard is also used over subfloors to increase the fire resistance of the assembly.

Lime

Lime is produced by burning limestone, marble, coral, or shells in a kiln at 2000°F (1100°C) to remove the carbonic acid gas; the purest lime (97%) is used in plaster and is a finishing lime.

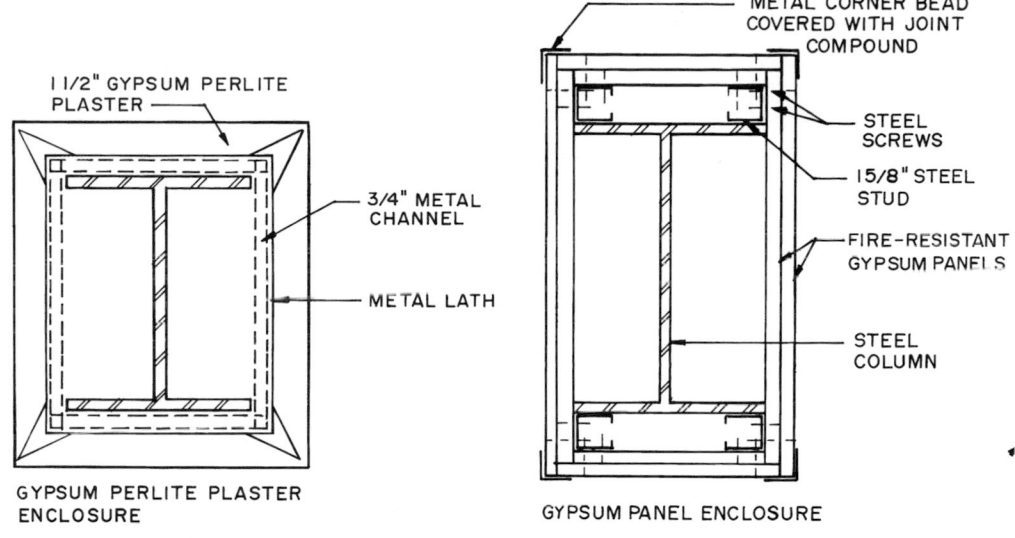

9-7 *Gypsum panels are used to provide fire protection to structural members.*

The lime produced by the kiln is called quicklime. It has the capacity to **slake** or **hydrate** when allowed to soak up to two or three times its weight in water. Slaking or hydration is a chemical reaction that causes the temperature of the mix to rise rapidly. This mixture is allowed to cool and sit three weeks before it can be used. Contractors buy lime already slaked so they do not have this waiting period.

The slaked quicklime has been transformed by the chemical reaction with water into hydroxite of lime, a fine dry powder. This powder is the lime product that is used in plaster. There are two types of hydroxite lime powder, Type N (normal) and Type S (special). Type N must be soaked 12 to 16 hours before it can be added to the plaster. Type S can be added to the plaster as soon as it is mixed with water.

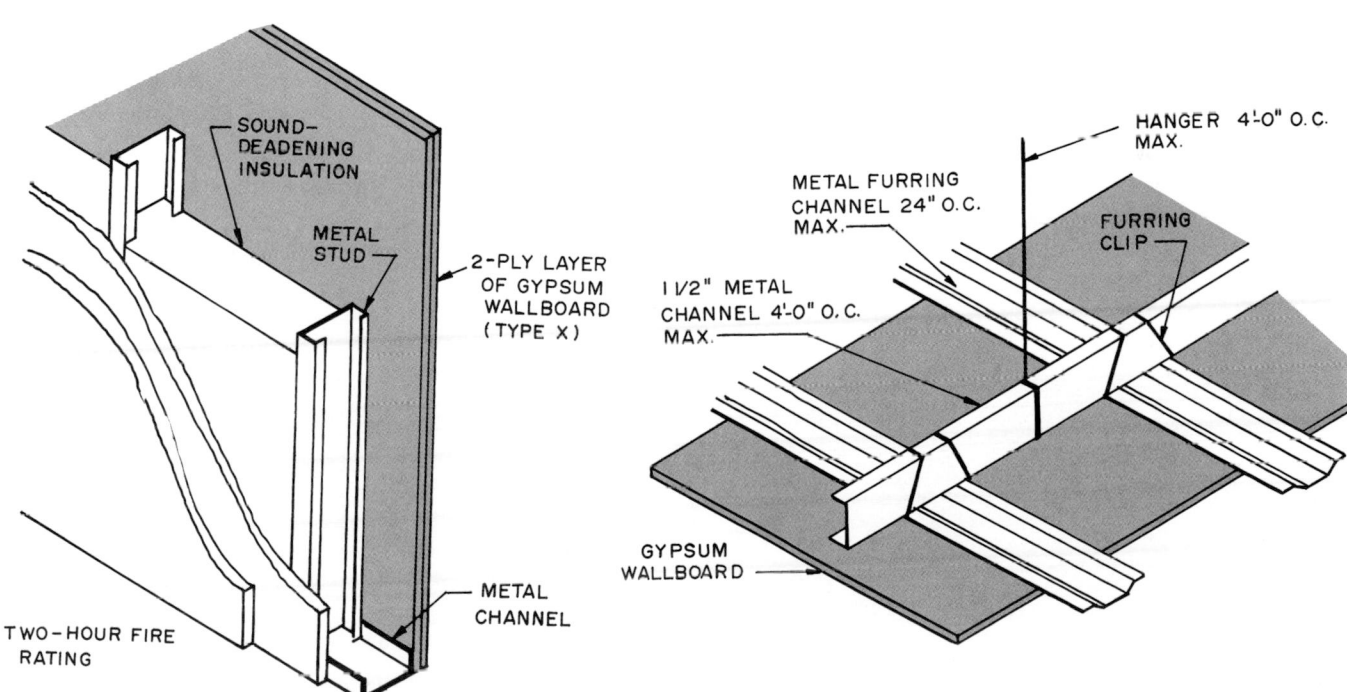

9-8 *A typical interior wall with a two-ply gypsum wallboard finish.*

9-9 *Gypsum wallboard can be secured to suspended steel channels forming a fire-resistant ceiling.*

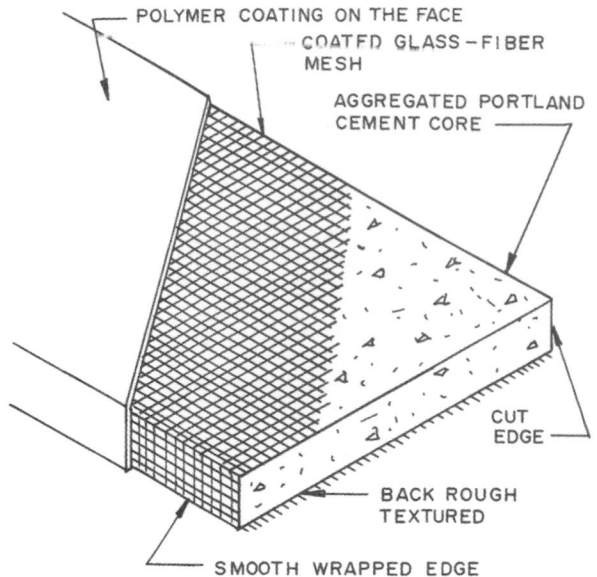

POLYMER COATING ON THE FACE
COATED GLASS–FIBER MESH
AGGREGATED PORTLAND CEMENT CORE
CUT EDGE
BACK ROUGH TEXTURED
SMOOTH WRAPPED EDGE

9-10 *Cement board is used where the wall or ceiling will be exposed to water or high humidity.*

Cement Board

Cement board is a product made with an aggregated portland cement core reinforced with polymer-coated, glass fiber mesh embedded in both surfaces (see 9-10). While it is not a gypsum product it is used for some of the same applications. Cement board is a durable, fire-resistant, and water-resistant panel used on bearing and nonbearing walls on interior and exterior surfaces. It can be attached to wood or steel framing with special screws for each application. Holes must be drilled to accept fasteners. Galvanized roofing nails can be used to secure it to wood framing. Joints can be covered with an open glass-fiber mesh tape. The panels can be finished with a portland cement mortar containing dry latex polymers. These materials are supplied by the manufacturer and their recommendation should be carefully followed.

Cement board may be used in areas with high humidity, such as baths, kitchens, and pools. It is also used for soffits, fences, and chimney enclosures. It is used as a substrate for the application of ceramic tile, slate, and quarry tile on interior surfaces. It can be used as a heat shield behind stoves and heaters.

ACOUSTICAL MATERIALS

Acoustics is the science of sound, including the generation, transmission, and effects of sound waves. Acoustical materials are those used to control sound. They are used to (1) reduce the **levels** of sound within an area, as in a room, by absorption and (2) to control the **transmission** of sound from one area to an adjacent area caused by the transmission of vibrations through the building structure.

Sound

Sound is the sensation produced by the stimulation of the organs of hearing by vibrations transmitted through the air. This includes vibrations traveling in the air at a speed of about 1130 ft/sec (345m/sec) at sea level and mechanical vibrations transmitted through an elastic medium, as steel. Sound is the movement of air molecules moving in a wavelike motion. A sound wave produces changes in the atmospheric pressure above and below the existing static pressure. This deviation in atmospheric pressure is called **sound pressure.**

Sound waves have a frequency, which is the number of times per second the sound pressure cycles above and below atmospheric pressure. If the frequencies and intensity of sound waves are within certain limits they can be heard by the human ear.

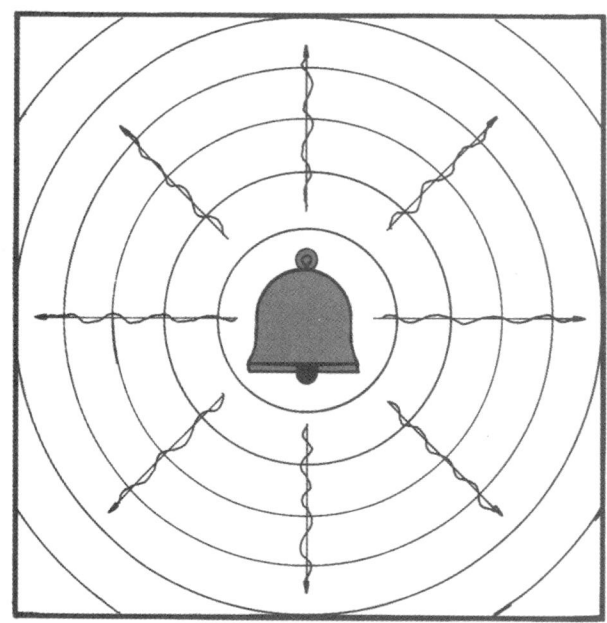

9-11 *Sound travels in all directions from a generating source.*

SOUND WAVES

Sound waves move in a spherical direction. That means that they move from the source in all three dimensions (see 9-11). There is a delay in the time a sound is created and it is heard. Since it takes audio sound about a second to travel 1130 ft/sec (345m/sec) in the air a per-

Table 9-8 SOUND PRESSURE LEVELS OF SELECTED SOUNDS

Sound Levels (decibels)	Source of Sound	Sensation
140	Near a jet aircraft	Deafening
130	Artillery fire	Threshold of pain
120	Elevated train, rock band, siren	Threshold of feeling
110	Riveting, air-hammering	Just below threshold of feeling
80–100	Power mower, thunder close by, symphony orchestra, noisy industrial plant	Very loud
60–80	Noisy office, average radio or TV, loud conversation	Loud
40–60	Conversation, quiet radio or TV, average office	Moderately loud
20–40	Average residence, private office, quiet conversation	Quiet
0–20	Whisper, normal breathing, threshold of audibility	Very faint

son 2260 ft. (690m) away will hear the sound 2 seconds after it is generated. When you see lightning in the distance you do not hear the thunder until several seconds later due to the differing speeds of sound and light. Sound travels at different speeds in various materials. In wood it travels about 11,000 ft/sec (3 355m/sec) and in steel about 16,000 ft/sec (4 880m/sec).

The **frequency** of sound is the number of **cycles** of alike waveforms per second. Sound travels in sine waves as shown in 9-12. Wave length equals the **velocity** of the sound (1130 ft/sec in air) divided by the **frequency** (number of cycles). For example a sound having a frequency of 15 cps would have a wave length of 1130/15 or 75 ft. The frequency, cycles per second, is expressed in units of hertz (Hz). A hertz is a unit of frequency equal to one cycle per second (CPS) so in the above example the frequency is expressed as 15 hertz (15 CPS = 15 Hz). The frequency of sound determines the pitch. Low, deep sounds have low frequencies while high pitched sounds have high frequencies. The human ear can receive sounds ranging from a low of about 20 Hz to a high of about 20,000 Hz (or 20 KHZ).

The frequency of sound to be considered usually varies rapidly and constantly. For example, the sound produced by music from a radio will often range from low to high frequencies rapidly and often. Frequency from some other source, as a dishwasher, will be more limited in range.

SOUND INTENSITY

Sound intensity depends upon the strength of the force that sets off the sound vibrations. Sound intensity is measured in decibels (dB). The decibel scale for normal applications ranges from 0 dB, which is just below the lowest audible sound (about 20 Hz), to 120 dB, which can produce a feeling of vibration in the ear. Intensities above this can cause actual damage to human hearing and structural damage to buildings. Representative sound levels in decibels for selected situations are in Table 9-8.

The intensity of sound will vary in much the same way as described for frequency. Music could produce very high decibel levels while a dishwasher will produce lower, steady level decibel intensity.

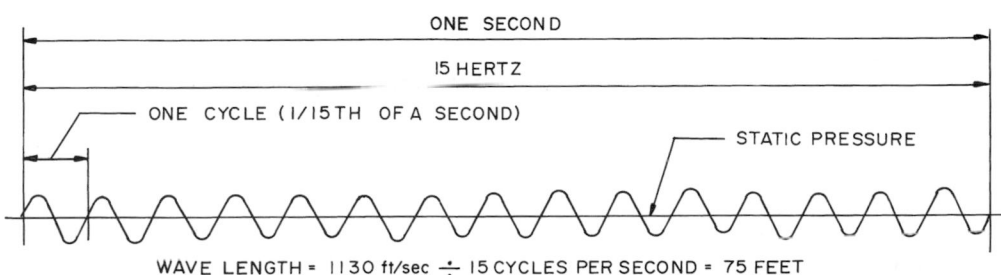

WAVE LENGTH = 1130 ft/sec ÷ 15 CYCLES PER SECOND = 75 FEET

9-12 *The sine wave audio frequency.*

Sound Control

Sound generated within a space, as a room, is controlled by acoustical materials that have sound absorbing properties and those that reduce sound transmission through assemblies of materials, as a wall or floor.

Sound-absorbing materials absorb some of the sound waves striking them and reflect the rest back into the area. The acoustical engineer selects materials to control the sound in the manner desired. Some materials, as a plaster wall, reflect most of the sound striking them back into the room. As the sound hits other walls it is again reflected (see 9-13). Sound-absorbing materials are porous and have openings into which the vibrating air particles move in and out causing friction, which generates heat. Some of the sound energy is lost as heat that may be reflected back into the room or transmitted through the material.

SOUND TRANSMISSION CLASS (STC)

Sound transmission class is a single number rating indicating the effectiveness of a material or an assembly of materials to reduce the transmission of airborne sound through the unit. The larger the STC number the more effective the material is as a sound transmission barrier. Materials are tested following the specifications in ASTM E90-70.

▼ Typical building code transmission control requirements for residences, apartments, and hotels follow:
1. All separating walls and floor to ceiling assemblies provide a STC of 50.
2. All penetrations in assemblies, as for piping and electrical, be sealed to maintain the STC 50 rating.
3. Entrance doors and their seals have a STC rating of 26 or more.
4. Floor-ceiling assemblies have an impact isolation class (IIC) rating of 50. This is covered in the next section.

Manufacturers of materials used to reduce sound transmission will indicate the STC rating for each material and the various thicknesses. Refer to 9-27, 9-28, and 9-29 for STC ratings for several assemblies of materials.

IMPACT ISOLATION CLASS (IIC)

Impact isolation class is a single number giving an approximate measure of the effectiveness of floor construction to provide iso-

Table 9-9 NOISE REDUCTION COEFFICIENTS FOR SELECTED MATERIALS

Material	NRC
Unpainted brick wall	0.02–0.05
Painted brick wall	0.01–0.02
Glazed clay tile	0.01–0.02
Concrete wall	0.01–0.02
Lightweight concrete block	0.45
Heavyweight concrete block	0.27
Standard plaster wall	0.01–0.04
Gypsum wall	0.01–0.04
Acoustical plaster wall	0.21–0.75
Glass	0.02–0.03
Fiberglass	0.50–0.95
Wood panel	0.10–0.25
Mineral wool	0.45–0.85
Acoustical tile	0.55–0.85
Carpeting	0.45–0.75
Vinyl floor covering	0.01–0.05

lation against the sound transmission from impacts. Impacts include walking, skidding, or dropping items on the floor, which set up vibrations which radiate to the area below. Floors having low IIC ratings can have improved performance by applying carpet or other sound-absorbing materials on them. Refer to 9-28 and 9-29 for several design suggestions. IIC ratings of 45 to 65 are common for floor/ceiling assemblies in multi-family dwellings.

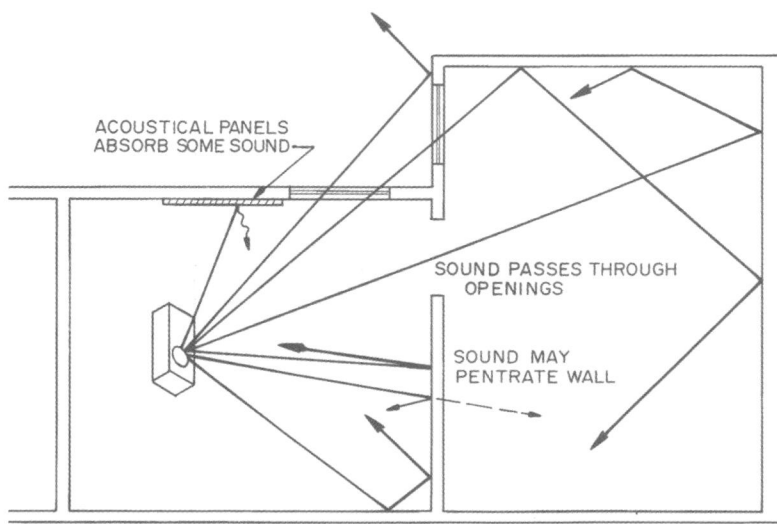

ACOUSTICAL PANELS ABSORB SOME SOUND

SOUND PASSES THROUGH OPENINGS

SOUND MAY PENTRATE WALL

9-13 *Sound is reflected when it strikes a hard, nonabsorbent surface.*

NOISE REDUCTION COEFFICIENT (NRC)

The noise reduction coefficient (NRC) is an indication of the amount of airborne sound energy absorbed by a material. The single number rating is the average of the sound absorption coefficients of an acoustical material at frequencies of 250, 500, 1000, and 2000 Hz. The larger the NRC number the greater the efficiency of the materials to absorb sound. Some typical examples are in Table 9-9.

Sound-Controlling Materials

The control of sound transmission and the absorption of sound can be accomplished using a wide variety of materials. Some standard construction materials, as brick or concrete block, are used to control sound transmission. Other materials are especially designed for acoustical purposes. Many of these materials are discussed in detail in other chapters.

FLOOR COVERINGS

The installation of **carpet** and **vinyl composition floor covering** reduces both impact noise and dampen air-borne noise. Carpeting is much more effective than vinyl floor covering and is even better when installed over a cushion.

ACOUSTICAL PLASTER

Acoustical plaster is composed of a plaster made with perlite or vermiculite aggregate. While it may be applied with a brush it usually is sprayed on the surfaces. It has the advantage of uniformly covering curved and irregular shapes. The plaster is applied in several layers with the finished job about ½ in. (12mm) thick. It has a NRC of about 0.21 to 0.75.

Another form of spray applied acoustical wall and ceiling covering that is not a plaster is composed of cellulosic fibers in a bonding agent. It will bond to almost any surface and provides acoustical control and thermal insulation.

ACOUSTICAL CEILING TILE & WALL PANELS

There are a variety of ceiling tiles and wall panels available. They are made from various materials such as wood fibers, sugarcane fibers, mineral wool, gypsum, fiberglass, aluminum, and steel. Most ceiling tiles have some form of perforations in their surface. Drilled or punched holes in a variety of patterns are common. Some tiles have slots, fissures, or striations (see 9-14). Other types use a sculptured, irregular molded surface. Following are brief descriptions of some of those available.

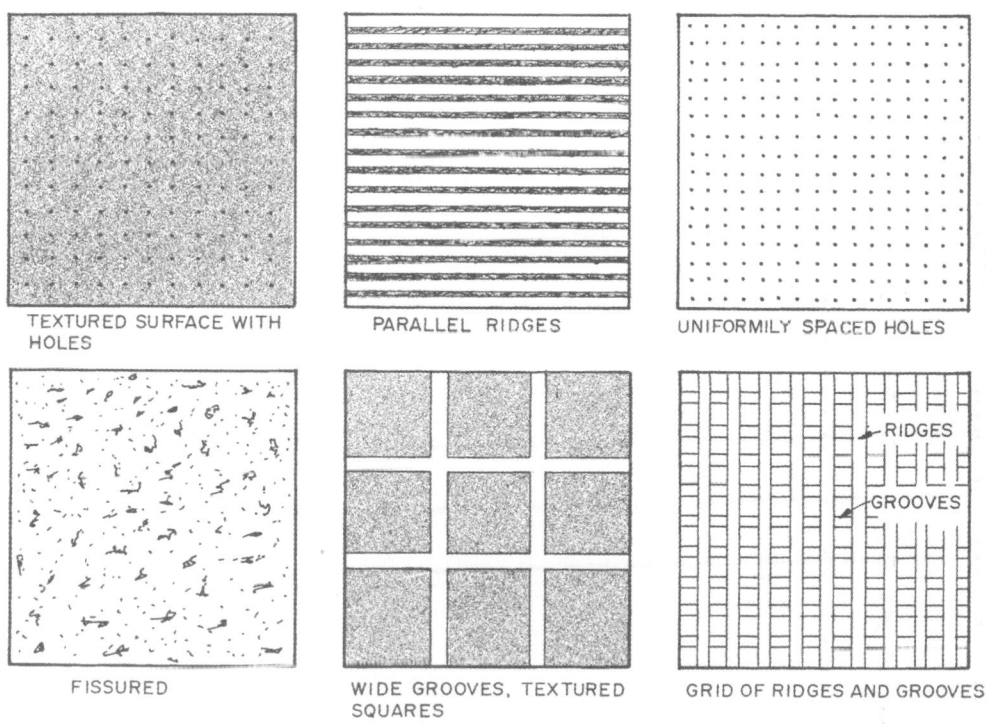

TEXTURED SURFACE WITH HOLES

PARALLEL RIDGES

UNIFORMILY SPACED HOLES

FISSURED

WIDE GROOVES, TEXTURED SQUARES

GRID OF RIDGES AND GROOVES

RIDGES

GROOVES

9-14 *A few of the acoustical surface treatments used on ceiling tile and panels.*

Wood fiber and cane fiber ceiling tiles are available in thicknesses ranging from ⅝ in. (16mm) to 1½ in. (38mm). Tile sizes commonly available include 12 × 12 in. (305 × 305mm), 12 × 24 in. (305 × 610mm), 24 × 24 in. (610 × 610mm), and 24 × 48 in. (610 × 1220mm).

Ceiling tiles and wall panels are also made from **molded mineral fibers** and are cast having a wide choice of surface textures. Some manufacturers use **mineral fiber** or **fiberglass** sound deadening panels covered with a fabric. This provides an attractive finished ceiling or wall panel. This type will have STC ratings of 35 to 40, NRC ratings of .60 to .80, and flame spread of 0-25.

Some ceiling tile and wall panels can resist damage due to moisture and bumps. One type is a **ceramic ceiling tile** made from mineral fibers in a ceramic bond. It is fire resistant and for use in areas of high humidity. It has STC ratings of 40-44, NRC ratings of .50 to .60, and flamespread 0. Another type uses **perforated aluminum** or **steel tiles.** One type applies a vinyl coating over a perforated aluminum tile backed by a mineral fiber substrate. The sound passes through the holes and is absorbed by the mineral fiber. Metal tiles have STC ratings of 35 to 45, NCR ratings of 0.60 to 0.70, and flamespread of 0. Another type that resists damage is made by bonding wood fibers into a porous panel resembling particleboard. It resists moisture and heavy blows.

Lightweight panels are made that have a vinyl covering over a laminate of high density molded glass fiber bonded to a core of 1 or 2 in. (25 to 50mm) sound absorbing fiberglass (see 9-15). Wall panels made from these materials are applied to walls in halls, restaurants,

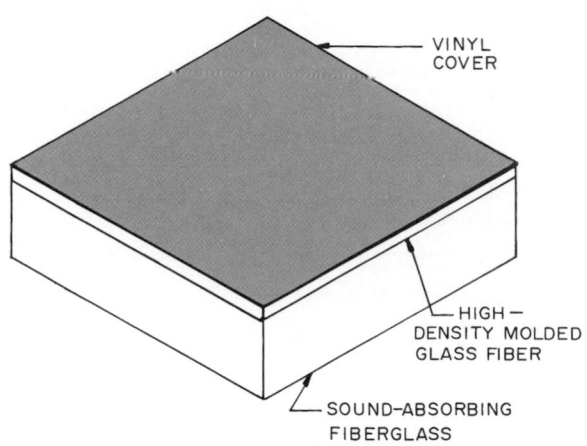

9-15 *A fabric covered acoustical panel with a tough layer that will resist damage from impact.*

gymnasiums, offices, and other places where sound control is necessary (see 9-16). One type uses a perforated, zinc-coated steel or aluminum panel bonded to a 2in. (50mm) thick fiberglass panel. They are mounted several inches off the wall (see 9-17).

Another type of ceiling panels used are **baffles.** Baffles are acoustical panels hung from the ceiling to reduce airborne sound in a large space such as a factory, restaurant, or auditorium (see 9-18). One type uses a light weight panel of vinyl covered fiberglass. Panels are 1 in. (25mm) thick, 10 in. (254mm) high, and from 2 to 4 ft. (610 to 1220mm) long (see 9-19). Another type uses a flame-resistant polyethylene cover over a 1½ in. (38mm) fiberglass core. It is flexible like a blanket and is hung from the ceiling. Another type of baffle uses rigid wood fiber sound deadening panels. They can be joined to form an attractive grid on the ceiling (see 9-20).

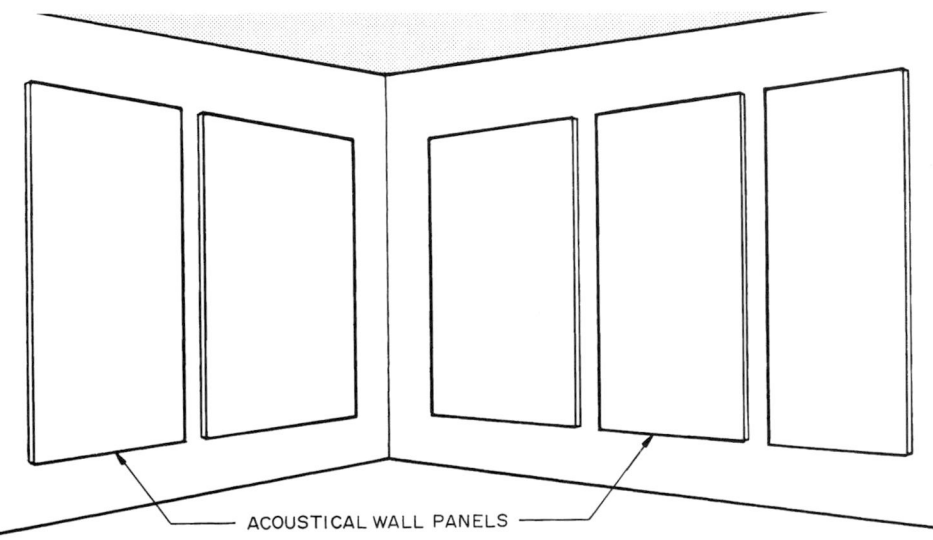

9-16 *Acoustical wall panels of molded glass fiber are used to control interior sound reverberations.*

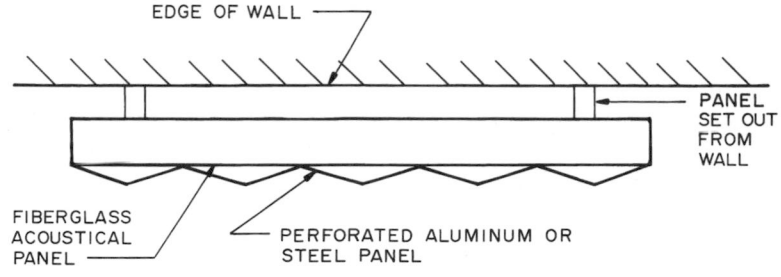

EDGE OF WALL

PANEL SET OUT FROM WALL

FIBERGLASS ACOUSTICAL PANEL

PERFORATED ALUMINUM OR STEEL PANEL

9-17 *A perforated metal acoustical panel.*

9-18 *Acoustical materials are used on ceiling baffles to reduce sound reverberations. Notice also the use of acoustical wall panels.*
Courtesy Metal Building Interior Products Co.

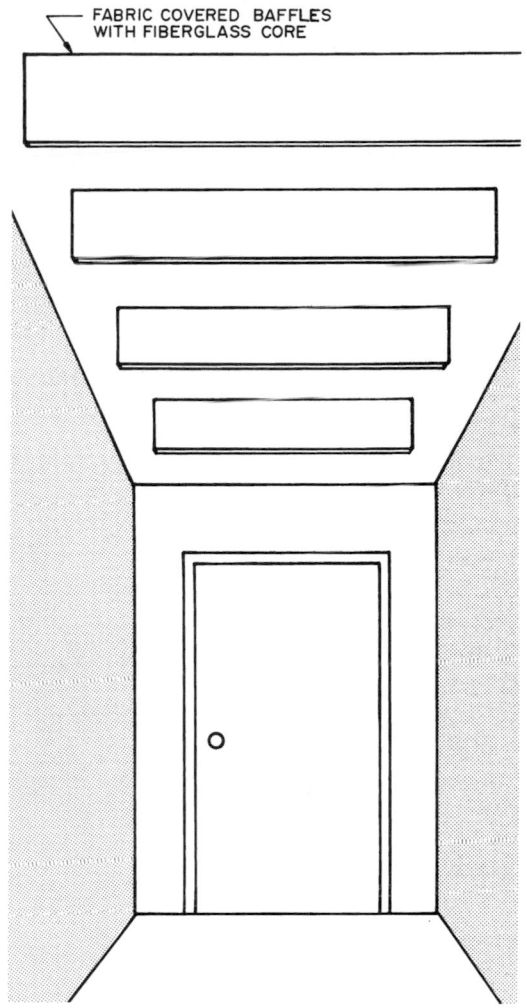

FABRIC COVERED BAFFLES WITH FIBERGLASS CORE

9-19 *Baffles at the ceiling used to control sound in a hall.*

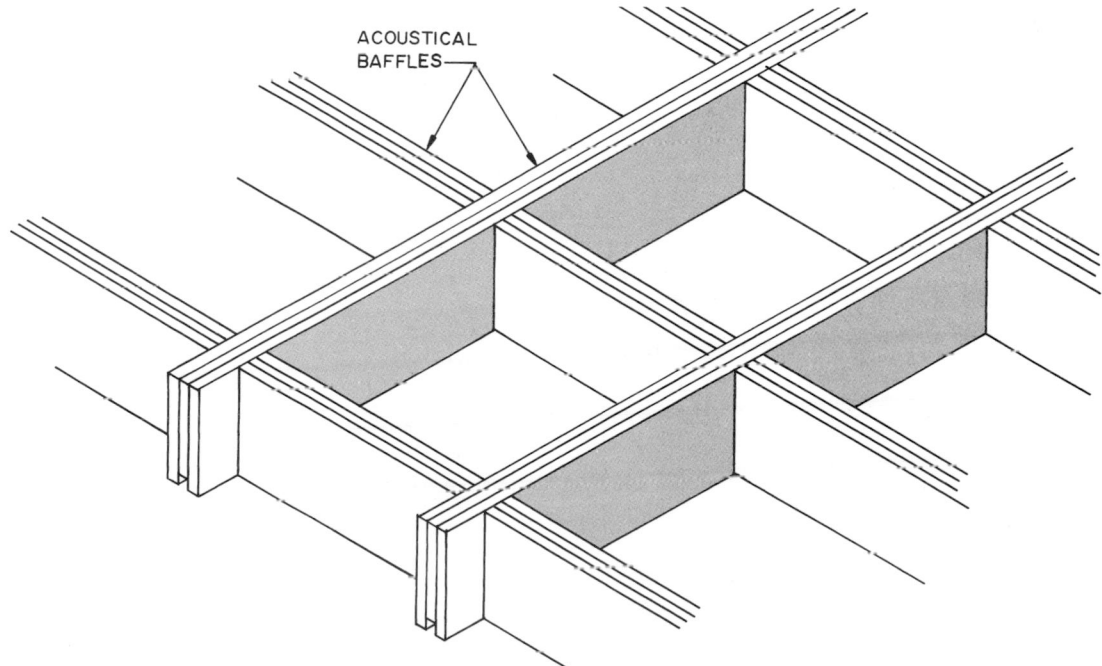

ACOUSTICAL BAFFLES

9-20 *Decorative baffles in a grid mounted on the ceiling.*

Sculptured acoustical wall units made from a high density molded fiberglass layer bonded to a sound absorbing glass fiber blanket are used to absorb sound and provide a decorative feature. They are covered with a wide range of fabrics. Typically they are in round, octagonal, and triangular shapes (see 9-21).

Some ceiling tile are designed to be glued to a sound ceiling substrate or can be nailed or stapled to the ceiling. Tile to be installed with an adhesive has nut size daubs of adhesive placed on the back. The tile is pressed against the ceiling substrate and slid into place (see 9-22).

If ceiling tile are to be installed to the bottom of floor joists without a substrate, wood 1 × 3 in. (25 × 75mm) strips are nailed perpendicular to the joists at 12 in. (305mm) on center spacing. The ceiling tiles are nailed or stapled through their tongues to the wood strips (see 9-23).

Ceiling tiles and panels are made with a variety of edge shapes producing different appearances. Some of these are shown in 9-24.

The suspended ceiling system is widely used in commercial construction and finds some use in residential work. It consists of a grid of metal runners hung from overhead on wires into which acoustical ceiling panels fit. Light fixtures are also designed to fit into this grid. A typical system is in 9-25. The space between the metal grid and the overhead structure can be used to run mechanical and electrical systems.

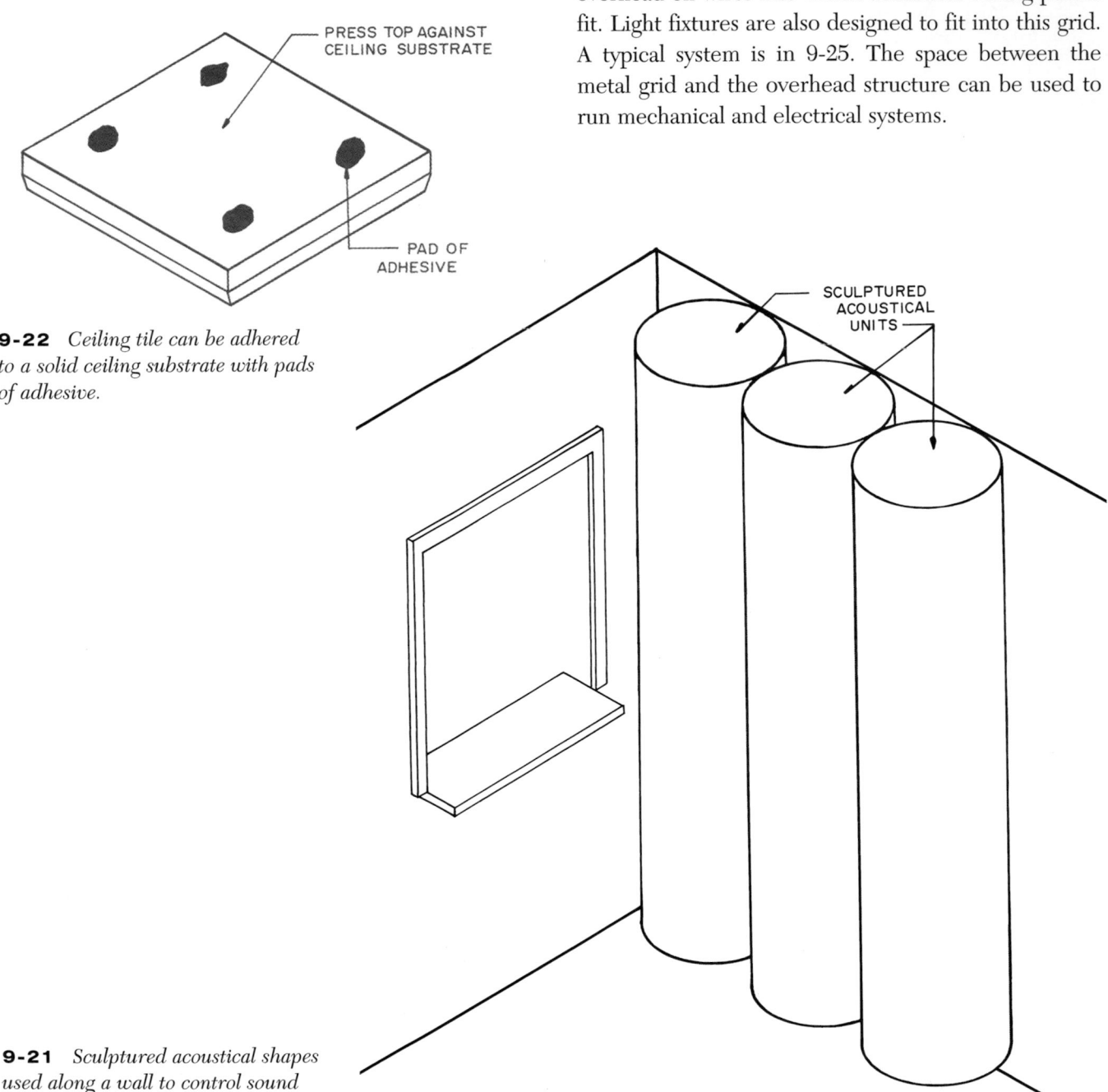

PRESS TOP AGAINST CEILING SUBSTRATE

PAD OF ADHESIVE

9-22 *Ceiling tile can be adhered to a solid ceiling substrate with pads of adhesive.*

SCULPTURED ACOUSTICAL UNITS

9-21 *Sculptured acoustical shapes used along a wall to control sound and serve as a decorative feature.*

9-23 *(Right) Ceiling tile can be installed by nailing or stapling to wood furring strips.*

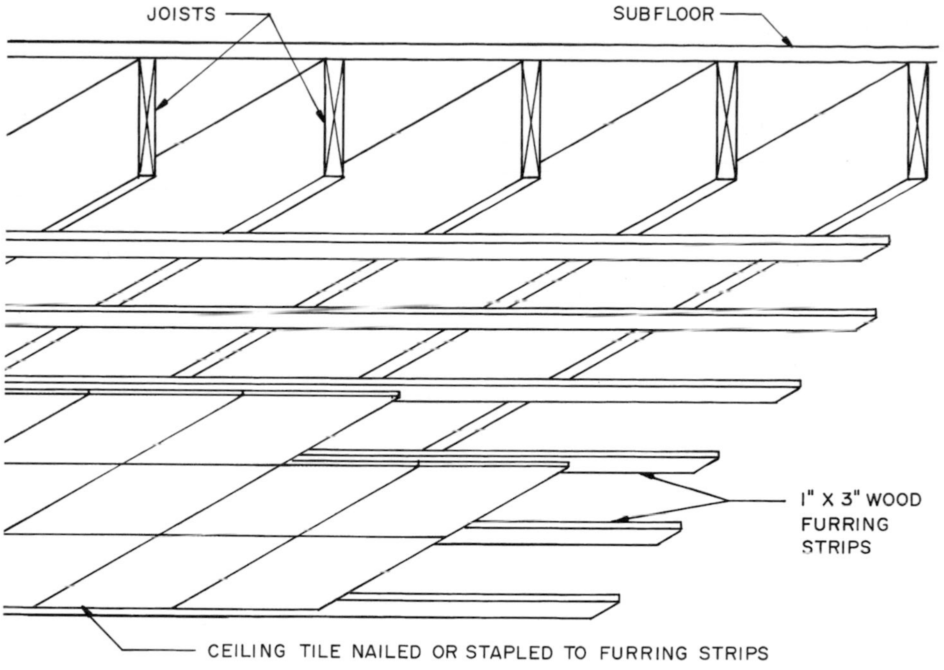

JOISTS   SUBFLOOR

1" X 3" WOOD FURRING STRIPS

CEILING TILE NAILED OR STAPLED TO FURRING STRIPS

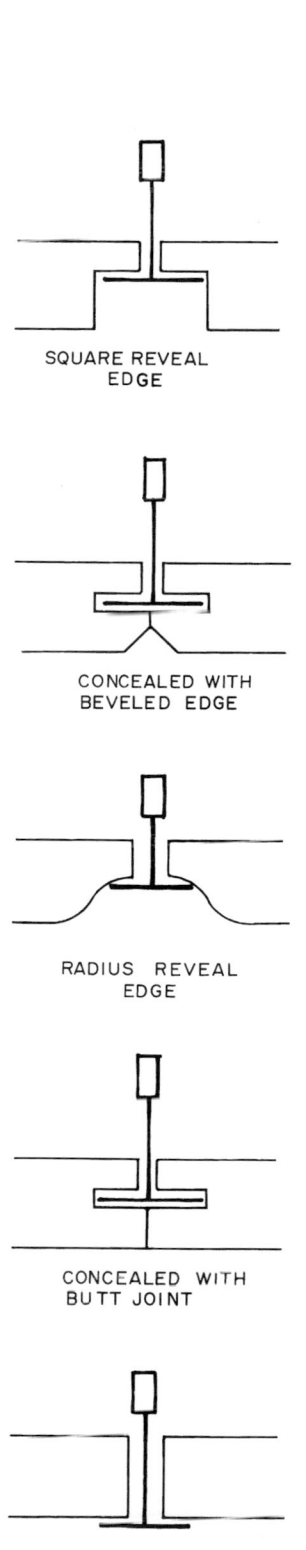

SQUARE REVEAL EDGE

CONCEALED WITH BEVELED EDGE

RADIUS REVEAL EDGE

CONCEALED WITH BUTT JOINT

SQUARE EDGE LAY-IN PANEL

9-24 *Typical panel edges for drop-in ceiling tiles and panels.*

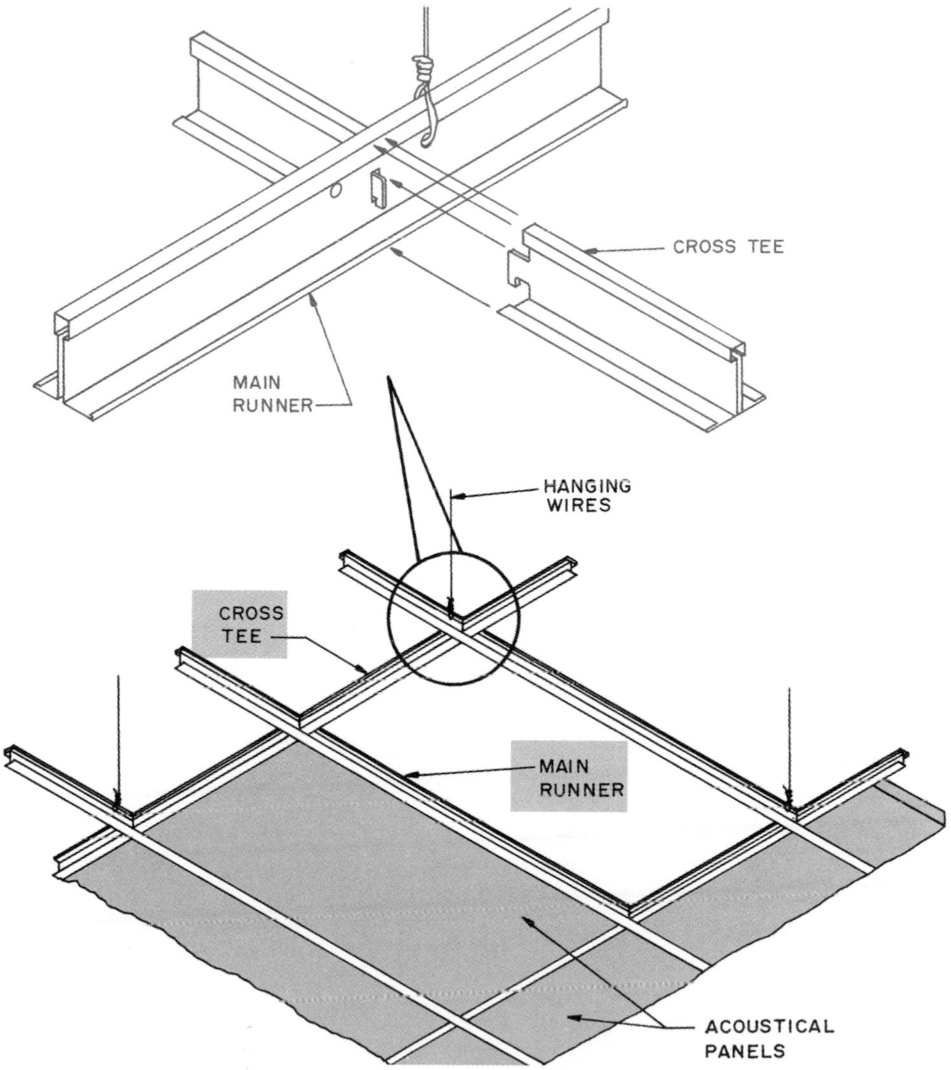

CROSS TEE

MAIN RUNNER

HANGING WIRES

CROSS TEE

MAIN RUNNER

ACOUSTICAL PANELS

9-25 *A view looking up from below at a suspended metal grid with acoustical panels resting on the flanges.*

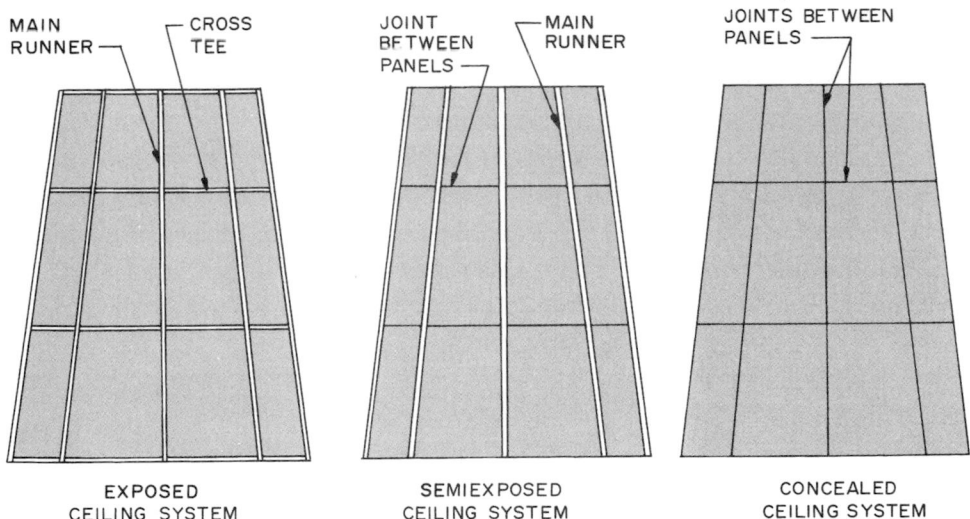

9-26 *Types of suspended ceiling systems.*

EXPOSED CEILING SYSTEM — MAIN RUNNER, CROSS TEE

SEMIEXPOSED CEILING SYSTEM — JOINT BETWEEN PANELS, MAIN RUNNER

CONCEALED CEILING SYSTEM — JOINTS BETWEEN PANELS

There are a number of grid systems available but basically they fall into three types—exposed grid, semi-exposed grid, and concealed grid. The manufacturer produces acoustical tiles with the edges designed to fit their grid producing the type of exposure desired. **Exposed systems** have the main runner and cross runner exposed.

This produces a square or rectangular grid appearance on the ceiling. The acoustical units rest on the flange of the T-shaped runner, as shown earlier in 9-24. **Semi-exposed systems** have the main runner exposed and the cross runner concealed. This gives a series of long parallel lines in the ceiling. The concealed system has no metal runners showing (see 9-26).

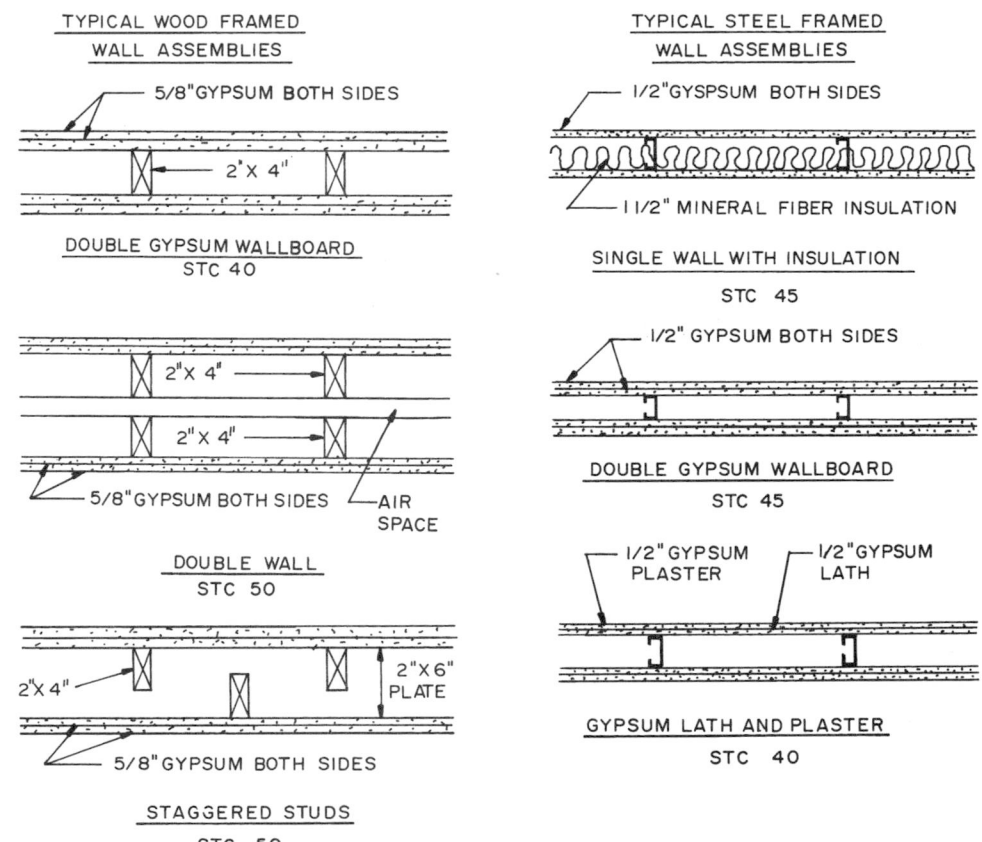

TYPICAL WOOD FRAMED WALL ASSEMBLIES

5/8" GYPSUM BOTH SIDES
2" x 4"
DOUBLE GYPSUM WALLBOARD
STC 40

2" x 4"
2" x 4"
5/8" GYPSUM BOTH SIDES — AIR SPACE
DOUBLE WALL
STC 50

2" x 4"
2" x 6" PLATE
5/8" GYPSUM BOTH SIDES
STAGGERED STUDS
STC 50

TYPICAL STEEL FRAMED WALL ASSEMBLIES

1/2" GYPSUM BOTH SIDES
1 1/2" MINERAL FIBER INSULATION
SINGLE WALL WITH INSULATION
STC 45

1/2" GYPSUM BOTH SIDES
DOUBLE GYPSUM WALLBOARD
STC 45

1/2" GYPSUM PLASTER — 1/2" GYPSUM LATH
GYPSUM LATH AND PLASTER
STC 40

9-27 *Examples of STC ratings for typical wall assemblies.*

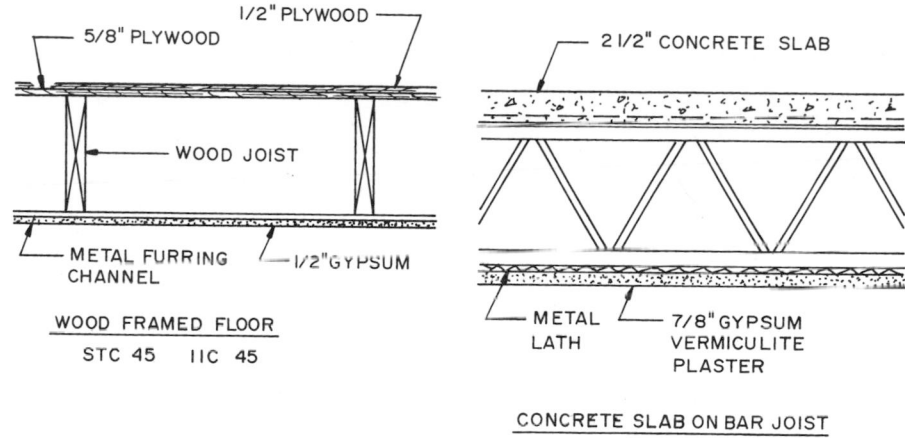

9-28 *STC and IIC ratings for typical floor-ceiling assemblies.*

WOOD FRAMED FLOOR
STC 45   IIC 45

CONCRETE SLAB ON BAR JOIST
STC 45   IIC 70

Sound Barriers

A wide range of material can be used to block the transmission of sound through an assembly, as a wall Many of these, as in brick, concrete, concrete block, and gypsum are discussed in earlier chapters. These units provide mass. Mass blocks the transmission of sound. For example, a 4 in. (100mm) brick or concrete masonry unit has an STC of about 40 A 4 in. (100mm) concrete floor has an STC of 44.

Lead sheet material is used in commercial buildings to block sound transmission. For example, a lead sheet can be laid on top of a suspended ceiling or hung from the bottom of the floor above blocking the transmission of sound that has penetrated the ceiling. Lead is available in sheets and foils as thin as 0.0005 in. (0.013mm). It is used in walls, doors, and other areas where a thin but effective sound barrier is needed.

Acoustical sealant is available as a pumpable material and is applied to all openings where sound may penetrate. For example, space around electrical boxes and pipes that pierce a wall or floor must be sealed. Openings, even very small, reduce the effectiveness of an otherwise efficient sound barrier.

Construction Techniques

There are ways to construct walls, floors, and ceiling to reduce sound transmission. Those that follow show some of the basic ways in common use.

Several wall assemblies are shown in 9-27. It shows three frequently used methods for increasing the STC rating of wood- and metal-framed walls. All of these could have sound deadening insulating batts installed to increase the STC value.

Several floor-ceiling assemblies are shown in 9-28 and 9-29. The floating floor technique shown is as used in commercial construction. The same type of construction can be used with wood-framed floors as used in residential work.

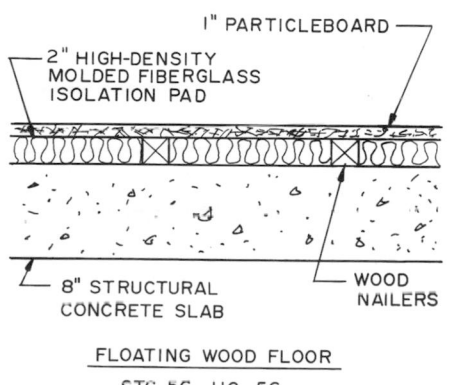

9-29 *STC and IIC ratings for typical floating floor construction.*

FLOATING WOOD FLOOR
STC 56   IIC 56

FLOATING CONCRETE SLAB
STC 73   IIC 70

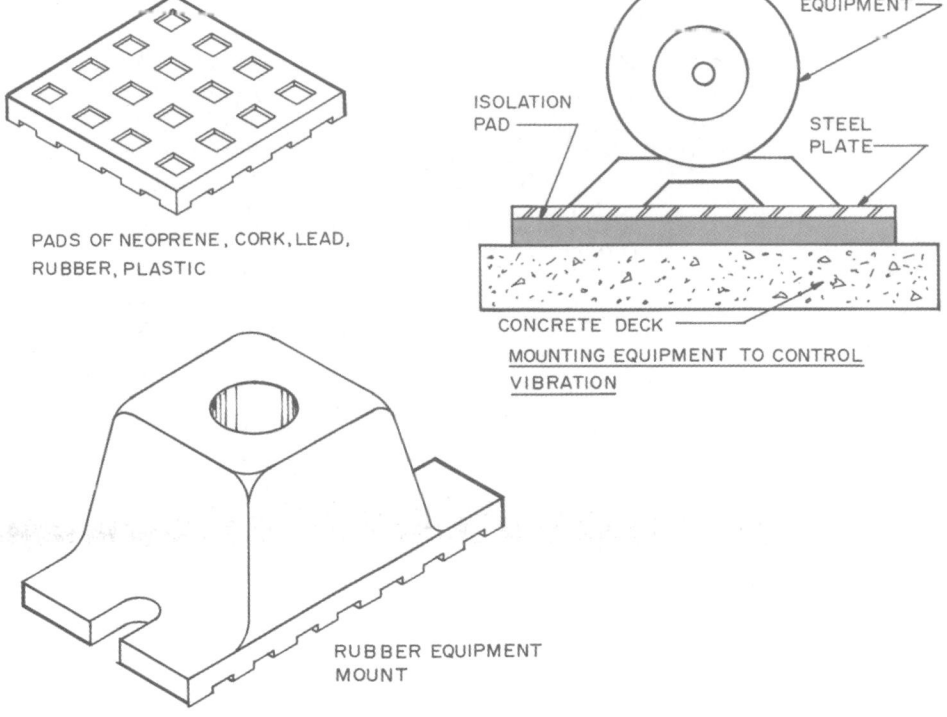

9-30 *Various pads used to damped the sound produced by vibrating mechanical equipment.*

Controlling Sound from Vibrations

Mechanical equipment is mounted on the structural frame of buildings. This produces vibration sounds that travel for great distances to other parts of the building. Vibrating water and sewer pipes create sounds that need to be controlled. Much of the sound can be dampened by mounting the mechanical equipment on various types of **isolator pads.** Rubber, neoprene, cork, and fiberglass pads are used for small pieces of equipment or larger pieces located in a basement area (see 9-30). For heavier duty units mounted within the building or on the roof, a steel spring mounting is used. The base of the spring pad must itself be isolated from the structure with an isolator pad to prevent the transmission of audible high frequency vibration through the spring to the structure.

INTERIOR WALLS, PARTITIONS & CEILINGS

Interior walls and partitions may be bearing or non-bearing and constructed with wood or steel studs or masonry. Various types of finish materials can be applied over the structure. Walls and partitions divide a building into areas having various types of occupancy and hide electrical and mechanical systems that may be run inside the wall cavity. They also provide privacy, security, sound control, insulation, and fire and smoke protection.

Ceilings provide an attractive finished surface as well as hide electrical and mechanical systems suspended above them. They also provide sound control and fire protection.

Interior finish on walls, partitions, and ceilings may be a gypsum board product, plaster, solid wood, plywood or hardboard panels, some type of fiberboard or ceramic tile.

Interior Walls & Partitions

Interior walls and partitions may be bearing or nonbearing. In most cases they are nonbearing and serve to divide the open space into usable areas and enclose openings such as stairs, shafts, and exits. The building structural system carries the loads of roof and floors, interior walls and partitions, furniture and other items to be placed in the building. The requirements an interior wall or partition must meet are specified by the building code. In Table 9-10 are minimum fire resistive requirements of walls, partitions, and opening protective devices as specified in the Standard Building Code.

Table 9-10 MINIMUM FIRE RESISTANCE OF WALLS, PARTITIONS & OPENING PROTECTIVES[1](hrs)

Protectives	Walls & Component	Opening Partitions
SHAFT ENCLOSURES (including stairways, exits & elevators)		
4 or more stories	2	1½B
less than 4 stories	1[2]	1B[2]
all refuse chutes	2	1½B
WALLS & PARTITIONS		
fire walls[3]	4	3A
within tenant space	See 704.2.3	
tenant space (see also 704.3)	1	¾C
horizontal exit	2	1½B
exit access corridors[4,5]	1	20 min.
smoke barriers	See 409.1.2	
refuse and laundry chute access rooms	1	¾C
incinerator rooms	2	1½B
refuse and laundry chute termination rooms	1	¾C
hazardous occupancy control areas	1	¾C
high-rise buildings	See 412	
covered mall buildings	See 413	
assembly buildings	See Note 2	
bathrooms & restrooms	See Note 6	
OCCUPANCY SEPARATIONS[7]	Required Fire Resistance	
	4	3A
	3	3A
	2	1½B
	1	¾C
EXTERIOR WALLS[8]	All	¾E

Notes:

See Standard Building Code for references.

1. Table 600 may require greater fire resistance of walls to insure structural stability.

2. All exits and stairways in Group A and H occupancies shall be 2 hours with 1½ hour B door assemblies.

3. See also 503.1.2.

4. See 704.2.3.

5. See 409 for sprinkled Group 1- buildings.

Table 9-11   MINIMUM INTERIOR FINISH CLASSIFICATION

	Unsprinklered			Sprinklered		
Occupancy	Exits[1]	Exits Access	Other Spaces	Exits[1]	Exit Access	Other Spaces
A	A	A	B	B	C	C
B	B	B	C	C	C	C
E	A	B	C	B	C	C
F	C	C	C	C	C	C
H		Sprinklers required		B	C	C
I Restrained	A	A	C	A	A	C
I Unrestrained		Sprinklers required		B	B	B[3]
M	B	B	C	C	C	C
R[2]	B	B	C	C	C	C
S	C	C	C	C	C	C

Notes:

1. In vertical exitways of buildings three stories or less in height of other than Group I Restrained, the interior finish may be Class B for unsprinklered buildings and Class C for sprinklered buildings.

2. Class C interior finish materials may be used within a dwelling unit.

3. Rooms with 4 or less persons require Class C interior finish.

   1. Class A Interior Finish, Flamespread 0–25, Smoke Developed 0–450. Any element thereof when so tested shall not continue to propagate fire.

   2. Class B Interior Finish. Flamespread 26–75, Smoke Developed 0–450.

   3. Class C Interior Finish. Flamespread 76–200, Smoke Developed 0–450.

Standard Building Code© 1994. Used with permission from the Standard Building Code, Southern Building Code Congress International, Inc.

Nonbearing nonfire resistant partitions—are those not used as fire separation walls. They must meet the code requirements specifying the combustibility of materials for the type of construction in which they are to be used. In Table 9-11 are minimum interior wall finish classifications based upon the occupancy of the building for materials other than floor finish and floor covering. They are based on the acceptable flamespread and smoke development classifications.

Fire-separation walls or **fire partitions**—are installed to provide enclosure of shafts, floor openings, exits, and subdividing an area. The construction and selection of materials are specified by codes for the type of construction involved. They are used to control the spread of fire between areas on a floor. The fire separation wall or fire partition should have a fire resistance rating equal to code requirements specified for the occupancy of the fire areas that were separated (see Table 9-12). They must extend from the top of the fire resistant floor to the bottom of the fire resistant floor or roof slab above and must be securely fastened to them. The size of openings in fire separation walls are restricted by code and must be protected by fire doors, windows, or shutters.

Fire walls—are fire-resistance-rated walls that restrict the spread of fire and extend continuously from the foundation to or through the roof. If it stops at the roof the roof assembly must be of noncombustible material. If it extends above the roof the height above the roof is specified by the code. The fire wall should be able to remain intact even when the construction on either side collapses due to fire damage. Fire-resistance ratings of fire walls will usually be 4 hours. The wall must be smoke tight where it meets the exterior wall. The size of openings in fire walls is restricted by code and requires protection by fire doors or wire glass.

Table 9-12 OCCUPANCY SEPARATION REQUIREMENTS[1]

Large or Small Assembly	2 hour
Business	1 hour
Educational	2 hour
Factory-Industrial	2 hour
Hazardous	See 704.1.4
Institutional	2 hour
Mercantile	1 hour
Residential	1 hour
Storage, Moderate Hazard S1	3 hour
Storage, Low Hazard S2	2 hour
Automobile Parking Garages[2]	1 hour
Automobile Repair Garages	2 hour

Note:

See Standard Building Code for references.

1. The minimum fire resistance of construction separating any two occupancies in a building of mixed occupancy shall be the higher rating required for the occupancies being separated.

2. See 411.2.6 for exceptions.

Standard Building Code© 1994. Used with permission from the Standard Building Code, Southern Building Code Congress International, Inc.

Party walls—are fire walls used jointly by two parties under an easement agreement. They are erected on an interior lot line dividing two parcels of land each which is a separate real estate entity. It is actually a common wall between two buildings that meet on a lot line.

Smoke barriers—are fire-resistant continuous membranes used to resist the movement of smoke. They have a fire-resistance rating specified by the building code. They form a continuous membrane from one outside wall to the other and from the floor slab to the roof or floor slab above. Doors in smoke barriers must meet special code requirements and have door closers that cause them to close when smoke detectors sense smoke. They typically have a one-hour minimum fire resistance requirement.

Interior wall, partitions, and ceilings are also subject to specifications relating to the transfer of sound into adjacent areas and the control of reverberating sound within the rooms.

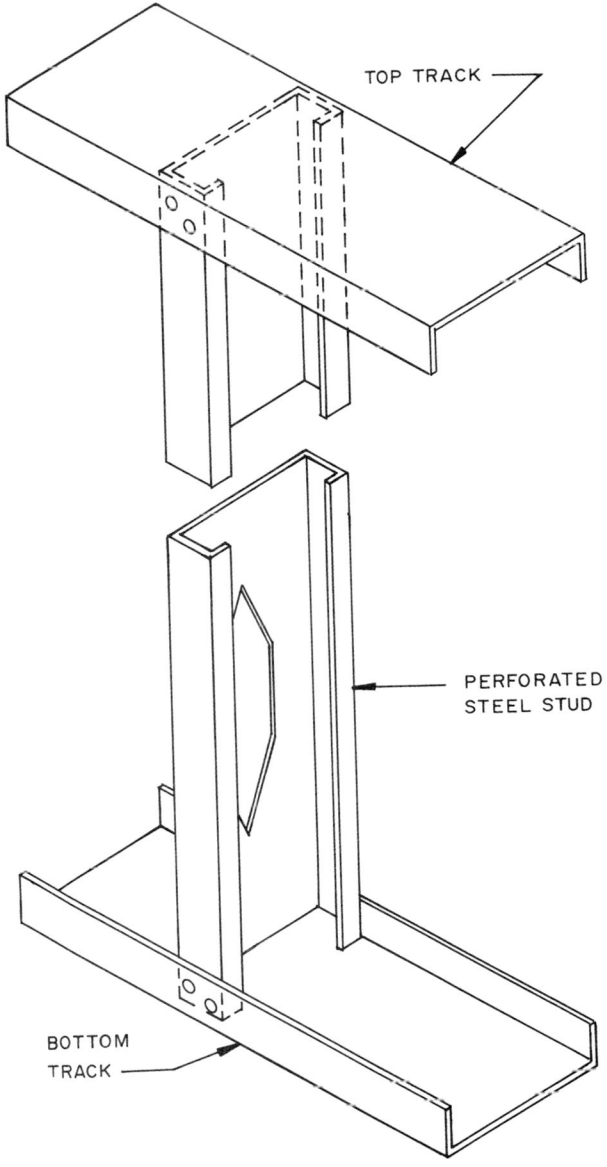

9-31 *Light gauge steel studs are used for interior walls and partitions.*

METAL STUD INTERIOR NONBEARING GYPSUM WALL SYSTEMS

Metal studs are formed from light-gauge steel and are available in a range of sizes. Some are manufactured from heavier gauge steel and are used for bearing walls. Other types are designed for use in curtain wall construction. Metal studs for nonbearing interior walls and partitions are typically made from 18-gauge steel sheet and may have solid or perforated webs (see 9-31).

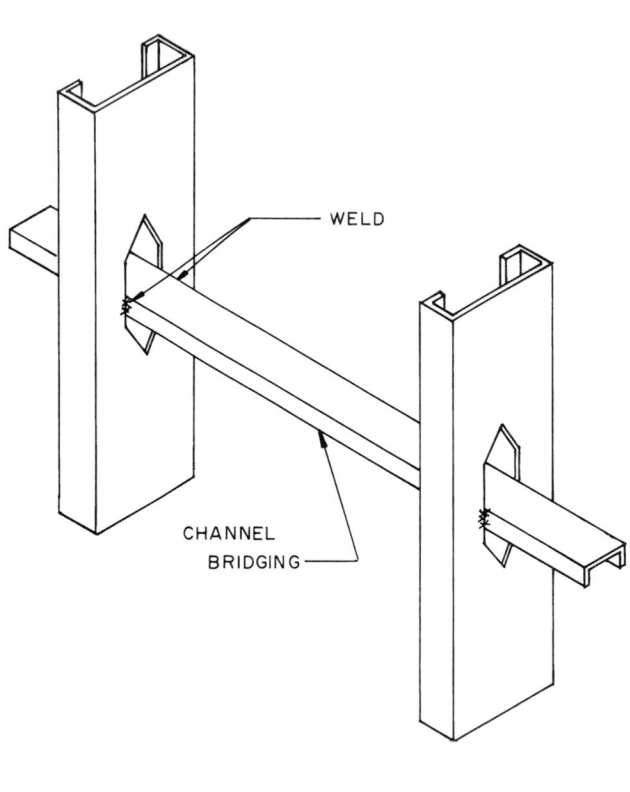

9-32 *One type of bridging used with penetrated steel studs.*

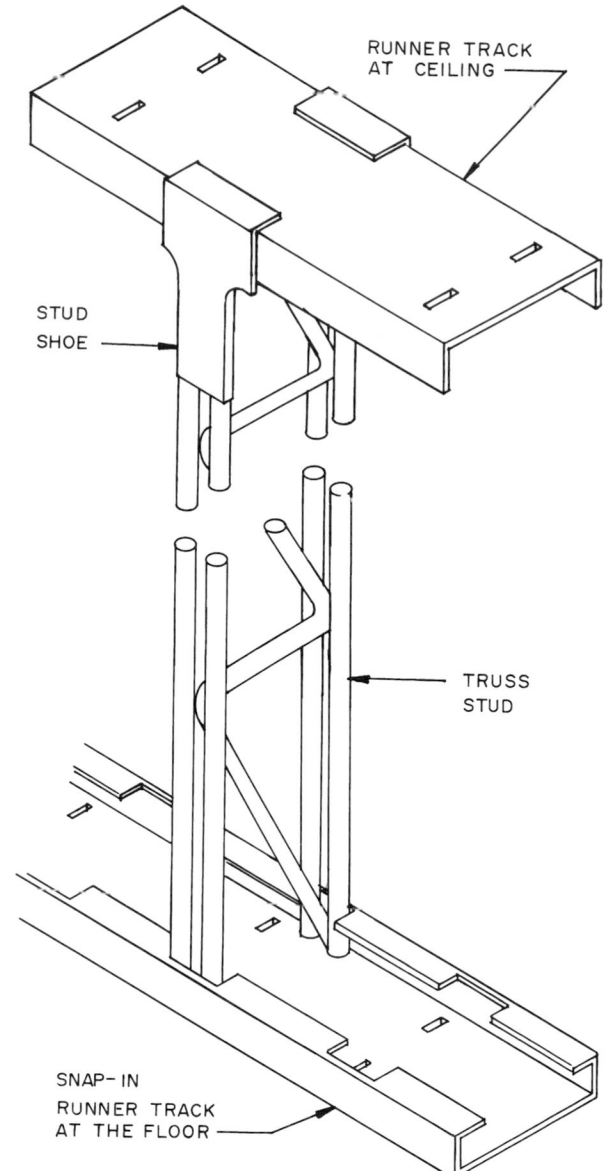

9-33 *A truss stud is set in snap-in runners and secured to the top runner with a stud shoe.*

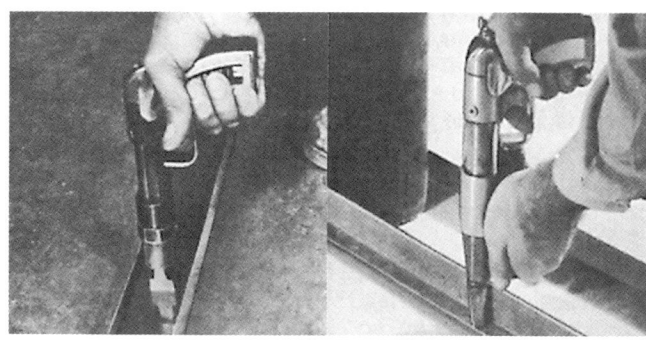

9-34 *The metal runner is fastened to the concrete floor with steel fasteners driven by a gun that uses a small charge of gunpowder.*

Courtesy USG Corporation

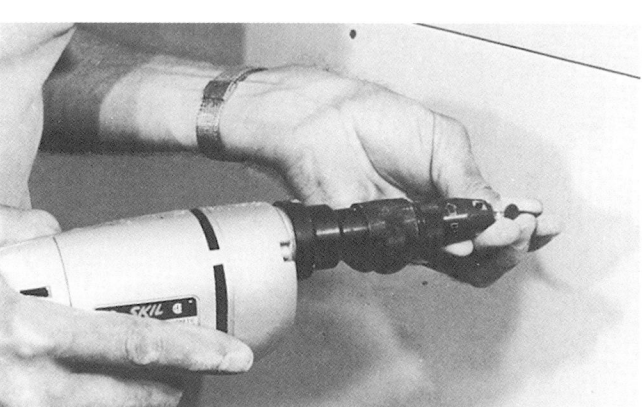

9-35 *Gypsum wallboard is secured to wood and metal studs with screws driven with a power screwdriver.*

Courtesy USG Corporation

The punch outs in the web facilitate installation of bridging, pipe, and electrical conduit (see 9-32). Another type of metal stud uses round rod bent on a diagonal truss design between vertical cords (see 9-33). These studs are used primarily for walls covered with gypsum lath or metal lath. The lath is joined to the studs with clips. The studs are secured to floor and ceiling runner tracks supplied by the manufacturer of the stud. The solid and perforated web studs are secured to the runners with self-drilling self-tapping screws. They are made of hardened steel and are driven with an electric screwdriver. The tracks are secured to the concrete floor and overhead material with power driven fasteners. The gun uses a gunpowder cartridge to drive the steel fastener through the track into the concrete floor (9-34). The sheets are carefully positioned and screwed to the metal studs (9-35).

TYPICAL WALL ASSEMBLIES

The actual assembly of gypsum wallboard walls will depend upon the desired sound transmission level, thermal insulation, and building code fire-resistance specifications. In 9-36 is a typical assembly for a **cavity-type wall** that, depending upon the circum-stance,s could serve as a fire, separation, or party wall. This is a continuous vertical, nonbearing wall assembly with metal studs and furring, sound attenuation, fire blankets, gypsum liner panels, and water-resistant fire-resistant gypsum facing panels. The studs used are C-H steel studs with steel E-studs at the end of the wall. A wall with this construction can have a two-hour fire rating.

A typical assembly for a fire-rated partition designed to offer effective sound control and a high fire rating is in Fig. 9-37. It uses steel studs, gypsum lath on steel furring, and a gypsum finish coat. A sound attenuation blanket is in the wall cavity.

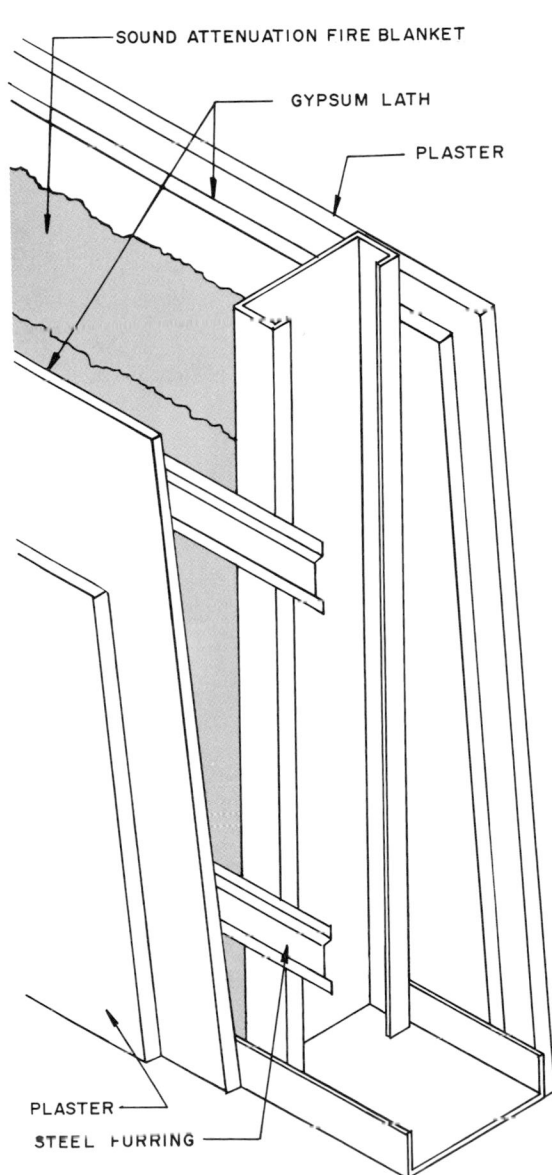

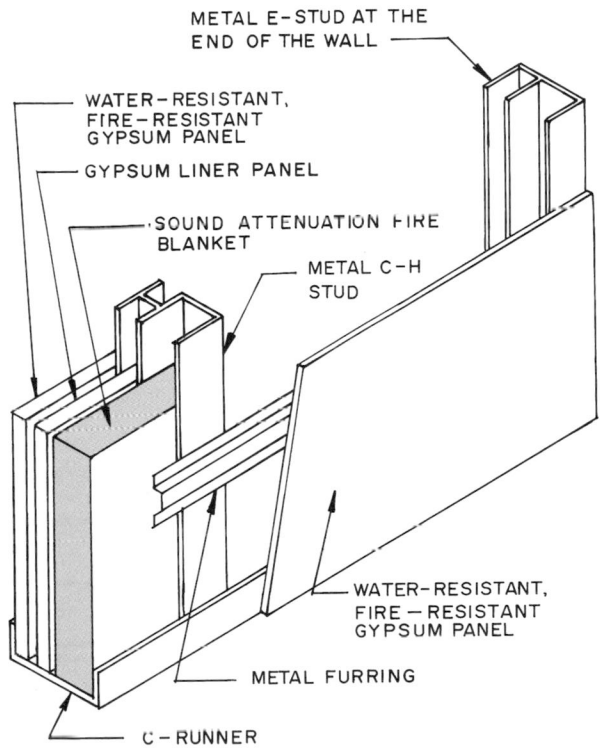

9-36 *A cavity-type separation wall utilizing a gypsum liner panel and a sound attenuation fire blanket.*

9-37 *A sound attenuation steel-framed wall with fire-resistant properties.*

Fire-resistant drywall partitions are used to enclose shafts in multistory buildings (9-38). Typical shafts include those for elevators, mechanical equipment, stairwells, and air returns.

A typical assembly for a nonbearing, fire-resistant gypsum board partition for enclosing shafts, air ducts, and stairwells is in 9-39. It uses a C-H stud system, a gypsum panel liner, and a single layer of fire-resistant gypsum board on each side. This assembly will give a two-hour fire rating. Gypsum fire-resistance wall construction is considerably lighter than masonry fire-resistant walls. Additional layers of fire-resistant gypsum board will increase the fire rating. In 9-40 are shown several steel-frame partition systems with gypsum board, lathe, and plaster and veneer plaster cladding. These contain no fire-resistant insulation or gypsum liner panels. The manufacturers of gypsum products have a wide range of wall designs for various fire ratings.

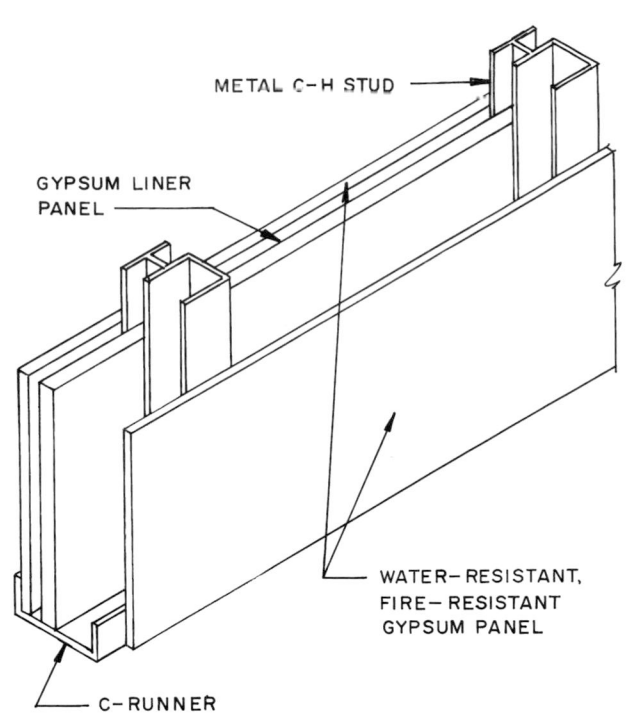

9-39 *Typical construction of a single layer, fire-resistant, nonbearing, cavity shaft wall as used on shafts, air ducts, and stairwells.*

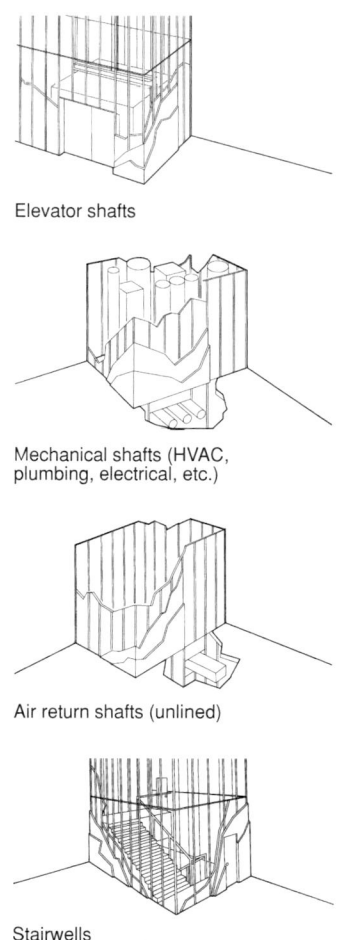

Elevator shafts

Mechanical shafts (HVAC, plumbing, electrical, etc.)

Air return shafts (unlined)

Stairwells

9-38 *These are examples of commonly found shaft walls that must meet fire code requirements.*

Courtesy USG Corporation

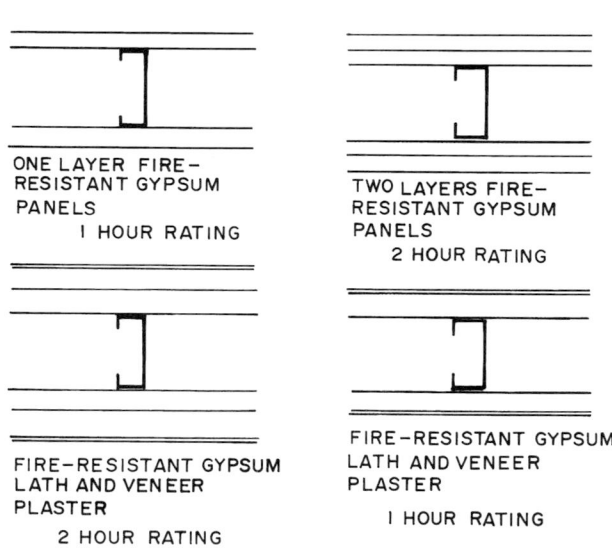

ONE LAYER FIRE-RESISTANT GYPSUM PANELS
I HOUR RATING

TWO LAYERS FIRE-RESISTANT GYPSUM PANELS
2 HOUR RATING

FIRE-RESISTANT GYPSUM LATH AND VENEER PLASTER
2 HOUR RATING

FIRE-RESISTANT GYPSUM LATH AND VENEER PLASTER
I HOUR RATING

9-40 *Typical fire ratings for various types of metal-framed gypsum wallboard and plaster partitions. Consult the manufacturer for specific data for recommended stud sizes and panel thicknesses.*

INSTALLING GYPSUM WALLBOARD

Gypsum wallboard is fastened to wood studs and ceiling joists with special nails or power-driven screws (9-41) and to metal studs and metal furring with self-drilling self-tapping screws (9-42). Panels can be installed with the long edge perpendicular to or parallel with the studs. Perpendicular application is usually used because it reduces the amount of joint to be taped (see 9-43). Perpendicular application also places the strongest dimension of the panel across the studs. Generally the ceiling panels are installed first and are run perpendicular or parallel with the joists depending upon which method produces the fewest joints. Panels may be single or double nailed. Single-nailed ceilings are usually spaced 7 in. (178mm) O.C. and 8 in. (203mm) O.C.

on side walls. Screws are spaced 12 in. (305mm) on the ceiling and 16 in. (406mm) on the sidewalls. To reduce the possibility of nail popping after the wall is finished the panel can be double nailed. The fasteners are driven until the head is set in a shallow dimple but the paper covering is not broken. Nails are driven with a drywall hammer that has the nailing face the required shape and diameter.

Gypsum panels are often installed in a double layer. The first layer is nailed or screwed to the framing and the second layer is bonded to it with an adhesive plus a few nails or screws. The adhesive is applied to the back of the panel as shown in 9-44. This produces a wall with greater sound attenuation and fire resistance and reduces the possibility of nail popping.

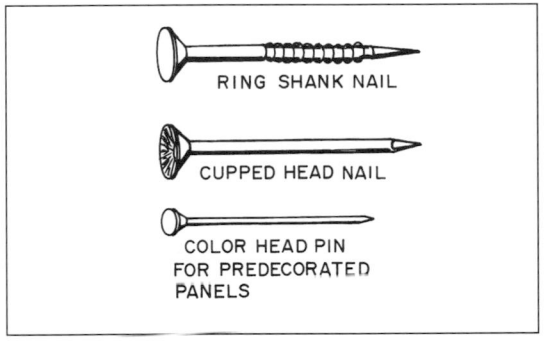

9-41 *Nails used to secure gypsum wallboard to wood studs.*

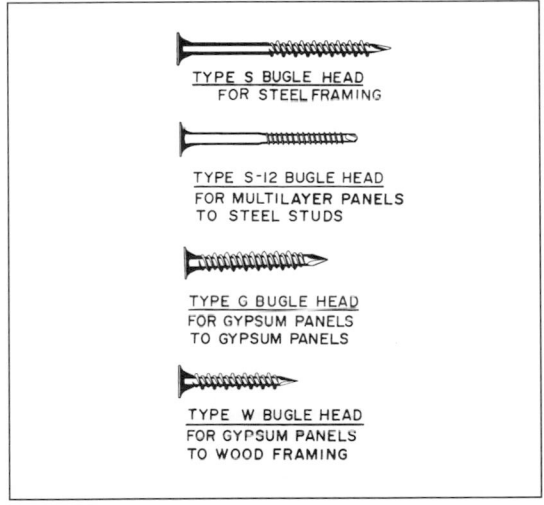

9-42 *Screws used to secure gypsum wallboard to wood and metal studs and for joining panels.*

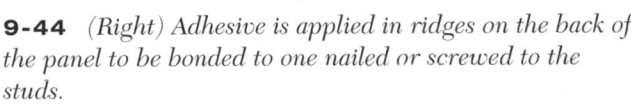

9-44 *(Right) Adhesive is applied in ridges on the back of the panel to be bonded to one nailed or screwed to the studs.*

Courtesy USG Corporation

9-43 *Gypsum wallboard panels are often installed with the long edge perpendicular to the studs.*

Courtesy USG Corporation

Gypsum panels are cut by scoring the surface along the edge of a metal T-square with a utility knife as shown in 9-45. The panel is bent breaking the core and the paper is scored on the back separating the parts of the panel.

The joints between the panels are covered with layers of joint compound and tape. The long edges are made with a taper allowing the compound and tape to fill it flush with the surface. The short ends of the panels are not tapered and produce a slight bulge if taped. Therefore, end joints are to to avoided when possible (9-46). The joint is covered with a thin layer of compound and the tape is pressed into it with a finishing knife (9-47). A thin skim coat is applied over the tape and the compound is allowed to dry. A layer of joint compound is placed over all nail and screw heads. Additional layers of compound are applied and sanded. Each layer is wider than the one before and is feathered out. The tape and joint compound can be mechanically applied to the wall and ceiling joints as shown in 9-48. The compound may be hand sanded to a final finish.

Internal corners are finished with paper or fiberglass tape bonded in joint compound and finished as described for joints. External corners are finished with metal or plastic corners nailed or screwed to the studs and covered with joint compound. Other finishing accessories include edge trim, control joints, and trim for round corners such as found on arches.

Door frames on walls with metal studs can be installed as shown in 9-49. These details show the installation of metal and wood door jambs.

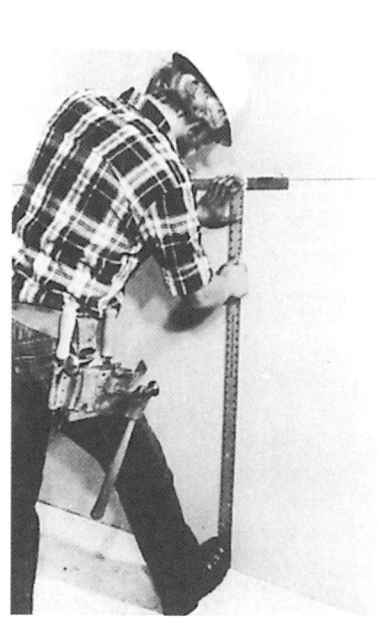

1. Score the panel along the straigthedge with a utility knife.

2. Bend the panel forward, breaking the core

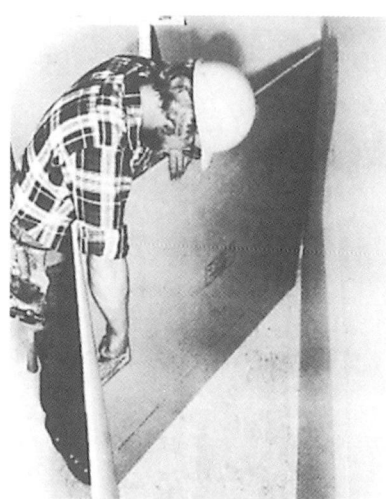

3. Cut the paper on the back of the panel.

4. Pull the parts of the panel apart.

9-45 Gypsum wallboard is cut by scoring the panel along a straightedge, bending it, breaking the core, and cutting the paper on the back side.
Courtesy USG Corporation

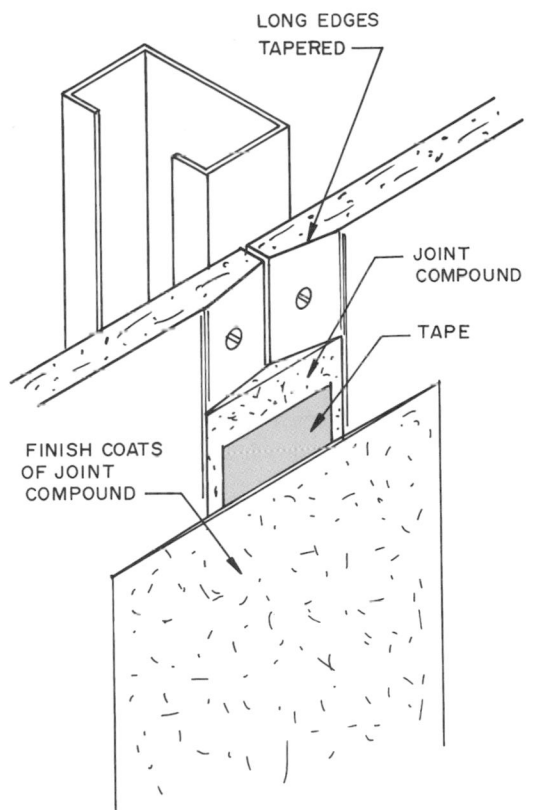

9-47 *The tape is pressed into the first layer of joint compound.*

Courtesy Georgia-Pacific Corporation

9-46 *The joints between panels are covered with layers of joint compound and tape.*

9-48 *This taping tool applies the tape to the joint after the first coat of joint compound is in place.*

Courtesy USG Corporation

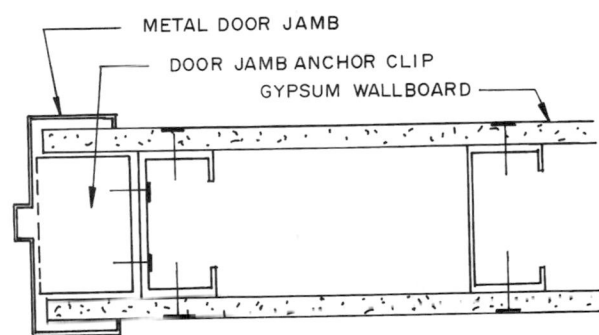

METAL DOOR JAMB INSTALLATION

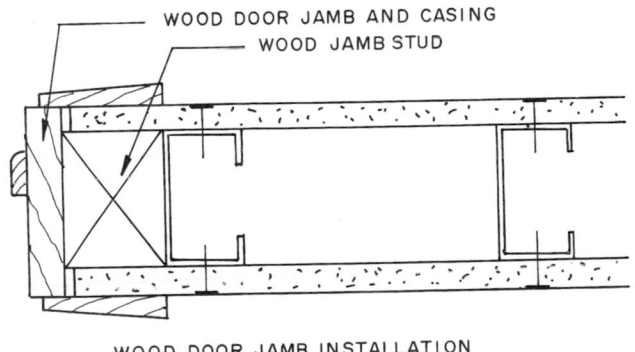

WOOD DOOR JAMB INSTALLATION

9-49 *Typical installation details for metal and wood door jambs.*

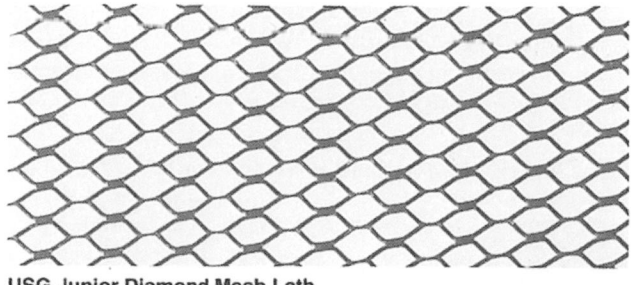

USG Junior Diamond Mesh Lath

USG Paper-Backed Metal Lath

USG 4-Mesh Z-Riblath

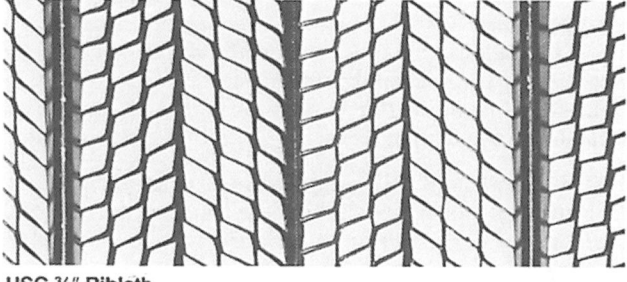

USG ⅜" Riblath

USG Self-Furring Diamond Mesh Lath

9-50 *Types of metal lath.*
Courtesy USG Corporation

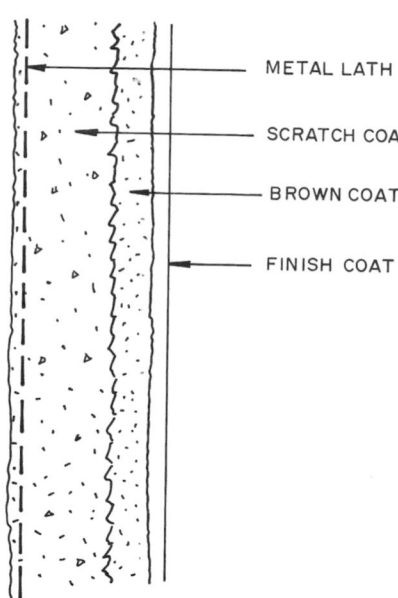

METAL LATH

SCRATCH COAT

BROWN COAT

FINISH COAT

9-51 *This is a three-coat plaster wall finish over metal lath.*

PLASTER WALL FINISHES

Plaster is applied over expanded metal lath, gypsum lath, or veneer plaster base.

Metal lath—is made by slitting sheets of thin metal alloy and stretching it to form a diamond-shaped mesh or by punching and forming the openings in the metal sheet. Typical types are riblath, diamond-mesh lath, sheet lath, and wire lath (see 9-50). They are usually wired to steel studs or secured with self-tapping screws.

Gypsum plaster lath—is a rigid, fire-resistant base upon which gypsum plasters are applied. The gypsum lath is available in several thicknesses and as two different products. One is a standard gypsum lath that is used for nailing or stapling to wood studs or screwed to metal studs and furring. A second type has the addition of a fire-resistant core.

Veneer plaster base—has a gypsum core and is faced with paper. It is used as the base for veneer plaster.

Plaster over expanded metal lath is a three-coat system (see 9-51). The first coat (the scratch coat) is applied to the metal lath with enough pressure to force it into the mesh and form a good key. It is cross raked leaving a flat but textured surface. After the scratch

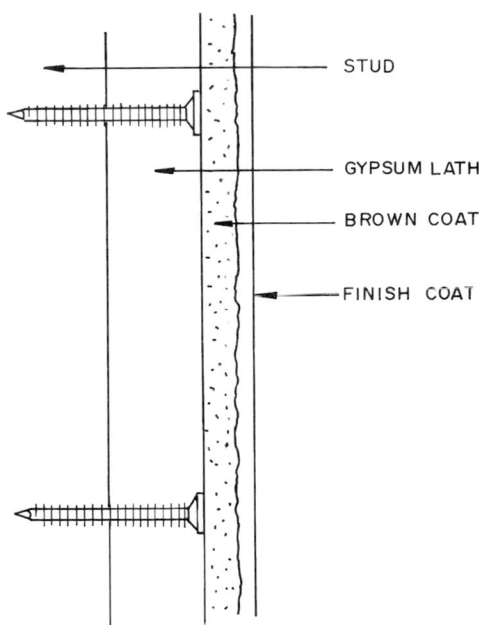

9-52 *This is a two-coat plaster wall finish over gypsum lath.*

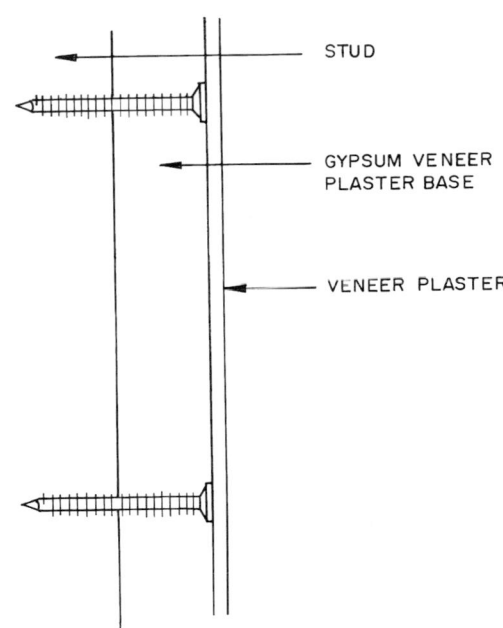

9-53 *Veneer plastic is a single-coat plaster applied over a gypsum veneer plaster base.*

coat has hardened the second coat (the brown coat) is applied over the scratch coat. It is leveled and allowed to harden. The third coat (the finish coat) is applied over the brown coat and the surface is finished to the desired texture.

Plaster over gypsum lath is a two-coat system; however, three coats may be used (see 9-52). The first coat is applied with pressure so it bonds to the lath and provides a level coat with a rough surface. The second coat is applied over the first coat and is finished to the desired texture.

Veneer plaster base is secured to the studs and is covered with one or two coats of veneer plaster (see 9-

53). The first coat is a thin layer that is covered by a second thin coat before the first coat has hardened. The second coat is troweled to the desired texture. Generally veneer plaster is completely hardened within 24 hours.

Plaster is installed at door openings as shown in 9-54. Other methods of assembly are used depending on the weight of the door and the method of securing the door frame to the steel wall stud. The door frame is grouted at the anchors to increase rigidity of the frame and improve resistance to frame rotation. The grouting may be located only at each frame anchor or be full grouted.

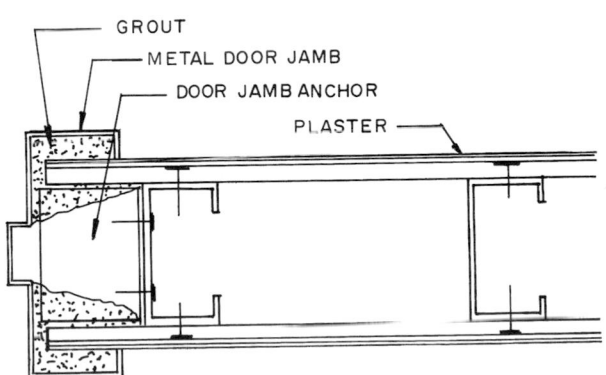

9-54 *One way to install metal door frames when the finish wall is plaster.*

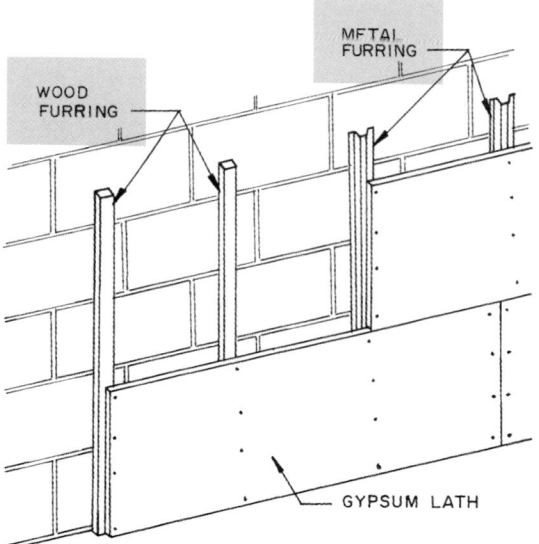

FUR THE WALL WITH WOOD OR METAL FURRING

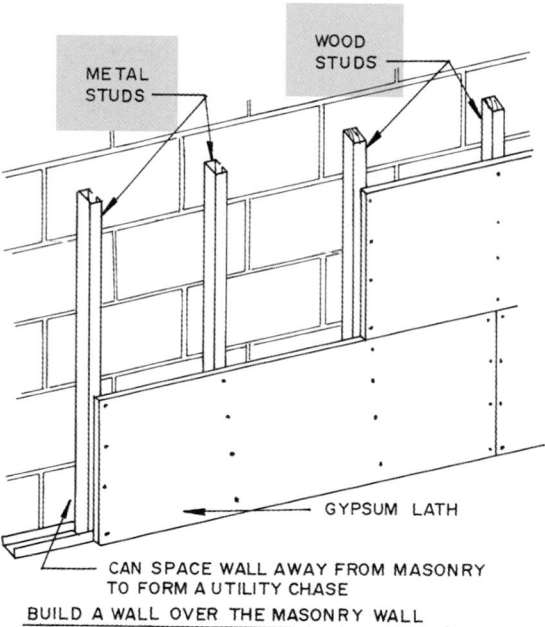

BUILD A WALL OVER THE MASONRY WALL

PLASTER OVER MASONRY & CONCRETE WALLS

Plaster may be applied directly to masonry, concrete, and clay tile walls. Concrete block walls are porous and provide a satisfactory base for plaster. Clay tile and brick walls may be used as a base for plaster walls. However, they must be sufficiently porous to provide suction for the plaster and be scored to increase the mechanical bonding. Smooth surface glazed or semiglazed tile cannot be covered. It requires the wall to be furred out and a metal or gypsum plaster base be installed. Monolithic concrete walls, although porous, usually require an application of a bonding agent to produce an adhesive bond necessary for direct application of gypsum plasters.

Generally exterior masonry walls are not plastered directly to the masonry because of the likely water seepage and condensation, which will wet the plaster.

Masonry concrete and clay tile walls can have gypsum lath installed by securing metal furring to the surface. On exterior walls this provides an air space, keeping the plaster base and plaster away from the masonry wall. Furring can be shimed to help bring a wall with minor irregularities into a flat surface. The furring may be placed horizontally or vertically (see 9-55).

PLASTERING

Plaster base may be metal lath or gypsum lath. The ends of the lath must rest on a stud and are nailed to it. Gypsum lath may also be secured by nailing or with self-drilling, self-tapping screws that are driven with a power screwdriver (see 9- 56).

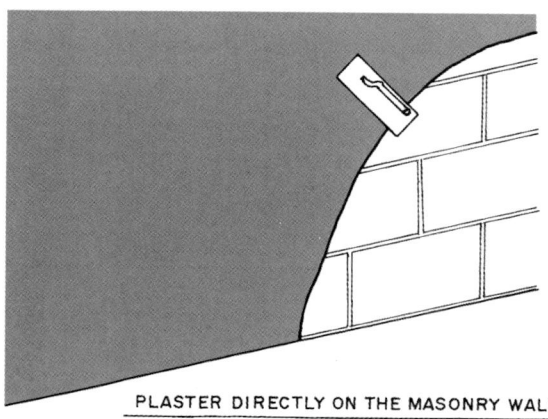

PLASTER DIRECTLY ON THE MASONRY WALL

9-55 *Masonry walls can be covered with a plaster finish.*

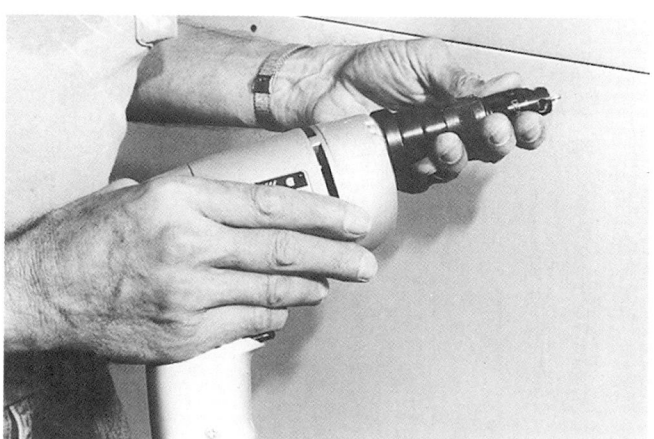

9-56 *Gypsum plaster lath is secured to studs with screws driven with a power screwdriver.*
Courtesy USG Corporation

Plaster may be applied to the base with a hand trowel as shown in 9-57. The plasterer holds in one hand a square flat tool called a **hawk**. The hawk holds a supply of plaster, which is removed by a trowel and applied to the wall. The gypsum lath base has metal clips supporting the ends of the panels because they do not rest upon a vertical support. Small wire clips are used to tie the gypsum panels to the truss-type metal studs.

END CLIP

9-57 *The plasterer is hand troweling the scratch coat to the gypsum lath base that is secured to metal studs with wire clips. The large metal clips support the ends of the panels that do not rest on a stud.*

Courtesy USG Corporation

9-58 *The plasterer on the right is applying a scratch coat by spraying it on the base. The plasterer in the center is hand troweling the brown coat while the person on the left is leveling the coat by pulling a darby across it. After it has been leveled it may be troweled again to smooth the surface.*

Courtesy USG Corporation

Plaster may also be applied by a spraying process as shown in 9-58. Various accessories are used with plaster finishes as described for drywall. Typical among these are casing beads, external corner beads, flexible corner beads, and expansion joints.

SOLID GYPSUM PARTITIONS

Solid gypsum partition systems may be studless or use channel studs. The solid studless partition has a vertical core of metal mesh lath. The lath specifications vary with the height of the wall. The lath is connected to the ceiling with a runner that is attached to the overhead and a metal runner at the floor that is fastened to the floor (see 9-59). The lath is covered with plaster

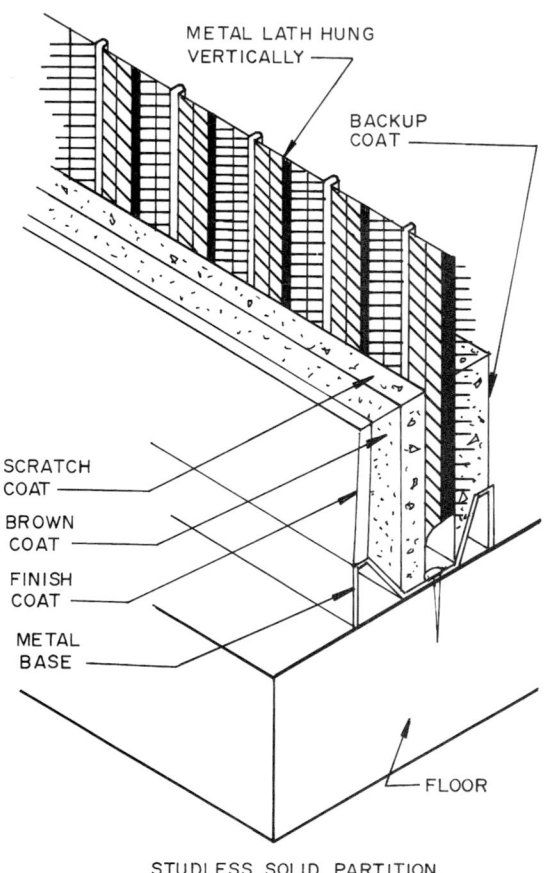

9-59 *Typical construction of a studless solid gypsum partition built around vertical metal lath.*

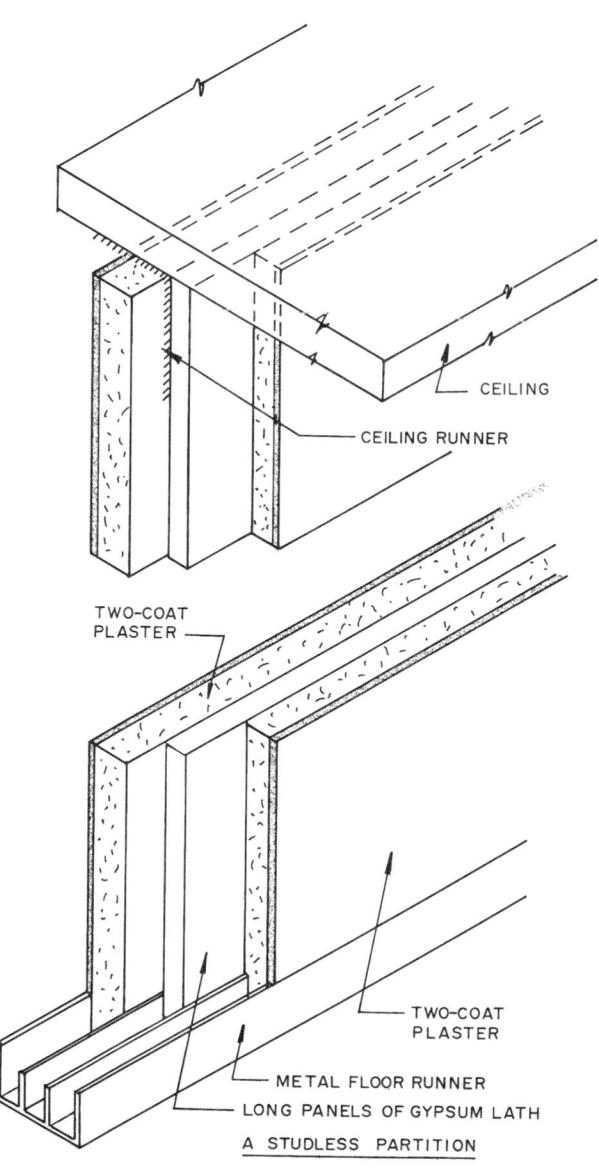

9-60 *A studless solid partition built around a gypsum lath core set in metal tracks.*

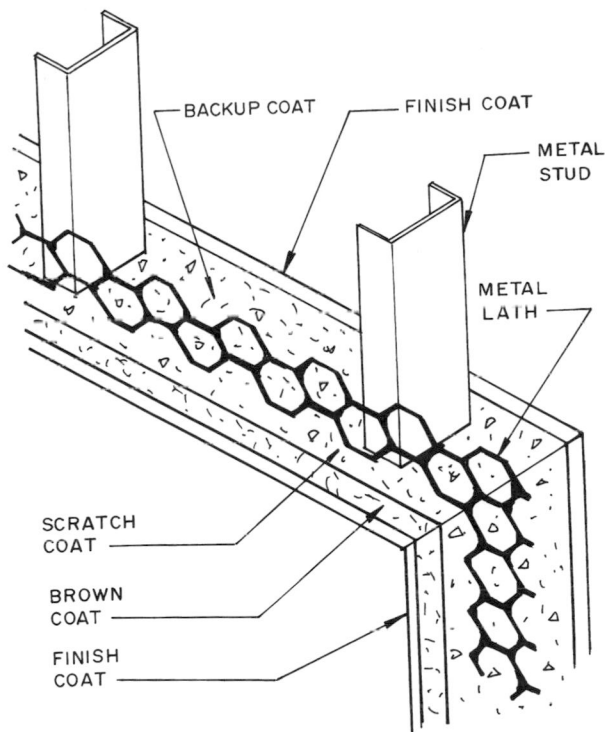

9-61 *Typical construction details for a solid gypsum partition built with metal C-studs and a metal lath core.*

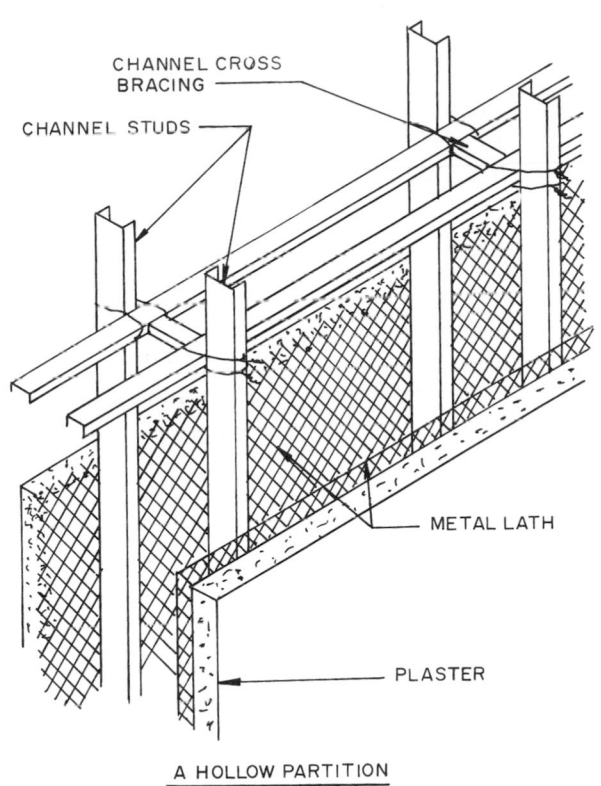

A HOLLOW PARTITION

9-62 *A double channel hollow partition can be used to provide space to run utilities.*

scratch, brown and finish coats. Another assembly for a studless partition is in 9-60. Long length gypsum lath is secured to runners at the overhead and floor. It is plastered in the conventional manner.

Solid gypsum partitions are also assembled using C-studs and metal lath. The studs are secured to the overhead and floor with metal runners. Metal lath is tied to the studs and plaster scratch, brown, and finish coats are applied (see 9-61).

Similar construction is used to assemble a double channel stud hollow partition. Vertical studs are secured to the overhead and floor with metal runners. Wire mesh is tied to each row of studs and conventional plastering procedures are used (see 9-62). The studs are stiffened and braced with horizontal channels. This type of partition is useful for running electrical and mechanical system components.

STRUCTURAL CLAY TILE PARTITIONS

Structural clay tile partitions are made with hollow clay masonry units that are glazed on one or both sides. The glazes provide a smooth surface impervious to penetration by water and are available in a wide range of colors (see 9-63). They resist abrasion, wear, and mildew and can be routinely scrubbed and sanitized.

9-63 *Partitions and walls are constructed with glazed structural ceramic tile units providing a fire-resistant wall that is easily cleaned and resists damage.*
Courtesy Stark Ceramics, Inc.

CERAMIC TILE WALL FACING

Ceramic tiles used to finish walls and partitions are available in a wide range of sizes and colors. They provide a hard, water-resistant facing and are durable and easily cleaned. Ceramic tile is generally bonded to walls faced with gypsum or cement panels designed to support the tile and resist damage if moisture penetrates the joints between the tiles. Ceramic tile may be bonded to concrete or masonry that has a surface prepared by sandblasting or scarifying to provide a true uncontaminated surface. Ceramic tile may also be applied to wall surfaces consisting of metal lath covered with a cement mortar bed (see 9-64). Portland cement mortar is used for setting tile in a thick bed. Other mortars and adhesives are used for thin beds.

WOOD & RECONSTITUTED WOOD WALL FINISHES

Interior wall finish may be some form of solid wood paneling or a reconstituted wood panel such as plywood or hardboard. They are available finished, including some types with vinyl wall coverings. Details about these products is in Division 6. The use of wood interior wall finishes may be severely limited or prohibited by code regulations for certain building types and occupancies.

Ceiling Construction

The ceiling serves a variety of functions. It provides a finished appearance to the overhead area, contributes to the acoustical treatment of the room, and provides a light reflecting surface to enhance illumination. Some types provide a space below the floor in which electrical and mechanical systems are run out of view. The ceiling may support lights, provide a measure of fire protection to the floor or roof above, and permit the use of sprinkler heads in a fire control system. Ceilings may be flat, horizontal surfaces, curved, or constructed on any of a wide variety of sloping surfaces that provide for the architectural enhancement of the area.

The finish materials may be plastic, metal, plaster, gypsum panels, wood, or one of the many fibrous panels available. The floor or roof decking may be exposed and used as the finished ceiling. This could be concrete, steel, or wood. The types of ceilings include suspended, contact, and exposed.

SUSPENDED CEILINGS

Suspended ceilings are possibly the most widely used in commercial construction. They offer considerable advantages because they allow electrical and mechanical systems to be hidden and are available with a variety of finished surface materials (see 9-65).

Suspended ceilings are hung from the floor or roof overhead by a series of wires which support a metal grid that carries the finished ceiling. The length of the wires can vary, thus enabling a flat horizontal ceiling to be installed below a structural floor or roof that has members of various sizes (see 9-66). Typically the lighting system is set in the grid and is flush with the ceiling (see 9-67). Some types of panels provide acoustical properties and others provide a degree of fire protection below the floor or roof above.

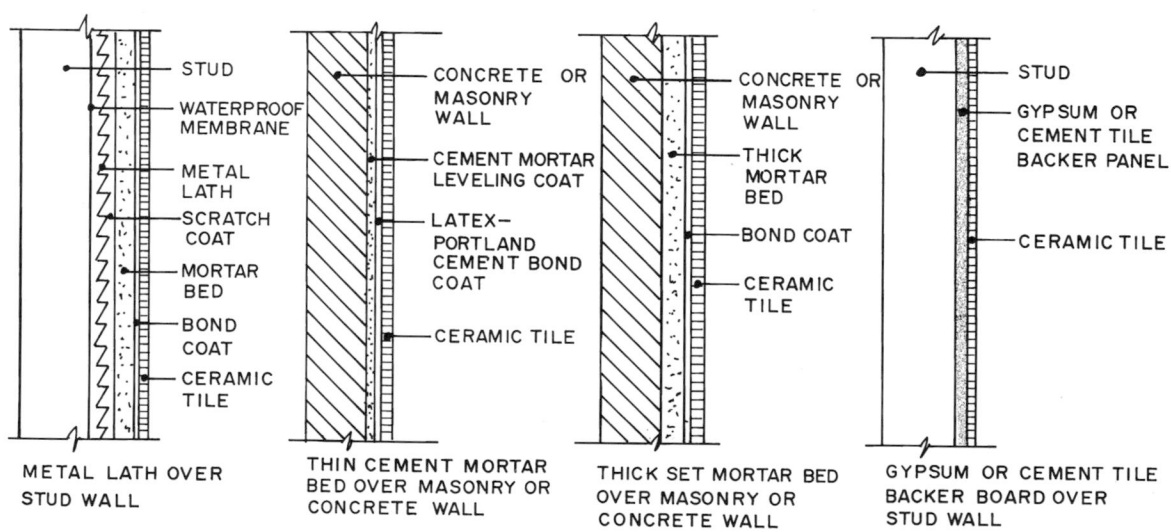

9-64 *Some of the constructions used when building a wall to have a ceramic tile face.*

9-65 *The exposed metal grid on this suspended ceiling is a neutral color permitting it to blend with the acoustic ceiling panels.*

Courtesy The Celotex Corporation

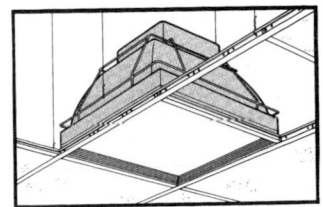

A high-intensity discharge lamp.

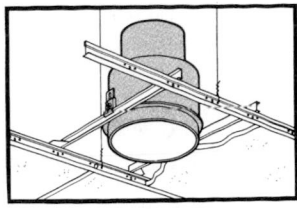

An incandescent fixture.

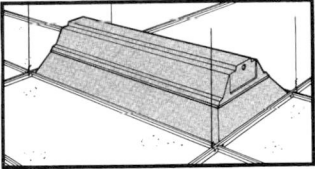

A vaulted flourescent light fixture.

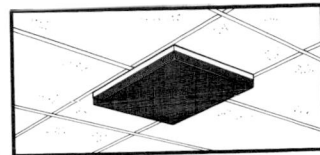

A surface-mounted fixture.

9-67 *Some of the types of light that are used with suspended ceilings.*

Courtesy Chicago Metallic Corporation

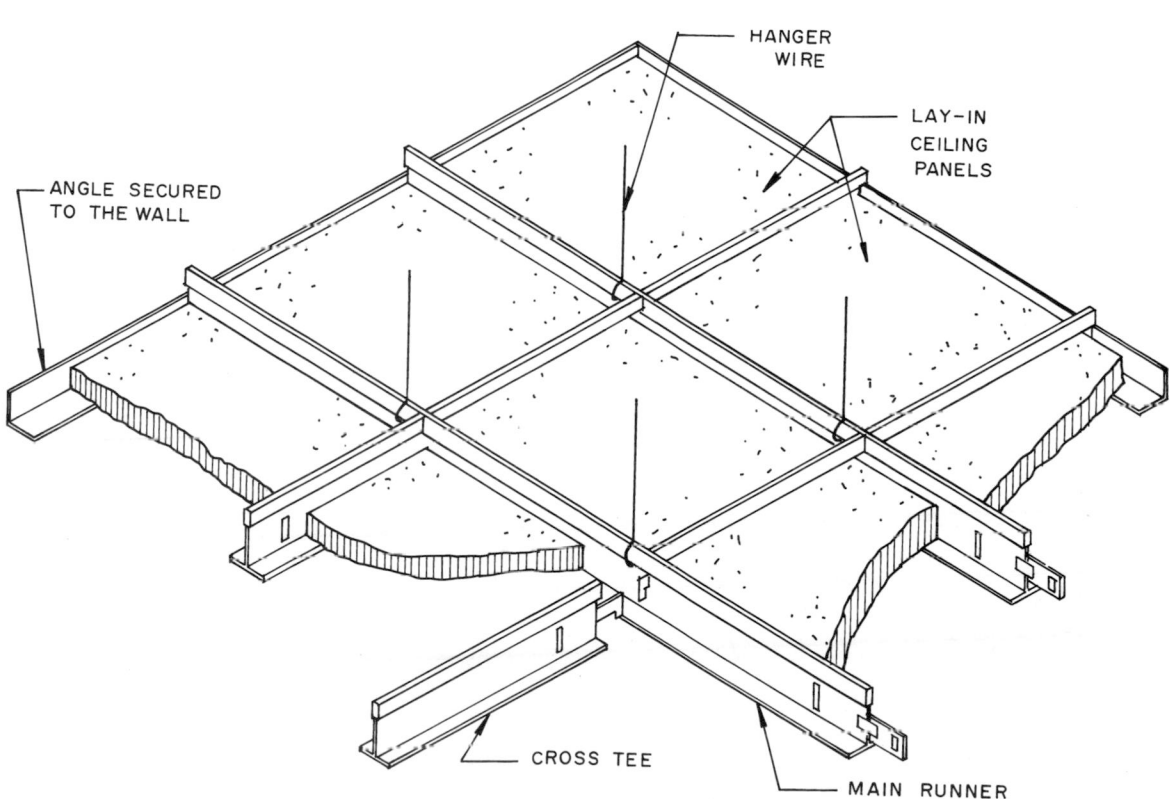

9-66 *A suspended ceiling is hung from the overhead by wires. The wire carries the main runners that are joined by cross tees forming a metal grid. In this design ceiling panels are laid into the grid forming the ceiling.*

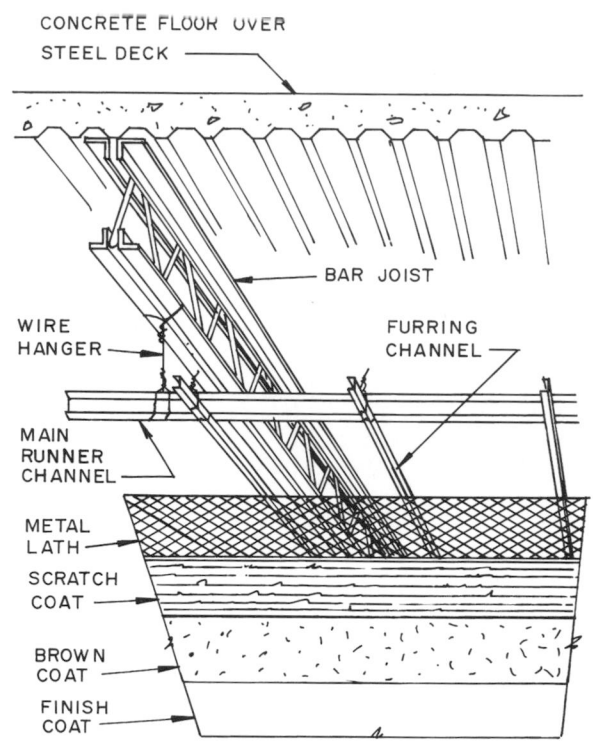

CONCRETE FLOOR OVER STEEL DECK

BAR JOIST

WIRE HANGER

FURRING CHANNEL

MAIN RUNNER CHANNEL

METAL LATH

SCRATCH COAT

BROWN COAT

FINISH COAT

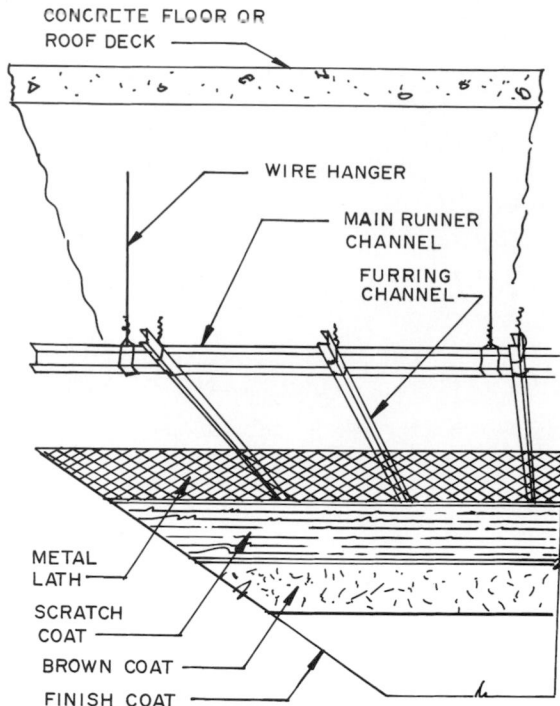

CONCRETE FLOOR OR ROOF DECK

WIRE HANGER

MAIN RUNNER CHANNEL

FURRING CHANNEL

METAL LATH

SCRATCH COAT

BROWN COAT

FINISH COAT

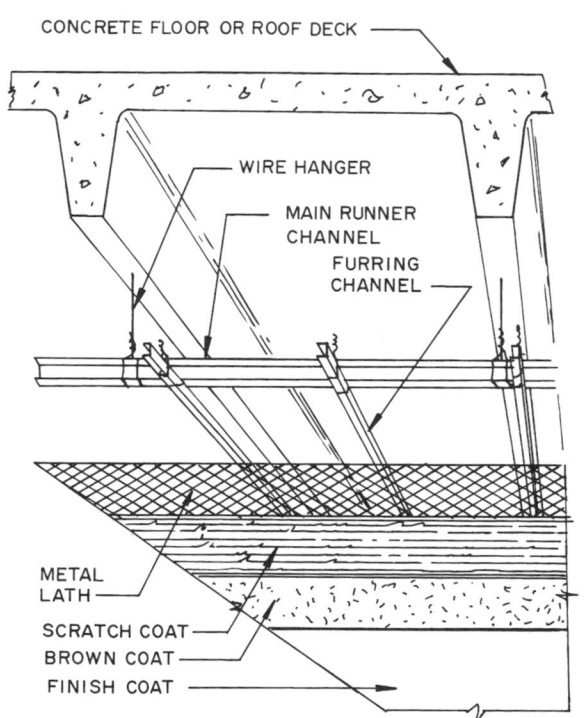

CONCRETE FLOOR OR ROOF DECK

WIRE HANGER

MAIN RUNNER CHANNEL

FURRING CHANNEL

METAL LATH

SCRATCH COAT

BROWN COAT

FINISH COAT

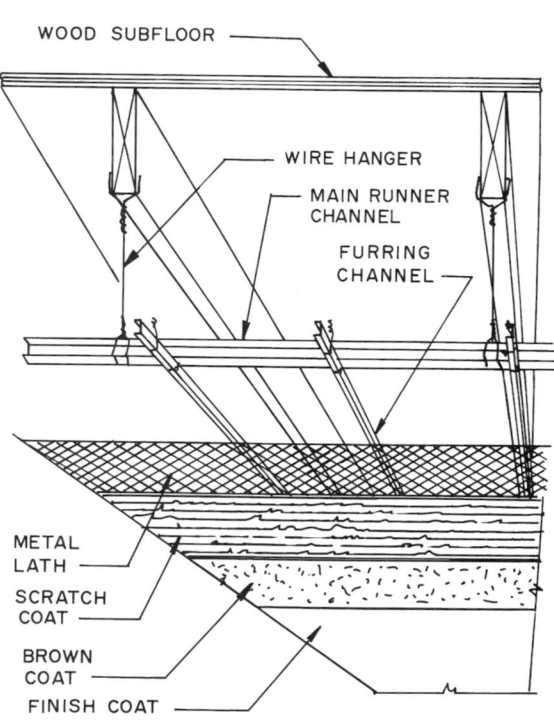

WOOD SUBFLOOR

WIRE HANGER

MAIN RUNNER CHANNEL

FURRING CHANNEL

METAL LATH

SCRATCH COAT

BROWN COAT

FINISH COAT

9-68 *A suspended plaster ceiling has the main channel hung from the overhead by wires and metal furring channels are wired to the main channel. The plaster coats are as described for finished walls.*

Suspended plaster ceilings are constructed by hanging channel runners from the overhead to which furring channels are tied with wire. The spacing is determined by the conditions. The wire or metal lath is wired to the furring. This construction produces a hard, durable, highly fire-resistant ceiling. Typical construction details are in 9-68.

There are various types of suspended acoustical systems available using a suspended metal grid into which lightweight panels and lights are placed. The grids are assembled with some form of locking connection as shown in 9-66. The drop-in panels may have an exposed grid or concealed grid. These systems also can accommodate heat ducts, sprinkler heads, and other required penetrations. Some grid systems have recessed panels and a large number of surface textures are available. Ceiling panels are available in several materials such as aluminum, fiberglass reinforced polymer gypsum cement, gypsum panels faced with vinyl and fabrics, fiber glass, and wood fibers.

FURRED CEILINGS

Furred ceilings are constructed using metal furring strips secured to the metal or wood structural system (see 9-69). Concrete overhead floors and roofs often have a C-channel hanger hung below them and the furring strips are wired to the hanger. The metal lath is wired to the furring strip (see 9-70). The spacing and size of the hangers depends upon the design of the system.

CONTACT CEILINGS

Ceilings can be attached directly to the floor or ceiling joists, bar joists, or other structural members. In wood frame construction gypsum wallboard panels are screwed directly to the joists as shown in Division 6. Plaster ceilings can have the metal lath secured directly to the steel, wood, or concrete structural system. The ceilings are then finished with several layers of plaster as described for plaster wall construction.

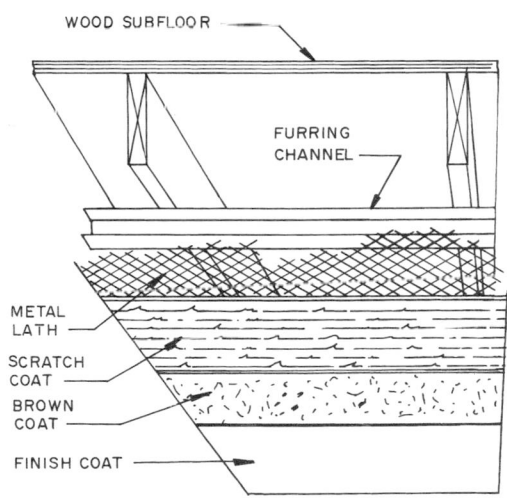

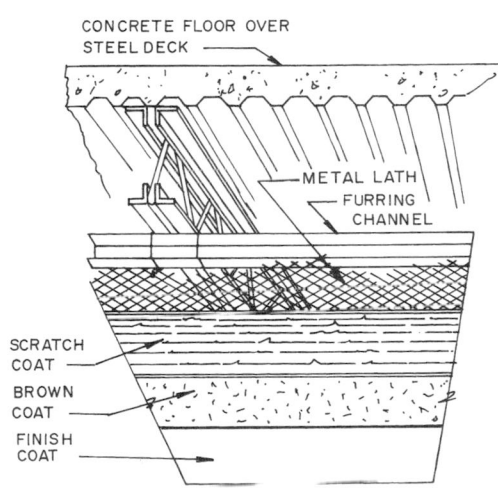

9-69 *Furred ceilings have the furring channels secured to the floor or roof structure.*

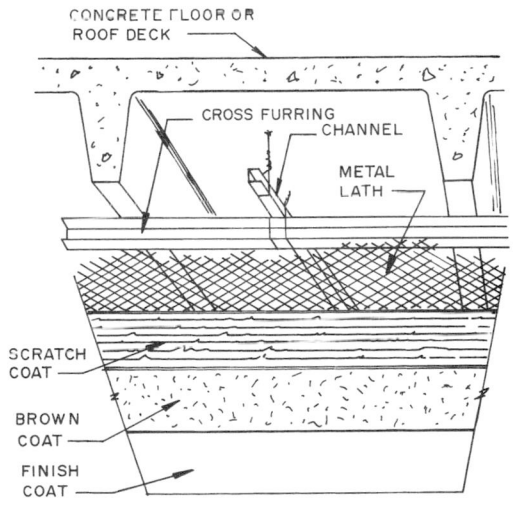

9-70 *This furred ceiling has a C-channel supporting the cross furring.*

FLOORING

Finish flooring materials are installed over substrate of various kinds and in a variety of situations. The designer must consider factors such as concrete slabs on grade or above grade, wood substrate, cellular steel floors with a concrete slab, the use of radiant heating, the traffic loads on the floor, required maintenance, fire resistance, building code requirements, acoustical requirements, color, and texture. The finish flooring must meet the prescribed conditions at an acceptable cost. Some areas require special conditions be observed such as a need for sanitation or unique situations as found in hospitals.

9-71 *Hardwood strip flooring protected with durable coatings provides a beautiful finished floor.*

Courtesy Harris-Tarkett, Inc.

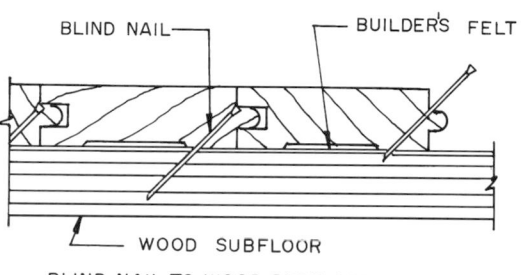

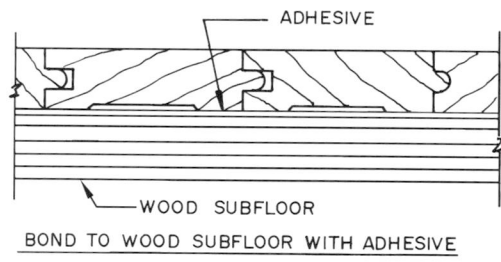

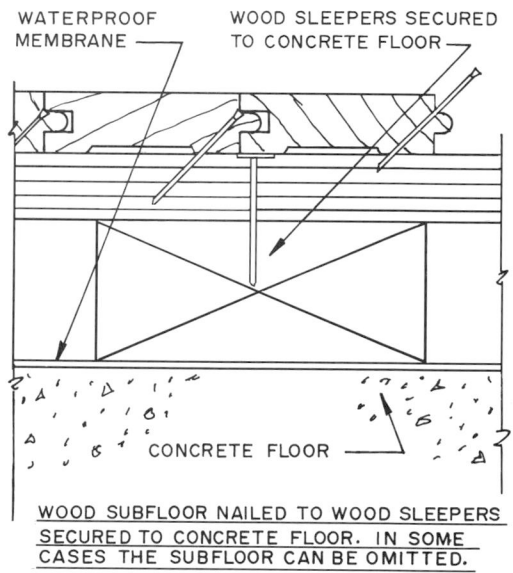

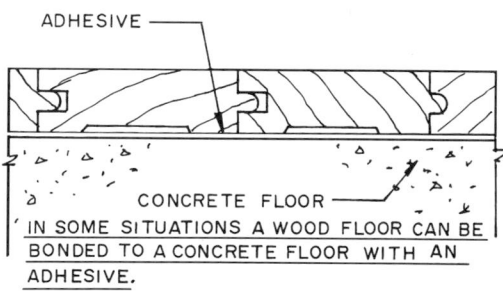

9-72 *Several ways wood flooring can be installed.*

Building Codes

Building codes have sections devoted to floor finish. They are related to rooms or enclosed spaces, vertical exits and passageways, and corridors providing access to exits. The requirements can vary depending upon the classification and occupancy of the building. The most commonly used finish flooring materials such as wood, vinyl, terrazzo, and clay tile do not present unusual hazards and are generally not subject to building code interior finish requirements. Carpeting does present a hazard and must meet the requirements specified by U.S. Department of Commerce standard DOC FF1 or the National Fire Protection Association Standard NFPA 253.

Wood Flooring

The various types of wood flooring are described in detail in Division 6. Both hard and softwoods are used to form strip flooring and various types of parquet and heavy wood block flooring (see 9-71). Some types have a factory applied finish while others are sanded and fin-ished after installation. Some types are nailed to a wood subfloor while others are bonded with a mastic adhesive. Many are available as a very thin veneer and cannot be sanded.

Standard wood strip flooring is manufactured in several thicknesses. Several methods for installing it are shown in 9-72. The recommendations of the manufacturer for method, nails, and adhesives should be observed. Parquet flooring is generally installed with adhesives. It is available in a wide variety of species of wood and patterns (see 9-73).

An example of other wood flooring products include a durable hardwood plank or parquet flooring that is impregnated with acrylic. This prefinished laminated product is ⅜ in. (915 mm) thick. It meets slip resistance standards set forth by the Americans with Disabilities Act of 1991 and has a Class B Flame Spread (ASTM E-84).

Still another product is composed of a hardwood veneer covered with a 20 mil vinyl top layer. The wood veneer is bonded to a base made up of several layers of vinyl and fiberglass (see 9-74).

9-73 *Parquet flooring has been widely used for many years and provides a wide variety of patterns.*
Courtesy Harris-Tarkett, Inc.

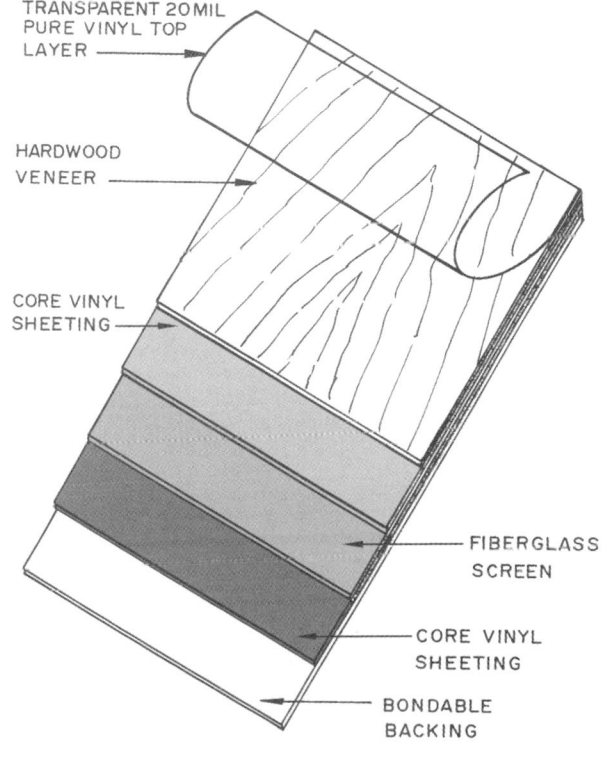

TRANSPARENT 20 MIL PURE VINYL TOP LAYER

HARDWOOD VENEER

CORE VINYL SHEETING

FIBERGLASS SCREEN

CORE VINYL SHEETING

BONDABLE BACKING

9-74 *This GenuWood™ flooring is a laminated product with a hardwood veneer and a pure vinyl top layer.*
Courtesy PermaGrain Products, Inc.

Resilient Flooring

Resilient flooring in a **polyvinyl chloride (PVC) material** is available in sheets and individual tiles. Sheet vinyl has a top vinyl layer and a composition backing. Vinyl composition tiles are composed of vinyl resins, plasticizers, stabilizers, fibers and pigments (see 9-75), and they are available in 9, 12, 18, and 36 in. (22.8, 30, 45.7 and 91.4 mm) squares and some rectangular shapes and thickness of ⅛ and ³⁄₃₂ in. (2.4 and 3.2 mm). Sheet vinyl flooring is available in rolls 6, 9, and 12 ft. (1.8, 2.7 and 3.7 m) wide and up to 50 ft. (15.2 m) long and a wide range of thicknesses from about 0.069 to 0.224 in. (1.75 to 5.69 mm). Also available are solid vinyl commercial vinyl (see 9-76) and **PVC** resilient flooring. Solid vinyl floor covering offers the maximum wear potential. PVC flooring is used where a heat-welded surface is required. Plastic materials are discussed in detail in Division 6. Vinyl composition floor coverings are tested for fire spread and smoke produced. Data is available from the manufacturer. They can be installed over wood or concrete subfloors with a mastic recommended by the manufacturer.

9-75 *Resilient vinyl composition tile is available in a wide range of patterns and colors.*

9-76 *This inlaid commercial vinyl flooring has a long life and provides an attractive, durable, easy to clean surface.*
Courtesy Azrock Industries, Inc.

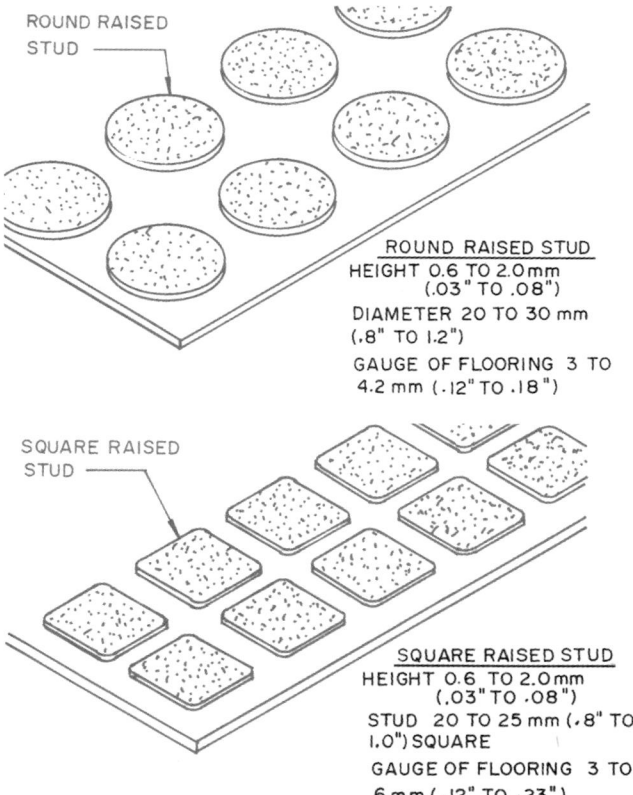

ROUND RAISED STUD

ROUND RAISED STUD
HEIGHT 0.6 TO 2.0mm (.03" TO .08")
DIAMETER 20 TO 30 mm (.8" TO 1.2")
GAUGE OF FLOORING 3 TO 4.2 mm (.12" TO .18")

SQUARE RAISED STUD

SQUARE RAISED STUD
HEIGHT 0.6 TO 2.0mm (.03" TO .08")
STUD 20 TO 25 mm (.8" TO 1.0") SQUARE
GAUGE OF FLOORING 3 TO 6 mm (.12" TO .23")

9-78 *Typical stud sizes and shapes used on rubber flooring tiles.*

9-77 *Rubber finish flooring is available in tiles and sheets.*

Asphalt resilient flooring is available in tiles usually 9 or 12 in. (2745 or 3660 mm) square. They are a composition of an asphalt binder for standard tile or a resinous binder for greaseproof tile, inert fillers, fibers, and pigments for color. They are available in four grades, A,B,C, and D based on color and usually ⅛ or 3/16 in. (3 or 4.8 mm) thick. They are durable and fire resistant and are bonded to wood or concrete subfloors with mastic recommended by the manufacturer.

Rubber flooring is a form of resilient flooring available in tiles and sheets (see 9-77). Tile sizes vary by the manufacturer but 12 and 36 in. (305 and 915 mm) square tiles are common. Sheets range from 36 to 50 in. (915 to 1270 mm) wide. Flooring is typically ⅛ and 3/16 in. (3 and 4.7 mm) thick. They are comfortable for walking, wear well, and resist damage from oils, solvents, alkalis, acids, and other chemicals. Most types have some form of grided surface that aids traction and a roughened or recessed back grid. Some typical designs are shown in 9-78. They are available in a wide range of colors.

Installation techniques depend upon the situation and manufacturer's recommendations should be observed. Typically they are bonded to concrete or wood subfloors with an approved adhesive. Generally they are not adhesive attached to below-grade slabs. On-grade slabs require the use of a moisture resistant adhesive. Manufacturers also have available other rubber flooring products such as stair tread, base, entrance mats that are set in a recess in the floor, and long runners that are not bonded to the subfloor.

Clay Tile

Clay floor tile includes quarry tile, paver tile, and ceramic tile. Information on these is found in Division 4.

Quarry tiles—are unglazed and made from shales and fire clays. They are very durable, clean easily, resist damage due to freezing, thawing, and abrasion, and they used on floors subject to heavy and extra heavy use. Proper installation is essential. Some types are made with an abrasive grain surface to provide slip resistance. While various manufacturers have a variety of sizes and shapes, square, rectangular and hexagon are most common. Various trim units such as stair nosing and cove base units are available (see 9-79).

Paver tiles—may be glazed or unglazed and are much like ceramic tiles but larger. Typical sizes are 4 × 4 in. (102 × 102 mm), 4 × 8 in. (102 × 203 mm), and 8 in. (203 mm) hexagon shape. They are weather resistant and resist abrasion. Typical uses include residential and moderate duty commercial floors such as shopping centers and restaurants and heavy duty floors such as a commercial kitchen (see 9-80). A variety of trim materials are available.

9-80 *Paver tiles are weather resistant and resist abrasion.*

Courtesy American Olean Tile Company

9-79 *Quarry tile provides a clean, durable finish floor.*

Ceramic mosaic tiles—are available glazed and unglazed and in squares, rectangles, and hexagon shapes. They are small tiles typically 1 or 2 in. (25.4 or 50.8 mm) square or hexagon shapes or 1 × 2 in. (25.4 × 50.8 mm) rectangles. They are used for interior floors, countertops, walls and swimming pools, exterior floors and walls with special installation procedures. Some types have slip-resistant surfaces (see 9-81). They have available in a variety of trim materials.

9-81 *Ceramic mosaic tile enable the floor to be laid in a variety of colors and geometric patterns.*

Courtesy American Olean Tile Company

9-82 *Textured glazed ceramic tile floors are used in areas subject to moderately heavy use.*

Glazed ceramic floor tiles—are available with smooth and textured surfaces. The textured surface provides greater slip resistance. They are available in 4 and 8 in. (101 and 202 mm) squares, 8 in. (202 mm) hexagon, and 4 × 8 in. (101 × 202 mm) sizes. They are used for interior residential and moderate duty commercial floors, such as in restaurants and shopping malls (see 9-82).

MORTARS & ADHESIVES FOR TILE FLOORS

Clay floor tile can be installed using a variety of bonding agents. Portland cement mortar is the only one recommended for a thick bed. Epoxies and furans do not contain portland cement and are more expensive.

Portland cement mortar—is a mixture of portland cement, hydrated lime, sand, and water. It can be used to provide a leveling bed up to 1¼ in. (32 mm) thick. The tiles may be set into this bed while it is still plastic. After it cures tiles may be bonded to the leveling bed with a thin-set coat such as dry-set or latex portland cement mortar.

Dry-set mortar—is a mixture of portland cement, resinous materials, sand, and water. It is a thin-set mortar (about 3/32 in. or 2.4 mm thick). It may be applied over the hardened thick set portland cement mortar bed and used to bond the tiles in place.

Epoxy mortar—contains no portland cement but is a mix of epoxy resin and a hardener. It has high bond strength, and resists impact and chemical attack. It is a thin coat adhesive.

Latex portland cement mortar—is a mixture of portland cement, a latex additive, and sand. It is a thin coat application usually about ⅛ in. (3 mm) thick.

Organic adhesives—harden by evaporation. They are applied with a notched trowel leaving a thickness of about 1/16 in. (1.6 mm). They are not used on exterior applications or interior uses exposed to considerable water.

Furan mortar—is a mixture of furan resin and a hardener. It has high resistance to chemicals.

TILE GROUTS

Grouts typically are a mixture of portland cement, fine sand, and lime. They generally come premixed requiring the addition of water. Latex, furan, epoxy, or silicone rubber can be added to influence the properties. The manufacturer should be consulted to get the most desirable grout for the situation at hand.

Brick floor laid with minimum mortar joint.

Brick floor laid without mortar.

Brick floor laid with wide mortar joint.

9-83 *Bricks provide an attractive durable finish flooring inside and outside the building.*

Brick Flooring & Brick & Concrete Pavers

Brick flooring and brick pavers are used to produce durable finish flooring. Brick pavers are thinner than standard brick ranging from ½ in. to 2¼ in. (12 to 57 mm) thick and are usually preferred to standard brick. Pavers are also available in concrete and asphalt. Glazed brick is also used for a finished floor. Brick flooring is used both indoors and outside as on a patio (see 9-83)

Some types are suitable for heavy use on industrial floors. The various types and sizes of brick are discussed in Division 4.

Bricks can be laid with the large flat face exposed or on edge exposing a narrow face. Brick pavers are laid with the flat face exposed. While there are various ways to lay them the most common are shown in 9-84. Exterior brick patios are often laid on a bed of sand with no mortar between the units. This is not satisfactory when heavy loads are to be placed on the floor. Floors subject to heavy loads are laid over a concrete slab. Concrete pavers are also available.

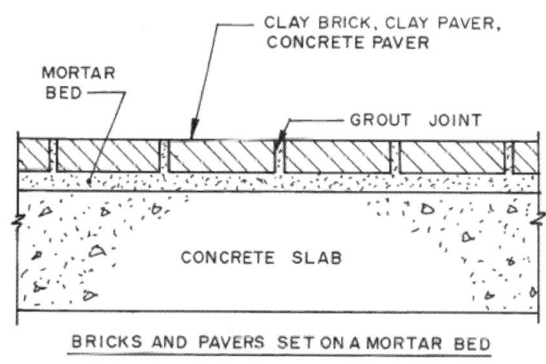

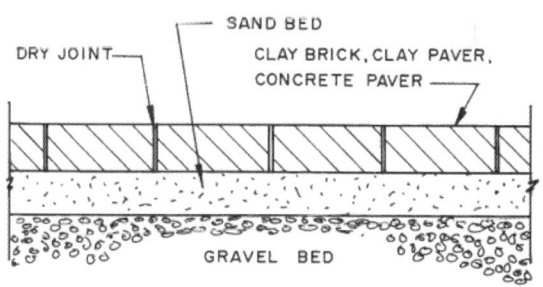

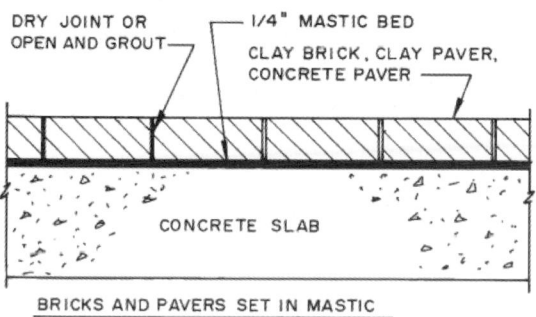

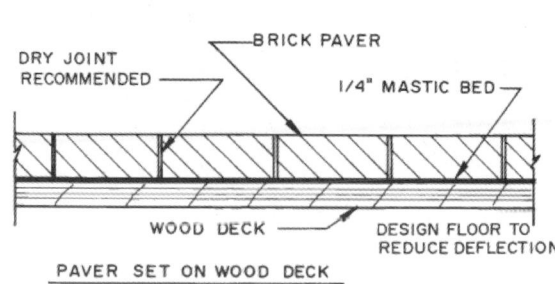

9-84 *Examples of typical floor construction using clay brick, clay pavers, or concrete pavers.*

Concrete Floors

Finished concrete floors are widely used and provide a durable and economic solution to many building floor situations. The properties of concrete and placing techniques are discussed in Division 3.

The concrete floor surface may be finished in several ways. The slab can be poured, leveled and screeded and the surface smoothed with a wood float. A smoother surface can be produced by smoothing the wood float surface with a steel trowel. These are done before the concrete has firmly set.

9-85 *Exposed aggregate concrete floors provide a durable and attractive finished surface.*

9-86 *The concrete floor in an industrial plant shown above was treated with ArmorSeal®.*

Courtesy The Sherwin-Williams Company

Various **textured surfaces** can be had by techniques, such as brushing the surface with a broom after it has been troweled but not set. An **exposed aggregate** finish is made by leveling the slab, embedding aggregates in the surface with a float to get a level surface, and, after the concrete has begun to set, flushing the concrete off the aggregate with water leaving it exposed and forming the finished surface (see 9-85). A high quality smooth finished surface can be produced by covering the base concrete with a 1 in. (25 mm) thick specially formulated concrete layer. This is applied before the base concrete has hardened.

Colored surfaces are produced by adding color pigments, generally metallic oxides, to the concrete mix as the batch is prepared. Another technique is to pour and finish the floor and spread a dry shake coloring material prepared especially for this purpose. The dry shake is spread evenly over the surface before it has hardened and floated to smooth the surface and even out the color.

Other finishes include painting, applying an abrasive aggregate to the surface to produce a nonslip finish, adding metallic aggregate to the surface to produce a more wear resistant finish, and using various chemical coatings to harden the surface (see 9-86).

Heavy duty concrete floors are used in many industrial applications. They are usually built as two-course construction. The reinforced floor base is poured and tapped with a durable concrete layer made with special abrasion-resistant aggregate, such as emery or iron filings.

Terrazzo

Terrazzo is a matrix consisting of marble or granite chips and portland cement and water or a synthetic resin. It is placed as a top layer over a concrete underbed, steel decking, or a wood subfloor.

Portland cement terrazzo is divided into sections by 1¼ in. (32 mm) high metal or plastic dividing strips, which reduce the possibility of cracking. The three methods for casting portland cement terrazzo are monolithic, bonded, and sand cushion (see 9-87).
Monolithic terrazzo—is a ⅝ in. (16 mm) thick layer placed as the topping on a green concrete slab.
Bonded terrazzo—is a topping 1¾ in. (44.5 mm) or thicker. The cured concrete slab is cleaned and coated with a neat portland cement, a concrete underbed of about 1 in. (25 mm) is laid. Divider strips are placed on the underbed and pushed into it. The terrazzo topping is leveled over the underbed.

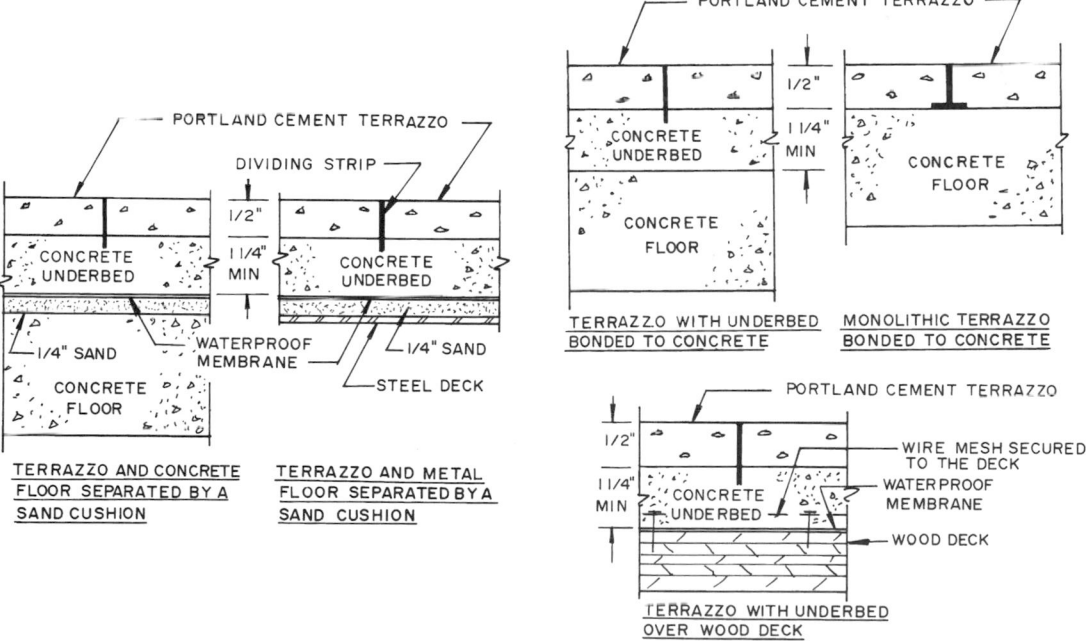

9-87 *Typical Portland cement terrazzo floor installations over various types of floor decks.*

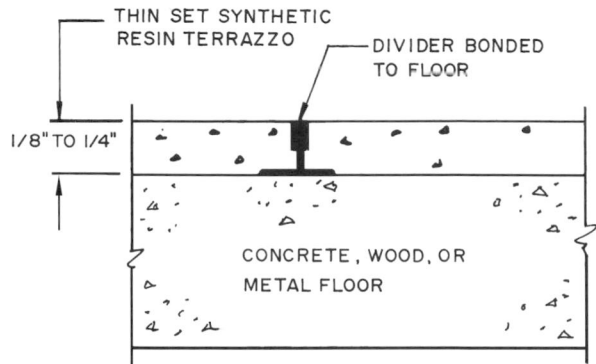

THIN SET SYNTHETIC
RESIN TERRAZZO

DIVIDER BONDED
TO FLOOR

1/8" TO 1/4"

CONCRETE, WOOD, OR
METAL FLOOR

9-88 *A typical synthetic resin terrazzo floor construction detail.*

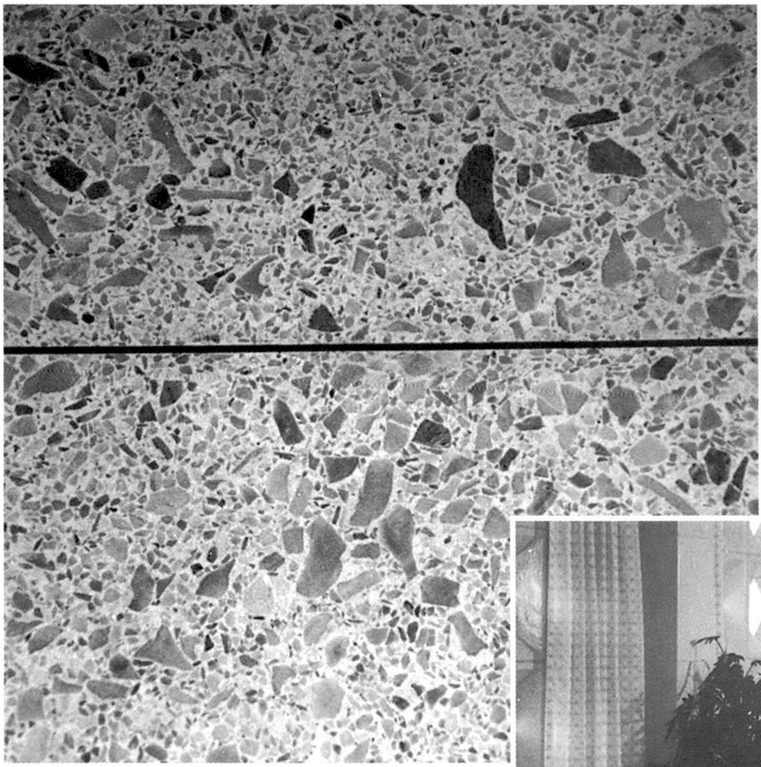

Close up of the finished terrazzo floor shown to the right. Note the metal divider strip.

9-89 *The glass block wall enhances the terrazzo floor. This is a very durable floor that is easily cleaned and presents a multicolored finish.*

Courtesy Pittsburgh Corning

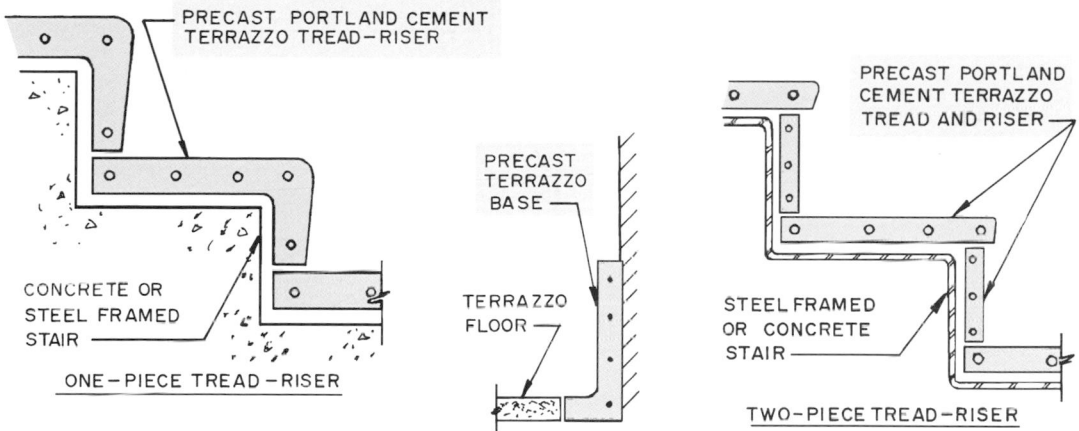

9-90 *Precast terrazzo units are often used for stair treads and risers.*

Sand cushion terrazzo—has a ½ in. (12.7 mm) bed of dry sand laid over the concrete slab. This is covered with a waterproof membrane and the concrete underbed is laid. Dividing strips are placed on the underbed and the terrazzo topping is poured and leveled. The sand cushion isolates the slab from the floor slab protecting it from damage due to possible movement of the floor slab.

Synthetic resin matrix terrazzo—should be installed following the directions of the manufacturer of the resin. It is a thin coat material usually ranging from ⅛ to ¼ in.(3 to 6mm) thick. The stone chips used in synthetic resin terrazzo are much smaller than those used in portland cement terrazzo. Dividing strips are bonded to the wood, metal, or concrete underbed with an adhesive (see 9-88).

After the terrazzo topping has hardened it is ground smooth and polished. The polishing brings out a gloss and the color of the chips. A nonslip surface is sometimes ground having less gloss. A protective sealer is applied over the finished floor (see 9-89).

Another form of terrazzo is as precast units. Synthetic resin terrazzo is cast in square and rectangular shapes in thickness from ⅛ to ¼ in. (3 to 6 mm). Precast stair treads, risers, window stools, wall base, and other shapes are available cast in portland cement terrazzo (see 9-90).

Stone Floor Coverings

Stone finish flooring is typically slate, marble, or granite (see 9-91). When used on the interior of a building they are laid over a wood or a concrete slab in the same manner as brick pavers.

Marble

Slate

9-91 *This marble floor is used on an entrance hall and is laid without mortar joints between the pieces.*

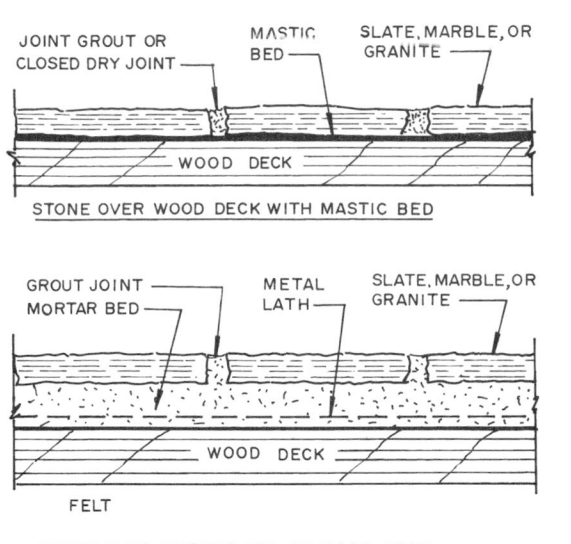

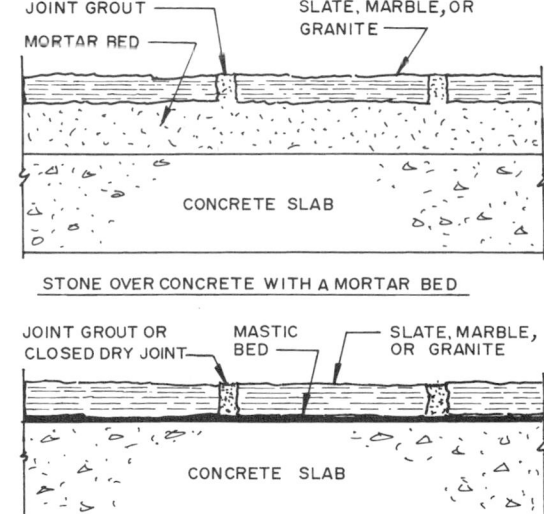

JOINT GROUT OR CLOSED DRY JOINT — MASTIC BED — SLATE, MARBLE, OR GRANITE

WOOD DECK

STONE OVER WOOD DECK WITH MASTIC BED

GROUT JOINT MORTAR BED — METAL LATH — SLATE, MARBLE, OR GRANITE

WOOD DECK

FELT

STONE OVER MORTAR BED ON WOOD DECK

JOINT GROUT MORTAR BED — SLATE, MARBLE, OR GRANITE

CONCRETE SLAB

STONE OVER CONCRETE WITH A MORTAR BED

JOINT GROUT OR CLOSED DRY JOINT — MASTIC BED — SLATE, MARBLE, OR GRANITE

CONCRETE SLAB

STONE OVER CONCRETE WITH MASTIC BED

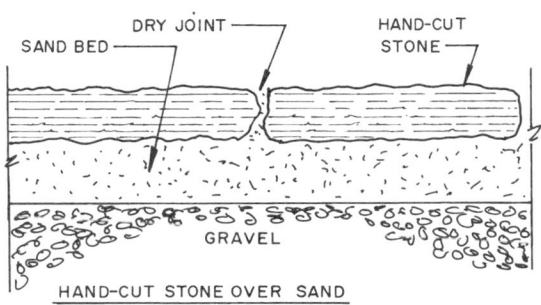

SAND BED — DRY JOINT — HAND-CUT STONE

GRAVEL

HAND-CUT STONE OVER SAND

9-92 *Various ways stone flooring may be laid for interior and exterior applications.*

9-93 *This flooring is made using cast marble tiles formed with marble aggregate cast in a plastic resin.*
Courtesy PremaGrain Products, Inc.

9-94 *This concrete garage floor has been given an epoxy primer and a single coat of Coguard™ which is a chemical product (urethane) that cures, producing a durable, weather-resistant coating.*
Courtesy Environmental Technologies Company

Sand laid stone is usually used for exterior work and the joints are left open. Interior stone floors may have the joints filled with joint grout or be set tight with no joint grout (see 9-92).

Another stone floor covering material is a cast stone tile flooring formed by impregnating and encapsulating the stone aggregate with high technology resins that create a very dense surface. It meets Underwriters Laboratory requirements for slip resistance (see 9-93).

Other Flooring Applications

Industrial buildings are generally built with a concrete slab. The finish flooring used is mainly to protect the slab from corrosive materials and the movement of heavy traffic. The concrete can be treated with chemicals to harden the surface (see 9-94) with a flooring such as wood blocks standing on end so the end grain forms the surface or asphalt mastic floors can be used (see 9-95). The entire floor or at least certain work areas may need treatment for slip resistance which could be incorporated in the concrete topcoating or be moveable rubber mats.

Some floors must have a low electrical resistance to prevent sparks from static electricity. This **nonsparking conductive flooring** is used in areas such as where volatile vapors are present. Other flooring materials must be grease resistant, slip resistant, fire-resistant, acid and alkali-resistant, static discharging, or X ray protective.

Various types of seamless floor coverings are available. Some are formed from plastic, such as an acrylic, latex epoxy, resins, methy methacrylate, and polyacrylate resins, and are spread over a concrete or wood subfloor. A wide range of flooring for athletic use is available. This includes coatings for showers, saunas, basketball, tennis and volleyball courts, tracks, and locker rooms. Materials used include elastomers, rubbers, polymers, and urethane products (see 9-96).

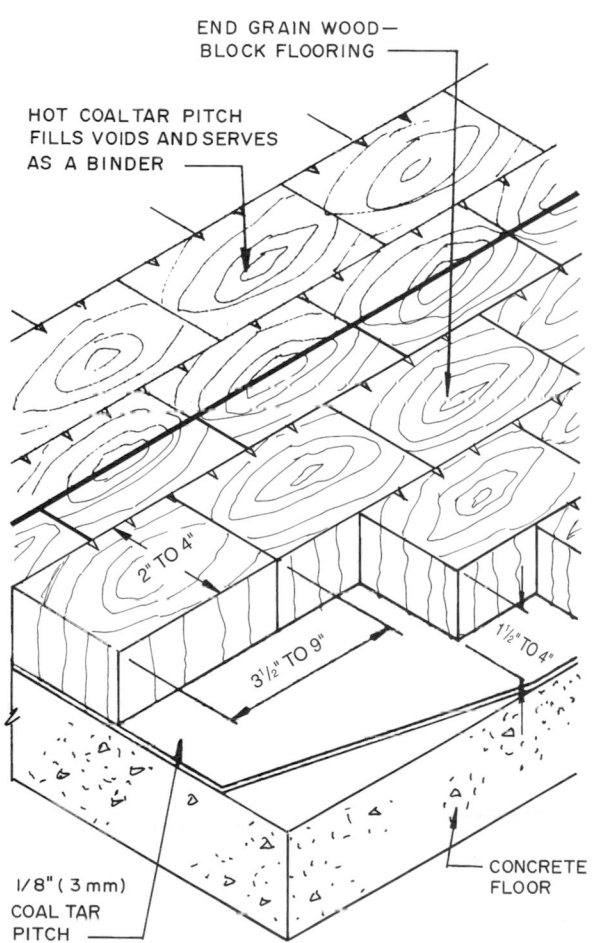

9-96 *Typical poured-in-place basketball floors are made from a two-component polyurethane elastomer*

9-95 *(Left) End-grain wood block floors are used in industrial areas where heavy wear is expected.*

CARPET

Carpet is a widely used floor covering in residential and commercial construction. It competes well with other floor covering products in terms of cost, durability, and maintenance. In addition it is available in a wide range of fibers, textures, and colors. It is made in widths of 6, 8, 12, and 25 ft. and wider in some cases, resulting in a material that produces few seams and rapidly covers large floor areas.

While carpeting is often selected for its beauty and comfort it also provides sound deadening properties by reducing impact noise and the reduction of sound reflection within a room.

Pile Fibers

Wool, a natural fiber, has been a major fiber in carpet construction for many years. The development of synthetic fibers has enabled the carpet manufacturer to produce products with a wider range of characteristics. Often carpet piles are a blend of fibers. For example, a wool carpet may have a blend of 10 to 30 percent nylon fibers to increase the toughness. Commonly used synthetic fibers are acrylic, modacrylic, nylon, polyester, and polypropylene (olefin).

WOOL FIBERS

Wool used in carpets is a natural fiber imported into the United States from wool producing countries, such as Australia, New Zealand, and Scotland. The product varies with the different breeds of sheep. The wool fibers range from 3½ to 7 in. (89 to 170mm) long. The wool pile is made by spinning the fibers into yarn. Wool is relatively unaffected by weak alkalis and organic solvents. It has good resistance to abrasion and aging and resists damage from mildew and sunlight. Generally it is more expensive than synthetic fibers.

ACRYLIC FIBERS

This synthetic fiber is similar to wool in texture and abrasion resistance. It is composed of more than 85 percent by weight of acrylonitrile. It has good resistance to mildew, aging, sunlight, moths, and chemicals. It is manufactured in a wide range of colors.

MADACRYLIC FIBERS

This synthetic fiber is a member of the acrylic group but contains at least 35 percent acrylonitrile but less than 85 percent as exists in acrylic fibers. It is soft, resilient, quick drying, and abrasion and flame resistant. It resists acids, alkalis, sunlight, and mildew.

NYLON FIBERS

Nylon is a synthetic fiber that is a petrochemical product produced as a continuous filament but it is often cut into short lengths varying from 1½ to 6 in. (38 to 152mm). The short fibers are spun into yarn much the same as with wool fibers. It is a strong fiber that has antistain properties and can be dyed. It is resistant to mildew, aging, and abrasion as well as being resilient and low in moisture absorbency.

POLYESTER FIBERS

Polyester fibers are synthetic fibers that have a high tensile strength, good resistance to mineral acids, and excellent mildew, aging, and abrasion resistance. Prolonged exposure to sunlight may cause some loss of strength. They are not as durable as nylon fibers.

POLYPROPYLENE FIBERS (OLEFIN)

Polypropylene is a synthetic fiber that is a class of olefin that consists of 85 percent propylene by weight. It has the lowest moisture absorption rate of all fibers in use. It resists mildew, abrasion, aging, sunlight, and common solvents. It is not as resilient as nylon, but is less expensive. Olefin is very lightweight, strong, and soil resistant.

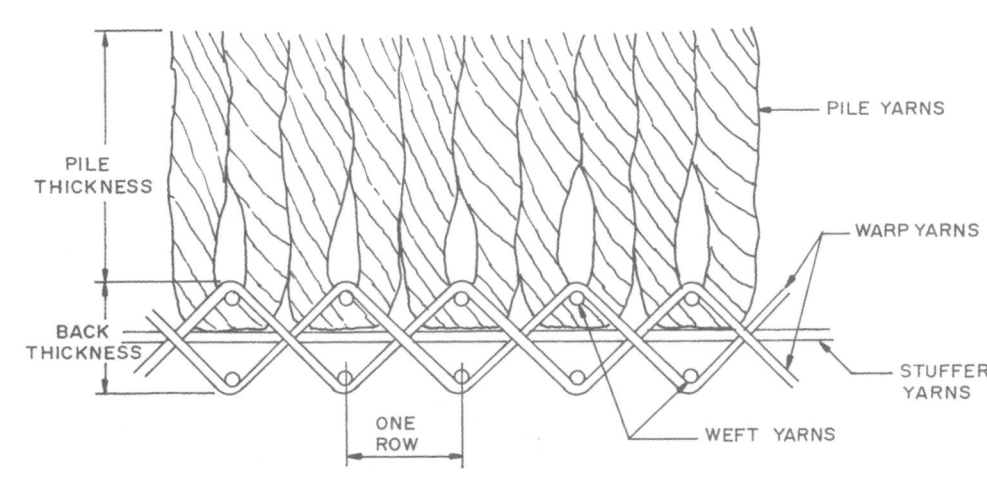

9-97 *Terms used to identify the parts of carpeting.*

Carpet Construction

While the construction of carpets varies somewhat the details in 9-97 will help show the terms used. The **pile yarns** form the exposed top surface that takes the wear and abrasion. Pile yarns are made from wool or one or more synthetic fibers. The **backing yarns** consist of **weft yarns** (running the width of the carpet) and **warp yarns** (running the length of the carpet). **Stuffer yarns** run lengthwise and give strength and stability to the carpet. The warp yarns pass over the weft yarns pulling the pile yarns into place.

WOVEN CARPETS

Woven carpets include Axminster, loomed, velvet, and Wilton construction.

Axminster Construction—is woven on a loom that inserts each tuft of pile yarn individually providing the possibility of varying the color of each tuft, which gives great flexibility for design. It is usually a cut pile that is even in height. The backing is heavily ridged and sized making it so stiff that it must be rolled lengthwise. A typical construction detail is in 9-98.

Loomed Construction—has the pile yarn bonded to a $^3/_{16}$ in. (5mm) thick rubber cushion. It has a low-loop single level pile. The pile back has a water-proof coating to which the rubber cushion is bonded. A typical construction detail is in 9-99.

Velvet Construction—is shown in 9-100. The pile, stuffer, and weft yarns are held together with the double warp yarns. The setting of the loom establishes the height of the piles and they are at this point loops. The wire that sets the height has a knife edge that cuts the pile yarns to the desired height. A textured surface can be produced by setting the wires to form and cut loops of different heights. Patterned designs cannot be produced by this loom but tweed effects can be had by varying the colors of the yarn.

Wilton Construction—is produced on a loom like the one used for velvet construction that has a mechanism to feed yarns of various colors. The various color yarns are selected by a series of punched cards. The cards regulate which color yarn is looped over the loop forming wires. These become the visible pile tufts and the other colored yarns are buried in the body of the carpet. This produces a multicolored product with a stiff backing (see 9-101).

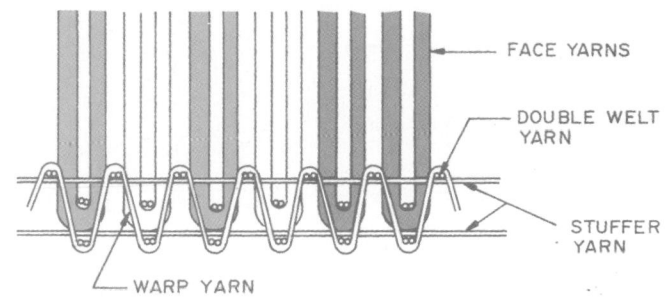

9-98 *Construction details of Axminster carpet.*

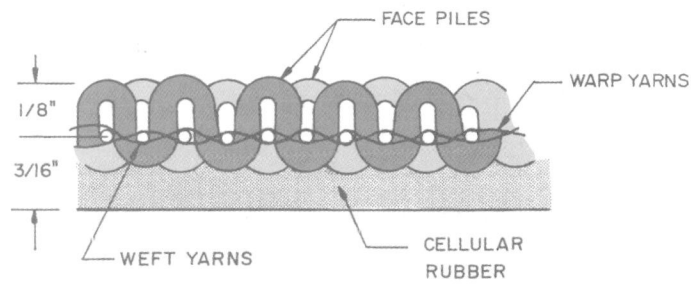

9-99 *Loomed carpet construction bonds the pile yarn to a rubber cushion.*

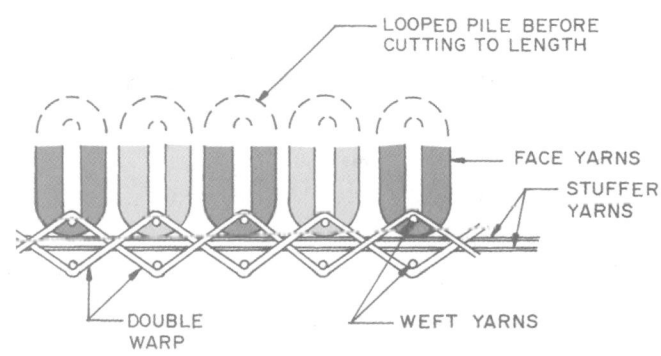

9-100 *Velvet construction has double warp yarns.*

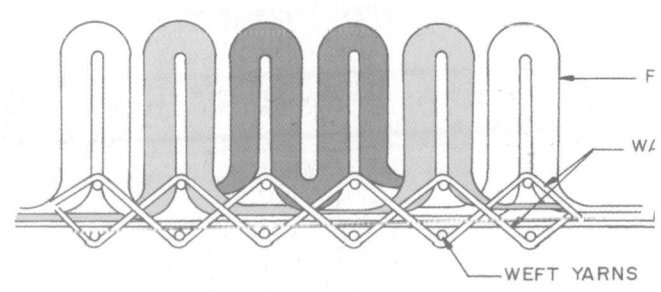

9-101 *Wilton construction is used to produce carpets containing multicolored yarns.*

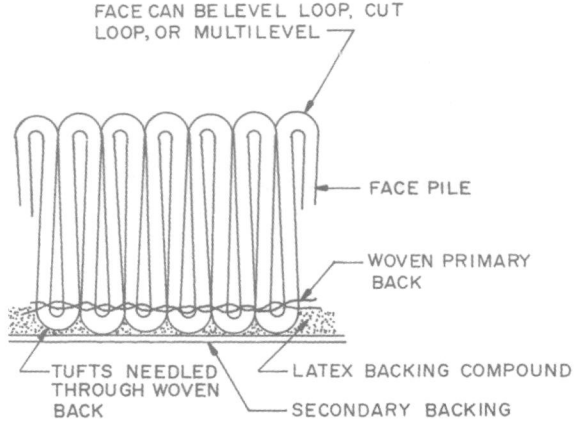

FACE CAN BE LEVEL LOOP, CUT
LOOP, OR MULTILEVEL

FACE PILE

WOVEN PRIMARY
BACK

TUFTS NEEDLED
THROUGH WOVEN
BACK

LATEX BACKING COMPOUND

SECONDARY BACKING

9-102 *Tufted carpet is made with stitching pile yarns inserted through a backing material and a second backing bonded over it.*

Wilton looms can produced sculptured and embossed textures by varying the pile height and a combination of cut and uncut loops. Uncut loop Wilton weaves are also available.

TUFTED CONSTRUCTION

Tufted construction involves stitching the pile yarn through a backing material, which is much like sewing on a power sewing machine. The back is then coated with a latex bonding agent, which holds the tufts to the base and bonds a second backing over it. The first backing is usually cotton, polypropylene olefin, or jute. The second backing is usually a coarse jute fabric (see 9-102). Tufted carpets are available in level loop, multilevel, cut, uncut, and a mixture of cut and uncut piles. Carved and textured surfaces are produced.

KNITTED CONSTRUCTION

Knitted carpet is made by looping together the pile yarn, stitching, and backing in a single operation. A coat of latex is spread on the back to bond these together and stiffen the carpet (9-103). They are available in solid colors and tweeds and single or multilevel uncut pile.

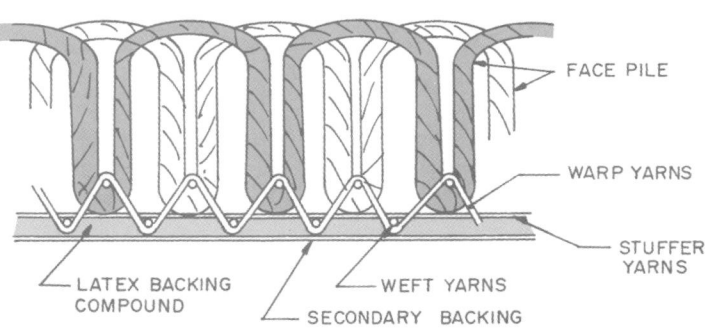

FACE PILE

WARP YARNS

STUFFER
YARNS

LATEX BACKING
COMPOUND

WEFT YARNS

SECONDARY BACKING

9-103 *Knitted carpet construction involves looping the pile yarns, stitching, and backing in a single operation.*

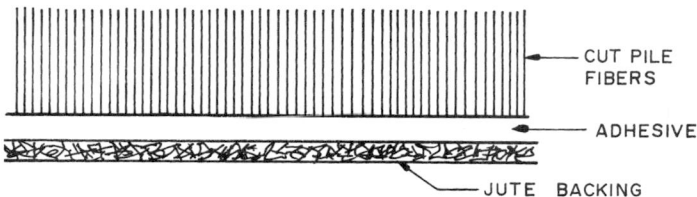

9-104 *Flocked carpet construction electrostatically sprays short strands of pile yarn onto an adhesive coated backing.*

FLOCKED CONSTRUCTION

Flocked construction involves electrostatically spraying short strands of pile yarn onto an adhesive coated backing sheet. The pile strands become vertically imbedded in the adhesive, a second backing is laminated to the first, and the unit is cured (see 9-104).

FUSION-BONDED CONSTRUCTION

Fusion-bonded carpet is made by bonding the pile yarn between two parallel sheets of backing. These are coated with vinyl adhesive making a sandwich-like product with the pile yarn in the center. After the adhesive hardens the pile yarn is cut in the middle forming two pieces of cut pile carpet. A secondary backing is often bonded to the first backing sheet (9-105).

Quality Specifications

Carpet is made in various qualities suitable for different locations. The Federal Housing Administration carpet quality specifications include pile yarn weight, pile density, and pile thickness. ASTM tests indicate how to measure these.

FLAMMABILITY REQUIREMENTS

Building codes specify flammability requirements for carpets in various locations, such as in commercial buildings. The flame resistance of carpets is most commonly ascertained by the Flooring Radiant Panel Test. The carpet specimen is mounted on the floor of the test chamber and exposed to intense radiant heat from above. The rate of flame spread is measured.

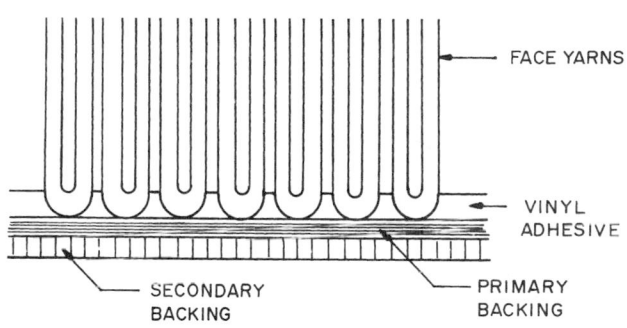

9-105 *Fusion-bonded carpet construction involves bonding pile yarn to backing sheets coated with a vinyl adhesive.*

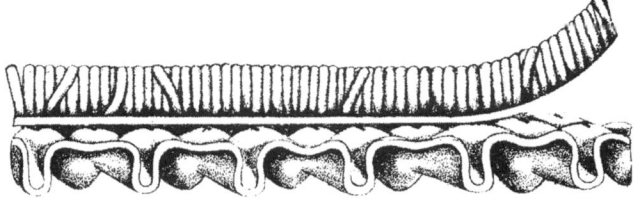

1. Carpet installed over separate cushion. This can be done by stretching the carpet in over tack strip, or by gluing the cushion to the floor and the carpet to the cushion.

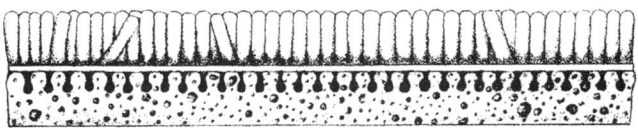

2. Carpet with attached cushion.

3. Direct glue down; carpet cemented directly to the floor without any cushion.

9-106 *Types of carpet cushion.*
Courtesy Cushion Carpet Council

Carpet Cushions

Generally carpet is installed over a cushion because this increases the life of the carpet, absorbs considerable traffic noise, has insulation value, and provides a soft, resilient floor covering. There are three types of carpet installation (9-106). The carpet can be installed over a separate cushion, the carpet can be made with a foam cushion attached to the back, or the carpet may be laid without a cushion.

Carpet cushions are available in three basic materials—urethane, fiber, and rubber. They are manufactured in Class 1, light and moderate traffic, and Class 2, heavy duty traffic. These recommendations meet U.S. Government requirements for cushion used in FHA financed housing. Although rubber cushion is more expensive, it lasts longer, resists mildew and mold, and is nonallergenic.

DIVISION 10

SPECIALTIES
CSI MASTERFORMAT™

Products listed as specialties include a vast spectrum of items used to provide a limited special function. Following are brief descriptions of some of these.

VISUAL DISPLAY BOARDS

Various types of display board are shown in 10-1.

Chalkboards are available with the exposed surface made from slate, porcelain enamel on steel, aluminum, or glass. The glass units use tempered float or plate glass coated with a colored, vitreous glaze that contains a fine abrasive. Chalkboards may be permanently mounted on a wall or be portable. Some have panels that slide horizontally or vertically. They are available in sizes up to 6 × 12 ft.(1830 to 3660 mm). Natural slate chalkboards are typically from ¼ to ⅜ in. (6 to 10 mm) thick, widths to 4 ft. (1220 mm), and random lengths to 6 ft.(1830 mm). Other chalkboards are available in widths to 4 ft.(1220 mm) and lengths to 10 ft. (3050 mm).

Marker boards may have a vinyl sheet surface or porcelain enamel on steel. The substrate may be particleboard, fiberboard, sheetrock, hardboard, plywood, or a plastic honeycomb. Special ink markers are used to write on the surface. Some are a form of watercolor and wipe off the marker board with a wet cloth. Another type is semipermanent ink and requires a special liquid cleaner.

Pegboard panels are usually ¼ in. (6 mm) thick hardboard with holes spaced 1 in. (25 mm) on center. They typically have an aluminum frame. Special wire hanging clips are available.

Bulletin boards or **tackboards** typically have a cork surface laminated to a fiberboard substrate. Some have a woven fabric facing and can hold paper with pins or the hook and loop fasteners that grip into the fabric.

Various types of manual and motorized projection screens are available. They may be integrated into a larger unit with chalkboards, bulletin boards and marker boards.

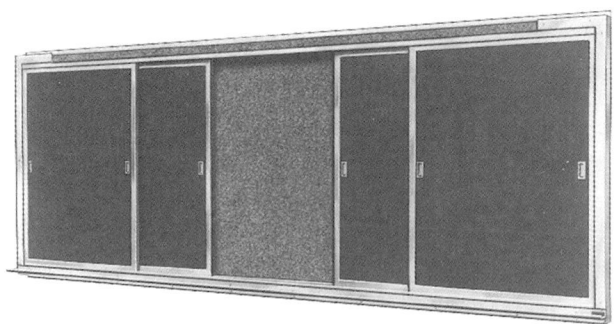

A chalkboard with sliding panels that uncover a tackboard below.

A display case.

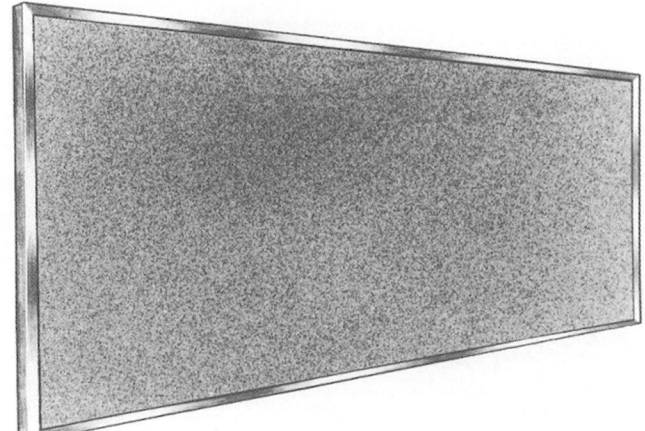

A bulletin board or tackboard.

10-1 *Some of the visual display units and chalkboards in common use.*
Courtesy Claridge Products and Equipment, Inc.

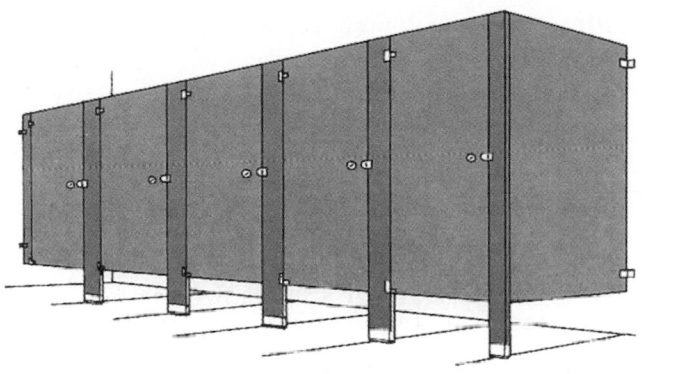

Floor-supported compartments.

Ceiling-hung units.

Overhead braced compartments.

10-2 *Toilet enclosures may be supported by the ceiling, wall, and floor.*

Courtesy Global Steel Products Corporation

COMPARTMENTS & CUBICLES

These products include enclosures such as metal, plastic laminate, and stone toilet compartments, shower and dressing compartments and various cubicles. They include dividers, screens and curtains for showers, hospital cubicles, toilets, and urinals. The toilet enclosures may be supported by the floor, ceiling, and wall (see 10-2). The toilet enclosures that are hung from the ceiling make cleaning the floor easier. Ceiling hung units require a structural steel ceiling support.

Metal toilet compartments may be steel with a baked enamel finish or stainless steel. Plastic laminate compartments have a 1/16 in. (1.6 mm) plastic laminate sheet adhered to particleboard or high-pressure plastic laminate core. Stone compartments are typically marble.

Cubicle enclosures are hung from an overhead track. The track path can be varied to suit the conditions. It can be mounted directly to the ceiling or suspended (see 10-3). Ceiling tracks are also used to support intravenous systems.

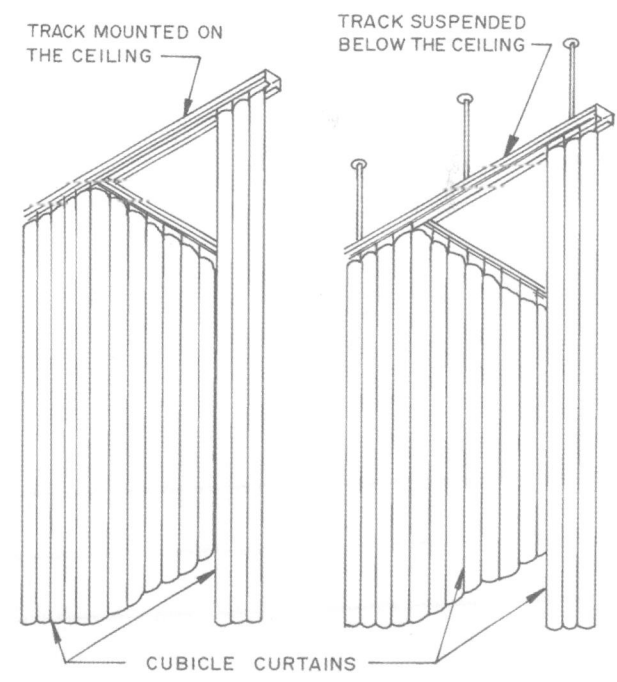

10-3 *Cubicle curtains may be hung from ceiling-mounted or suspended tracks.*

LOUVERS, VENTS, GRILLES & SCREENS

Some types of grille, screen, louver, and vent are special items. Louvers and other items for ventilation that are not an integral part of the mechanical system are speciality items. This includes items such as operable, stationary, and motorized metal wall louvers, louvered equipment enclosures, door louvers, and various types of wall and soffit vents. For example, a curtain wall louver is a specialty product. Interior and exterior grilles and screens made of any material are used for air distribution, sun screen, and other functions in addition to ventilation purposes (see 10-4).

Grilles—are open gratings or barriers used to cover, conceal, protect, or decorate an opening. They can be steel, aluminum, cast iron, or wood (see 10-5).

Gratings—of various types are used to cover **trench frames** which control surface water flow. These are typically cast iron. Street and parking lot construction have to be designed to carry away surface water from rain and melting snow.

Curb inlets—and trench frames are used as intakes for this purpose (see 10-6).

Tree grates—are another form of grating that are used both inside and outside a building. They provide protection yet permit the tree to receive water (see 10-7).

10-4 *These metal louvers are part of the building ventilation system and are designed to be architecturally pleasing.*

10-5 *These cast-metal grilles provide security and are a major architectural feature.*
Courtesy Lawler Foundry Corporation

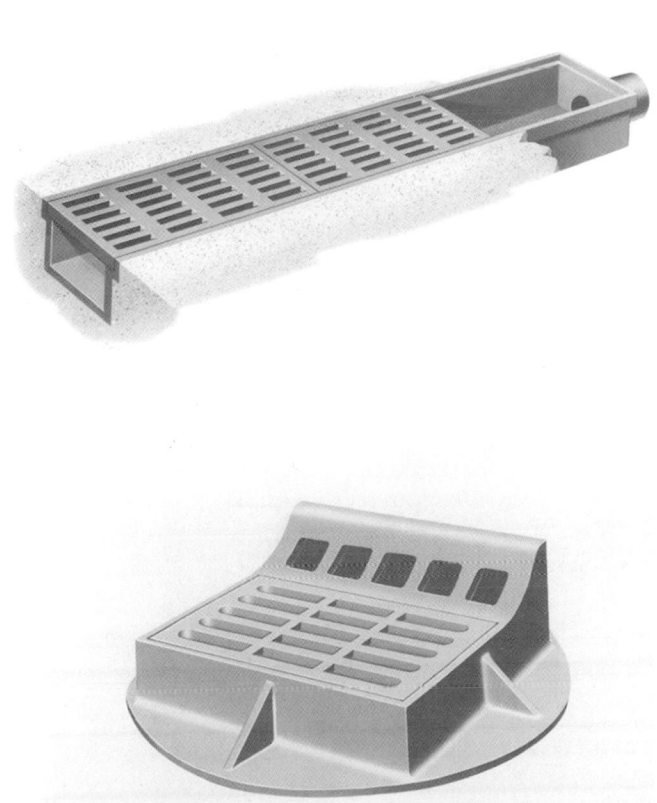

10-6 *A cast-iron trench frame (top) and this heavy-duty curb inlet frame and grate are used to control surface water.*
Courtesy Neenah Foundry Company

10-7 *A cast-iron tree guard protects the tree roots from surface damage yet permits rain to reach them.*
Courtesy Neenah Foundry Company

SERVICE WALL SYSTEMS

Service wall systems incorporate services such as clocks, fire hoses, drinking fountains, fire extinguisher cabinets, telephones, and waste receptacles (see 10-8).

WALL & CORNER GUARDS

Wall and corner guards are protective devices made from metal, plastic, rubber, and other materials that will withstand impact and for exterior use exposure to the weather. In 10-9 are cast-iron wheel guards used to protect the sides of a large opening. Rubber bumpers are secured to walls, such as at a truck loading dock, to protect the wall and truck from damage.

Other types of corner guards are used to protect wall corners that may be damaged by passing traffic such as moving carts and forklifts. These are typically protected with molded rubber or plastic guards installed on the finished corners.

ACCESS FLOORING

Access flooring is free-standing flooring made in modular units and raised above the basic floor system to provide a space to run mechanical and electrical services. The systems commercially available have some type of metal adjustable pedestals upon which the modular floor panels rest. The floor panels are typically covered with vinyl composition floor covering (see 10-10), although carpet could be used.

10-8 *Drinking fountains are a part of the service wall system.*

Courtesy EBCO Manufacturing Company

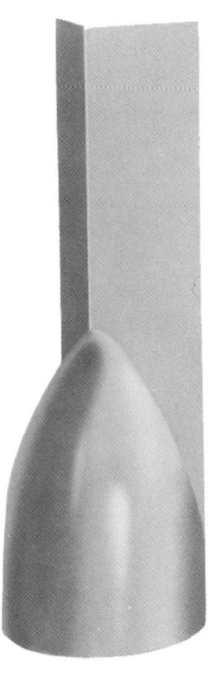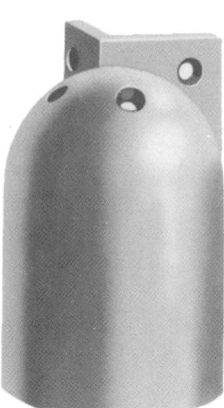

10-9 *These cast-iron wheel guards will withstand severe impact and abrasion as they protect the sides of garage door openings.*

Courtesy Neenah Foundry Company

10-10 *This access floor permits computer and electrical runs to be located below the floor and easily accessed for changes as needed.*

Reproduced with permission from *The Building Systems Integration Handbook*,
Richard Rush, Editor, Butterworth-Heinmann Publishers, Newton, MA

PEST CONTROL

Various mechanical, chemical, and electrical pest repellent systems and protective devices are available. These include control of rodents, birds, and insects.

FIREPLACES & STOVES

Manufactured fireplaces and stoves including dampers, metal chimneys, and fireplace screens and doors are specialties.

Manufactured fireplaces are typically metal-framed fire-box-and-vent units designed to accommodate **gas-fired** logs. They may be vented vertically through a metal pipe or horizontally through the wall. Some types do not require venting. They are delivered completely assembled (10-11).

A wide range of **wood-burning stoves** are also available. Some are placed in the room and become not only a source of heat but an attractive part of the decor. Others are large units with ducts running to several rooms.

10-11 *This free-standing top-vented gas-fired stove provides an attractive focal point in a room.*

Courtesy Heat-N-Glo Fireplace Products, Inc.

MANUFACTURED EXTERIOR SPECIALTIES

These items include stock and custom-designed steeples, spires, cupolas, and weather vanes.

Steeple—is an ornamental construction usually ending in a spire (see 10-12). It is usually erected on a roof or tower and may hold a clock or bells.

Spire—is a tall, pointed, pyramidal roof built upon a tower or steeple.

Cupola—is a light roofed structure mounted on the roof and is typically circular or a polygon. In addition to the architectural beauty, it often serves as a roof vent (see 10-13).

FLAGPOLES

This specialty area includes complete flagpole assemblies including the required accessories. They are typically made from wood, metal, and fiberglass. They may be ground-set or wall-mounted (see 10-14).

10-12 *This beautiful steeple contains a clock and is crowned with a tall spire.*

10-13 *This cupola has louvered sides so it can serve as a roof vent.*

10-14 *A flagpole flies the stars and stripes.*

IDENTIFYING DEVICES

Identifying devices include directories, direction signs, bulletin boards, and various letters, signs, and plaques used to communicate or identify. These may be fixed devices or be lighted, computerized, or some type of electronic device.

Directories—usually have replaceable letters and a locked glass door (see 10-15). They are available as wall-mounted or free-standing units.

Bulletin boards—may have a cork face or a cork face covered with nylon or vinyl fabric. The substrate is typically a fiberboard panel.

Various types of **accessibility sign** have become standardized for use across the country. Many building codes require accessible elements to be identified by the International Symbol of Accessibility (see 10-16). The International Symbol of Accessibility is posted to indicate all elements and spaces of accessible facilities. This includes facilities such as parking spaces reserved for individuals with disabilities, accessible passenger loading zones, accessible entrances, and accessible toilet and bathing facilities (see 10-17). In addition, signs are used to identify volume control telephones, text telephones, and assistive listening systems.

One example of code-regulated signs is the identification of exits and signs pointing the direction to move to get to an exit. Exit signs are illuminated. These can point to stairs, elevators, or doors exiting directly from the building to the outside. Signs are also used to identify refuge areas and the direction to move to reach them. A refuge area is an area that has protection against smoke and fire. The maximum occupant capacity for assembly rooms, restraints, and other areas where numbers of people gather is indicated by signs posted in the area. Control of parking and identification of facilities for the handicapped are other major parts of any sign system related to a building (see 10-18). These and many other signs are required by the building codes.

Other **symbol signs** are widely used to graphically present a message and help communicate with people using various languages. These are found extensively in places such as airports, hotels, and hospitals, and as highway traffic control signs (see 10-19).

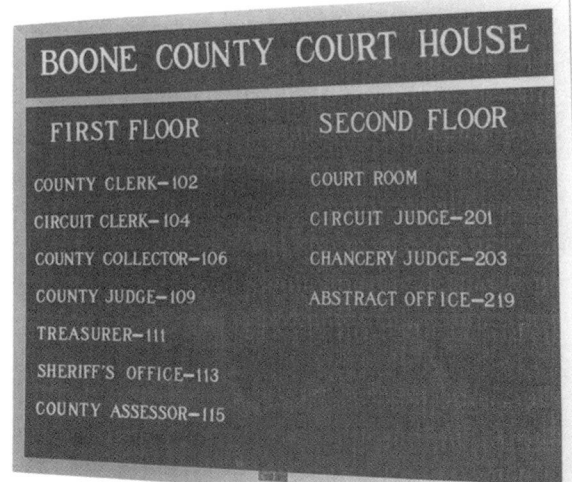

10-15 *This building directory gives room locations for the offices in the courthouse.*
Courtesy Claridge Products and Equipment, Inc.

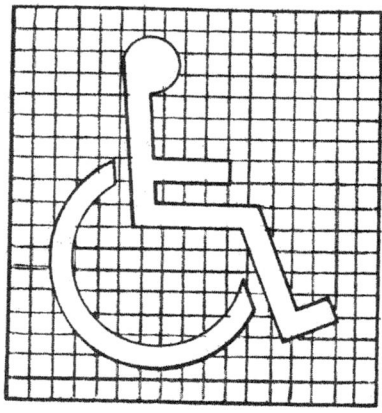

Proportions
International Symbol of Accessibility.

Display Conditions
International Symbol of Accessibility

10-16 *The International Symbol of Accessibility.*

International TDD Symbol

International Symbol of Access for Hearing Loss

10-17 *Symbols used to identify telephones and assistive listening devices for those with hearing impairments.*

10-18 *Various signs are used to control parking and identification of facilities accessible by the handicapped.*

10-19 *Some of the symbols used to communicate graphically.*

PEDESTRIAN CONTROL DEVICES

The products used for pedestrian control include portable posts and railings, rotary gates, turnstiles, electronic detection and counting systems, and items to limit access (see 10-20). These are strong, durable products that can stand exposure to the weather.

LOCKERS & SHELVING

Lockers provide temporary storage such as for airports (10-21) and athletic facilities (10-22). Some are coin operated. All types of open manufactured metal and wood storage shelving used for general storage are specialty items. This includes open shelving, various manufactured shelving systems, and mobile storage systems.

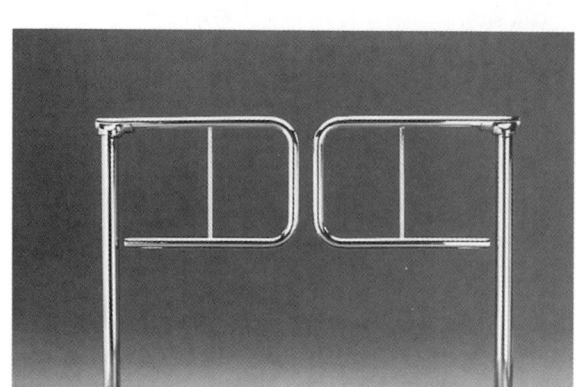

A self-closing double gate.

A plastic-coated welded steel barrier.

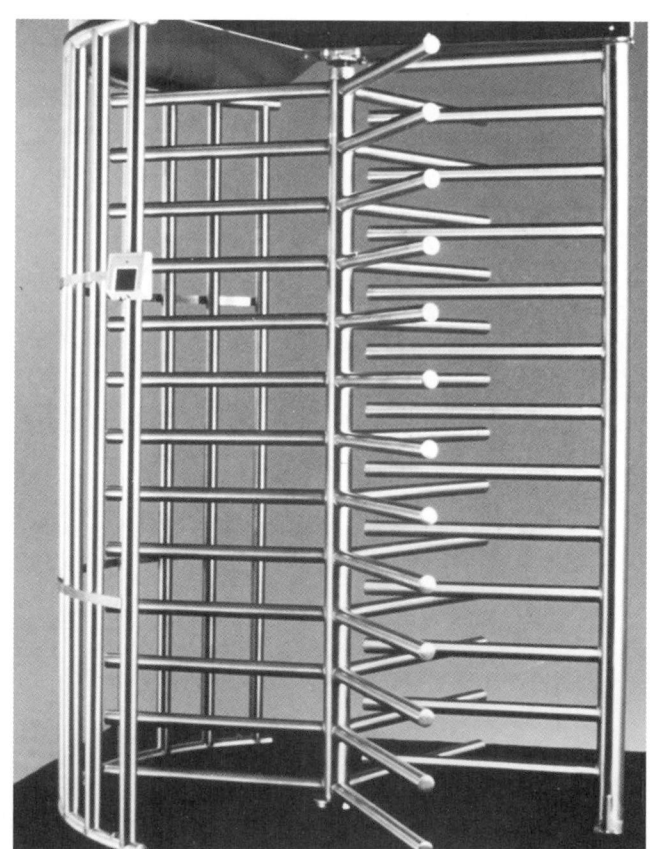

A 7-ft.-high nonpenetrable security turnstile provides total passageway security.

10-20   *These are strong, durable products for pedestrian control.*
Courtesy Alvardo Manufacturing Company

10-21 *Extensive banks of lockers are used in public places such as airports. These are coin-operated units.*

Courtesy American Locker Company, Inc.

10-22 *(Right) Single-tier lockers are widely used in schools and athletic facilites.*

Courtesy Republic Storage Systems Co., Inc., Canton, Ohio

FIRE-PROTECTION SPECIALTIES

This includes all portable fire-fighting devices and their storage facilities and excludes items directly connected to an installed fire protection system. Typical items include fire extinguishers, extinguisher cabinets, fire blankets and their storage cabinets, and any fire-extinguishing units on wheels (see 10-23).

10-23 *(Right) A typical fire extinguisher stored in a protective cabinet.*

10-24 *This covered walkway provides safe access to buildings on the opposite sides of a major thoroughfare.*

Courtesy Lin-El, Inc.

PROTECTIVE COVERS

Various types of awning, canopy, marquee, covered walkway, and sheltered bus and car stop structure are available as standard manufactured units or may be custom-designed and built. They are made from any of the commonly used materials that will withstand exposure to the weather (see 10-24).

POSTAL SPECIALTIES

Postal specialty products include all types of postal service facility and device such as view windows, letter slots, letter boxes, collection boxes, and chutes. In large commercial buildings these can become an extensive system used to speed the collection of mail (see 10-25).

10-25 *Post office boxes typically used in U.S. post offices.*

EXTERIOR PROTECTION DEVICES FOR OPENINGS

An extensive range of protection devices are manufactured, including products such as sun control screens, shutters, louvers, screens to provide security, and panels to provide insulation and protection against storms. These may be fixed or manually or electrically movable (see 10-26).

DEMOUNTABLE PARTITIONS

Demountable (movable) partitions include various dividers, screens, enclosures, and partitions. They are available as free-standing post-and-panel units, hung from the wall or ceiling, or secured to the floor and ceiling. They may be solid, offering privacy and some sound control or open mesh or screen panels, giving an area security but permitting visual examination. Some types use folding gates or doors. They are made of a variety of materials including wood, plastic laminates, and metal (see 10-27).

OPERABLE PARTITIONS

Operable partitions are those manually or power operated, providing the moving partition to enclose or open up an area. They are hung from a track on the ceiling and are supported by a steel floor track (see 10-28). Some are top-hung and have a rubber seal at the floor. They include folding panel partitions, accordion folding partitions, and sliding and coiling partitions. Folding and accordion doors are not in this classification.

10-26 *These windows have steel detention protection grids, providing maximum security yet permitting light to enter the building.*
Courtesy William Bayley Company

10-27 *Low demountable partitions provide privacy yet use the general illumination and heating and air-conditioning system.*
Courtesy O'Brien Partitions Co., Inc.

10-28 *Operable partitions are used to divide large rooms into several smaller rooms.*

10-29 *A round floor-mounted telephone kiosk.*

Courtesy Sherron Division, Redyref Pressed and Welded, Inc.

TELEPHONE SPECIALTIES

Telephone specialties include manufactured telephone enclosures that may be completely enclosed, supported on the floor (see 10-29), or wall-hung (see 10-30). Telephone directory units and shelving are usually wall-hung and are available in many sizes. Telephone enclosures for outdoor use are weather-tight and have lighting for use at night. Some companies will manufacture the enclosures to the architect's design.

10-30 *A bank of wall-hung telephones with privacy dividers.*

Courtesy American Specialties, Inc.

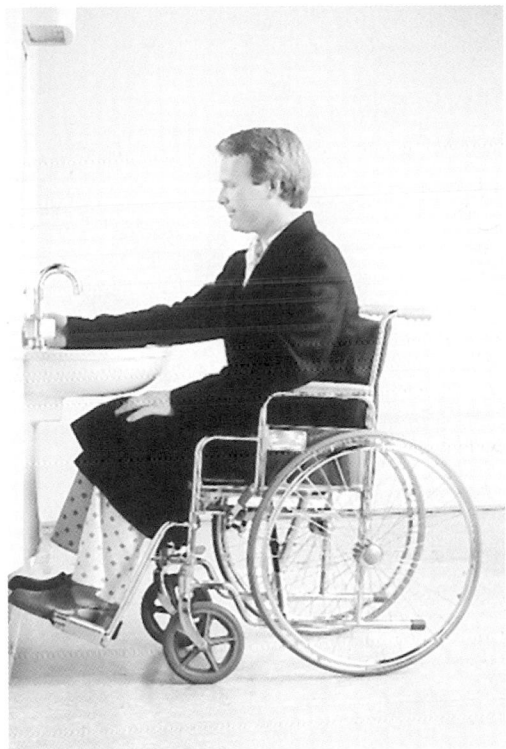

10-31 *This wash center provides barrier free access.*

Courtesy Bradley Corporation

TOILET & BATH ACCESSORIES

Toilet and bath accessories include all items manufactured for use in connection with toilets and baths. These include products used in most commercial restrooms plus special units designed for use in hospitals and areas where accessibility and security are important (see 10-31). Items in this specialty area include items such as mirrors, air fresheners, electric dryers, grab bars (see 10-32), waste receptacles, and dispensers for soap, toilet tissue, towels, razor blades, lotion, and paper cups. Hospital products include special cabinets and bedpan storage units. Security items include mirrors that resist theft and breakage and specially designed soap and tissue dispensers.

SCALES

Any type of weighing device is considered a specialty item.

10-32 *This wall unit provides the required grab bar for the handicapped plus the toilet paper and seat cover supply.*

Courtesy Bobrick Washroom Equipment, Inc.

10-33 *This type of wardrobe, often used in hospitals and nursing homes, provides clothes hanging space plus shelf storage.*

WARDROBE & CLOSET SPECIALTIES

These specialties are units used to store clothing such as hats and coats. Among these are wardrobes providing hanging space as well as several drawers (see 10-33). Other units typically used in hospitals and dormitories combine a wardrobe with a vanity area and lavatory. Wardrobes are often constructed with a plastic laminate over a substrate such as particleboard. Racks are generally steel. A wide variety of racks, wall-mounted and floor-supported, are used to store coats, hats, and packages.

DIVISION 11

EQUIPMENT
CSI MASTERFORMAT™

11010 Maintenance Equipment

11020 Security & Vault Equipment

11030 Teller & Service Equipment

11040 Ecclesiastical Equipment

11050 Library Equipment

11060 Theater & Stage Equipment

11070 Instrumental Equipment

11080 Registration Equipment

11090 Checkroom Equipment

11100 Mercantile Equipment

11110 Commercial Laundry & Dry Cleaning Equipmentt

11120 Vending Equipment

11130 Audiovisual Equipment

11140 Vehicle Service Equipment

11150 Parking Control Equipment

11160 Loading Dock Equipment

11170 Solid Waste Handling Equipment

11190 Detention Equipment

11200 Water Supply & Treatment Equipment

11280 Hydraulic Gates & Valves

11300 Fluid Waste Treatment & Disposal Equipment

11400 Food Service Equipment

11450 Residential Equipment

11460 Unit Kitchens

11470 Darkroom Equipment

11480 Athletic, Recreational & Therapeutic Equipment

11500 Industrial & Process Equipment

11600 Laboratory Equipment

11650 Planetarium Equipment

11660 Observatory Equipment

11680 Office Equipment

11700 Medical Equipment

11780 Mortuary Equipment

11850 Navigation Equipment

11870 Agricultural Equipment

11900 Exhibit Equipment

Division 11 groups equipment related to residential and commercial construction. The level-two sections are shown on the previous page. A few of these are discussed in the following pages.

MAINTENANCE EQUIPMENT

Maintenance equipment includes free-standing and built-in equipment used to maintain the building. This includes products such as window washing systems, vacuum systems, floor and wall cleaning equipment, and various types of housekeeping cart. Central vacuum systems have a central powered vacuum unit (see 11-1) connected to pipes run through the wall to sealed outlets in the various rooms. The flexible plastic hoses with various end attachments are connected to the outlet when desired to vacuum an area (see 11-2).

Window washing systems for high-rise buildings have hoists mounted on the roof, and the equipment that washes the windows on the exterior is controlled by the equipment on the roof. The system usually rides on tracks on the roof. Some systems lower scaffolding from the roof from which workers wash the windows.

SECURITY & ECCLESIASTICAL EQUIPMENT

Security and vault equipment is designed to store money or other valuables and includes vault doors (see 11-3), day gates, security and emergency systems, safes, and safe deposit boxes.

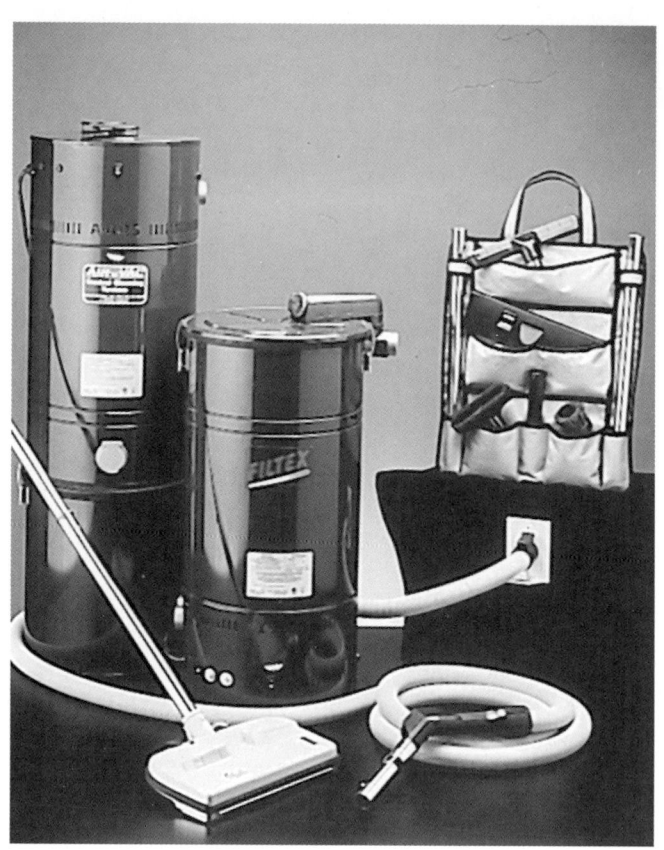

11-1 *(Right) A central power vacuum unit that operates the system throughout the entire house.*
Courtesy M & S Systems, Inc.

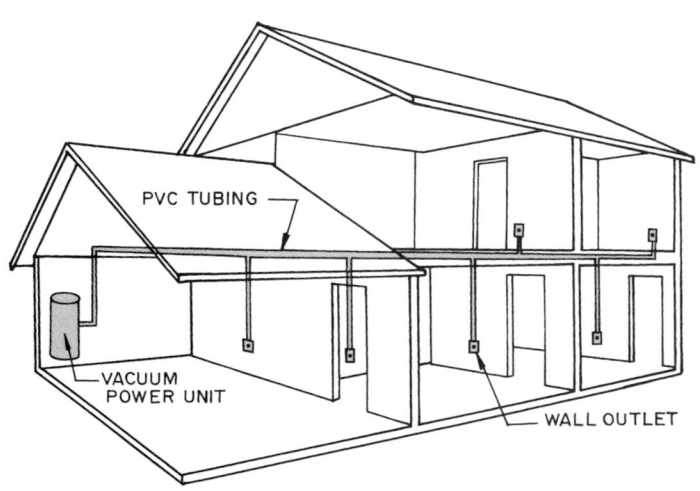

11-2 *A central power vacuum system operates off a power unit that is usually located in a garage, basement, or utility room. The tubing runs to each room and has an inlet connection for the vacuum hose.*

PVC TUBING

VACUUM POWER UNIT

WALL OUTLET

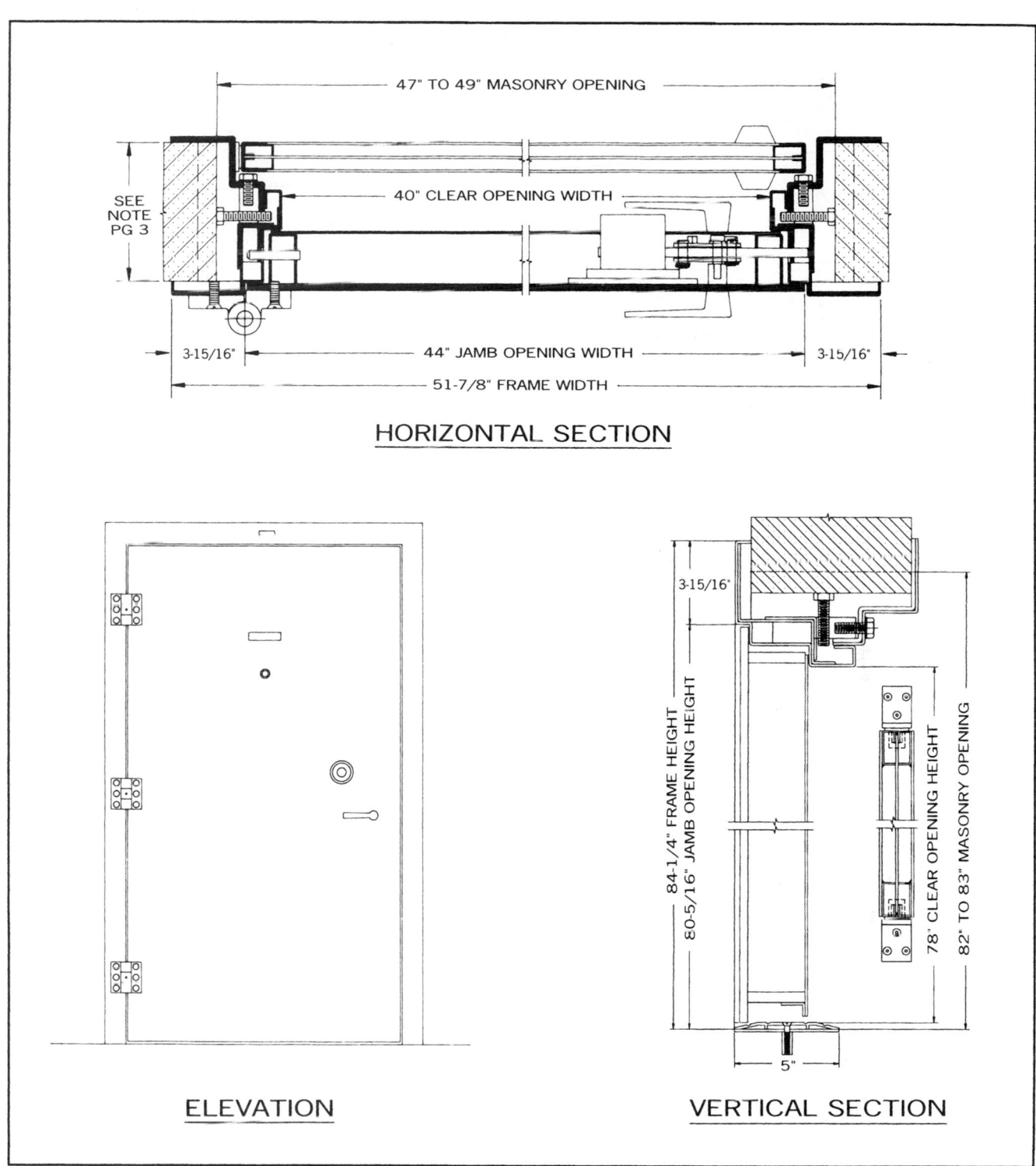

11-3 *Security vault doors provide maximum security for valuables. Various types are made for different levels of security. The one in this illustration rates GSA Class 5. It is rated to provide 20 man-minutes against manipulation of the lock, and against radiological attack.*

Courtesy Overly Manufacturing Company

11-4 *Automatic teller machines are widely used to facilitate the banking needs of clients.*

Courtesy Branch Banking and Trust Company

Teller and service equipment is also designed for use where money transactions occur and includes service and teller windows, package transfer units, automatic banking systems (see 11-4), and teller equipment systems (see 11-5).

Ecclesiastical equipment includes items related to churches, such as baptisteries and various chancel items (see 11-6). This will vary according to the denomination and desires of the congregation. Some items are stock while others are custom built. All types of materials are used including stone, wood, and metal (see 11-7).

11-6 *The minister delivers the sermon from this custom-built pulpit.*

11-5 *Teller counters are used to provide individual service to customers in banks and related businesses.*

Library equipment includes book storage and retrieval devices and includes book theft protection equipment, library stack systems, study carrels, and book depositories (see 11-8).

MERCANTILE EQUIPMENT

This includes equipment used in retail and service stores including display cases, cash registers and checking equipment, food processing equipment, and special items as in specialty shops such as barber and beauty shops (see 11-9).

SOLID WASTE HANDLING EQUIPMENT

Waste handling equipment includes units to collect, shred, compact, and incinerate solid waste. Some of the systems manufactured include package incinerators, compactors, storage bins, pulping machines, pneumatic waste transfer systems, and various chutes and storage collectors. One such system is shown in Division 14.

11-8 *Libraries use large shelving units to store their inventory of publications.*

11-7 *The offering plates are kept on this stand beside the altar.*

11-9 *Retail display fixtures are designed to exhibit a wide variety of merchandise of different sizes, weights, and values.*

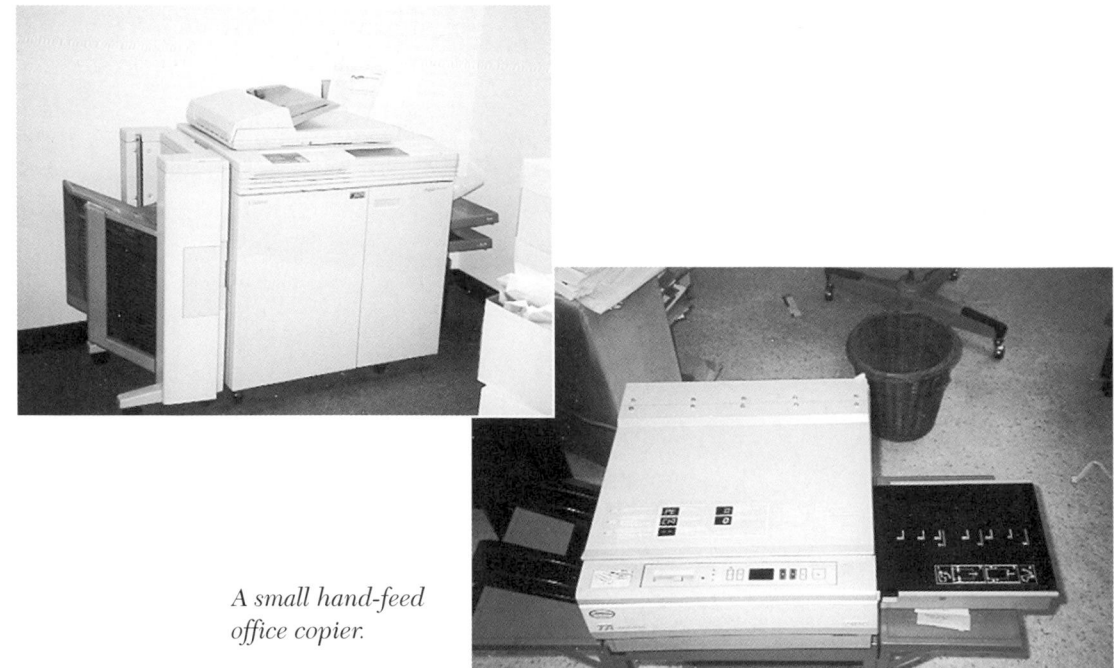

A large automatic-feed production copier that will collate.

A small hand-feed office copier.

11-10 *Copiers are available from small hand-feed units to large automatic-feed, multiple-page units that collate and print in color.*

OFFICE EQUIPMENT

The **office equipment** section includes products such as copiers (see 11-10), computers (see 11-11), word processors, printers, fax machines, and modems. It does not include any of the office furniture such as desks, files, and chairs. Office furniture is detailed in Division 12. Office equipment is manufactured by a number of different companies. Since some of these products must communicate with each other, care should be exercised when different brands are mixed to be sure of compatibility.

A laser computer printer.

A personal computer.

11-11 *The personal computer has a large storage capacity and can support an extensive range of applications, produce graphics, display information, and play music. A laser printer allows the output of hard copy.*

RESIDENTIAL EQUIPMENT & UNIT KITCHENS

Residential equipment includes all appliances that may be built in or free standing such as stoves, washers, dryers, freezers, microwaves, dishwashers, compactors, refrigerators, surface cooking units, ovens, range hoods, and indoor barbecue units. These are manufactured by a number of different companies and if brands are mixed it is important to consider the possible difference in color (see 11-12).

Unit kitchens are manufactured units that combine a refrigerator, cooking unit, and sink plus cabinets into one unit. They are delivered totally assembled for rapid installation (see 11-13).

This section does not include kitchen cabinets. They are detailed in Division 12.

FLUID WASTE TREATMENT & DISPOSAL EQUIPMENT

This section includes a wide range of equipment used to treat and dispose of fluid waste. Such equipment includes an oil/water separator, fluid pumping stations, sewage and sludge pumps, scum removal equipment, aeration equipment, and package sewage treatment plants. The selection of the units is a major part of the engineering design of a waste treatment system. The units must be durable and dependable.

11-13 *Unit kitchens are compact single units that fit along one wall.*
Courtesy Dwyer Products Corporation

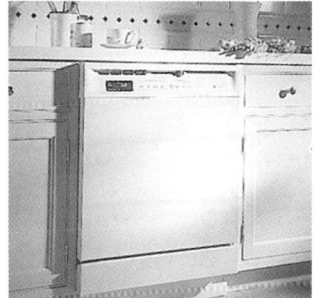

11-12 *Kitchen appliances include a wide range of electric and gas units in a range of sizes and colors.*
Courtesy General Electric Appliances

MEDICAL EQUIPMENT

The number of items of **medical equipment** used for health care for humans and animals is extensive. It includes items such as sterilizing equipment, examination and treatment equipment, optical and dental equipment (see 11-14), and items used in operating rooms and radiology facilities (see 11-15). Technical changes and improvements occur rapidly in this area, and those planning these facilities must be constantly alert for new equipment. The items must be of the highest quality.

Laboratory equipment may be preassembled by the manufacturer or designed to be assembled on site. It includes the many items used in laboratories for testing, research, and developing new products. Typical items include fume hoods, incubators, sterilizers, refrigerators, and various types of emergency safety appliance designed for use in laboratories. This includes a range of special service fittings and the accessories to use them.

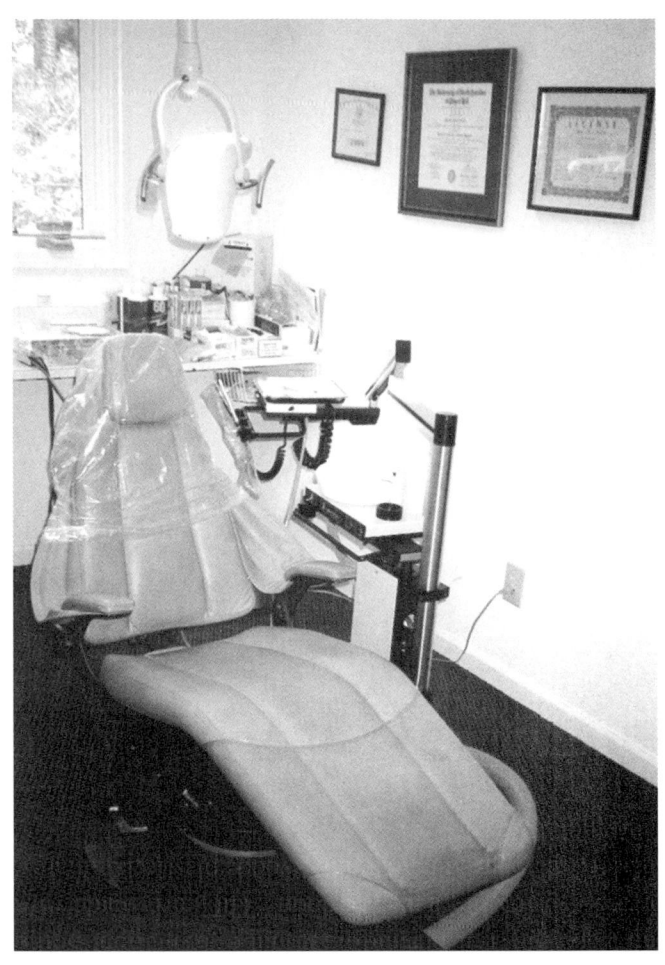

11-14 *(Right) Typical equipment in a dental office.*

11-15 *Medical facilities utilize a large number of small testing, treatment, and examination equipment.*

DIVISION 12

FURNISHINGS

CSI MasterFormat™

Courtesy Hausman Industries, Inc.

Furnishings include products that are decorative in nature such as artwork, rugs, and window treatments and other items of a useful nature such as casework, furniture for commercial buildings, and multiple seating. The interior decorator plays a major role in the selection of these items and sometimes becomes involved in the design of special products. The architect also is an important part of the overall design process.

ARTWORK

Photo murals are a popular form of interior artwork. They are available in a wide range of sizes and scenes. Murals painted on the wall are much more expensive but frequently become important art items.

Mosaic murals are formed by bonding a stone, such as marble, clay tiles, or glass pieces, together on heavy paper forming the desired design. The sheets are set against a mortar setting bed on the wall. When the setting bed has hardened, the paper is soaked off the face (see 12-1).

▼ All types of wall decorations are used, including:
  ▼ **paintings**
  ▼ **prints**
  ▼ **fabric wall hangings**
  ▼ **tapestries** of various types (see 12-2)

Some high-quality **woven rugs** are used as wall hangings (see 12-3).

Sculpture

Sculpture is another form of artwork that can be used extensively in interior and exterior locations. Some types are carved from stone or made from cast stone (see 12-4) while others are cast in bronze, lead, and aluminum (see 12-5).

Reconstructed sculpture is made by welding various shapes of metal. It often has other materials, such as stone, integrated into the piece.

Relief artwork is usually a panel-type piece with the images raised above the surfaces producing a three dimensional appearance. This gives the piece depth.

12-1 *This mosaic mural was formed using small ceramic tiles and covers the entire wall of this restaurant.*
Courtesy American Olean Tile Company

12-2 *Paintings are extensively used for interior decoration.*

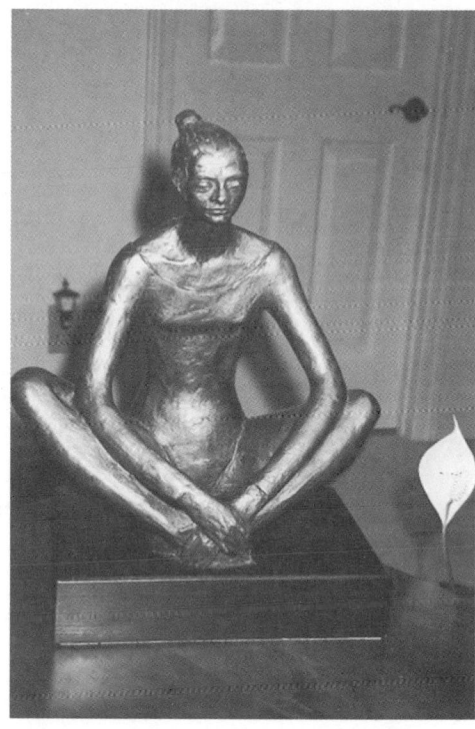

12-4 *This ballerina was made from cast stone.*

12-3 *Woven rugs and tapestries are popular wall decorations.*

12-5 *Cast-metal sculpture has been used for many years to produce striking art objects. This is Heloise, wed to the priest Pierre Abelard (lived 1079–1142).*

Stained-Glass Work

Stained-glass windows are a mosaic of pieces of translucent glass cut to the desired shape. Sometimes the glass requires additional detailing, shading, or texturing. This is accomplished by painting it with special mineral pigments and firing it in a kiln. The two types of stained glass are leaded and faceted.

Leaded glass—windows have the pieces of glass joined together with H-shaped lead strips called "cames" (see 12-6). Large windows require round bracing bars be wired to the leaded glass. This provides bracing yet allows for thermal movement. Exterior leaded glass is pressed into a bed of glazing sealant or tape (see 12-7).

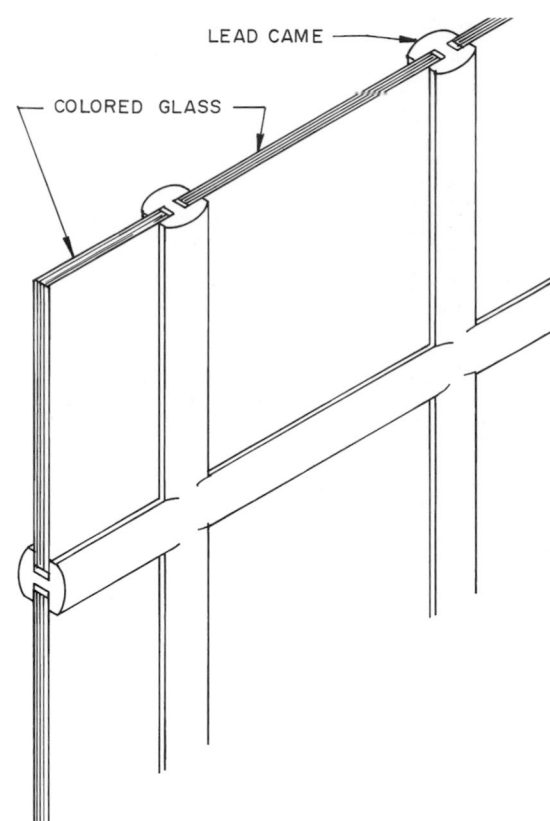

LEAD CAME

COLORED GLASS

12-6 *(Right) Leaded glass is formed by joining glass pieces cut in various sizes and shapes with lead cames.*

12-7 *These dramatic leaded glass windows contribute to the atmosphere of the church interior.*
Courtesy Rambusch, Cunningham Werdnigg, photographer

12-8 *This faceted glass wall provides privacy, admits light, and is a major decorative feature.*

Faceted glass—uses colored glass slabs 1 in. (25 mm) thick. The glass is cut or faceted to reflect the rays of light in many directions. The glass is made in a wide range of translucency, opacity, and transparency. This influences the color and degree of reflected light (see 12-8).

The faceted glass is set into a reinforced concrete or epoxy resin matrix much like bricks (see 12-9). This forms a panel 1 in. (25 mm) thick through. The light can pass through the faceted glass pieces bound together by the matrix. The panels are mounted in wood or metal frames or in a masonry wall.

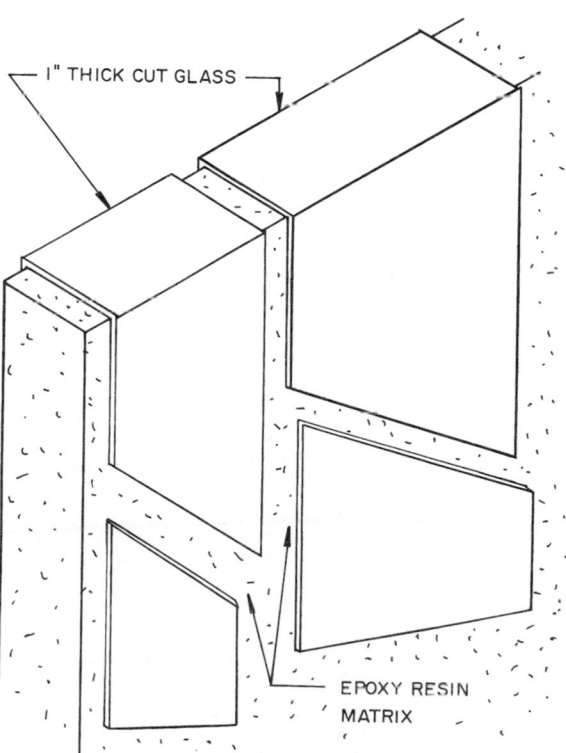

12-9 *Faceted glass walls are formed by bonding 1-inch-thick glass cut in various shapes and bonded with an opaque epoxy resin or reinforced concrete.*

1" THICK CUT GLASS

EPOXY RESIN MATRIX

Ecclestical Artwork

Ecclesiastical artwork includes pieces of religious significance to specific denominations. They can take the form of the types discussed in this chapter. Specific pieces could include altar pieces and religious symbols (see 12-10).

FABRICS

This section relates to fabrics, leathers, and furs for upholstery, curtains, draperies, and wall hangings and the various fabric treatments and fillers. For example, fire codes may require these products be fireproofed or their use will not be permitted (see 12-11).

12-10 *(Right) Ecclesiastical artwork reflects the beliefs of the denomination.*

12-11 *Draperies, curtains, and other wall hangings in commercial buildings must be noncombustible or be maintained flame-resistant.*

12-12 *Residential casework includes kitchen and bath appliances.*
Courtesy General Electric Appliances

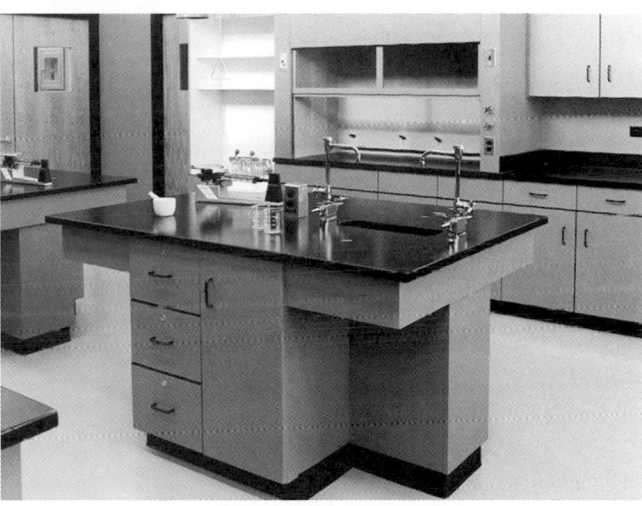

12-13 *Laboratory casework frequently contains utility connections such as water, waste removal, electricity, gas, and computer connections.*
Courtesy Stevens Industries, Inc.

MANUFACTURED CASEWORK

Manufactured casework includes stock cabinets manufactured from wood, steel, and plastic laminates. It includes the countertops, sinks, and any accessories or fixtures mounted on the countertops. For residential applications it includes both kitchen and bath casework (see 12-12). Medical applications include casework designed for dental, hospital, and optical uses, and nurses' stations, and veterinary casework. All types of laboratory casework used in medical, research, and other laboratories are part of this section (see 12-13). Various manufacturers produce specialized casework, which has a limited market. Included are casework for banks, dormitories, ecclesiastical uses, hotels, motels, schools, restaurants, and various display units (see 12-14).

12-14 *Commercial casework includes stock and custom-built units for commercial applications. This installation has storage, chalkboards, and bulletin boards.*
Courtesy Stevens Industries, Inc.

WINDOW TREATMENT

Interior window treatment can serve to be decorative and practical by blocking the intrusion of sunlight and providing privacy.

Metal blinds—with narrow horizontal slats (often called venetian blinds) can be adjusted to block the sun or opened to permit a view to the outside. They can be raised to the top of the window so they are completely out of view (see 12-15).

Vertical blinds—are made with fabric slats hanging from the top of the window. They can be adjusted to be open or closed (see 12-16).

Shutters—made of wood or plastic are also used on the interior of the window. Some have adjustable louvers that will let in a little light. They permit very little view to the exterior.

Shades—are manufactured including insulating, lightproof, translucent, and woven wood or plastic types. They roll up to provide a view and are pulled down to block the view. An accordian type with a fiberglass fabric is also available (see 12-17).

Curtains—of various types can serve the same purpose as shades. Some have metal tracks across the top that permit the drapes to be open or closed over the windows. Other units remain on each side or over the top of the window and serve a decorative purpose. Again, fire codes must be observed when selecting shades and curtains.

12-15 *Blinds with horizontal metal slats can be adjusted to block the sun or be opened to permit the entrance of light.*

12-16 *Vertical blinds are widely used in residential and commercial construction.*

12-17 *Various types of shade are available for blocking the sun and providing privacy.*

FURNITURE & ACCESSORIES

Furniture under CSI MasterFormat™ Division 12, includes all types of free-standing units for commercial and residential use. The listing of areas included in open office furniture includes office partitions, storage, work surfaces, shelving and light fixtures. General furniture includes all types from residential through commercial areas such as classroom, hotel, library (see 12-18), medical, office (see 12-19), and restaurant furniture. Furniture accessories such as clocks, ashtrays, lamps, waste receptacles, and desk accessories are included (see12-20).

12-18 *Library carrels provide a quiet study area.*

RUGS & MATS

This section includes loose rugs and mats, gratings and foot grilles. Carpet covering the floor area is included in Division 9. Typical products include chair pads, floor mats, entrance tiles, and floor runners. They are made from a variety of materials including rubber, carpet, vinyl, aluminum, and stainless steel.

12-19 *Various types of upholstered furniture are specified under CSI MasterFormat™ Division 12.*

12-20 *Clocks are one type of furniture accessory.*

12-21 *Multiple seating includes products used in any area where large numbers of people gather.*
Courtesy Stevens Industries, Inc.

MULTIPLE SEATING

Multiple seating includes seating for an area where audiences gather such as a school, restaurant, theater, church, stadium, or auditorium (see 12-21). It includes fixed, portable, and telescoping seating. Booths and various table and seat modules as found in restaurants are included.

INTERIOR PLANTS & PLANTERS

Extensive use of live and artificial plants is common in all types of commercial building. This section includes the plants, plant holders, and any landscaping accessories and maintenance materials required (see 12-22).

12-22 *An extensive variety of live and artificial plants are used for interior decoration.*

DIVISION 13

SPECIAL CONSTRUCTION

CSI MasterFormat™

Special construction encompasses an extensive array of structures, systems, and assemblies serving a limited specific purpose. Construction ranges from complex facilities, such as nuclear reactors to simpler projects such as site-constructed incinerators and specialized digestion tank covers and assemblies. Following are discussions of a few of these sections.

AIR-SUPPORTED STRUCTURES

These are structures with single and multiple wall enclosures made of a flexible material that is pneumatically supported. The structures are pre-engineered and prefabricated for delivery to the site. Typical materials include fiberglass fabric that is teflon coated and polyester that is vinyl or neoprene coated.

Air-supported structures form an envelope that encloses a pressurized space. Some are unreinforced membranes in which the membrane is the primary structural element (see 13-1). Another type uses a reinforced membrane which is reinforced by a network of cables or webbing forming the primary structural system and the membrane spans the area between the cables or webbing. Air locks are used as entrances and emergency exits are counter-pressure balanced, are self-closing, and have panic hardware. The membrane is light admitting and must meet local fire codes.

13-2 *This shows the installation of an integrated ceiling, which when completed will cover the vaulted ceiling and include lighting, a sound system, and a sprinkler system.*
Courtesy Chicago Metallic Corporation

INTEGRATED ASSEMBLIES

Integrated ceilings are a type of integrated assembly. They are pre-engineered and prefabricated for assembly on the site. Some integrated systems include lighting, air supply registers, ducts, air return grilles, fire sprinkler systems, and communication linkage. These systems are coordinated with CSI Master-Format™ Divisions 15 and 16 (see 13-2).

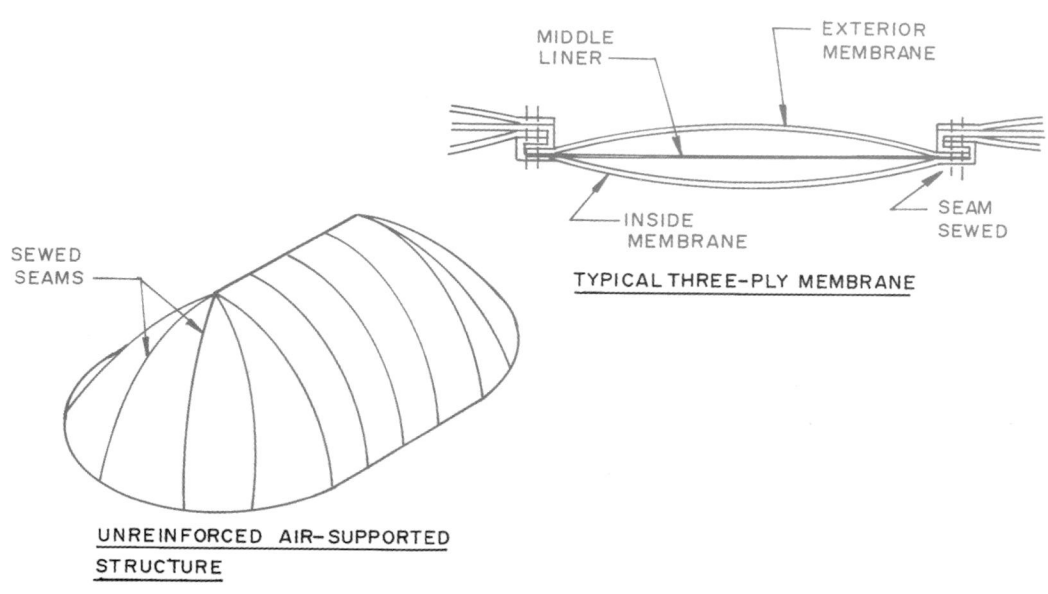

13-1 *This is an unreinforced air-supported structure constructed with a three-ply membrane.*

SPECIAL-PURPOSE ROOMS

There are many rooms designed and constructed to meet specific performance requirements, such as fire protection, thermal control, or sound control. The specific rooms in the CSI MasterFormat™ also include athletic rooms, sound-conditioned rooms, clean rooms, cold-storage rooms, insulated rooms, shelters and booths, planetariums, prefabricated rooms, saunas, steam rooms, and vaults.

NUCLEAR REACTORS

The design and construction of nuclear reactors is a very demanding project and is regulated by a number of government agencies. The use of nuclear fuels in high-temperature high-concentration process to generate electric power is widespread. In 13-3 a fuel bundle is being inspected and readied for placement in a storage pool. The massive construction required when building a nuclear reactor is shown in 13-4. See Division 16 for additional information.

PRE-ENGINEERED STRUCTURES

All types of pre-engineered prefabricated buildings and structures are classified as "Special Constructions" even if they are erected on temporary foundations. Some of the structures included are metal building systems, glazed structures such as a greenhouse, portable buildings, grandstands and bleachers, and cable-supported, fabric, and log structures. Information on some of these can be found in Division 5.

WASTE & UTILITIES SECTIONS

A number of Special Constructions occur in the area of waste treatment and systems involved with utilities.

Filter underdrains and media include the piping and filters used in water and fluid waste treatment. Filter media include anthracite, charcoal, diatomaceous earth, and mixed and sand media.

Digester Covers and Appurtenances include special tank covers and assemblies used on digestion tanks.

13-3 *Fuel bundles are being inspected and prepared for placement in the storage pool.*
Courtesy Tennessee Valley Authority

13-4 *Massive concrete construction is used in construction of nuclear reactors with some walls many feet thick.*
Courtesy Bureau of Reclamation, U.S. Department of the Interior

Oxygenation Systems include site-assembled piping systems and related equipment for the dissolution and mixing of gaseous oxygen in liquid waste. This includes the oxygen generators, oxygen storage, and the dissolution system.

Utility Control Systems include the operating and monitoring systems for water supply, wastewater, and electrical power generation plants. It includes metering devices, display panels, control panels, and sensing and communicating equipment.

CONTROL & INSTRUMENTATION

Process & Industrial Control Systems are a section in Special Construction. The types of control vary with the industry and can include mechanical, electrical, fluid, pneumatic, and computer controlling devices. These systems may be totally automated or semiautomated. They can monitor situations such as temperatures or pressures, regulate the system so it operates at the required speed and quality and control safety within a manufacturing, processing or assembly plant.

Instruments listed as **Recording Instruments** are installed to measure and record various occurrences such as seismic information, stresses in structures, and meteorological information, such as solar and wind energy.

Transportation Control Instrumentation systems are used to monitor and control the various aspects of transportation systems such as airport control, railway control, subway control, or transit vehicle control.

13-5 *Solar panels in a field arrangement providing services to a large industrial building.*
Courtesy Solar Development, Inc.

SOLAR ENERGY SYSTEMS

Solar energy systems may be active or passive.

Active solar systems are assembled from manufactured components that utilize solar energy for conversion to thermal energy and electrical power. The components include air and liquid flat plate collectors, concentrating collectors, and vacuum tube collectors (see 13-5). Manufacturers have complete solar systems available. One example is a solar water heater (see 13-6). A schematic of a warm-water flat-plate solar heating system is in see 13-7. The flow is controlled by valves that are activated by thermostats. The liquid is moved by electricity-operated pumps.

Passive solar systems utilize natural means to collect, store, and distribute the heat through the building. Typical systems are shown in 13-8 and 13-9.

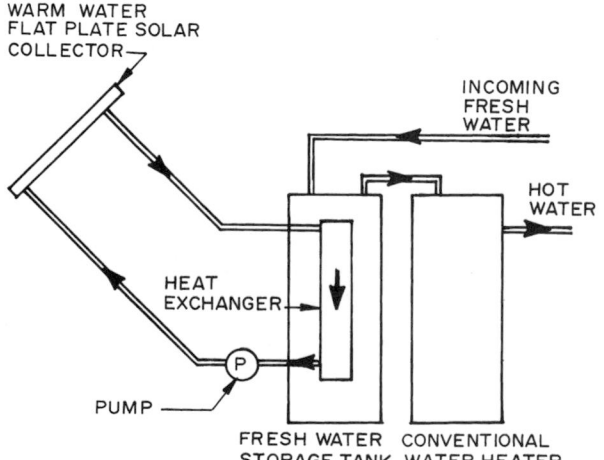

13-6 *A simplified schematic of a solar hot water heater. The controls have been omitted. The solar collector heats the fluid that runs through the heat exchanger. It heats the water in the storage tank before it goes to the conventional water heater.*

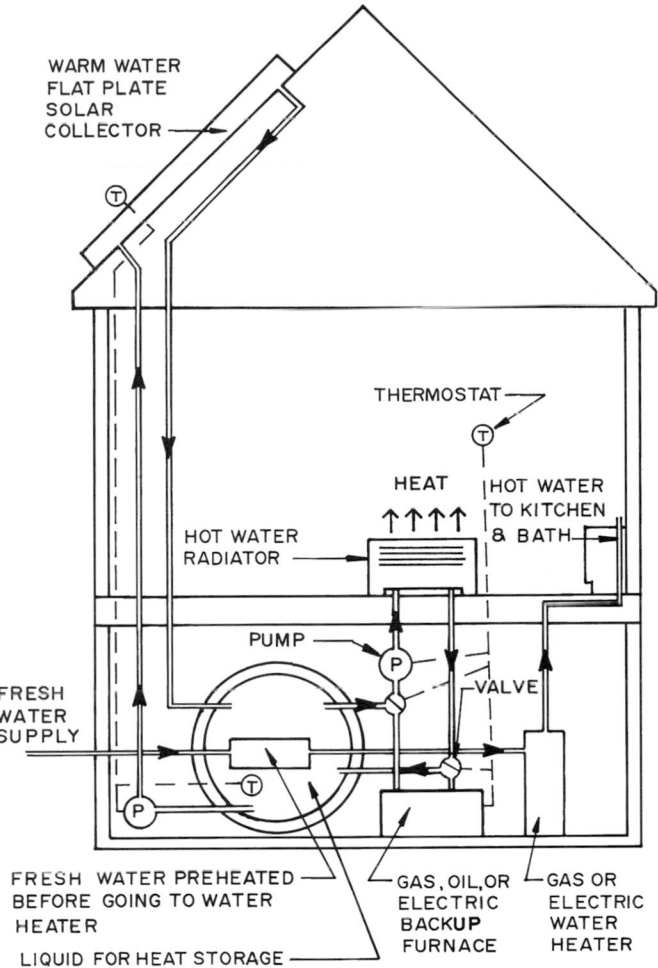

WARM WATER
FLAT PLATE
SOLAR
COLLECTOR

THERMOSTAT

HEAT
↑ ↑ ↑ ↑

HOT WATER
TO KITCHEN
& BATH

HOT WATER
RADIATOR

PUMP

VALVE

FRESH
WATER
SUPPLY

FRESH WATER PREHEATED
BEFORE GOING TO WATER
HEATER

LIQUID FOR HEAT STORAGE

GAS, OIL, OR
ELECTRIC
BACKUP
FURNACE

GAS OR
ELECTRIC
WATER
HEATER

13-7 *This schematic shows an active solar heating system using a warm water flat plate collector. Notice the heated liquid storage tank is also used to preheat the fresh water going to a conventional water heater.*

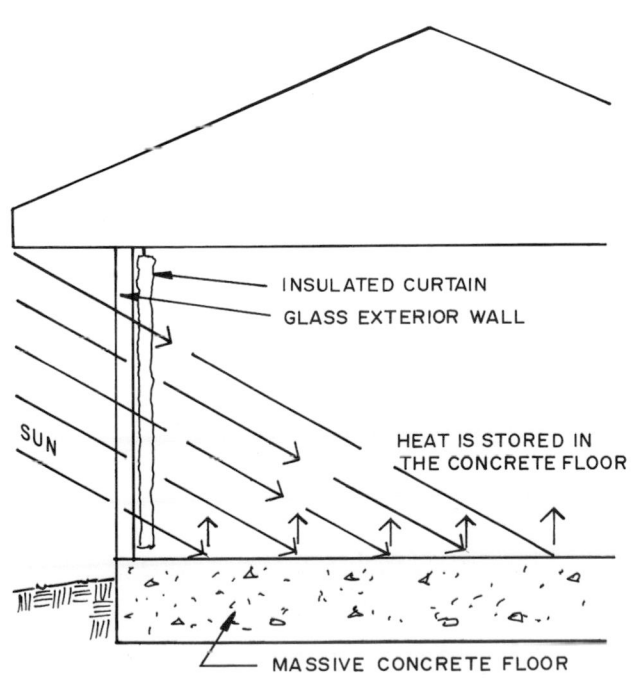

INSULATED CURTAIN
GLASS EXTERIOR WALL

SUN

HEAT IS STORED IN
THE CONCRETE FLOOR

MASSIVE CONCRETE FLOOR

13-8 *This passive solar heating design utilizes a massive concrete floor for heat storage.*

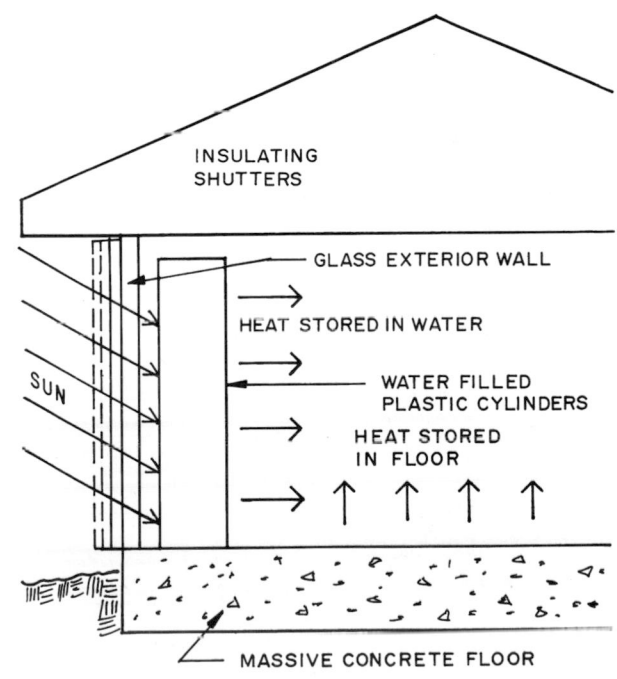

INSULATING
SHUTTERS

GLASS EXTERIOR WALL

HEAT STORED IN WATER

WATER FILLED
PLASTIC CYLINDERS

SUN

HEAT STORED
IN FLOOR

MASSIVE CONCRETE FLOOR

13-9 *This type of passive solar heating uses plastic cylinders filled with water and a massive concrete floor for heat storage.*

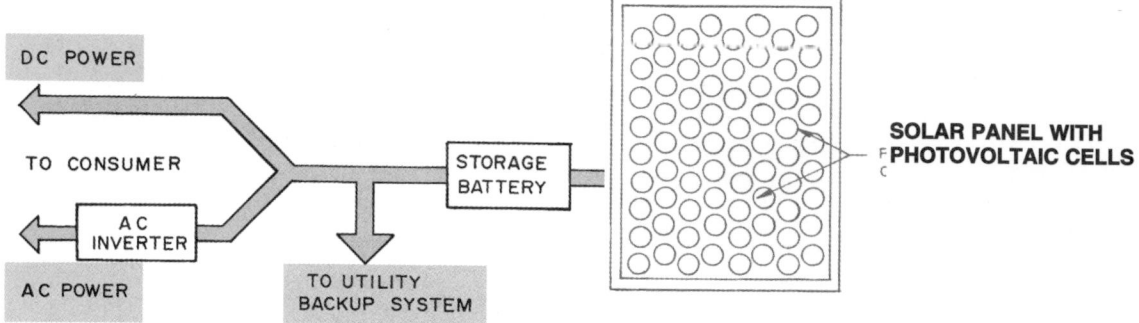

13-10 *A schematic of a typical solar photovoltaic electric power producing system.*

Solar photovoltaic collectors provide electricity when under the influence of light. Solar photovoltaic cells are thin, flat semiconductors that convert light energy into direct-current electricity. They find many uses, especially in remote areas where an array of cells can be used to operate electrical equipment, such as a pump or lights, without a battery. They can be grouped in large numbers to add additional electrical power to an electric company grid (see 13-10). Solar cells may be connected to both a direct load for immediate use and to charge a battery. The battery supplies the needed power when the sun is not shining. A typical solar photovoltaic system is also shown in 13-10. The system can also be connected with the local utility as a backup.

WIND ENERGY CONVERSION SYSTEMS

Another source of electrical energy is derived from fabricated systems using wind to drive wind turbines. These are tall steel towers with large propellors con-

nected to an electricity generating device mounted on the top. They are placed in groups on sites known to have steady dependable winds. The power is sold to the local electric utility and fed into their power grid for delivery to the consumer. A typical system is shown in 13-11. The turbine delivers DC electricity that can be converted to AC by an inverter. Usually part of the power generated is used to charge batteries that are used when the wind is not blowing. This must be done with DC power. The system can also be connected to the local utility as a backup.

COGENERATION SYSTEMS

Cogeneration systems are systems that use fossil fuel, geothermal, wind, or solar energy to produce electricity and heat as a system separate from the local utility. Cogeneration systems are connected into the local electrical utility network so that they can buy extra electricity when it is needed and sell power to the utility when it is developed.

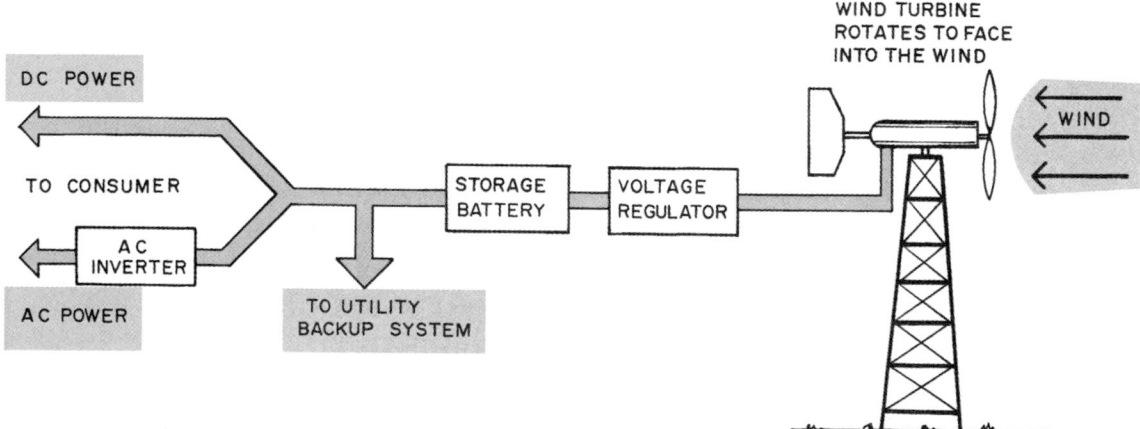

13-11 *A schematic of a typical wind energy conversion system with an AC converter that converts the DC power generated into AC.*

Specifications:

1. Audio sensors require junction box.
   Box shall be mounted flush with ceiling. Audio sensors shall not be mounted more than 25'-0" from any wall surface or more than 50'-0" apart. Minimum of one audio sensor per each enclosed area.
2. Waterproof floor outlet with bell nozzle.
3. Main junction box (12" x 12" x 4") complete with cover and located in accessible plan in equipment room or work area.
4. Outlet box (2"W x 3"H x 2-1/4"D) for transformer by Mosler. Mount as illustrated in main junction box.
5. Status control.
6. Alarm control cabinet must be in vault whenever possible.

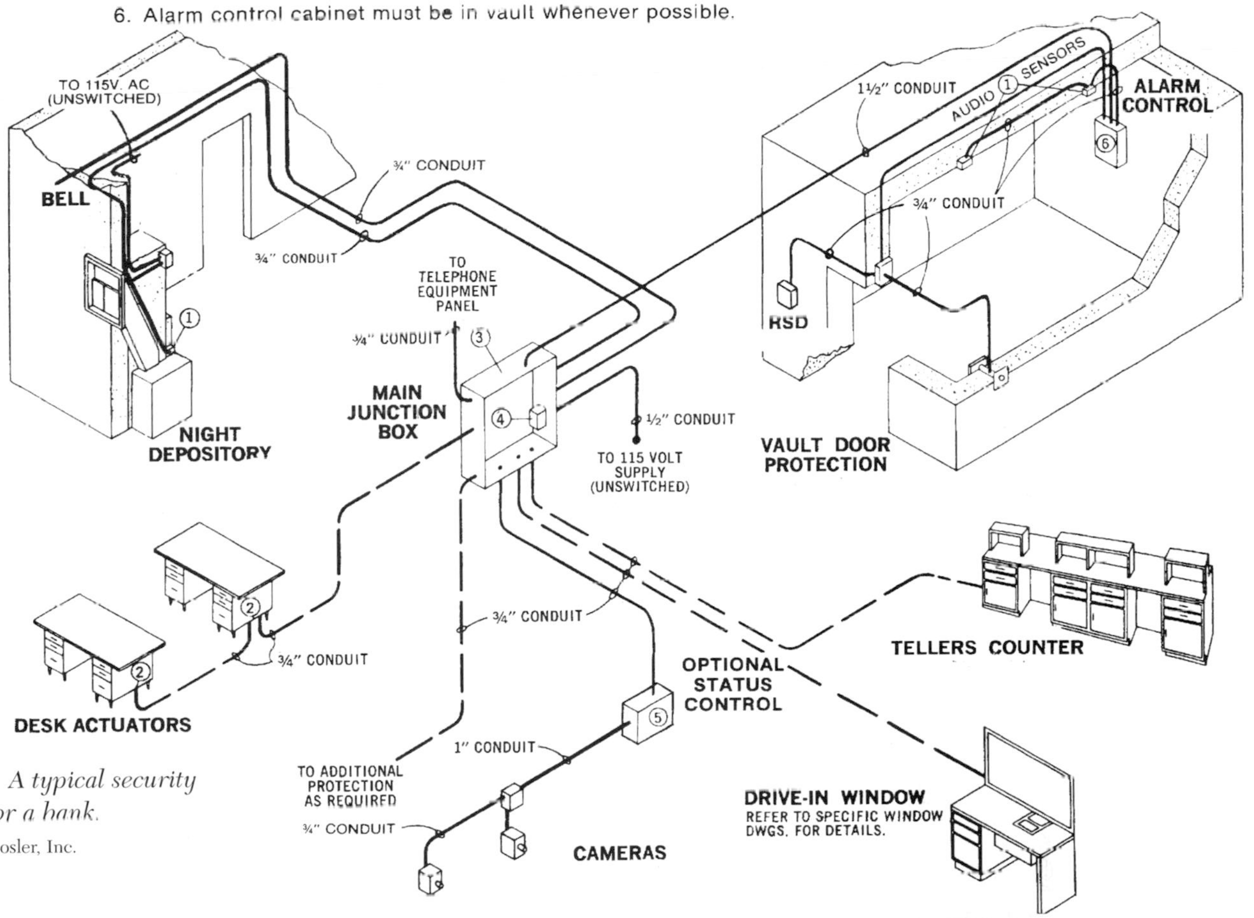

TO 115V. AC (UNSWITCHED)

BELL

¾" CONDUIT

¾" CONDUIT

①

NIGHT DEPOSITORY

TO TELEPHONE EQUIPMENT PANEL

¼" CONDUIT

③

MAIN JUNCTION BOX

④

½" CONDUIT

TO 115 VOLT SUPPLY (UNSWITCHED)

1½" CONDUIT

AUDIO ① SENSORS

ALARM CONTROL

⑥

¾" CONDUIT

RSD

VAULT DOOR PROTECTION

②
②

DESK ACTUATORS

¾" CONDUIT

¾" CONDUIT

¾" CONDUIT

OPTIONAL STATUS CONTROL

TELLERS COUNTER

TO ADDITIONAL PROTECTION AS REQUIRED

¾" CONDUIT

1" CONDUIT

⑤

CAMERAS

DRIVE-IN WINDOW
REFER TO SPECIFIC WINDOW DWGS. FOR DETAILS.

13-12 *A typical security system for a bank.*
Courtesy Mosler, Inc.

BUILDING AUTOMATION SYSTEMS

Building automation systems monitor and control various types of equipment, special purpose device, conveying system, and mechanical and electrical system. The automation system integrates all the systems within a building. Among those commonly found are systems to monitor and control energy and environmental systems, communications, security, clocks, alarms, detection devices, and door controls. Devices for moving people and materials, such as moving walks, moving ramps, elevators, and escalators are also automated.

One such system is shown in 13-12. This illustration shows a possible security arrangement for a bank. It uses audio sensors, waterproof outlets, and an alarm control cabinet, which is usually located in the vault.

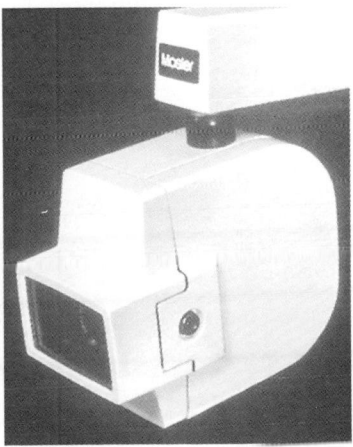

13-13 *Security cameras provide wide coverage and record the activities within their area of surveillance.*
Courtesy Mosler, Inc.

Alarm actuators are at the tellers' windows, drive-in windows, and employees' desks, and surveillance cameras are placed in strategic locations (see 13-13).

FIRE SUPPRESSION & SUPERVISORY SYSTEMS

Fire detection and alarm systems may be automatic such as a fire detector, heat/smoke detector (13-14), or a water flow switch that indicates an automatic sprinkler system has been activated. A fire flame detector is used where the materials that may burn, such as gasoline, do not generate smoke at first. They detect the presence of either infrared or ultraviolet radiation. Some systems also have manual pull stations that are activated by someone who sees a fire start (see 13-15). Various types of fire alarm are in use. The most common are horns and bells (see 13-16) and flashing lights (see 13-17). These are often combined into one unit providing sound and light signals. Fire suppression systems include automatic sprinklers (see 13-18), water fog generators that are used on highly flammable solids or liquids, and a liquid foaming agent introduced into the water in a sprinkler system. Several types of automatic gas suppression systems are available of which the use of carbon dioxide is common. A high-expansion foam is sometimes used in small compartmentalized areas. The foam must completely cover the area to put out the fire. See Division 15 for additional information on fire suppression systems.

SPECIAL SECURITY CONSTRUCTION

Security construction includes the various systems and equipment to achieve a high level of security. This includes things such as crash barriers, blast resistant products, gun ports, bullet-resistant security gates and doors, identification, and monitoring systems. A full alarm system uses closed-circuit television cameras that may be controlled by guards or automatically pan an area and zoom in when needed.

13-14 *One type of fire detector that senses smoke and sounds an alarm.*

13-15 *Two types of manually activated fire alarm station.*

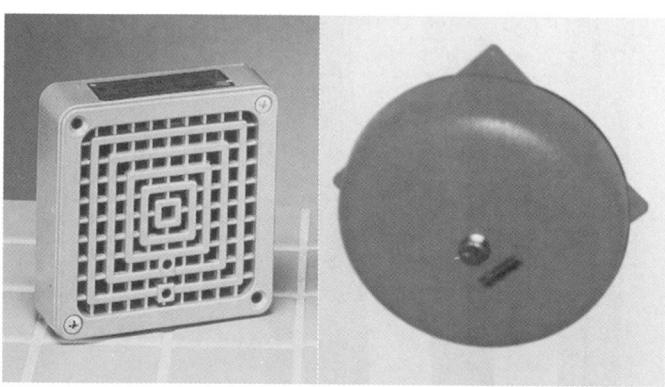

Horn   *Bell*

13-16 *Horns are used to provide a loud warning when the fire detection system senses a fire.*

Courtesy Federal Signal Corporation

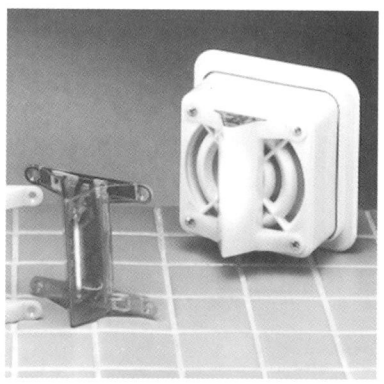

13-17 *This fire warning horn has a light attached, providing a signal for the hearing impaired.*

Courtesy Federal Signal Corporation

13-18 *One type of sprinkler head on an automatic sprinkler system.*

DIVISION 14

CONVEYING SYSTEMS

CSI MASTERFORMAT™

Courtesy Dover Elvevator Systems

There are many different types of conveying system, each designed for a special purpose. Some move materials and equipment while others move people. They range from a simple dumbwaiter used to carry mail between floors in a building to large, complex elevator systems carrying people or heavy materials and equipment.

CONVEYING-SYSTEM CODE STANDARDS

Elevators, escalators, moving walks, and dumbwaiters are designed and installed following the code requirements of the *American National Standard Safety Code for Elevators, Dumbwaiters, Escalators, and Moving Walks*, ANSI/ASME A17.1, and local building codes. Standard elevator sizes and shapes have been developed by National Elevator Industries, Inc. (NEII). They have standards such as *Elevator Engineering Standard Layouts* and *Suggested Minimum Passenger Elevator Requirements for the Handicapped*. These standards establish certain basic rules that require specific compliance and that should be coordinated with the design criteria of the complete elevator installation. Following are some examples.

The **hoistway design** criteria specify that the hoistway be enclosed its entire height with fire-resistant materials such as masonry, drywall, or concrete. Pits must also be of noncombustible material and waterproofed to prevent entry of groundwater. The top of the hoistway must be enclosed with a concrete or metal floor.

The hoistway must be designed to prevent the accumulation of gases and smoke in case of a fire. No windows are permitted and the hoistway and machinery space must be free of any pipes or ducts. Clear inside dimensions for hoistways for standard size elevators are specified in the National Elevator Industry standard, *Elevator Engineering Standard Layout*.

The **elevator machine room** must be enclosed with fire-resistive enclosures and doors. The only machinery allowed in the room is that required for the operation of the elevator. Since maintenance is required, a permanent and easy access to the machine room is required.

The **electrical equipment and wiring** must conform to the National Electrical Code, ANSI/NFRA 70. All main electrical feeds are installed outside the hoistway. The only electrical equipment allowed in the hoistway are those things directly connected with the elevator. The machine room should have permanent lighting and natural or mechanical ventilation.

Among the **escalator code requirements** pertaining to work provided by others related to the escalator are such things as requiring the floor openings for escalators to be sealed against the passage of heat, smoke, and flame in accordance with Life Safety Code ANSI/NFRA No. 101 and local codes. The bottom sides of the escalator and machinery spaces must be enclosed with fire-resistant materials. The machinery spaces must have ventilation and electric lighting. The stair treads on the escalator must have adequate lighting.

Buffers are required in the pit. Buffers are energy-absorbing units located in the pit at the bottom of the hoistway. They absorb any impact from a car that may descend below the normal lowest level.

The code requirements include factors such as required tests, emergency condition operation, venting, opening protection, signals, and signs.

ELEVATOR SYSTEMS

An elevator consists of a hoisting mechanism connected to a car or platform that slides vertically on guides on the sides of a fire-resistant hoistway. It is used to move passengers and materials between floors of a multistory building.

> **Passenger elevators**—are designed to transport people between floors.
> **Freight elevators**—carry materials between floors and only allow the operator or others needed to handle the materials being transported.
> **Hospital elevators**—have special cars large enough to transport patients on stretchers or beds and needed attendants. They can also serve as passenger elevators.

Some local codes require buildings over two stories high to have elevators to accommodate disabled persons. Elevators are not considered part of the system of egress for code purposes because they might not operate as required during an emergency, such as a fire or earthquake. People could get stranded between floors. If there is a power failure the elevator will automatically return to the lowest landing and the doors will open, allowing the passengers to exit. In the event of a fire the elevator fire service will be activated by the building's smoke alarm or by the fire service keyswitch located in the hall. When this happens all car calls are

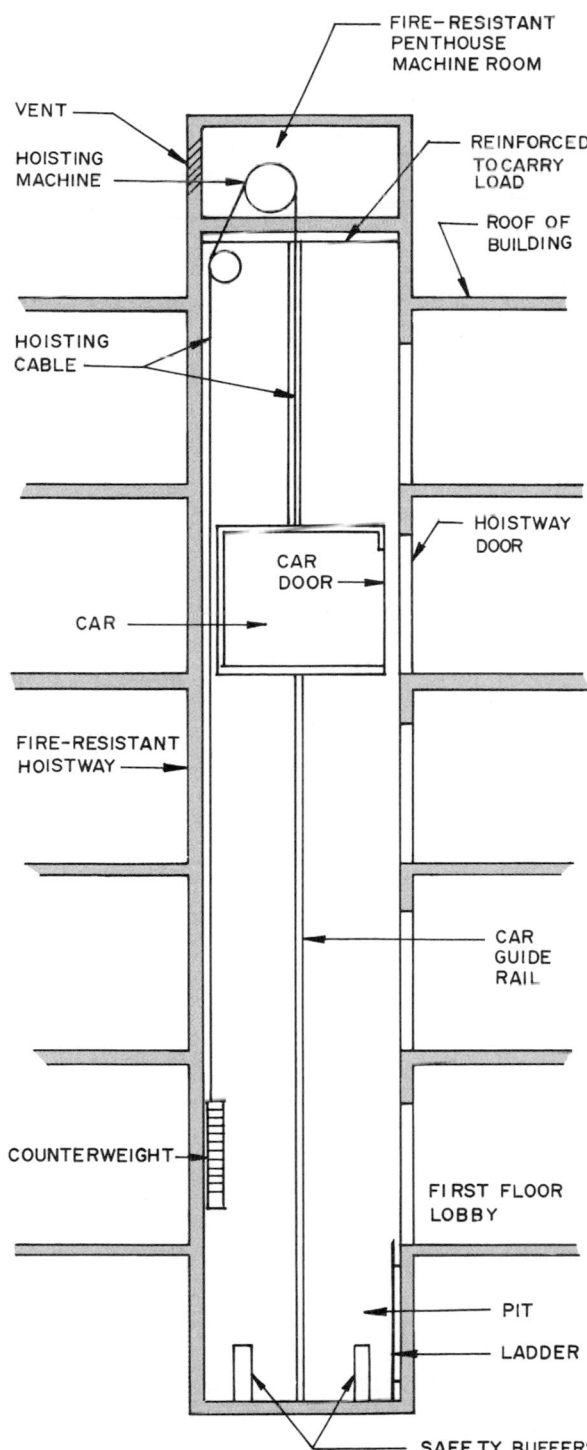

VENT

HOISTING MACHINE

HOISTING CABLE

FIRE-RESISTANT PENTHOUSE MACHINE ROOM

REINFORCED TO CARRY LOAD

ROOF OF BUILDING

HOISTWAY DOOR

CAR DOOR

CAR

FIRE-RESISTANT HOISTWAY

CAR GUIDE RAIL

COUNTERWEIGHT

FIRST FLOOR LOBBY

PIT

LADDER

SAFETY BUFFERS

14-1 *A traction type electric elevator generally has the hoisting machine at the top of the hoistway.*

canceled and the car returns to the main floor and the doors open. Since elevators are vital for use by fire-fighters and other emergency personnel, they may be reactivated for use.

Elevator Hoistways

The hoistway is a vertical fire-resistant enclosed shaft in which the elevator moves. It has a pit at the bottom and openings at each floor. The openings are protected by doors controlled by the operating system. Some types require a penthouse on the roof above the shaft. Codes typically require the shaft have a 2-hour fire rating and opening be protected with doors having a ½-hour fire rating. A metal or concrete deck is required on the top of the hoistway.

HOISTWAY DOORS

The doors of the hoistway are controlled by the operating system and must have a 1½-hour fire rating. The codes specify the types of door that are acceptable and the opening requirements. Passenger elevators generally use horizontal sliding doors; however, swinging doors may be used. Vertical sliding doors are used on freight elevators.

The doors are closed automatically when the car leaves the **landing zone.** The landing zone is an area 18 in. (490 mm) above or below the landing floor. The elevator car will not move unless all the doors on all levels are closed and locked. Doors have locks that prohibit them from being opened by someone on the landing side. Emergency access is available for maintenance and rescue personnel.

MACHINE ROOMS

Machine rooms are part of the hoistway and provide a fire-resistant enclosure for the installation of required hoisting machinery, controls, pumps, and hydraulic oil storage. Since control systems are computerized, the area should be air conditioned to control the temperature. Nothing not involved with the operation of the elevator is permitted in the machine room. The size of the machine room will vary depending on the type of equipment. Machine rooms for traction-type elevators are generally in a penthouse on the roof over the hoistway (see 14-1). The floor must be designed to carry the weight of the machinery plus loads required for lifting and lowering the car. In some cases traction elevator machine rooms can be located in the pit at the bottom of the shaft.

Hydraulic elevator machine rooms are usually located to the side of the hoistway (see 14-2). They contain the hydraulic equipment and controls.

VENTING HOISTWAYS

During a fire the elevator hoistway may become a vertical flue for gases and smoke and be used to carry away these noxious materials by venting the hoistway at the top. However, building designs utilizing other means of containing and venting smoke and gases from the site of the fire to prevent them from moving to unaffected floors are being used. Codes may require the hoistway to be vented, but other design considerations for controlling smoke must be considered as the hoistway is designed.

Elevator Cars

The elevator car is a platform enclosed by walls and a roof that is designed to carry passengers, materials, packages, and many special uses such as for hospital beds and stretchers. The car contains lighting, controls, venting units, handrails, telephones, and various types of wall, ceiling, and floor finish materials. **Passenger elevators** are finished to be attractive yet durable. **Freight elevator** cars must have durable, abrasion resistant interior finishes. Typically they have steel floors and walls.

The doors on the car are automatically operated and cannot be opened while the car is moving or is outside the landing zone. The car will not move if the door is open. Elevator cars are available that open to the front and to the front and rear. Double-entrance cars are widely used on hospital and freight elevators. Elevator car doors may be center-opening, two-speed sliding doors, or single sliding doors (see 14-3). The center-opening doors provide the quickest entry and exit. They are always used on high speed systems. Two-speed sliding doors provide the widest possible opening but are slower to open and close than center-opening doors. Single sliding doors are the most economical and the slowest. They move right or left depending upon the car design, and the opening width is limited by the width of the car.

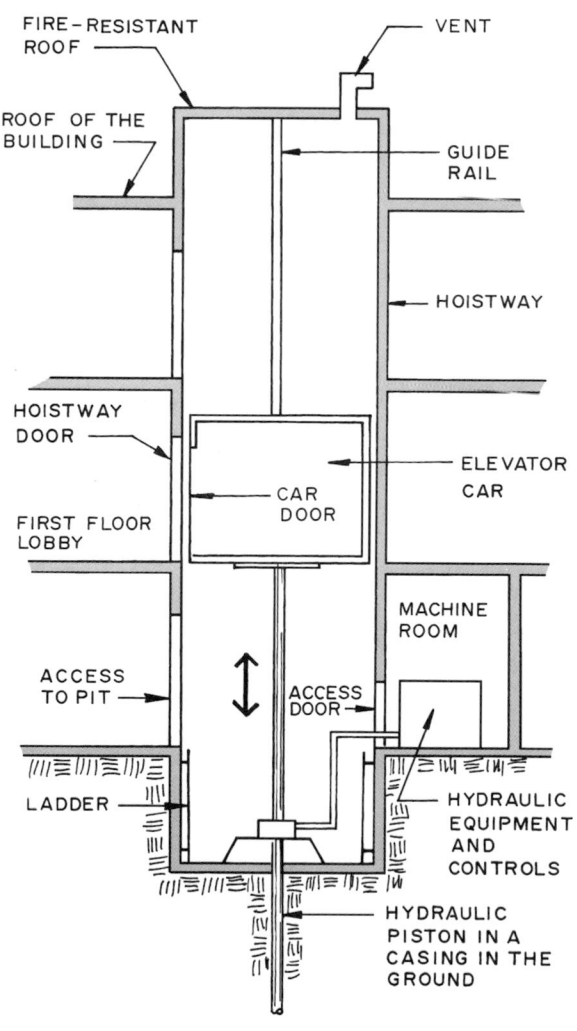

14-2 *A hydraulic elevator typically has the machine room on the side of the hoistway.*

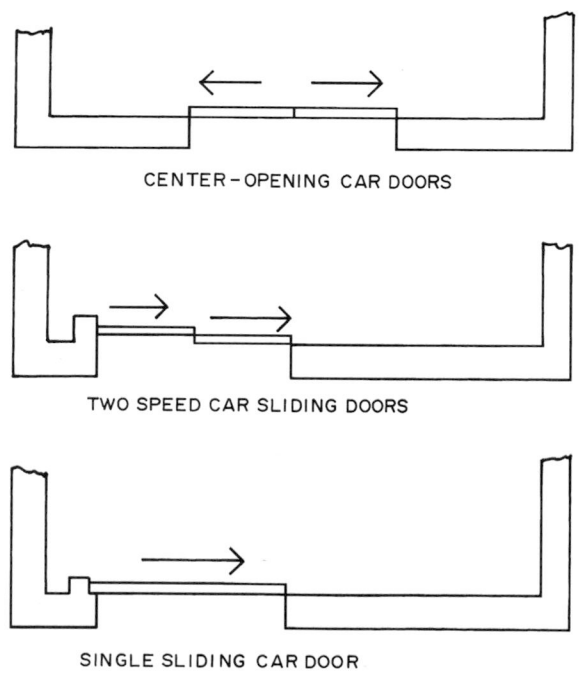

14-3 *Typical types of car door action.*

The car contains a panel with several controls. It must be located so it can be reached by a person in a wheelchair. It has an alarm button sounding an alarm outside the hoistway and an emergency stop switch. The telephone provides a means of communication with someone outside the hoistway. The panel has buttons used to indicate the floor desired and to hold the door open when the computer tries to close them (see 14-4).

The capacity of a car is indicated by the maximum number of people it is designed to carry. The capacity is the total load in pounds divided by 150. For example, an elevator rated at 1500 lbs. would be marked to carry a maximum of 10 people.

Car sizes vary with the different manufacturers, but typical sizes for passenger elevators range from 6'-0" × 4'-0" (1.8 × 1.2 m) to 8'-0" × 6'-0" (2.4 × 1.8 m). Door widths are typically 3-0" to 4'-0" (.9 to 1.2 m) and 7'-0" to 9'-0" (2.1 to 2.7 m) high. Hospital elevations are in the range of 9'-0" × 7'-0" (2.7 × 2.1 m). Freight elevators range from 5'-0" × 7'-0" (1.5 × 2.1 m) to 12'-0" × 16'-0" (3.7 × 4.9 m). They have clear door widths from 5'-0" to 12'-0" (1.5 to 3.7 m).

Electric Passenger Elevators

Electric passenger elevators suspend the car from wire ropes and use weights to counterbalance the car. The car is guided by vertical guide rails. The electrically driven hoisting mechanism may be located at the top or bottom of the shaft. The two types in common use are geared and gearless traction mechanisms. A geared electric passenger elevator is shown in 14-5.

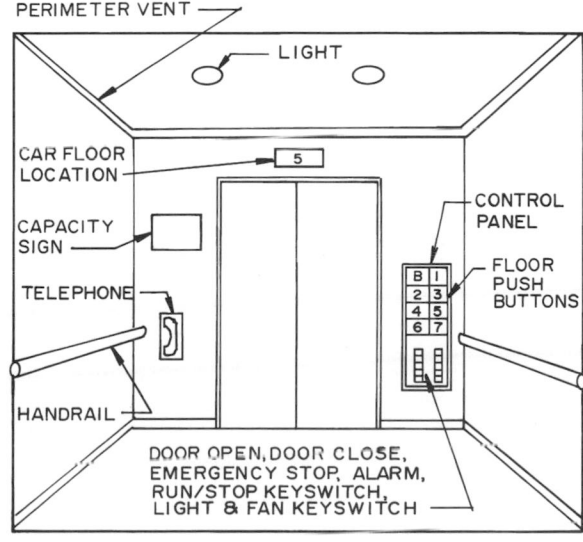

14-4 *Typical elevator car controls.*

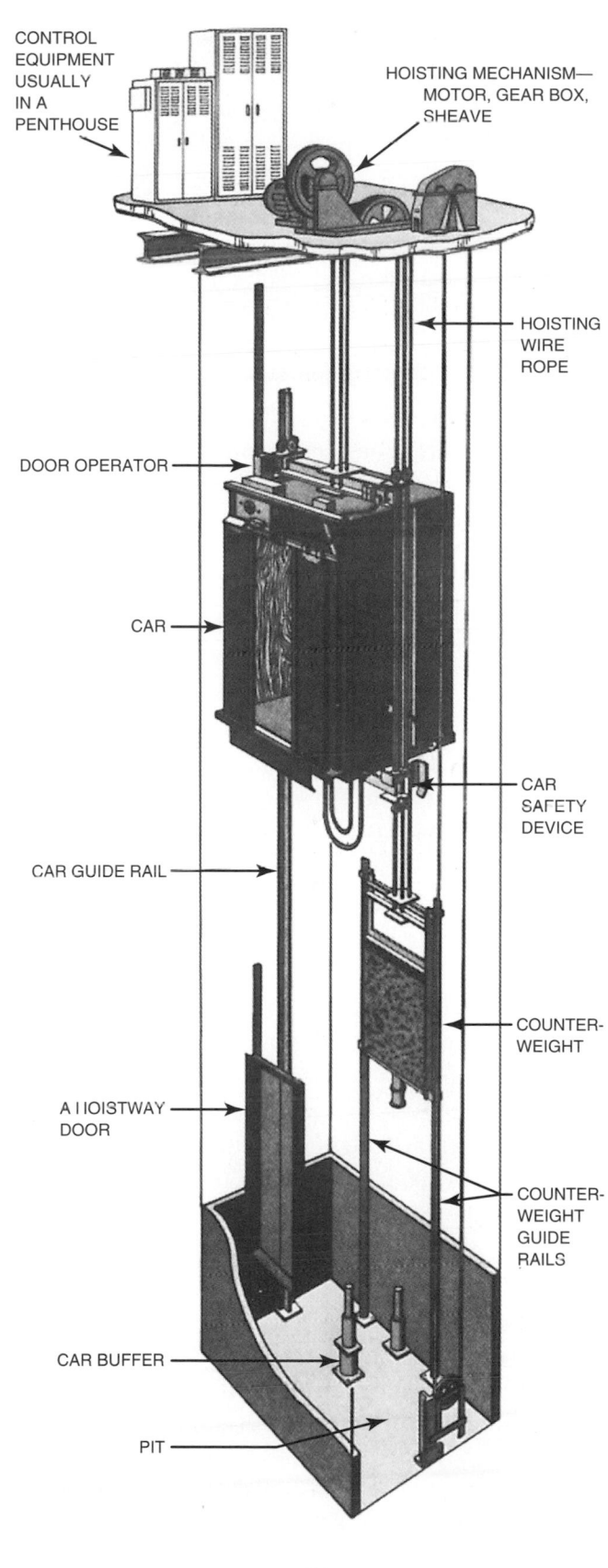

14-5 *This is a geared electric passenger elevator that uses an electric motor to drive a gear box that regulates the speed of the traction sheave.*

TRACTION DRIVING MECHANISMS

Electric elevators are powered by traction machines consisting of an electric motor connected to a driving sheave. This may be a direct connection or through a series of gears. A wire rope runs through grooves in the face of the sheave or the traction is provided by friction.

Gear-driven traction—machines provide slower rising speeds and are used when slower speeds are desired. The gearing may be through a helical gearbox or with a worm gear.

Gearless direct drive—machines provide high speeds and are usually used on high-rise buildings. They have the traction sheave connected directly to the motor.

GEARED TRACTION ELEVATORS

This hoisting mechanism uses a worm gear to drive a large spur gear that is connected to the traction sheave (refer to 14-5, on the previous page). A sheave is a pulley that has a grooved rim for retaining a wire rope used to transmit force to the rope. The wire rope runs over the traction sheave and is moved as it rotates. This type of drive is used when low speeds and high lifting capacity are required. The speed can be changed by varying the size of the spur gear, which changes the gear ratio.

Typical car rise speeds for geared traction **passenger elevators** range from 350 to 500 ft./min. (106 to 152 m/min.). The range of lifting capacity is typically from 2000 to 4500 lbs. (900 to 2025 kg). The hoisting mechanism is usually housed in a penthouse on the roof.

Geared traction **freight elevators** typically have a rise from 50 to 200 ft./min. (15 to 60 m/min.) and carry loads up to about 20,000 lbs. (9072 kg).

GEARLESS TRACTION ELEVATORS

A gearless traction elevator is set up much like the geared traction elevator. However, the traction sheave and brake are mounted directly on the motor shaft. The speed of rotation of the traction sheave (which contains the wire rope) is the same as the speed of the motor. The speed can be varied by using a DC electric motor built to run at the speed required. Car rise speeds can range from 500 to 1200 ft./min. (152 to 366 m/min.). Gearless elevators are also run using a motor generator drive. The range of lifting capacity is typically from 2000 to 7000 lbs. (900 to 3150 kg). The hoisting mechanism is usually housed in a penthouse on the roof.

ELECTRIC ELEVATOR CONTROL

The elevator control system regulates the starting, stopping, safety devices, speed of movement, direction of movement, acceleration, and deceleration of the car. Two types of control are in general use multivoltage and variable-voltage, variable-frequency control (VVVF).

Multivoltage control is used with machines using DC motors. It controls the speed by varying the voltage supplied to the motor armature. It provides a smooth regulation of speed and is widely used on passenger elevators.

VVVF control is used to control AC motors and produces a smooth overall operation of the car. It is more efficient than multivoltage controls and DC motor operation and is being increasingly used instead of DC controls and motors.

CAR SAFETIES

Car safeties are used to stop the movement of the car and hold it in position. A governor monitors the speed of the car, and if it exceeds the safe car speed a car safety is activated. A car safety is a device that applies brake shoes against the guide rails, stopping the car. The gov-

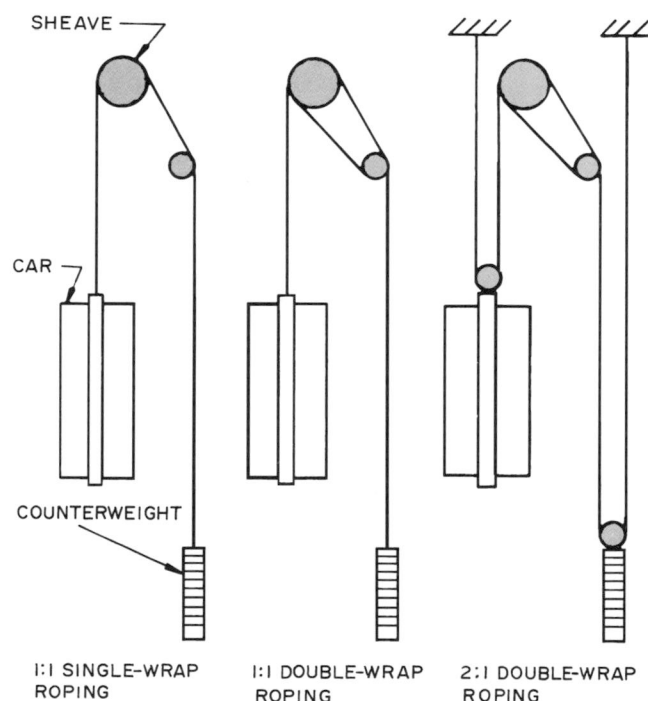

14-6 *Several types of roping system used on electric traction elevators.*

ernor also switches off the electrical power to the motor. Some types also activate a brake shoe to the motor drive shaft.

ROPING

Traction-type machine cars must be suspended from at least three hoisting ropes (ANSI 17.1). The wire rope consists of steel strands laid helically around a hemp core. Each steel strand is made up of steel wires wrapped helically around a steel wire core.

The roping of the traction-type elevation greatly effects the speed of the car and the loads on the hoisting wires. Typical roping systems are shown in 14-6.

Single-wrap roping has the wire pass over the driving sheave one time and on to the counterweight. The deflection sheave moves the wire clear of the moving car. This provides a 1:1 ratio, which means it will move the car 1 ft. (305 mm) for every 1 ft. (305 mm) of travel along the circumference of the sheave.

Double-wrap roping has the wire wrapped twice around the driving sheave and a secondary sheave and on to the counterweight. Double-wrap roping can be used to provide a 1:1 or 2:1 ratio. The 1:1 double-wrap ratio causes less wear on the wire but increases the load on the sheave. The 2:1 double-wrap roping is used for heavy-load elevators. This system moves the car 1 ft. (305 mm) for every 2 ft. (605 mm) of rope travel along the circumference of the sheave. Other roping systems are in use.

COUNTERWEIGHTS

Counterweights are steel plates that slide on vertical steel guides. As the elevator rises they lower and as the elevator lowers they rise. Counterweights generally equal the weight of the unloaded car and hoisting wires plus a percentage of the load capacity of the elevator. Since they rely on gravity, they provide help in raising the car, thus reducing somewhat the power required.

OPERATING SYSTEMS

Elevator operating systems control the operation of the elevator. They range from simple to complex, automatic systems.

Typical **operator-controlled systems** include the car-switch and signal systems. The **car-switch system** is operator controlled. The operator controls the direction of travel and initiates when it is to move. The **sig-**

nal system is operator controlled. The operator presses buttons indicating the floor stops and then presses a start button. The car automatically stops at each floor for which a button was pressed.

Automatic operating systems—do not require an operator. Signals are initiated by the passengers or by an automatic operating device. Typical automatic operating systems include single, selective collective, and group.

Single automatic systems—start when the passenger pushes a button indicating the floor desired (refer to 14-4). The car moves automatically to that floor, stops, and the doors open. The car is called to a floor by someone wanting to board by pressing a button on the wall by the doors to the elevator.

Selective collective automatic system—has two call buttons, up and down. When the "up" button in the car or on the hall wall has been pressed, the car moves up but stops at every floor where the hall wall "up" button has been pressed. After it reaches the top floor, it descends, stopping at each floor where a hall wall button was pressed or for which a floor button in the car was pressed (see 14-7).

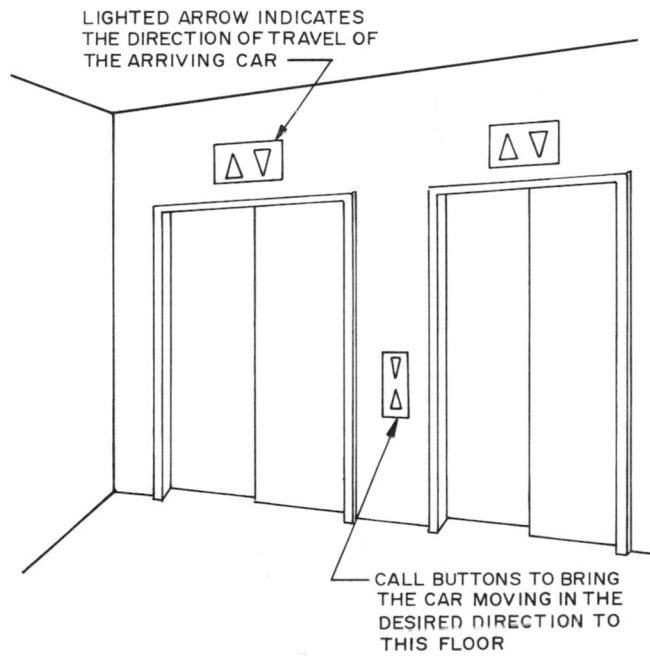

14-7 *Call buttons on the wall bring the elevator to the floor. The overhead direction arrows indicate the direction the arriving elevator is moving.*

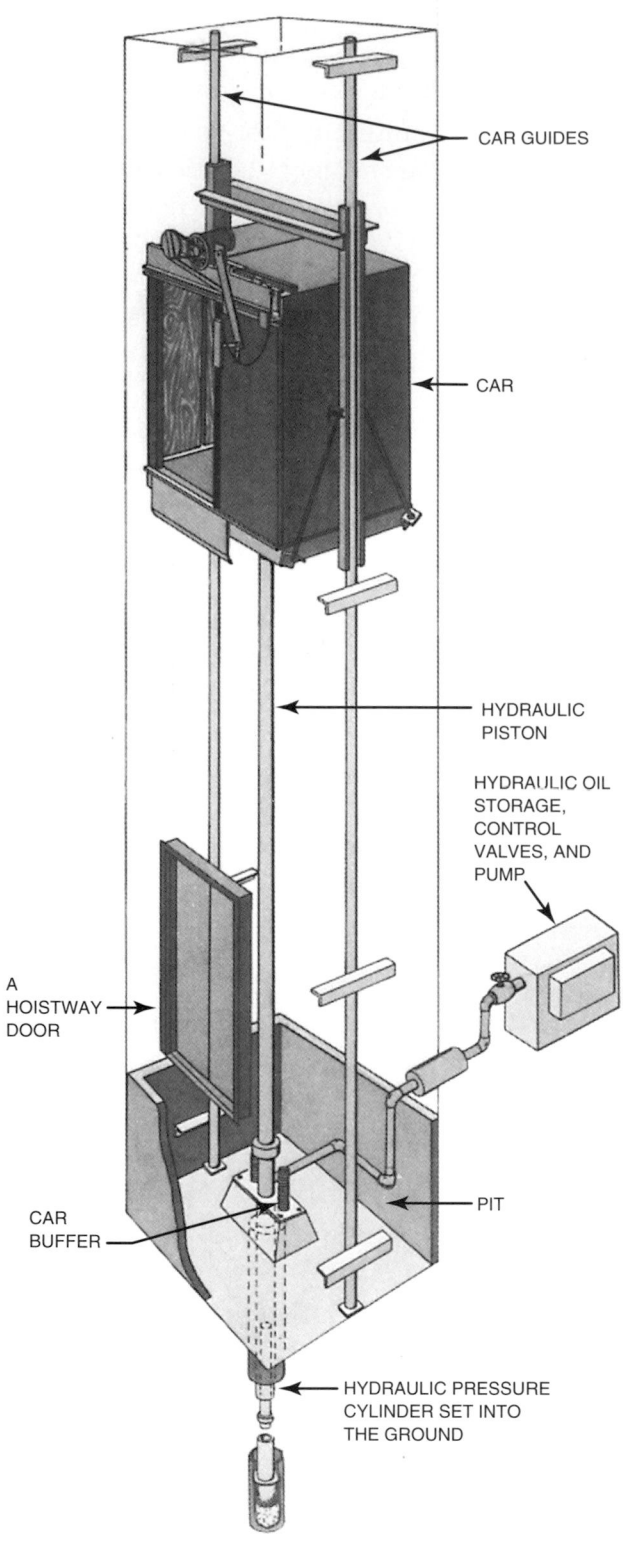

CAR GUIDES

CAR

HYDRAULIC PISTON

HYDRAULIC OIL STORAGE, CONTROL VALVES, AND PUMP

A HOISTWAY DOOR

CAR BUFFER

PIT

HYDRAULIC PRESSURE CYLINDER SET INTO THE GROUND

14-8 *A hydraulic elevator moves the car with a long piston controlled by a hydraulic pump and valves.*

Courtesy Otis Elevator Company

Group automatic systems—control the operation of several cars all serving the same floors using a supervisory control system. The control system automatically decides which cars to send to various floors and coordinates the flow of traffic. The cars are dispatched from the first floor on predetermined intervals and whichever car is nearest a floor where a passenger is waiting stops to make the pickup.

Supervisory control system—is able to make adjustments in car flow and direction to handle times when requirements vary, such as late afternoon in a high-rise office building when many workers are trying to leave at about the same time.

While systems vary it may have a database utilizing traffic information to decide which pattern of car utilization will produce the shortest waiting time. For example, in peak "up" traffic times the cars may make fewer stops per round trip, returning to the lobby more often. A car may be assigned to serve only one group of floors and return quickly to the lobby. Another car can serve another group of floors. This reduces the stops for each car and gets them back to the lobby quickly for another trip. The reverse can occur when the down trips become heavily loaded. It is necessary that the control system display in the lobby indicate which floors each elevator will serve.

Hydraulic Elevators

Hydraulic elevators are used for low-rise installations. The car is mounted on top of a piston that slides inside a hydraulic pressure cylinder. The hydraulic oil is pumped by an electric pump to the bottom of the piston, forcing it to rise and moving the car with it. When the car reaches the desired floor, the pump is automatically stopped and the oil under pressure holds the car at that level. The car is lowered by releasing oil from the pressure cylinder back into the storage tank. The installation requires that the pressure cylinder be sunk into the ground a distance equal to the length of the cylinder (see 14-8). Hydraulic elevators do not require a penthouse on the roof to house the hoisting mechanism.

The speed of rise of **hydraulic passenger elevators** ranges from 100 to 150 ft./min. (30 to 46 m/min.). The range of lifting capacity is typically 1500 to 4000 lbs. (675 to 1800 kg). **Hydraulic freight elevators** rise at speeds of 50 to 100 ft./min. (15 to 30 m/min.) and carry loads from 3000 to 12,000 lbs. (1350 to 5400 kg). Since gravity is at work, the down movement is faster than the rise.

14-9 *Freight elevators are designed to carry the anticipated loads and resist actions from forklifts and other devices used to load them.*
Courtesy Dover Elevator Systems

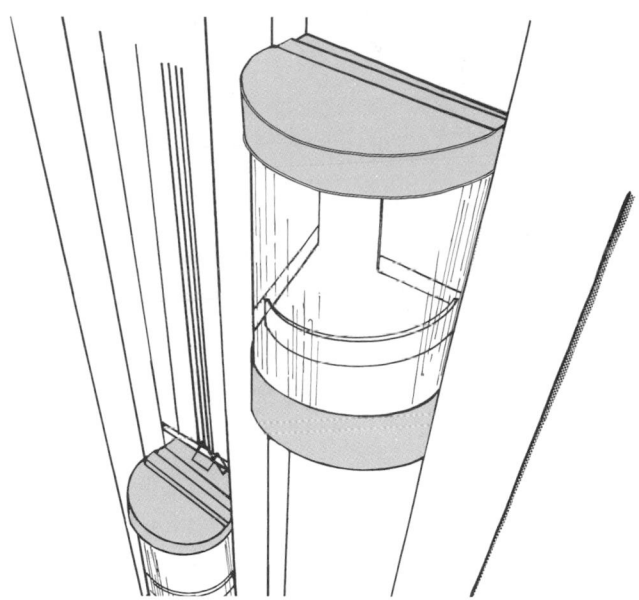

14-10 *Observation elevators are glass enclosed and travel outside of a hoistway or in a hoistway open on one side.*

Freight Elevators

Freight elevators are designed to carry general freight, loads carried to industrial trucks, motor vehicles, and heavy concentrated loads. Hydraulic elevators may be used in low-rise buildings and electric in mid- and high-rise buildings. There are three classes of freight elevator specified in the *American National Standard Safety Code for Elevators, Dumbwaiters, Escalators, and Moving Walks,* ANSI/ASME A17.1.

Class A—is limited to carrying a maximum of one-quarter of the rated load. It is for general freight hand loaded and unloaded with a lightweight hand truck.

Class B—handle only motor vehicles.

Class C—elevators are designed to handle heavy concentrated loads.

> **Class C1**—elevators can carry materials plus industrial trucks (forklifts, pallet trucks, etc).
>
> **Class C2**—elevators are used where materials are loaded by industrial trucks but the trucks are not carried with the materials.
>
> **Class C3**—elevators carry heavy concentrated loads other than trucks.

While there are some premanufactured freight elevators available, many have to be custom designed. The size of the car and the loading capacity vary widely with industrial requirements. The platform is designed to carry the anticipated load plus extra weight and movement when it is being loaded. For example, a truck loading materials can cause sideways movement of the car and shock if the load is not slowly lowered to the floor. If the load is placed off center in the car it produces eccentric loading, which must be considered as the system is designed.

The doors on the car are often an open steel mesh and lift vertically. The door to the hoistway is typically a vertical bi-parting door. They are solid doors and have a locking system that will not allow the doors to open until the car arrives at the floor (see 14-9).

Generally freight elevators are automatically operated as described for passenger elevators. Some operator-controlled systems are available.

Freight elevators are designed and manufactured in accordance with ASME A17.1.

Observation Elevators

Observation elevators move up a hoistway secured to the exterior of a building. The cars are designed to have considerable glass in the walls so a view of the surrounding area is available as the elevator rises. The hoisting machinery is placed out of view. The elevators may be geared, gearless, or hydraulic. The lower end of the hoistway (at the ground) must be enclosed with a safety barrier (14-10).

Residential Elevators

Residential elevators are available in a range of sizes including cars large enough to carry a wheelchair (see 14-11). While designs vary, they may use hydraulic or the electric traction system similar to that described for passenger elevators. The car in the electric system is guided by channel guides. The doors have safety locking devices preventing them from opening until the car is at the floor. The hydraulic unit uses a piston to raise the car.

The cars are steel reinforced and available with a variety of interior wall, floor, and ceiling finishes. Car sizes range from 36 in. (10,980 mm) to 42 in. (12,810 mm) square and 36 × 48 in. (10,980 to 14,640 mm) for wheelchairs. Most handle loads up to 450 lbs. (202.5 kg).

14-11 *This elevator is used in commercial and residential applications. It is large enough to handle a wheelchair. It is built into the wall structure.*

Courtesy Access Industries, Inc.

AUTOMATED TRANSFER SYSTEMS

Automated transfer systems provide for vertical distribution of materials as well as horizontal transfer systems. These are widely used in hospitals, clinics, office buildings, hotels, and manufacturing plants. Among these are dumbwaiters, tote box and cart transfer systems, and horizontal tote box transfer systems.

Dumbwaiters

A dumbwaiter is a mechanism used to raise and lower a small car vertically within a building. It is used in hospitals, restaurants, libraries, and office buildings to move mail, supplies, and materials (such as food, medicine, books) from one floor to another. The size of the cars is controlled by local and national codes. Standard heights are 3, 3½, and 4 ft. (915, 1067, and 1220 mm) and the maximum platform size is 9 sq. ft. (.837 m^2). The units are designed to carry loads ranging from light-duty lifts carrying 25 to 50 lbs. (11.25 to 22.50 kg) to heavy duty types carrying up to 500 lbs. (225 kg). They may be manually or electrically powered.

Manually operated units—have an endless rope connected to a large pulley that is connected by gears to a pulley connected to the hoisting mechanism. These have automatic braking mechanisms. They are usually limited to two-story applications.

Electrically-powered dumbwaiters—are used on buildings of any height. They are available as drum type, traction type, or hydraulic as explained in the section on passenger elevators. The drum type has a maximum height of rise of 40 ft. (12.2 m) while the traction-type height is unlimited.

Electric dumbwaiters have a speed of 50 to 150 ft./min. (15.25 to 45.75 m/min.). The higher speeds are typically used in buildings over 50 ft (15 m) high. Some types permit the standard speed, as 50 ft./min. (15 m/min) to be reduced to as low as 25 ft./min. (7.6 m/min.) when carrying fragile items. The cars available can open on the front, front and rear, or front and side. Standard doors are bi-parting; however, slide-up, slide-down and swinging doors are available. Doors may be power or manually operated.

Dumbwaiters may be counter loading or floor loading. Floor loading enables carts to be moved directly on the dumbwaiter (see 14-12).

The electric traction and drum dumbwaiters ride on vertical rails secured to each floor with brackets forming the vertical structure. The lifting mechanism is typically located on the top of the shaft.

Cart & Tote Box Transfer Systems

Cart and tote box transfer systems use high-performance dumbwaiter and elevator lift equipment that may be used in connection with a horizontal transfer system. The systems available will lift up to 1000 lbs. (450 kg).

Carts—are various sizes but may be up to 30 in. (9150 mm) wide, 55 in. (16,775 mm) deep, and up to 65 in. (19,825 mm) high (see 14-13).

Tote boxes—are about 15 in. (4575 mm) wide, 20 in. (6100 mm) long and 10 in. (3050 mm) deep. They can be carried on standard dumbwaiters. Small carts with weights not exceeding the capacity of a standard dumbwaiter can be carried by them. Tote box systems use counter-high doors.

Horizontal transfer systems move tote boxes and other items into the vertical transfer system. Typically items, such as mail, are placed in the tote box, which is placed on the conveyor. The conveyor system is installed in front of the vertical transfer system door. The boxes are placed on the conveyor and the operator presses the desired floor destination button. When a car arrives at the receiving floor the door opens automatically and a car transfer device loads the tote box into the car of the tote box transfer system. The door closes and the car moves up or down to the desired floor, where the door opens and is automatically unloaded onto a conveyor.

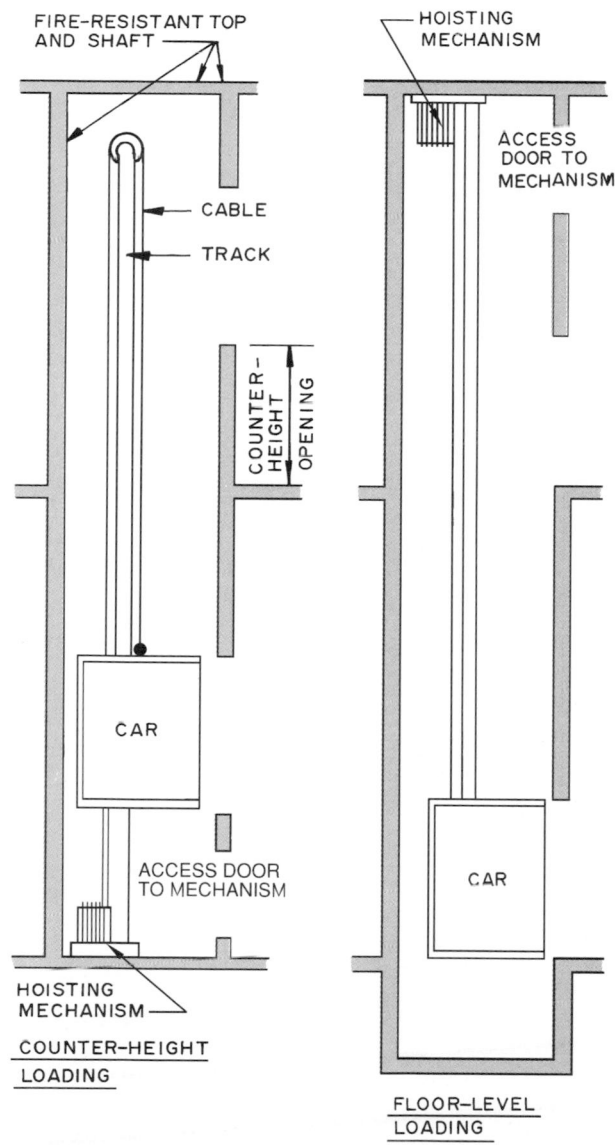

14-12 *Dumbwaiters may have floor-level or counter-height loading.*

14-13 *The cart transfer system uses floor-level loading and carts designed to fit into the dumbwaiter.*
Courtesy Matot, Inc.

14-14 *This porch lift is designed for exterior use on commercial and residential buildings.*

Courtesy Access Industries, Inc.

WHEELCHAIR LIFTS & STAIR LIFTS

Wheelchair lifts are used on interior and exterior locations. They are made up of a steel platform with steel sides and a front gate that lowers to form a ramp to help load the wheelchair (see 14-14). The ramp has a rubber skid proof surface. Wheelchair lifts are operated by an electric motor and have an automatic stop switch activated whenever the person in the wheelchair releases the control starting movement. The lift will not operate until all entry and exit doors are closed.

Another type of wheelchair lift moves the passenger and wheelchair up the stair to another level. This system uses a platform with side enclosures to move the wheelchair up the stair riding on a rail system fastened to the wall or stair treads. It can travel up a multilevel straight stair. Two lifts are used when it is necessary to turn a 90° or 180° corner.

Stair lifts are used to move people who do not use a wheelchair but have difficulty climbing stairs. The lift consists of a chair that runs along a track that may be mounted on the stair or along the wall of the stair well. The chair can move around corners and across stair landings (see 14-15). The rail can continue past the top of the stair, permitting the passenger to get off the chair a safe distance from the stairway. The lifts are electrically powered and a series of gears and shafts provide the motion.

ESCALATORS

Escalators are inclined, continuous, power-driven stairways used to move passengers up or down between floors. When planning a building the escalator should be located where it will be most accessible to traffic. Adequate space must be allowed at each landing since people will be concentrated in this area (see 14-16). An alternate method for moving between floors, such as stairs parallel with the escalator, is necessary when codes require it. This provides access should the escalator fail or be under repair. Escalators may be used as a required means of egress if they meet all of the requirements for an emergency egress stairway. This includes such things as having the floor openings enclosed, providing fire and smoke protection, and having an approved sprinkler system. Escalators not serving on a required exit should have the floor openings enclosed or protected by one of the following systems: a **partial enclosure** that uses self-closing doors; an automatic self-closing rolling shutter, a system of high-velocity **water-spray nozzles;** or an **automatic water curtain** with an air-exhaust system.

14-15 *A chair lift moves a person up a stair while comfortably seated.*

Courtesy Access Industries, Inc.

14-16 *This escalator has a large open area at each landing to provide needed space as people crowd on and off.*
Courtesy Otis Elevator Company

Escalators can move large numbers of people much faster than elevators. Typical speeds are 90 to 100 ft/min. (27 to 30 m/min.) and most can move 2000 to 4000 or more people per hour. However, they are seldom used to move passengers over more than 5 or 6 stories.

Escalator Components

An escalator has a welded-steel truss structural frame. The stair is a series of moving steps of cast metal, grooved to provide safe footing. The treads and risers are secured to a continuous chain that is moved by an electric-driven geared unit. Each side of the stair has a solid balustrade covering the ends of the stair and supports a handrail. The handrail moves at the same speed as the stair. The escalator has electronic control devices for operation and emergency situations (see 14-17).

Standards & Safety

Escalator standards are available in the publication *American National Standard Safety Code for Elevators, Dumbwaiters, Escalators, and Moving Walks*, ANSI/ASME A17.1 and the Life Safety Code of the National Fire Protection Association. In Canada the requirements are in CAN 3B-44.

14-17 *Escalators are built using heavy structural components and drive mechanisms that can stand constant year-round operation.*
Courtesy Dover Elevator Systems

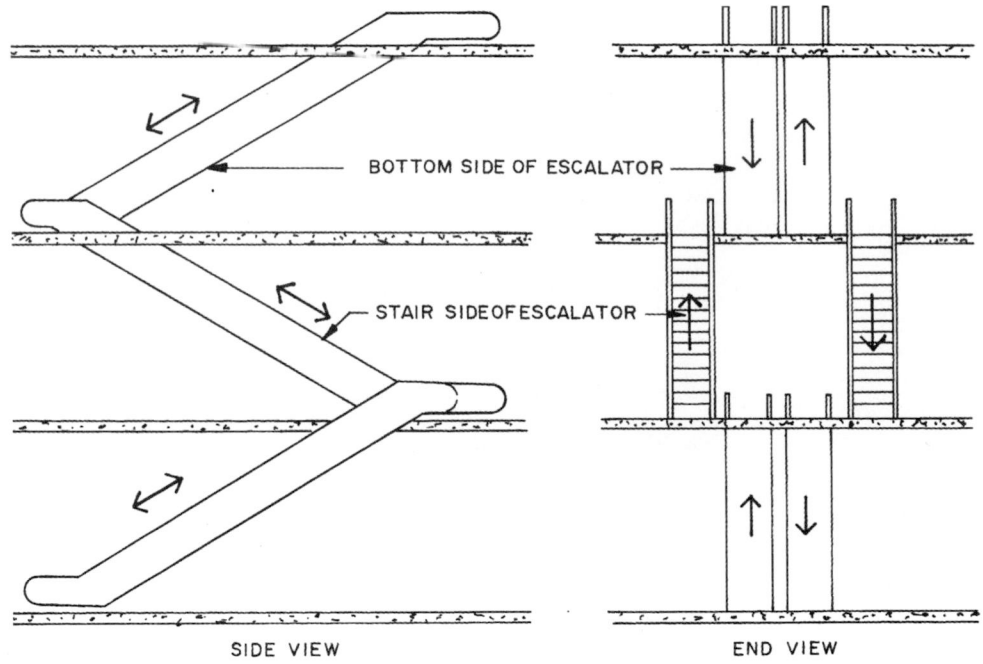

SIDE VIEW   END VIEW

BOTTOM SIDE OF ESCALATOR

STAIR SIDE OF ESCALATOR

14-18 *Escalators may be installed in a parallel arrangement.*

The safety features for escalators included by the codes and the various manufacturers include emergency stop buttons, broken step and drive chain switches, a brake that is electronically released when a power failure occurs, a switch to prevent the speed from exceeding the design speed, step lights, switches to stop the escalator if a handrail breaks, switches to control the handrail speed, a switch to shut it down if a step breaks, smooth balustrades protecting the sides of the step, and landing plates that protect against items' getting caught as the steps flow under the floor.

Installation Examples

Escalators are often installed in pairs with one for the up direction and one heading down. They may be located in a parallel arrangement (see 14-18) or in a parallel crisscross pattern (see 14-19).

Escalator Sizes

The stair sizes are established in ANSI/ASME A17.1. The width of escalator treads available is typically 24, 32, and 40 in. (0.6, 0.8, and 1.0 m). They are built on an angle of 30° and have various maximum rises depending upon the design. Rises of 20 to 30 ft. (6.1 to 9.2 m) are common for a single unit. The installation in 14-20 shows design sizes for one style of escalator.

MOVING WALKS & RAMPS

Moving walks are horizontal conveyor belts designed to move people. They may have a slight rise or fall seldom exceeding 5° (see 14-21). **Moving ramps** move people up or down an incline with a maximum slope of 12°. They frequently connect with moving walks. They are used where large numbers of people need to be moved over long distances, such as in an airport terminal.

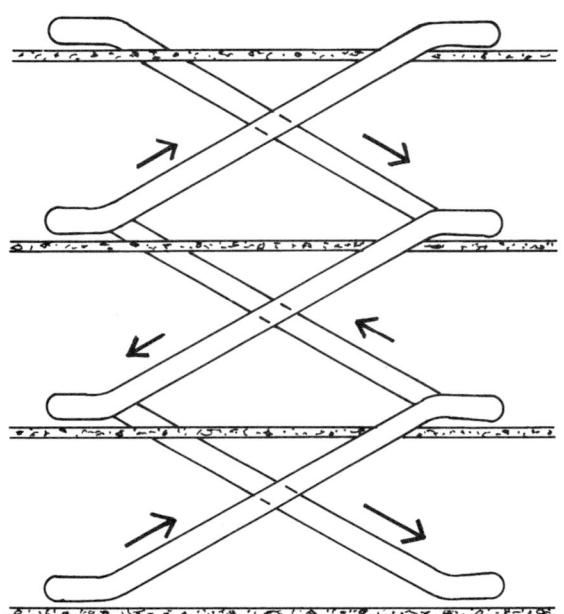

14-19 *This is a typical crisscross escalator installation.*

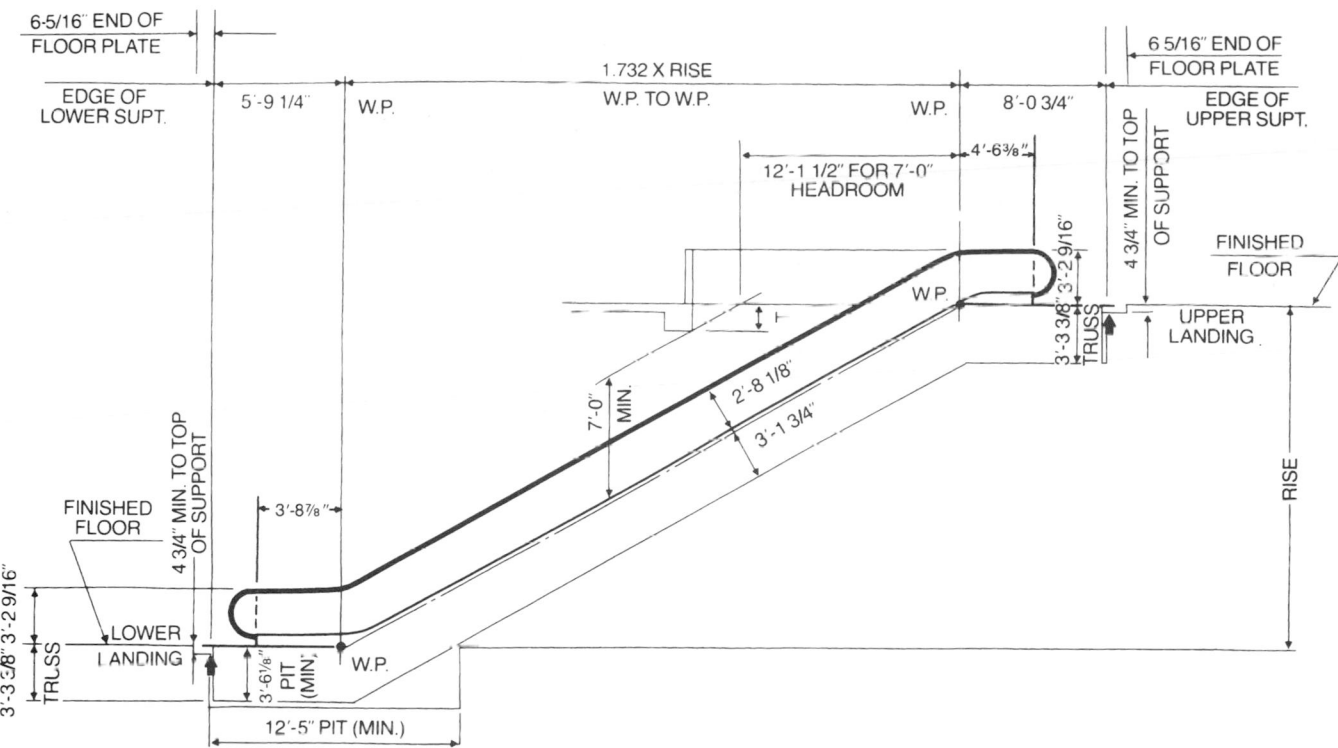

14-20 *Typical escalator design sizes.*
Courtesy Otis Elevator Company

14-21 *The pedestrians' view as they approach a moving walk.*
Courtesy Otis Elevator Company

Typical widths of the moving flexible rubber-covered endless belt for moving walks and ramps are 24, 32, and 40 in. (0.6, 0.8, and 1.0 m). The 24 in. belt accommodates one adult, the 32 in. belt provides room for an adult and a child or an adult with a shopping cart, and the 40 in. accommodates two adults or one adult with luggage. The system has a mechanical and electrical safety system and controls similar to that used on escalators and is electrically driven.

SHUTTLE TRANSIT

Shuttle transit systems provide horizontal transportation for distances beyond the practical limits of moving walks. They have many applications, such as linking office or retail areas to remote parking facilities or moving passengers between terminals in a large airport. Business and industrial parks can use them to provide rapid internal transportation (see 14-22). The system operates much like the elevator. The cars may be dispatched on a regular schedule or on call (see 14-23). Each boarding station has a sign showing the scheduled arrival times. The cars are unmanned and operate on automatic controls. The system uses standard elevator gearless traction machine drives and cable equipment. The steel cable is attached along the side of the shuttle. The steel guide rails and power rails are located adjacent to the running surface. The verti-

14-22 *This Otis Shuttle is one of a number of people movers used in areas where people must be moved over distances farther than they would like to walk. People are moved rapidly and comfortably.*

Courtesy Otis Elevator Company

cal load is supported by a cushion of air developed between pads on the bottom of the car and the guideway running surface.

MATERIAL CONVEYORS

Material conveyors are used to move items, such as packages, luggage, parts, aggregate, and concrete within a building or on the construction site. They include belt, roller, and segmented moving surfaces. The belt conveyor may be flat or troughed and both are powered by an electric motor.

Flat belt conveyor—is used to move items such as packages and manufactured parts within a building to stations where the items can be loaded or unloaded from the moving belt. The system can turn corners using special power belt curves. It is able to move items horizontally or down small inclines.

Troughed conveyor belt—runs on rollers forming a U-shape and is used to move dry loose materials.

Roller conveyors—may have solid steel rollers across the unit or use a series of individual wheels sometimes referred to as skate wheels.

   Solid roller unit—may be gravity operated or power operated. It is used for medium- and heavy-duty work and may be permanently installed or be able to be dismounted and moved.

   Skate wheel conveyor—is used for light-duty use such as unloading a delivery truck having light packages. It is gravity operated and usually is a portable unit.

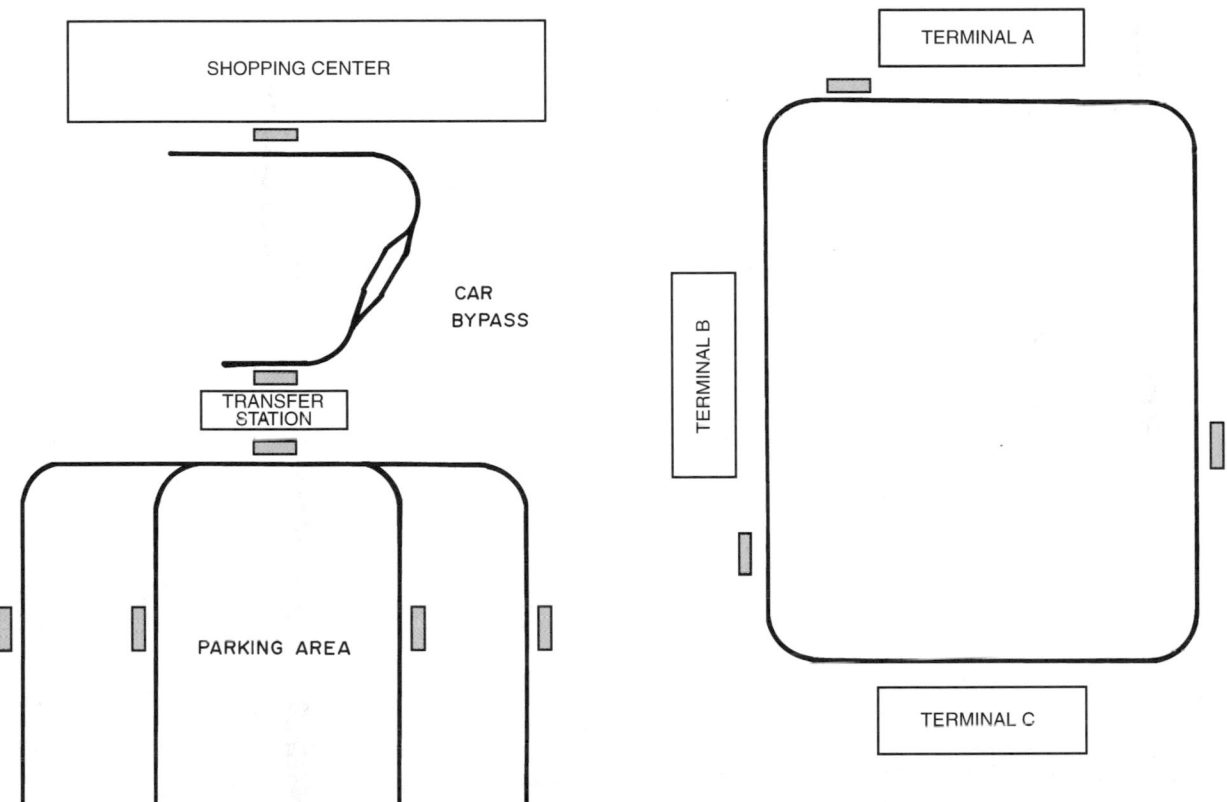

14-23 *The Otis Shuttle can connect widely spaced buildings or areas, such as a shopping center and parking or terminals at an airport. It operates unmanned, on automatic controls, and follows a schedule.*

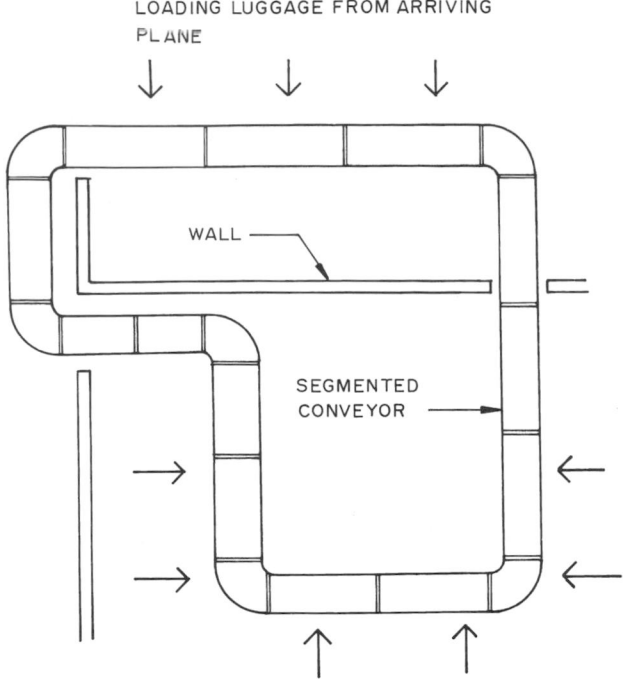

LOADING LUGGAGE FROM ARRIVING PLANE

WALL

SEGMENTED CONVEYOR

ARRIVING PASSENGERS PICK UP LUGGAGE

14-24 *Segmented conveyor belts can move materials around corners.*

Segmented conveyor—has a moving surface made up of flat sections joined with hingelike connectors (see 14-24). The airport luggage conveyors are usually this type. They handle heavy loads and are power driven. Some types can be used to move solid and loose waste materials.

Portable belt conveyors—are used to move materials on the construction site. The unit in 14-25 is set up to move concrete from a ready-mix truck. It uses a troughing-type conveyor and is powered by a gasoline engine.

As shown in 14-26, a loading hopper has been attached to the setup in 14-25 to receive the concrete. This conveyor can move aggregate and other materials. It is available in a wide range of lengths and carrying capacities.

14-25 *This high capacity portable belt conveyor will move large amounts of concrete, aggregate, and other materials.*
Courtesy Morgen Manufacturing Company

14-26 *A hopper is mounted on the end of the conveyer to receive the concrete.*
Courtesy Morgen Manufacturing Company

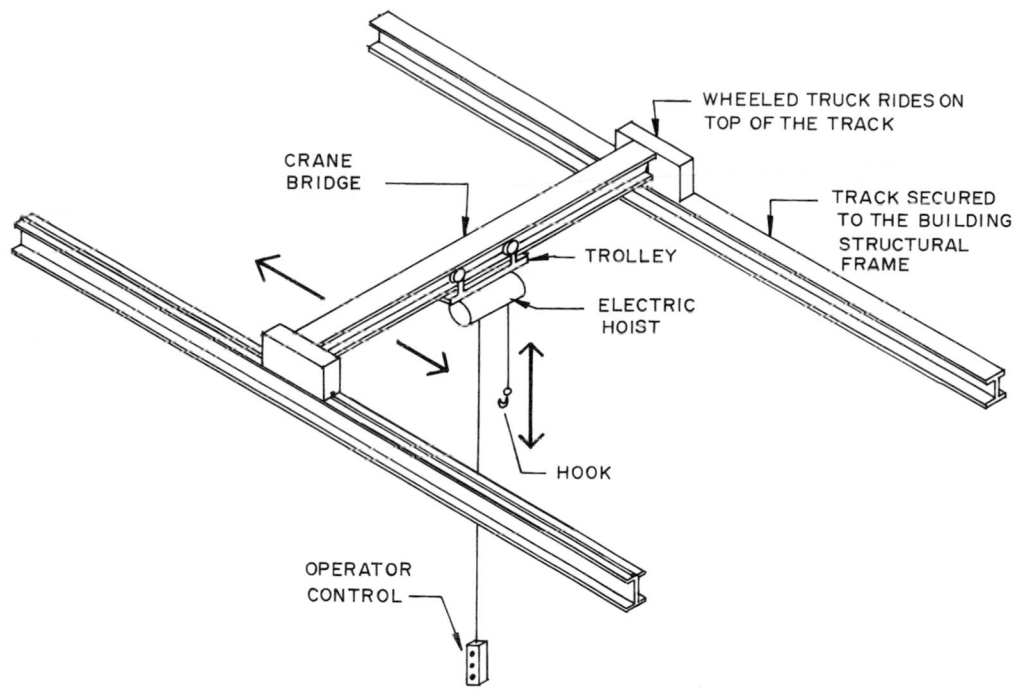

14-27 *An overhead crane runs on the top of the track.*

CRANES & HOISTS

Cranes and hoists are used to move materials and heavy items within a building and on outdoor locations. The choice of the type to use depends upon the applications required and is an important part of the design of the structure. Overhead, monorail and underhung cranes run on rails generally supported by the structural frame of the building, so the load imposed by the crane must be carefully calculated. Some types use a structural system independent from the building structure.

Overhead cranes—move along the top of fixed overhead rails. They are often used for specific jobs such as moving steel members from storage to the fabrication area and moving the finished members to a storage or shipping area (see 14-27). The lifting action is produced by a hoist that is part of the trolley. The trolley moves along a steel track. Large overhead cranes may have the operator seated in a cab connected to the crane bridge.

Monorail—has the hoist slung below a single steel track as shown in 14-28.

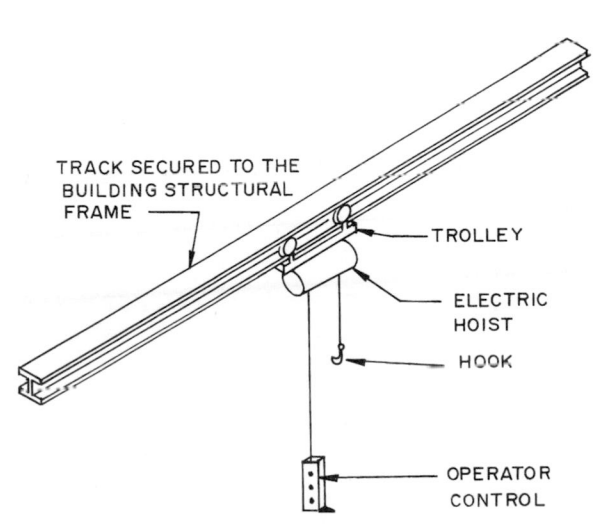

14-28 *A monorail crane moves the hoist along a single track.*

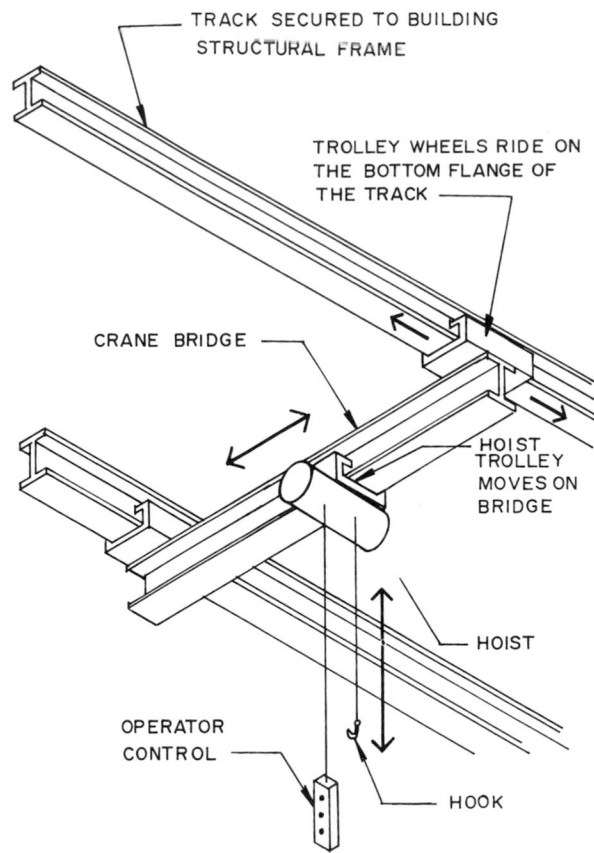

TRACK SECURED TO BUILDING STRUCTURAL FRAME

TROLLEY WHEELS RIDE ON THE BOTTOM FLANGE OF THE TRACK

CRANE BRIDGE

HOIST TROLLEY MOVES ON BRIDGE

OPERATOR CONTROL

HOIST

HOOK

14-29 *An underhung crane has the crane bridge trolleys riding on the lower track flange.*

The underhung crane—is much like the overhead crane except it travels along the bottom flange of the track (see 14-29). A variation is a wall-mounted jib crane. These cranes are powered by electric motors and generally controlled by an operator on the floor. The control cable extends from the hoist to the floor. The operator can control the hoist to lift or lower a load and the trolley to move along the track.

Gantry crane—has an overhead track and trolley that runs on a horizontal track supported by legs that run on rails secured to the floor. It is used by industries that must move very heavy loads. It may operate inside or outside the building and does not require support from the structural system of the building (see 14-30).

On large cranes the operator is seated in a cab located below the bridge. A diesel-powered **mobile gantry** is shown in 14-31. It moves on wide-base radial tires so is more versatile than the track-bound gantry. It is available in a wide range of structural configurations. The cab is located to give the operator excellent visibility and is reinforced with steel guards. It also has a heater, defroster, windshield wipers, tinted glass, and a dome light. The control panel includes a load weight indicator so the weight being lifted is known. It is available with two-wheel and four-wheel drive.

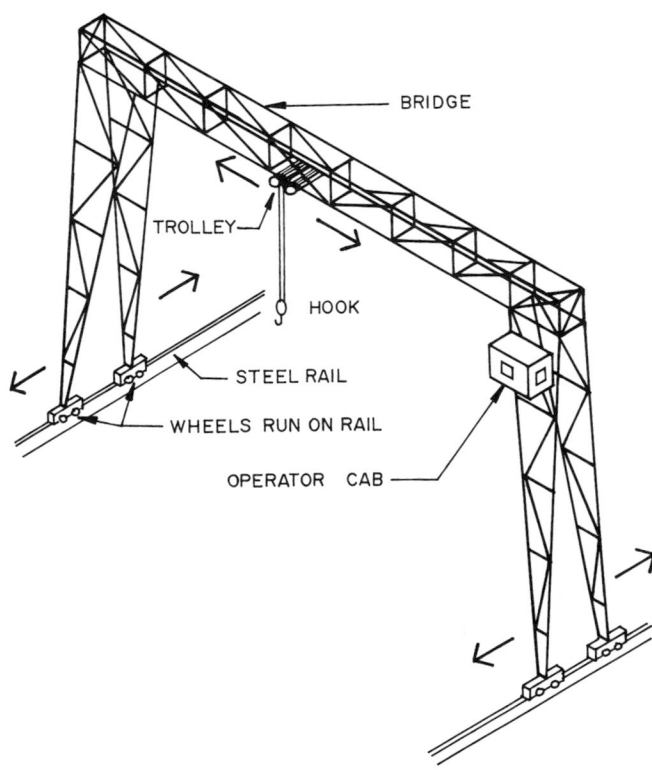

BRIDGE

TROLLEY

HOOK

STEEL RAIL

WHEELS RUN ON RAIL

OPERATOR CAB

14-30 *This heavy-duty gantry crane rides on rails and can lift loads of several hundred tons.*

14-31 *A mobile gantry crane.*
Courtesy Shuttlecraft, Inc.

DIVISION 15

MECHANICAL

CSI MASTERFORMAT™

15050 Basic Mechanical Materials & Methods

15100 Building Services Piping

15200 Process Piping

15300 Fire Protection Piping

15400 Plumbing Fixtures & Equipment

15500 Heat-Generation Equipment

15600 Refrigeration Equipment

15700 Heating, Ventilating & Air-Conditioning Equipment

15800 Air Distribution

15900 HVAC Instrumentation & Controls

15950 Testing, Adjusting & Balancing

FIRE-PROTECTION SYSTEMS

As a building is designed and consideration is given to the fire-protection system, it becomes integrally involved with the design of the plumbing, mechanical, communications, and signaling systems. As the fire-protection system is designed it must provide for early detection of a fire and give adequate warning. The means of exiting the building is a part of the planning and is regulated by codes. This is especially critical in multistory buildings. The design should consider compartmentalization of the building, smoke control, and the type of fire control system to be used. A source of emergency electrical power is required.

Fire-Protection Standards

The standards of the **National Fire Protection Association** (NFPA) cover the wide range of fire-protection requirements. The standards are typically adopted by local building inspection departments and become part of their code. The codes are published as individual books, such as *Fire Protection Systems and the National Fire Alarm Code*. Some publications are the result of a joint effort with the Society of Fire Protection Engineers. The entire code is available as a multi-volume set titled *National Fire Codes*. NFPA also publishes the *National Electrical Code* and the *Life Safety Code*.

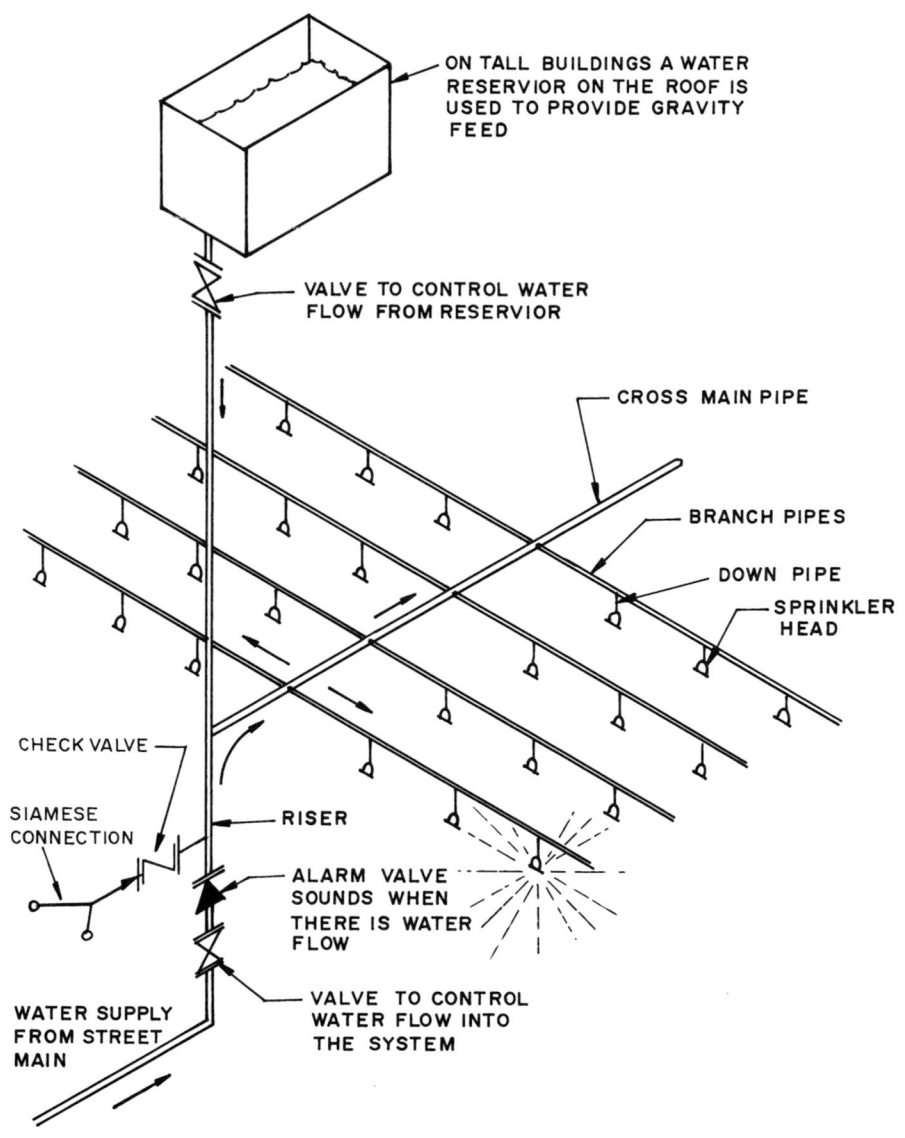

15-1 *A simplified illustration of an automatic waterline-suppression sprinkler system.*

15-2 *An exterior alarm on an automatic water sprinkler system.*

15-3 *A Siamese connection is used by the fire department to connect a supplemental source of water to the building fire-suppression system.*

15-4 *An adjustable pendent sprinkler head used in residential and commercial systems.*

Courtesy Central Sprinkler Company

The Underwriters Laboratories, Inc. (UL) tests and approves fire-protection equipment and issues reports in its Fire Protection Equipment List.

▼ Other codes (see also Division 1) include:
BOCA National Fire Prevention Code, Building Officials and Code Administrators.
Standard Fire Prevention Code, Southern Building Code Congress International, Inc.
Uniform Fire Code, International Conference of Building Officials.

Fire-Suppression Systems

Fire-suppression systems include various types of sprinkler system, foam and fog extinguishing system, gas system, and chemical system. Manual water extinguishing systems are also used.

AUTOMATIC WATER SPRINKLER SYSTEMS

A typical automatic fire-suppression sprinkler system uses a water supply from the city water system. Tall buildings require a backup supply such as a storage tank on the roof (see 15-1). The inflow is controlled by a valve. When water flow is detected, an alarm valve is activated (see 15-2). The riser has a Siamese connection (see 15-3) that is located outside the building and is used by the fire department to pump additional water into the system from an outside source such as a secondary water supply or a street hydrant. Each area on each floor that is protected by sprinklers will have a similar system. In large buildings separate risers to various sections will be used.

Sprinkler heads of various designs are available. One type is shown in 15-4. They may be an upright type or a pendent type (see 15-5).

Automatic water sprinkler systems include wet-pipe, dry-pipe, deluge, preaction, and various types of water fog and liquid foam. These use some type of sprinkler head or nozzle.

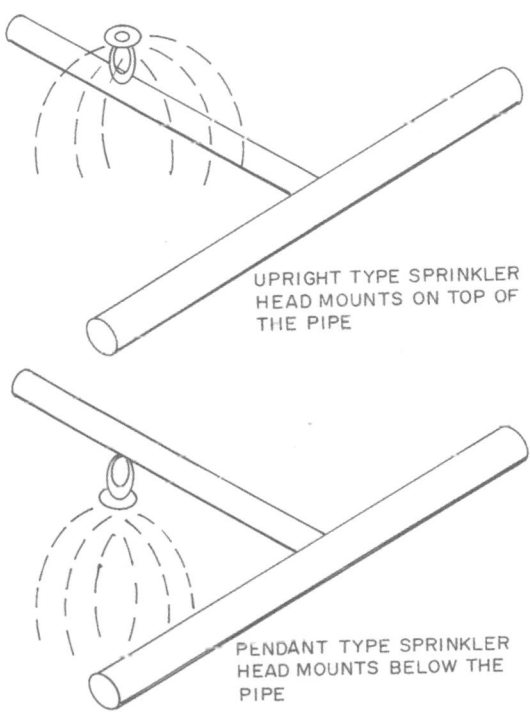

UPRIGHT TYPE SPRINKLER HEAD MOUNTS ON TOP OF THE PIPE

PENDANT TYPE SPRINKLER HEAD MOUNTS BELOW THE PIPE

15-5 *Sprinkler heads may be pendent or upright types.*

Wet-pipe systems keep water under pressure in the system of pipes at all times. When sprinkler heads are activated by heat from a fire, the water is immediately released. This is the most widely used water system. Only the sprinkler heads over the fire area are activated, thus limiting damage from water. If used in areas where the pipes may freeze, the system is filled with an antifreeze and water mixture.

Dry-pipe system pipes are maintained with air or nitrogen under pressure. When a sprinkler head opens due to heat from a fire, the air pressure is released, causing the dry pipe to open, filling the pipes with water and on through the open sprinkler heads. Since the pipes are dry this system is widely used in areas subject to freezing temperatures. Normally the upright sprinkler head system is used because the pendent heads may hold water and freeze.

The deluge system is designed to deliver as much water as possible as quickly as possible. It keeps the sprinkler heads or spray nozzles open at all times and the pipes are dry. It wets down the entire area because all the sprinklers are open. The system is activated by a sensitive fire-detection system that opens the deluge valve rather than by the activating of each of the sprinkler heads.

Preaction systems have the sprinkler heads closed and the pipes dry. They use a fire-detection system that is more sensitive than typical sprinkler heads. This detection system opens the preactive valve, allowing water to flow only to the sprinkler heads opened by heat from the fire. The accidental opening of a sprinkler head will not cause the system to discharge water because the pipes are dry. This provides areas with sensitive equipment or expensive items from damage due to an accidental opening of a sprinkler head.

Water fog systems use a standard sprinkler piping system and have spray heads or nozzles instead of the sprinkler heads. They are used in areas where highly flammable materials are stored. The fog tends to cool the material, keeping it below ignition temperature.

FOAM FIRE-SUPPRESSANT SYSTEMS

Foam fire-suppressant systems are used on fires, such as a gasoline fire, where water systems are ineffective. Foams are masses of air or gas filled bubbles formed by mechanical or chemical techniques.

Chemical foam is formed by a reaction between water and several chemicals producing a foam filled with bubbles produced by carbon dioxide.

Air foam is formed with water and a chemical. It is moved through pipes and hoses and sprayed through discharge nozzles.

High-expansion foam is formed by passing air through a screen that is constantly wetted with a chemical solution and a small amount of water. It is moved to the fire area in large ducts and generally totally fills the compartment with the fire. Someone trapped in a compartment filled with foam will have a difficult time (see 15-6).

GAS FIRE-SUPPRESSION SYSTEMS

Gas fire-suppression systems have the advantage of being able to flood an area, suppressing the fire with little harm to the contents. This is especially important in areas with sensitive equipment, such as computer rooms, commercial aircraft, telephone exchanges, and libraries. Gas used for fire suppression is usually carbon dioxide or Halon 1301.

Carbon dioxide covers the fire with a blanket of heavy gas that reduces the oxygen content of the sur-

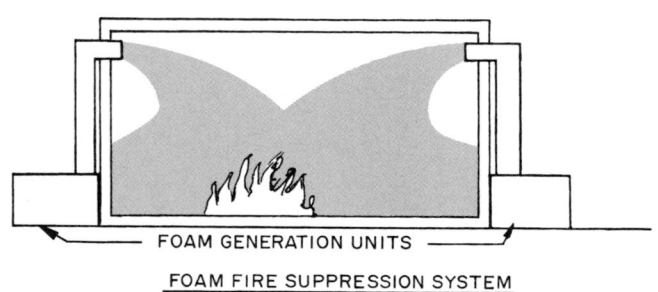

FOAM GENERATION UNITS

FOAM FIRE SUPPRESSION SYSTEM

15-6 *Foam systems produce large amounts of air- or gas-filled bubbles, filling the compartment with foam from large-diameter ducts.*

rounding air so the combustion is extinguished. It is used in small compartmentalized areas such as an electrical cabinet. It is not used in areas where people are present.

Carbon dioxide is stored in liquid form under great pressure and is released as a gas that cools and smothers the fire. A fire sensing system activates a control valve that releases the gas through a series of pipes with nozzles directed at the potential point of a fire.

Halon 1301 extinguishes a fire by interfering with the combustion process, thus preventing combustion from occurring. Low concentrations of Halon 1301 extinguish flames rapidly, while higher concentrations completely inert the atmosphere, preventing further burning or explosion. Halon 1301 is stored as a gas in cylinders under pressure (see 15-7). It vaporizes as it enters the fire area through discharge nozzles. The vapor diffuses into the surrounding atmosphere and leaves no residue. Most metals, rubber, and plastics are not affected by Halon 1301.

DRY-CHEMICAL SYSTEMS

Dry-chemical systems do not penetrate the burning material but remain on the surface and smother the fire. Chemicals frequently used include sodium bicar-bonate, potassium chloride, and monoammonium phosphate. They are effective on flammable liquid, electrical, and ordinary combustible materials. They have a high tolerance for extreme weather conditions and temperatures.

Automatic systems pipe chemicals directly to an area where a fire might occur. These tend to be a limited area such as a piece of equipment that because of the materials being used is a constant fire hazard. Dry chemicals are also widely used in handheld fire extinguishers.

MANUAL FIRE-SUPPRESSION SYSTEMS

Manual fire-suppression systems have a water piping system on each floor of the building to which fire hose stations are connected. When water pressure is not enough to supply stations in the upper floors of a building or will not supply water in adequate quantity, standpipes are added to the system. A standpipe is a pipe or tank on the roof of a building that stores a supply of water. It provides an extra supply when normal water pressures fail. The standpipe system can include pumps to increase water pressure (see 15-8, on the following page). The standpipe also provides a reserve for the potable water needed for daily use in the building.

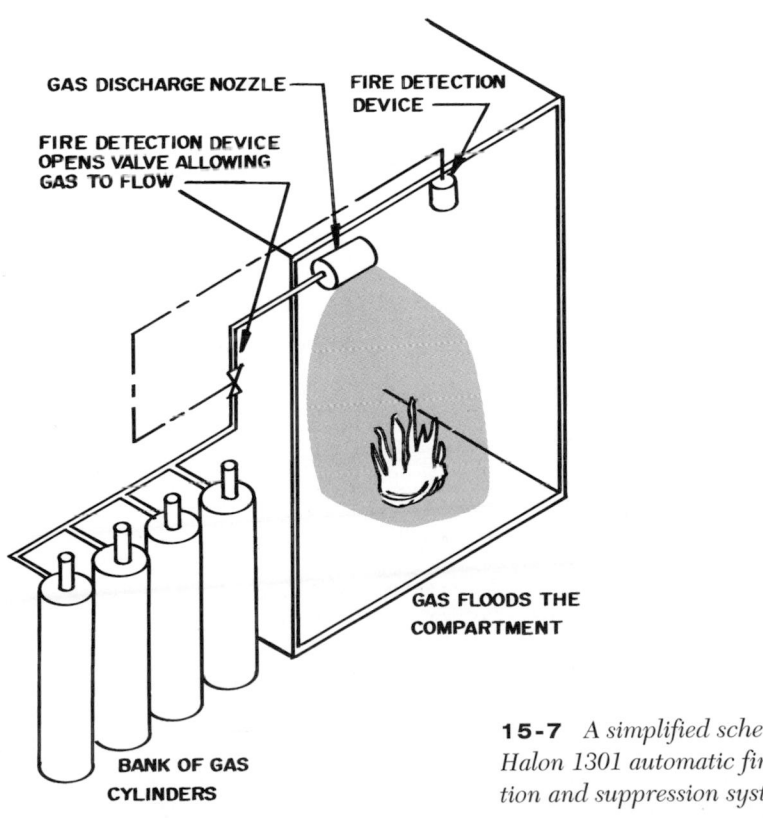

GAS DISCHARGE NOZZLE

FIRE DETECTION DEVICE

FIRE DETECTION DEVICE OPENS VALVE ALLOWING GAS TO FLOW

GAS FLOODS THE COMPARTMENT

BANK OF GAS CYLINDERS

15-7 *A simplified schematic of a Halon 1301 automatic fire-detection and suppression system.*

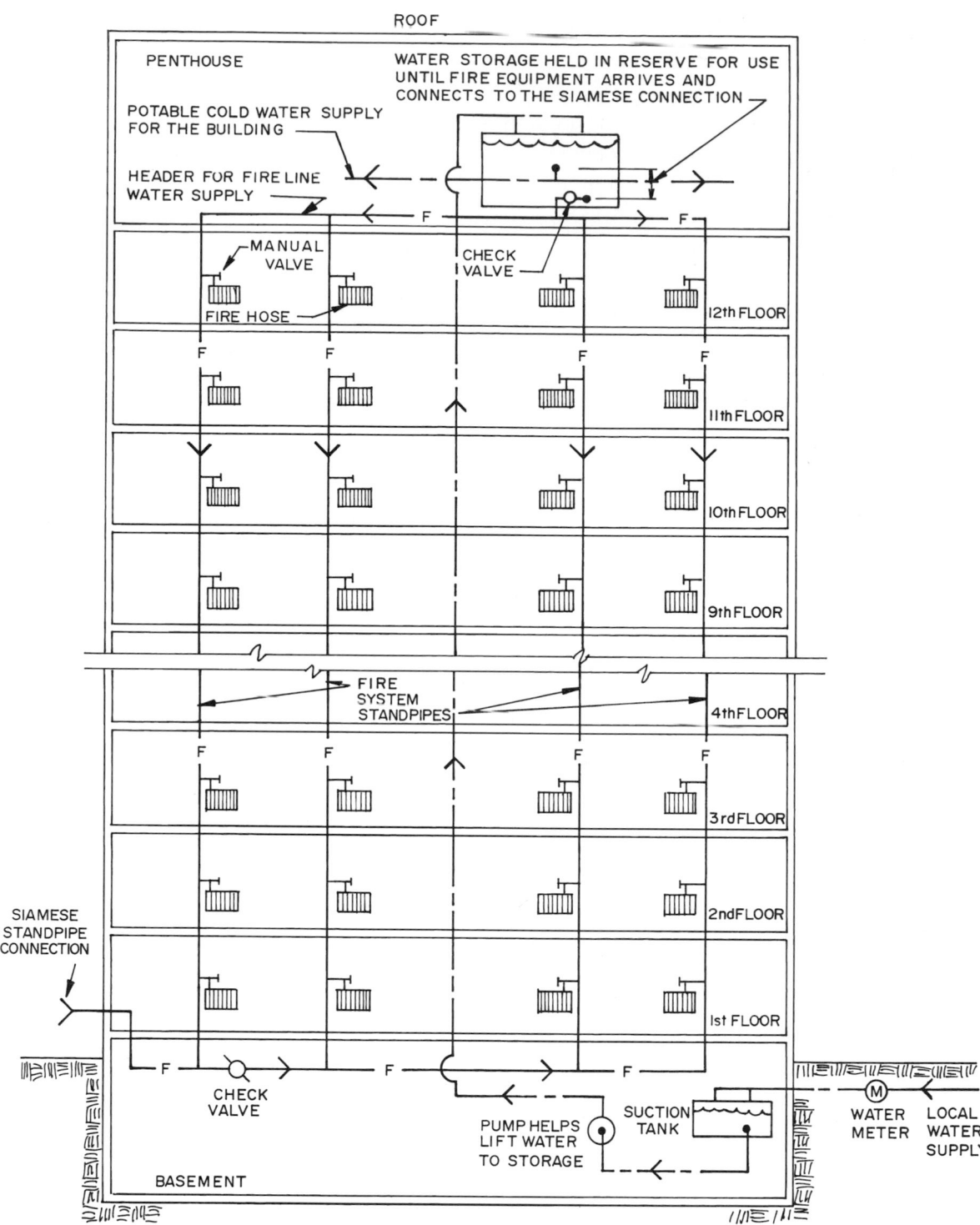

15-8 *A simplified schematic of a manual fire-suppression system utilizing standpipes. A water storage tank on the top of the building supplies water to the standpipes in addition to that from the local water supply. The standpipes supply water to the fire hose stations. The storage tank provides a short-term gravity-feed extra supply to assist until the fire department can connect to the Siamese connection.*

Table 15-1   PIPE & TUBING USED FOR WATER SUPPLY SYSTEMS

Type	Connections	Diameters	Special Qualities
Galvanized steel	Threaded	⅛–4 in. (3–102 mm)	Strong, long life
Welded steel	Threaded	⅛–4 in. (3–102 mm)	Strong, long life
Copper tube	Soldered, brazed	¼–6 in. (6–152 mm)	Corrosion resistant
Red brass	Threaded, brazed	⅛–6 in. (3–152 mm)	Corrosion resistant
Plastic	Solvent-joined, heat-fused, serrated inserts	¼–6 in. (6–152 mm)	Lightweight, corrosion resistant

PLUMBING SYSTEMS

A typical building plumbing system will include potable hot and cold water distribution systems and a sanitary disposal system. There are many other plumbing systems designed to suit a special need, such as compressed air systems, fuel oil systems, natural gas systems, and oxygen gas systems. All of these are designed and installed as specified by plumbing codes.

Plumbing Codes

The materials, plumbing design, and installation procedures are strictly regulated by local plumbing codes. Local governments will usually adopt one of the model codes.

▼ Model codes include the following:
> **Uniform Plumbing Code** published by the International Association of Plumbing and Mechanical Officials.
> **National Standard Plumbing Code** published by the National Association of Plumbing, Heating, and Cooling Contractors.
> **Southern Standard Plumbing Code** published by the Southern Building Code Congress International, Inc.
> **BOCA National Plumbing Code** published by the Building Officials and Code Administrators International.

Piping, Tubing & Fittings

Water is piped through buildings to a number of locations where it serves various purposes. Potable (drinkable) water is delivered to sinks, lavatories, and water heaters. Typically it is also run to toilets (water closets)

and exterior hose bibbs. While these do not require potable water, a less pure and less expensive supply, known as gray water, is not generally available (refer to 15-27). These pipes are for most situations hidden in walls, floors, or ceilings. They can be exposed in areas where their being seen is not important. However, they must be protected from freezing in all locations. The system must provide valves to control the flow of water into the building, within parts of large buildings, and to the individual fixtures.

The various types of pipe used for potable water systems are detailed in Table 15-1. Decisions on which to use are made based on cost, length of service expected, and the quality of the water. Some waters are loaded with minerals and are very corrosive. If hot water is to be carried, the pipe must withstand temperatures of 180°F (82°C) or higher. In special applications the effect of the liquid or gas to be carried must be considered. Gasoline, liquified petroleum gas, and other such materials will have a substantial effect on pipes.

Plastic, copper, and steel pipe are all made with the same diameters. However, some types are made in larger diameters than others. They are also made with a range of wall thicknesses. The thicker the pipe wall, the greater the pressure it can carry.

Galvanized steel pipe has been used for many years. It is strong, can be used for hot and cold water, and has a long life. The pipe and fittings are joined with threaded connections. It is available in three grades—standard-weight, extra-strong, and double extra-strong.

Gas piping within the building is **black steel pipe** meeting the requirements of ASTM A53 or ASTM A106. It uses malleable-iron or steel fittings. Some codes permit the use of copper or brass pipes if the type of gas to be carried will not cause them to corrode.

Table 15-2 COPPER TUBING TYPES, USES & SIZES

Type	Color	Uses	Nominal Diameter
K	Green	Potable water, fire protection, solar, fuel/fuel oil, HVAC, snow melting	Straight lengths ¼–12 in. (6–305 mm) Coils ¼–2 in. (6–50.8 mm)
L	Blue	Potable water, fire protection, solar, fuel/fuel oil, HVAC, snow melting	Straight lengths ¼–12 in. (6–305 mm)
M	Red	Potable water, fire protection, solar, fuel/fuel oil, HVAC, snow melting	Straight lengths ¼–12 in. (6–305 mm)
DWV	Yellow	Drain, waste, vent, HVAC, solar	Straight lengths 1¼–8 in. (32–203 mm)
ACR	Blue / Soft ACR not marked	Air-conditioning, refrigeration, natural gas, liquified petroleum gas	Straight lengths ⅜–4⅛ in. (9.5–105 mm)
OXY, MED, OXY/MED, OXY/ACR, OXY/MED	K-Green L-Blue	Medical gas	Straight lengths ¼–8 in. (6–203 mm)
G	Yellow	Natural gas, liquified petroleum gas	Straight lengths ⅜–1⅛ in. (9.5–28.5 mm) Coils ⅜–⅞ in. (9.5–22 mm)

Some types of plastic pipe are permitted for underground outside installation. They must be given a protective coating to help resist corrosion from elements in the soil.

Welded steel pipe is made by rolling a flat hot steel strip into a circular shape and butt-welding the edges as they are pressed together. They are available in the same grades as galvanized steel pipe. The outside diameter is the same for all three grades and the difference in the thickness of the wall is taken up on the inside of the pipe. It is joined by threaded fittings.

Red brass pipe is used for water lines especially if the water contains corrosive elements. It is manufactured with threaded fittings and plain ends that are brazed to socket-type fittings.

Copper water tubing is an excellent hot and cold water distribution material. It is resistant to corrosion and does not rust. The flexible type is easy to bend into various shapes, reducing the number of fittings needed. The joints fit together in a socket arrangement and are secured by soldering. The solder used must not contain lead because it will contaminate the water, causing those who use it to have health problems. Copper water tubing has a high coefficient of expansions. When carrying heated liquids it could expand enough in length to cause damage to the pipe unless provision is made to allow for it.

The types of copper tubing are shown in Table 15-2. Copper water tubing is available in three types, K, L, and M. Type K has the heaviest wall and has a green stripe along its length. Because of its strength it can be used for underground runs. Type L has a medium wall and a blue stripe and type M has a light wall and a red stripe. Both are used for potable water, heating and air-

Table 15-3 MAJOR TYPES & USES OF PLASTIC PIPE

| Type | Condition | Connections | Maximum Operating Temperature | | Typical Uses |
			°F	°C	
Acrylonitrile butadiene styrene (ABS)	Rigid	Threaded, serrated fittings, solvent	100 pressure 180 nonpressure	38 82.9	Cold water, waste, vent, sewer, drain, conduit, gas
Chlorinated polyvinyl chloride (CPVC)	Rigid	Threaded couplings, serrated fittings, solvent	180 at 100 psig (type 11)	82.9	Cold and hot water, chemical piping
Polyethylene (PE)	Flexible	Serrated fittings, fusion in socket, butt fusion	100 pressure 180 nonpressure	38 82.9	Cold water, gas, waste, chemicals
Polypropylene (PP)	Rigid	Mechanical couplings, butt fusion, socket fusion	100 pressure 180 nonpressure	38 82.9	Chemical piping, chemical drainage
Polyvinyl chloride (PVC)	Rigid	Solvent, threading, mechanical couplings, serrated fittings	100 pressure 180 nonpressure	38 82.9	Cold water, gas, waste, vents, drains, sewers, conduit
Styrene rubber plastic (SRP)	Rigid	Solvent, serrated fittings, elastomer seal	150 nonpressure (not used under pressure)	66	Sewage disposal field, storm drainage, soil drainage

conditioning, and fuel and fuel oil systems. Type ACR is used for air-conditioning and refrigeration systems, DWV (yellow stripe) is used for drain, waste, and ventilation systems and several other types identified as OXY and MED are used for medical gas applications.

Copper tubing is manufactured in hard and soft tempers. The soft-tempered pipe is easily bent, reducing the need for many connections. The hard-tempered pipe requires soldered connectors to turn corners. Copper tubing is available in diameter from ¼ to 8 in. (6 to 203 mm).

Plastic pipe is manufactured in several synethetic resins. The resin used greatly influences the strength and use of the pipe. In Table 15-3 are the commonly used resins for producing plastic pipe. Notice that most are not acceptable for hot water piping. Plastic pipe is lightweight, flexible, and available in long lengths.

Various types are joined with solvent cement, elastomeric seals for bell-end piping and fittings, serrated insert fittings secured with stainless steel clamps, heat fusion (used on plastics for which there is no solvent), and threaded fittings, which can connect to threaded metal pipe.

Plastic piping and sanitary piping are manufactured in diameters from ½ in. (12.7 mm) to 48 in. (1219 mm) in diameter and in lengths to 20 ft. (6 m). The wall thicknesses are identified by a schedule number such as Schedule 40, 80, or 120. The larger the number, the thicker the pipe wall.

Pipes made from **glass, nickel, silver,** and **chrome** are used for special applications. They are corrosion resistant. The metal pipes usually have threaded connections, while the glass pipe has a neoprene gasket secured with a stainless steel strap over a stainless steel sleeve.

PIPE CONNECTIONS

Frequently used methods for connecting pipe are shown in 15-9. Water pipe uses butt welded, socket, and threaded fittings. The bell and spigot used on sewer lines and the flanged connection is used in areas such as petrochemicals and power generation piping.

VALVES

A wide variety of valves are available for controlling the flow of water, oil, gas, and various chemicals. Some of the frequently used types are in 15-10.

Pressure regulator valve—limits the water pressure, preventing damage to piping and equipment.

Water hammer arrestor—has a hydraulic piston that absorbs shock waves produced by sudden changes in water flow and reduces banging in the pipes.

Backflow preventer—keeps water from backing in the system, and the **expansion tank** absorbs excess pressure by absorbing the extra volume of water cre-

ated when water is heated and it expands. It has a rubber bladder that flexes against the water pressure.

T & P valve—is a safety relief valve that senses a buildup of temperature or opens to release the excess pressure, preventing an explosion.

Float valves—are used to control water levels in tanks. As the water level rises, the float rises and shuts off the water when a set capacity has been reached.

Strainers—have a 20-mesh screen that collects dirt and debris. They may be on the main line or on lines to a particular piece of equipment.

Saddle valves—are installed on piping that has water under pressure.

Stop & waste valve—has been commonly used to shut off water to the building on the owner's side of the meter. It is a gate-style valve. Now the more dependable **ball valve** is used for shutoffs.

Globe valves—are used to control the flow of hot and cold water, oil, and gas.

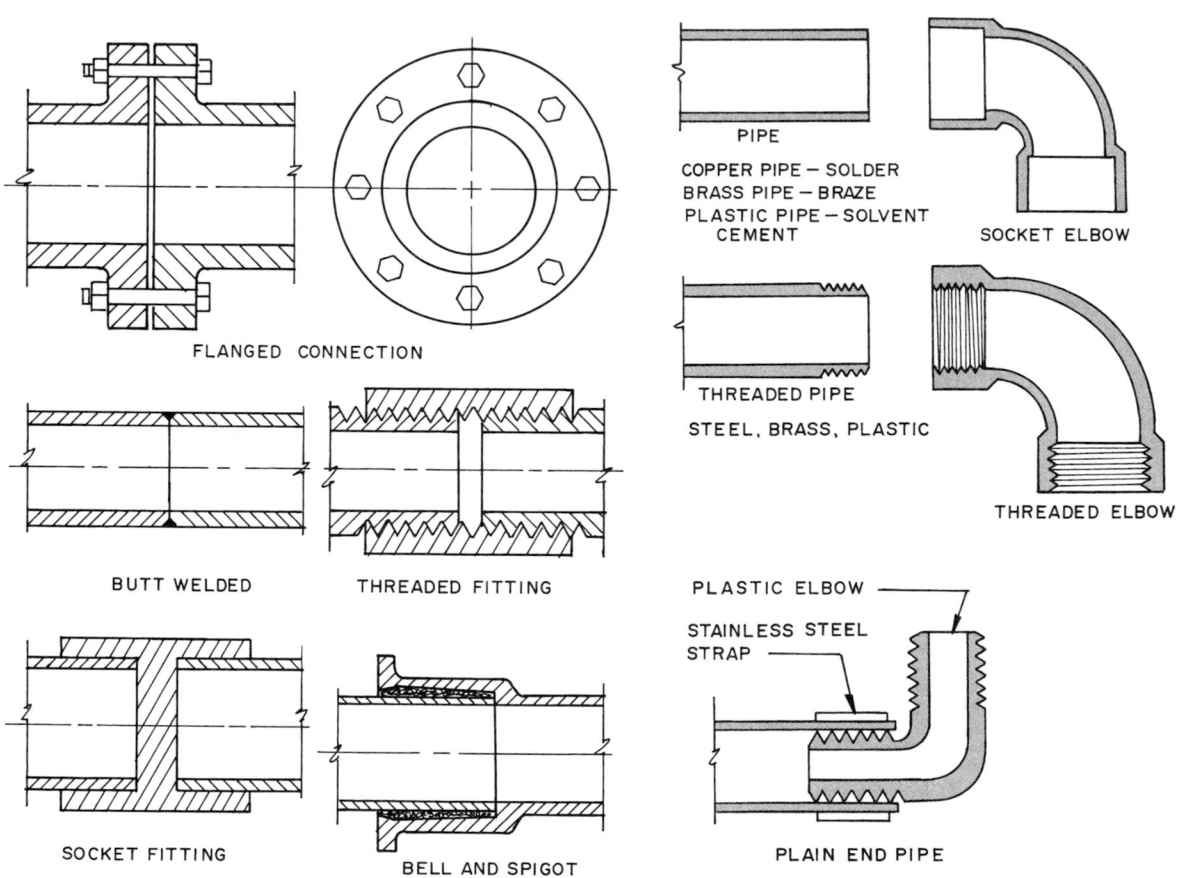

FLANGED CONNECTION

BUTT WELDED   THREADED FITTING

SOCKET FITTING   BELL AND SPIGOT

PIPE

COPPER PIPE — SOLDER
BRASS PIPE — BRAZE
PLASTIC PIPE — SOLVENT CEMENT

SOCKET ELBOW

THREADED PIPE
STEEL, BRASS, PLASTIC

THREADED ELBOW

PLASTIC ELBOW
STAINLESS STEEL STRAP

PLAIN END PIPE

15-9 *Various types of pipe and fitting connection.*

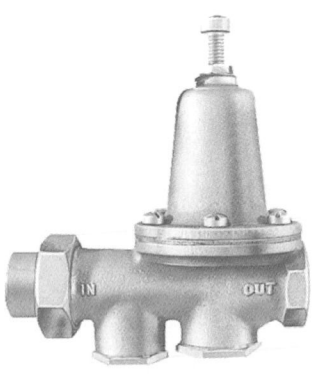

Water pressure reducing valve

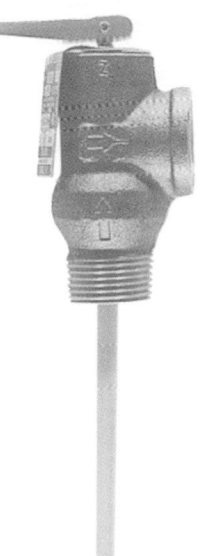

Check valve

Strainer

Ball valve

Relief valve

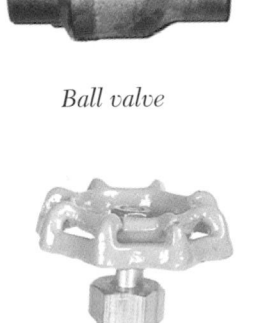

Stop and waste valve

WATER HAMMER
ARRESTOR
NO. 15

Water hammer
arrestor

Gate valve

Float valve

Expansion tank

15-10 *Typical valves used in water distribution systems.*
Courtesy Webster Valve Company

Table 15-4   MINIMUM FLOW & PRESSURE REQUIRED BY TYPICAL PLUMBING FIXTURES

Fixture	Flow Pressure (psi)	(kPa)	Flow Rate (gpm)	(L/s)
Ordinary basin faucet	8	55	2.0	0.13
Self-closing basin faucet	8	55	2.5	0.16
Sink faucet, ⅜ in. (9.5 mm)	8	55	4.5	0.28
Sink faucet, ½ in. (12.7 mm)	8	55	4.5	0.28
Bathtub faucet	8	55	6.0	0.38
Laundry tub faucet, ½ in. (12.7 mm)	8	55	5.0	0.32
Shower	8	55	5.0	0.32
Ball cock for closet	8	55	3.0	0.19
Flush valve for closet	15	103	15–40	0.95–2.52
Flushometer valve for urinal	15	103	15.0	0.95
Garden hose (50 ft., ¾-in. sill cock) (15 m, 19 mm)	30	207	5.0	0.32
Garden hose (50 ft., ⅝-in. outlet) (15 m, 16 mm)	15	103	3.33	0.21
Drinking fountain	15	103	0.75	0.05
Fire hose, 1½ in. (38 mm), ½-in. nozzle (12.7 mm)	30	207	40.0	2.52

Reproduced from Manual of Individual Water Supply Systems, U.S. Environmental Protection Agency

PIPE INSULATION

Cold water pipes will form condensation on the exterior when warm humid air hits them. The condensation drips down inside a wall cavity, ceiling or floor and wets the insulation and penetrates drywall or other wall finish. The pipes are covered with preformed fiberglass or foam insulation that is usually ½ to 1 in. (12 to 25 mm) thick. The insulation is fitted around the pipe and taped. Hot water pipes are insulated to reduce the loss of heat to the cooler atmosphere.

Pipes carrying chilled water, brine, refrigerant, domestic hot water, commercial hot water and steam, steam condensate, and hot water heating systems require the use of minimum insulation. If hot and cold piping run parallel they should be 6 to 8 in. (152 to 203 mm) apart to reduce the chance of heat and cooling exchange.

Potable Water Supply

Water for public and privately owned central water systems may be obtained from rivers, lakes, ponds, surface runoff, and wells. Surface-collected water tends to contain a number of contaminates, which need to be removed to produce potable water. These quality problems include **hardness** caused by calcium and magnesium salts, **color** caused by manganese or iron, **corrosion** caused by acidity in the water, **pollution** caused by sewage and organic matter, **odor and taste** caused by organic matter, and **turbidity** caused by silt and other suspended matter. The water treatment plant processes the water to correct these problems.

Water treatment includes processes such as **screening** the water at the intake, **sedimentation** (which allows particles to drop in a settling basin), **coagulation** (removing suspended matter with a chemical such as hydrated aluminum sulfate) in a settling basin, **filtration** (removing suspended particles and some bacteria through a filter such as sand, diatomaceous earth, and chlorine), **disinfection** (removing harmful organisms using bromine, iodine, ozone, or heat treatment), **softening** (removing calcium and magnesium), and **aeration** (exposing water to the air, as by spraying, to improve taste and color).

In areas without a central water supply the water is obtained from wells. These are usually drilled, driven, or jetted. A **drilled well** is dug with a steel auger that drills into the earth into which a pipe is placed.

A **driven well** is formed by putting a steel point on the end of sections of pipe and driving them into the earth. A **jetted well** has a well point from which a high-pressure stream of water is pumped, opening a hole for the well pipe to move behind it into the earth.

The water from wells tends not to require the extensive treatment needed for surface water. It must be tested for purity and treated as required such as to reduce hardness, iron, or manganese.

WATER PIPE SIZING

The size of the required water supply pipe depends upon the available pressure, the vertical and horizontal distances involved (friction in the pipe), the number of fittings (tees, elbows), and the demand of the use at the fixture. Experience has allowed values to be determined for the flow of various fixtures under recommended minimum pressures. The values of all of these are considered as the pipe diameter is calculated. Several fixtures on one water line do not require the water flow to increase directly for each because not all of the fixtures will likely be used at the same time. Typical flow pressures and flow rates are shown in Table 15-4.

POTABLE WATER DISTRIBUTION SYSTEMS

Codes specify that all fixtures in a building must be supplied with potable water at adequate pressures. Potable water in residential buildings generally uses the water pressure from the central water system or from the pump on a well to supply the fixtures (see 15-11). The residential system shown is an **upfeed system** with the water moving up from the source into the building and to the fixtures. The piping used must have the proper size to carry the required amount of water needed by each fixture. The circulation pipes and risers usually feed more than one fixture, so the total expected flow must be figured.

The water enters through a meter and a shutoff valve. In cold climates the meter will be inside the building and the water service line will be below the frost line. The depth varies in different geographical areas. The water continues to a water softener if needed and then to a water heater and horizontal circulation lines. The water moves to fixtures in the upper stories through risers and to individual fixtures through small-diameter pipe connectors. The hot water runs from the water heater through a similar piping system.

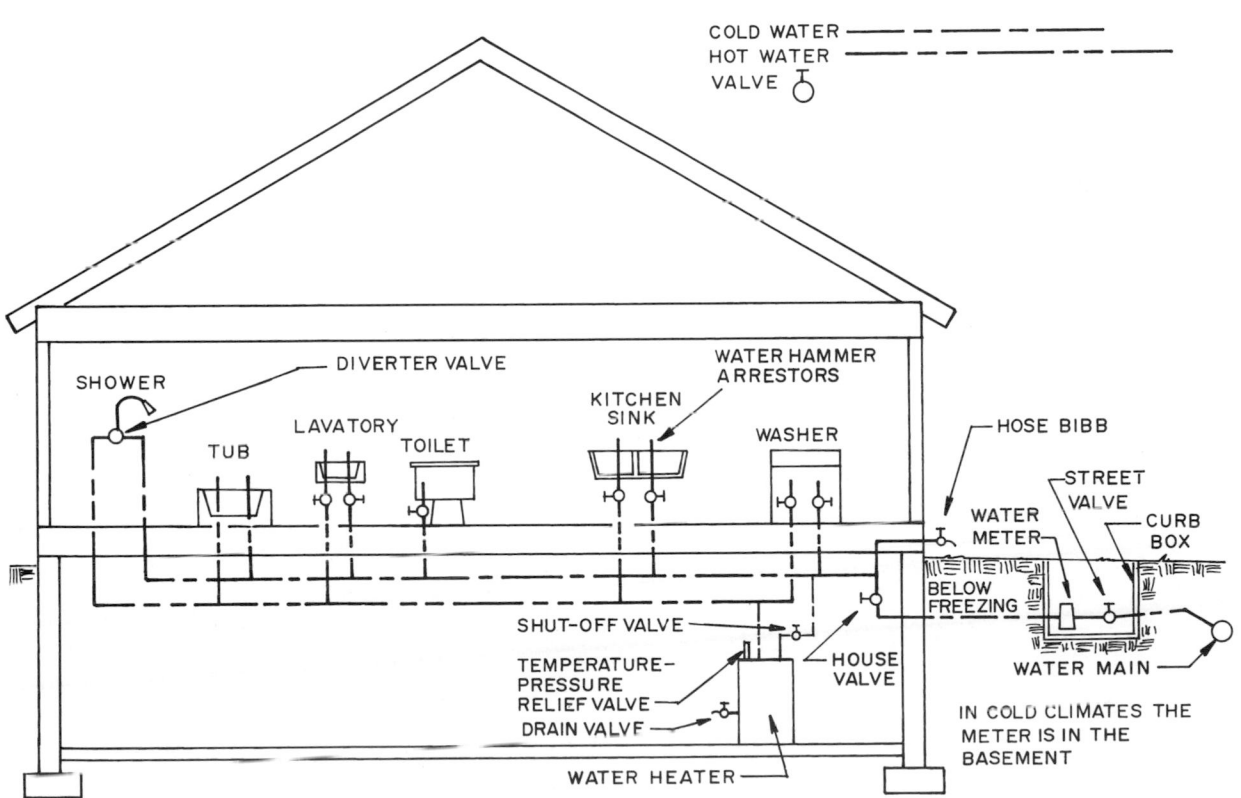

15-11 *Components of a typical residential potable water distribution system.*

Low multistory buildings may also use an **upfeed system** if the demand on water is not excessive (see 15-12). A series of pumps in the basement provides the needed pressure to raise the water to the desired height and maintain it as fixtures are used. This system is useful on multistory buildings that are not high enough to warrant the expense of a rooftop water storage system. The system will have two or more pumps. As the demand for water increases a second or third pump will come on line, increasing the flow and maintaining the pressure. It should be noted that the supply available from the central water system main should be large enough so that the demands of the building can be met without reducing water service to neighboring buildings. Notice that this system does not have a reserve water supply and a power failure can greatly reduce the water supply available.

Tall multistory buildings will use a **down-feed potable** hot and cold water distribution system. Very tall buildings will usually be divided into zones each having its own pumps and storage tanks (see 15-13). The water from the central water system is pumped from a street main or a water storage suction tank to a roof storage tank with one or more pumps. This storage tank holds a reserve for the fire-suppression system and a supply of potable water for fixtures. The water flows from a header pipe to downfeed risers from which branch water lines run to fixtures on each floor (see 15-14).

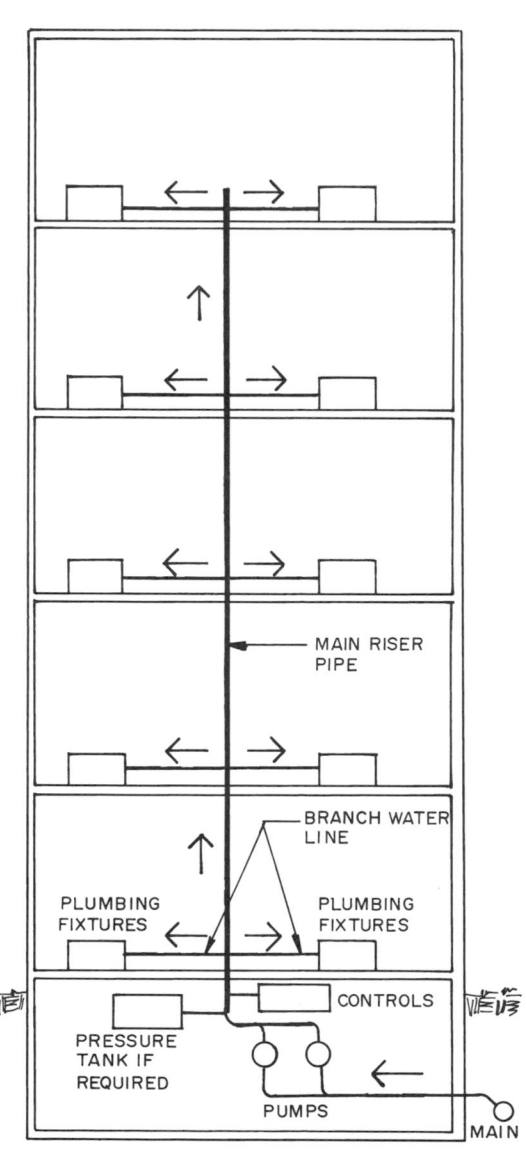

15-12 *This is an upfeed potable water distribution system for a low-rise building.*

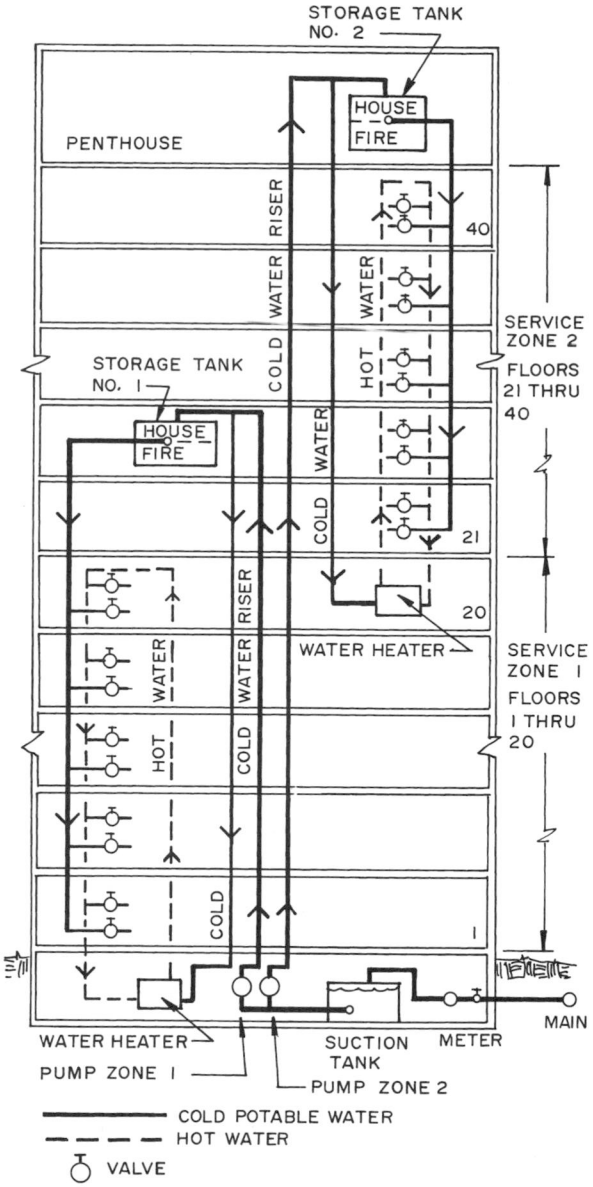

15-13 *A simplified schematic of a two-zone down-feed hot and cold water distribution system.*

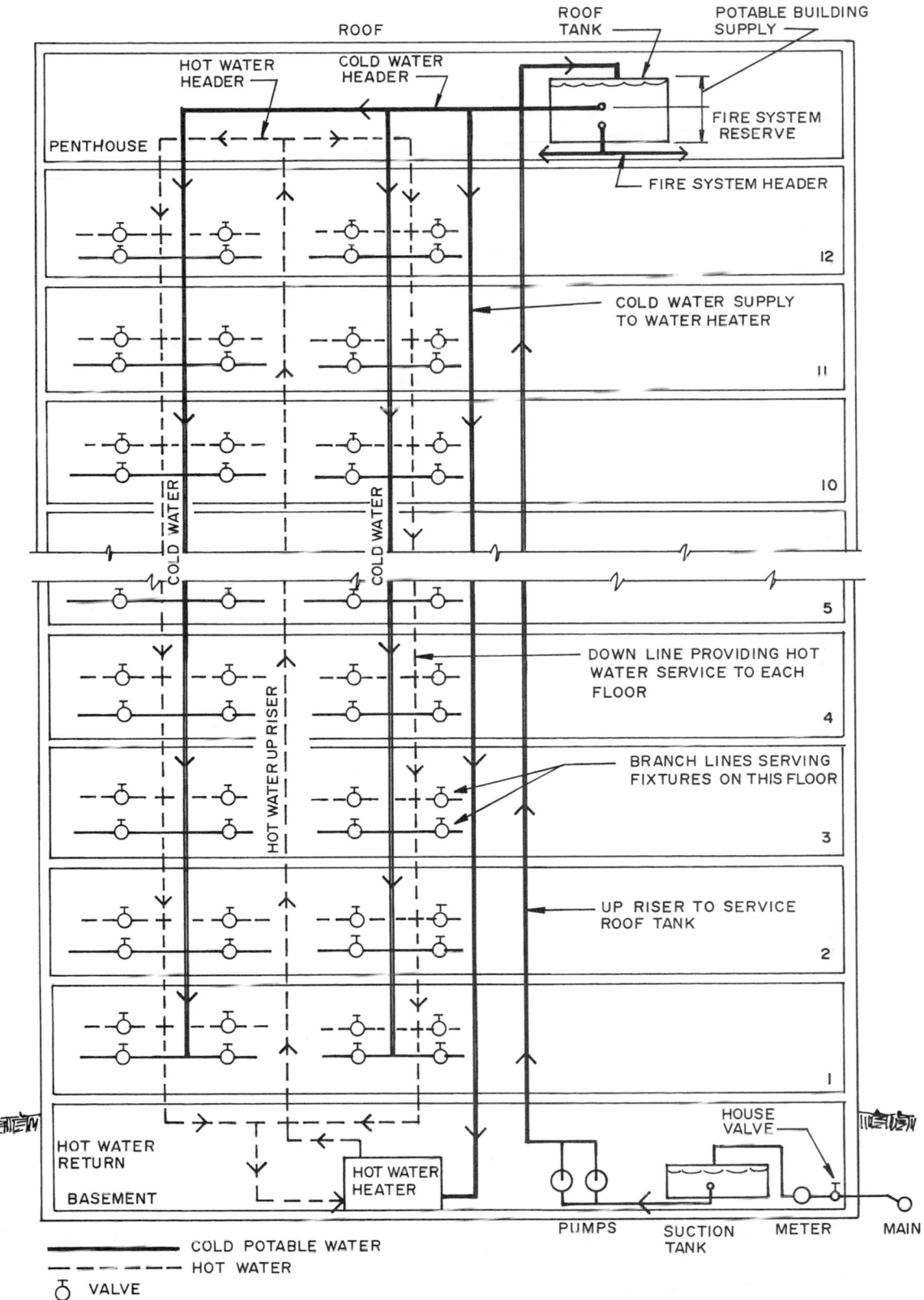

ROOF
ROOF TANK
POTABLE BUILDING SUPPLY
HOT WATER HEADER
COLD WATER HEADER
PENTHOUSE
FIRE SYSTEM RESERVE
FIRE SYSTEM HEADER
12
COLD WATER SUPPLY TO WATER HEATER
11
COLD WATER
10
COLD WATER
5
DOWN LINE PROVIDING HOT WATER SERVICE TO EACH FLOOR
HOT WATER UP RISER
4
BRANCH LINES SERVING FIXTURES ON THIS FLOOR
3
UP RISER TO SERVICE ROOF TANK
2
1
HOT WATER RETURN
BASEMENT
HOT WATER HEATER
HOUSE VALVE
PUMPS   SUCTION TANK   METER   MAIN

——————— COLD POTABLE WATER
— — — — — HOT WATER
⌀ VALVE

15-14 *A simplified schematic of a down feed potable hot and cold water distribution system typically used in high-rise buildings.*

Table 15-5 CAST-IRON SOIL PIPE USES & SIZES

Type	Uses	Nominal Diameter
Hubless soil pipe, standard & extra heavy	Sanitary & storm drains, waste, vents	2–15 in. (50.8–381 mm)
Hub-type soil pipe, service & extra heavy	Sanitary & storm drains, waste, vents	2–15 in. (50.8–381 mm)

Sanitary Piping Systems

The sanitary piping system removes the water and other waste materials discharged at the various fixtures. The system carries the material down through the building to the building drain from which it is discharged through the building sewer and onto the public sewer system or septic tank. The waste material flows from the building at a level below the lowest fixture and flows by gravity through the building sewer. If the public sewer is not below the sloped building sewer waste will not drain and pumps will be used to lift the waste and move it to the public sewer.

Sanitary piping systems are designed to carry the waste materials using pipe diameters that are sized to carry the design flow rapidly and not clog the pipes, create annoying water noises, and produce minimum pressure variations where fixtures connect to waste pipes and waste pipes connect to branch soil pipes and where branch soil pipes connect to the soil stack.

The system relies on gravity flow so it is not under pressure. The effluent flow increases in speed as it procedes down the piping system. Friction within the pipe limits the speed and is a factor when sizing pipe.

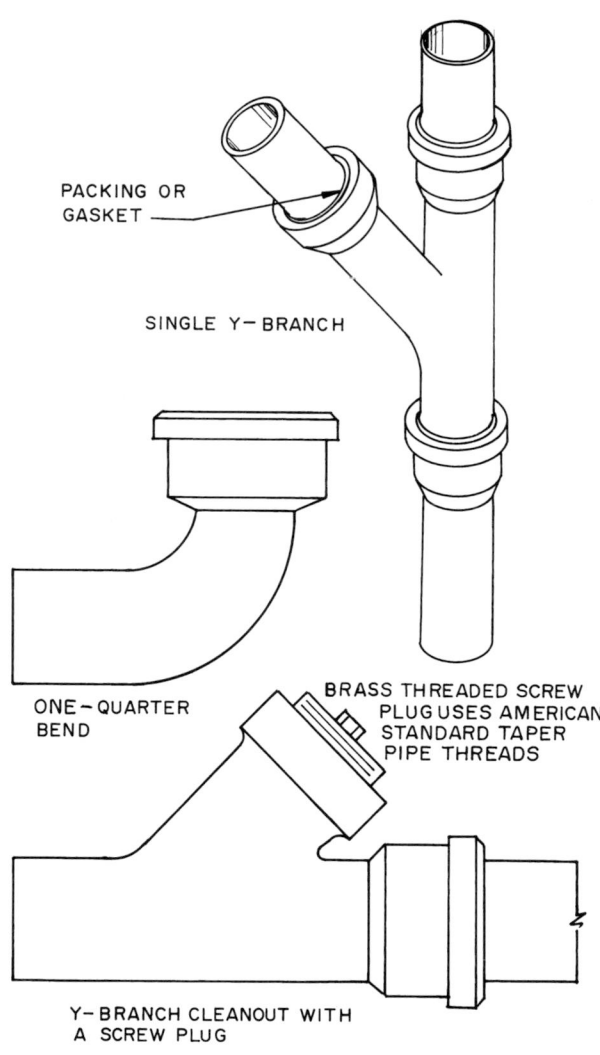

PACKING OR GASKET

SINGLE Y-BRANCH

ONE-QUARTER BEND

BRASS THREADED SCREW PLUG USES AMERICAN STANDARD TAPER PIPE THREADS

Y-BRANCH CLEANOUT WITH A SCREW PLUG

15-15 *Typical hub type cast iron soil pipe fittings.*

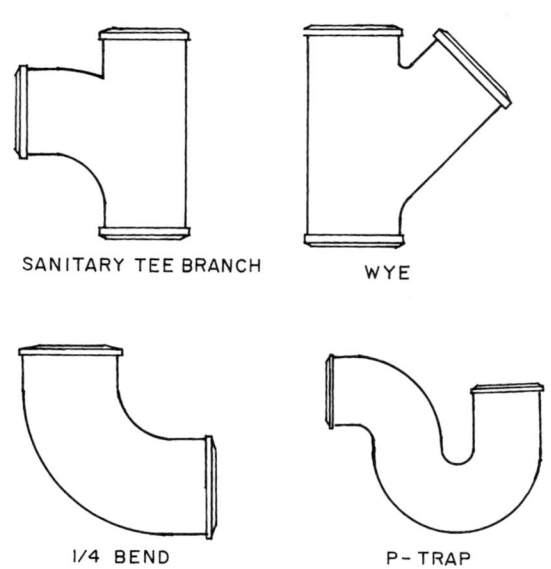

SANITARY TEE BRANCH

WYE

1/4 BEND

P-TRAP

15-16 *Typical hubless type cast-iron soil pipe fittings.*

Plumbing codes specify the required slope for horizontal piping. The requirement varies with pipe diameter but is typically from ⅛ to ½ in. (3 to 12.7 mm). Flow velocity increases with the amount of slope and high velocities in horizontal pipes can increase siphonage. The capacity of a pipe should be increased by using a larger diameter pipe rather than increasing the slope.

SANITARY PIPING

Piping used for sanitary systems may be cast iron, copper, plastic, lead, glass, and clay. They are used for various drainage, waste and vent installations (DWV).

Cast-iron soil pipe and fittings are gray iron castings suitable for installation and service for storm drain, sanitary, waste, and vent piping.

Hub-type cast-iron soil pipe and fitting specifications are detailed in ASTM A 74-87. They are available in Extra Heavy and Service classifications in diameters shown in Table 15-5. The pipes and fittings must be coated with a material to protect the surface.

Hubless cast-iron soil pipe and fittings are specified in ASTM A 888-90. Neither type are intended to be used on pressure systems. The selection of the proper design size allows the free air needed for gravity drainage. Both types are available in a wide range of fittings. A few are shown in 15-15 and 15-16. The hub type joints may be sealed with lead and oakum or with a neoprene compression gasket while hubless connections use a gasket and a stainless steel casing and retaining clamps (see 15-17).

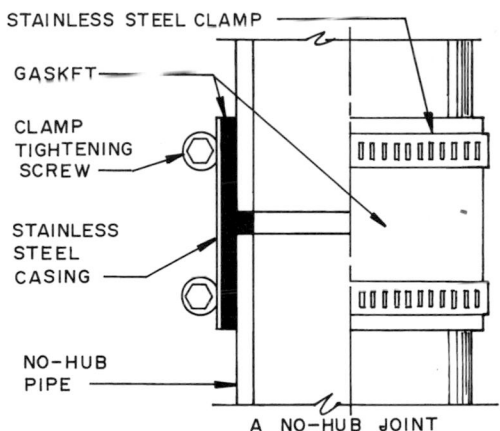

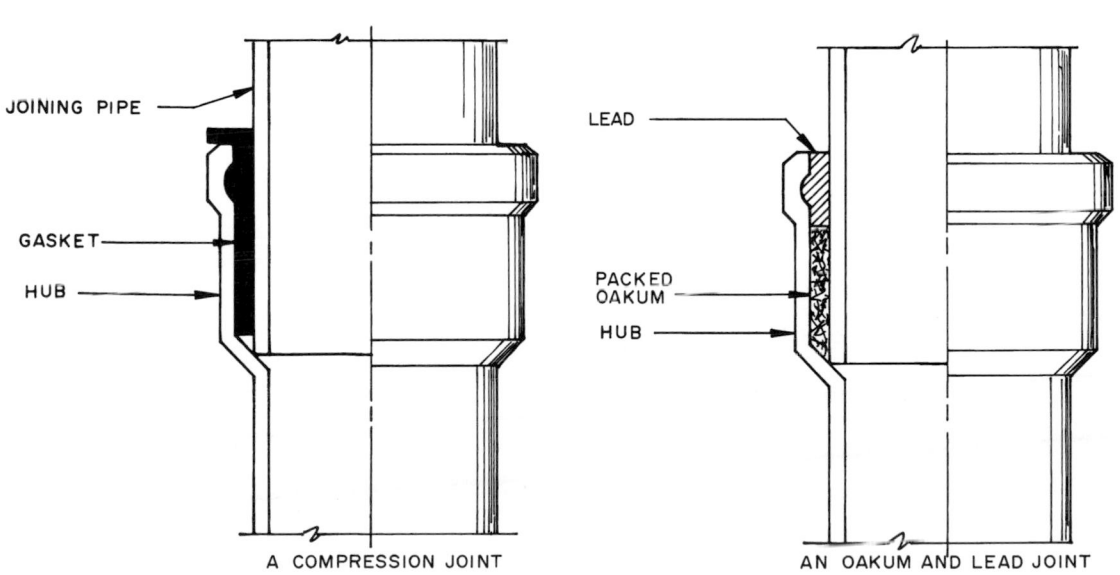

15-17 *Joints used to join hub-type and hubless cast-iron soil pipe.*
Courtesy Cast Iron Pipe Institute

Table 15-6 DRAINAGE FIXTURE UNITS
FOR SELECTED PLUMBING FIXTURES

Plumbing Fixture	Drainage Fixture Unit Value
Automatic clothes dryer	3
Bathtub w/ or w/out overhead shower	2
Clinic sink	6
Dental unit	1
Drinking fountain	½
Dishwasher, domestic	2
Floor drain, 2 in. waste pipe	3
Kitchen sink, 1½ in. trap	2
Kitchen sink with dishwasher	3
Lavatory, 1¼ in. waste pipe	1
Shower stall, residential	2
Urinal, stall, washout	4
Water closet, tank operated	4
Water closet, valve operated	6

From *National Standard Plumbing Code,* National
Association of Plumbing, Heating, and Cooling Contractors

Copper tube used for sanitary waste systems are classified as DWV (Drainage, Waste, Vent). (The available diameters are in Table 15-2.) It is used for all parts of drainage plumbing in residential, low-rise, and high-rise buildings including soil and vent stacks, and soil, waste, and vent branches. In high-rise buildings expansion must be considered as the system is designed. Changes of one degree F will cause the DWV pipe to change 0.001 in. (0.025 mm) for a 10 ft. section.

Joints in copper DWV tubing are made by properly cleaning and fluxing the joining parts and allowing the solder to flow between them by capillary action. Brazed connections are used if greater strength is needed.

Plastic pipe suitable for DWV systems include acrilylonitrile butadiene styrene (ABS), styrene rubber plastic (SRP), and polyvinyle chloride (PVC). All three can be used for sewer systems while ABS and PVC can be used for drain, waste, and vent systems. Local plumbing codes should be checked to verify the approved uses. Plastic pipe and fittings have identification symbols on each piece as shown in 15-18. The pipe and fittings are joined with a solvent cement.

Lead and glass drain pipe is used for the disposal of special liquids such as in a chemical plant or research laboratory. They resist attack by many chemicals. Lead pipe is joined by welding the joints. Glass pipe is connected with a gasket and stainless steel clamp as shown for hubless pipe in 15-17, on the previous page.

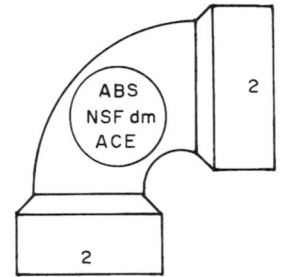

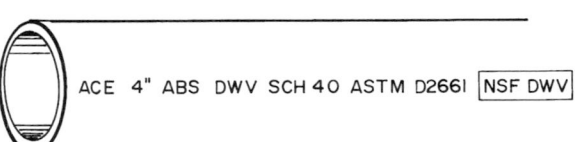

ACE 4" ABS DWV SCH 40 ASTM D2661 NSF DWV

ACE	The name of the manufacturer.
4 in.	Diameter of the pipe.
ABS	Acrylonitrile-Butadiene-Styrene, the material.
DWV	Suitable for drainage waste and vent.
SCH 40	Schedule 40. This identifies the wall thickness of the pipe.
ASTM D2661	"Standards Number" assigned by the American Society for Testing and Materials.
NSF DWV	Tested by the National Sanitation Foundation Testing Laboratory. The pipe meets or exceeds the current standards for sanitary service.

15-18 *Identification symbols used on plastic pipe.*
Courtesy Plastic Pipe Institute

Clay pipe is used for waste disposal lines that occur outside of a building. Their use is carefully controlled by building codes. The pipe is made from clay and burned in a kiln much like brick. They are available in diameters from 4 to 36 in. (102 to 914 mm) and have a variety of fittings. They use a hub joint which is filled with a packing compound. Clay pipe is impervious to acids and alkalines and does not deteriorate with age. It is not used where it might be subjected to shock or loads.

ASCERTAINING WASTE PIPE SIZES

The diameter of the pipes used in a sanitary piping system depends upon the amount of waste they are to carry. Plumbing codes typically establish the number of drainage fixture units running through each pipe size (Table 15-6). A **fixture unit** is a measure of the probable discharge into the drainage system by the various plumbing fixtures. It is expressed in units of cubic volume per minute. The value for a particular fixture depends upon the volume rate of discharge, the time duration of a single discharge, and the average time between successive discharges. A fixture unit is generally equal to 7½ gallons of flow.

INDIRECT WASTES

Indirect waste such as wastes from food handling equipment, dishwashers, sterilizers, and commercial laundries are discharged with an indirect waste pipe. The waste from the indirect waste pipe is not discharged directly into the building sanitary piping system but into a fixture connected to the building sanitary system. This provides an air gap between the two systems. An air gap is used when it is required to prevent one system from accidently backing up into another. Toxic and corrosive waste discharges must be automatically diluted with water or chemically neutralized before being introduced into the building sanitary system.

RESIDENTIAL SANITARY PIPING

A typical simple sanitary piping system for a residence is in 15-19. The waste from the fixtures drains through waste pipes or branch soil pipes into a soil stack. The waste pipes are sized to carry the flow from the fixture. The branch soil pipe is sized to carry the flow from all the fixtures flowing into it. The soil stack extends below the building and connects to the building drain. The building sewer connects to the central sewer system or a septic tank.

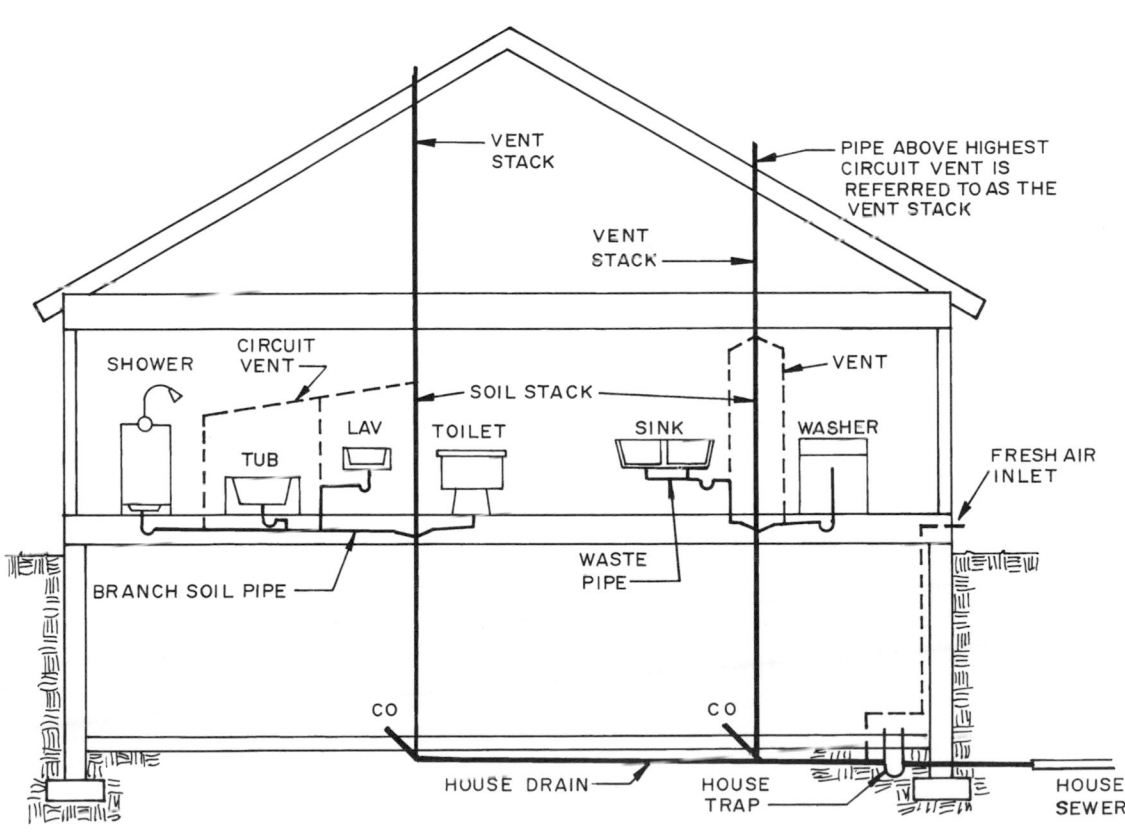

15-19 *A simplified schematic of a typical residential sanitary piping system.*

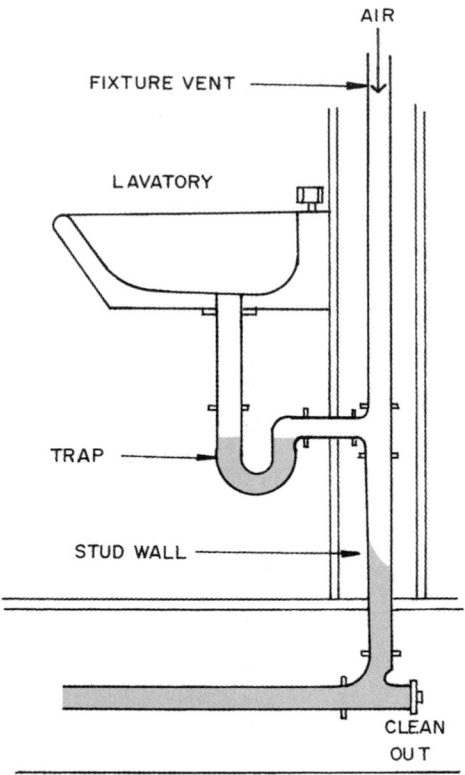

15-20 *Fixtures are vented to keep water in the trap under atmospheric pressure so the water in it is not siphoned out when a nearby fixture discharges waste in the system.*

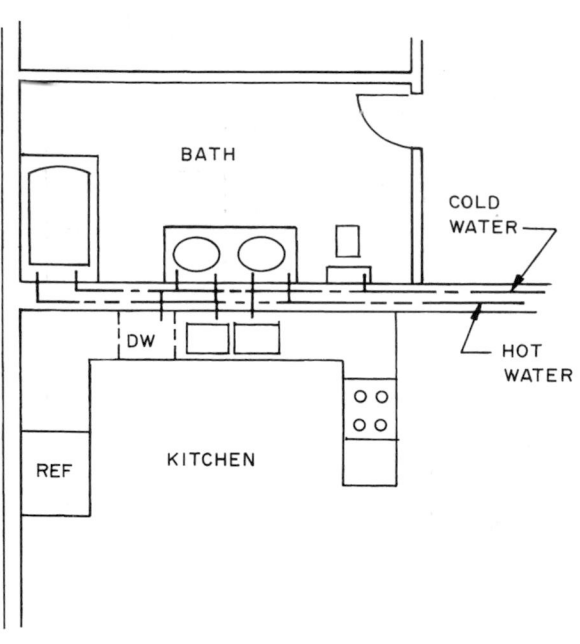

15-22 *Plumbing costs can be reduced by clustering the plumbing fixtures.*

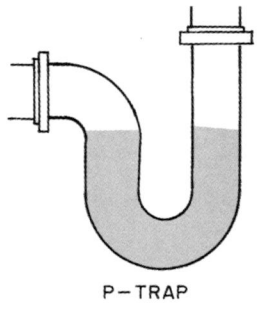

P—TRAP

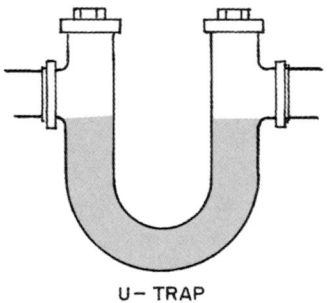

U—TRAP

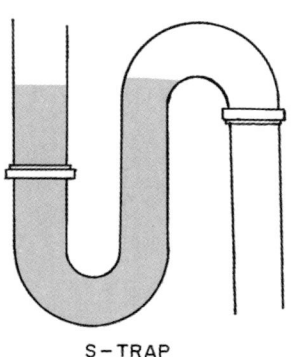

S—TRAP

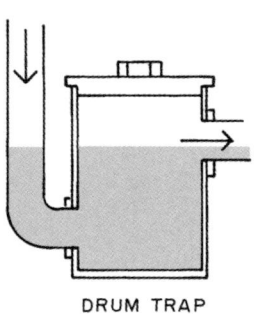

DRUM TRAP

15-21 *Some of the commonly used traps.*

Each fixture has a vent pipe connected to a vent stack running through the roof. The vent pipe keeps the water in the trap of the fixture under atmospheric pressure thus eliminating the chance of it being siphoned out when another fixture is used. For example, a trap in a lavatory could have the water siphoned out if a connecting toilet is flushed if the waste system is a closed installation (see 15-20). Some fixtures, such as a toilet, have a trap built into the unit.

Various types of traps are used. Lavatories typically have a p-trap. A tub may have a drum trap (see 15-21). In order to keep plumbing costs down the designer attempts to place fixtures on each side of a wall. This reduces the piping needed and long horizontal runs to reach out of the way fixtures (see 15-22). Typically the wall containing the plumbing will use at least 2 × 6 in. studs.

MULTISTORY BUILDING SANITARY PIPING SYSTEMS

In multistory buildings it is common to design plumbing chases on each floor directly above the other in which water and waste disposal systems can run the height of the building. Heating, air-conditioning, and fire suppressing systems can also use these chases. This consolidates the plumbing, reduces costs, and makes the planning of the uses of the space on each floor more flexible. Plumbing fixtures are typically located next to the sides of the vertical chase. However horizontal runs to locations within the floor area can be made from the chase (see 15-23). In other cases pipe risers can be located inside a space around a structural column. This enables fixtures to be placed in various locations within the building (see 15-24).

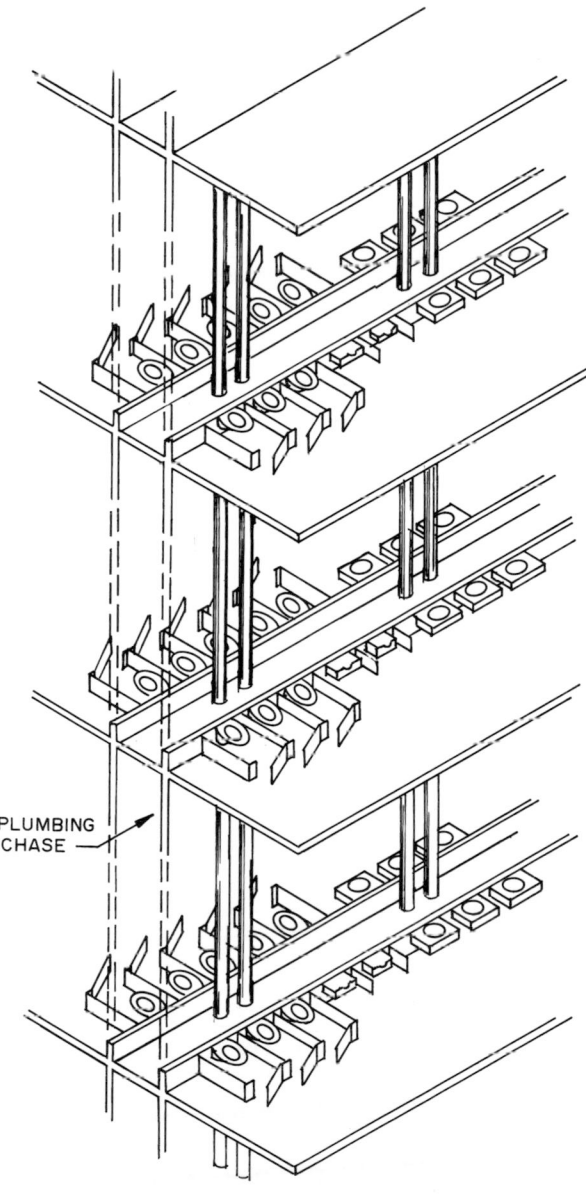

15-23 *A plumbing chase is used to stack fixtures on floors above each other.*

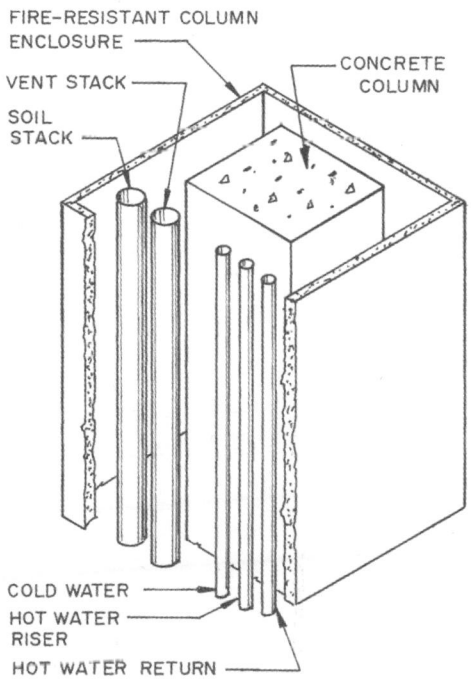

15-24 *A column enclosure can be used to run pipe risers from floor to floor.*

A schematic of a small multistory building sanitary piping system is in 15-25. The fixtures for the restrooms with the related piping are detailed on the second floor with notes indicating they are the same on all other floors. The soil stack, vent stack, and building drain are shown for the entire building. Notice the venting of multiple fixtures to horizontal circuit vents that connect to the vent stack. In some cases the fixture waste pipes connect directly to the soil stack while in other situations they feed into a horizontal branch soil pipe which connects to the vertical soil stack. The circuit vents, waste pipes, and branch soil pipes slope toward the soil stack providing gravity flow. The pipe sizes are calculated depending upon the flow as explained earlier.

The system in 15-25 requires two sets of pipes, a vent system and a waste disposal system. This two-pipe system is most commonly used in the United States. The venting controls the possibility of siphoning water from the traps allowing sewer gas to enter the building. Another system, the Sovent system, is used in Europe and Africa and has been approved for use in the United States by some of the national plumbing codes.

The Sovent System

The Sovent system is a vertical cast iron drainage and waste system conveying wastes from the upper levels of a building to the base of each Sovent stack. The stack begins just above the lowest deaerator fitting and continues up to just above the highest fixture connection

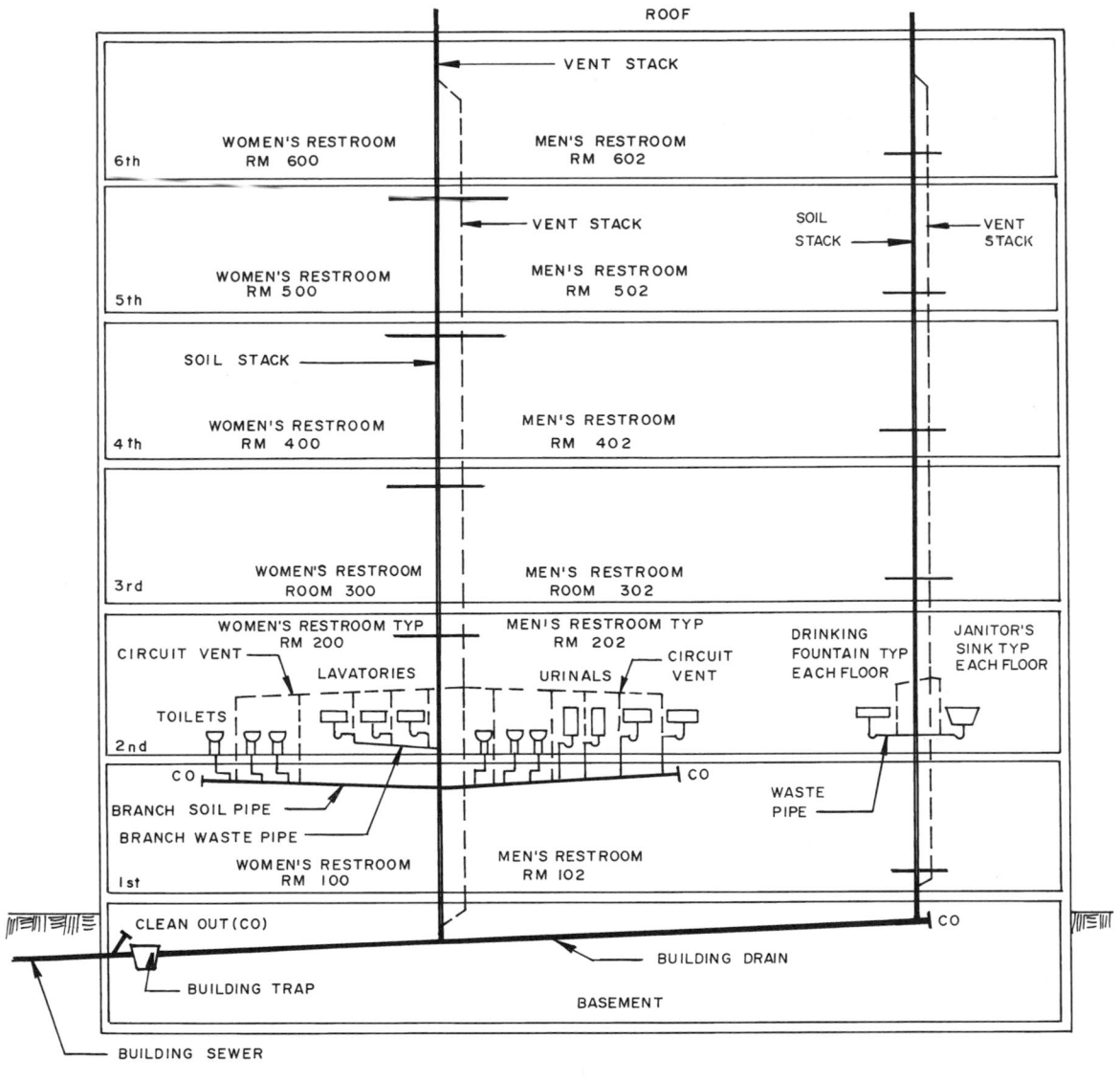

15-25 *A simplified schematic of a sanitary piping system for a low-rise multistory building.*

(see 15-26). This includes horizontal stack offset located at intermediate levels. The stack uses traditional fittings and pipe made from approved drain, waste, vent (DWV) materials. The choice is controlled by the manufacturer and codes. The Sovent stack penetrates the roof to the atmosphere much like traditional systems.

Waste flowing in a vertical pipe will cling to the interior wall and moves down the pipe in a swirling motion. This leaves an open airway in the center needed for flow to continue. As the falling waste gathers speed it meets air resistance and flattens out and may form a complete blockage of the pipe. This blockage is elimi-

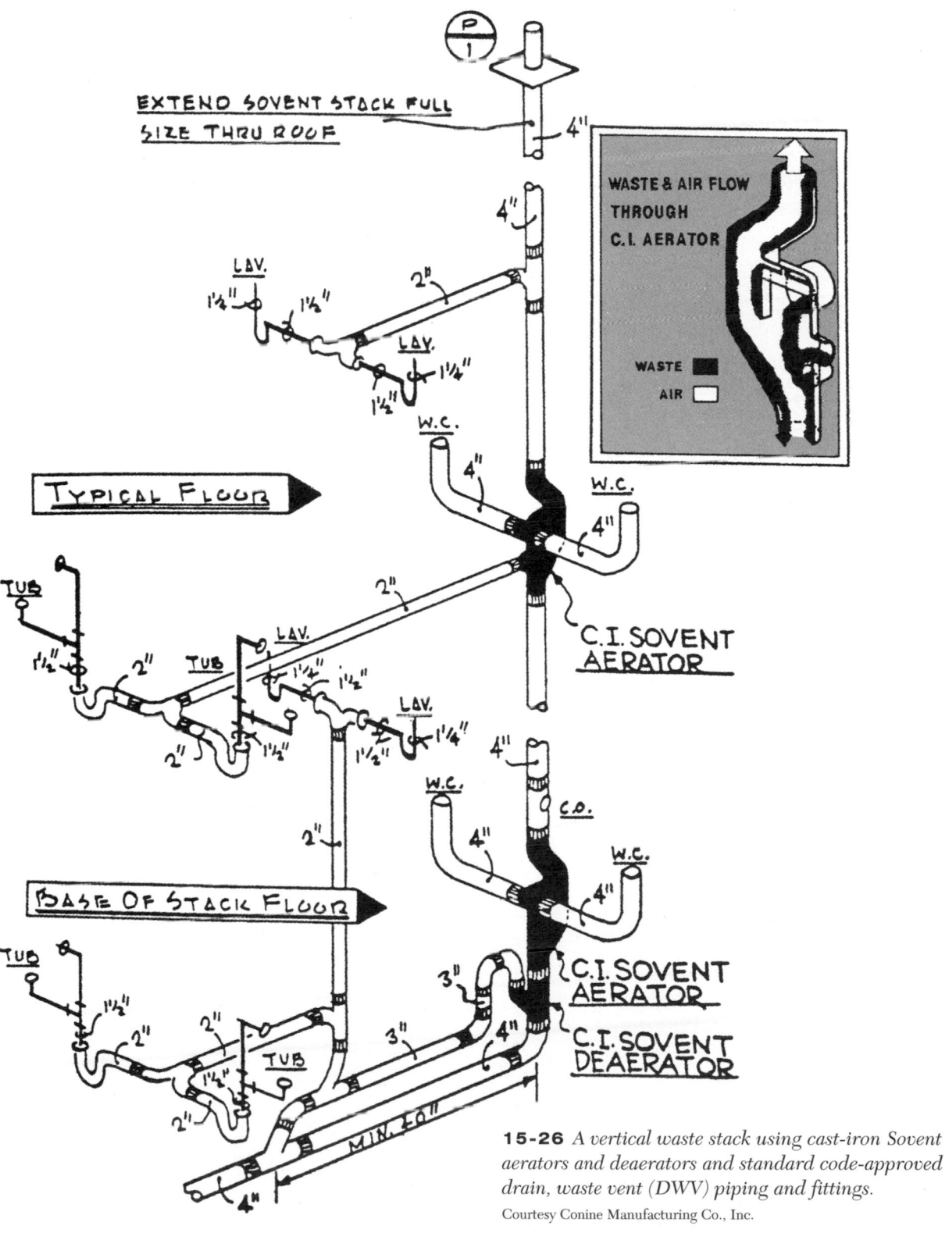

15-26 *A vertical waste stack using cast-iron Sovent aerators and deaerators and standard code-approved drain, waste vent (DWV) piping and fittings.*
Courtesy Conine Manufacturing Co., Inc.

nated by the Sovent aerator fitting. The waste enters the offset chamber, which breaks any attempt of the waste to form a plug and reduces its velocity. As the waste exits the offset chamber it clings to the interior pipe surfaces leaving an open center air space. It then enters the mixing chamber. Here the incoming waste from horizontal branches hits the baffel, drops, and is mixed with the down falling waste. The waste flows down the stack to the lowest level that contains the deaerator fitting.

The deaerator fitting is designed to handle the pressure fluctuations that occur when falling water suddenly turns horizontal. Here the velocity is slowed while additional waste continues to pile behind it causing a wave (hydraulic jump) to form and possible development of positive and negative pressures. The water strikes the nosepiece, which reduces its velocity and allows the air and waste to separate and continue to flow down the horizontal pipe. The pressure-relief line provides an outlet for pressures developed and eliminates any chance of the systems developing a vacuum, thus pulling the water out of the traps at each fixture.

Nonpotable Water Systems

National awareness of the need for water conservation has produced a need for new technologies and alternate methods of handling water and wastewater. In a building where large amounts of water are used, producing a considerable flow of wastewater, recyling this flow helps conserve the water supply and reduces the load on wastewater disposal facilities.

A schematic for an in-building wastewater treatment and recycling facility is shown in 15-27. Potable water is delivered to the drinking fountains, lavatories, and other uses where it is needed (1). This is a completely separate system from the nonpotable system. The toilet uses nonpotable water. The wastewater enters a pretreatment trash trap (2) and a sump (3) provides temporary storage in case of a mechanical failure of the system. Usually this will hold several days supply of wastewater. A tank truck can be used to haul away excess wastewater if necessary. The wastewater proceeds automatically through the various processes, including biological treatment, filtering, color removal, disinfection, and ozone treatment, producing treated water which is kept in a storage reservoir (12). The water is moved under pressure to the piping serving the toilets (13).

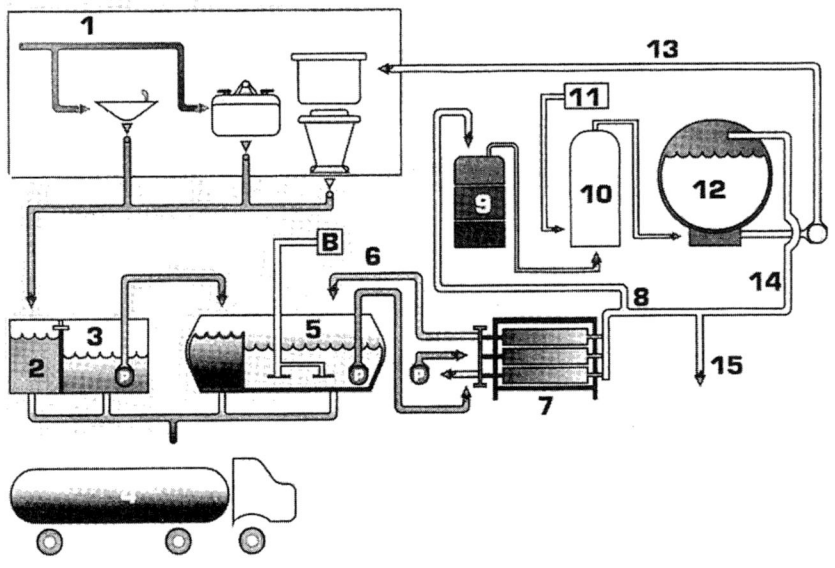

15-27 *A schematic for an in-building wastewater treatment and recycling facility.*

Courtesy SFA Enterprises, Inc. and John Irwin, Thetford Systems, Inc.

1. Potable water
2. Trash removal
3. Storage
4. Annual sludge removal
5. Biological treatment
6. Solids recycle
7. Membrane filtration
8. Filtered water
9. Color removal—activated carbon
10. Disinfection
11. Ozone
12. Treated water storage
13. Nonpotable flush water
14. Overflow
15. Low-volume, highly treated discharge to sewer or on-site soil absorption system

It is vital that when the potable and nonpotable systems are designed they are kept separate and have no chance of accidentally becoming connected. All piping must be clearly marked "nonpotable water supply" or color coded and marked with colored tape. Valves, wall outlets, and other possible places of attachment are marked with warning tags.

The National Sanitation Foundation (NSF) has established a standard for recycled water quality in the publication, *Certification Standard No. 1.*

Roof Drainage

Areas of multistory buildings such as flat roofs or balconies collect water during a storm and require a drainage system. This system is separate from the sanitary sewer and must drain into the public storm sewer system (see 15-28).

Flat roofs and other such surfaces are usually sloped toward interior roof drains. The roof drains are raised cast iron, plastic, or aluminum strainers set in the center of a sloped area. The water drains through downpipes, which may be hidden internally or mounted on the exterior of the building. The downpipes connect a storm drain below the building, which connects to the storm sewer outside the building. Another technique is to slope the roof toward scuppers and drain the surface water through them into downpipes and on to the storm sewer.

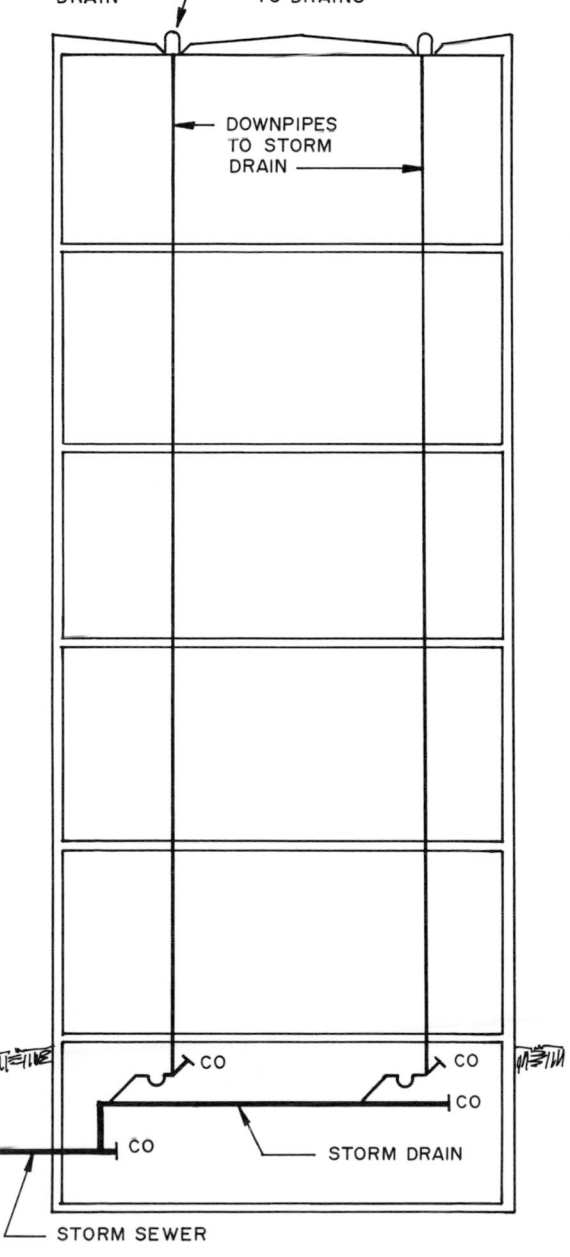

15-28 *A simplified schematic showing a storm drain system for a roof on a multistory building.*

Table 15-7 MINIMUM PLUMBING FACILITIES

Type of Building or Occupancy	Water Closets	Urinals	Lavatories	Bathtubs or Showers	Drinking Fountains
Assembly places (theaters, etc.)	1 per 1–15 2 per 16–35 3 per 36–55 Over 55, 1 per additional 40	1 per 50 males	1 per 40	—	1 per 75
Hospitals					
Individual room	1 per room	—	1 per room	1 per room	—
Ward room	1 per 8 patients	—	1 per 10 patients	1 per 20 patients	—
Restaurant	1 per 1–50 2 per 51–150 3 per 151–300	1 per 1–150 males	1 per 1–150 2 per 151–200 3 per 201–400	—	—
Worship place, assembly area	1 per 300 males 1 per 150 females	1 per 300 males	1 per toilet room	—	1 per 75
Worship place, educational & activities	1 per 250 males 1 per 125 females	1 per 250 males	1 per toilet room	—	1 per 75

Reprinted from the *Uniform Plumbing Code*™ with the permission of the International Association of Plumbing and Mechanical Officials © copyright 1994.

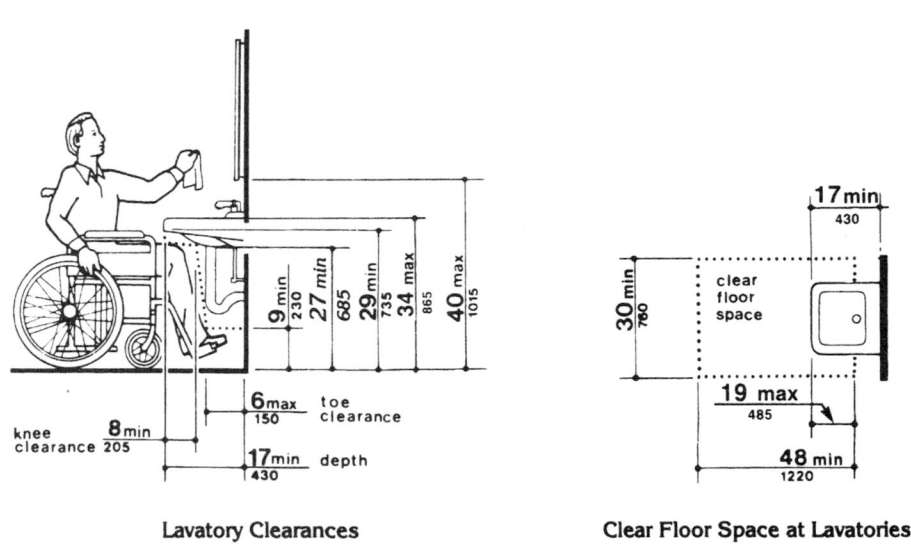

Lavatory Clearances   **Clear Floor Space at Lavatories**

15-29 *Lavatories must be mounted with the top surface not more than 34 in. (865 mm) above the floor. They must be positioned so a person in a wheelchair will have the required knee room. Any exposed pipes must be located out of the way and any sharp edges must be covered.*

From *Americans with Disabilities Act*, U.S. Architectural and Transportation Barriers Compliance Board, Washington, D.C.

Plumbing Fixtures

One decision to be made as a plumbing design is prepared is how many fixtures are required. Minimum requirements are specified in the various plumbing codes. An example of some selected types of building occupancy requirements as specified by the Uniform Plumbing Code are in Table 15-7. These are minimum recommendations and known conditions may warrant the use of additional fixtures.

Plumbing fixtures require a steady supply of clean water to assist with the discharge of waste materials. They are under constant wear from water, bacteria, and other harmful elements. Therefore they must be made from durable materials having a smooth, nonporous surface. Typical materials include stainless steel, copper, brass, enameled cast iron, vitreous china, molded plastics, gel-coated fiberglass, and acrylic-faced fiberglass.

Special fixtures and design specifications are available in the publication, *Americans with Disabilities Act*. It records heights, spacings, grab bar requirements, and other features required for all aspects of a buildings design to accommodate those with physical disabilities. Recommendations relating to both fixture placement for lavatories (see 15-29) and toilets (15-30) illustrate some of the requirements.

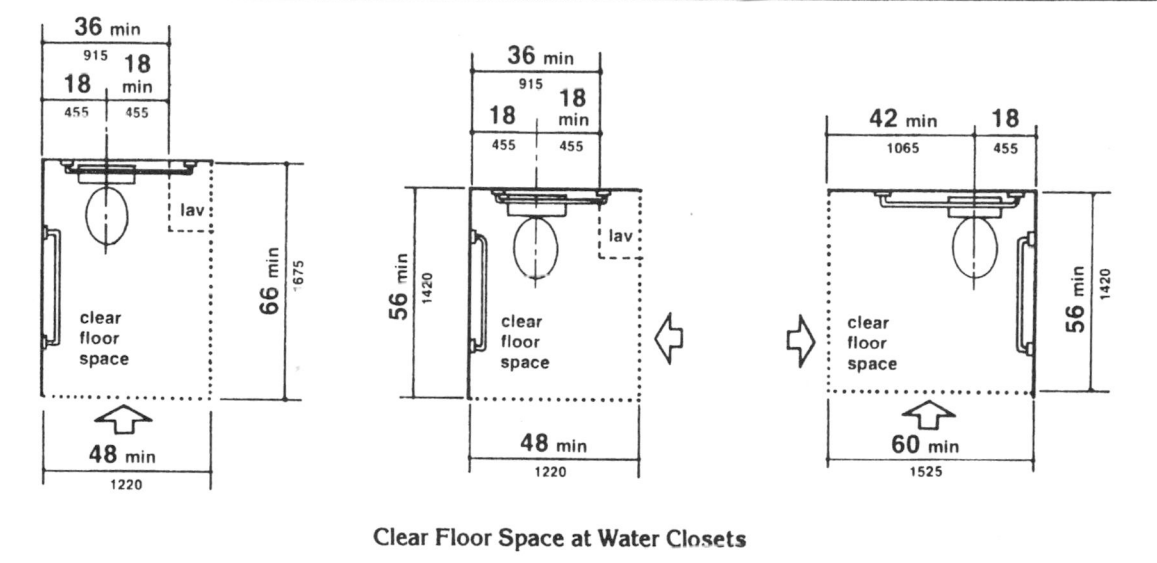

Clear Floor Space at Water Closets

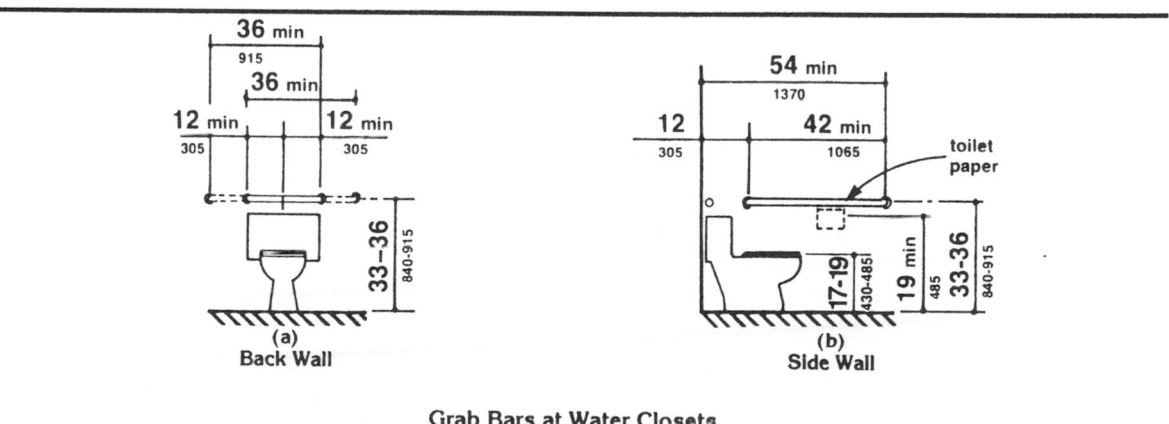

Grab Bars at Water Closets

15-30 *Water closets require a minimum of 48 inches (1220 mm) clear of other items. The water closet must be between 17 and 19 inches (430 and 485 mm) high. Grab bars to the side and rear are required.*

From *Americans with Disabilities Act*, U.S. Architectural and Transportation Barriers Compliance Board, Washington, D.C.

Water closets (toilets) are available that are wall hung or floor mounted. Floor mounted tank type are common in residential buildings. Wall hung water closets are widely used in industrial and public restrooms because they make it easier to keep the floor clean around them. They do require special wall construction to carry the heavy weight involved (see 15-31). Residential water closets typically use a flush tank while commercial establishments use a high pressure flushing system. Water closets have various methods for flushing. Some are quiet and use less water than others.

Urinals are wall hung units used in men's restrooms in public facilities to reduce the number of water closets required. They are less expensive to install and take little space (see 15-32). Some types have privacy shields between units. They may have a manual high pressure flush valve or be connected to an automatic high pressure flushing system. Sometimes one is set lower than the others.

Lavatories in residences are typically mounted in a base cabinet. The lavatory may be set into a top covered with plastic laminate or ceramic tile. Others have a molded plastic top with the lavatory bowl and top one integral piece (see 15-33). This reduces the problems that occur around the stainless steel edge of those set into the top. Another popular lavatory is a pedestal type. The lavatory is actually wall hung and the pedestal covers up the plumbing below (see 15-34). Wall hung residential lavatories often have decorative metal legs instead of a pedestal.

Lavatories in public facilities are typically wall hung units though some use lavatories set into a wall hung counter top without a cabinet base below. These usually have self-closing faucets which prevent someone from letting the water run after they leave. One type of faucet has an infrared control that turns on the water when hands are placed below the faucet and turns it off when the hands are removed. This saves on water use and reduces the cost of hot water.

15-31 *This wall hung toilet has an exposed flush valve.*
Courtesy Universal-Rundle Corporation

15-32 *A wall hung urinal with an exposed flush valve.*

15-34 *A pedestal lavatory is decorative and takes up very little space.*

15-35 *A round freestanding stainless steel wash fountain. It measures 28 inches (711 mm) from the floor to the top of the rim.*
Courtesy Bradley Corporation

15-33 *(Right) This molded-plastic bathroom lavatory has the top and bowl cast as a single unit.*

Washfountains are used in restrooms and dressing rooms where a large number of people will appear and need to wash up at the same time such as in an industrial plant. Washfountains are typically half round wall hung units or round free-standing units. Large units can accommodate up to eight people at one time (see 15-35).

Kitchen sinks are available in a range of types with the most typical being some form of two-bowl unit. However single- and triple-bowl units are available. The one bowl is for general use and the second bowl will contain the garbage disposal. They are installed in plastic laminate or ceramic tile covered tops.

Service sinks are used by janitors to clean mops and other cleaning activities. They are usually wall hung and deep. Mop service basins are floor mounted units about 1 ft. (305 mm) high.

Bathtubs are available in a range of sizes and designs. Some have two sides closed and fit in a corner. Others have one side closed and fit between end walls (see 15-36). Square and rectangular units are available. Whirlpool tubs have pumps that circulate the water forming a whirlpool effect. Some bathtubs made from gel-coated fiberglass have the tub and wall enclosure formed as a single unit. This leaves no cracks to form mold and is easy to clean (see 15-37).

Showers widely used are gel-coated fiberglass units that are in one piece. They are free from any joints making them one solid, easy to clean unit. Other types are available that have a cast terrazzo base and walls are covered with ceramic tile or have a prefabricated enclosure made from galvanized-bonderized steel with an enamel finish or molded plastic panels.

Various types of **drinking fountains** are available. Some are wall-hung and protrude into the room. Others are recessed or semi-recessed into the wall. Free-standing and some recessed drinking fountains have a water-cooling system in the base producing temperature-controlled water (see 15-38). Units that provide access for the handicapped are wall mounted and have an electrically activated push button valve on the front that requires only a light touch. Some pedestal types are designed for outdoor use and may have a foot operated valve that is frost-proof. Stainless steel is the major material used on the surfaces exposed to water. Drinking fountains require a sanitary drain as well as a source of potable water.

15- 36 *This bathtub fits between end walls and has one side exposed.*
Courtesy Universal-Rundle Corporation

15-37 *This fiberglass molded tub, shower, and wall covering unit has smooth, easy to keep clean junctions between the various sections.*
Courtesy Universal-Rundle Corporation

15-38 *Drinking fountains are available as free-standing, semi-recessed, and recessed units.*
Courtesy EBCO Manufacturing Company

HEATING, VENTILATING, AIR-CONDITIONING (HVAC) & REFRIGERATION

The design of a heating, ventilation, and air-conditioning (HVAC) system involves many factors. A solution for one building may be insufficient for another structure. Solutions will vary depending upon the occupancy. For example, an office building will give major consideration to human comfort while a manufacturing plant will have a range of other considerations. The design of the building must be considered. For example, factors such as type and amount of glazing, insulation, air infiltration and exfiltration, heat and cooling requirements due to machinery, industrial processes, solar load, materials used for walls, ceilings, floors, and roofs and their coefficient of thermal conductivity, and space within the building provided to house mechanical units and access various parts of the structure with ducts, pipes, and other parts of a system are considered. Some other factors such as a need to control humidity, removal of chemicals, noxious gases, or dust, availability and cost of fuel, cost of the various systems and the expected maintenance expenses and possibility of down time and climatic factors, illustrate a few additional things a HVAC engineer has to consider as a system is designed.

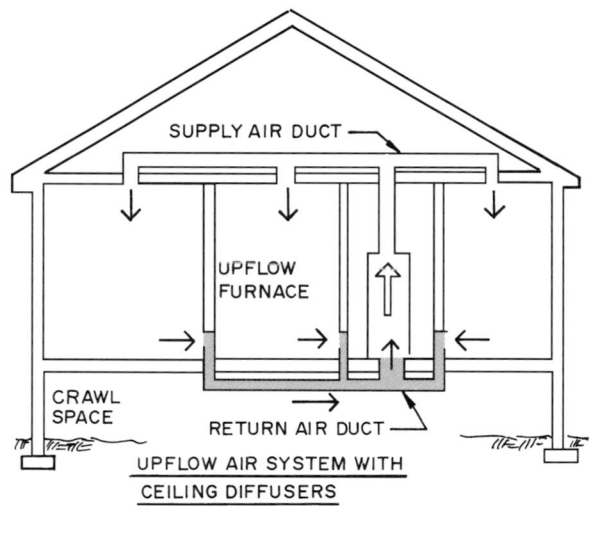

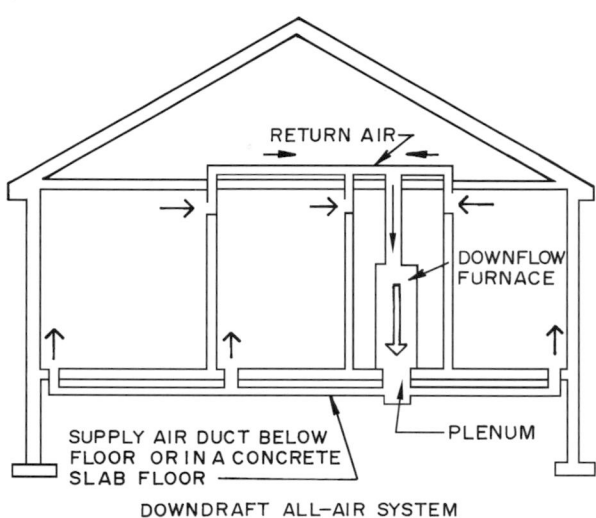

15-40 *Down flow warm air heating systems are used in a building with a crawl space or concrete floor.*

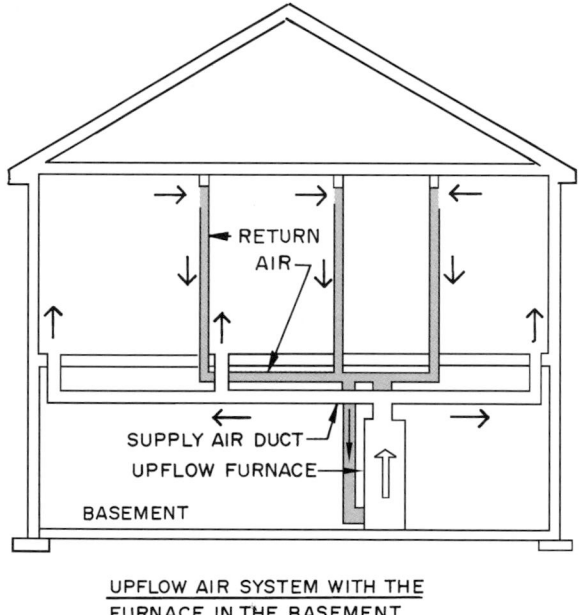

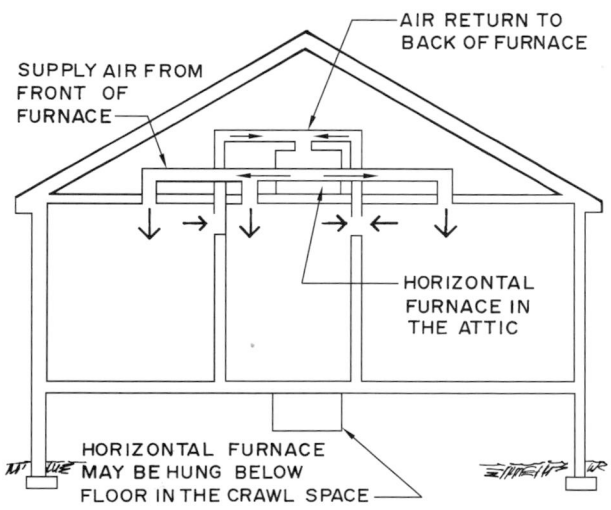

15-39 *Typical up-flow warm air heating systems.*

15-41 *Horizontal warm air furnaces are hung below the floor or in the attic.*

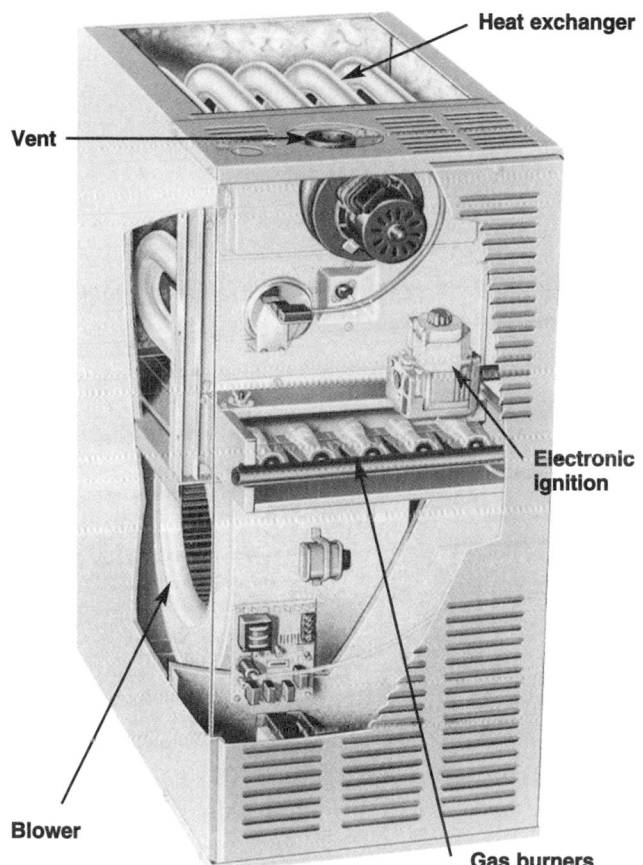

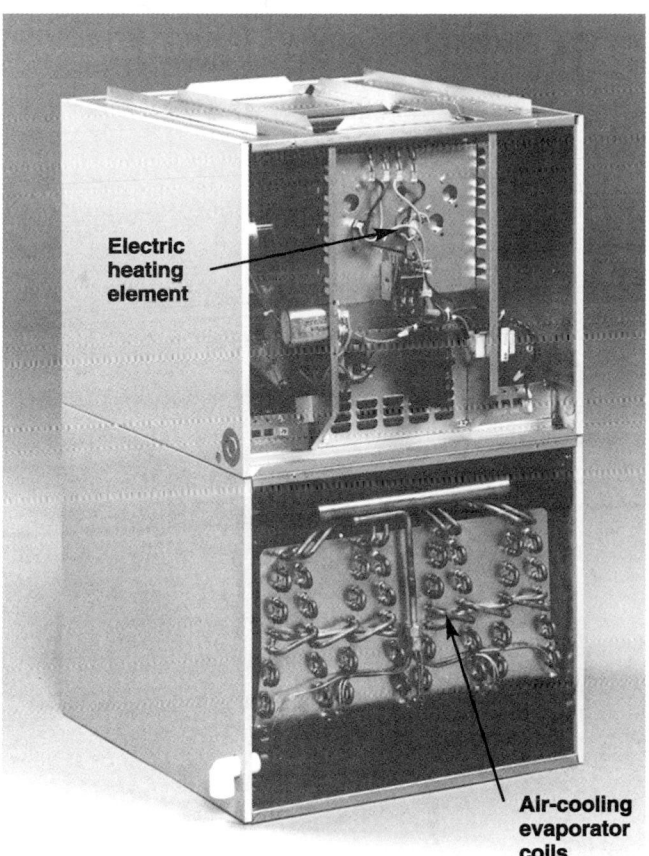

15-42 *This gas fired warm air furnace can be converted from the factory shipped up-flow/horizontal configuration to the down-flow mode.*
Courtesy Lennox International, Inc.

15-43 *This electric central heating/cooling furnace has options of 5 to 25KW and up to 5 tons of cooling capacity.*
Courtesy Rheem Manufacturing Company

Fuel & Energy Used for HVAC

The major **fuels** include butane, coal, fuel oil, natural gas, propane, solid waste and wood. Butane, coal, fuel oil, propane, solid waste, and wood require some form of storage be used. The **energy sources** include electricity, geothermal, ground heat, and solar.

Warm Air Heating Equipment

Warm air systems basically have some type of heat generating device (furnace), controls, and a distribution system of ducts. The furnace may be fired by oil, gas, or electricity and some solar systems use warm air distribution. An airconditioning unit can be installed as a part of the furnace so the system can heat and cool the building.

WARM AIR FURNACES

Warm air furnaces use blowers to move the heated air through ducts. These include upflow, downflow, and horizontal types. An **upflow warm air furnace** moves air out the top of the unit into the duct system. It is used when the furnace is in a basement or on the floor when the ducts are to be run in the attic (see 15-39). A **downflow warm air furnace** moves the air out the bottom of the unit and are typically used when the ducts are in a concrete slab floor or below the floor in a crawl space (see 15-40). **Horizontal warm air furnaces** are mounted in the attic or hung below the floor joists as shown in 15-41. Each of these may have an air-cooling unit which uses the furnace blower and ducts to provide cool air to the building spaces.

In 15-42 is shown the internal construction of a gas-fired warm air furnace. The blower operates on two speeds and also serves to circulate cool air if an air-cooling unit is mounted on top of the furnaces. They are available in a wide range of sizes and output (BTUH)

Warm air furnaces may also use electric resistance coils to heat the air that is then distributed through ducts by a blower. The furnace is similar to that shown for gas furnaces. A major problem is the high cost of electricity in many areas (see 15-43).

Oil-fired furnaces require a large storage tank outside the building from which the fuel oil is pumped to the burner in the furnace. The location and method of placement is regulated by codes. The oil tank construction must meet Underwriters Laboratory (UL) specifications. Tanks may be placed above or below ground. If below ground they should be below the frost line. Some codes permit oil storage tanks inside the building.

The oil-fired furnace is much like the gas-fired furnace (see 15-44) except it has an oil burner unit having a nozzle that breaks the oil into a fine spray that is ignited by a high voltage ignition using spark electrodes to ignite the vapor.

HEAT PUMPS

A heat pump is a machine that can heat or cool a building using forced air through ducts. Basically it is a machine taking heat from one source, such as the outside air, and transferring it to another area, such as the air inside of a building. The commonly available types include air-to-air, water-to-water, and air-to-water. Ground source heat pumps are finding increasing use.

15-44 *This oil-fired furnace is available in up-flow, down-flow, and horizontal units with a BTU/hr heating capacity up to 150,000.*

Courtesy Rheem Manufacturing Company

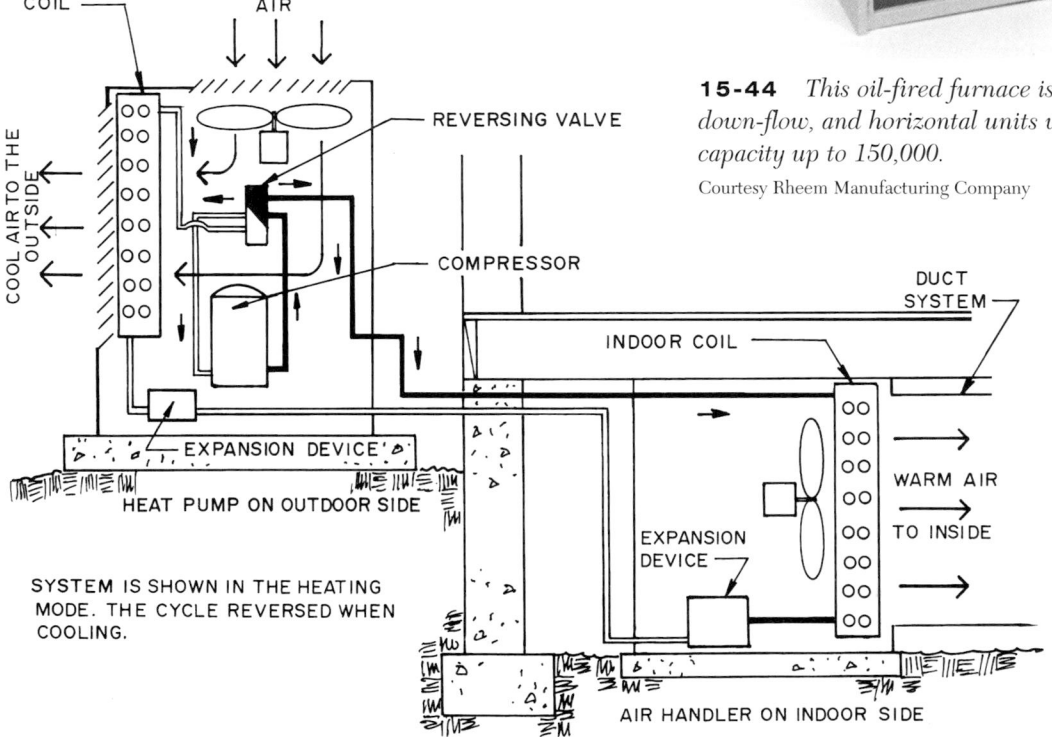

15-45 *The heating cycle of an air-to-air electric heat pump.*

The most frequently used type is an air-to-air unit. The unit has a compressor similar to that used in a refrigerator. A refrigerant (a gas) is circulated in coils. When in a heating mode the coils in the outdoor unit enable the refrigerant to absorb heat from the outdoor air. The refrigerant is moved to the compressor where it is compressed raising its temperature. It moves to indoor coils where indoor air is blown over them by a blower removing the heat and moving it through ducts into the building (see 15-45). When in the cooling mode the reverse happens. Heat is absorbed by the inside coils and dispersed to the outside air by the outside coils.

The groundwater heat pump (GWHP) is a water-to-air system. A water-to-air system pumps water from a well to the heat pump. The well water typically maintain a constant temperature, which, while it will vary with the geographic area, will typically fall between 55 to 65°F (13 to 18°C). The refrigerant in the heat pump coils absorbs the heat in the water and discharges it into the building through an air handling unit and ducts. It can reverse the process and remove heat from inside the building and discharge it to the water which is disposed of in a disposal well.

Water-to-water heat can be used to heat water and at the same time cool water. In an operation where both are required at the same time the heat pump can remove heat from a liquid source, as some fluid in a manufacturing operation that needs chilling, and move the heat to another liquid, such as producing hot water for cleaning operations.

The **ground-loop heat pump** is used in areas where groundwater is not readily available, too costly to acquire, or of poor quality for heat pump use. The system does avoid the cleaning requirements required of groundwater systems due to minerals in the water. The system may use an earth coil in a horizontal loop or be installed vertically in drilled holes. The earth coil system transfers heat from or to a water/antifreeze solution circulated through plastic pipes buried in the earth. The heat from the earth is transferred to the refrigerant in a heat exchanger and compressed and distributed by a blower through ducts. In the cooling mode the reverse occurs with indoor heat being transferred to the water/antifreeze solution, which disperses it to the cooler ground.

Air-Conditioning with Duct Systems

Duct systems can also be used to air condition a building. The heat pump previously discussed provides both heating and cooling modes. When oil, gas, or electric warm air furnaces are used a cooling coil unit is placed on the furnace and connected to an outside air conditioner as shown in 15-46. They are connected by pipes carrying the refrigerant from the compressor in the outdoor unit to the indoor coil. The furnace blower moves air over the cold coils and through the ducts into the building. The refrigerant absorbs heat from the air blown over the indoor coil. The refrigerant moves to the outdoor air-conditioning compressor where the heat it has absorbed is transferred to the atmosphere and the refrigerant moves back to the interior coil to repeat the process.

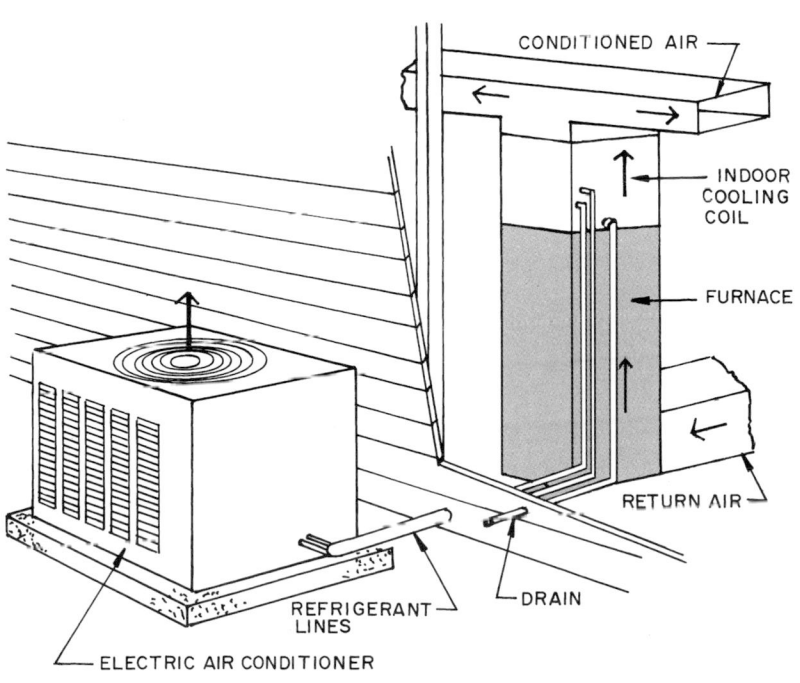

15-46 *An electric air conditioner unit is installed out of doors and supplies the chilled refrigerant to cooling coils on the furnace.*

All-Air Distribution Systems

Procedures for designing all-air duct distribution systems can be found in the ASHRAE Handbook-Fundamentals. All-air duct systems may be low or high velocity with the high velocity design requiring smaller diameter pipes and high pressures. The designer must consider how the size of the ducts influences performance, outside air requirements, supply air temperatures, airflow, air changes, zoning humidity control, heat gain and heat loss of the space to be conditioned, control of noise in the system, possible use of heat recovery devices, and many special factors such as exist in hospitals, manufacturing plants, and laboratories.

The systems in common use include single duct, multizone, reheat, variable air volume, and dual duct. Various combinations of these can be used to meet design requirements.

SINGLE ZONE SYSTEMS

Single zone systems use one air handling unit (AHU) to supply an entire building or a portion of a building that is considered a single zone. The furnace and air handling unit can be installed outside of or within the space to be conditioned. A return air duct system is usually required. A typical example is a small residence. A large residence may have two or more heat and cooling sources using two or more single zone systems. A simple schematic is in 15-47. It should be noted that this system supplies air at a constant rate. Therefore room temperature is varied by changing the air temperature.

REHEAT SYSTEMS

A reheat system is a variation of the single duct single zone and multizone systems. It supplies a single source of preconditioned or recirculated air at a constant rate through ducts to several zoned areas. The air is processed through the air handling unit where it can be filtered, humidified, and cooled. The temperature of the air is that required for the coolest zone. The cool air is then sent through ducts to each zone. The duct to each zone has a reheat coil which heats the incoming air to the temperature required for that zone. The reheat can be an electric resistance unit or a hot water or steam coil. This system permits the simultaneous cooling and heating of areas with different requirements.

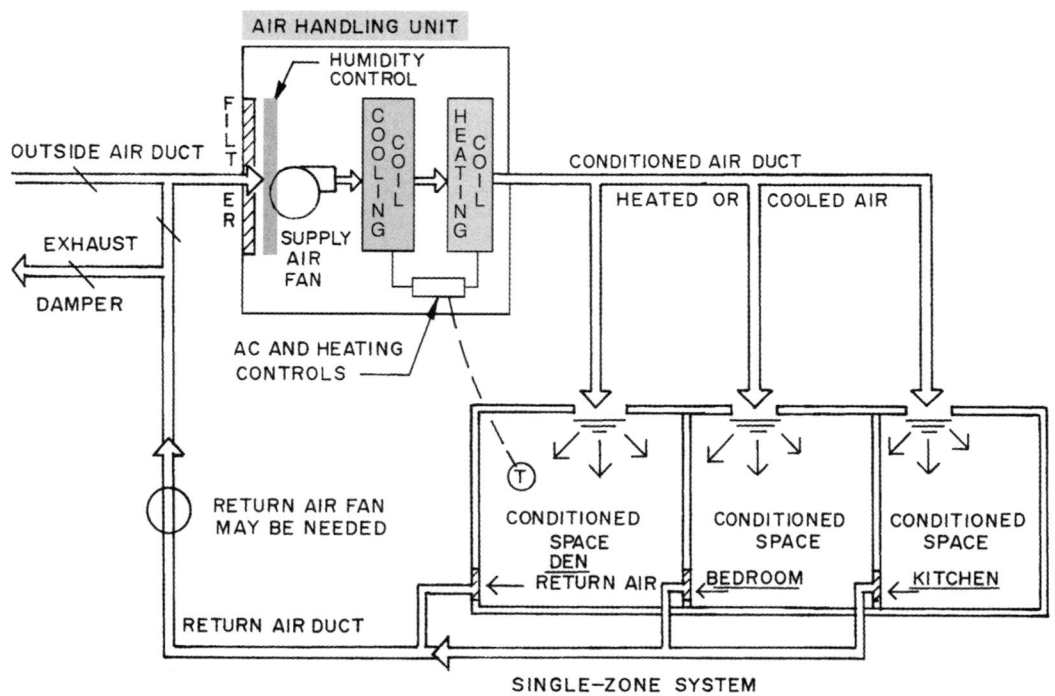

15-47 *A single zone heating/cooling duct system for a small building that has a single source of heating and cooling.*

VARIABLE AIR VOLUME SYSTEMS

Variable air volume (VAV) systems control the temperature in a space by supplying air at a constant temperature and varying the **quantity of air** supplied rather than changing the temperature of the air as is done in the single duct, reheat, and multiple zone systems. The air from the air handling unit is moved through single ducts to each zone where a variable air volume terminal is located. This terminal varies the air supply to the space while the air temperature is held constant. Variable air volume systems are not useful in situations where the control of humidity is important.

DUAL DUCT SYSTEMS

Dual duct systems move conditioned air in the air handling unit to the spaces by two parallel ducts. One duct carries cold air and the other warm air. At each space mixing valves combine the warm and cold air in the proportion needed to meet the air temperature requirements of the space. These may be constant volume or variable volume air systems. Constant volume systems may use a reheat. Variable air volume systems may use a single fan or dual supply fans.

Additional details on these systems can be found in the *ASHRAE Handbook, HVAC Systems and Equipment,* American Society of Heating, Refrigerating and Air-Conditioning Engineers.

Types of Duct System

The air distribution system receives heated or cooled air from the furnace. It moves into the air handling unit where it may be filtered, humidified, or dehumidified. The air handling unit has a fan, filters, humidifiers, coils, and dampers. From the air handling unit the conditioned air is moved through ducts to the diffusers in the various rooms to be heated or cooled and a separate system of ducts moves air from these spaces back to the furnace for reconditioning, possibly exhausting some of the air and adding fresh air brought in from the outside. All of this is accomplished by a series of electrical controls.

DUCTS

Warm air distribution systems use ducts to move the heated air from the furnace to the diffusers (outlets) in the various rooms. The design of the ducts and selection of materials is vital to a properly functioning system. The designer must carefully calculate the sizes required, ascertain the air velocity to use, and calculate the pressure.

A standard for duct design is published by the Air Conditioning Contractors of America (ACCA) *Manual D, Duct Design for Residential Winter and Summer Air Conditioning.* The Sheet Metal and Air Conditioning Contractors National Association also has publications relating to duct design and installation.

DUCT CLASSIFICATION

Duct systems are regulated by various laws, building codes, local ordinances, and standards. These must be considered by the engineer as the duct system is designed. Projects built for the federal government will have standards issued by various agencies such as the General Services Administration and the Federal Construction Council.

▼ Duct construction is classified in terms of the pressure and use.
  Commercial duct systems include HVAC systems for applications such as educational, business, general factory, and mercantile structures.
  Industrial duct systems include those used for industrial exhaust and air pollution control.
  Residential ducts are specified by local building codes. An often used source for multifamily dwellings is National Fire Protection Association Standard 90A. Supply ducts may be galvanized steel, aluminum, or other materials rated by Underwriters Laboratory Standard 181. Rigid and flexible fiberglass supply ducts must meet the standard, Fibrous Glass Duct Construction Standards of the Sheet Metal and Air Conditioning Contractors National Association.

▼ Commercial ducts are also usually regulated by NFPA Standard 90A and UL181. This classifies ducts into two groups:
  Class O—zero flame spread, zero smoke spread
  Class 1—25 flame spread, 50 smoke developed

Class O ducts are of iron, steel, aluminum, concrete, masonry, or clay tile. Class I ducts include many of the flexible and rigid fiberglass ducts manufactured.

▼ Industrial ducts are specified by NFPA Standard 91. These are used for duct systems that might convey flammable vapors or air containing various particles.

▼ Partical-conveying ducts are available in four classifications.

Class 1—Nonparticulate applications such as
makeup air, general ventilation, and gaseous
emission control.

Class 2—Moderately abrasive particles in the
air such as sanding or buffing.

Class 3—Highly abrasive material in low concentration such as handling sand or abrasive
cleaning.

Class 4—Highly abrasive particles in high concentration.

Abrasive ratings are specified in Round Industrial
Duct Construction Standards by the Sheet Metal and
Air Conditioning Contractors National Association
(SMACNA).

Industrial ducts are generally galvanized steel,
uncoated carbon steel, or aluminum. Aluminum is not
used if the air contains abrasive particles. Those carrying corrosive vapors must have appropriate protective
coatings.

FORCED-AIR DUCT SYSTEMS

The duct systems used in residential and small commercial buildings include the perimeter loop, perimeter radial, and extended plenum.

Perimeter loop is typically used with concrete slab
floors and a downflow furnace. However it could
be used in a building with a basement or crawl
space (see 15-48). The perimeter duct is placed in
the thickened edge of the slab as shown. It is
essential that the edge and bottom of the slab be
insulated. Registers are placed along the perimeter
duct as needed. The return ducts in this system
would be in the attic.

Perimeter radial system is also used in concrete
slab construction but can be used in basements
and crawl spaces. It uses a downflow furnace (see
15-49). A variation of this system uses an upflow
furnace with the radial ducts in the attic.

Extended plenum system may have the furnace in
the basement or first floor. It can also be used with
horizontal furnaces in the crawl space or attic. The
plenum is extended to provide the needed airflow
to feed the ducts running from it to outlets.

If the furnace is on the first floor the plenum
could be in the attic and the ducts run over the
ceiling joists with diffusers running through the
ceiling into the room (see 15-50). Horizontal warm
air furnaces are commonly placed below the floor
or in the attic.

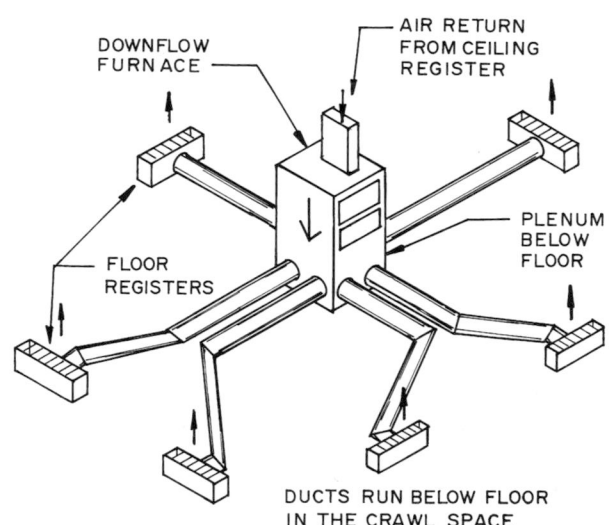

15-48 *A perimeter loop warm air duct system circulates
air through a continuous duct system fed by several horizontal ducts.*

15-49 *A perimeter radial duct system extends individual
ducts to each space to be heated and cooled.*

Humidifiers & Dehumidifiers

Control of the relative humidity in a space is an important factor in the overall conditioning of the environment. The requirements vary depending upon the occupancy. Humidity is the water vapor within a space. Relative humidity is a ratio of the weight of water vapor actually in the air to the maximum possible weight of water vapor the air could contain when at the same temperature. It is expressed as a percentage. For example, if the relative humidity is 100 percent the air can hold no more water vapor and an increase in water vapor will cause moisture to condense and form water drops. Human comfort depends a great deal upon the relative humidity of the air, typically indoor relative humidity should be kept between 30 to 60 percent. Low humidity causes drying of the membranes of the nose, throat, skin, and hair. Furniture, cabinets, interior trim, and other wood products can shrink and check if the relative humidity is too low. Likewise high relative humidity can cause doors and drawers to swell and stick. Heating the air in the winter removes moisture so a humidifier is used to increase the relative humidity. In the summer the air in many geographic areas has a high relative humidity. The air-conditioning system must dehumidify (remove water vapor) the air.

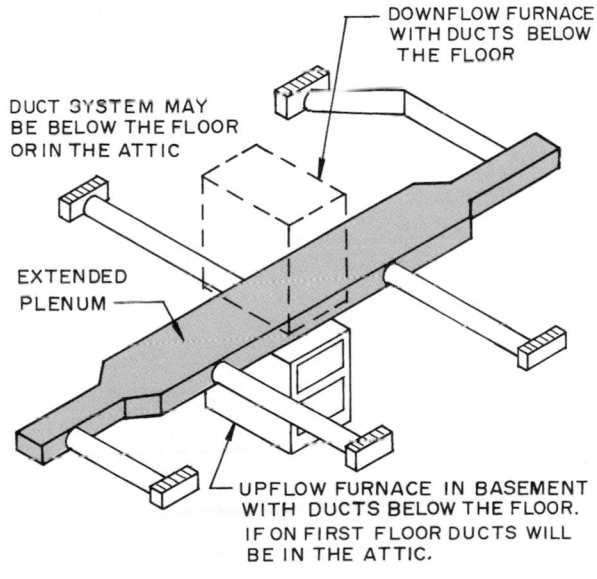

15-50 *The extended plenum runs from the furnace and individual ducts are taken off it to the rooms to be heated and cooled.*

HUMIDIFICATION EQUIPMENT

Humidifiers increase the amount of water vapor in the air. They operate in two different ways. One type adds heat and as the air flows through the humidifier wet section it picks up moisture and the air is then distributed to the room. The other type takes heat from the air flow.

DESICCANT DEHUMDIFICATION

Dehumidification involves the removal of water vapor from the air, gases, or other fluids by some mechanical or chemical means. A dehumidifier is a device used to remove moisture from the air thus reducing the relative humidity.

Desiccation is the use of a desiccant for removing moisture from a material. A **desiccant** is any absorbent liquid or solid that is used to remove water or water vapor from a material. **Absorbent materials** will extract substances from a liquid or gas medium with which it is in contact. Absorbent materials have the ability to have molecules of gases, liquids, or solids to adhere to its surfaces without changing the absorbent material chemically or physically.

Dehumidification is typically accomplished in residences by the cooling of air during the air-conditioning mode. For commercial and industrial applications other methods are available.

There are many industrial applications where dehumidification is required. Among these are the need to maintain a dry atmosphere in a warehouse, producing dry air when needed to aid in the drying of a material in an industrial process, to dry natural and liquified gas, and to lower the relative humidity in a plant manufacturing products using hygroscopic materials, such as wood.

Dehumidification equipment uses both solid and liquid desiccant materials. **Desiccant** materials are those that are absorbent or adsorbent and are used to remove water or water vapor from a material. They may be liquid or solid.

Hydronic (Hot Water) Heating Systems

Hydronic heating systems are used in residential and commercial installations. A system consists of a boiler fueled by oil or natural gas, and a system of pipes, radiators, pumps, and controls. Three types of systems are typically used in residential and small commercial buildings. These are one-pipe, two-pipe direct return, and two-pipe reverse return.

ONE-PIPE SYSTEMS

One-pipe systems use a single loop of pipe to circulate water at 180°F (82.9°C) to radiators and to return the cooler water back to the boiler to be reheated and recirculated as shown in 15-51. Notice that radiator 1 receives the water directly from the boiler from which it flows into radiator 2. The water temperature at radiator 2 will be lower than at radiator 1. This loss can be compensated for by having a larger No. 2 radiator and the loss continues through the remainder of the radiators on the loop. This system is difficult to keep in balance.

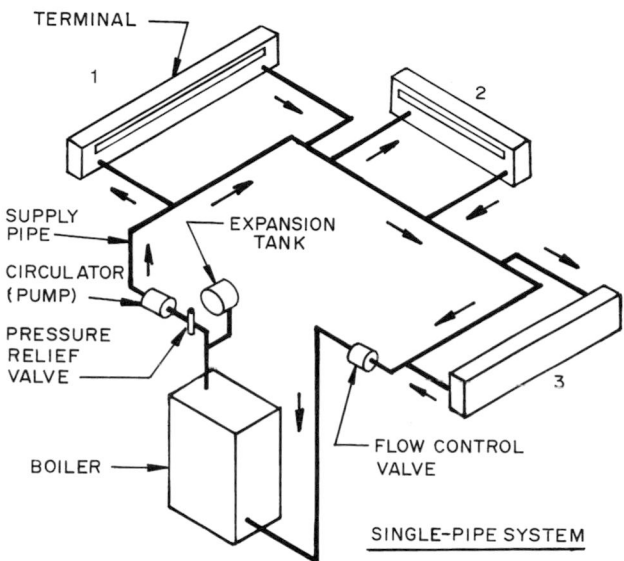

15-51 *A single-pipe hydronic system carries the water from the boiler through each terminal and back to the boiler.*

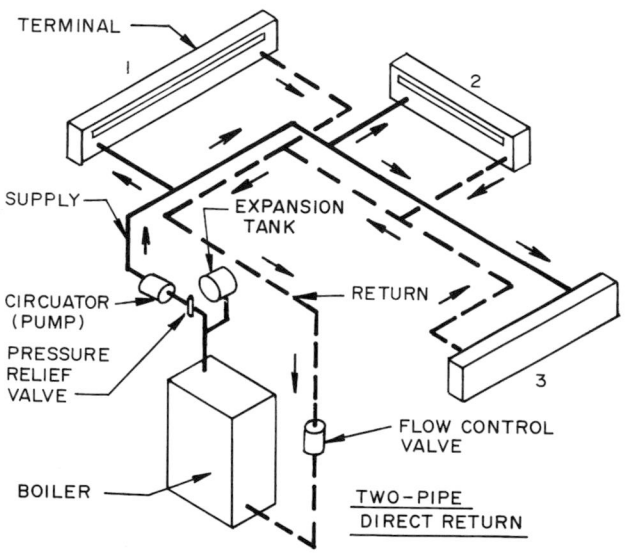

15-52 *The two-pipe hydronic direct return hot water system has a separate return line for the cooled water which flows in a direction opposite the supply line.*

TWO-PIPE DIRECT RETURN SYSTEM

All the radiators in this system receive hot water directly from the boiler through the hot water supply line. The cool water from each radiator is returned directly to the boiler through a separate loop of pipe. The last radiator tends to receive less water because of higher pipe resistance to water flow. It has the longest hot water supply and cool water return lines. This is overcome by increasing the size of the hot water supply line and sizing the circulation pump to provide the needed flow at the radiators on the end of the circuit (see 15-52). Balancing devices are used to regulate the flow of hot water through each radiator. In general this system is not widely used because of the difficulty in getting a balanced distribution.

TWO-PIPE REVERSED RETURN SYSTEM

This system is much like the two-pipe direct return except the return water flows in the same direction as the hot water. This provides a system having about the same pipe resistance on all radiators. For example, radiator 1 has the shortest hot water supply line but the longest return line. The reverse is true for radiator 3. This system is typically used in larger buildings with longer pipe runs (see 15-53).

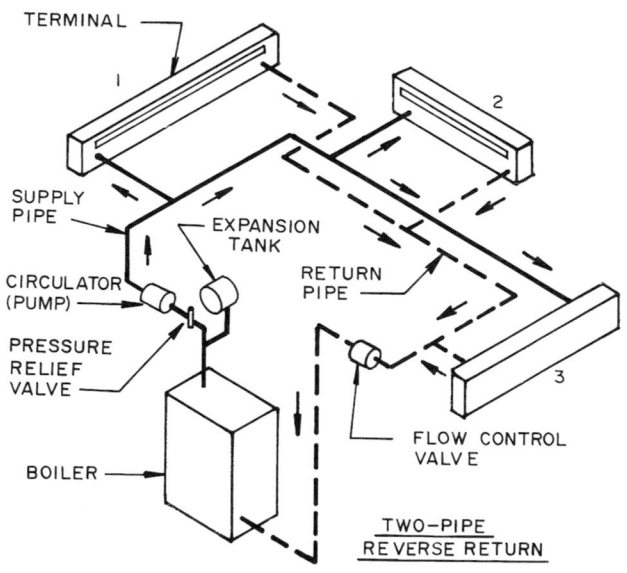

15-53 *The two-pipe reverse return hydronic hot water system has the return water flowing in the same direction as the hot water supply providing about the same pipe resistance to water flow for each terminal.*

MULTIZONE TWO-PIPE SYSTEMS

A multizone system enables the temperatures in various parts of a building to be controlled separately. Each zone has a complete two-pipe system fed from a central boiler. The flow of hot water is individually controlled to each zone. This permits some zones to be kept at lower temperatures when not in use resulting in a saving of energy costs. Large multistory and multiuse buildings will use multizone two-pipe hydronic systems (see 15-54).

HYDRONIC CONTROLS

The water temperature in the boiler (180°F or 83°C) is controlled by a thermostat immersed in the water in the boiler that regulates the operation of the oil or gas burner and maintains the proper water temperature.

A room thermostat is used to start and stop the circulation pumps regulating the supply of water to the radiators. The boiler is often used to supply hot water for domestic use. In this case it must be sized to meet both heating and domestic water demands.

Pipe Systems for Water Heating & Cooling

Water distribution systems can be used for heating with hot water and cooling with chilled water. The systems are usually closed systems and use circulators (pumps) to move the water through the system and the terminal units. The hot water is supplied by a boiler and the cold water by a chiller. A chiller is a refrigerating machine used to remove heat from the water to be circulated for cooling air in a building. Three-pipe and four-pipe distributions are used.

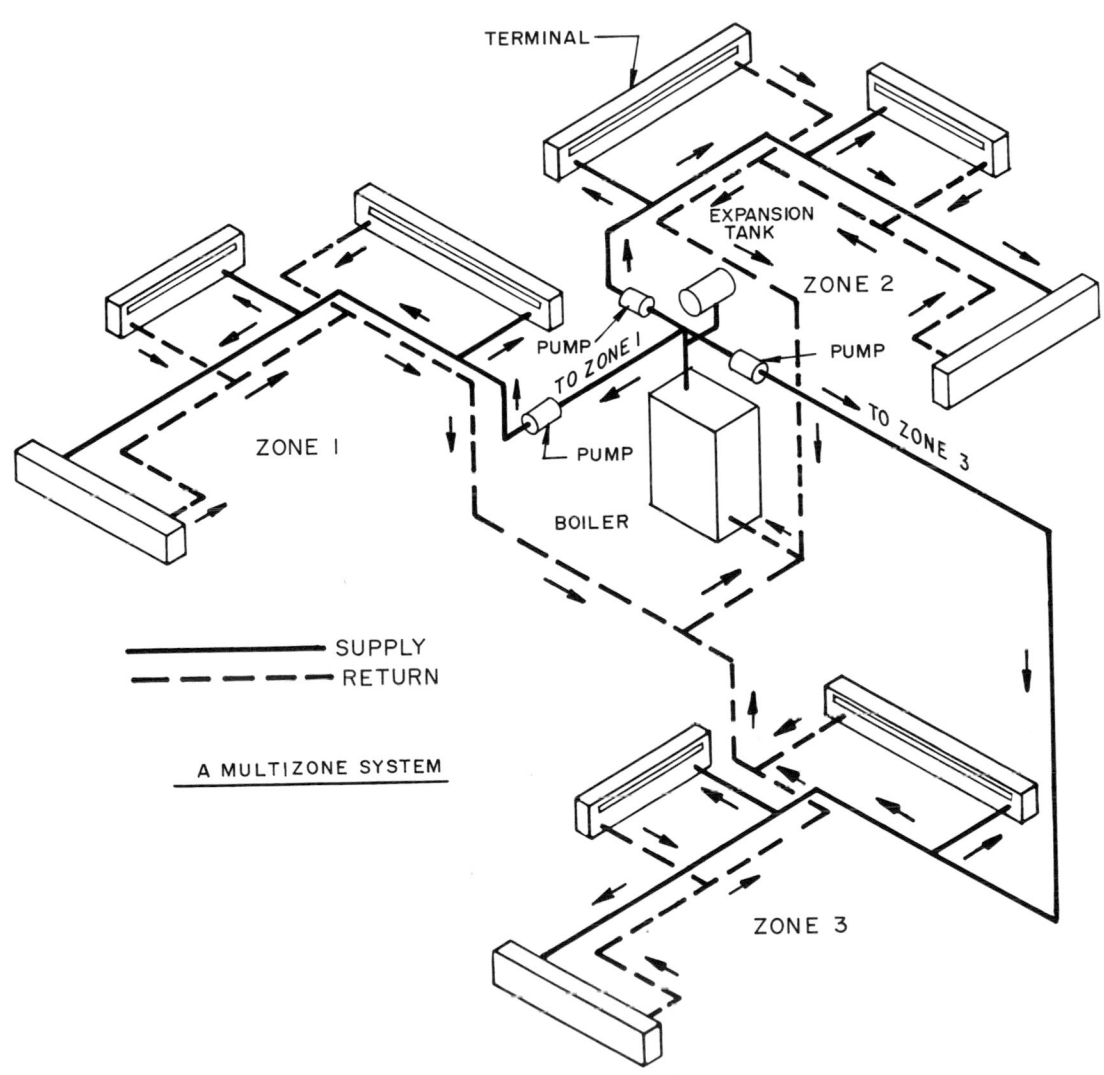

15-54 *A multizone two-pipe hydronic system is used to provide different temperatures in each zone.*

THREE PIPE SYSTEMS

Three-pipe systems provide heating and cooling supply to the terminal units by running a heating supply pipe to each terminal. A third pipe is the return in which the hot and chilled water are mixed and returned to the boiler and chiller, which results in warmed chilled water going into the chiller and cooled warm water going into the boiler. This results in increased cost to reheat and rechill the water before it is recirculated. This is not economical and is not widely used.

FOUR-PIPE SYSTEMS

Four-pipe systems are used when the system provides both heating and cooling modes. The system provides each terminal with separate hot water and chilled water supply and return lines. This provides heating and cooling as required any time either is needed. Systems that use the same coil for heating and cooling control the flow with two valves on each of the heating and cooling supply at each terminal. Others put separate heating and cooling coils in the terminal unit (see 15-55).

Steam Heating Systems

A steam heating system has a boiler or other steam generating device, a piping system, radiators or convectors, and controls. The boiler is usually oil or gas fired but coal, wood, waste products, solar, nuclear, electrical energy, or cogeneration sources can also be used.

Steam delivers considerably more heat per pound (0.45k) than a pound of water but when it becomes vapor (steam) it expands much more than hot water. Therefore a steam system requires larger diameter pipes than hot water systems. Steam produces high pressures, which force it through the piping system without the use of pumps as are used in hot water systems. The pipes must also be sized to allow gravity flow of the condensed water back to the boiler without interfering with the flow of the steam. Engineers use the design data for pipe sizes available in the ASHRAE Handbook (American Society of Heating, Refrigerating and Air-Conditioning Engineers). Steam space heating systems are usually classified as **low pressure.**

It is more difficult to control the temperatures of steam than hot water so hot water is more widely used for space heating systems. If a building requires steam for an industrial process the steam supply is generally used for space heating. This is usually **high-**

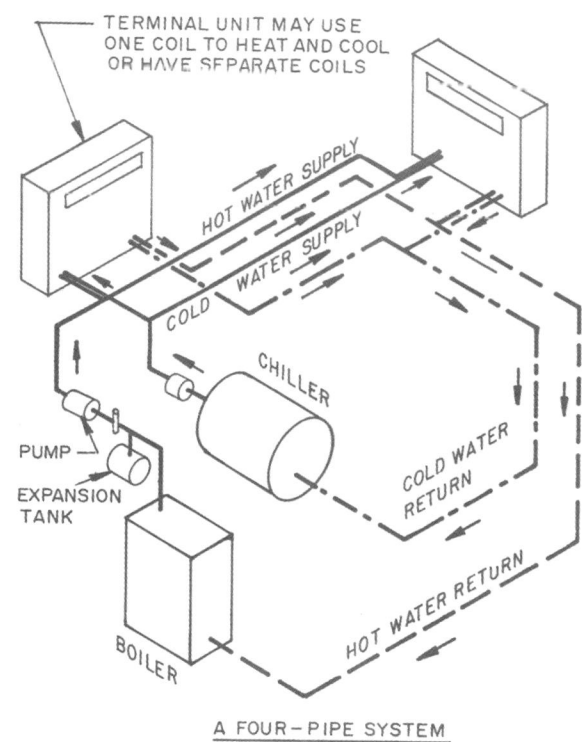

15-55 *A four-pipe hydronic system providing both heating and cooling modes.*

pressure steam and requires the use of pressure-reducing valves in series to get the pressure to that acceptable for space heating.

The two types of steam heating systems are one-pipe and two-pipe.

ONE-PIPE STEAM HEATING SYSTEM

The one-pipe system uses a single pipe supplying steam to the radiators and also serves as the return line for the condensed water to flow back to the boiler. This system is only used in small buildings and is not widely used.

TWO-PIPE STEAM HEATING SYSTEMS

A **two-pipe gravity return system** is shown in 15-56. This system has separate piping for the steam ssupply and the return of the condensate to the boiler. **Thermostatic traps** are on the outlet line of each radiator or other terminal unit and keep the steam within the unit until it has dispersed its latent heat. Then the trap opens, the condensate flows out the return line, and additional steam enters the terminal unit. Gravity flow two-pipe systems are used only in small systems.

Two-pipe steam systems use either gravity or mechanical returns. Gravity returns are used in small systems. Most systems use higher steam pressures to supply steam to the terminal units and a condensate pump or vacuum pump to return the condensate to the boiler.

A **two-pipe vacuum system** is similar to the two-pipe mechanical return system but has a vacuum pump added to give an additional pressure difference. It circulates condensate removing gas that is not condensable and discharging it to the atmosphere creating a vacuum on both the steam supply line and the condensate return line. The vacuum return system is used on larger buildings because it requires a lower steam pressure and can fill the system with steam rapidly.

Boilers

Boilers are used to transfer heat from a fuel source to a fluid, such as water. The liquid is contained in a cast iron, steel or copper pressure vessel which transfers heat to the water to produce hot water or steam (see15-57). Boilers are constructed according to the ASME *Boiler and Pressure Vessel Code*. The Hydronics Institute publishes the *Testing and Rating Standard for Heating Boilers*.

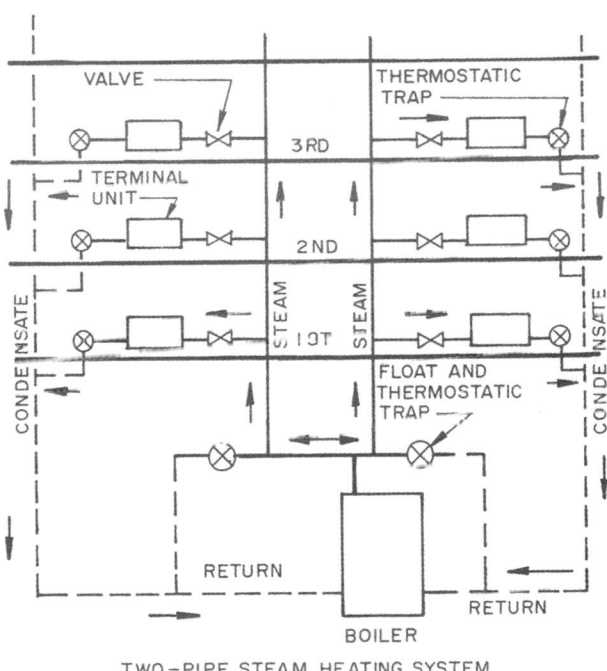

15-56 *Two-pipe steam heating systems have separate piping lines for the steam supply and the return condensate.*

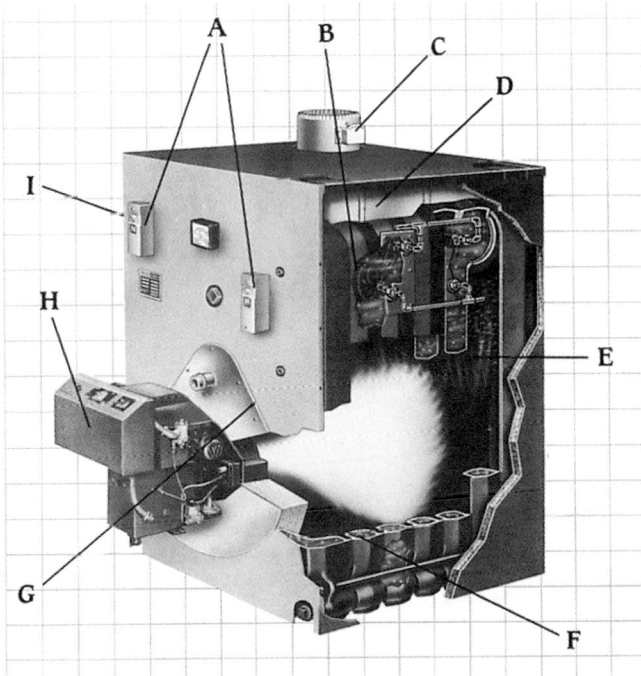

PF-5 Series Features

A. Front-mounted controls for easier adjustment and maintenance

B. Tankless heater for optimum domestic hot water

C. Adjustable lock-type damper for improved efficiency

D. Aluminized steel flue canopy for long life

E. Cast-iron vertical flue design for maximum heat

F. Wet base thermal pump construction for improved circulation

G. Burner mounting plate with flame observation port

H. Four manufacturer burner options to best fit your needs

I. Left side cleanout for easy entrance to all flue surfaces

15-57 *This is an oil-fired cast-iron wet base boiler. The letter "F" points to the cast-iron boiler sections that are assembled forming the boiler sections on all sides of the heat source.*
Courtesy Burnham Corporation, Hydronics Division, Lancaster, PA

▼ Boilers are classified by working pressure, temperature, fuel, size, and whether they are steam or water boilers.

Steam boilers are used for space heating, and auxiliary uses such as in a commercial laundry or industrial processes where steam is required. Steam is typically used in large commercial and industrial buildings.

Hot water boilers are used for space heating and domestic hot water supply. They are typically used in residential and small commercial buildings and steam systems in larger multi-story buildings.

▼ Boilers are classified as high-pressure or low-pressure.

High-pressure boilers are referred to as **power boilers** and produce steam pressures above 15 psi and hot water pressures and temperatures above 160 psi and 250°F (122°C). They are typically of steel construction and use firetube or watertube design.

Low-pressure boilers are referred to as **heating boilers** and are limited to a maximum steam pressure of 15 psi and a maximum hot water pressure and temperature of 250°F (122°C). They are made from cast iron, steel, or copper.

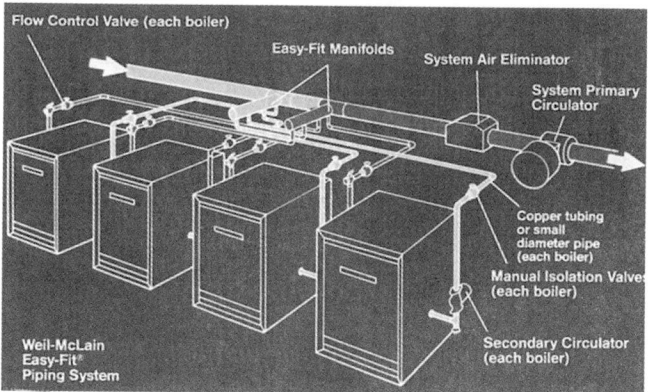

15-58 *Gas-fired boilers installed in series.*
Courtesy Weil-McLain

Large commercial hot water and steam heating systems use large boilers often installed in series as shown in 15-58. Typical boiler connections for multiple boiler installations are controlled by local codes.

Terminal Equipment for Hot Water & Steam Heating Systems

The terminal units used on hot water and steam heating systems include natural convection units, forced convection units, and radiant panels. Those used on high-pressure steam systems have heavier construction than those used for hot water and low-pressure steam. These include cast-iron and steel radiators, convectors, finned tube units, and baseboards.

NATURAL CONVECTION TERMINAL UNITS

Natural convection heating devices include various types of terminal units which may be wall hung, recessed or in the form of a baseboard. These devices distribute heat by using natural air circulation. This occurs because heated air rises and cool air settles producing a natural circulation. The cool air enters the unit below the finned heating tube and rises as it is heated, passing out through a grille or other opening at the top of the unit. They are available in various styles and sizes.

Finned-tube units have metal fins secured to a metal tube and are a convection type heater. However, some radiant heat is produced. If placed so human contact may be possible they have an enclosing cover; otherwise, they need not be covered. The fins are usually aluminum or copper and are secured to copper tubing, which conducts steam or hot water from the boiler. The fins become hot and the air flows by them and is heated.

Baseboard units are located along the wall where it meets the floor. The heating element may be a finned tube, cast-iron, or aluminum unit. The unit is enclosed in a metal enclosure with openings at the floor and near the top of the enclosure providing for natural air circulation.

Steel radiators are another heating device that heats by convection and radiation. These are in the form of panels typically mounted on a wall or ceiling. However they can be freestanding. They are heated with hot water circulating through tubes which feed flat hollow panels or a series of tubes. It works with low temperature hydronic systems and, although not as hot as conventional hot water radiators, are still fairly hot to the touch. The panels are made of heavy-gauge steel.

FORCED-CONVECTION HEATING UNITS

Forced-convection units are used for spaces that are to be heated and cooled. However, they can be used for heating or cooling only as is done with natural convection units. The units available include unit ventilators, unit heaters, induction units, fan-coil units, and large central air-handling units. Forced-convection units use some form of fan or blower to produce air movement over a cooling or heating coil and move the heated or cooled air through the space to be conditioned.

Fan-coil units have a fan, filter, and heating and cooling coils. Some have separate heating and cooling coils while others use the same coil for the heating water in the winter and the chilled water in the summer. If used on a two-pipe system, the entire system must be set for heating or cooling. The hot water in the pipes must be drained before the chilled water is introduced. In a large building this can take several days. If used with three- or four-pipe systems each fan-coil unit has separate heating and cooling coils so each fan coil unit can provide heating or cooling when required.

Unit ventilators are much like the fan-coil unit except they have an opening through the outside wall, which allows it to bring in outside air. The opening is covered with a decorative louvered grille. The control system regulates dampers to control the influx of outside air. These are typically used in rooms to be occupied by large numbers of people where frequent air changes may be necessary. This helps meet code requirements for room ventilation.

Induction units are much like fan-coil units. However, the movement of the air in the room is through the use of high-pressure air that is piped through a nozzle behind the coil causing room air to circulate through the coil and out into the room. The nozzle and pressurized air replaces the fan in the fan-coil unit. They are usually mounted on a outside wall below a window.

Unit heaters have a fan that circulates air over some type of heat exchange surface, such as a hot water coil. They are enclosed in a metal case and usually are suspended from the ceiling. Most common uses are in large open industrial plants or businesses, such as an auto repair shop. They can have electric heating elements, be fired by natural gas or propane, or be part of a steam or hot water heating system. They provide a rapid flow of heated air. Unit heaters are also used to temper cold outside air that is introduced.

Chilled-Water Cooling Systems for Large Buildings

The most commonly used cooling system uses chilled water to remove heat from the air in a building. This involves the use of mechanical equipment, including a means of refrigeration, cooling coils, and heat exchangers. The chilled-water system is a closed circuit of piping in which water is recirculated from a chiller where it is chilled and moved through other equipment used to remove heat from the water and circulate it through terminal units in the space to be cooled and back to the chiller.

The three types of liquid chillers include (1) centrifugal, (2) absorption, and (3) positive displacement.

CENTRIFUGAL CHILLERS

Centrifugal chillers are driven by an electric motor, steam or gas turbine, or an internal combustion engine that may be diesel or natural gas fueled.

They use a **vapor-compression refrigeration system.** The major parts of a mechanical vapor-compression chilled-water system include a compressor, condenser, evaporator, and an expansion valve. An air-cooled refrigerant condenser is a device in which heat removal from the refrigerant is accomplished entirely by heat absorption by means of air flowing over the condensing surfaces.

A **compressor** is a machine that mechanically compresses the refrigerant. An evaporator is that part of a refrigerating system in which the refrigerant is evaporated to absorb heat from the contacting heat source. The expansion valve reduces the temperature and pressure of the refrigerant as it passes through it.

The **vapor compression refrigeration cycle** used by centrifugal chillers is illustrated in 15-59. This is a direct expansion (DX) cycle. The refrigerant in vapor form is compressed by the cylinder in the compressor. This causes it to become hot and at a high pressure. The refrigerant moves to the condenser where a fan belows outside air through the condenser coil removing much of the heat in the refrigerant causing it to condense into a warm liquid, which is still at a high pressure. The pressure pushes the liquid refrigerant toward the evaporator. On the way it passes through a thermal expansion valve imposing a pressure drop causing it to expand, which reduces the temperature. The cool liquid refrigerant passes through the cooling coil of the evaporator where it absorbs heat from the air passing through the coil. When the refrigerant leaves the evaporator it is a vapor where it moves to the compressor, compressed, and the cycle is repeated.

A centrifugal chiller is shown in 15-60.

ABSORPTION CHILLER REFRIGERATION SYSTEM

An absorption chiller can be described as a refrigerating machine that uses heat energy and absorption imput to generate chilled water. Absorption cooling uses an evaporated refrigerant that is frequently water. The mechanical compression stage used in centrifugal chillers is not used but instead uses a process in which an absorbant such as lithium bromide solution is used. This solution pulls vapor off evaporator coils creating a cooling reaction. A source of heat such as a gas burner, low-pressure steam, or hot water is used to regenerate the absorbent solution by separating it from the absorbed vapor. Since a source of heat is used to operate an absorption chiller it is especially economical to operate when a building has waste steam or other heat sources that can be used to provide a major amount of the heat required.

The absorption chiller is a quiet operating unit having few moving parts, generating little vibration, and is lightweight compared with compression type chillers.

An absorption chiller is shown in 15-61.

Double-Effect Absorption Chillers

A gas-fired double-effect absorption chiller-heater is shown in 15-62. Units of this type are used in commercial applications where chilled water is used for cooling and hot water for heating using a central air-conditioning system. The condenser is water cooled during the cooling period and the heat is rejected through a cooling tower. The double-effect absorption cooling cycle has two generators. One is heated by natural gas and the other by the hot semiconcentrated refrigerant vapor.

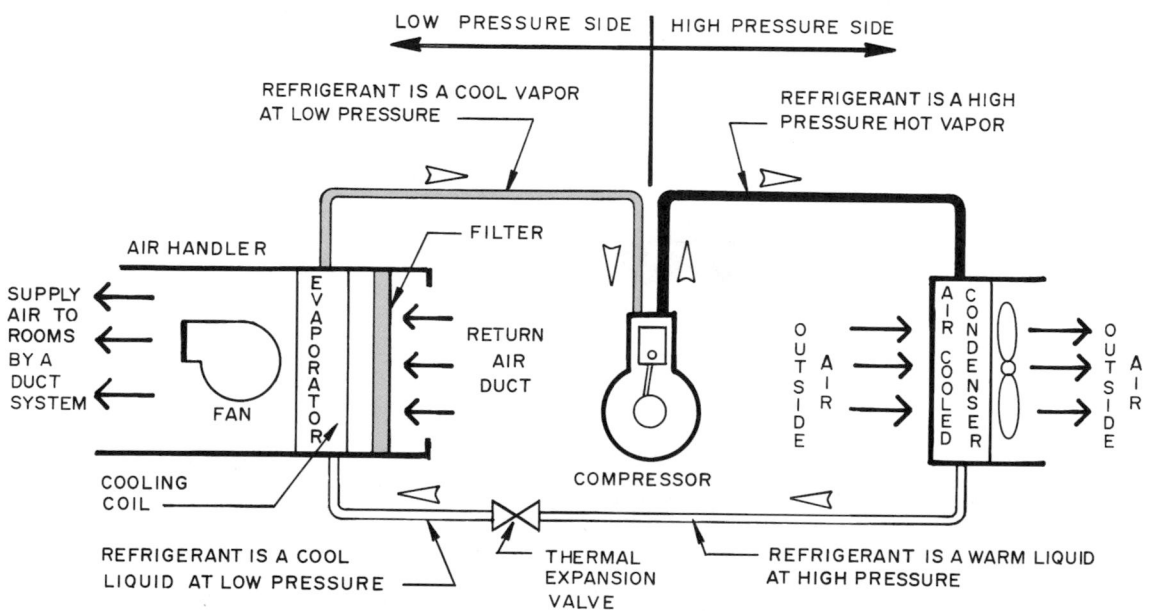

15-59 *A simplified illustration of an air-cooled direct expansion (DX) vapor compression refrigeration cycle used in centrifugal chillers.*

POSITIVE-DISPLACEMENT CHILLERS

Positive-displacement chillers use scroll, reciprocating and rotary screw compressors, and typically use electric motor drives.

Screw compressor—uses a rotary motion to drive two intermeshing helical rotors to produce compression.

Scroll compressor—is a positive-displacement compressor in which the reduction of the internal volume of the chamber in which compression occurs is produced by a rotating scroll within a stationary scroll. A scroll is an involute spiral.

Reciprocating compressor—is a positive-displacement compressor in which the change in the volume of the compression chamber is produced by the reciprocating movement of a piston.

Positive-displacement compressors—increase the pressure of the refrigerant vapor by reducing the volume of the compression change in which it has been injected using a mechanical means such as a piston. A water-cooled package chiller with a rotary screw compressor that uses microcomputer controls is shown in 15-63.

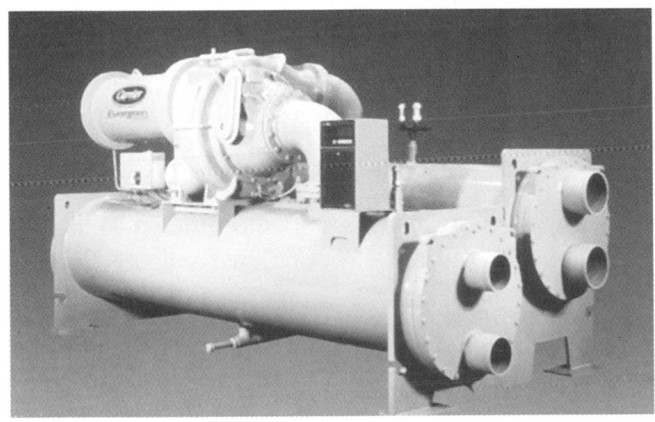

15-60 *This centrifugal chiller is designed to accommodate non-ozone-depleting HFC-134a refrigerant. This is an environmentally responsible, nonrestricted refrigerant used in this high efficiency chlorine-free chiller. The chiller is designed for use in buildings 300,000 sq. ft. (27,600m²) or larger.*

Courtesy Carrier Corporation

15-62 *A gas-fired direct-fired double-effect absorption chiller/heater.*

Courtesy American Yazaki Corporation

15-61 *A direct-fired double effect absorption chiller/heater.*

Courtesy The Trane Company

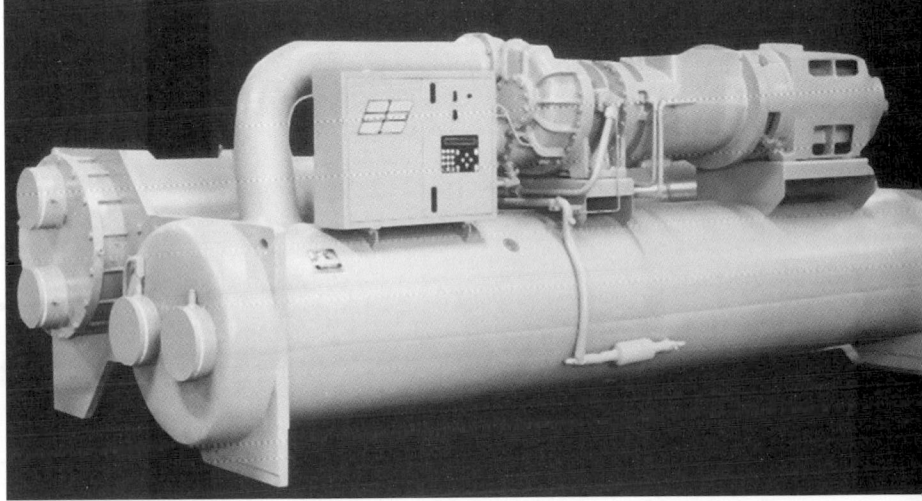

15-63 *This is a water-cooled package chiller with a rotary screw compressor that uses highly efficient microcomputer controls.*

Courtesy Dunham-Bush, Inc.

Radiant Heating & Cooling

Radiant panel systems use panels on the walls, ceiling, or in the floor whose surface temperatures can be controlled. These are usually operated by electrical resistance units or circulating water or air. Radiant energy is transmitted through the air in straight lines, does not heat the air, but does raise the temperature of solid objects upon which it falls. The objects obtain the heat by absorption. It can be reflected off a surface.

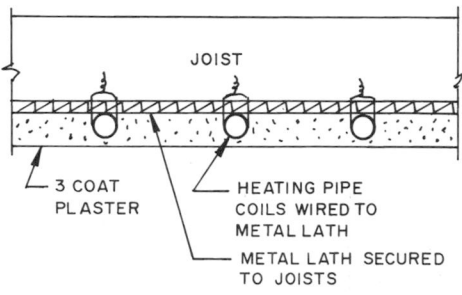

COILS IN PLASTER BELOW METAL LATH

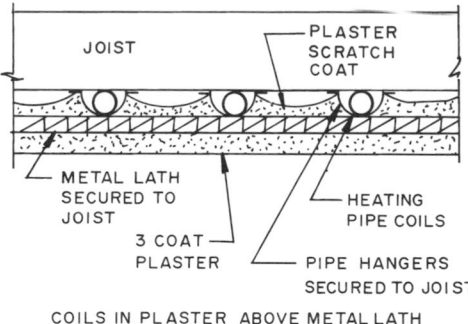

COILS IN PLASTER ABOVE METAL LATH

15-64 *Typical radiant heat ceiling piping panel installations.*

PIPING IN CEILINGS

There are several piping systems used for radiant heating and cooling in ceilings. Hydronic ceiling panels can use a two-pipe or four-pipe distribution system similar to that discussed for hot water systems. The design of the system is critical to a successful end result and must be done by a qualified engineer. Typically the panels are installed with the pipes embedded above or in a finished plaster ceiling as shown in 15-64.

The suspended ceiling has the pipes tied to an overhead supporting member with metal lath and plaster below. When wood or metal joists are used the pipes are secured to them with metal pipe hangers and the metal lath and finished plaster are placed below. The coils may be embedded in the plaster coat by securing the metal lath above the pipes and plastering over them. Other types of finished ceilings can be used but plaster is most common.

Another type of ceiling panel consists of copper tubing bonded to flat metal panels, which will be the exposed finished ceiling. The individual panels are hung between the channels of a suspended ceiling as shown in 15-65. This system is used to heat and cool the interior air. Several panel designs are available but most use copper tubing and aluminum panels.

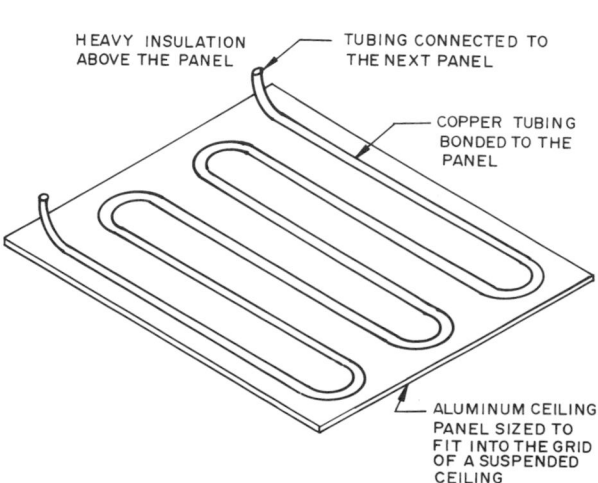

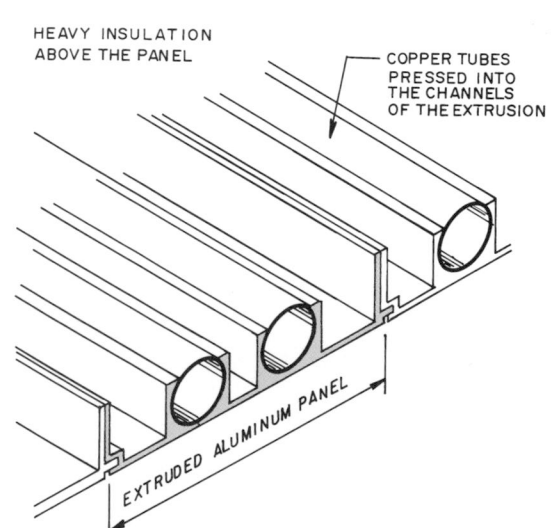

15-65 *Two types of hydronic metal ceiling panels.*

PIPING IN FLOORS

Radiant heating piping may be embedded in concrete floors or placed above or below a wood subfloor. A detail for concrete slab heating is in 15-66. Plastic, ferrous, and nonferrous pipe and tube may be used. They are arranged in continuous coils or have header coils. Usually 1½ to 4 in. (38 to 101 mm) of concrete covers the pipe.

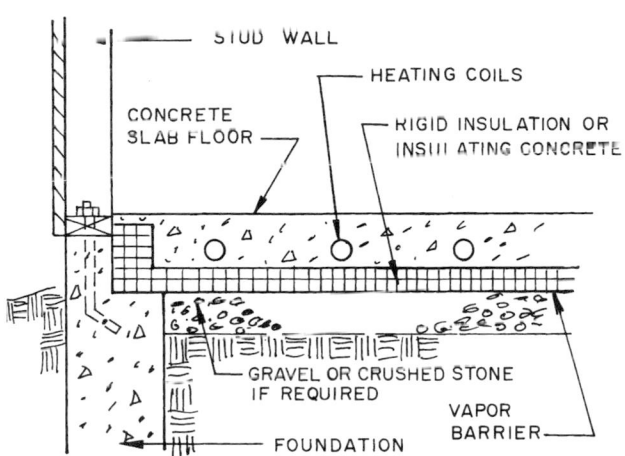

15-66 *A typical detail for radiant heating pipes embedded in a concrete slab.*

The edges of the slab must be fully insulated with rigid insulation. Sometimes insulating concrete is used.

Piping may be placed on top of a wood subfloor and covered with 1 to 2 in. (25 to 50 mm) of concrete or gypsum underlayment. Gypsum products designed specifically for floor heating are available. Concrete should be of structural quality.

Electric Heating Systems

The types of electrically heated systems include (1) factory assembled panels that mount on walls or ceiling, (2) fabrics and wall covering material containing resistance heating wires, and (3) various types of electric resistance cables that may be embedded in concrete or laminated in drywall ceilings or in plaster.

One type of ceiling panel is made to fit into the grid formed by the structural members of suspended ceilings (see 15-67). The panels are available with various constructions such as conductors embedded in a panel (may be a gypsum panel) or some form of laminated panel.

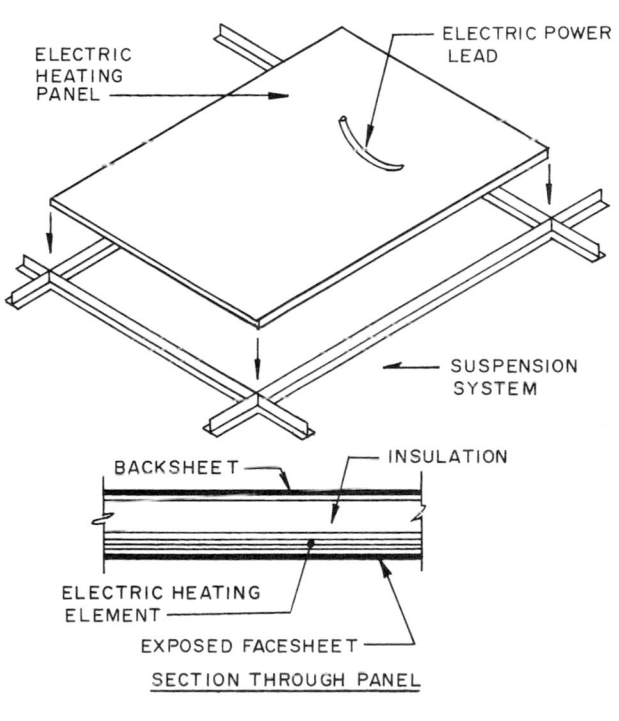

15-67 *An electric heating ceiling panel supported by the structural grid of a suspended ceiling system.*

Another ceiling heating system uses electric heating cable stapled to a ceiling covering such as gypsum board or plaster lath or other fire-resistant material. The wires are covered by coats of plaster. If metal lath is used it must first be covered with a brown coat of plaster so a nonelectrical conducting surface is available (see 15-68). Electrical heating cable can also be laid in concrete slabs.

The floor is laid in two pours. The first pour is 3 in. (76 mm) or more of insulating concrete. The cable is laid on this slab and is fastened in place by stapling into the slab or using special nail anchors. These hold the cables so the spacing is maintained when the second layer of concrete is poured. The second layer is usually 1½ to 2 in. (38 to 51 mm) thick and must not be insulating concrete. A finish floor can be laid over this slab.

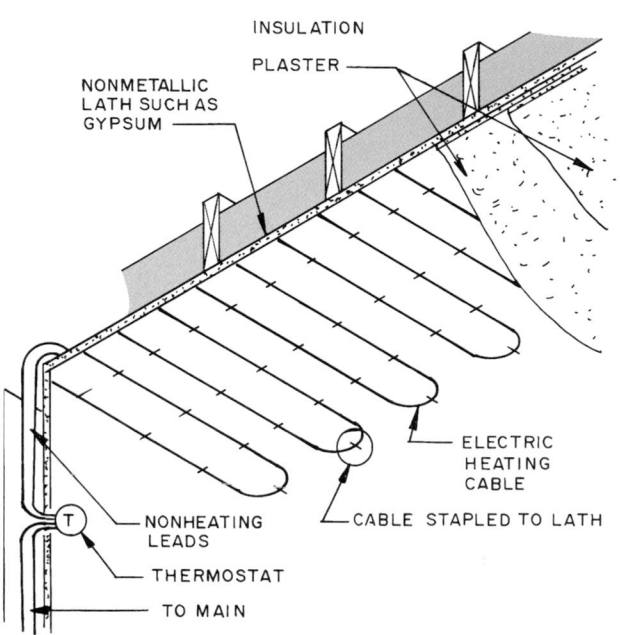

15-68 *An electric ceiling heating system composed of heating cable secured to lath and covered with plaster.*

DIVISION 16

ELECTRICAL

CSI MasterFormat™

16050 Basic Electrical Materials & Methods

16100 Wiring Methods

16200 Electrical Power

16300 Transmission & Distribution

16400 Low-Voltage Distribution

16500 Lighting

16700 Communications

16800 Sound & Video

Courtesy Square D Company

The modern building depends heavily on electricity to make it functional and habitable. Electricity provides the power for the lighting; runs motors for heating, ventilating, and air-conditioning; provides power to operate the many devices brought into the building; powers elevators, escalators, and other conveying systems; supplies the power to operate the communications, fire, and security systems; and is used to operate a vast array of manufacturing machinery.

ELECTRICAL LOADS

The loads put on an electrical system vary widely depending upon the occupancy. All buildings have extensive lighting requirements. This is a major interior wiring system. Many functions, such as air-conditioning and heating, require the use of electric motors. Other systems, such as refrigeration and ventilation, have motors, compressors and other electricity consuming devices. The range of appliances and other electrical devices is extensive and produce varying loads as periods of demand occur. Industrial plants have heavy electrical demands to operate machinery, perform manufacturing operations, melt, fuse, and otherwise process materials. Internal transportation (elevators, escalators, moving walks, conveyors) all require electric power. Internal communications and controls are electrically operated and many things would not function without them. Then there are hundreds of special equipment items such as those found in hospitals, computer centers, and radio and television studios. The determination of electrical loads and internal systems is a major part of an adequately designed building.

Electricity

Electric current can be defined as the flow of electrons along a conductor, such as a copper wire. It is produced by a generator or battery, which forces electrons to follow the conductor to a consuming device, such as a light, and back to the producing source (see 16-1) forming a loop or continuous circuit. The flow of electric current in a circuit resembles the flow of a fluid in a hydraulic circuit. In a hydraulic system a pump puts the fluid under pressure (pounds per square inch), the fluid flow is measured in a quantity such as gallons. The fluid meets some resistance as it enters a fixture, and the flow is controlled by a valve. An electric circuit has a battery or generator to produce electricity and the electromotive force (volts) to move it along the conductor in quantities measured in amperes, finds resistance (ohms) when it enters a fixture and the flow is controlled by a switch. Therefore the units used to identify the factors related to the flow of electric current are:

AMPERE

A unit of the rate of flow of electric current. An electromotive force of 1 ohm results in a current flow of 1 ampere. The symbol for ampere is A.

VOLT

The unit of electromotive force (pressure) that causes electric current to flow along a conductor. The symbol for volt is V.

OHM

The unit of electrical resistance of a conductor. The symbol for ohm is the Greek capital letter omega (Ω) although R is used in most formulas.

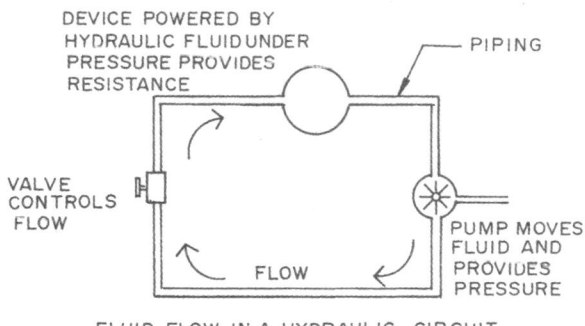

FLUID FLOW IN A HYDRAULIC CIRCUIT

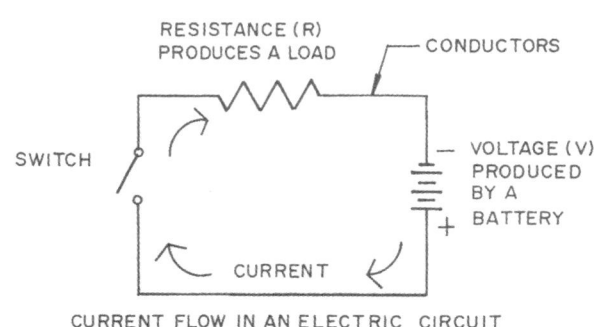

CURRENT FLOW IN AN ELECTRIC CIRCUIT

16-1 *A comparison of electrical and hydraulic circuits reveals similarities. Electric switches and hydraulic valves control flow, electric current, and hydraulic fluid flow in the circuit. Power is supplied by an electric device, such as a battery and a hydraulic pump, electric wire and hydraulic piping form the circuit, and resistance to flow occurs in both circuits.*

OHM'S LAW

Amperes, volts, and ohms are related to each other and a variance in one will affect the others. This relationship is identified as Ohm's Law. The relationship for each measure is shown by the following formula in which I is the electric current (amperes) or intensity of electron flow, R is resistance (ohms), and E is the electromotive force or volts.

$$I = \frac{E}{R} \qquad amperes = \frac{volts}{ohms}$$

$$R = \frac{E}{I} \qquad ohms = \frac{volts}{amperes}$$

$$E = I \times R \quad volts = amperes \times ohms$$

Conductors

Electric current flows along materials called **conductors.** The commonly used conductors for electric wiring are copper and aluminum. Copper is a better conductor than aluminum so if aluminum wire is used it must have a larger diameter to carry the same amount of current. Good conductors have a low resistance to the flow of electricity. Materials that have a very high resistance to electrical flow get hot because of the friction generated by the flow of electrons. These materials, such as nichrome, are used in applications such as electric heaters where the production of heat is wanted. Other materials do not conduct electricity and are called insulators.

Glass, ceramics, and plastics are examples. They are used on electrical devices where it is necessary to provide protection from the electricity. For example, the switch lever on a light switch is a nonconductor type material.

For electricity to do work it must flow through a circuit. It flows from the positive connection to the negative connection as shown in 16-2. The negative electrons move to the positive pole and on through the circuit to the consuming device and back to the negative pole. A switch can be put in the circuit to interrupt the flow when desired. The switch when closed completes the circuit.

AC & DC Current

The two types of electric current are direct current (DC) and alternating current (AC). **Direct current** has a constant flow in one direction while **alternating current** varies periodically in value and directions, first flowing in one direction in the circuit and then flowing in the opposite direction. Each complete repetition is called a **cycle** and the number of repetitions per second is called the frequency. The frequency at which this occurs is measured in **hertz** (Hz). In the United States the frequency for alternating current is 60 cycles per second or 60 hertz.

Power & Energy

Energy is the term used to express **work** and is expressed in units of kilowatt hours, footpounds, Btu, joules, or calories. **Power** is the rate at which energy is used and is expressed in terms of watts, kilowatts, and other units shown in Table 16-1 on the following page. Since power is the rate at which energy is used time is a factor. The relationship between power and energy is shown by the following equation:

$$Power = \frac{energy}{time}$$

$$Energy = power \times time$$

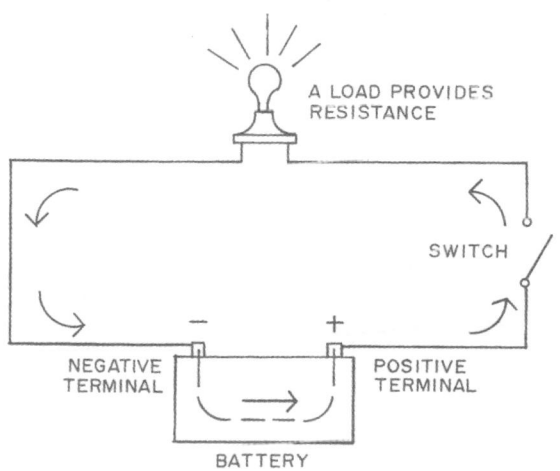

A LOAD PROVIDES RESISTANCE

SWITCH

NEGATIVE TERMINAL

POSITIVE TERMINAL

BATTERY

16-2 *Electricity flows through a circuit from the positive connection to the negative connection.*

Table 16-1 UNITS OF POWER & ENERGY

English System	Metric System
Units of Power[a]	
Horsepower (hp)	Joule per second (J/s)
Btu per second (Btu/s)	Calorie per second (cal/s)
Watt (W)	Watt (W)
Kilowatt (kW)	Kilowatt (kW)
Units of Energy[b]	
Btu	Calorie (cal)
Foot-pound (ft-lb)	Joule (J)
Kilowatt-hour (kWh)	Kilowatt-hour (kWh)

[a]The rate at which work is done
[b]The amount of work done

The unit of electric power in electric circuits is expressed in watts (W) or kilowatts (kW), which is 1000 watts. One kilowatt = 1000 watts. One watt hour of energy represents one watt of power used for one hour. A **watt** is one ampere flowing under an electromotive force of one volt. The power (watts) into an electrical device having a resistance of R (ohms) in which the current is I (amperes) is found by the equation:

$$W = I^2 \times R$$

$$\text{watts} = \text{amperes}^2 \times \text{ohms}$$

It can be seen therefore that power is related to current (amperes), electromotive force (voltage), and resistance (ohms).

ELECTRICAL CODES

The design and installation of electrical systems are strictly regulated by electrical codes. The National Electrical Code (NEC) of the National Fire Protection Association (NFPA) specifies the safety principles that must be followed. Local governments may have their own electrical codes which typically include all National Electrical Code requirements but may contain additional regulations. The Underwriters Laboratories, Inc. test and certify electrical devices of all types. Approved items carry the UL label.

ELECTRIC POWER SOURCES

Most electrical current is produced by some type of generator. These include hydroelectric, nuclear, and fossil-fueled electrical generators.

Alternating current is produced by an a-c generator also called an alternator. The alternator is powered by any of the four sources of energy mentioned above. The normal alternator frequency is 60 Hz (hertz).

Hydroelectric Power Generation

Electric power is generated by hydroelectric generation plants using the energy of falling water to turn a generator and produce electricity. The water turns the turbine which drives the generator that produces the power. Transformers step up the voltage to the requirements needed to transmit it over electric lines to various destinations. A typical installation will have many turbines, generators, and transformers.

Fossil Fuel-Powered Generation

Fossil fuels used to produce steam to produce electricity include coal, oil, and natural gas. Some experience using the burning of waste materials is underway. The steam produced by burning the fuels is used to drive steam turbines, which turn a generator and produce electricity. Since fossil fuels are a nonrenewable resource and are becoming increasingly expensive alternate sources, such as nuclear fission, are becoming commonly used. Some kinds of coal and oil fuels cause air pollution problems and require exhaust fumes at the stack be monitored and cleaned if necessary (see 16-3).

Nuclear-Powered Generation

In 16-4 is an overall photo of the Harris Nuclear power plant located at New Hill, North Carolina and operated by Carolina Power and Light. The facility occupies a 10,723 acre site and has a 4,100 acre lake. It is capable of producing 900,000 kilowatts of electricity and is licensed to operate for 40 years.

Nuclear-powered electric generation plants are similar to fossil-fueled plants in that they produce steam to run a steam turbine that drives the generator. The energy is produced in the nuclear reactor by fission. Fission is a process in which the center or nucleus of certain atoms are split when struck by a subatomic particle called neutron. The products of fission fly apart at high speed and generate heat as they collide with surrounding matter. The fission reaction is controlled in

16-3 *The Mayo fossil-fuel electric generating plant located at Roxboro, NC. This coal-burning electric power generating plant was built in 1983 and can produce up to 750,000 kw. At full capacity it takes 73 rail cars to supply the 7,300 tons of coal per day required to run the plant.*

Courtesy Carolina Power and Light Company

16-4 *The Harris nuclear electric generating plant located at New Hill, N.C. required 24 million pounds of reinforced steel, 500,000 cubic yards of concrete, and 2,000 miles of power and control cable. It is capable of producing 900,000 kilowatts of electricity.*

Courtesy Carolina Power and Light Company

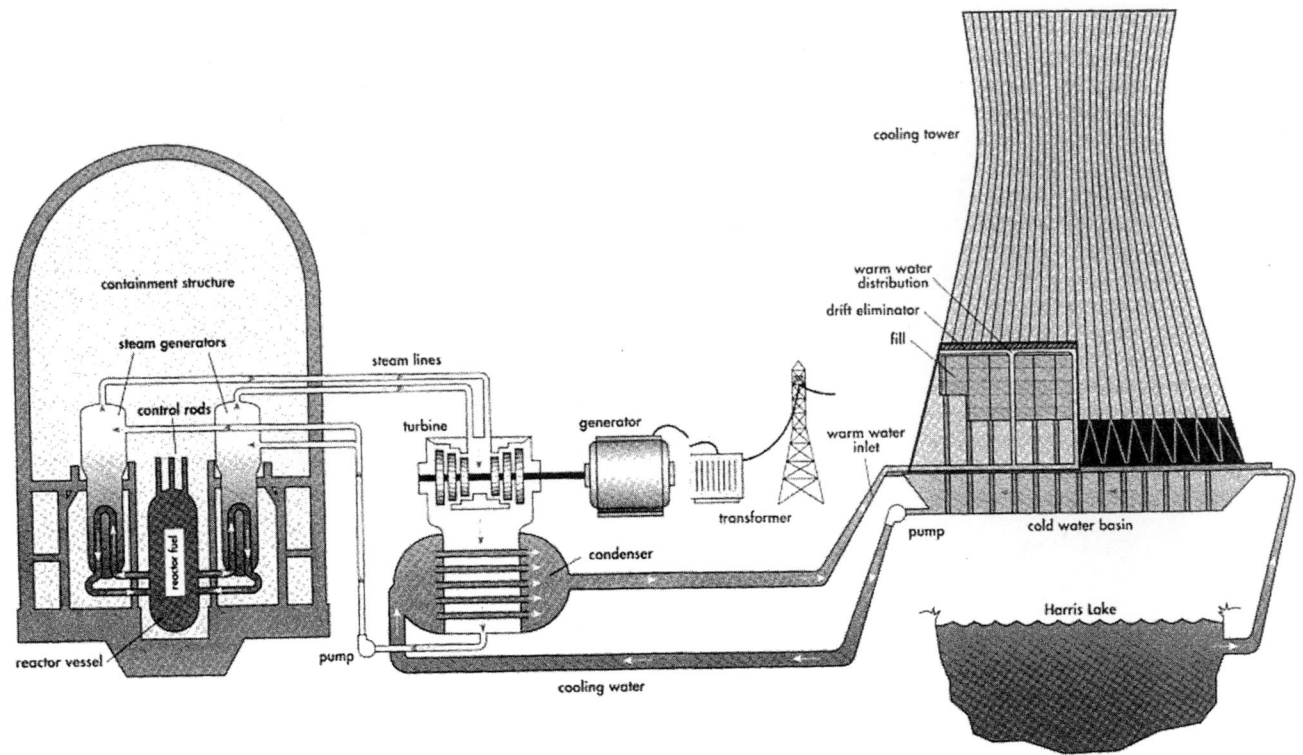

16-5 *A simplified illustration showing how a nuclear electric generating plant operates.*
Courtesy Carolina Power and Light Company

the nuclear core of the reactor, which consists of fuel rods in a chemical form of plutonium or uranium and thorium. Heat energy is produced by the fission reaction of the nuclear fuel. The heat is removed by a coolant and used to produce steam to drive a steam turbine which drives the electrical generator. 16-5 shows how the Harris nuclear plants work. To ensure the safety in the nuclear plant the concept of "defense-in-depth" is employed. There are several layers of protection each of which is independent of the others so if one should fail, others will continue to protect the plant, the workers, and the general public.

On-Site Power Generation

On-site power generation is used when a utility system is not available or unable to provide reliable service. Some systems are used to provide additional service during peak periods. Some facilities, such as a hospital, will have an on-site back-up power generation system which is used when the utility system fails, as during a storm.

The methods used to generate electricity on-site include wind turbines, solar photovoltaic cells, thermal sources, and cogeneration.

WIND TURBINES

Wind turbines utilize prevailing winds to drive a propellor mounted on a generator that is mounted on a tower. These are often built in a group and may be tied into the utility power system (see 16-6).

SOLAR PHOTOVOLTAIC CELLS

Solar photovoltaic cells converts some of the sunlight that falls on them into electricity. They are set in a large grouping and are connected together. Often they are connected to a battery and charge the battery when the sun is shining and use the battery power at night or on cloudy days.

THERMAL SOURCES

Thermal sources use an engine or turbine coupled to the shaft of a generator to produce electricity. Internal combustion engines run on fuels such as diesel fuel, methane gas, or natural gas (see 16-7). These are typically used as an emergency back-up source of electricity to a vast range of commercial facilities such as hospitals, banks, computer centers, retail stores, schools, and wastewater treatment plants. The system includes the generator set, transfer switches, and paralleling switchgear.

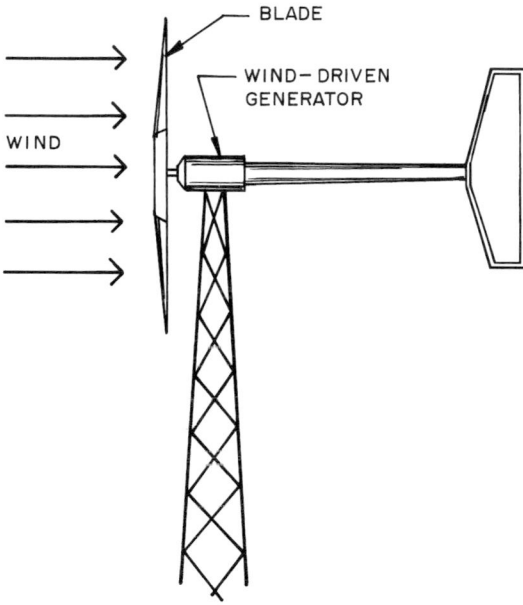

16-6 *Wind-driven generators are one source of electricity.*

16-7 *Emergency standby generator may be powered by diesel or gas engines. They are part of a total system, including the generator set, transfer switches, and paralleling switchgear.*
Courtesy Onan Corporation

Turbines are either gas or steam powered. Gas turbines burn various types of gaseous and liquid fuels such as natural gas or fuel oil. Steam turbines are driven by a source that produces a large quantity of high-pressure steam. The revolving turbine drives the generator producing electricity. The steam is produced by a boiler that is fired with a fuel such as coal, solid waste, natural gas, or oil.

COGENERATION

Cogeneration involves the utilization of normally wasted heat energy produced by a reliable, steady source to generate electricity on-site or to use it to heat or cool the building. For example, if an internal combustion engine or a turbine is being used to produce on-site electricity, the heat produced by these power sources is reclaimed and can be used to heat water or produce steam, which can be used to heat the building or provide cooling using an absorption chiller. An auxiliary conventionally fired boiler is used in the system to provide extra hot water or steam if needed.

Salvaged heat may also be used to produce steam used to drive a turbine producing a supplemental on-site supply of electricity. This is especially effective in areas where high-pressure steam is required for an industrial process such as food processing or pulp and paper manufacturing, which provides a steady, high temperature source of wasted heat (see 16-8).

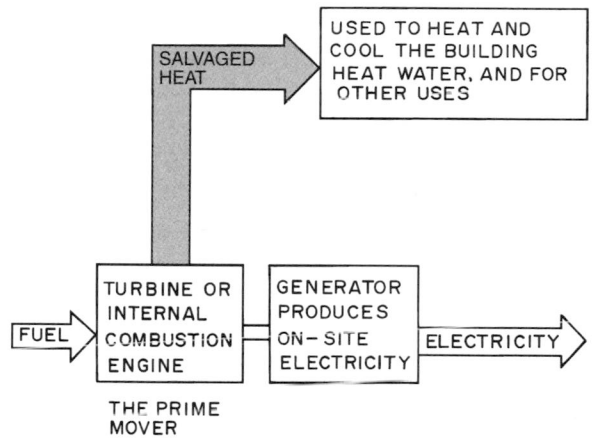

Cogeneration utilizes heat that normally would be wasted to produce on-site electricity or to heat or cool a building.

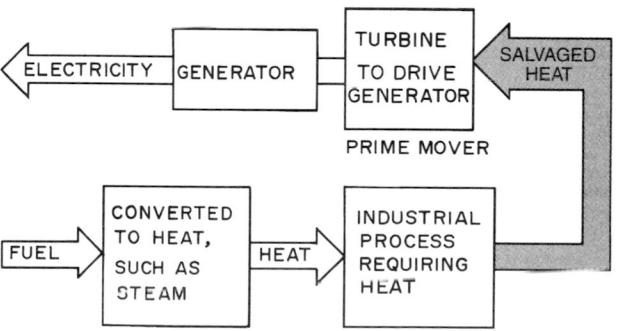

16-8 *Cogeneration can utilize heat generated by processes within a building to produce steam that can drive a turbine producing on-site electricity.*

ELECTRICAL EQUIPMENT

Electrical Power Conductors

A major consideration in designing and installing an electric power system is to safely transmit the power to the end use. Electric power systems have the potential for causing fires, property damage, and human injury and death. Electrical codes are strict and inspection during construction must be thorough. The key to the situation is to isolate the electrical conductors as they pass through the building until they reach the point of use, such as an electric light.

CABLES

Electric cable design and use are strictly regulated by codes such as the National Electrical Code published by the National Fire Protection Association. The types of conductor (see 16-9) rated by this code follow:

Type AC—Insulated conductors are wrapped in paper and enclosed in a flexible, spiral-wrapped, metal covering. It has an internal copper bonding strip in contact with the metal covering providing a means of grounding. It is used only in dry locations. This wire is referred to as BX cable.

Type ACL—It has the same insulation and covering as Type AC but has lead-covered conductors making it useful in wet applications.

Type ACT—The individual copper conductors have a moisture-resistant fibrous covering and run inside a spiral metal sheath.

Type MC—Insulated copper conductors are sheathed in a flexible metal casing. If it has a lead sheath it can be used in wet locations. It is a heavy-duty industrial feeder cable.

Type MI—The conductors are mineral insulated and sheathed in a gastight and watertight metal tube. It can be used in hazardous locations and underground. It can be fire rated.

Type NM or NMC—A nonmetallic-sheathed cable used in protected areas. It is also called Romex. NM has flame-retardant and moisture-resistant outer casing and is restricted to interior use. NMC is also fungus resistant and corrosion resistant and can be used on exterior applications.

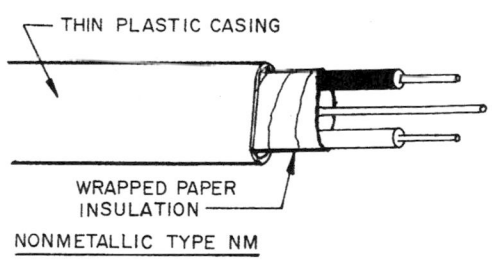

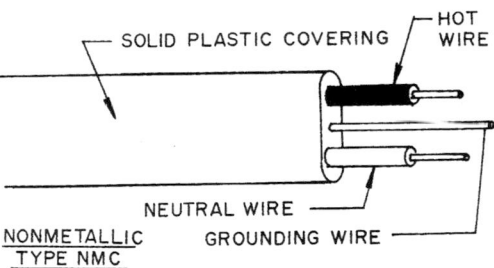

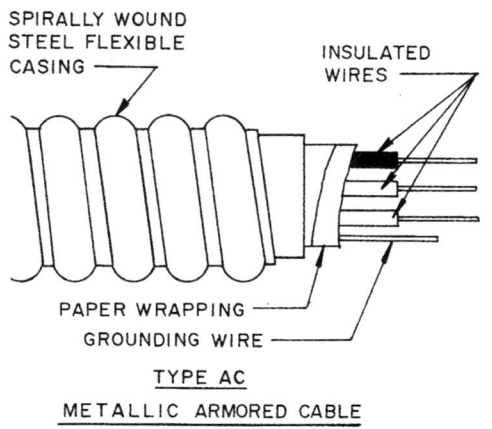

16-9 *Several of the commonly used electrical conductors.*

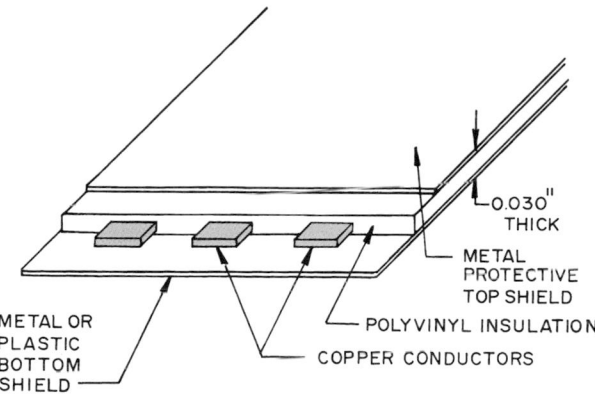

16-10 *A typical under-carpet wiring system. It uses a flat power cable that runs under the carpet. Outlets are placed as needed along the cable.*

Type SE or USE—A moisture-resistant, fire-resistant insulated cable that has a braid of armor providing protection against atmospheric corrosion.

Type USE—has a lead covering permitting it to be used underground. This is used as the service entrance cable.

Type SE—is used for service entrance wiring or general interior use. Type USE is used for underground service entrance wiring.

Type SNM—The conductors are in a core of moisture-resistant, flame-resistant, nonmetallic material. This assembly is covered with a metal tape and a wire shield and sheathed in an extruded nonmetallic material impervious to oil, moisture, fire, sunlight, corrosion, and fungus. It can be used for hazardous applications.

Type UF—The conductors are enclosed in a sheath resistant to corrosion, fire, fungus, and moisture and can be directly buried in the earth.

FLAT CONDUCTOR CABLES

Flat conductor cables consist of copper cables formed and flat embedded in a plastic sheathing A typical cable is .030 in. (0.78 mm) thick and around 2½ in. (63.5 mm) wide (see 16-10). The conductor is placed on the subfloor, a shielding material is placed over it, and the carpet is installed. These are used for standard 120V electric power, and cables used in communication and data systems. Outlets are installed as required by making connections through the carpet. Codes generally require it be covered with carpet squares so it is easily accessible.

CABLE BUS & BUSWAYS

Busways are bare conductors run in a metal trough. The conductors are mounted on insulators keeping them clear of the trough. Electrical connections are made to the conductors with various types of plug-ins (see 16-11).

16-11 *These lights are mounted on busways secured to the ceiling. The lights can be placed anywhere along the busway.*

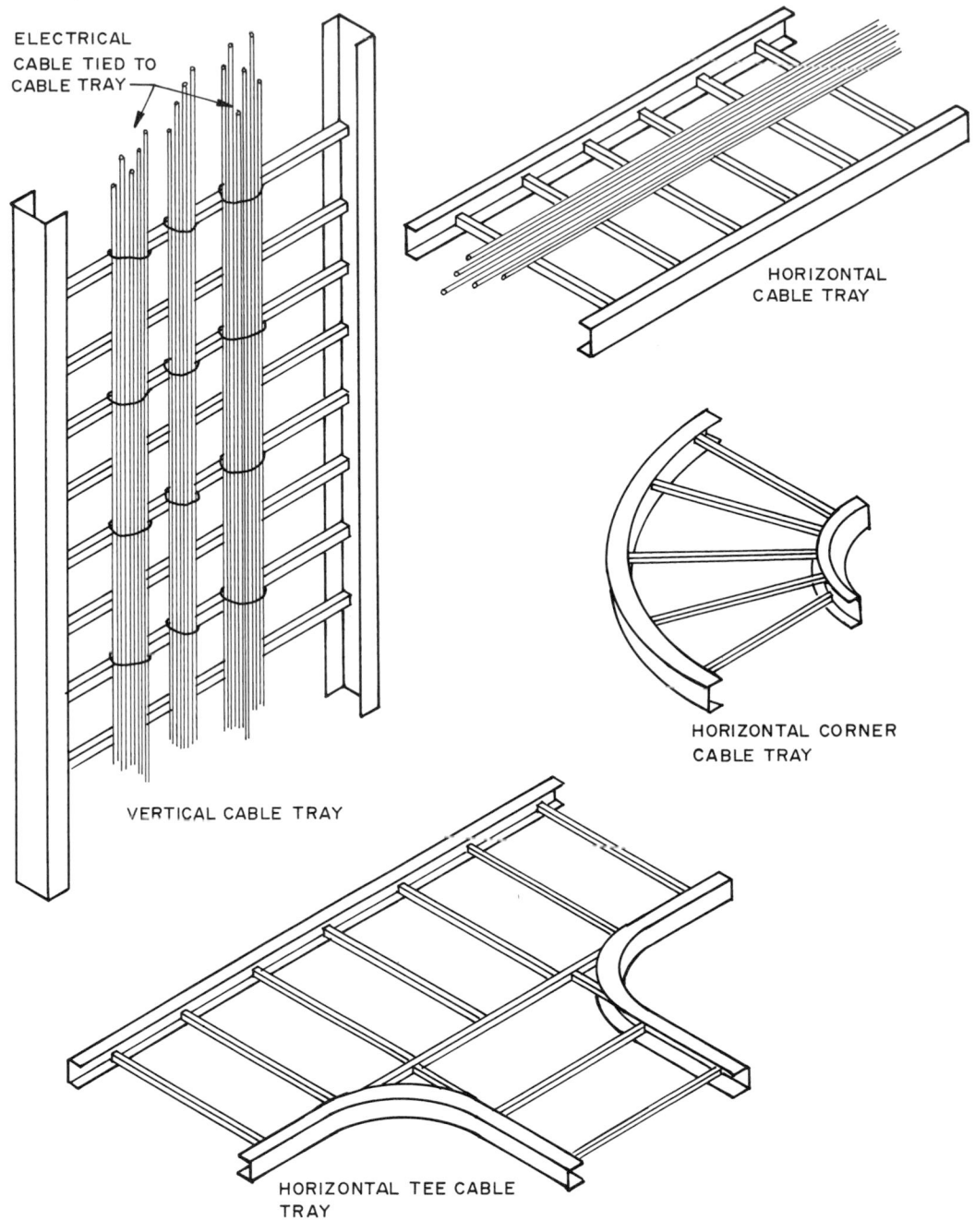

ELECTRICAL
CABLE TIED TO
CABLE TRAY

HORIZONTAL
CABLE TRAY

VERTICAL CABLE TRAY

HORIZONTAL CORNER
CABLE TRAY

HORIZONTAL TEE CABLE
TRAY

16-12 *Some typical cable trays.*

Raceways

Raceways are used to support, enclose, and protect electrical wires.

CABLE TRAYS

Often **raceways** or **cable trays** are open-faced metal channels used to provide support for electric wires that have adequate insulation and do not require extra protection. Open raceways only support the wires. Wires in this system are open to inspection and modifications (see 16-12).

CONDUIT

Conduit is a form of closed raceway. They support insulated electric wire and provide protection. Conduit does not have conductors inside when it is installed. The wires are run in later (see 16-13). One type is a steel pipe available in three thicknesses. The heavy-wall conduit is referred to as **rigid steel conduit** (RSC). The intermediate-wall thickness type is called **intermediate metal conduit** (IMC), and the thin-wall type is called **electric metallic tubing** (EMT).

Flexible metal conduit called Greenfield is used for short runs such as connecting an electric unit, as a furnace, to the power source. It is available as a watertight conduit.

Rigid nonmetallic conduit is also available in **polyvinyl chloride** (PVC) and **high-density polyethylene** (PE).

Conduit can be run inside walls, ceilings below and through floors, and in concrete slabs. Codes regulate the use and locations of the various types.

The rigid metal conduit is available with inside diameters ranging from ½ to 4 in. (12.7 to 101.6 mm). The thin-wall conduit can be bent to form curved corners. Junctions and sharp turns are made with metal fittings. The steel heavy-wall conduit can be used in concrete slabs. Conduit made from aluminum is also available in the same sizes. It is light weight, nonsparking, weathers well, and is easy to work. If embedded in concrete it may cause cracking. If buried in the earth it should be coated with asphalt or another type of protective coating. Metal and nonmetal conduit can be left exposed to view when appearance is not important.

Other raceways are made in rectangular shapes and are intended to be surface mounted and exposed to view. They are painted to match the wall and ceiling and can have electric outlets, telephone connections, and computer and communications wiring connections (see 16-14).

16-13 *This plastic conduit protects the feeder cables into the meter pan.*

(Above) Baseboard

(Right) Wall-mounted

16-14 *Rectangular raceways are used for surface-mounted wiring installations.*
Courtesy Wiremold Company

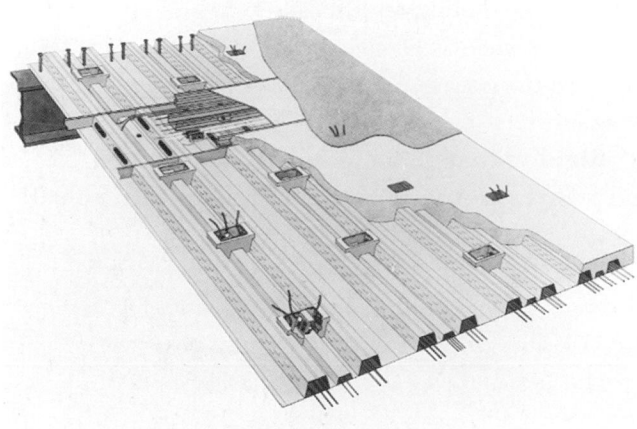

16-15 *This cellular composite steel floor, Robertson Q-Floor, provides channels for running wiring and punchouts for the location of electrical outlets into the building. The outlet boxes are set in place before the concrete floor is poured.*

Courtesy H.H. Robertson

Metal raceways are also part of cellular steel floor decking. The decking serves as a substrate to support the cast-in-place concrete floor. The cells in the decking are used to carry various electrical and communications wires (see 16-15). Perpendicular to the cells in the steel decking are metal ducts spaced as required having outlets through which wires may be pulled to provide power in that area. These are typically spaced on a grid permitting electric power to be available in a number of places in the floor.

Meters

The amount of electric power used is measured in watt-hours (Wh). Since the amount increases rapidly it is reported in kilowatt-hours (kWh). One kWh is equal to 1000 watt-hours (Wh). The amount used is measured by a kilowatt-hour meter (see 16-16). Three-wire meters are typically used for residential applications.

16-16 *A meter is used to measure the kilowatt hours of electricity used.*

16-17 *The incoming power is run into the meter pan. Power enters the building electrical system after the utility company installs the meter.*

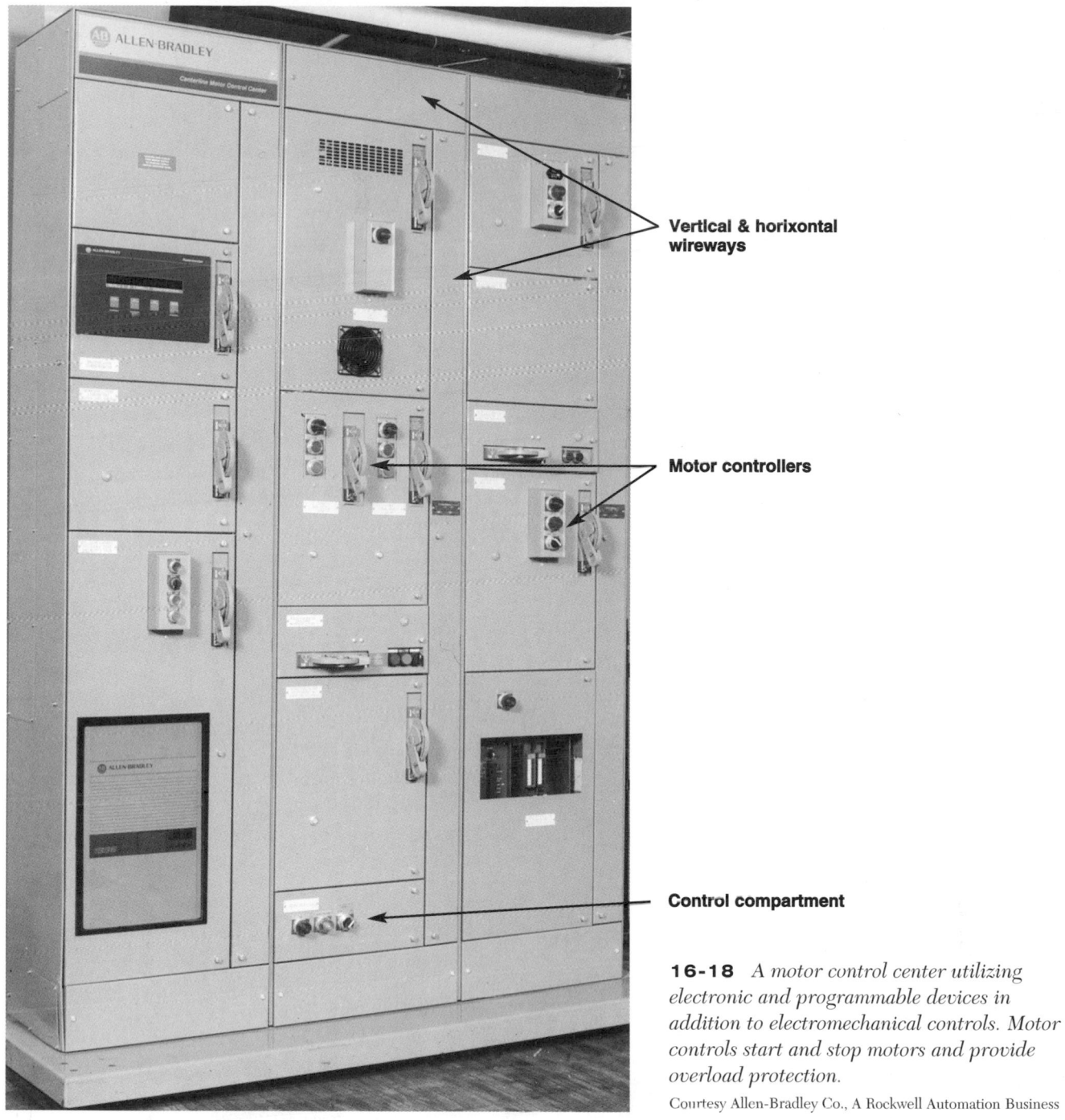

Vertical & horixontal wireways

Motor controllers

Control compartment

16-18 *A motor control center utilizing electronic and programmable devices in addition to electromechanical controls. Motor controls start and stop motors and provide overload protection.*

Courtesy Allen-Bradley Co., A Rockwell Automation Business

Commercial and industrial applications having heavy demands and higher voltage requirements typically use four-wire meters.

The owner of the building provides the meter pans and any current transformers needed to step-down the voltage within the building. The electric utility supplies the meter (see 16-17).

Typically the meter is placed outside the building so the utility staff have ready access. They can be inside the building if easy access is provided.

Motor Control Centers

Motor controllers are used to start and stop motors and to protect them from overloads. They have a disconnect switch that must be located within sight of the controller and the motor. The overload protection is typically a heat-operated relay that opens the circuit when line temperatures rise. The motor circuit is closed after the overload situation has been corrected by pressing a reset button (see. 16-18).

16-19 *A liquid-cooled transformer used on a commercial building.*

Transformers

Transformers are used to change (transform) alternating current from one voltage to another. The voltage coming into the transformer is the primary voltage and that leaving the transformer is the secondary voltage. For example, a transformer is used to step down a primary 4160V current received from the utility distribution system to a secondary voltage such as 480V as it enters the building. Another transformer in the building's vault (or closet) could step this down to the 120V power required for use within the building. Primary voltages include 4160V, 7200V, 12,470V, and 13,200V while secondary voltages include 480V, 277V, 240V, 208V, and 120V.

Transformers may be dry (air cooled) or liquid cooled. Transformers generate heat, which must be removed to prevent overheating. Dry transformers remove heat by circulating air through spaces in the transformer. Liquid-cooled transformers circulate an oil through coolers that absorb the heat and transfer it to the outside air. Mineral oil is the lowest cost liquid used but it is flammable and cannot be used in all locations. A number of liquid coolants, such as silicone liquids have low flammability and are widely used. Liquid transformers are used on large installations (see 16-19).

Transformers may be located indoors or outdoors. Dry-type outdoor installations must be weatherproof enclosures and larger sizes must be kept away from combustible materials. Liquid-type transformers on or adjacent to buildings where combustion is possible must have a means for protecting against fires caused be excess heat or oil leaks.

Dry-type transformers installed indoors do not present the fire hazard of oil-type transformers. They need a fire-resistant barrier between them and combustible material and require adequate ventilation. Large sizes are usually placed in a fire-resistant vault.

Oil-type transformers placed indoors are generally required to be in a vault. The vault is fire resistant and must meet codes (see 16-20).

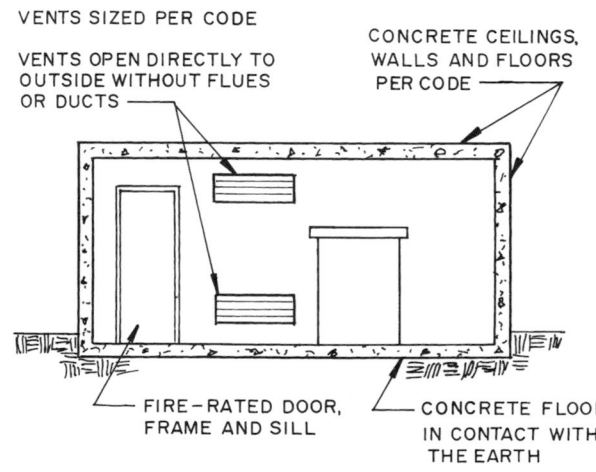

VENTS SIZED PER CODE

VENTS OPEN DIRECTLY TO OUTSIDE WITHOUT FLUES OR DUCTS

CONCRETE CEILINGS, WALLS AND FLOORS PER CODE

FIRE-RATED DOOR, FRAME AND SILL

CONCRETE FLOOR IN CONTACT WITH THE EARTH

16-20 *A typical fire-resistant vault for oil-type transformers installed inside a building.*

The oil-type transformer used in vault walls, ceiling, and floor must have a fire rating to meet the code. They also must be ventilated and have the doors, frames, and sills also meet fire ratings. No valves, piping, ducts, or other fittings not connected with the installation of the transformer are permitted in the vault.

Some of the smaller sizes of liquid-cooled transformers that use a nonflammable fluid may be installed indoors and outdoors in any location.

Switches

Switches are used to open an electrical circuit and interrupt the flow of current. A typical switch as used on low power applications, such as lighting, has some means of physically opening and closing the circuit such as manually moving a lever or pushing a button, using an electric coil, a motor, or a spring. These move metal contacts which separate opening the circuit or close completing the circuit. Solid-state switches also interrupt the circuit but do so by electronically creating a conducting or nonconducting condition. There are no moving parts. They are classified by the National Electric Manufacturers Association (NEMA) and the Underwriters Laboratories (UL).

Switches have different actions to enable them to perform specific functions. These are illustrated by the simple knife switch illustrations in 16-21.

▼ Switching actions and functions for typical throw-switch applications.

Single-pole, single-throw (SPST) switch— is moved to make or break a connection between two contact points. When in contact the circuit is complete. When open the circuit is broken.

Single-pole, double-throw (SPDT) action— is used to control a unit, as a light, from two locations. The light can be turned on and off from either location. This requires a SPDT switch at both locations.

Double-pole, single-throw (DPST) action— is similar to the SPST action. However, it opens both wires in the circuit. They are used when neither of the wires are grounded such as switches used on 240V motors and appliances. It can also be used to control two circuits at the same time. Both circuits are either open or closed simultaneously.

Double-pole, double-throw (DPDT) action—can be used to control more than one circuit at a time. It can be used to reverse the direction of rotation of a DC motor by reversing the polarity.

SINGLE-POLE, SINGLE-THROW
KNIFE SWITCH

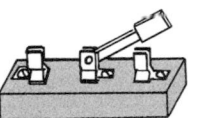

SINGLE-POLE, DOUBLE-THROW
KNIFE SWITCH

DOUBLE-POLE, SINGLE-THROW
KNIFE SWITCH

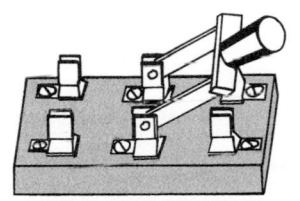

DOUBLE-POLE DOUBLE-THROW
KNIFE SWITCH

16-21 *Typical switch actions are illustrated by these knife switches.*

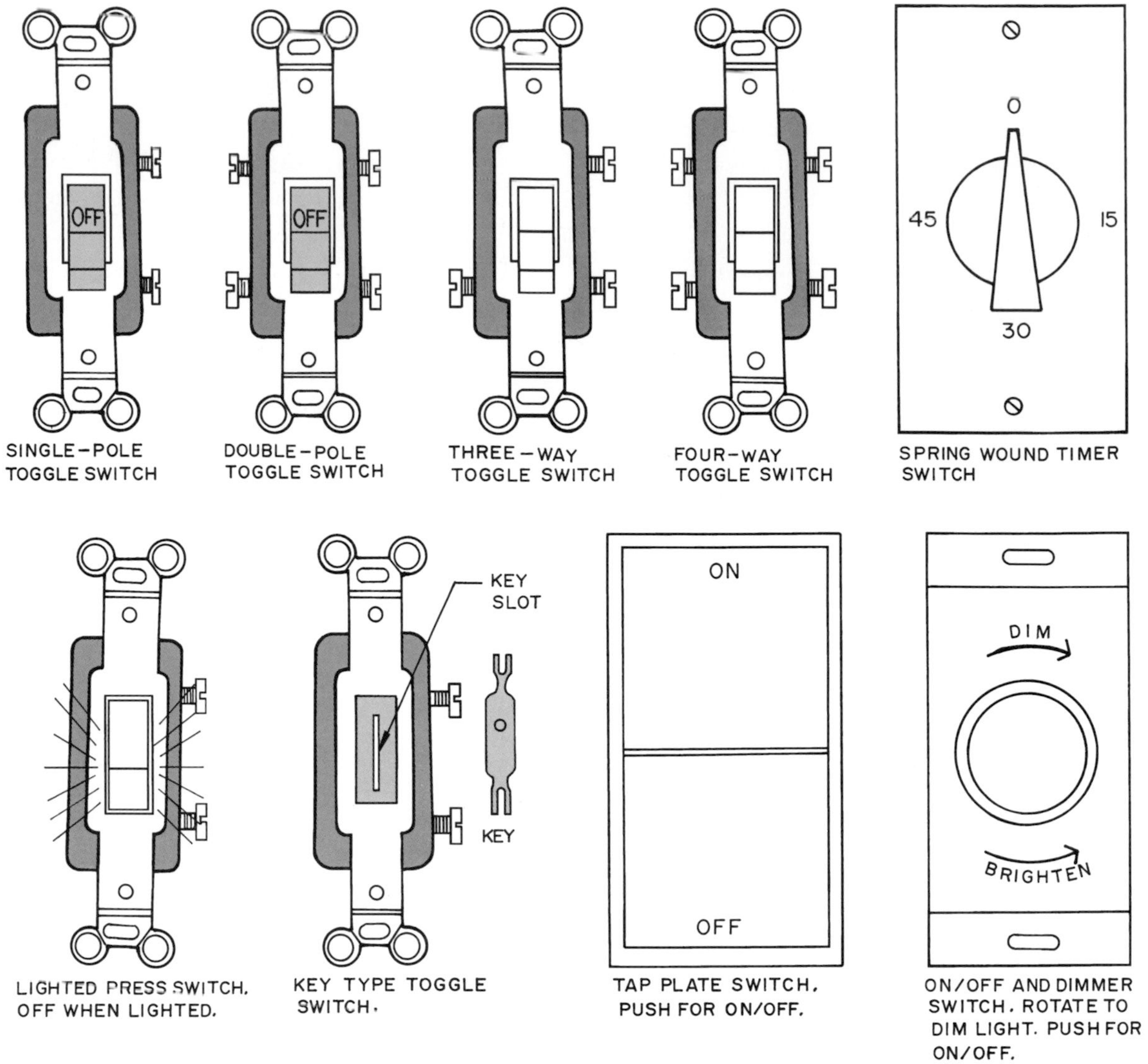

SINGLE—POLE
TOGGLE SWITCH

DOUBLE—POLE
TOGGLE SWITCH

THREE—WAY
TOGGLE SWITCH

FOUR—WAY
TOGGLE SWITCH

SPRING WOUND TIMER
SWITCH

LIGHTED PRESS SWITCH.
OFF WHEN LIGHTED.

KEY
SLOT

KEY

KEY TYPE TOGGLE
SWITCH.

ON

OFF

TAP PLATE SWITCH.
PUSH FOR ON/OFF.

DIM

BRIGHTEN

ON/OFF AND DIMMER
SWITCH. ROTATE TO
DIM LIGHT. PUSH FOR
ON/OFF.

16-22 *These are examples of switches typically used to control circuits with loads up to 30A.*

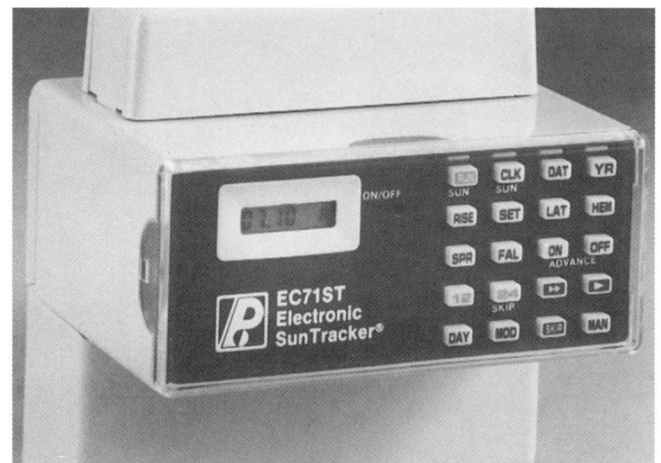

16-23 *A programmable time switch that permits the independent programming of 8 to 20A, 240V circuits. It can provide multiple ON–OFF daily switching on a 7-day week and special 365-day functions.*
Courtesy Paragon Electric Company

Switches that control circuits with loads up to 30A, such as lighting, are shown in 16-22. They are rated to carry various loads which typically run 15, 20, or 30A at 120V. They may be toggle, push, keyed, rocker, touch, tap-plate, or rotary type. They are available as SPST, DPST, DPDT, SPDT. They are also available as three-way and four-way switches. Three-way switches are used to control a light from two different locations. Four-way switches control a light from three different locations. Other switches can operate as timers allowing the unit, such as a fan, to operate for a set time after which it automatically shuts off. This type uses a spring-wound timer. Keyed switches provide security because a key is needed to operate it. Programmable switches are solid-state switches that can be programmed to switch a circuit on and off at preset times (16-23).

Switches described above do not contain a fuse. However fusible switches are available (see 16-24). The fuses and switches are enclosed in a metal box that can be closed and locked. The switch is manually operated with a handle on the outside of the box.

OTHER SWITCHES

▼ There are many other switches designed to serve special purposes.

Service disconnect—will disconnect all electric power to a building except for specific emergency equipment requirements. Generally these are located outside the building. They may be part of a switchboard.

Contactor—is a form of switch. It uses an electromagnet to close the contact blocks. It is operated from a remote location and can be activated by various devices such as pushbutton or thermostat.

Remote-control switch—is a form of contactor that remains latched until its electric circuit is energized. In this way the coil is energized only when the circuit is to be opened. It is used on applications where the circuit must be kept closed (as in a light installation) for long periods of time.

Time-controlled switches—use some type of timer such as an electronic timer or a miniaturized low-speed motor, which rotates a disc that contacts the switch to open and close the circuit (see 16-25).

16-24 *This is a heavy-duty fused three-pole three-wire industrial switch.*

Courtesy Siemens Energy and Automation, Inc.

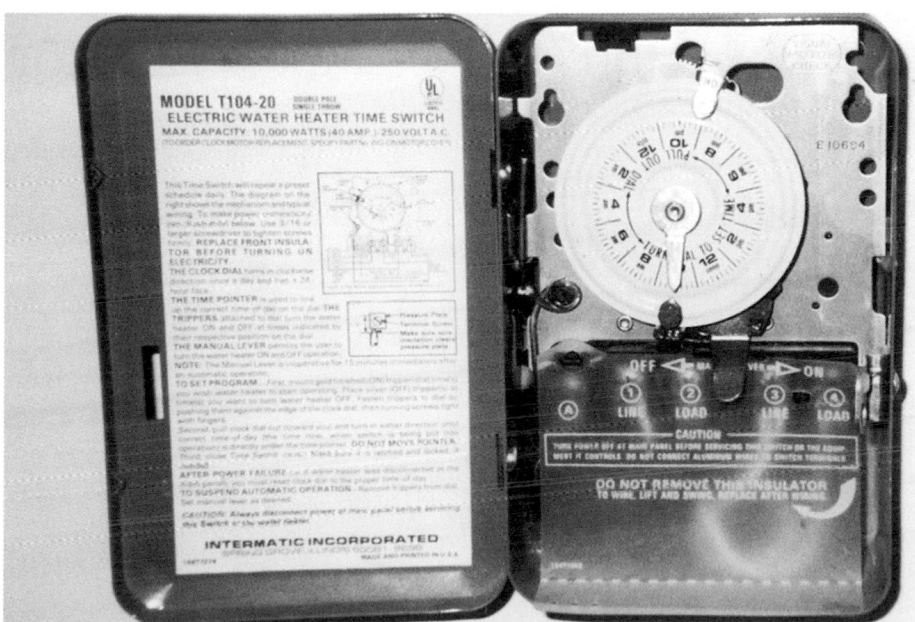

16-25 *Time switches permit setting the clock so the circuit is open and closed as frequently as necessary. This switch operates on a 24-hour period and repeats the settings every 24 hours.*

16-26 *An 800-amp 3-pole automatic transfer switch with a control panel. When the normal flow of electricity fails it will automatically transfer to the emergency service.*
Courtesy Automatic Switch Company

Automatic transfer switch—is a double throw switch that switches to a source of emergency power when the normal electrical service is interrupted (see 16-26). A typical application would be in the power supply for a hospital operating room.

Isolation switches—are opened only after the current flow in the circuit has been interrupted by another regular use switch. They are not used to interrupt the flow of current in a circuit.

OVERCURRENT PROTECTION DEVICES

Circuit breakers are electromechanical units disconnecting a circuit automatically whenever the current attains an established value (see 16-27) that would cause overheating and a possible fire in the circuit. Examples of occurrences that could cause a circuit breaker to open the circuit would be excess current flow caused by overloading the circuit or a short circuit. Circuit breakers are available in a molded case that is inserted in the circuit at the panel or a large air-type breaker that is not in a protective case.

16-27 *A circuit breaker disconnects a circuit automatically whenever there is excess current flow, a short circuit, power surge, or overheating. This is a single-pole 50A frame plug-in circuit breaker. It also provides ground fault protection.*
Courtesy Square D Company

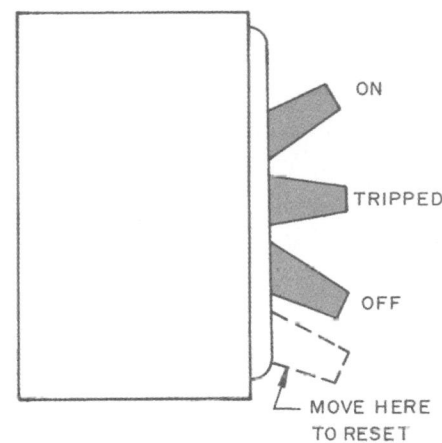

16-28 *Positions of the handle on a circuit breaker. To restore power move the lever to the "reset" position and back to the "on" position.*

ON

TRIPPED

OFF

MOVE HERE
TO RESET

When these circuit breakers open a circuit they are reset after the deficiency has been corrected by moving the switch handle (see 16-28). Circuit breakers are mounted in a circuit breaker box or a panelboard.

Fuses are also used to protect circuits from overloads, short circuits, and power surges. They have an internal metal link that will melt when overheated thus breaking the circuit. The two types of fuses are cartridge and plug (see 16-29).

The two types of cartridge fuses are **knife-blade** and **ferrule**. The blades on the knife-blade type slip into metal clips on the panelboard while the ferrule type fits the copper rings on each end into clips. This connects the incoming power at the service entrance to the bus bars in the panelboard. Most cartridge fuses are discarded after they have blown. However, some types have replaceable links.

Ferrule type cartridge fuses are rated from 10 to 60 amperes and are generally used to protect currents for individual appliances such as an electric stove. Knife-blade cartridge fuses are used for service over 60 amperes and are used in the service entrance between the incoming power line and the circuits in the panelboard.

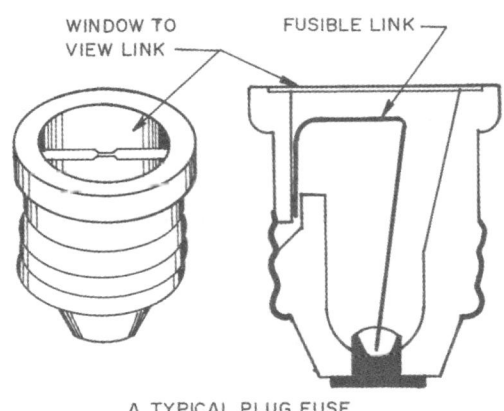

WINDOW TO
VIEW LINK

FUSIBLE LINK

A TYPICAL PLUG FUSE

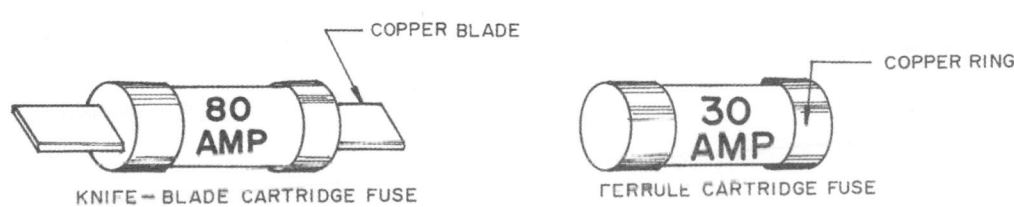

COPPER BLADE

80
AMP

KNIFE-BLADE CARTRIDGE FUSE

COPPER RING

30
AMP

FERRULE CARTRIDGE FUSE

16-29 *Commonly used fuses.*

Plug fuses are available in 15-amp, 20-amp, 25-amp, and 30-amp sizes. They are used to protect individual circuits requiring small current requirements, such as a series of lights. They are installed in a fuse box or panelboard (see 16-30) and serve the same function as a circuit breaker. However, when they are blown they are discarded and replaced with a new fuse when the deficiency has been corrected.

▼ There are several varieties of plug fuse.

Standard plug fuse—will blow when the fusible link is overheated.

Time-delay plug fuse—has a fusible link that will melt immediately only when a short circuit occurs. If there is an overload the link softens but does not break. If the overload is quickly removed, such as a momentary load when starting a large electric motor, it will not break. However, if the overload continues the link will melt.

Type S plug fuse—functions in the same manner as a time-delay fuse. However, the threaded base is too small to be screwed into the fuse panel. An adapter base is required. The adapters are threaded for fuses of different ampere capacities. The adapter is screwed into the threaded fuse socket on the panelboard. Since the adapter will only accept a plug fuse designed to fit it a fuse of the wrong amperage cannot be screwed in place.

Switchgear & Switchboards

Switchgear and switchboards are freestanding units consisting of an assembly of fuses and/or circuit breakers, switches, and other line components.

Switchboards—are large units such as those used in commercial buildings. All the components are contained within an insulated metal case and the devices are controlled with insulated handles mounted on the face of the panel. The assembly is used for low-voltage (600 volts or less) installations (see 16-31).

Switchgear—are switching and interrupting devices with associated control, regulating, metering, and protective devices used for high-voltage (over 600V) service typically assembled in an insulated metal structure. However, the term switchgear also describes individual switching units not assembled in a panel.

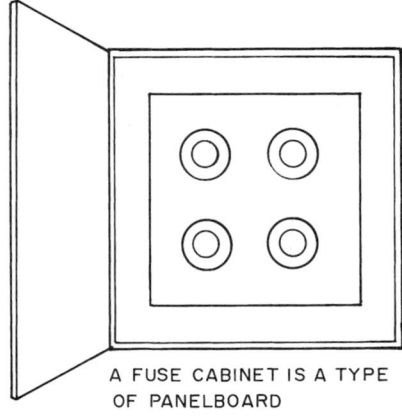

A FUSE CABINET IS A TYPE OF PANELBOARD

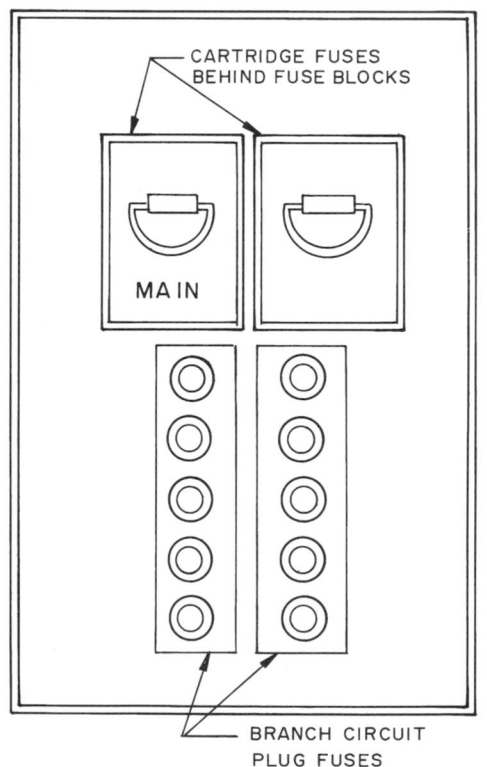

CARTRIDGE FUSES BEHIND FUSE BLOCKS

MAIN

BRANCH CIRCUIT PLUG FUSES

A LARGE FUSED PANELBOARD

16-30 *Plug-type fuses are installed in a fuse box or a panelboard.*

Switchboards and switchgear used in commercial and industrial buildings are commonly placed in a basement vault designed specifically to house the switchgear. It must be adequately ventilated and meet the requirements of the National Electrical Code. The design of the vault should include adequate entrances and exits so equipment can be installed and removed and workers in the room can exit quickly in an emergency. They must be located in permanently dry conditions.

Switchgear installed outdoors may be housed inside a small building, or use switchgear housed in a weatherproof metal case (see 16-32).

Panelboards

A panelboard receives a large amount of electrical power from the public utility and distributes it in smaller amounts through a number of circuits. In small buildings it serves the same basic purpose as a switchboard but on a smaller scale. A panelboard is typically used in residential and small commercial construction to distribute power to each of the circuits within the building after receiving input from the public utility line (see 16-33).

16-31 *A switchboard is a single electric control panel or assembly of panels on which are mounted switches, overcurrent devices, and instrumentation enclosed in an insulated metal structure with the devices controlled by handles on the front of the panel.*

Courtesy Square D Company

16-32 *Switchgear installed outdoors must be installed in a weatherproof enclosure.*

16-33 *This panelboard has the main breaker at the bottom and two vertical rows of circuit breakers.*

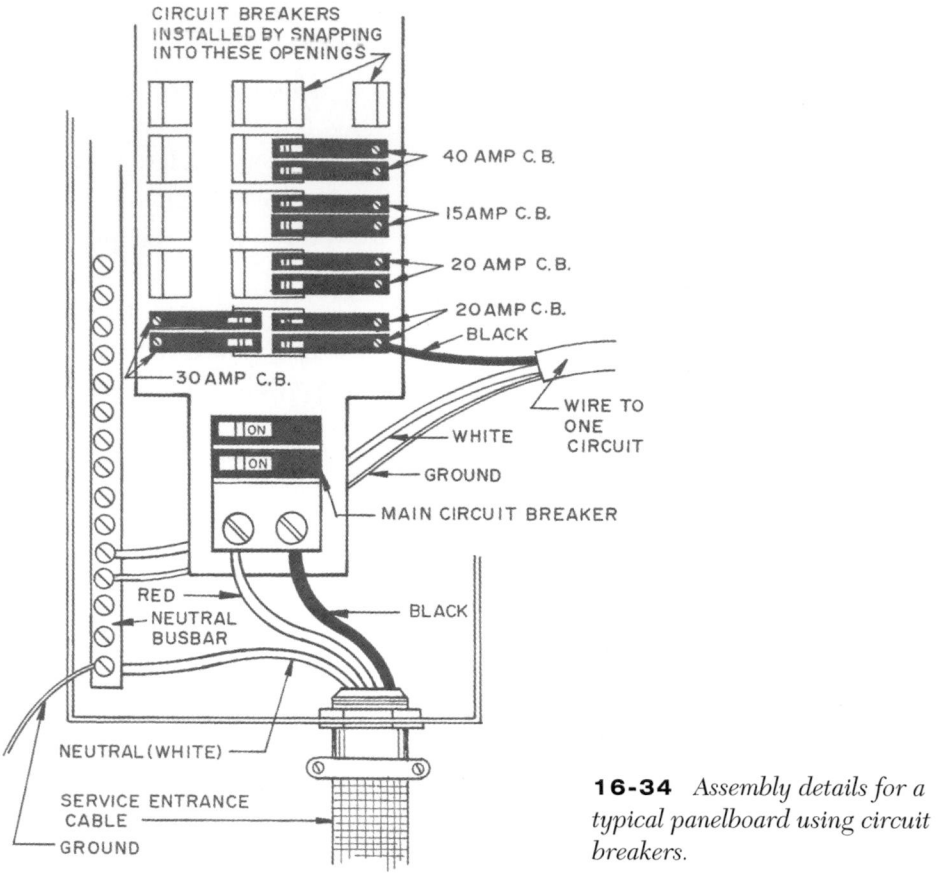

CIRCUIT BREAKERS
INSTALLED BY SNAPPING
INTO THESE OPENINGS

40 AMP C.B.

15 AMP C.B.

20 AMP C.B.

20 AMP C.B.
BLACK

30 AMP C.B.

WIRE TO
ONE
CIRCUIT

WHITE

GROUND

MAIN CIRCUIT BREAKER

RED
NEUTRAL
BUSBAR

BLACK

NEUTRAL (WHITE)

SERVICE ENTRANCE
CABLE

GROUND

16-34 *Assembly details for a typical panelboard using circuit breakers.*

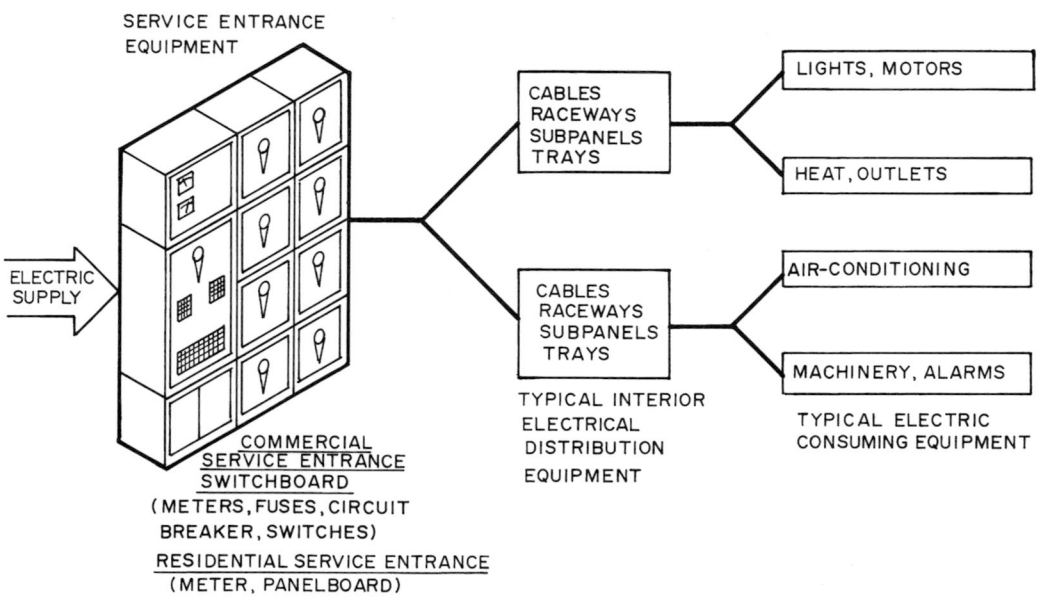

SERVICE ENTRANCE
EQUIPMENT

ELECTRIC
SUPPLY

COMMERCIAL
SERVICE ENTRANCE
SWITCHBOARD
(METERS, FUSES, CIRCUIT
BREAKER, SWITCHES)

RESIDENTIAL SERVICE ENTRANCE
(METER, PANELBOARD)

CABLES
RACEWAYS
SUBPANELS
TRAYS

CABLES
RACEWAYS
SUBPANELS
TRAYS

TYPICAL INTERIOR
ELECTRICAL
DISTRIBUTION
EQUIPMENT

LIGHTS, MOTORS

HEAT, OUTLETS

AIR-CONDITIONING

MACHINERY, ALARMS

TYPICAL ELECTRIC
CONSUMING EQUIPMENT

16-35 *A typical electrical system for a commercial or industrial building includes the service entrance equipment, distribution equipment, and the equipment that consumes the electrical power.*

▼ A typical panelboard using circuit breakers is illustrated in 16-34.

Incoming service entrance cable—connects to the neutral bus bar and the main circuit breaker.

Main—controls the power flow from the utility into the service panel. The white wire is the neutral, connected to the neutral bus bar.

Red and black wires—connect to the main and through it to the snap-on connections for the circuit breaker.

Circuit breakers—control the power to each circuit.

Panelboards have locked metal covers so no live parts are exposed unless the door is opened. Medium-duty panelboards are used for lighting and general purpose electrical outlets. Heavy-duty panels are used for distribution of power in industrial applications.

In large construction panelboards are located on the various floors and are fed from the service switches and switchgear. The lights and outlets in an area of a floor are controlled from a panelboard on that floor.

BUILDING ELECTRICAL DISTRIBUTION SYSTEMS

Interior building electrical systems distribute power as needed to the various sections of the building and transmit information through an internal communications system. The basic standard for the design and installation is the National Electrical Code published by the National Fire Protection Association. The Underwriters Laboratory certifies electrical equipment and materials through the use of testing specifications and procedures.

The electrical power system of a commercial or industrial building includes the service entrance equipment, interior distribution system, and the equipment that uses the electricity (see 16-35). The service entrance includes equipment such as meters, switches, panelboards, switchgear, switchboards, fuses, and circuit breakers. The distribution system includes trays, wiring, raceways, wireways, conduit, and busways. The third part of the interior electrical system, the items to use the electricity include things such as lights, motors, heaters, industrial equipment, and communications equipment. These are the loads considered when the system is designed.

The following examples of building electrical distribution systems show generalized examples. The actual design of a system varies considerably depending upon the requirements of the building and the equipment to be installed and requires the services of an experienced electrical engineer.

The Service Entrance

The service entrance is the system bringing the electrical power from the lines of the electrical utility into the building.

A typical above-ground service entrance for a residential or small commercial building is shown in 16-36.

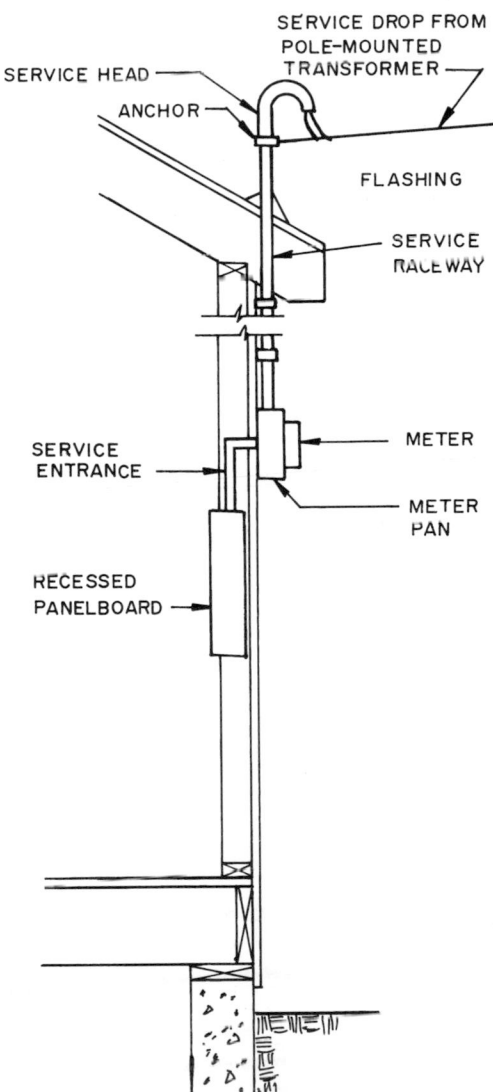

16-36 *Typical above-ground service entrance for residential and small commercial buildings.*

A typical below-ground service entrance for a residential or small commercial building is in 16-37. The service from the public utility may be overhead or underground. The meter is generally on the exterior of the building. This enables the utility to read the meter and install or remove it without entering the building. From the service entrance panelboard branch circuits are run to various parts of the building for lights, outlets, appliances, furnace, and other required services. Each circuit has a circuit breaker overload device in the panelboard. Service can be run to a subpanel in an area some distance from the panelboard where branch circuits can be taken from it. If the building has several occupants, as a duplex, individual meters may be set up for each apartment.

A service entrance for commercial buildings can take various forms depending upon the requirements of the occupants. It requires a service switch or circuit breaker, which disconnects power to the entire building. This is typically located outside the building but exceptions can be made. The power enters through a meter installation, which includes a meter pan, meter cabinet, and a current transformer in a cabinet or vault.

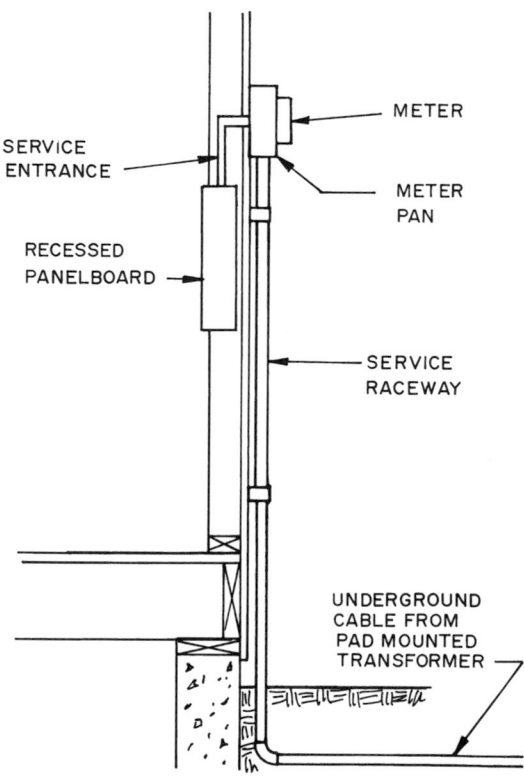

16-37 *Typical below-ground service entrance for residential and small commercial buildings.*

LIGHTING

The use of natural lighting is a major consideration when a designer is working on the design of the illumination for a building interior. Most spaces have natural and artificial light integrated so they flow into the space and work together to provide the type and degree of illumination required. Natural lighting can be controlled by various shading devices and the type of glazing used.

Illumination Design Considerations

As the lighting system is designed the activities in each area must be carefully analyzed. Certainly the purpose of the lighting system is to enable the occupants to have a pleasant atmosphere and be visually able to perform required tasks. All of this must be accomplished using as little electrical power as possible.

When designing a lighting system the ceiling height, reflectance values of the ceiling, walls, floors and furniture, and the footcandle requirements all relate to each other. The choice of the type of luminary, its spacing, how it is mounted, and its height above the area to be lighted must be considered.

TASK LIGHTING

Task lighting involves providing an adequate level of light for a person to perform work of a specific nature, such as electronic assembly on a production line, without undue eye strain. It should be possible to work at a specific task without having the same high levels of lighting in the entire room. The tasks performed, as in a manufacturing plant, need to be identified, their location within the building and the area involved specified, the number of people performing the task, their relationship (nearness) to each other, and any patterns of movement of the occupants performing the tasks established.

The surface upon which the task is to be performed will influence lighting decisions. The most common situation involves some form of horizontal surface while inclined and vertical work surfaces are frequently found. The widespread use of computers presents a multiple problem with a horizontal keyboard, vertical rack to hold the copy being set and a monitor which produces light directly at the operator. The reflective values of the surface also require consideration because highly reflective work surfaces can be very disturbing.

GENERAL ILLUMINATION

General illumination provides a degree of light over a large area, such as a room in which specific task lighting luminaires are also present. General illumination provides a way to ease visual discomfort caused when looking from a high-level lighted area (as a task) out across a large area (a room). General illumination reduces the difference in brightness levels thus maintaining visual comfort. Since the level of lighting intensity is less the luminaires can be widely spaced and use less electricity. The designer attempts to get a balance between the brightness of a task-oriented area and the larger general area of the room.

SELECTIVE LIGHTING

Selective lighting is used to focus light upon a specific object or area such as light focused on a painting or on a display case containing jewelry. The general illumination becomes the background and the specific lighting focuses visual attention upon the object.

GLARE

Glare refers to light that is intense and offending to the viewer. It may be from luminaires that shine directly into the eyes of the viewer causing visual discomfort. This can be reduced by lowering the brightness of the luminaires or by relocating or shielding them if possible.

Reflected glare is often more frequently found than direct glare shining in the viewer's eyes. It occurs when the light is reflected off a surface. It can be controlled by selecting surfaces with low reflective coefficients, reducing the brightness of the luminaire or changing the angle of the reflecting surface.

REFLECTION

The **reflection coefficient** of walls and ceiling is an important part of lighting planning. The luminous intensity of a luminaire can be reduced or enhanced by the reflective potential of the surrounding ceiling, walls, and floor. Dark colors absorb light while light colors reflect light. Therefore, where maximum use of the luminous intensity of a luminaire is enhanced by light-colored ceilings and walls. Bright colors, as red or yellow, are often not used because they tend to provide a glare, causing visual discomfort. Manufacturers of wall and ceiling materials and paints generally have had them tested and can give their **reflectance coefficient.** For example, if half the incident light on a sur-

face is reflected back the reflectance coefficient is 50% or 0.50. The other 50% is absorbed or transmitted through the material. The amount transmitted is called **luminous transmittance.** The rate of flow of light energy through a surface is called **luminous flux.**

Levels of Illumination

Various activities require different levels of illumination. For example, drafting activities require about 200 footcandles while a corridor only needs 10 to 20 footcandles. Recommended levels of illumination have been established by the Illuminating Engineering Society of North America. Adequate illumination depends upon the footcandles (quantity) plus the brightness, distribution, and color of the light (quality).

Room Cavity Ratio

The size of the space to be lighted affects the way light is distributed throughout the space. High ceilings reduce the efficiency of lighting. Open spaces can be more efficiently lighted than small rooms or partitioned space. The relation between the space proportions, room height, room perimeter, and the height of the work surface to be illuminated divided by the floor area is called the room cavity ratio.

Measuring Illuminance Levels

The **level of illuminance** is commonly measured with an illuminance meter, often referred to as a light meter (see 16-38). It has a photoelectric panel connected to a microammeter and an electronic control circuitry. It is calibrated to read in footcandles or lux. Some types have a remote sensor. The meter is held so the photoelectric panel is parallel to the plane of the area to be

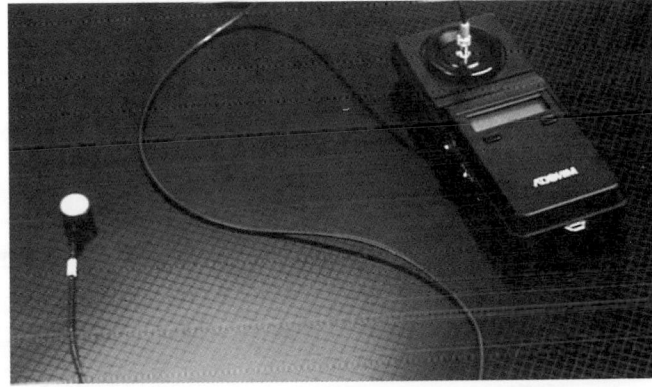

16-38 *The level of illuminance is measured with an illuminance (light) meter.*

Courtesy Minolta Corporation

tested. For example, to measure ceiling illuminance it is held parallel with the ceiling while to measure wall illuminance it is held parallel with the wall. It is important to follow the manufacturers instructions so accurate data are recorded.

The **measure of illuminance** is a measure of the **quantity** of illumination. One lumen of luminous flux uniformly imposed on one square foot of area results in an illuminance of one footcandle. To totally evaluate the illumination available it is necessary to get a measure of the luminance of quality of illumination.

Lighting Systems

The most frequently used lighting system is possibly a **direct** system, however **semi-direct, indirect, semi-indirect, general-diffuse** or **direct–indirect** find

many applications (see 16-39). The exact classification of a lighting system depends upon the amount of light from the luminaire directed up and down.

▼ The classification of lighting systems.

   Direct lighting systems project most of the light down to the floor or work surface and is the most efficient system. A light color reflective floor will reflect some light back toward the ceiling, reducing the darkness at the ceiling somewhat.

   Indirect lighting systems project most of the light up on to the ceiling and some on to the walls. Here it is reflected back to the floor and work surfaces. This system requires that ceilings and walls have a high reflectance

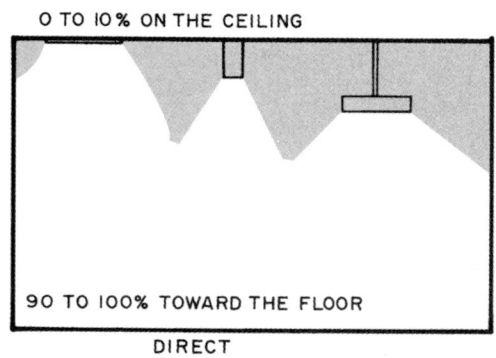

0 TO 10% ON THE CEILING

90 TO 100% TOWARD THE FLOOR

DIRECT

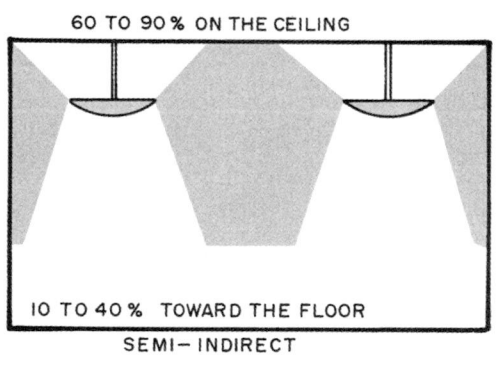

10 TO 40% ON THE CEILING

60 TO 90% TOWARD THE FLOOR

SEMI–DIRECT

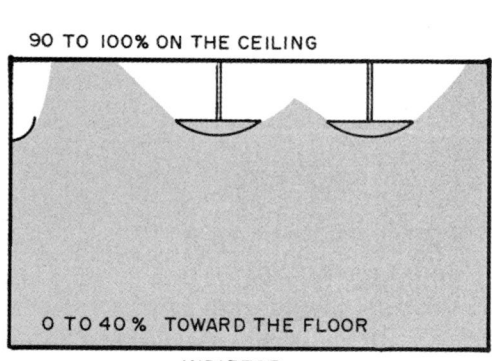

90 TO 100% ON THE CEILING

0 TO 40% TOWARD THE FLOOR

INDIRECT

60 TO 90% ON THE CEILING

10 TO 40% TOWARD THE FLOOR

SEMI–INDIRECT

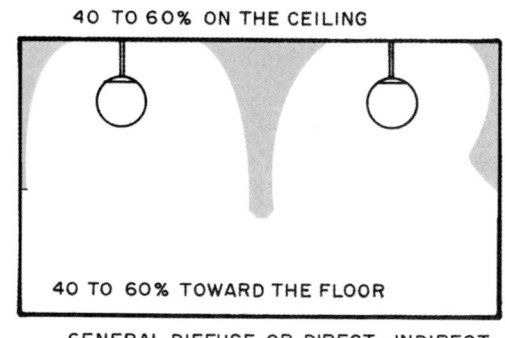

40 TO 60% ON THE CEILING

40 TO 60% TOWARD THE FLOOR

GENERAL DIFFUSE OR DIRECT–INDIRECT

16-39 *The distribution of light can be varied by using these types of lighting systems.*

coefficient. It produces an illumination that is rather uniform, free of glare and diffuse. It is rather inefficient and requires that lumanaires have more powerful lamps than direct lighting.

Semi-indirect lighting systems allow more light to project down than indirect. They do this by using a diffuser that is translucent, permitting some light to project down yet reflecting most of it up to the ceiling. This produces a diffuse low glare illumination.

General-diffuse systems (also called direct-indirect) project about an equal amount of light up and down. It produces a light ceiling and upper wall area yet projects light down to the floor and work surfaces. This system typically uses a globe diffuser, which allows light to project in all directions.

Types of Lamp

A variety of lamps are available and each has particular advantages and disadvantages. Some are useful only for special applications.

INCANDESCENT

An incandescent lamp has a filament joined to a metal base with copper lead-in wires that enclosed in a glass bulb. The filament is a tungsten wire that resists the flow of electricity and in doing so gets hot and glows, producing light. The metal base screws into a socket which is part of the luminaire. Most incandescent lamps are filled with argon and nitrogen, which makes possible the use of a high-filament temperature. The melting point of tungsten is 3655K (6170°F). Lamps filled with krypton gas have a longer life than argon and nitrogen lamps and cost more.

TUNGSTEN-HALOGEN LAMPS

Tungsten-halogen lamps (quartz-iodine) are a type of incandescent lamp. They have a tungsten filament and are filled with a halogen, such as iodine or bromine, and an inert gas, which reduces the evaporation of the filament. The bulb is made from quartz because it will withstand higher temperatures than glass. The halogen additive in the lamp reacts chemically with any tungsten deposited on the bulb and redeposits it on the tungsten filament, improving the efficiency of the lamp. An exmaple with a reflector is shown in 16-40.

16-40 *A tungsten-halogen quartz-iodine lamp installed in a reflector to provide indirect lighting.*

Tungsten-halogen lamps operate at high temperatures and even though the quartz envelope can withstand high temperatures the lamps could explode. Therefore manufacturers recommend some type of shielding or protective screen be used to contain flying fragments from unprotected quartz lamps.

FLUORESCENT LAMPS

Fluorescent lamps have a long circular glass tube that is sealed at each end. A mixture of an inert gas, such as argon and low pressure mercury vapor, is in the tube. A cathode is in each end from which electrons are produced that start and maintain operation of a mercury arc, which produces an ultraviolet arc. The ultraviolet arc is absorbed by phosphors coating the inside of the tube and cause it to fluoresce (radiate) light.

Fluorescent lamps will not operate directly off 120V alternating current because it will not cause the arc discharge. The luminaire has a **ballast** that provides the starting and operating voltages. Standard lamps have a starter that preheats the cathodes, producing the high voltage arc needed to start the lamp. Rapid start lamps have the same basic construction as standard lamps but have a circuit that keeps the lamp electrodes constantly preheated by means of low voltage windings that are a part of the ballast.

High output (HO) lamps and **very high output** (VHO) lamps require special ballasts. They are used where high output is required in a limited area such as a merchandise display or an outdoor sign. They generate considerable heat and this must be considered when they are used. They will function in cold situations where standard lamps will not light.

Instant-start fluorescent lamps use a high-voltage transformer to generate an arc, lighting the lamp without the preheating delay in standard lamps. They have a single pin at each end. The high-voltage start greatly reduces the life of the lamp, but it will start in temperatures below that of rapid-start lamps.

Since fluorescent lamps are much more efficient than incandescent lamps lower wattage lamps can be used.

HIGH-INTENSITY DISCHARGE LAMPS (HID)

High-intensity discharge lamps produce light by passing an electric arc through a metallic vapor that is confined in a sealed quartz or ceramic tube. The high-intensity discharge lamps available include mercury vapor, high- and low-pressure sodium, and metal-halide. With appropriate color correction they can be used in many indoor and outdoor situations.

Mercury Vapor Lamps (M-V)

Mercury vapor lamps are available in clear, white, white deluxe, and color corrected. The clear lamp produces a blue-green light, which causes distortion of other colors. Color correction is accomplished by coating the outer bulb with phosphors, which make the lamp useful for some indoor applications.

Mercury vapor lamps use mercury and argon gas, which helps in starting since mercury has a low vapor pressure at room temperature. Energizing the circuit moves a starting voltage across the space between the starting electrode and the main electrode. This creates an argon arc that vaporizes the mercury. A mercury vapor lamp takes 3 to 6 minutes to reach its full warmup and maximum output. A ballast is used to start the lamp and control the arc. After the lamp is turned off it must cool and the pressure allowed to lower before it can be relighted, which takes from 3 to 8 minutes depending on the construction of the lamp.

Metal-Halide Lamps

Metal-halide lamps are another type of high-intensity discharge lamp. They are designed much like the mercury vapor lamp but have halides of metals, such as sodium, indium, or thallium, added. These salts produce a light that is radiated at frequencies other than the colors radiated by mercury, producing a lamp color that is better than mercury vapor lamps. It has about the same starting delay and restart times as the mercury vapor.

It should be noted that metal-halide lamps carry the same safety warning as mercury vapor and that they tend to explode and must be encased in an approved enclosing fixture. Some types have a plastic coating making them safe to use in an open fixture. They also are marked for their proper burning position. They are designed to be installed with their base up, base down, or horizontally. If installed in the wrong position their operating characteristics will be limited.

High- and Low-Pressure Sodium Lamps

High-pressure sodium (HPS) lamps are arc discharge lamps with sodium under high pressure in the glass arc tube. This produces a light with a yellow tint. It is a highly efficient lamp superior to all other types and in areas where the color is not objectionable can replace mercury vapor and metal-halide lamps.

Low-pressure sodium lamps, referred to as SOX, produce a deep yellow light that is usually not acceptable for general interior lighting. It is widely used for street, highway, and parking lighting. It has a long life and is the most economical type of lighting.

APPENDIX A

PROFESSIONAL & TECHNICAL ORGANIZATIONS

DIVISION 2. SITE WORK

American Association of State Highway
and Transportation Officials System
449 Capitol St. NW, Suite 225
Washington, DC 20001

American Planning Association (APA)
1313 E. 60th St.
Chicago, IL 60637

American Society of Civil Engineers
(ASCE)
345 E. 47th St.
New York, NY 10017

American Society of Landscape
Architects (ASLA)
4401 Connecticut Ave. NW
Washington, DC 20008

American Water Works Association
(AWWA)
6660 W. Quincy Ave.
Denver, CO 80235

ASFE: Professional Firms Practicing
in the Geosciences
8811 Colesville Road
Silver Spring, MD 20910

Asphalt Institute (AI)
P.O. Box 14052
Lexington, KY 40512–4052

Canadian Society of Landscape
Architects (CSLA)
1339 Fifteenth Ave. SW, Apt. 310
Calgary, Alberta, Canada T3C 3V3

National Sanitation Foundation
International (NSFI)
P.O. Box 130140
Ann Arbor, MI 48113

Urban Land Institute (ULI)
624 Indiana Ave. NW, Suite 400
Washington, DC 20004

DIVISION 3. CONCRETE

American Concrete Institute (ACI)
P.O. Box 9094
Farmington Hills, MI 48333–9094

American Society of Concrete
Construction (ASCC)
1902 Techny Ct.
Northbrook, IL 60062

Architectural Precast Association (APA)
1850 Lee Rd., Suite 230
Winter Park, FL 32789

Concrete Reinforcing Steel Institute
(CRSI)
933 North Plum Grove Rd.
Schaumberg, IL 60173

National Concrete Masonry Association
(NCMA)
2302 Horsepen Rd.
Herndon, VA 22071

National Precast Concrete Association
(NPCA)
10333 N. Meridian St., Suite 272
Indianapolis, IN 46290

Portland Cement Association (PCA)
5420 Old Orchard Rd.
Skokie, IL 60077–1083

Post-Tensioning Institute (PTI)
1717 W. Northern Ave., Suite 114
Phoenix, AZ 85021–5471

Precast/Prestressed Concrete Institute
(PCI)
175 W. Jackson Blvd., Suite 1859
Chicago, IL 60604

Tilt-Up Concrete Association (TCA)
P.O. Box 204
Mt. Vernon, IA 52310

Wire Reinforcement Institute (WRI)
2911 Glencagle Dr.
Findlay, OH 45840

DIVISION 4. MASONRY

Brick Institute of America (BIA)
11490 Commerce Park Dr.
Reston, VA 22091–1525

Building Stone Institute (BSI)
P.O. Box 507
Purdys, NY 10578

Cast Stone Institute (CastSI)
1850 Lee Rd., Suite 230
Winter Park, FL 32789

Indiana Limestone Institute of America
(ILIA)
Stone City Bank Bldg., Suite 400
Bedford, IN 47421

International Masonry Institute (IMI)
823 Fifteenth St. NW, Suite 1001
Washington, DC 20005

Marble Institute of America (MIA)
33505 State St.
Farmington, MI 48335

The Masonry Council (MC)
2619 Spruce St.
Boulder, CO 80302

Masonry Institute of America (MasIA)
2550 Beverly Blvd.
Los Angeles, CA 90057

National Concrete Masonry Association
(NCMA)
2302 HorsePen Rd.
Herndon, VA 22071

National Lime Association (NLA)
200 N. Glebe Rd., Suite 800
Arlington, VA 22203

DIVISION 5. METALS

Aluminum Anodizers Council (AAC)
1000 N. Rand, Suite 214
Wauconda, IL 60084

Aluminum Association (AA)
900 Nineteenth St. NW, Suite 300
Washington, DC 20006

American Hot Dip Galvanizers
Association
1133 Fifteenth St. NW
Washington, DC 20005

American Institute for Hollow
Structural Sections (AIHSS)
929 McLaughin Run Road, Suite 8
Pittsburgh, PA 15017

American Institute of Steel Construction (AISC)
One Wacker Drive, Suite 3100
Chicago, IL 60601–2011

American Iron and Steel Institute (AISI)
1101 Seventeenth St. NW,
Suite 1300
Washington, DC 20036

American Zinc Association
1112 Sixteenth St. NW, Suite 240
Washington, DC 20036

Copper Development Association (CDA)
260 Madison Ave.
New York, NY 10016

International Lead Zinc Research Organization, Inc.
2525 Meridian Parkway
Research Triangle Park, NC 27709–2306

Iron and Steel Society (ISS)
410 Commonwealth Dr.
Warrendale, PA 15086

Lead Industries Association, Inc.
295 Madison Ave.
New York, NY 10017

Metal Lath/Steel Framing Association
600 S. Federal St., Suite 400
Chicago, IL 60605

National Association of Architectural Metal Manufacturers (NAAMM)
600 S. Federal St., Suite 400
Chicago, IL 60605

National Ornamental and Miscellaneous Metals Association (NOMMA)
804-10 Main St., Suite E
Forest Park, GA 30050

Nickel Development Institute (NDI)
15 Toronto St., Suite 402
Toronto, Ontario, Canada M5C 2E3

Sheet Metal and Air-Conditioning Contractors National Association (SMACNA)
P.O. Box 221230
Chantilly, VA 22022–1230

Society of Automotive Engineers, Inc.
400 Commonwealth Drive
Warrendale, PA 15096–0001

Specialty Steel Industry of North America
3050 K St. NW
Washington, DC 20007

Steel Deck Institute (SDI)
P.O. Box 9506
Canton, Ohio 44711

Steel Joist Institute (SJI)
1205 Forty-Eighth Ave. N, Suite A
Myrtle Beach, SC 29577

Steel Structures Painting Council (SSPC)
4516 Henry St.
Pittsburgh, PA 15213–3728

Structural Insulated Panel Association (SIPA)
1511 K Street NW, Suite 600
Washington, DC 20005

Wire Reinforcement Institute
2911 Gleneagle Drive
Findlay, OH 45840–2908

DIVISION 6. WOOD & PLASTICS
Wood Organizations

American Fiberboard Association & American Hardboard Association
1210 W. Northwest Highway
Palatine, IL 60067

American Forest and Paper Association
1111 Nineteenth St. NW, Suite 800
Washington, DC 20036

American Institute of Timber Construction
7012 South Revere Parkway, Suite 140
Englewood, CO 80112

American Lumber Standards Committee
P.O. Box 210
Germantown, MD 20875–0210

American Wood Council
1111 Nineteenth St. NW, Suite 800
Washington, DC 29936

American Wood Preservers Association
P.O. Box 286
Woodstock, MD 21163–0286

American Wood Preservers Institute
Tysons International Building,
1945 Old Gallows Road, Suite 550
Vienna, VA 22182

APA—The Engineered Wood Association
P.O. Box 11700
Tacoma, WA 98411

Architectural Woodwork Institute
13924 Braddock Rd., Suite 100
Centerville, VA 22020

California Redwood Association (CRA)
405 Enfrente Dr., Suite 200
Novato, CA 94949

Canadian Wood Council
1730 St. Laurent Blvd., Suite 350
Ottawa, Ontario, Canada K1G 5L1

Cedar Shake and Shingle Bureau
515 One Hundred Sixteenth Ave. NE, Suite 275
Bellevue, WA 98004–5294

Cultured Marble Institute (CMI)
1735 N. Lynn St., suite 950
Arlington, VA 22209

Fine Hardwood Veneer Association
260 S. First St., Suite 2
Zionsville, IN 46077

Forest Products Laboratory
One Gifford Pinchot Dr.
Madison, WI 53705

Forest Products Society
2801 Marshall Ct.
Madison, WI 53705

Hardwood Council
P.O. Box 525
Oakmont, PA 15139

Hardwood Manufacturers Association
400 Penn Center Blvd., Suite 530
Pittsburgh, PA 15235

Hardwood Plywood & Veneer Association
P.O. Box 2789
Reston, VA 22090

Maple Flooring Manufacturers Association (MFMA)
60 Revere Drive, Suite 500
Northbrook, IL 60062

The Modular Building Systems Council and The Panelized Building Systems Council Divisions of The Building Systems Council National Association of Home Builders
1201 Fifteenth St. NW
Washington, DC 20005

National Association of Home Builders
1201 Fifteenth St. NW
Washington, DC 20005

National Hardwood Lumber Association (NHLA)
P.O. Box 34518
Memphis, TN 38184–0518

National Oak Flooring Manufacturers Association (NOFMA)
P.O. Box 3009
Memphis, TN 38173–0009

National Particleboard Association
(NPA)
18928 Premiere Ct.
Gaithersburg, MD 20879

Oak Flooring Institute
P.O. Box 3009
Memphis, TN 38173–0009

Southern Forest Products Association
(SFPA)
P.O. Box 641700
Kenner, LA 70064-1700

Southern Pine Association
P.O. Box 641700
Kenner, LA 70064–1700

Structural Board Association
45 Sheppard Ave. East, Suite 412
Willowdale, Ontario,
Canada M2N 5W9

Structural Insulation Board Association
1511 K St. NW, Suite 600
Washington, DC 20005

Western Wood Products Association
(WWPA)
Yeon Building
522 SW Fifth Ave.
Portland, OR 97204

Wood Protection Council (WPC)
1201 L Street NW, Suite 400
Washington, DC 20005

Plastics Organizations

Decorative Laminate Products
Association (DLPA)
600 S. Federal St., Suite 400
Chicago, IL 60605

Plastics Institute of America
(PIA)
227 Fairfield Road
Fairfield, NJ 07004

Society of Plastics Engineers
(SPE)
14 Fairfield Drive
Brookfield, CT 06804–0403

The Society of the Plastics Industry
(SPI)
1275 K Street NW, Suite 400
Washington, DC 20005

Vinyl Institute (VI)
65 Madison Ave.
Morristown, NJ 07960

Vinyl Siding Institute (VSI)
355 Lexington Ave.
New York, NY 10017

DIVISION 7. THERMAL & MOISTURE
Protection Adhesive and Sealant
Council (ASC)
1627 K Street NW, Suite 1000
Washington, DC 20006–1707

American Society of Heating,
Refrigerating and Air-Conditioning
Engineers (ASHRAE)
1791 Tullie Circle NE
Atlanta, GA 30329

Asphalt Institute (AI)
P.O. Box 14052
Lexington, KY 40512–4052

Asphalt Roofing Manufacturers
Association (ARMA)
6288 Montrose Road
Rockville, MD 20852

Associated Foam Manufacturers (AFM)
P.O. Box 246
Excelsion, MN 55331

Cellulose Insulation Manufacturers
Association
136 South Keowee St.
Dayton, OH 45402

Exterior Insulation Manufacturers
Association (EIMA)
2759 State Road 580, Suite 112
Clearwater, FL 34621

Institute of Roofing and Waterproofing
Consultants (IRWC)
4242 Kirchoff Rd.
Rolling Meadows, IL 60008

Insulation Contractors Association
of America
P.O. Box 26237
Alexandria, VA 22313

Mineral Insulation Manufacturers
Association (MIMA)
44 Canal Center Plaza, Suite 310
Alexandria, VA 22314

National Insulation and Abatement
Contractors Association (NIACA)
99 Canal Center Plaza
Alexandria, VA 22314

National Roofing Contractor
Association (NRCA)
10255 W. Higgins Rd., Suite 600
O'Hare International Center
Rosemont, IL 60018

National Tile Roofing Manufacturers
Association (NTRMA)
P.O. Box 947
Eugene, OR 97440

Noise Control Association (NCA)
680 Rainier Lane
Port Ludlow, WA 98365

North American Insulation Manufacturers Association (NAIMA)
44 Canal Center Plaza
Alexandria, VA 22314

Perlite Institute (PI)
88 New Dorp Plaza
Staten Island, NY 10306–2994

Polyisocyanurate Insulation Manufacturers Association (PIMA)
1001 Pennsylvania Ave. NW
Washington, DC 20004

Roof Coatings Manufacturers
Association (RCMA)
60 Revere Drive, Suite 500
Northbrook, IL 60062

Roof Consultants Institute (RCI)
7424 Chapel Hill Rd.
Raleigh, NC 27607

Roofing Industry Educational
Institute (RIEI)
14 Inverness Dr. E., Bldg H,
Suite 110
Englewood, CO 80112–5608

Roofing Products Division/Rubber
Manufacturers Association
(RPD/RMA)
1400 K Street NW
Washington, DC 20005

Sealant, Waterproofing and Restoration
Institute (SWRI)
3101 Broadway, Suite 585
Kansas City, MO 64111

Single Ply Roofing Institute
20 Walnut St., Suite 8
Wellesley Hills, MA 02181

Society of the Plastics Industry (SPI)
1275 K Street NW, Suite 400
Washington, DC 20005

Specialty Steel Industry of North
America
3050 K St. NW
Washington, DC 20007

Vermiculite Association (VA)
600 S. Federal St., Suite 400
Chicago, IL 60605

DIVISION 8. DOORS & WINDOWS
Aluminum Extruders Council (AEC)
1000 N. Rand Rd., Suite 214
Wauconda, IL 60084

Aluminum Fenestration Products
Association (AFPA)
1000 N. Rand Rd., Suite 214
Wauconda, IL 60084

American Architectural Manufacturers
Association (AAMA)
1540 E. Dundee Rd., Suite 310
Palatine, IL 60067

Architectural Translucent Skylight &
Curtain Wall Association
(ATSCWA)
7120 Stewart Ave.
Wausau, WI 54401

Builders Hardware Manufacturers
Association (BHMA)
355 Lexington Ave., 17th Floor
New York, NY 10017

Flat Glass Marketing Association (FGMA)
Glass Tempering Association (GTA)
Laminators Safety Glass Association
(LSGA)
3310 SW Harrison St.
Topeka, KS 66611

National Fenestration Rating Council
(NFRC)
1300 Spring St., Suite 120
Silver Spring, MD 20910

National Glass Association (NGA)
8200 Greensboro Dr., Suite 302
McLean, VA 22102

National Wood Window and Door
Association (NWWDA)
1400 E. Touhy Ave.
Des Plaines, IL 60018

Sealed Insulating Glass Manufacturers
(SIGM)
401 N. Michigan Ave.
Chicago, IL 60611–4267

Steel Door Institute (SDI)
30200 Detroit Rd.
Cleveland, OH 44145

Steel Window Institute
1300 Summer Ave.
Cleveland, OH 44115

Vinyl Window and Door Institute
(VWDI)
355 Lexington Ave.
New York, NY 10017

DIVISION 9. FINISHES
Acoustical Society of America (ASA)
500 Sunnyside Blvd.
Woodbury, NY 11797

American Fiber Manufacturers
Association (AFMA)
1150 Seventeenth St. NW, Suite 310
Washington, DC 20036

Architectural Spray Coaters
Association (ASCA)
230 W. Wells, Suite 311
Milwaukee, WI 53203

Audio Engineering Society
(AES)
60 East 42nd St., Rm. 2520
New York, NY 10165

Carpet and Rug Institute (CRI)
P.O. Box 2048
Dalton, GA 30720

Carpet Cushion Council (CCC)
P.O. Box 546
Riverside, CT 06878

Ceilings and Interior Systems
Construction (CISC)
579 W. North Ave., Suite 301
Elmhurst, IL 60126

Facing Tile Institute
P.O. Box 8880
Canton, OH 44711

Foundation of the Wall & Ceiling
Industry (FWCI)
307 E. Annandale, Suite 200
Falls Church, VA 22042

Gypsum Association (GA)
810 First St. NE, Suite 510
Washington, DC 20002

International Institute of Lath &
Plaster (IILP)
820 Transfer Road
St. Paul, MN 55114

Maple Flooring Manufacturers
Association (MFMA)
60 Revere Dr., Suite 500
Northbrook, IL 60062

Materials and Methods Standards
Association (MMSA)
P.O. Box 350
Grand Haven, MI 49417

National Council of Acoustical
Consultants
66 Morris Ave., Suite 1A
Springfield, NJ 07081–1409

National Oak Flooring Manufacturers
Association (NOFMA)
P.O. Box 3009
Memphis, TN 38173–0009

National Paint and Coatings Association
(NPCA)
1500 Rhode Island Ave. NW
Washington, DC 20005

National Terrazzo and Mosaic
Association (NTMA)
3166 Des Plaines Ave., Suite 132
Des Plaines, IL 60018

National Tile Contractors Association
(NTCA)
P.O. Box 13629
Jackson, MS 39236

National Wood Flooring Association
(NWFA)
233 Old Meramec Station Rd.
Manchester, MO 63021

Painting and Decorating Contractors of
America (PDCA)
3913 Old Lee Highway, Suite 33B
Fairfax, VA 22030

Resilient Floor Covering Institute
(RFCI)
966 Hungerford Dr., Suite 12B
Rockville, MD 20850

Rubber Manufacturers Association
(RMA)
1400 K Street NW
Washington, DC 20005

Tile Council of America (TCA)
P.O. Box 1787
Clemson, SC 29633

Wallcovering Manufacturers Association & Wallcovering Information
Bureau (WMA and WIB)
401 N. Michigan Ave., Suite 2200
Chicago, IL 60611

Wood and Synthetic Flooring Institute
(WSFI)
4415 W. Harrison St., Suite 242C
Hillside, IL 60162

DIVISION 11. EQUIPMENT
American Society for Industrial
Security (ASIS)
1655 N. Fort Meyer Dr., Suite 1200
Arlington, VA 22209

National Burglar and Fire Alarm
Association (NBFAA)
7101 Wisconsin Ave., Suite 1390
Bethesda, MD 20814

National Crime Prevention Council
(NCPC)
1700 K Street NW, 2nd Floor
Washington, DC 20006–3817

National Crime Prevention Institute (NCPI)
University of Louisville, Belknap Campus, Burhans Hall
Louisville, KY 40292

Security Industry Association (SIA)
1801 K Street NW, Suite 1203L
Washington, DC 20006–1301

DIVISION 12. FURNISHINGS

American Society of Furniture Designers (ASFD)
521 S. Hamilton St.
P.O. Box 2688
High Point, NC 27261

Business and Institutional Furniture Manufacturers Association (BIFMA)
2335 Burton St. SE
Grand Rapids, MI 49506

Contract Furnishings Council (CFC)
1190 Merchandise Mart
Chicago, IL 60654

DIVISION 13. SPECIAL CONSTRUCTION

Acoustical Society of America (ASA)
500 Sunnyside Blvd.
Woodbury, NY 11797

American Solar Energy Society (ASES)
2400 Central Ave., Suite G-1
Boulder, CO 80301

Audio Engineering Society (AES)
60 East 42nd St., Room 2520
New York, NY 10165

Building & Fire Research Laboratory National Institute of Standards & Technology
Building 226, Rm 216
Gaithersburg, MD 20899

Building Seismic Safety Council (BSSC) National Institute of Building Sciences
1201 L Street NW, Suite 400
Washington, DC 20005

Fire Equipment Manufacturers Association (FEMA)
1300 Summer Ave.
Cleveland, OH 44115

Florida Solar Energy Center (FSEC)
300 State Rd. 401
Cape Canaveral, FL 32920

Industrial Fabrics Association International (IFAI)
345 Cedar St., Suite 800
St. Paul, MN 55101–1088

International Fire Code Institute (IFCI)
5360 S. Workman Mill Rd.
Whittier, CA 90601

Metal Building Manufacturers Association (MBMA)
1300 Summer Ave.
Cleveland, OH 44115

National Council of Acoustical Consultants (NCAC)
66 Morris Ave., Suite 1A
Springfield, NJ 07081

National Fire Laboratory (NFL)
Institute for Research in Construction
Ottawa, Ontario, Canada K1A 0R6

National Fire Protection Association (NFPA)
1 Batterymarch Park
Quincy, MA 02269

National Fire Sprinkler Association, Inc. (NFSA)
P.O. Box 1000
Patterson, NY 12563

National Pool and Spa Institute (NPSI)
2111 Eisenhower Ave.
Alexandria, VA 22314

National Renewable Energy Laboratory (NREL)
1617 Cole Blvd.
Golden, CO 80401

Northeast Sustainable Energy Association (NSEA)
23 Ames St.
Greenfield, MA 01301

Society of Fire Protection Engineers (SFPE)
One Liberty Square
Boston, MA 02109

Solar Energy Industries Association (SEIA)
122 C St. NW
Washington, DC 20001

DIVISION 14. CONVEYING SYSTEMS

Conveyor Equipment Manufacturing Association
1000 Vermont Ave. NW
Washington, DC 20005

National Association of Elevator Contractors (NAEC)
1298 Wellbrook Circle NE
Conyers, GA 30207

National Elevator Industry, Inc.
185 Bridge Plaza North
Fort Lee, NV 07024

DIVISION 15. MECHANICAL

Air Conditioning & Refrigeration Institute
4301 N. Fairfax Dr., Suite 425
Arlington, VA 22203

Air Conditioning Contractors of America
1513 Sixteenth St. NW
Washington, DC 20036

Air Movement and Control Association (AMCA)
30 W. University Dr.
Arlington Heights, IL 60004

American Boiler Manufacturers Association
950 N. Glebe Rd., Suite 160
Arlington, VA 22203

American Gas Association (AGA)
1515 Wilson Blvd.
Arlington, VA 22209

American Society of Heating, Refrigerating & Air-Conditioning Engineers (ASHRAE)
1791 Tullie Circle NE
Atlanta, GA 30329

American Society of Mechanical Engineers (ASME)
345 E. 47th St.
New York, NY 10017

American Society of Plumbing Engineers (ASPE)
3617 Thousand Oaks Blvd., Suite 210
Westlake Village, CA 91362

American Society of Sanitary Engineers (ASSE)
P.O. Box 40362
Bay Village, OH 44140

American Water Works Association (AWWA)
6666 W. Quincy Ave.
Denver, CO 80235

Cast Iron Soil Pipe Institute (CISPI)
5959 Shallowford Rd., Suite 419
Chattanooga, TN 37412

Cooling Tower Institute (CTI)
P.O. Box 73383
Houston, TX 77273

Gas Research Institute
8600 West Bryn Mawr Ave.
Chicago, IL 60631

Heat Exchange Institute
1621 Euclid Ave., Suite 1230
Cleveland, OH 44113

The Hydronics Institute, Inc. (HI)
P.O. Box 218
Berkeley Heights, NJ 07922–0218

Institute of Heating & Air Conditioning
Industries (IHACI)
606 N. Larchmont Blvd., Suite 4A
Los Angeles, CA 90004

International Association of Plumbing
& Mechanical Officials (IAPMO)
5360 South Workman Mill Rd.
Whittier, CA 90601

International Society for Indoor Air
Quality & Climate
Box 22038, Sub 32
Ottawa, Ontario, Canada K1V 0W2

National Association of Plumbing,
Heating, & Cooling Contractors
(NAPHCC)
P.O. Box 6808
Falls Church, VA 22046

The National Burglar & Fire Alarm
Association
7101 Wisconsin Ave.
Bethesda, MD 29814

National Fire Sprinkler Association
Robin Hill Corp. Park, Route 22
P.O. Box 1000
Patterson, NY 12563

National Propane Gas Association
(APGA)
1600 Eisenhower Lane, Suite 100
Lisle, IL 60521

National Sanitation Foundation
International (NSFI)
P.O. Box 130140
Ann Arbor, MI 48113

Plastic Pipe and Fittings Association
(PPFA)
800 Roosvelt Rd., Bldg. C, Suite 20
Glen Ellyn, IL 60137

Plastic Pipe Institute
1275 K St. NW
Washington, DC 20005

Plumbing Manufacturers Institute
(PMI)
800 Roosevelt Rd., Bldg. C, Suite 20
Glen Ellyn, IL 60137

Security Industries Association
1801 K St. NW, Suite 1203L
Washington, DC 20006

Sheet Metal & Air Conditioning Contractors Association International
(SMACNA)
4201 Lafayette Center Drive
Chantilly, VA 22022–1209

Solar Energy Industries Association
(SEIA)
122 C St. NW
Washington, DC 20001

Waste Material Management Division
U.S. Department of Energy
1000 Independence Ave. SW
Washington, DC 20585

Water Quality Association (WQA)
4151 Naperville Rd.
Lisle, IL 60532

DIVISION 16. ELECTRICAL

American Nuclear Society
555 North Kensington Ave.
LaGrange Park, IL 60525

American Public Power Association
2301 M St. NW, Suite 300
Washington, DC 20037

Edison Electric Institute (EEI)
701 Pennsylvania Ave. NW
Washington, DC 20004

Electric Power Research Institute
207 Coggins Drive
P.O. Box 23205
Pleasant Hill, CA 94523

Illuminating Engineering Society of
North America (IES)
120 Wall St., 17th Floor
New York, NY 10005

International Association of Lighting
Designers (IALD)
18 E. 16th St., Suite 208
New York, NY 10003

Lighting Research Center (LRC)
Rensselaer Polytechnic Institute
Greens Building, Room 115
Troy, NY 12180

Lighting Research Institute (LRI)
120 Wall St., 17th Floor
New York, NY 10005

National Electrical Manufacturers
Association (NEMA)
2000 L St. NW
Washington, DC 20036

National Lighting Bureau (NLB)
2101 L St. NW
Washington, DC 20037

Underwriters Laboratories (UL)
333 Pfingsten Rd.
Northbrook, IL 60062

Information Sources for Recycled
& Energy-Efficient Materials
Alliance to Save Energy
1725 K Street NW, Suite 914
Washington, DC 20006

American Council for an Energy-
Efficient Economy
1001 Connecticut Ave. NW
Washington, DC 20036

American Solar Energy Society
2400 Central Ave., Suite G-1
Boulder, CO 80301

American Wind Energy Association
122 C St. NW, Suite 400
Washington, DC 20001

Association of Energy Engineers
4025 Pleasantdale Road, Suite 420
Atlanta, GA 30340

The Boston Society of Architects
52 Broad St.
Boston, MA 02109–4301

Center for Resourceful Building
Technology
P.O. Box 3413
Missoula, MT 59806

Energy Efficiency & Renewable
Energy Clearing House
P.O. Box 3048
Merrifield, VA 22116

Environmental Resource Guide
John Wiley & Sons, Inc.
605 Third Ave.
New York, NY 10158–0012

EnviroSafe Products
81 Winent Place
Staten Island, NY 10309

National Energy Foundation
5160 Wiley Post Way, Suite 200
Salt Lake City, UT 84116

Passive Solar Industries Group
1090 Vermont Ave. NW, Suite 1200
Washington, DC 20005

Resource Conservation Technology
2633 North Calvert St.
Baltimore, MD 21218

Technical Information Program
National Renewable Energy
Laboratory
1617 Cole Blvd.
Golden, CO 80401–3393

Wastebusters, Inc.
1390 Richmond Terrace
Staten Island, NY 10310

BUILDING CODE AGENCIES

Building Officials and Code Administrators International (BOCA)
4051 West Fossmore Rd.
Country Club Hills, IL 60478

Council of American Building Officials (CABO)
5203 Leesburg Pike, Suite 708
Falls Church, VA 22041

International Association of Plumbing & Mechanical Officials
20001 Walnut Drive South
Walnut, CA 91789

International Conference of Building Officials (ICBO)
5360 Workmen Mill Rd.
Whittier, CA 90601–2298

National Conference of States on Building Codes & Standards (NCS-BCS)
505 Huntmar Park Dr., Suite 210
Herndon, VA 22070

Southern Building Code Congress International (SBCC)
900 Montclair Rd.
Birmingham, AL 35213–1206

OTHER ORGANIZATIONS

American Association of State & Highway Transportation Officials
444 N. Capitol St. NW, Suite 249
Washington, DC 20001

American Institute of Architects
1735 New York Ave. NW
Washington, DC 20006

American National Metric Council
1735 N. Lynn St., Suite 950
Arlington, VA 22209–2022

American National Standards Institute (ANSI)
11 West 42nd St., 13th Floor
New York, NY 10036

American Society for Testing & Materials (ASTM)
1916 Race St.
Philadelphia, PA 19103

American Society of Civil Engineers
345 East 47th Street
New York, NY 10017–2398

Building & Fire Research Laboratory
National Institute of Standards & Technology
Bldg. 226, Room B216
Gaithersburg, MD 20899

Canada Mortgage & Housing Corporation
700 Montreal Road
Ottawa, Ontario, Canada K1A 0P7

Canada Standards Association (CSA)
178 Rexdale Blvd.
Rexdale, Ontario, Canada M9W 1R3

Canadian Construction Materials Centre (CCMC), Institute for Research in Construction, National Research Council Canada
Ottawa, Ontario, Canada K1A 0R6

Canadian Home Builders Association (CHBA)
150 Laurier Ave. West, Suite 200
Ottawa, Ontario, Canada K1P 5J4

Construction Specifications Institute (CSI)
601 Madison St.
Alexandria, VA 22314–1791

Department of Housing & Urban Development (HUD)
451 Seventh Street SW
Washington, DC 20410

Environmental Hazards Management Institute
10 Newmarker Road
P.O. Box 932
Durham, NH 03824

Environmental Protection Agency
401 M Street NW
Washington, DC 20460

Federal Specifications
General Services Administrations, Specification Activity Office,
Bldg. 197 Washington Navy Yard
2nd and M Streets SE
Washington, DC 20407

Institute for Research in Construction and the National Research Council of Canada
Montreal Road Building M-24
Ottawa, Ontario, Canada K1A 0R6

National Association of Home Builders (NAHB)
1201 Fifteenth Street NW
Washington, DC 20005–2800

National Fire Protection Association
1 Batterymarch Park
Quincy, MA 02269

National Institute of Building Science (NIBS)
1201 L Street NW
Washington, DC 20005

National Institute of Science & Technology
U.S. Department of Commerce
Bldg. 101, Room 813
Gaithersburg, MD 20899

National Institute of Standards & Technology
U.S. Department of Commerce
Bldg. 202, Rm. 204
Gaithersburg, MD 20899

National Research Center of the National Association of Home Builders
400 Prince Georges Center Blvd.
Upper Marlboro, MD 20772–8731

National Standards of Canada
350 Sparks St.
Ottawa, Ontario, Canada K1R 7S8

Occupational Safety and Health Administration (OSHA)
Francis Perkins Department of Labor Building
200 Constitution Ave. NW
Washington, DC 20210

Small Homes Council-Building Research Council (SHC-BRC)
University of Illinois
1 East St. Mary's Road
Champaign, IL 61820

Underwriters Laboratories of Canada (ULC)
7 Crouse Rd., Scarborough, Ontario, Canada M1R 3A9

Underwriters Laboratories, Inc.
333 Pfingsten Rd.
Northbrook, IL 60062–2096

U.S. Metric Association
10245 Andasol Ave.
Northridge, CA 91325

METRIC INFORMATION

Tables giving metric equivalents for common fractions, two-place decimal inches, and millimeter-to-decimal-inches are printed inside the front and rear covers for ready reference.

BASE SI UNITS

Quantity	Unit	Symbol
Length	Meter	m
Mass	Kilogram	kg
Time	Second	s
Electric current	Ampere	A
Thermodynamic temperature	Kelvin	K
Amount of substance	Mole	mo
Luminous intensity	Candela	cd

DERIVED METRIC UNITS WITH COMPOUND NAMES

Physical Quantity	Unit	Symbol
Area	Square meter	m^2
Volume	Cubic meter	m^3
Density	Kilogram per cubic meter	kg/m^3
Velocity	Meter per second	m/s
Angular velocity	Radian per second	rad/s
Acceleration	Meter per second squared	m/s^2
Angular acceleration	Radian per second squared	rad/s^2
Volume rate of flow	Cubic meter per second	m^3/s
Moment of inertia	Kilogram meter squared	$kg \bullet m^2$
Moment of force	Newton meter	N•m
Intensity of heat flow	Watt per square meter	W/m^2
Thermal conductivity	Watt per meter Kelvin	W/m•K
Luminance	Candela per square meter	cd/m^2

SUPPLEMENTARY SI UNITS

Quantity	Unit	Symbol
Plane angle	Radian	rad
Solid angle	Steradian	sr

SI PREFIXES

Multiplication Factor			Prefix	Symbol
1 000 000 000 000 000 000	=	10^{18}	exa	E
1 000 000 000 000 000	=	10^{15}	peta	P
1 000 000 000 000	=	10^{12}	tera	T
1 000 000 000	=	10^9	giga	G
1 000 000	=	10^6	mega	M
1 000	=	10^3	kilo	k
100	=	10^2	hecto	h
10	=	10^1	deka	da
0.1	=	10^{-1}	deci	d
0.01	=	10^{-2}	centi	c
0.001	=	10^{-3}	milli	m
0.000 001	=	10^{-6}	micro	m
0.000 000 001	=	10^{-9}	nano	n
0.000 000 000 001	=	10^{-12}	pico	p
0.000 000 000 000 001	=	10^{-15}	femto	f
0.000 000 000 000 000 001	=	10^{-18}	atto	a

METRIC UNIT TO IMPERIAL UNIT CONVERSION FACTORS[a]

Metric Units		Imperial Equivalents
Length		
1 millimeter (mm)	=	0.0393701 inch
1 meter (m)	=	39.3701 inches
	=	3.28084 feet
1 kilometer (km)	=	0.621371 mile
Length/Time		
1 meter per second (m/s)	=	3.28084 feet per second
1 kilometer per hour (km/h)	=	0.621371 mile per hour
Area		
1 square millimeter (mm²)	=	0.001550 square inch
1 square meter (m²)	=	10.7639 square feet
1 hectare (ha)	=	2.47105 acres
1 square kilometer (km²)	=	0.386102 square mile
Volume		
1 cubic millimeter (mm³)	=	0.0000610237 cubic inch
1 cubic meter (m³)	=	35.3147 cubic feet
	=	1.30795 cubic yards
1 milliliter (mL)	=	0.0351951 fluid ounce
1 liter (L)	=	0.219969 gallon
Mass		
1 gram (g)	=	0.0352740 ounce
1 kilogram (kg)	=	2.20462 pounds
1 tonne (t) (5 1,000 kg)	=	1.10231 tons (2000 lb.)
Force		
1 newton (N)	=	0.224809 pound-force
Stress		
1 megapascal (MPa)	=	145.038 pounds-force psi
Loading		
1 kilonewton per sq. meter	=	20.8854 pounds-force psf
1 kilonewton per meter	=	68.5218 pounds-force per ft.
Miscellaneous		
1 joule (J)	=	0.00094781 Btu
1 joule (J)	=	1 watt-second
1 watt (W)	=	0.00134048 electric hp

IMPERIAL UNIT TO METRIC UNIT CONVERSION FACTORS[b]

Imperial Units		Metric Equivalents
Length		
1 inch	=	25.4 mm
	=	0.0254 m
1 foot	=	0.3048 m
1 mile	=	1.60934 km
Length/Time		
1 foot per second	=	0.3048 m/s
1 mile per hour	=	1.60934 km/h
Area		
1 square inch	=	645.16 mm²
1 square foot	=	0.0929030 m²
1 acre	=	0.404686 ha
1 square mile	=	2.58999 km²
Volume		
1 cubic inch	=	16387.1 mm³
1 cubic foot	=	0.0283168 m³
1 cubic yard	=	0.764555 m³
1 fluid ounce	=	28.4131 mL
1 gallon	=	4.54609 L
Mass		
1 ounce	=	28.3495 g
1 pound	=	0.453592 kg
1 ton (2000 lb.)	=	0.907185 t
Force		
1 pound	=	4.44822 N
Stress		
1 psi	=	0.00689476 MPa
Loading		
1 psf	=	0.0478803 kN/m2
1 plf	=	0.0145930 kN/m
Miscellaneous		
1 Btu	=	1055.06 J
1 watt-second	=	1 J
1 horsepower	=	746 W

Notes:

1. 1.0 newton = 1.0 kilogram × 9.80665 m/s² (International Standard Gravity Value)

2. 1.0 pascal = 1.0 newton per square meter

[a]Multiply the number of metric units by the imperial (English) equivalent to convert a measurement from metric units to imperial units.

Notes:

1. 1.0 newton = 1.0 kilogram × 9.80665 m/s² (International Standard Gravity Value)

2. 1.0 pascal = 1.0 newton per square meter

[b]Multiply the number of imperial (English) units by the metric equivalent to convert a measurement from imperial units to metric units.

WEIGHTS OF BUILDING MATERIALS

Brick & Block Masonry	lb./ft.²	kg/m²
4" brickwall	40	196
4" concrete brick, stone or gravel	46	225
4" concrete brick, lightweight	33	161
4" concrete block, stone or gravel	34	167
4" concrete block, lightweight	22	108
6" concrete, stone or gravel	50	245
6" concrete block, lightweight	31	152
8" concrete block, stone or gravel	55	270
8" concrete block, lightweight	35	172
12" concrete block, stone or gravel	85	417
12" concrete block, lightweight	55	270

Concrete	lb./ft.³	kg/m³
Plain, slag	132	2155
Plain, stone	144	2307
Reinforced, slag	138	2211
Reinforced, stone	150	2403

Lightweight Concrete	lb./ft.³	kg/m³
Concrete, perlite	35–50	561–801
Concrete, pumice	60–90	961–1442
Concrete, vermiculite	25–60	400–961

Structural Clay Tile	lb./ft.2	kg/m²
4" hollow	23	368
6" hollow	38	609
8" hollow	45	721

Structural Facing Tile	lb./ft.²	kg/m³
2" facing tile	14	68.6
4" facing tile	24	118
6" facing tile	34	167

Stone Veneer	lb./ft.³	kg/m³
2" granite, ½" parging	30	481
4" limestone, ½" parging	36	577
4" sandstone, ½" parging	49	785
1" marble	13	208

Wood	lb./ft.²	kg/m2
Ash, white	40.5	198
Birch	44	202
Cedar	22	108
Cypress	33	162
Douglas fir	32	157
White pine	27	132
Pine, southern yellow	26	127
Redwood	26	127
Plywood, ½"	1.5	7.4

Metals	lb./ft.³	kg/m³
Aluminum	165	2643
Copper	556	8907
Iron, cast	450	7209
Steel	490	7850
Steel, stainless	490–510	7850–8170

Wall, Ceiling & Floor	lb./ft.²	kg/m²
Acoustical tile, ½"	0.8	3.9
Gypsum wallboard, ½"	2	9.8
Plaster, 2" partition	20	98
Plaster, 4" partition	32	157
Plaster, ½"	4.5	22
Plaster on lath	10	49
Tile, glazed, ⅜"	3	14.7
Tile, quarry, ½ "	5.8	28.4
Terrazzo, 1"	25	122.5
Vinyl composition floor tile	1.4	69
Hardwood flooring, ²⁵/₃₂"	4	19.6
Flexicore 6", lightweight concrete	30	14.7
Flexicore 6", stone concrete	40	196
Plank, cinder concrete, 2"	15	73.5
Plank, gypsum, 2"	12	58.8
Concrete reinforced, stone, 1"	12.5	61.3
Concrete reinforced, lightweight, 1"	6–10	29.4–49
Concrete plain, stone, 1"	12	58.8
Concrete plain, lightweight, 1"	3–9	14.7–44.1

Suspended Ceilings	lb./ft.²	kg/m²
Acoustic plaster on gypsum lath	10–11	49–54
Mineral fiberboard	1.4	6.9

Residential Assemblies	lb./ft.²	kg/m²
Wood framed floor	10	49
Ceiling	10	49
Frame exterior wall, 4" studs	10	49
Frame exterior wall, 6" studs	13	64
Brick veneer of 4" frame	50	245
Brick veneer over 4" concrete block	74	363
Interior partitions with gypsum both sides (allowance per sq. ft. of floor area—not weight of material)	20	320

Partitions	lb./ft.²	kg/m²
2 × 4 wood studs, gypsum wallboard 2 sides	8	39.2
4" metal stud, gypsum wallboard 2 sides	6	29.4
6" concrete block, gypsum wallboard 2 sides	35	171.5

Roofing	lb./ft.²	kg/m²
Built-up	6.5	31.9
Concrete roof tile	9.5	46.6
Copper	1.5–2.5	7.4–12.3
Steel deck alone	2.5	12.3
Shingles, asphalt	1.7–2.8	8.3–13.7
Shingles, wood	2–3	9.8–14.7
Slate, ½"	14–18	68.6–88.2
Tile, clay	8–16	39.2–78.4

Glass	lb./ft.²	kg/m²
¼" (6.3 mm) plate or float	3.3	16.2
½" (12.7 mm) plate or float	6.6	32.3
¹⁄₃₂" (0.79 mm) sheet	2.8	13.7
¼" (6.3 mm) sheet	3.5	17.2
⅛" (3.2 mm) double strength	1.6	7.8
⁷⁄₃₂" (5.6 mm) sheet	2.85	14.0
¼" (6.3 mm) laminated	3.30	16.2
½" (12.7 mm) laminated	6.35	31.1
2" (50.2 mm) bullet resistant	26.2	128.4
1" (25.4 mm) insulating, ½" (12.7 mm) air space	6.54	32.0
¼" (6.3 mm) wired	3.5	17.1
⅜" (9.5 mm) wired	5.0	24.4
3⅞" × 5¾" square, (98.0 × 146 mm) glass block	16.0	78.4

NAMES & ATOMIC SYMBOLS OF SELECTED CHEMICAL ELEMENTS

Name	Atomic Symbol	Name	Atomic Symbol
Aluminum	Al	Gallium	Ga
Antimony	Sb	Germanium	Ge
Argon	Ar	Gold	Au
Arsenic	As	Hafnium	Hf
Barium	Ba	Helium	He
Beryllium	Be	Holmium	Ho
Bismuth	Bi	Hydrogen	H
Boron	B	Illinium	Il
Bromine	Br	Indium	In
Cadmium	Cd	Iodine	I
Calcium	Ca	Iridium	Ir
Carbon	C	Iron	Fe
Cerium	Ce	Krypton	Kr
Cesium	Cs	Lanthanum	La
Chlorine	Cl	Lead	Pb
Chromium	Cr	Lithium	Li
Cobalt	Co	Lutecium	Lu
Copper	Cu	Magnesium	Mg
Dysprosium	Dy	Manganese	Mn
Erbium	Er	Masurium	Ma
Europium	Eu	Mercury	Hg
Fluorine	F	Molybdenum	Mo
Gadolinium	Gd	Neodymium	Nd

Name	Atomic Symbol	Name	Atomic Symbol
Neon	Ne	Silver	Ag
Nickel	Ni	Sodium	Na
Niobium	Nb	Strontium	Sr
Nitrogen	N	Sulfur	S
Osmium	Os	Tantalum	Ta
Oxygen	O	Tellurium	Te
Palladium	Pd	Terbium	Tb
Phosphorus	P	Thallium	Tl
Platinum	Pt	Thorium	Th
Polonium	Po	Thulium	Tm
Potassium	K	Tin	Sn
Praseodymium	Pr	Titanium	Ti
Protactinium	Pa	Tungsten	W
Radium	Ra	Uranium	U
Radon	Rn	Vanadium	V
Rhenium	Re	Xenon	Xe
Rhodium	Rh	Ytterbium	Yb
Rubidium	Rb	Yttrium	Y
Ruthenium	Ru	Zinc	Zn
Samarium	Sm	Zirconium	Zr
Scandium	Sc		
Selenium	Se		
Silicon	Si		

COEFFICIENTS OF THERMAL EXPANSION[a] FOR SELECTED CONSTRUCTION MATERIALS

Material	Multiply by 10^{-6} in./in./°F	Multiply by 10^{-6} mm/mm/°C
Concrete		
Normal weight concrete	5.5	9.8
Gypsum		
Gypsum panels	9.0	16.2
Gypsum plaster	7.0	12.6
Wood fiber plaster	8.0	14.4
Masonry		
Brick (varies some)	3.0	5.6
Concrete masonry units	5.2	9.4
Marble	7.3	13.1
Granite	4.7	8.5
Limestone	4.4	7.9
Metal		
Iron, gray cast	5.7	10.5
Iron, malleable	5.6	10.5
Steel, carbon (ASTM A285)	5.6	10.5
Steel, high strength (ASTM A141)	6.4	11.7
Steel, stainless (type 201)	8.7	15.7
Steel, stainless (type 405)	6.0	10.8
Nickel (211)	7.4	13.3
Copper (CA110)	9.4	16.5
Bronze, commercial	10.2	19.3
Brass, red	10.4	19.9
Aluminum, wrought	12.8	23.0
Polymer, Thermosetting		
Phenolics	45.0	81.0
Urea-melamine	20.0	36.0
Polyesters	45.0	75.5
Epoxies	33.0	72.0

Material	Multiply by 10^{-6} in./in./°F	Multiply by 10^{-6} mm/mm/°C
Polymer, Thermoplastic		
Polyethylene, high density	70.0	120.0
Polypropylene	50.0	90.0
Polystyrene	38.0	68.5
Polyvinyl chloride (PVC)	30.0	54.0
Acrylonitrile-butadiene-styrene (ABS)	50.0	90.0
Acrylics	40.0	72.0

Wood

For most hardwoods and softwoods the parallel-to-grain thermal expansion values range from 1.7×10^{-6} to 2.5×10^{-6} in./in./°F or 3.1×10^{-6} to 4.5×10^{-6} mm/mm/°C.

Linear expansion coefficients across the grain are proportional to wood density. They range from 5 to 10 times greater than the parallel-to-grain coefficients and, therefore, are of more concern.

Material	Multiply by 10^{-4} in./in./°F	Multiply by 10^{-4} mm/mm/°C
Foam Insulation		
Polystyrene	3.5	6.3
Polyurethane	2.7	4.9

Material	Multiply by 10^{-7} in./in./°F	Multiply by 10^{-7} mm/mm/°C
Glass		
Glass, soda lime window sheet	47.0	85.0
Glass, soda lime plate	48.0	87.0

[a] The change in dimension of a material per unit of dimension per degree change in temperature.

GLOSSARY

Additional definitions are available in construction dictionaries such as the following:

Harris, C. M., *Dictionary of Architecture and Construction*, McGraw-Hill, New York, 1987.

Kennedy, F., *The Wiley Dictionary of Civil Engineering and Construction*, John Wiley & Sons, Somerset, N.J., 1990.

Philbin, T., *The Illustrated Dictionary of Building Terms*, McGraw-Hill, New York, 1996.

Putnam, R., *Builders Comprehensive Dictionary*, Prentice-Hall, Englewood Cliffs, N.J., 1984.

abrasion resistance Resistance to being worn away by rubbing or friction.

absorption The process in which a liquid or mixture of liquid and gases is drawn into the pores of a porous solid material.

absorptivity The relative ability to absorb sound and light.

accelerator An admixture used to speed up the setting of concrete.

access floor A freestanding floor raised above the basic floor.

acoustical ceiling A ceiling of fibrous tiles, panels, or other sound-deadening material.

acoustical glass A glazing unit used to reduce the transmission of sound through the glazed opening by bonding a soft interlayer between the layers of glass.

acoustical plaster Calcined gypsum mixed with lightweight aggregates.

acoustics The science of sound generation, sound transmission, and the effects of sound waves.

acrylic A transparent thermoplastic made from esters of acrylic acid.

active solar system A solar heating system that uses mechanical means to move and store solar energy.

additive Materials mixed with the basic plastic resin to alter its properties.

adhesion The ability of a coating to stick to a surface.

adhesive A substance used to hold materials together by surface attachment.

admixture A material other than portland cement, aggregate, and water that is added to concrete to alter its properties.

adsorbent A material that has the ability to cause molecules of gases, liquids, or solids to adhere to its surfaces without changing the adsorbent physically or chemically.

agglomeration A process that bonds ground iron-ore particles into pellets to facilitate handling.

aggregate Inert granules such as crushed stone, gravel, and expanded minerals mixed with portland cement and sand to form concrete.

air, combustion Air used to provide for the combustion of a fuel.

air-conditioner A mechanical device used to provide air-conditioning.

air-conditioning The process of treating air to control simultaneously its humidity, cleanliness, and temperature and to provide distribution within a building.

air-dried lumber Wood dried by exposing it to the air.

air-entrained cement A portland cement with an admixture that causes a controlled quantity of stable, microscopic air bubbles to form in the concrete.

air-entrained concrete Concrete with an admixture added that produces millions of microscopic air bubbles in the concrete.

air gap An unobstructed vertical distance between the lowest opening of pipe that supplies a plumbing fixture and the level at which the fixture will overflow.

air-supported structures Structures with the enclosing envelope made from a flexible material that is pneumatically supported.

alkali A substance such as lye, soda, or lime that can be destructive to paint.

alkyd Synthetic resin modified with oil for good adhesion, gloss, color retention, and flexibility.

alloy A metallic material composed of two or more chemical elements one of which is a metal.

alloying element Any substance added to a molten metal to change its mechanical or physical properties.

alternating current An electric current that varies periodically in value and direction by flowing first in one direction and then in the opposite direction.

alumina A hydrated form of aluminum oxide from which aluminum is made.

ambient temperature The temperature of the surrounding air.

ampere A basic SI unit that measures the rate of flow of electric current.

anaerobic bonding agents Bonding agents that set hard when not exposed to oxygen.

angle of repose The angle of the sloped surface of the sides of an excavation.

annealing Heating a metal to a high temperature followed by controlled cooling to relieve internal stresses.

anodizing An electrolytic process that forms a permanent, protective oxide coating on aluminum.

APA performance-rated panels Plywood manufactured to the structural specifications and standards of APA - The Engineered Wood Association.

arc resistance The total elapsed time in seconds an electric current must arc to cause a part to fail.

architectural terra-cotta Clay masonry units made with a textured or sculptured face.

asphalt Dark brown to black hydrocarbon solids or semisolids having bituminous constituents that gradually liquify when heated.

autoclave A high-pressure steam room that rapidly cures green concrete units.

awning window A window that pivots near the top edge of the sash and projects toward the exterior.

Axminster construction Carpet formed by weaving on a loom that inserts each tuft of pile individually into the backing.

backfill Earth filled in around a foundation wall to replace earth removed for construction of the foundation.

bagasse Crushed sugar cane or beet refuse from sugar making.

ballast An electrical device to provide the starting voltage and operating current for fluorescent, mercury, and other electric discharge lamps.

balloon frame A system of framing a wood-framed building in which all vertical structural members (studs) of the exterior bearing walls and partitions extend the full height of the frame, from the bottom plate to the top plate, and support the floor joists and roof.

bank measure The volume of soil in situ in cubic yards.

batch The amount of concrete mixed at one time.

bauxite Ore containing high percentages of aluminum oxide.

beam A straight horizontal structural member whose main purpose is to carry transverse loads.

bearing pile A pile that carries a vertical load.

bearing plate A steel plate placed under a beam, column, or truss to distribute the end reaction from the beam to the supporting member.

bearing wall A wall that supports a roof, floor, or ceiling.

bedrock The hard, solid rock formation at or below the surface of the earth.

bending moment The moment that produces bending on a beam or other structural member.

bending stress A compressive or tensile stress developed by applying nonaxial force to a structural member.

beneficiation A process of grinding and concentration that removes unwanted elements from iron-ore before the ore is used to produce steel.

Bentonite clay An absorptive clay that swells several times its dry volume when saturated with water.

binder Film-forming ingredient in paint that binds the suspended pigment particles together.

bitumen A generic term describing a material that is a mixture of predominantly hydrocarbons in solid or viscous form. It is derived from coal and petroleum.

bituminous coatings Coatings formulated by dissolving natural bitumens in an organic solvent.

bleeding Excess water that rises to the surface of concrete shortly after it has been poured.

blocking Wood pieces inserted between joists, studs, rafters, and other structural members to stabilize the frame, provide a nailing surface for finish materials, and block the passage of fire between the members.

board foot The measure of lumber having a volume of 144 in^3.

boards Lumber less than 2 in. (50.8 mm) thick and 1 in. (25.4 mm) or more wide.

boiler A closed vessel used to produce hot water or steam.

bond beam A continuous reinforced beam formed from horizontal masonry members bonded with reinforced concrete.

bond breaker A material used to prevent adjoining materials from adhering.

bonding agent A compound that holds materials together by bonding the surfaces to be joined.

box beam A structural member of metal or plywood whose cross section is a closed rectangular box shape.

box sill A type of sill used in frame construction in which the floor joists butt and are nailed to a header joist and rest on the sill.

branch A pipe in a plumbing system into which no other branch pipes discharge and that discharges into a main or submain.

branch circuit The electrical wiring between the overcurrent protection device and the connected outlets.

branch circuit, appliance An electric circuit supplying energy to outlets to which appliances are to be connected.

branch circuit, general purpose An electric circuit that supplies energy to a number of outlets for lighting and small appliances.

branch circuit, individual An electric circuit that supplies energy to only one piece of equipment.

branch interval A length of soil or waste stack 8 ft. or more in height (equal to one story) within which the horizontal branches from one floor or story of a building are connected to a stack.

branch, plumbing A horizontal run of waste piping that carries waste material to a vertical riser.

branch vent A vent connecting one or more individual vents into a vent stack or stack vent.

breaking strength The point at which a material actually begins to break.

Brinell hardness A measure of resistance of a material to indention.

Brinell hardness number A measure of Brinell hardness that is obtained by dividing the load in kilograms by the area of the indention given in square millimeters.

British thermal unit (Btu) The amount of heat required to raise the temperature of 1 lb. of water by 1°F.

British thermal unit per hour (Btu/h) The rate of heat flow per hour.

brittleness The characteristic of a material that tends to crack or break without appreciable plastic deformation.

brown coat The second coat of plaster in a three-coat plaster finish.

buffer, elevator Energy-absorbing units placed in the elevator pit.

building code A set of legal regulations that ensure a minimum standard of health and safety in buildings.

building drain The lowest horizontal piping of a plumbing drainage system that receives the discharge from soil, waste, and other drainage pipes within the building and carries the waste to the building sewer.

building sewer Horizontal piping that carries the waste discharge from the building drain to the public sewer or septic tank.

built-up roof membrane A continuous, semiflexible roof membrane built up of plies of saturated felts, coated felts, fabrics, or mats that have surface coats of bitumens. The last ply is covered with mineral aggregates, bituminous materials, or a granular-surface roofing sheet.

bulletin board A surface used to display announcements and other material usually attached with tacks.

burning Curing bricks by placing them in a kiln and subjecting them to a high temperature.

bus A rigid electric conductor enclosed in a protective busway.

busway A rigid conduit used to protect a bus running through it.

BX cable A cable sheathed with spirally wrapped metal strip identified as Type AC.

cable tray A ladderlike metal frame open on the top used to support insulated electrical cables.

caisson A watertight structure within which work can be carried out below the surface of water.

calcareous clays Clays containing at least 15 percent calcium carbonate.

calcined gypsum Ground gypsum that has been heated to drive off the water content.

camber An arching in a structural member due to tension applied to the steel framing.

candela A metric unit of luminous intensity that closely approximates candlepower.

candlepower A term used to express the luminous intensity of a light source. It is the same magnitude as a candela.

cant strip A triangular molding secured over the joint between a wall and a roof deck.

capillary action The movement of a liquid through small openings of fibrous material by the adhesive force between the liquid and the material.

capillary break A groove in a member used to create an opening that is too wide to be bridged by a drop of water, thus eliminating the passage of water by capillary action.

car, elevator The load-carrying unit of an elevator, consisting of a platform, walls, ceiling, door, and a structural frame.

car safeties, elevator Devices used to stop a car and hold it in position should it travel at excessive speed or go into a free fall.

carbohydrates Organic compounds that form the supporting tissue of plants.

carbon steel Any steel for which no minimum content for alloying agents is specified but for which the carbon content is the element used to determine its properties.

cast-in-place concrete Concrete members formed and poured on the building site in the locations where they are needed.

cast-in-place piles Concrete piles cast in a hollow metal shell driven into the earth or an uncased hole.

casting A metal part produced by pouring a molten metal into a mold.

cast iron A hard, brittle metal made of iron that contains a high percentage of carbon.

caulking A resilient material used to seal cracks and prevent leakage of water.

cavity wall A masonry wall made up of two wythes of masonry units separated by an air space.

cellulose An inert carbohydrate that is the chief constituent of the cell walls of plants, wood, and paper.

Celsius temperature The temperature scale used with the SI system in which the boiling point of water is 100°C and freezing is 0°C.

cement 1. A powder with adhesive and cohesive properties that sets into a hard, solid mass when mixed with water. 2. Bonding agent made from synthetic rubber suspended in a liquid.

cement board A panel product made with an aggregate portland cement core reinforced with polymer-coated glass-fiber mesh on each side.

cementitious materials Materials that have cementing properties.

cement-lime mortar Mortar made with the addition of slaked lime to the cement.

central service core A fire-resistant vertical shaft through a multistory building used to route electrical, mechanical, and transportation systems.

ceramic A class of products made of clay fired at high temperatures.

ceramic glaze A compound of metallic oxides, chemicals, and clays fused to a material at high temperature, providing a hard, smooth surface.

chalkboard A surface that can be written on with chalk and from which the chalk can be easily removed.

chase A recessed area in a wall for holding pipes and conduit that passes vertically between floors.

chemical strengthening A process for strengthening glass that involves immersing the glass in a molten salt bath.

chiller A refrigerating machine composed of a compressor, a condenser, and an evaporator, used to transfer heat from one fluid to another.

chord, bottom A horizontal or inclined structural member forming the lower edge of a truss.

chord, top A horizontal or inclined structural member forming the top edge of a truss.

circuit breaker An electrical device used to open and close a circuit by nonautomatic means or to open a circuit by automatic means at a predetermined overcurrent without damage to itself.

cladding The external finish covering the base material on a wall.

Class A,B,C roofing Classification of roofing materials by their resistance to fire when tested in accordance with ASTM E108.

clay A very cohesive material made up of microscopic particles (less than 0.00008 in. or 0.002 mm).

clay tile A unit made from fired and sometimes glazed clay and used as a finish surface on floors and walls.

cleanouts Openings in the waste piping system that permit cleaning obstructions from the pipe.

clear coating A transparent protective and/or decorative film.

clear span The horizontal distance between the interior edges of supporting members.

coal tar Tar produced through the destructive distillation of coal during the conversion of coal to coke.

coal tar pitch A dark brown to almost black hydrocarbon material derived by distilling coke-oven tar.

coating A paint, varnish, lacquer, or other finish used to create a protective and/or decorative layer.

coefficient of heat transmission The total amount of heat that passes through an assembly of materials, including any air spaces and surface air films. It is expressed in Btu per hr., per ft.², per °F temperature difference between the inside and outside air.

coefficient of thermal conductance The amount of heat, expressed in Btu, that can pass through a specified thickness of a material per hr., per ft.², per °F temperature difference between the surfaces.

coefficient of thermal expansion The total amount of heat, expressed in Btu, that passes by conduction through a 1 in. thickness of a homogeneous material per hr., per ft.², per °F, which is measured as the temperature difference between the two surfaces of the material.

cofferdam A temporary watertight enclosure around an area of water-bearing soil or an area of water from which water is pumped allowing construction to take place in the water-free area.

cogeneration systems Systems using fossil fuel, geothermal energy, wind, or solar energy to produce electricity and heat.

cohesion When referring to soils, it is the sticking together of soil particles whose forces of attraction exceed the forces that tend to separate them.

cohesion The molecular forces between particles within a body which acts to unite them.

cohesionless soil A soil that when unconfined has little or no cohesion when submerged and no significant strength when air dried.

cohesive soil A soil that when unconfined has considerable cohesion when submerged and considerable strength when air dried.

cold-rolled steel Steel rolled to the final desired shape at a temperature at which it is no longer plastic.

compaction Compressing soil to increase its density.

compartment A small area within a larger area enclosed by partitions.

composite materials Materials made by combining several layers of different materials.

composite panels Panels having a reconstituted wood core bonded between layers of solid veneer.

compression The condition of being shortened (compressed) by force.

compression test A test used to determine the behavior of materials under compression.

compressive strength The maximum stress a material can withstand before it is crushed.

compressive stresses Stresses created when forces push on a member and tend to shorten it.

compressor A mechanical device for increasing the pressure of a gas.

concentrated load Any load that acts on a very small area of a structure.

concrete A solid, hard material produced by combining portland cement, aggregates, sand, and water and sometimes admixtures.

concrete masonry Factory manufactured concrete units, such as concrete brick or block.

concrete pump A pump that moves concrete through hoses to the area where it is to be placed.

condensate A liquid formed by the condensation of a vapor.

condensation The process of changing from a gaseous to a liquid state.

condensation point The temperature at which a vapor liquefies if the latent heat is removed at standard or a stated pressure.

condenser A heat-exchanger unit in which a vapor has some heat removed, causing it to form a liquid.

conduction, thermal The process of heat transfer through a material to another part of that material or to a material touching it.

conductivity, electric A measure of the ability of a material to conduct electric current.

conductor, electric Wire through which electric current flows.

conduit A steel or plastic tube through which electrical wires are run.

consolidation The process of compacting freshly placed concrete in a form.

control joint A groove formed in concrete or masonry structures to allow a place where cracking can occur, thus reducing the development of high stresses.

controller An electric device or a group of devices used to govern the electric power delivered to the equipment to which it is connected.

convection The process of carrying heat from one spot to another by movement of a liquid or gas. The heated liquid or gas expands and becomes lighter, causing it to rise while the cooler, heavier dense liquid or air settles.

convector A unit designed to transfer heat from hot water or steam to the air by convection.

cooling tower A heat-transfer device in which the atmospheric air cools warm water flowing through the tower, usually by evaporation.

corrosion The deterioration of a metal or of concrete by chemical or electrochemical reaction caused by exposure to the weather.

covalent bonding A process in which small numbers of atoms are bonded into molecules.

creep Permanent dimensional deformation occurring over a period of time in a material subjected to constant stress at elevated temperatures.

creep strength The stress that produces a given size change when constantly applied for a given period of time at a specific temperature.

creep test A test to determine the creep behavior of materials subjected to constant stress at a constant temperature.

cricket A small false roof used to divert water from behind a projection above the roof, such as a chimney.

cubicle A very small enclosed space often large enough for just one person.

cupola A small roofed structure built on a roof, usually to vent the area below the roof.

curb A low wall of wood or masonry extending above the level of the roof and surrounding an opening in the roof.

cure To chemically cross-link polymer chains by heating and/or adding a chemical agent.

curing Protecting concrete after placing so that proper hydration occurs.

curing agent Part of a two-part compound that, when added to the second part, sets up the curing action. Also referred to as the catalyst.

current, electric The flow of electrons along a conductor.

curtain wall An exterior wall of lightweight construction that is nonload-bearing, supporting only its own weight.

cushion A layer of resilient material applied to a floor over which a carpet is to be laid.

damper A movable vane used to vary the volume of air passing through a duct, inlet, or outlet.

damping capacity The ability of a material to absorb vibrational energy.

darby A tool used to level concrete in a form after it has been screeded.

dead load A permanent load that provides steady pressure on the building structure, such as roofing materials.

decibel A unit for measuring sound energy or power.

deflection Displacement of a member from its static position as a result of forces acting on that member.

deformation A change in shape of a member caused by a load or force acting on it without a breach of the continuity of its parts.

dehumidification The removal of water vapor from the air.

dehumidifier A cooling, absorption, or adsorption device used for removing moisture from the air.

dehydration The removal of water vapor from any substance.

delamination The separation of the plies in a laminate or plies from a base material.

desiccant Any absorbent, adsorbent, liquid, or solid that removes water or water vapor from a material.

desiccation The process of evaporating or removing water vapor from a material.

dew point The temperature at which air must be cooled to reach a given pressure and water content for it to reach saturation (100 percent relative humidity).

dewatering Pumping subsurface water from an excavation to maintain dry and stable working conditions.

diaphragm A horizontal roof or floor structural element designed to resist lateral loads and transmit them to shear walls (vertical resisting elements).

dielectric strength The maximum voltage a dielectric (nonconductor) can withstand without fracture.

diffuser A circular, square, or rectangular air distributing outlet, usually in the ceiling, that has members to discharge supply air in several directions, mixing the supply air with the secondary air in the room.

dimension lumber Lumber from 2 in. (50.8 mm) up to but not including 5 in. (127 mm) thick and 2 in. (50.8 mm) or more wide.

direct current Electricity that flows in one direction.

double glazing Two parallel sheets of glass with an air space between.

double-hung window A window having two vertically sliding sashes.

double tees T-shaped precast floor and roof units that span long distances unsupported.

dressed lumber Lumber having one or more sides planed smooth.

drypack A stiff granular grout.

dry-press process The process used to make bricks when the clay contains 10 percent or less moisture.

duct A hollow tube through which air is circulated.

ductile Capable of being stretched or deformed without fracturing (plastic deformation).

ductility A measure of the capability of a material to be stretched or deformed without breaking.

durability of a coating The ability of a coating to hold up against destructive agents, such as weathering and sunlight.

dynamic load Any load that is nonstatic.

E value The ratio of stress to strain.

eaves The lower part of a roof that projects over the exterior wall.

EER (Energy Efficiency Ratio) The Btu output divided by the input in watts. The higher the EER the more efficient the equipment.

efflorescence A white soluble salt deposit on the surface of concrete and masonry, usually caused by free alkalies leached from the mortar by moisture moving through it.

effluent In plumbing, the liquid discharge from a waste disposal system.

elastic deformation The ability of a material to return to its original position after a load has been removed.

elastic limit The greatest stress a material can withstand without permanent deformation upon the release of the stress.

elasticity The property of a material that causes it to return to its original shape upon removal of a deforming load.

elastomer A macromolecular material that returns to its approximate initial dimensions and shape after being subjected to substantial deformation.

elastomeric Having the properties of an elastomer.

electric conduction The ability of a material to conduct an electric current.

electric current The movement of electrons in an electric conductor.

electric current, alternating An electric current that reverses the direction of flow periodically.

electric current, direct An electric current that does not reverse its polarity.

electric power The rate of generating, transferring, or using electric energy. It is expressed in watts (W) and kilowatts (kW).

elevator A hoisting and lowering mechanism equipped with an enclosed car that moves between floors in a building.

elongation Drawing out to a greater length when under load or expansion due to temperature increases.

enamel A classification of paints that dry to a hard flat semigloss or gloss finish.

epoxy finish A clear finish having excellent adhesion qualities, abrasion and chemical resistance, and water resistance.

epoxy resin A class of synthetic thermosetting resins derived from certain special types of organic chemicals.

equilibrium The state of being equally balanced.

equilibrium moisture content The moisture content at which wood neither gains nor loses moisture when surrounded by air at a specified relative humidity and temperature.

erection plan An assembly drawing showing where each structural steel member is located on the building frame.

escalator A continuous moving stair used to move people up and down between floors.

evaporator That part of a refrigerating system in which the refrigerant is evaporated, allowing it to absorb heat from the contacting heat source.

expansion joint A joint used to separate two parts of a building to allow expansion and contraction movement of the parts.

exposed aggregate finish A finished concrete surface in which a coarse aggregate is exposed to view.

extruding A process in which a billet of material is shaped into a strip having a uniform cross section by forcing the material through a die.

fabric A cloth made by weaving, knitting, or felting fibers.

face brick Brick made or selected to produce an attractive exterior wall.

faceted glass window A window made by bonding 1 in. (25.4 mm) thick glass pieces with an epoxy resin matrix or reinforced concrete.

Fahrenheit temperature The temperature scale on which at standard atmospheric pressure the boiling point of water is 212°F, the freezing point is 32°F, and absolute zero is 2459.69°F.

fan coil unit The fan and heat exchanger for cooling and heating that are assembled in a common cabinet.

fascia The finish board covering the edges of rafters at the eaves.

fatigue A condition that occurs in a material when it is subject to fluctuating or cyclic strains and stresses that lead to permanent deformation.

fatigue limit The number of cycles of loading of a specified type that a specified material can withstand before failure.

fatigue strength A measure of the ability of a material or structural member to carry a load without failure when the loading is applied a specified number of times.

fatigue test A test to determine the behavior of a material under fluctuating stresses.

felt A sheet material made using a fiber mat that has been saturated and topped with asphalt.

fenestration An area that allows light to pass into a building, commonly referring to glazed windows. Also, the arrangement of windows in an exterior wall.

ferrous Iron-based metallic materials.

fiber saturation point The moisture content of wood at which the cell walls are saturated but there is no water in the cell cavities.

fiberboard A panel made from vegetable fibers and binding agents.

fibered gypsum A neat gypsum with cattle hair or organic fillers added.

filler Inert material added to a plastic resin to alter the strength and working properties and to lower the cost.

finish coat The third or final coat of gypsum plaster.

finish floor The flooring that is left exposed to view.

finish lime A hydrated lime used in finish coats of plaster and in ornamental plasters.

finish plaster The topcoat of plaster on a wall or ceiling.

firebrick A brick made from special clays that will withstand high temperatures.

fireclays Deep mined clays that withstand heat.

fire endurance The time during which a material or an assembly of materials provides resistance against the passage of fire.

fireproofing Material used to protect various members from damage due to fire.

fire-rated partition A partition assembly that has been tested and given a rating indicating the length of time it will resist a fire in hours.

fire resistance The capacity of a material or assembly of materials to withstand fire or give protection from it.

fire-resistant gypsum A gypsum product that has increased fire-resistance properties due to the addition of fire-resistant materials in the gypsum core.

fire-stop A member used to close openings between studs, joists, and other members to retard the spread of fire through openings between them.

fire wall A construction of noncombustible materials that subdivides a building or separates adjoining buildings to retard the spread of fire.

flame spread Flaming combustion that occurs along the surface of a material.

flames spread rate The rate at which flames will spread across a surface of a material.

flame spread rating A numerical designation given to a material to indicate its comparative ability to restrict flaming combustion over its surface.

flammability The ability of a material to resist burning.

flash point The temperature at which a flammable material will suddenly break into flame.

flash set Very rapid setting of the cement in concrete.

flashing A thin impervious material used to prevent water from penetrating the joints between building elements.

float A flat hand tool used to smooth the surface of freshly placed concrete after it has been leveled with a darby.

float process A glass manufacturing process in which the molten glass ribbon flows through a furnace supported on a bed of molten metal.

flocked construction Carpet formed by electrostatically spraying short strands onto an adhesive-coated backing material.

flood coat A heavy coating of asphalt poured and spread over a surface.

fluorescence The emission of visible light from a substance as a result of the absorption of radiation of short wavelengths.

flux A mineral added to molten iron to cause impurities to separate into a layer of molten slag on top of the iron.

flying formwork Large sections of formwork for pouring concrete slabs that are lifted from story to story by a crane in an assembled condition.

footcandle The unit of illumination equal to 1 lumen per square foot.

footing The lowest, widest part of the foundation that distributes the load over a broad area of the soil.

footlambert A unit for measuring brightness or luminance. It is equal to 1 lumen per square foot when brightness is measured from the surface.

formwork Temporary construction used to contain and give shape and support to concrete as it cures.

foundation The lower part of a building, which transfers structural loads from the building to the soil.

framed connections Connections joining structural steel members with a metal, such as an angle, that is secured to the web of the beam.

framing plan A drawing showing the location of structural members.

freezing cycle day A day when the temperature of the air rises above or falls below 32°F or 0°C.

freezing point The temperature at which a given substance will solidify (freeze).

frequency The number of cycles per second of current or voltage in alternating current, of a sound wave, or of a vibrating solid expressed in hertz.

frost line The maximum depth in the earth to which the soil can be expected to freeze during a severe winter.

frost point The temperature at which frost forms on exposed, chilled surfaces.

fuel contributed A rating of the amount of combustible material in a coating.

furring Strips applied over a surface to increase thickness or to provide a base for the attachment of other material.

fuse An overcurrent protection device that opens an electric circuit when the fusible element is broken by heat due to overcurrent passing through it.

fusion-bonded construction Carpet formed by bonding pile yarn between two sheets of backing material and cutting the pile yarn in the center, forming two pieces of carpet.

galling The wearing or abrading of one material against another under extreme pressure.

galvanic corrosion Corrosion that develops by galvanic action when two dissimilar metals are in contact in the atmosphere.

gearless traction elevator An elevator with the traction sheave connected to a spur gear that is driven by a worm gear connected to the shaft of the electric motor.

girder A major structural member used to support beams.

glass An amorphous, noncrystalline solid made by fusing silica with a basic oxide.

glazed structural clay tile Hollow clay tile products with glazed faces typically used to build interior walls.

glued laminated lumber (glulam) A structural wood member made by bonding together laminations of dimension lumber.

glues Bonding agents made from animal and vegetable products.

grade (1) Related to soil, the elevation or slope of the ground. (2) In relation to lumber, a means of classifying lumber or other wood products based on specified quality characteristics.

grade beam A ground-level reinforced structural member that supports the exterior wall of a structure and bears directly upon columns or piers.

grade level The elevation of the soil at a specific location.

grade mark A stamp on a product, such as wood, plywood, or steel, indicating the product's quality.

grading Adjusting the level of the ground on a site.

gravel Hard rock material in particles larger than ¼ in. (6.4 mm) in diameter but smaller than 3 in. (76 mm).

green lumber Lumber having a moisture content more than 19 percent.

grille An open grate used to cover, conceal, protect, or decorate an opening.

ground A conducting connection between an electrical circuit and the earth or a conducting body that serves in place of the earth.

ground-fault circuit interrupter A device providing protection from electric shock by de-energizing a circuit within an established period of time when the current to ground exceeds a predetermined value that is less than that needed to activate a standard overcurrent protective device.

groundwater Water that exists below the surface of the earth and passes through the subsoil.

grout A viscous mixture of portland cement, water, and aggregate used to fill cavities in concrete. Also refers to a specially formulated mortar used to fill under the baseplates of steel columns and in connections in precast concrete.

gypsum Hydrous calcium sulfate.

gypsum backerboard A gypsum panel used as the base on which to bond tile or gypsum wallboard.

gypsum board A gypsum panel used for interior wall and ceiling surfaces. It contains a gypsum core and surfaces covered with paper.

gypsum lath A panel having a gypsum core and a paper covering providing a bonding surface for plaster.

gypsum plaster Ground gypsum that has been calcined and mixed with additives to control setting time and working qualities.

gypsum sheathing A gypsum panel with a water-repellent core. Used for sheathing exterior walls.

hardboard A general term used to describe a panel made from interfelted ligno cellulosic fibers consolidated under heat and pressure.

hardness A measure of the ability of a material to resist indention or surface scratching.

hardwood A botanical group of trees that have broad leaves that are shed in the winter. (It does not refer to the hardness of the wood.)

hardwood plywood Plywood with various species of hardwoods used on the outer veneers.

haunch A projection used to support a member, such as a beam.

header joist A structural member fastened between two parallel full-length framing members to support cut off members at openings.

heartwood The wood extending from the pith to the sapwood.

heat exchanger A device to transfer heat between two physically separated fluids.

heat loss The energy needed to warm outside air leaking into a building through cracks around doors, windows, and other places.

heat pump A heating/refrigerating system in which heat is taken from a heat source, such as the air, and given up to the space to be heated. For cooling it takes heat from the air in the space and gives it up outdoors.

heat-strengthened glass Glass that has been strengthened by heat treatment.

heat-treatable alloys Aluminum alloys whose strength characteristics can be improved by heat treating.

heat treating Heating and cooling a solid metal to produce changes in physical and mechanical properties.

heating value The amount of heat produced by the complete combustion of a unit quantity of fuel.

heavy timber construction A type of wood-frame construction using heavy timbers for the columns, beams, joists, and rafters.

hiding power The ability of a paint to hide the previous color or substrate.

hinge joint A joint that permits some action similar to a hinge and in which there is no appreciable separation of the joining members.

hip roof A roof consisting of four sloping planes that intersect forming a pyramidal shape.

hoistway, elevator A fire-resistant vertical shaft in which the elevator moves.

hollow clay masonry A unit whose core area is 25 to 40 percent of the gross cross-sectional area of the unit.

hollow concrete masonry Concrete masonry units that have open cores.

hollow-core door A door with face veneers on the outer surfaces, wood spacers around the edges, and a hollow interior supported with a honeycomb grid.

hollow-core slab A precast concrete structural slab that uses internal cavities to reduce its weight.

horizontal shear The tendency of the top wood fibers to move horizontally in relationship to the bottom fibers.

hot melt Adhesives that bond when they are heated to a liquid form.

humidifier A device used to add moisture to the air.

humidify Add water vapor to the air.

humidity The amount of water vapor within a given space.

hydrate The capacity of lime to soak up water several times its weight.

hydrated lime Calcium hydroxide made by burning calcium carbonate, which forms calcium oxide that can then chemically combine with water.

hydration A chemical reaction between water and cement that produces heat and causes the cement to cure or harden.

hydraulic elevator An elevator having the car mounted on top of a hydraulic piston that is moved by the action of hydraulic oil under pressure.

hydraulic mortar A mortar that is capable of setting and hardening under water.

hydronic heating system A system that circulates hot water through a system of pipes and convectors to heat a building.

hydronics The science of cooling and heating water.

hydrostatic pressure The pressure equivalent to that exerted on a surface by a column of water of a specified height.

hydroxide of lime The product produced by the chemical reaction during the slaking or hydrating of lime.

hygrometer An instrument used to measure humidity conditions of the air.

hygrometric expansion The expansion and contraction of materials in relation to their moisture content.

hygroscopic The ability to readily absorb and retain moisture from the air.

Hz The abbreviation for hertz, the unit of measurement of the frequency of electric current. It represents the number of cycles per second.

I joist A wood joist made of an assembly of laminated veneer wood top and bottom flanges and a web of plywood or oriented strandboard.

igneous rock Rock formed by the solidification of molten material to a solid state.

illuminance The density of luminous power in lumens per a specified area.

impact insulation class An index of the extent to which a floor assembly transmits impact noise from a room above to the room below.

impact noise Sound generated by impact on the floor or other parts of the building that is carried through the building.

impact strength The energy required to fracture a specimen when struck with a rapidly applied load.

impact test A test for determining the resistance of a specimen fracture from a high-velocity blow.

in situ Undisturbed soil.

incandescence The emission of visible light produced by heating.

independent footings Footings supporting a single structural element, such as a column.

ingot A mass of molten metal cast in a mold and solidified to be stored until used for forging or rolling into a finished product.

insulating glass A glazing unit used to reduce the transfer of heat through a glazed opening by leaving an air space between layers of glass.

insulation, electric A material that is a poor conductor of electricity.

interceptor A trapping device designed to collect materials that will not be able to be handled by a sewage treatment plant, such as grease, glass or metal chips, and hair.

intumescence The swelling of a fire-retardant coating when heated, which forms a low-density film that provides some resistance to the spread of flame on the surface.

iron A metallic element existing in the crust of the earth from which ferrous alloys, such as cast iron, are made.

jamb The vertical member forming the side of a door or window frame.

joint compound A plastic gypsum mixture used to cover the joints and fasteners in gypsum wallboard installations.

jointing Forming control joints in a concrete slab.

joist A horizontal structural member used to carry the floor and ceiling loads.

joule A meter-kilogram-second unit of work or energy.

jute A coarse fiber obtained from two East Indian tiliaceous plants.

Keene's cement A hard, high-strength, white, quick-setting finishing plaster made from burnt gypsum and alum.

kiln (1) A chamber with controlled humidity, temperature, and airflow in which lumber is dried. (2) A low-pressure steam room in which green concrete units are cured.

kiln dried Wood products dried in a kiln.

kinetic energy The energy of a body with respect to the motion of the body.

knitted construction Carpet formed by looping pile yarn, stitching, and backing together.

lacquer A fast-drying clear or pigmented coating that dries by solvent evaporation.

laminate A material made by bonding several layers of material.

laminate, wood A product made by bonding layers of wood or other material to a wood substrate.

laminated glass Glass panels that have outer layers of glass laminated to an inner layer of transparent plastic.

laminated veneer lumber A structural lumber manufactured from wood veneers so that the grain of all veneers runs parallel to the axis of the member.

lamp A general term used to describe the source of artificial light. Often called a bulb or tube.

landing zone, elevator The area 18 in. (5490 mm) above or below the landing floor

latent heat Heat involved with the action of changing the state of a substance, such as changing water to steam.

lateral force A force acting generally in a horizontal direction, such as wind against an exterior wall or soil pressure against a foundation wall.

lateral loads Loads moving in a horizontal direction, such as the wind.

latex A water-based coating, such as styrene, butadiene, acrylic, and polyvinyl acetate.

lath The base material for the application of plaster.

leveling plate A steel plate set in grout on top of a concrete foundation to create a level bearing surface for the base of a steel column.

lift-slab construction A method of building site-cast concrete buildings by casting all the floor and roof slabs in a stack on the ground and lifting them up the columns with a series of jacks and welding them in place.

light A pane of glass.

light-gauge steel structural members Load-bearing members formed from light-gauge steel rolled into structural shapes.

lighting fixture See luminaire.

lighting outlet An electrical outlet to which a light fixture is connected.

light meter See luminance meter.

lightweight steel framing Structural steel framing members made from cold-rolled lightweight sheet steel.

lignin An amorphous substance that penetrates and surrounds the cellulose strands in wood, binding them together.

lime A white to gray powder produced by burning limestone, marble, coral, or shells.

limestone A sedimentary rock consisting of calcium and magnesium.

lintel A beam spanning an opening in a wall.

liquid limit Related to soils, the water content expressed as a percentage of dry weight at which the soil will start to flow when tested by the shaking method.

live load Nonpermanent moving or movable external loads on a structure, such as furniture or snow.

load-bearing Carrying an imposed load.

loomed construction Carpet formed by bonding the pile yarn to a rubber cushion.

louver A unit composed of sloping vanes used to restrict the entry of rain into openings in exterior walls yet permit the flow of air through the opening.

low-E glass Low emissivity glass that has a thin metallic coating that selectively reflects ultraviolet and infrared wavelengths of the energy spectrum.

low-emissivity coating A surface coating used on glass that permits the passage of most shortwave electromagnetic radiation (light and heat) but reflects longer-wave radiation (heat).

low-slope roofs Roofs that are nearly flat.

lumber A product produced by harvesting, sawing, drying, and processing wood.

lumber, boards Lumber nominally less than 2 in. thick and 2 in. or more wide.

lumber, dimension Lumber cut and dressed to standard sizes.

lumber, dressed size The size of lumber after it has been cut to size and the surfaces planed.

lumber, machine stress-rated Lumber that has been mechanically tested to determine its stiffness and bending strength.

lumber, matched Lumber that is edge dressed to make close tongue-and-groove edge joints.

lumber, nominal size The size of lumber after it has been sawn to size but has not been surfaced.

lumber, patterned Lumber that is shaped to a pattern or to a molded form.

lumber, rough Lumber that has not been surfaced but may be sawn, edged, and trimmed.

lumber, shiplapped Lumber that is edge dressed to make a lapped joint.

lumber, shop and factory Lumber intended to be cut up and used in some manufacturing process.

lumber, structural Lumber that is intended for use where allowable strength or stiffness of the piece is known.

lumber grader A person who inspects each piece of lumber and assigns a grade to it.

lumen A unit for measuring the flow of light energy. See luminous flux.

luminaire A complete lighting unit consisting of one or more lamps plus elements needed to distribute light, hold and protect the lamps, and connect power to the lamps. Also called a lighting fixture.

luminance The luminous intensity of a surface of a given area viewed from a given direction.

luminance meter A photoelectric instrument used to measure luminance. Also called a light meter.

luminescence The emission of light not directly caused by incandescence.

luminous flux The rate of flow of light energy through a surface, expressed in lumens.

luminous intensity The force that generates visible light expressed by candela, lumens per steradian, or candlepower.

luminous transmittance A measure of the capacity of a material to transmit incident light in relation to the total incident light striking it.

lux A unit of illumination equal to 1 lumen per square meter.

makeup air Air brought into the building from the outside to replace air that has been exhausted.

malleability The characteristic of a material that allows plastic deformation in compression without rupture.

malleability The property of a metal that permits it to be formed mechanically, such as by rolling or forging, without fracturing.

marble A metamorphic rock formed largely of calcite, dolomite, or dense limestone.

marker board A surface that can be written on with a water-based or semipermanent removable ink.

masonry cement A hydraulic cement used in mortars to increase plasticity and water retention.

MasterFormat The trademarked title of a uniform system for indexing construction specifications published by the Construction Specifications Institute and Construction Specifications Canada.

mastic Trowelable bituminous adhesive compound.

mastic A doughlike compound available in many different formulations designed for use as sealants and adhesives.

mat foundation A large, single concrete footing equal in area to the area covered by the footprint of the building.

mat foundation A large reinforced concrete slab on the earth that covers the entire area of the building foundation.

mechanical action The bonding of materials by adhesives that enter the pores and harden, forming a mechanical link.

mechanical properties Properties exhibited by a material's reaction to applied forces, such as tensile strength and compressive strength.

melamine A white crystalline made from calcium cyanamide.

melting temperature The temperature at which a material turns from a solid to a liquid.

membrane A continuous, unbroken roof covering.

metal lath Perforated sheets of thin metal secured to studs that serve as the base for a finished plaster wall.

metamorphic rock Rock formed by the action of pressure and/or heat on sedimentary soil or rock.

meter, electric A device measuring and recording the amount of electricity passing through it in kilowatt-hours.

mildewcide An agent that helps prevent the growth of mold and mildew on painted surfaces.

mild steel Steel containing less than 0.3 percent carbon.

millwork Wood interior finish items manufactured in a factory, such as doors, windows, and cabinets.

modified bitumens A roofing membrane composed of a polyester or fiberglass mat saturated with a polymer-modified asphalt.

modular size A dimension that conforms to a given module, such as the 48 in. width of plywood panels.

module A repetitive dimension or other unit.

modulus of elasticity The property of a material that indicates its resistance to bending. It is the ratio of unit stress to unit strain.

modulus of rupture A measure of the ultimate load-carrying capacity of a structural member.

moisture barrier A membrane used to block the passage of water, and water vapor through an assembly of materials, such as a wall.

moisture content The amount of water contained in wood, expressed as a percentage of the weight of the wet wood to the weight of an oven-dry sample.

molding A wood strip material that has curved and shaped surfaces. It is used for decorative purposes.

moment A force that acts at a distance from a point and that tends to cause the body to rotate about that point.

moment of inertia The sum of the products of the mass and the square of the perpendicular distance to the axis of rotation of each particle in a body rotating about an axis.

monolithic concrete Concrete cast with no joints except construction joints; a continuous pour.

monomer An organic molecule that can be converted into a polymer by chemical reaction with similar molecules or organic molecules.

mortar A plastic mixture of cementitious materials, water, and a fine aggregate.

mortar flow A measure of the consistency of freshly mixed mortar related to the diameter of a molded truncated cone specimen after the sample has been vibrated a specified number of times.

mosaic A decoration made up of small pieces of inlaid stone, glass, or tile.

motor control A device that governs the electrical power delivered to one or more electric motors.

motor control center Controllers used to start and stop electric motors and protect them from overloads.

moving ramp A conveyor belt system used to move people or packages up or down an incline.

moving walk A conveyor belt system operating at floor level used to move people in a horizontal direction.

mullion Horizontal or vertical members between adjacent window or door units.

muntin Small horizontal and vertical bars between small lights of glass in windows and doors.

nail popping The loosening of nails holding gypsum board to a wall or ceiling. It produces a bulge in the surface of the gypsum panel.

natural fibers Fibers found in nature, such as wool and cotton.

neat plaster A gypsum plaster with no aggregates or fillers added. Sometimes called unfibered gypsum.

needle beam A steel or wood beam that is run through an opening in a bearing wall and used to support the wall and related loads as work on the foundation below the wall is performed.

noise reduction coefficient A single number indicated by the amount of airborne sound energy absorbed into a material.

nonbearing Refers to a structural part that does not carry a load.

noncalcareous clays Clays containing silicate of alumina, feldspar, and iron oxide.

nondestructive testing Methods of testing an item that do not destroy the item being tested.

nonferrous Metallic materials in which iron is not a principal element.

nonheat-treatable alloys Alloys that do not increase in strength when they are heat treated but that do gain strength by the addition of alloying elements.

nonprestressed units Concrete structural members in which the reinforcing steel is not subject to prestressing or posttensioning.

nylon A synthetic plastic made from coal, tar, and water.

oakum A caulking material made from hemp fibers treated with tar.

oil-based paint Paint composed of resins requiring solvent for reduction purposes.

opaque coatings Coatings that completely obscure the color and much of the texture of the substrate.

open-web joist A prefabricated steel truss made of welded members, used for floor and roof construction.

organic material A class of compounds comprising only those existing in plants and animals.

oriented strand board A panel made from wood strands that have their strand face oriented in the long direction of the panel.

outlet box A box that is part of the electrical wiring system that contains one or more receptacles.

overlaid plywood Plywood panels whose exterior surfaces are covered with a resin-impregnated fiber ply.

overload The operation of electrical equipment in excess of the normal full-loaded electrical rating or of a conductor carrying current in excess of its rated capacity.

oxidation A reaction between a material and oxygen in the atmosphere.

oxide layer In aluminum, a very thin protective layer formed naturally on aluminum due to its reaction to oxygen.

oxidize To convert an element into its oxide, such as rusting steel.

pan A metal form used to form the cavities between joists in cast-in-place concrete floors and roofs.

panelboard A panel that includes fuses or circuit breakers used to protect the circuits in a building from overloads.

panelized construction Construction that uses preassembled panels for walls, floors, and roof.

parallel strand lumber Lumber made from lengths of wood veneer bonded to produce a solid member.

parapet The top of an exterior wall that extends above the line of the roof.

parging The application of a portland cement plaster on masonry and concrete walls to make them less permeable to water.

particleboard A sheet product manufactured from wood particles and a synthetic resin or other binder.

partition A non-load-bearing interior wall.

party wall A wall that is common to two buildings and is on the boundary between them.

passive solar systems Solar systems that use natural means to store solar energy.

paver A thin brick used as the finished floor covering.

perlite A lightweight material made from volcanic rock.

perm The unit of vapor permeability.

phenol A class of acid organic compounds used in the manufacture of various resins, plastics, and wood preservatives.

phenolic A synthetic resin made by the reaction of a phenol with an aldehyde.

photovoltaic cells Thin, flat semiconductors that convert light energy into direct-current electricity.

physical properties The properties associated with the physical characteristics of a material, such as thermal expansion and density.

pier A column designed to support a load.

pig An ingot of cast iron.

pig iron A high-carbon-content iron produced by the blast furnace and used to produce cast iron and steel.

pigments Paint ingredients mainly used to provide color and hiding power.

pilaster A vertical projection from a masonry or concrete wall providing increased stiffening.

pile A wood, steel, or concrete column usually driven into the soil to be used to carry a vertical load.

pile cap A concrete slab or beam that covers the head of several piles, tying them together.

pile hammer A machine for delivering blows to the top of a pile, driving it into the earth.

pit, elevator The part of the hoistway that extends below the floor of the lowest landing to the floor at the bottom of the hoistway.

pitch The slope of a roof or other plane surface.

pitch Related to carpets, the number of tufts in a 27 in. width of carpet.

plaster A cementitious material, usually a mixture of portland cement, lime or gypsum, sand, and water. Used to finish interior walls and ceilings.

plaster base Any material suitable for the application of plaster.

plaster of paris A calcined gypsum mixed with water to form a thick, pastelike mixture.

plastic An organic material that is solid in its finished state but is capable of being molded or of receiving form.

plastic behavior The ability of a material to become soft and formed into desired shapes.

plastic deformation The deformation of a material beyond the point at which it will recover its original shape.

plastic limit Related to soils, the percent moisture content at which the soil begins to crumble when it is rolled into a thread ⅛ in. (3 mm) in diameter.

plasticity The ability of a material to be deformed into a different shape.

plasticizer Liquid material added to some plastics to reduce their hardness and increase pliability. Also, an additive to concrete and mortar to increase plasticity.

plate glass A high-quality glass sheet that has both surfaces ground flat and carefully polished.

platform frame A wood structural frame for light construction with the studs extending only one floor high upon which the second floor is constructed.

plenum The space above a suspended ceiling.

plenum chamber A chamber in a heating and air conditioning system that has air pressure higher than the surrounding air and that is connected to ducts supplying conditioned air to the rooms.

ply One of a number of layers in a layered construction.

plywood A glued wood panel made up of thin layers of wood veneer with the grain of adjacent layers at right angles to each other or of outer veneers glued to a core of solid wood or reconstituted wood.

plywood, cold-pressed Interior type plywood manufactured in a press without external applications of heat.

plywood, exterior Plywood bonded with a type of adhesive that is highly resistant to moisture and heat.

plywood, interior Plywood manufactured for indoor use or in locations in which it would be subject to moisture for only a brief time.

plywood, marine Plywood panels with the same glue as exterior plywood but with more restrictive veneer specifications.

plywood, molded Plywood that is glued to the desired shape either between curved forms or by fluid pressure applied with flexible bags or blankets.

plywood, postformed Panels formed when flat plywood sheets are reshaped into a curved configuration by steaming or the use of plasticizing agents.

plywood stressed-skin panel A structural panel constructed with outer skins of plywood applied over an internal frame of wood members forming a rigid panel.

pole construction Construction using large-diameter log poles in a vertical position to carry the loads of the floors and roof.

polycarbonate A polyester made by linking certain phenols through carbonate groups.

polyester A linear polymer made by linear linking of oxybenzoyl units.

polyethylene A thermoplastic resin made by polymerizing ethylene.

polyimide A polymer based on the combination of certain anhydrides with aromatic diamines.

polymer A chemical compound formed by the union of simple molecules to form more complex molecules.

polymerization A chemical reaction in which molecules of a monomer are linked together to form large molecules whose molecular weight is a multiple of that of the original substance.

polyolefin A polymer composed of open-chain hydrocarbons having double bonds.

polypropylene A polymer produced by the linking of repeated propylene monomers.

polystyrene A clear, colorless plastic resin made by polymerizing styrene.

polyurethane A thermoplastic or thermosetting resin derived by condensation reaction of a polyisocyanate and a hydroxyle.

polyvinyl chloride A thermoplastic resin derived by the polymerization of vinyl and acetate.

ponding The collection of water in shallow pools on the top surface of the roof.

porcelain A strong vitreous material bonded to metal at high temperature.

porcelain enamel An inorganic metal oxide coating bonded to metal by fusion at a high temperature.

portland cement A cementitious binder used in concrete, mortar, and stucco, obtained from pulverizing clinker consisting of hydraulic calcium silicates.

post, plank, and beam framing A wood-framing system using beams for horizontal structural members that rest on posts, forming the vertical members.

posttensioning A method used to place concrete under tension in which steel tenons are tensioned after the concrete has been poured and hardened.

potable water Water that is safe to drink and meets standards of the local health authority.

power The rate at which work is performed, expressed in watts or horsepower.

pozzolan A siliceous or siliceous and aluminus material blended with portland cement that chemically reacts with calcium hydroxide to form compounds possessing cementitious properties.

pozzolan cement A cement made from volcanic rock that contains considerable silica.

precast concrete Concrete cast in a form and cured before it is lifted into its intended position.

pressure-treated lumber Lumber that has chemicals forced into it under pressure to retard decay and provide resistance to fire.

prestressed concrete Concrete that has been pretensioned or posttensioned.

pretensioning A method used to place a concrete member under tension by pouring concrete over steel tendons that are under tension before the concrete is poured.

primer A base coat in a paint system. It is applied before the finish coats.

proportional limit The upper limit at which stress is proportional to strain.

purlin A horizontal structural member that spans beams, frames, or trusses and supports the roof deck or rafters or a joist supporting a roof deck.

quarry An open excavation in the earth from which building stone is removed.

quarry tile A large clay tile used for finished flooring.

quartersawing Sawing lumber so the hard annual rings are nearly perpendicular to the surface.

R-value The numerical value used to indicate the resistance to the flow of heat.

raceway An enclosed channel designed to carry wires and cables.

racking The distortion of a rectangular frame when subject to shear forces.

radiant heat Heat transferred by radiation.

radiation The transfer of heat through space by means of electromagnetic waves.

rafter Member of a roof structural frame that supports the sheathing and other roof loads.

rainscreen principle The principle that states that wall cladding can be made watertight by placing wind-pressurized air chambers behind the joints, which reduces the air pressure differentials between the inside and outside that could cause water to move through the joints.

rake The board along the sloping edge of a gable.

ready-mix concrete Concrete mixed in a central plant and delivered to the site by truck.

rebar Steel bar used to reinforce concrete.

receptacle A device installed in an electrical outlet box to receive a plug to supply electric current to portable equipment.

receptacle outlet An outlet box in which one or more receptacles are installed.

reduction (1) A process in which iron is separated from oxygen with which it is chemically mixed by smelting the ore in a blast furnace. (2) In regard to aluminum, the electrolytic process used to separate molten aluminum from the alumina.

reflectance The ratio of the intensity of light reflected by a material to the intensity of the incident light.

reflectance coefficient A measure stated as a percentage of the amount of light reflected off a surface.

reflective coated glass Glass having a thin layer of a metal or metal oxide deposited on the surface to reflect heat and light.

reflectivity The relative ability of a surface to reflect sound or light.

refractive index The ratio of the speed of light in a material to the speed of light in a vacuum.

refractory Nonmetallic ceramic material used where high temperatures (above 2700°F) are present, such as in furnace linings.

refrigerant The medium of heat transfer that absorbs heat by evaporating at low temperatures and pressure and giving up heat when it condenses at a higher temperature and pressure.

reinforced brick masonry Brick masonry construction that has steel reinforcing bars inserted to provide tensile strength.

relative humidity The ratio of the amount of water vapor present in the air to that which the air would hold if saturated at the same temperature.

remote-control circuit An electric circuit that controls any other circuit using a relay or other such device.

resilience The property of a material whereby it gives up its stored energy when the deforming force is removed.

resiliency The ability to regain the initial shape after being deformed.

resilient flooring Finished flooring made from a resilient material, such as polyvinyl chloride or rubber.

resin A natural or synthetic material that is the main ingredient of paint and binds the ingredients together.

resistance, electric The physical property of a conductor or electric-consuming device to resist the flow of electricity, reducing power and generating heat.

retaining wall A wall that bears against soil or other material and resists lateral and other forces from the material being held in place.

retarder An admixture used to slow the setting of concrete.

return air Air removed from a space and vented or reconditioned by a furnace, air conditioner, or other apparatus.

rigid frame A structural framework in which the beams and columns are rigidly connected (no hinged joints).

riser, plumbing A water supply running vertically through the building to supply water to the various branches and fixtures on each floor.

rivet A metal fastener used to hold metal plates by passing through holes in each and having a head formed on the protruding end.

rock A solid mineral material found naturally in large masses.

rock anchor A posttensioned cable or steel rod inserted into a hole drilled in rock and grouted in place.

romex A nonmetallic-sheathed electric cable of type NM or NMC.

roof drain A drain on the roof to carry away water to a downspout.

roof planks Precast gypsum concrete members used for decking roofs.

room cavity ratio A relationship between the height of a room, its perimeter, and the height of the work surface above the floor divided by the floor area.

rows Related to carpets, the number of tufts per inch.

saddle A ridge in a roof deck that divides two sloping parts, diverting water toward roof drains.

safing Fire-stopping material placed in spaces between the floor and curtain walls in high-rise construction.

sand Fine rock particles from 0.002 in. (0.05 mm) in diameter to less than 0.25 in. (6.4 mm) in diameter.

sandstone A sedimentary rock formed from sand.

sanitary sewage Waste material containing human excrement and other liquid wastes.

sanitary sewer A sewer to receive sanitary sewage without the infusion of other water such as rain, surface water, or other clear water drainage.

sapwood The wood near the outside of the log just under the bark.

scratch coat The first coat of gypsum plaster that is applied to the lath.

screed A tool used to strike off the surface of freshly poured concrete so it is flush with the top of the form.

screeding The process of striking off the surface of freshly poured concrete with a screed so it is flush with the top of the form.

sculpture A three-dimensional work of art.

scupper An outlet in a parapet wall for the drainage of overflow water from the roof to the outside of the building.

scuttle An opening through the ceiling and roof to provide access to the roof. It is covered with a waterproof cover. Also referred to as a roof hatch.

sealant A mastic used to seal joints and seams.

sealer A material used to seal the surface of a material against moisture.

seasoning Removing moisture from green wood.

seated connections Connections that join structural steel members with metal connectors, such as an angle, upon which one member, such as a beam, rests.

security glass Glass panels assembled with multiple layers of glass and plastic to produce a panel that will resist impact.

sedimentary rock Rock formed from the deposit of sedimentary materials on the bottom of a body of water or on the surface of the earth.

segregation The tendency of large aggregate to separate from the sand-cement mortar in the concrete mix.

seismic area A geographic area where earthquake activity may occur.

seismic load Forces produced on a structural mass by the movements caused by an earthquake.

semitransparent Coatings that allow some of the texture and color of the substrate to show through.

sensible heat Heat that causes a detectable change in temperature.

septic tank A watertight tank into which sewage is run and where it remains for a period of time to permit hydrolysis and gasification of the contents, which then flow from the tank and are absorbed into the soil.

service entrance The point at which power is supplied to a building and where electrical service equipment, such as the service switch, meter, overcurrent devices, and raceways, are located.

service equipment The equipment needed to control and cut off the power supply to the building, such as switches and circuit breakers.

service temperature The maximum temperature at which a plastic can be used without altering its properties.

shaft wall A fire-resistant wall that protects the elevator, stairwell, and vertical mechanical chases in high-rise construction.

shales Clays that have been subjected to high pressures, causing them to become relatively hard.

shear A deformation in which planes of a material slide so as to remain parallel.

shear panel A floor, wall, or roof designed to serve as a deep beam to assist in stabilizing a building against deformation by lateral forces.

shear plate connector A circular metal connector recessed into a wood member that is to be bolted to a steel member.

shear stress The result of forces acting parallel to an area but in opposite directions, causing one portion of the material to "slide" past another.

shear studs Metal studs welded to a steel frame that protrude up into the cast-in-place concrete deck.

sheathing A covering placed over exterior studs or rafters that serves as a base below the exterior cladding.

sheave A pulley over which the elevator wire hoisting rope runs.

sheeting Wood, metal, or concrete members used to hold up the face of an excavation.

shop drawings Related to steel construction, working drawings giving the information needed to fabricate structural steel members.

shoring Bracing used to temporarily hold a wall in position.

shrinkage limit Related to soils, the water content at which the soil volume is at its minimum.

siamese connection A connection outside a building to which firefighters connect an alternate water supply to boost the water used by the fire suppression system.

sill The wood member anchored to the top of a foundation to receive joists; the horizontal bottom part of a door or window frame.

silt Fine sand with particles smaller than 0.002 in. (0.05 mm) and larger than 0.00008 in. (0.002 mm).

single-ply roofing A roofing membrane composed of a sheet of waterproof material secured to the roof deck.

sintering A process that fuses iron-ore dust with coke and fluxes into a clinker.

site-cast concrete Concrete poured and cured in its final position.

site investigation An investigation and testing of the surface and subsoil of the site to record information needed to design the foundation and the structure.

site plan A drawing of a construction site, showing the location of the building, contours of the land, and other features.

skylight A roof opening that is covered with a watertight transparent cover.

slab-on-grade A concrete slab poured and hardened directly on the surface of the earth.

slag A molten mass composed of fluxes and impurities removed from iron ore in the furnace.

slake The process of adding water to quicklime, hydrating it and forming lime putty.

slip form A form designed to move upward as concrete is poured in it.

slump A measure of the consistency of freshly mixed concrete, mortar, or stucco.

slump test A test to ascertain the slump of concrete samples.

slurry A liquid mixture of water, bentonite clay, or portland cement.

slurry wall A wall built of a slurry used to hold up the sides of an area to be excavated.

smelting A process in which iron ore is heated, separating the iron from the impurities.

smoke barriers Continuous membranes used to resist the passage of smoke.

smoke developed rating A relative numerical classification of the fumes developed by a burning material.

soffit The undersurface of an eave.

soft-mud process A process used to make bricks when the clay contains moisture in excess of 15 percent.

softwood A botanical group of trees that have needles and are evergreen.

soil anchors Metal shafts grouted into holes drilled into the sides of an excavation to stabilize it.

soil stack A vertical plumbing pipe into which waste flows through waste pipes from each fixture.

soil vent That portion of a soil stack above the highest fixture waste connection to it.

solar energy Radiant energy originating from the sun.

solar screen A device used to divert solar energy from windows.

solid clay masonry A unit whose core does not exceed 15 percent of the gross cross-sectional area of the unit.

solvents Liquids used in paint and other finishing materials that give the coating workability and that evaporate, permitting the finish material to harden.

sound A sensation produced by the stimulation of the organs of hearing by vibrations transmitted through the air. It is the movement of air molecules in a wavelike motion.

sound transmission class A single number rating of the resistance of a construction to the passage of airborne sound.

span The distance between two supporting members.

span rating Number indicating the distance a sheet of plywood or other material can span between supports.

spandrel beam A beam running from column to column in outside walls that carries the curtain wall above it.

special units Concrete masonry units that are designed and made for a special use.

specific adhesion Bonding dense materials using the attraction of unlike electrical charges.

specific gravity The ratio of the weight of one cubic foot of a material to the weight of one cubic foot of water.

specifications A written document in which the scope of the work, materials to be used, installation procedures, and quality of workmanship are detailed.

spire A tall pyramidal roof built upon a tower or steeple.

split ring connector A ring-shaped metal insert placed in circular recesses cut in joining wood members that are held together with a bolt or lag screw.

spoil bank An area where soil from the excavation is stored.

spread foundation A foundation that distributes the load over a large area.

spreading rate The area over which a paint can be spread expressed in square feet per gallon.

square In roofing, 100 square feet of roofing material.

stabilizers Additives used to stabilize plastic by helping it resist heat, loss of strength, and the effect of radiation on the bonds between the molecular chains.

stain A solution of coloring matter in a vehicle used to enhance the grain of wood during the finishing operation.

stained-glass window A window made of small colored glass pieces joined with leaded canes.

standing seam The vertical seam formed when two sheets of metal roofing are joined.

static loads Any load that does not change in magnitude or position with time.

steam Water in a vapor state.

steam separator A device used to remove moisture from steam after it flows from the boiler.

steam trap A device used to allow the passage of condensate while preventing the passage of steam.

steel A malleable alloy of iron and a small carefully controlled carbon content.

steeple A towerlike ornamental construction, usually square or hexagonal, placed on the roof of a building and topped with a spire.

steep-slope roofs Roofs with sufficient slope to permit rapid runoff of rain.

stiff-mud process A process used to make bricks from clay that has 12 to 15 percent moisture.

stiffness Resistance to bending or flexing.

stone Rock selected or processed by shaping to size for building or other use.

strain The deformation of a body or material when it is subjected to an external force.

strain hardening Increasing the strength of a metal by cold-rolling.

strand-casting machine A machine that casts molten steel into a continuous strand of metal that hardens and is cut into required lengths.

stress Force applied to a structural member, assembly of materials, or a component per unit of its area.

stressed-skin panels An assembly with high-strength facing panels separated by wood spacing strips and bonded firmly to them.

stress-rated lumber Lumber that has its modulus of elasticity determined by actual tests.

structural sandwich construction A wood construction consisting of a high-strength facing material bonded to and acting integrally with a low-density core material.

stucco A portland cement plaster used as the finish material on building exteriors.

stud The vertical framing member in frame wall construction.

subfloor The structural floor joined to the joists that supports the finish flooring.

substrate A subsurface to which another material is bonded.

suction The rate at which clay masonry units absorb moisture.

suction rate The weight of water absorbed when a brick is partially immersed in water for one minute expressed in grams per minute or ounces per minute.

superplasticizer An admixture used with concrete to make the wet concrete very fluid without adding additional water.

supply air Conditioned air entering a space from a heating, ventilating, or air-conditioning (HVAC) unit.

surface burning rating The rating of interior and surface finish material providing indexes for flame spread and smoke developed.

surface clays Clays obtained by open-pit mining.

surface effect The effect caused by the entrainment of secondary air against or parallel to a wall or ceiling when an outlet discharges air against or parallel to the wall or ceiling.

suspended ceiling A finish ceiling hung from the overhead by a series of wires.

switch, bypass isolation A manually operated switch used in connection with a transfer switch to provide a means of directly connecting load conductors to a power source and for disconnecting the transfer switch.

switch, general-use A switch used for general electrical control.

switch, general-use snap A general-use switch built so it can be installed in a device box or on the box cover.

switch, isolating A switch used to isolate an electric circuit from the source of power.

switch, motor-control A switch used to interrupt the maximum operating overload current of a motor.

switch, transfer A device used to transfer one or more conductors from one power source to another.

switchboard A large single panel or an assembly of panels with switches, overcurrent protection devices, and buses that are mounted on the face or both sides of the panel.

switches Devices to open and close an electric circuit or to change the connection within a circuit.

synthetic fibers Fibers formed by chemical reactions.

synthetic materials Materials formed by the artificial building up of simple compounds.

T-sill A type of sill construction used in balloon framing in which the header joist is placed inside the studs and is butted by the floor joists.

tapestry A fabric upon which colored threads are woven by hand to produce a design.

tee Precast concrete or metal structural members in the shape of the letter T.

temper designation A specification of the temper or metallurgical condition of an aluminum alloy.

tempered glass Heat-treated glass that has great resistance to breakage and increased toughness.

tempering The reheating of hardened steel to decrease hardness and increase toughness.

tempering glass A process used to strengthen glass by raising the temperature of the glass to near the softening point and then blowing jets of cold air on both sides suddenly to chill it and create surface tension in the glass.

tendon A steel bar or cable in pre-stressed concrete used to impart stress in the concrete.

tensile bond strength The ability of a mortar to resist forces tending to pull the masonry apart.

tensile strength The maximum tensile stress that a material can sustain at the point of failure.

tensile stress The stress per unit area of the cross section of a material that resists elongation.

tension The condition of being pulled or stretched.

terra-cotta A hard unglazed clay tile used for ornamental work.

terrazzo A finish-floor material made up of concrete and an aggregate of marble chips that after curing is ground smooth and polished.

texture plaster A finish plaster used to produce rough, textured finished surfaces.

thermal break Material with a low thermal conductivity that is inserted between materials, such as metal with a high thermal conductivity, to retard the passage of cold or heat through the highly conductive material.

thermal bridge A thermal conducting material that conducts heat through an insulated assembly of materials.

thermal conductance (C) Thermal conductance is the same as thermal conductivity except it is based on a specified thickness of a material rather than on one inch as used for conductivity.

thermal conduction The process of heat transfer through a solid by transmitting kinetic energy from one molecule to the next.

thermal conductivity (k) The rate of heat flow through one square foot of material one inch thick expressed in Btu per hour when a temperature difference of one degree Fahrenheit is maintained between the two surfaces.

thermal convection Heat transmission by the circulation of a liquid or a heated air or gas.

thermal insulation A material that has a high resistance to heat flow.

thermal properties The behavior of a material when subjected to a change in temperature.

thermal radiation The transmission of heat from a hot surface to a cool one by means of electromagnetic waves.

thermal resistance (R) An index of the rate of heat flow through a material or assembly of materials. It is the reciprocal of thermal conductance (C).

thermoforming A process in which heated plastic sheets are made to assume the contour of a mold by using the force of air pressure, vacuum, or mechanical stretching.

thermoplastics Plastics that soften by heating and reharden when cooled without changing the chemical composition.

thermosetting plastics Cured plastics that are chemically cross-linked and when heated will not soften but will be degraded.

thermostat A temperature-sensitive instrument that controls the flow of electricity to units used to heat and cool spaces in a building.

throw The horizontal or vertical distance an airstream travels after leaving the air outlet before it loses velocity.

tieback anchors Steel anchors grouted into holes drilled in the excavation wall to hold the sheeting, thus reducing the number of braces required.

timber Wood structural members having a minimum thickness of 6 in. (140 mm).

timber joinery The joining of structural wood members using wood joints, such as the mortise and tenon.

tolerance The permissible deviation from a given dimension or the acceptable variation in size from the given dimension.

topcoat The final coat of paint.

torque A twisting or rotating action.

torsion strength The maximum stress a material will withstand before fracturing under a twisting force.

torsion test A test used to ascertain the behavior of materials subject to torsion.

toughness A measure of the ability of a material to absorb energy from a blow or shock without fracturing.

transformer An electrical device used to convert an incoming electric current from one voltage to another voltage.

trap Device used to maintain a water seal against sewer gases that back up the waste pipe. Usually each fixture has a trap.

tread The horizontal part of a stair upon which the foot is placed.

troweling Producing a final smooth finish on freshly poured concrete with a steel-bladed tool after the concrete has been floated.

truss An assembly of structural members joined to form a rigid framework, usually connected to form triangles.

truss-framed system An assembled truss unit made up of a floor truss, wall stud, and a roof truss.

truss plate A steel plate used to strengthen the joints in truss assemblies.

tufted construction Carpet formed by stitching the pile yarn through the backing material.

twist Warping in which one or more corners of a piece of wood twist out of the plane of the piece.

two-way concrete joist system Floor and roof construction that has two perpendicular systems of parallel intersecting joists.

two-way flat plate Reinforced concrete construction in which the main reinforcement runs in two directions and both surfaces are flat planes and it is supported by columns.

two-way flat slab Reinforced concrete floor or roof construction in which a two-way flat plate is supported by columns with drop panels or column capitals.

U The overall coefficient of heat transmission; that is, the combined thermal value of all the materials in an assembly of building materials, such as a wall. U is expressed in Btu-h per square foot of an area per degree Fahrenheit temperature difference.

U_o The average overall coefficient of heat transmission; that is, the average of the U-value of all the materials that make up the component assembly.

ultimate strength The maximum stress, such as tensile, compressive, or shear, that a material can withstand.

ultimate tensile strength The maximum tensile stress of a material up to the point of rupture.

ultrasonic testing A method of nondestructive testing of materials that uses high frequency sound vibrations to find defects in the material.

under-carpet wiring A flat, insulated electric wire that is run under the carpet.

underlayment Sheet material, such as hardboard, that is laid over the subfloor to provide a smooth, stiff surface for the finish flooring.

underpinning Placing a new foundation below the existing foundation.

unfibered gypsum A neat gypsum. It has no additives.

uniform load Any load that is spread out evenly over a large area.

unreinforced concrete Concrete placed without steel-reinforcing bars or welded-wire fabric.

valence The points on an atom to which valences of other elements can bond.

valley The intersection of two inclined surfaces.

vapor retarder A material having a high resistance to the passage of vapor through it.

varnish A transparent coating that dries on exposure to air, providing a protective coating.

vehicle The liquid portion of a paint composed mainly of solvents, resins, or oils.

velvet construction Carpet formed by joining the pile, stuffer, and weft yarns with double warp yarns.

veneer A thin sheet of material used to cover another surface.

veneer gypsum base A gypsum board product designed to serve as the base for the application of gypsum veneer plaster.

veneer plaster A thin layer of plaster applied over a special veneer gypsum base sheet.

vent stack That part of the soil stack above the highest vent branch.

vents, plumbing Pipes permitting the waste system to operate under atmospheric pressure. They allow air to enter and leave the system, preventing water in the traps from being siphoned off. If this occurs, sewer gases can enter the building.

vermiculite An insulation material or aggregate made of expanded mica.

vertical load A load acting in a direction perpendicular to the plane of the horizon.

vertical shear The tendency of one part of a member to move vertically in relationship to the adjacent part.

viscosity The resistance of a liquid to flow under an applied load or pressure.

vitrification A process of using high kiln temperatures to fuse the surface grains of clay products so they are impervious to the passage of water.

volatile organic compounds Compounds released to the atmosphere as a coating dries.

volt The unit of potential difference or electromotive force. One volt applied across a resistance of one ohm results in a current flow of one ampere.

voltage The force, pressure, or electromotive force that causes electric current to flow in an electric circuit.

waferboard A mat-formed panel made of wood wafers, randomly arranged and bonded with a waterproof binder.

waffle slab A concrete slab that has ribs running in two directions forming a wafflelike grid.

wainscot A protective or decorative finish wall covering applied to the lower part of an interior wall.

warp A variation in a board from a flat, plane condition.

waste pipe Horizontal plumbing pipes that connect a fixture to the soil pipe.

water-based coatings Coatings formulated with water as the solvent.

water-cement ratio In a concrete or mortar mixture, the ratio of the amount of water (minus that held by the aggregates) to the amount of cement used.

waterproofing Material used to make a surface impervious to the penetration of water.

water repellent Liquid that penetrates the pores of wood and prevents moisture from penetrating without altering the desirable qualities of the wood.

water retention The property of a mortar that prevents the rapid loss of water by absorption into the masonry units.

water-smoking A process used to drive off the remaining water from clay products before they are fired in the kiln.

water stop A rubber or plastic diaphragm placed across a joint in cast concrete to prevent the passage of water through the joint.

water-struck brick Brick made in a mold that was wetted before the clay was placed in the mold.

water table The level below the ground where the soil is saturated with water.

water-vapor permeability The rate of water-vapor transmission through a given area of flat material of a given thickness induced by a given vapor pressure difference between the two surfaces under specified temperature and humidity conditions.

water-vapor transmission rate The steady-state vapor flow in a given time through a given area of a body, normal to specified parallel surfaces, under specific conditions of temperature and humidity at each surface.

watt The unit of measurement of electrical power or rate of work. It is a pressure of one volt flowing at the rate of one ampere.

weatherability The ability of a plastic to resist deterioration due to moisture, ultraviolet light, heat, and chemicals found in the air.

weathered joint A mortar joint finished so the mortar slopes outward, allowing water to shed away from the joint.

weathering Changes in the strength, color, surface, or other properties of a material due to the action of the weather.

weathering index A value that reflects the ability of clay masonry units to resist the effects of weathering.

weathering steel A steel alloy that forms a natural self-protecting rust.

weep hole Small openings at the bottom of exterior cavity walls to allow moisture in the cavity to drain out.

welded-wire fabric A form of steel reinforcing made from wire strands welded where they cross, forming a mesh.

Wilton construction Carpet formed on a loom capable of feeding yarns of various colors.

wind load Any load on a building caused by pressure or suction developed by the wind.

wind uplift Upward forces on a building caused by negative air pressures produced under certain wind conditions.

winning A term used to describe the mining of clay.

wired glass Glass made with a wire grid embedded in it.

wood preservative Substance that is toxic to fungi, insects, borers, and other wood-destroying organisms.

workability (1) Describes the ease or difficulty with which concrete can be placed and worked into its final location. (2) In relation to mortar, the property of freshly mixed mortar that determines the ease and homogeneity with which it can be spread and finished.

working joints Joints in exterior walls that allow for expansion and contraction of materials in the wall.

wracking When a component, such as a wall, is forced out of plumb.

wrought products Products formed by any of the standard manufacturing processes, such as drawing, rolling, forging, or extruding.

yield point The point at which strains increase without a corresponding increase in stress.

yield strength The load at which a limited permanent deformation occurs.

zoning ordinances Local regulations that control the use and development of land.

SELECTED BIBLIOGRAPHY

Also consult Appendix A by division category for related professional and technical organizations that have numerous publications available.

DIVISION 1 GENERAL REQUIREMENTS

Clough, R.H. and G.A. Sears. *Construction Contracting;* New York: John Wiley & Sons, 1993.

Fisk, E.R. *Construction Project Administration;* Englewood Cliffs, NJ: Prentice-Hall, 1992.

Levitt, R.E. and N.W. Samelson. *Construction Safety Management;* New York: John Wiley & Sons, 1993.

Nunnally, S.W. *Construction Methods and Management;* Englewood Cliffs, NJ: Prentice-Hall, 1993.

Samuels, B.M. *Construction Law;* Englewood Cliffs, NJ: Prentice-Hall, 1996.

Trauner, T.J. *Managing the Construction Project: A Practical Guide for the Project Manager,* New York: John Wiley & Sons, 1992.

DIVISION 2 SITE CONSTRUCTION

Ambrose, J.E. *Building Construction;* New York: Chapman & Hall, 1992.

DeChiara, J., and L.E. Koppelman. *Time-Saver Standards for Site Planning;* New York: McGraw-Hill, 1984.

Merritt, F.E., and J.T. Ricketts. *Building Design and Construction Handbook;* New York: McGraw-Hill, 1994.

Nash, George. *Do-It-Yourself Housebuilding;* New York: Sterling Publishing Co., 1995.

William O. Fellers. *Materials Science, Testing, and Properties for Technicians;* Upper Saddle River, NJ: Prentice Hall, 1990.

DIVISION 3 CONCRETE

ACI Building Code Requirements for Structural Concrete and Commentary; Detroit: American Concrete Institute.

Jacobs, D.H. *Concrete: A Home Owner's Illustrated Guide;* Blue Ridge Summit, PA: TAB Book Co., 1992.

NAHB Research Foundation. *Residential Concrete;* Washington DC: National Association of Home Builders, 1983.

Ramsey, C.G, and J.R. Hoke, eds. *Architectural Graphic Standards;* New York: John Wiley & Sons, 2000.

DIVISION 4 MASONRY

Stitt, F.A. *Architect's Detail Library; ;* New York: John Wiley & Sons, 1997.

Division 6 WOOD & PLASTICS

WOOD

American Institute of Timber Construction. *Timber Construction Manual;* New York: John Wiley & Sons, 1994.

Canadian Wood Council. *Wood Reference Handbook;* New York: John Wiley & Sons, 1993.

Chen, W.F. *The Civil Engineering Handbook,* Boco Raton, FL: CRC Press, Inc., 1994.

Forest Products Laboratory. *Handbook of Wod and Wood-based Materials;* Washington, DC: Hemisphere Publishing Corp., 1989.

Merritt, F.S. and J.T. Ricketts. *Building Design and Construction Handbook;* New York: McGraw-Hill, 1994.

WOOD PRODUCTS

Canadian Wood Council. *Wood Building Technology;* New York: John Wiley & Sons, 1993.

Canadian Wood Council. *Wood Reference Handbook;* New York: John Wiley & Sons, 1993.

Forest Products Laboratory. *Wood Engineering Handbook;* Upper Saddle River, NJ: Prentice Hall, 1991.

Hiro, J.E. *Millwork Handbook;* New York: Sterling Publishing Co., 1993.

LIGHT-FRAME CONSTRUCTION

Canadian Wood Council. *Wood Building Technology;* New York: John Wiley & Sons, 1993.

Canadian Wood Council. *Wood Reference Handbook;* New York: John Wiley & Sons, 1993.

Lorre, E.N. *Residential Steel Framing Construction Guide;* Las Vegas, NV: Technical Publications, 1997.

Newman, M. *Structural Details for Wood Construction;* New York: McGraw-Hill, 1987.

Ramsey, C.G, H.R. Sleeper, and J. Ambrose, ed. *Residential and Light Construction;* New York: John Wiley & Sons, 1992.

Spence, W.P. *Carpentry & Bulding Construction;* New York: Sterling Publishing Co., 1999.

Spence, W.P. *Constructing Staircases, Balustrades & Landings;* New York: Sterling Publishing Co., 2000.

Spence, W.P. *Finish Carpentry;* New York: Sterling Publishing Co., 1995.

Spence, W.P. *Residential Framing;* New York: Sterling Publishing Co., 1993.

Tallon, R. *Graphic Guide to Frame Construction;* Newtown, CT: The Taunton Press, 1999.

HEAVY TIMBER CONSTRUCTION

American Institute of Timber Construction. *Timber Construction Manual;* New York: John Wiley & Sons, 1994.

Benson, T. *The Timber-Frame Home: Design, Construction, Finishing;* Newtown, CT: Taunton Press, 1997.

Canadian Wood Council. *Wood Building Technology;* New York: John Wiley & Sons, 1993.

Canadian Wood Council. *Wood Reference Handbook;* New York: John Wiley & Sons, 1993.

Merritt, F.S. and J.T. Ricketts. *Building Design and Construction Handbook;* New York: McGraw-Hill, 1994.

Spence, W.P. *Residential Framing;* New York: Sterling Publishing Co., 1993.

PLASTICS

Hornbostel, C. *Construction Materials: Types, Uses and Applications;* New York: John Wiley & Sons, 1992.

Ramsey, C.G, and J.R. Hoke, eds. *Architectural Graphic Standards;* New York. John Wiley & Sons, 2000.

Task Committee on Properties of Selected Plastics. *Structural Plastics Selection Manual;* New York: American Society of Civil Engineers, 1985.

Division 7   THERMAL & MOISTURE

NAHB Research Foundation. *Insulation Manual;* Washington DC: National Association of Home Builders, 1990.

Scharff, Robert. *Roofing Handbook;* New York: McGraw-Hill, 1995.

Spence, W.P. *Finish Carpentry;* New York: Sterling Publishing Co., 1995.

The NRCA Roofing and Waterproofing Manual; Rosemont IL: National Roofing Contractors Association.

Watson, J.A. *Commercial Roofing Systems;* Englewood Cliffs, NJ: Prentice-Hall, 1984.

Division 8   DOORS & WINDOWS

Exterior Wall Construction in High-Rise Buildings; Ottawa, Ontario, Canada: Canada Mortgage and Housing Corp.

Ramsey, C.G, and H.R. Sleeper, J.R. Hokke, eds. *Architectural Graphic Standards;* New York: John Wiley & Sons, 1991.

Spence, W.P. *Finish Carpentry;* New York: Sterling Publishing Co., 1995.

Spence, W.P. *Residential Framing;* New York: Sterling Publishing Co., 1993.

Division 9   FINISHES

ACOUSTICAL MATERIAL

Egan, M.D. *Architectural Acoustics;* New York: McGraw-Hill, 1988.

Rettinger, M. *Handbook of Architectural Acoustics and Noise Control;* New York: McGraw-Hill, 1988.

FLOORING

Spence, W.P. *Finish Carpentry;* New York: Sterling Publishing Co., 1995.

Todd, K. *Carpentry Layout;* Solana, CA:cCraftsman Book Co., 1988

GYPSUM, LIME & PLASTER

Gorman, J.R., S. Jaffe, W.F. Pruler, and J.J. Rose. *Plaster and Drywall Systems Manual;* New York: McGraw-Hill, 1988.

Gypsum Construction Handbook; Chicago: United States Gypsum Company.

Van Den Braden, F. and T.L. Hartsell, *Plastering Skills;* American Technical Publisher, Inc., 1984.

INTERIOR WALLS, PARTITIONS & CEILINGS

Gorman, J.R., S. Jaffe, W.F. Pruler, and J.J. Rose. *Plaster and Drywall Systems Manual;* New York: McGraw-Hill, 1988.

Gypsum Construction Guide, NC: Gold Bond Building Products, National Gypsum Company.

Gypsum Construction Handbook; Chicago: United States Gypsum Company.

Remodeler's Guide to Suspended Ceilings, Chicago: Chicago Metallic Corporation.

Spence, W.P. *Installing & Finishing Drywall;* New York: Sterling Publishing Co., 1998.

Spence, W.P. *Finish Carpentry;* New York: Sterling Publishing Co., 1995.

Spence, W.P. *Residential Framing;* New York: Sterling Publishing Co., 1993.

PROTECTIVE & DECORATIVE COATINGS

Weismantel, G.E., *Paint Handbook;* New York: McGraw-Hill, 1981.

Division 10 SPECIALTIES

Sweet's General Building and Renovation, Catalog File, Section 10, Specialties; New York: McGraw-Hill.

Division 11 EQUIPMENT

Sweet's General Building and Renovation, Catalog File, Section 11, Equipment; New York: McGraw-Hill.

Division 12 FURNISHINGS

DeChiara, J., J. Panero, and M. Zelnick. *Time-Saver Standards for Interior Design and Space Planning;* New York: McGraw-Hill, 1984.

Sweet's General Building and Renovation Catalog File, Section 12, Furnishings; New York: McGraw-Hill.

Division 13 SPECIAL CONSTRUCTION

Sweet's General Building and Renovation Catalog File, Section 13. Special Construction; New York. McGraw-Hill.

Division 14 CONVEYING SYSTEMS

Ambrose, J.E. B*uilding Construction and Design;* New York: Chapman & Hall, 1992.

American National Standard Safety Code for Elevators, Dumbwaiters, Escalators, and Moving Walks, New York: American National Standards Institute.

Merritt, F.S. and J.T. Ricketts. *Building Design and Construction Handbook;* New York: McGraw-Hill, 1994.

Stein, B. and J.S. Reynolds. *Mechanical and Electrical Equipment for Buildings;* New York: John Wiley & Sons, 1999

Division 15 MECHANICAL

Sweet's General Building and Renovation Catalog File, Section 12, Furnishings; New York: McGraw-Hill.

ASHRAE Handbook; Atlanta, GA: American Society of Heating, Refrigerating, and Air-Conditioning Engineers.

Levenhagen, J.I., and D.H. Spethmann; *HVAC Controls and Systems;* New York: McGraw-Hill, 1998S.

Spence, W.P. *Architectural Working Drawings: Residential and Commercial Buildings;* New York: John Wiley & Sons.

Stein, B., and J.S. Reynolds, *Mechanical and Electrical Equipment for Buildings;* New York: John Wiley & Sons, 1999.

Division 16 ELECTRICAL

Bradshaw, V, *Building Control Systems;* New York: John Wiley & Sons, 1993.

Kaufman, J.E. *IES Lighting Handbook Application Volume,* New York: Illuminating Engineering Society of America, 1987.

Merritt, F.S., and J.T. Ricketts. *Building Design and Construction Handbook;* New York: McGraw-Hill, 1994.

Stein, B. *Mechanical and Electrical Systems;* New York: John Wiley & Sons, 1996.

INDEX